Statistical Design and
Analysis of Experiments

Statistical Design and Analysis of Experiments

with Applications to Engineering and Science

ROBERT L. MASON

Statistical Analysis Section
Southwest Research Institute
San Antonio, Texas

RICHARD F. GUNST

Department of Statistical Science
Southern Methodist University
Dallas, Texas

JAMES L. HESS

Engineering Department
E. I. du Pont de Nemours & Co.
Wilmington, Delaware

WILEY

JOHN WILEY & SONS

New York • Chichester • Brisbane • Toronto • Singapore

Library of Congress Cataloging in Publication Data:

Mason, Robert L., 1946–
    Statistical design and analysis of experiments: with applications
to engineering and science/Robert L. Mason, Richard F. Gunst,
James L. Hess.
        p. cm.—(Wiley series in probability and mathematical
statistics. Applied probability and statistics)
    Includes bibliographies and indexes.
    ISBN 0–471–85364–X
    1. Experimental design. 2. Mathematical statistics. I. Gunst.
Richard F., 1947–   . II. Hess, James L. III. Title. IV. Series.
QA279.M373   1989
519.5—dc19   88-20893
                                                              CIP

Printed in the United States of America

10 9 8 7 6 5 4 3 2

To Carmen, Ann, Janis Sue

Preface

Statistical Design and Analysis of Experiments is intended to be a practitioner's guide to statistical methods for designing and analyzing experiments. The topics selected for inclusion in this book represent statistical techniques that we feel are most useful to experimenters and data analysts who must either collect, analyze, or interpret data. The material included in this book also was selected to be of value to managers, supervisors, and other administrators who must make decisions based in part on the analyses of data that may have been performed by others.

The intended audience for this book consists of two groups. The first group covers a broad spectrum of practicing engineers and scientists, including those in supervisory positions, who utilize or wish to utilize statistical approaches to solving problems in an experimental setting. This audience includes those who have little formal training in statistics but who are motivated by industrial or academic experiences in laboratory or process experimentation. These practicing engineers and scientists should find the contents of this book to be self-contained, with little need for reference to other sources for background information.

The second group for whom this book is intended is students in introductory statistics courses in colleges and universities. This book is appropriate for courses in which statistical experimental design and the analysis of data are the main topics. It is appropriate for upper-level undergraduate or introductory graduate-level courses, especially in disciplines for which the students have had or will have laboratory or similar data-collection experience. The focus is on the use of statistical techniques, not on the theoretical underpinnings of those techniques. College algebra is the only prerequisite.

This book stresses the strategy of experimentation, data analysis, and the interpretation of experimental results. It features numerous examples using actual engineering and scientific studies. It presents statistics as an integral component of experimentation from the planning stage to the presentation of the conclusions.

The topics selected for inclusion in this book constitute a compilation of many of the rich varieties and alternatives that are available. We have chosen topics according to one or more of three criteria: (i) techniques that are widely recognized as fundamental to the statistical design and analysis of experiments, (ii) selected new procedures that are increasingly recognized as valuable alternatives to or improvements of standard techniques, and (iii) procedures for which proper application does not require extensive training in statistical theory. In decisions regarding topics that are included in this book, we were guided by our collective experiences as statistical consultants and by our desire to produce a book that would be informative and readable. The topics selected can be implemented by practitioners and do not require a high level of training in statistics.

Statistical Design and Analysis of Experiments is divided into five sections. The Introduction, consisting of Chapters 1 and 2, presents an overview of many conceptual foundations of modern statistical practice. The distinctions between populations or processes and samples, parameters and statistics, and mathematical and statistical modeling are discussed.

Part I consists of Chapters 3 to 5 and presents elementary descriptive statistics and graphical displays. These chapters introduce the reader to the basic issues surrounding the statistical analysis of data. The informational content of simple graphical and numerical methods of viewing data is stressed.

Chapters 6 to 11 constitute Part II. They are the heart of the experimental-design portion of the book. Unlike many other statistics books, this book intentionally separates discussions of the design of an experiment (Part II) from those of the analysis of the resulting data (Parts III and IV). This separation is desirable for two reasons. First, readers benefit from the reinforcement of concepts by considering the topics on experimental design in close proximity to one another. In addition, alternatives to the various designs are easily cross-referenced, making the distinctions between the designs clearer. Second, concentrated attention is devoted to experimental-design issues without having these matters confused by the equally important issues relating to the analysis of data from the designs. All too often, texts devote a paragraph to the design of an experiment and several pages to the analysis of the resulting data. Our experiences with this approach are that the material on experimental design is slighted when designs and analyses are presented together. A much clearer understanding of proper methods for designing experiments is achieved by separating the topics.

Part III, consisting of Chapters 12 to 20, stresses the analysis of data from designed experiments. Analysis of data from each of the designs discussed in Part II is presented. Confidence-interval and hypothesis-testing procedures are detailed for single-factor and multifactor experiments. Statistical models are used to describe responses from experiments, with careful attention to the specification of the terms of the various models and their relationship to the possible individual and joint effects of the experimental factors.

Part IV consists of Chapters 21 to 28 and is devoted to regression modeling. Linear regression modeling using least-squares estimators of the model parameters is detailed, along with various diagnostic techniques for assessing the assumptions typically made with both regression and analysis-of-variance models. Identification of influential observations and alternative regression estimators are also discussed.

We are grateful to the Literary Executor of the late Sir Ronald A. Fisher, F. R. S., to Dr. Frank Yates, F. R. S., and to the Longman Group, Ltd., London, for permission to reprint part of Table XXIII from their book *Statistical Tables for Biological, Agricultural and Medical Research* (6th edition, 1974). We also are indebted to many individuals for contributing to this work. Colleagues read earlier versions of the manuscript and made many valuable suggestions on content and readability. We are especially grateful to Jack Benson, Neil Blaylock, Janet Buckingham, Ken Chatto, Don Marquardt, Jack Webster, Mike Wobser, and an anonymous reviewer. Several graduate students at Southern Methodist University criticized key portions of the book.

We were fortunate to receive excellent typing and word-processing support as the manuscript progressed from initial drafts to the final copy. We offer our thanks for this support to Sheila Crane, Joan Lexa, Judy Roberts, and Marilyn Smith.

Contents

Statistical Design and
Analysis of Experiments

Introduction

CHAPTER 1

Statistics in Engineering and Science

In this chapter we discuss the role of statistics in engineering and scientific experimentation. Two detailed examples introduce the uses of statistics in experimental work:

- *Chemical Analysis Variation Study—a demonstration of the use of statistical design and analysis for the identification of causes of bias and excess variability in experimental results.*
- *Radiocarbon Dating Example—an illustration of the complementary roles of basic knowledge of physical processes and data analysis in the modeling of a data-generating process.*

The term "scientific" suggests a process of objective investigation that ensures that valid conclusions can be drawn from an experimental study. Scientific investigations are important not only in the academic laboratories of research universities but also in the engineering laboratories of industrial manufacturers. *Quality* and *productivity* are characteristic goals of industrial processes, which are expected to result in goods and services that are highly sought by consumers and that yield profits for the firms that supply them. Recognition is now being given to the necessary link between the scientific study of industrial processes and the quality of the goods produced. The stimulus for this recognition is the intense international competition among firms selling similar products to a limited consumer group.

The setting just described provides one motivation for examining the role of statistics in scientific and engineering investigations. It is no longer satisfactory just to monitor on-line industrial processes in order to ensure that products are within desired specification limits. Competition demands that a better product be produced within the limits of economic realities. Better products are initiated in academic and industrial research laboratories, made feasible in pilot studies and new-product research studies, and checked for adherence to design specifications throughout production. All of these activities require experimentation and the collection of data. The definition of the discipline of statistics in Exhibit 1.1 is used to distinguish the field of statistics from other academic disciplines and is oriented toward the experimental focus of this text. It clearly identifies statistics as a scientific discipline, which demands the same type of rigor and adherence to

3

EXHIBIT 1.1

Statistics. Statistics is the science of problem-solving in the presence of variability.

---

basic principles as physics or chemistry. The definition also implies that when problem solving involves the collection of data, the science of statistics should be an integral component of the process.

Perhaps the key term in this definition is the last one. The problem-solving process involves a degree of uncertainty through the natural variation of results that occurs in virtually all experimental work.

When the term "statistics" is mentioned, many people think of games of chance as the primary application. In a similar vein, many consider statisticians to be "number librarians," merely counters of pertinent facts. Both of these views are far too narrow, given the diverse and extensive applications of statistical theory and methodology.

Outcomes of games of chance involve uncertainty, and one relies on probabilities, the primary criteria for statistical decisions, to make choices. Likewise, the determination of environmental standards for automobile emissions, the forces that act on pipes used in drilling oil wells, and the testing of commercial drugs all involve some degree of uncertainty. Uncertainty arises because the level of emissions for an individual automobile, the forces exerted on a pipe in one well, and individual patient reactions to a drug vary with each observation, even if the observations are taken under "controlled" conditions. These types of applications are only a few of many that could be mentioned. Many others are discussed in subsequent chapters of this book.

Figure 1.1 symbolizes the fact that statistics should play a role in every facet of data collection and analysis, from initial problem formulation to the drawing of final conclusions. This figure distinguishes two types of studies: experimental and observational. In experimental studies the variables of interest often can be controlled and fixed at predetermined values for each test run in the experiment. In observational studies many of the variables of interest cannot be controlled, but they can be recorded and analyzed. In this book we emphasize experimental studies, although many of the analytic procedures discussed can be applied to observational studies.

Data are at the center of experimental and observational studies. As will be stressed in Section 1.3, all data are subject to a variety of sources which induce variation in measurements. This variation can occur because of fixed differences among machines, random differences due to changes in ambient conditions, measurement error in instrument readings, or effects due to many other known or unknown influences.

Statistical experimental design will be shown to be effective in eliminating known sources of bias, guarding against unknown sources of bias, ensuring that the experiment provides precise information about the responses of interest, and guaranteeing

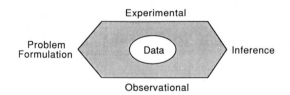

Figure 1.1 Involvement of statistics in scientific investigations.

that excessive experimental resources are not needlessly wasted through the use of an uneconomical design. Likewise, whether one simply wishes to describe the results of an experiment or one wishes to draw inferential conclusions about a process, statistical data-analysis techniques aid in clearly and concisely summarizing salient features of experimental data.

In the next two sections of this chapter examples are given that demonstrate the role of statistics in the experimental process. The examples are chosen to illustrate some of the many uses of statistics.

1.1 CHEMICAL ANALYSIS VARIATION STUDY

In most industrial processes there are numerous sources of possible variation. Frequently studies are conducted to investigate the causes of excessive variation. These studies could focus on a single source or simultaneously examine several sources. Consider, for example, a chemical analysis which involves different specimens of raw materials and which is performed by several operators. Variation could occur because the operators systematically differ in their method of analysis. Variation also could occur because one or more of the operators do not consistently adhere to the analytic procedures, thereby introducing uncontrolled variability to the measurement process. In addition, the specimens sent for analysis could differ on factors other than the ones under examination.

Table 1.1 Results of an Experiment to Identify Sources of Variation in Chemical Analyses[a]

Operator	Specimen	Combustion Run	Chemical Analysis	
			1	2
1	1	1	156	154
		2	151	154
		3	154	160
	2	1	148	150
		2	154	157
		3	147	149
2	3	1	125	125
		2	94	95
		3	98	102
	4	1	118	124
		2	112	117
		3	98	110
3	5	1	184	184
		2	172	186
		3	181	191
	6	1	172	176
		2	181	184
		3	175	177

[a]Adapted from Snee, R. D. (1983). "Graphical Analysis of Process Variation Studies," *Journal of Quality Technology*, **15**, 76–88. Copyright, American Society for Quality Control, Inc., Milwaukee, WI. Reprinted by permission.

In order to investigate sources of variability for a chemical analysis similar to the one just described, an experiment was statistically designed and analyzed to ensure that relevant sources of variation could be identified and measured. A test specimen was treated in a combustion-type furnace, and a chemical analysis was performed on it. In the experiment three operators each analyzed two specimens, made three combustion runs on each specimen, and titrated each run in duplicate. The results of the experiment are displayed in Table 1.1 and graphed in Figure 1.2.

Figure 1.2 highlights a major problem with the chemical analysis procedure. There are definite differences in the analytic results of the three operators. Operator 1 exhibits very consistent results for each of the two specimens and each of the three combustion runs. Operator 2 produces analytic results which are lower on the average than those of the other two operators. Operator 3 shows good consistency between the two specimens, but the repeat analyses of two of the combustion runs on specimen 5 appear to have substantially larger variation than for most of the other repeat analyses in the data set. Operator 2 likewise shows good average consistency for the two specimens, but large variation both for the triplicate combustion runs for each specimen and for at least one of the repeat analyses for the fourth specimen.

Thus the experimental results indicate that the primary sources of variation in this

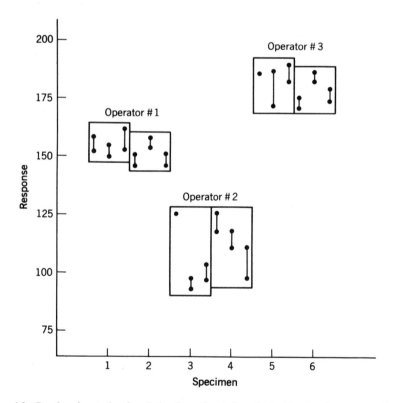

Figure 1.2 Results of a study of variation in a chemical analysis. (Combustion runs are boxed; duplicate analyses are connected by vertical lines.)

chemical analysis are the systematic differences (biases) among operators and, in some instances, the (random) inconsistency of the chemical analyses performed by a single operator. In reaching these conclusions statistics played a role in both the design of the experiment and the formal analysis of the results, the foregoing graphical display being one component of the analysis. The quality of this data-collection effort enables straightforward, unambiguous conclusions to be drawn. Such clear-cut inferences are often lacking when data are not collected according to a detailed statistical experimental design.

1.2 RADIOCARBON DATING EXAMPLE

In addition to designing experiments and collecting data according to sound statistical principles, statistical methodology is often used to fit theoretical models to data. In the next example we describe the relationship between two variables by using two alternative model-fitting methods. The use of either method implies specific assumptions about the physical process from which the data are obtained. The choice of method depends on each experimenter's understanding of underlying assumptions about each method, as well as an understanding of the physical processes from which the data are obtained. This example illustrates the critical link between model assumptions and model fitting.

Radiocarbon dating is an important technique for estimating the age of anthropological and geological artifacts. Radiocarbon dating is based on measuring the radioactive decay of carbon-14 atoms in the artifacts. The decay measurements are obtained by counting the number of beta particles (electrons) emitted over a fixed time period, say 100 minutes. By comparing the number of beta particles emitted from the sample with the number emitted from a reference source of recent origin, the age of the sample can be estimated, since it is known that the concentration of carbon-14 atoms in a sample diminishes by about 1% every 83 years.

Among the many factors which must be investigated when establishing laboratory procedures for carbon-14 dating is the energy window for which counts are made. A larger energy window results in larger counts than a smaller one. Figure 1.3 depicts a scattergram of 100-minute decay counts for a sample of charcoal using two channels of a single counter. Channel A of this counter has a larger energy window than does channel B. The 96 pairs of counts display an approximately linear relationship in Figure 1.3; consequently, one might postulate a linear relationship of the form

$$\text{channel-A count} = \alpha + \beta \cdot (\text{channel-B count}) + (\text{measurement error}).$$

One can obtain estimates for the constants α and β by many different curve-fitting techniques. Using a least-squares fitting procedure (Chapter 21) yields

$$\text{channel-A count} = 1129 + 0.85 \cdot (\text{channel-B count}).$$

The solid line in Figure 1.3 indicates that the least-squares fit is in good agreement with the data.

Superimposed on Figure 1.3 is a second fit to the data obtained by a maximum-likelihood curve-fitting technique (Section 28.1). This alternative equation, indicated by the dashed line, is

$$\text{channel-A count} = 753 + 1.01 \cdot (\text{channel-B count}).$$

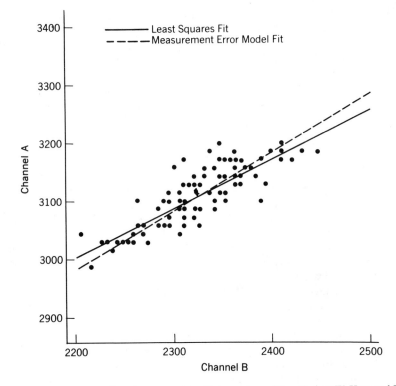

Figure 1.3   Linear fits to radiocarbon count data. Data courtesy of Drs. Herbert W. Haas and James Devine, Radiocarbon Laboratory, Southern Methodist University, Dallas, TX.

Qualitatively the two fits both seem to adequately represent the trend in the data. Either of the fitted equations could be used to represent the theoretical relationship (apart from measurement error) between the two variables. This illustrates the common phenomenon that without a theoretical basis to justify assumptions it is often difficult to select from among several alternative fitted models one that best represents a specific set of data.

The dashed line in Figure 1.3 is based on the assumption that the true theoretical relationship between the counts is a straight line but that both variables are measured with error:

$$\text{channel-A count} = \alpha + \beta \cdot (\text{channel-B count} + \text{channel-B measurement error})$$

$$+ \text{ channel-A measurement error.}$$

Hindsight suggests that if one of the channels of the counter is subject to measurement error, so is the other one. Least-squares fitting procedures generally require that there be no measurement error associated with the predictor variable in the model (here, the channel-B count). Thus the theoretical model containing two measurement errors is not only more faithful to the physical nature of the process under investigation, but it also dictates that methodologies other than least squares may be necessary for proper model fitting.

While we stress least-squares model-fitting procedures throughout Parts III and IV of this book, one reason for presenting this example was to emphasize that there are alternatives to these "traditional" techniques. These alternatives may be more appropriate when one critically examines the assumptions that underlie the use of the methodology and those of the data.

1.3 THE ROLE OF STATISTICS IN EXPERIMENTATION

Statistics is a scientific discipline devoted to the drawing of valid inferences from experimental or observational data. The study of variation, including the construction of experimental designs and the development of models which describe variation, characterizes research activities in the field of statistics. A basic principle that is the cornerstone of the material covered in this book is the following:

All measurements are subject to variation.

The use of the term "measurement" in this statement is not intended to exclude qualitative responses of interest in an experiment, but the main focus of this text is on designs and analyses which are appropriate for quantitative measurements.

Each of the examples presented in the two previous sections clearly exhibits the variation which is inherent in experimental work. The statistical procedures which were discussed enabled the goals of each study to be achieved in spite of the experimental variability. In the first example the use of a statistical design enabled the experimenters to unambiguously isolate the key sources of the excessive variation in the data. Once the major sources of variation were identified, action was taken to reduce the variation to more acceptable levels. In the second example, the two known sources of variation were modeled. The measurement-error model explicitly accounted for the variation due to each of the two channels. Estimation of the unknown constants in the measurement-error model produced a good fit to both the estimation data and a second data set of counts.

These two examples illustrate three general features of the statistical design and analysis of experiments. First, statistical considerations should be included in the project design phase of any experiment. At this stage of a project one should consider the nature of the data to be collected, including what measurements are to be taken, what is known about the likely variation to be encountered, and what factors might influence the variation in the measurements.

Second, a statistical design should be selected that controls, insofar as possible, variation from known sources. The design should allow the estimation of the magnitude of uncontrollable variation and the modeling of relationships between the measurements of interest and factors (sources) believed to influence these measurements.

Uncontrollable variation can arise from many sources. Two general sources of importance to the statistical design of experiments are experimental error and measurement error. Experimental error is introduced whenever test conditions are changed. For example, machine settings are not always exact enough to be fixed at precisely the same value or location when two different test runs call for identical settings. Batches of supposedly identical chemical solutions do not always have exactly the same chemical composition. Measurement errors arise from the inability to obtain exactly the same measurement on two successive test runs when all experimental conditions are unchanged.

Third, a statistical analysis of the experimental results should allow inferences to be

Table 1.2   Role of Statistics in Experimentation

Project Planning Phase

- What is to be measured?
- How large is the likely variation?
- What are the influential factors?

Experimental Design Phase

- Control known sources of variation
- Allow estimation of the size of the uncontrolled variation
- Permit an investigation of suitable models

Statistical Analysis Phase

- Make inferences on design factors
- Guide subsequent designs
- Suggest more appropriate models

drawn on the relationships between the design factors and the measurements. This analysis should be based on both the statistical design and the model used to relate the measurements to the sources of variation. If additional experimentation is necessary or desirable, the analysis should guide the experimenter to an appropriate design and, if needed, a more appropriate model of the measurement process.

Thus the role of statistics in engineering and scientific experimentation can be described using three basic categories: project planning, experimental design, and data analysis. These three basic steps in the statistical design and analysis of experimental results are depicted in Table 1.2.

Table 1.2 provides a partial explanation for the layout of this book. While it is beyond the scope of this text to delve in depth into the various modes of project planning, Chapter 2 and all of Part I contain introductory material that stresses various statistical concepts that are important to the successful planning of an experiment. Special emphasis is placed on understanding the omnipresence of variability in measurements through discussions of populations and samples, statistical modeling, and quantitative and graphical summarization of data.

Part II covers a variety of statistical designs. This material on the statistical design of experiments covers a wide range of alternative designs that may be considered appropriate, depending on decisions made at the planning phase of the study. Based on the design chosen for the study, one or more of the analytic techniques described in Parts III and IV may be used for modeling and inference.

REFERENCES

Text References

Snee, R. D., Hare, L. B., and Trout, J. R. (1985). *Experiments in Industry: Design, Analysis, and Interpretation of Results*, Milwaukee, WI: American Society for Quality Control.
   This collection of eleven case histories focuses on the design, analysis, and interpretation of scientific

experiments. The stated objective is "to show scientists and engineers not familiar with this methodology how the statistical approach to experimentation can be used to develop and improve products and processes and to solve problems."

Tanur, J. M., Mosteller, F., Kruskal, W. H., Link, R. F., Pieters, R. S., and Rising, G. R. (1972). *Statistics: A Guide to the Unknown,* San Francisco: Holden-Day, Inc.

This collection of 44 essays on applications of statistics presents excellent examples of the uses and abuses of statistical methodology. The authors of these essays present applications of statistics and probability in nontechnical expositions which for the most part do not require previous coursework in statistics or probability. Individual essays illustrate the benefits of carefully planned experimental designs as well as numerous examples of statistical analysis of data from designed experiments and observational studies.

Andrews, D. F. and Herzberg, A. M. (1985). *Data: A Collection of Problems from Many Fields for the Student and Research Worker,* New York: Springer-Verlag, Inc.

This book is a collection of 71 data sets with descriptions of the experiment or the study from which the data were collected. The data sets exemplify the rich variety of problems for which the science of statistics can be used as an integral component in problem-solving.

CHAPTER 2

Basic Statistical Concepts

In this chapter concepts are introduced that are fundamental to the use of statistics in experimental work. They include

- *distinguishing statistical samples from populations,*
- *relating sample statistics to population parameters, and*
- *characterizing deterministic and empirical models.*

Statistical methodology plays a key role in experimental research. Statistical analyses aid in the efficient design of experiments and in problem-solving and decision-making processes. Experimental data, in the form of a representative sample of observations, enable us to draw inferences about a phenomenon, population, or process of interest. These inferences are obtained by using sample statistics to draw conclusions about postulated models of the underlying data-generating mechanism. In the next three sections we discuss several concepts which are fundamental to an understanding of statistical inference.

2.1 POPULATIONS AND SAMPLES

All possible items or units which determine an outcome of a well-defined experiment are collectively called a "population" (see Exhibit 2.1). An item or a unit could be a measurement, or it could be material on which a measurement is taken. For example, in a study of geopressure as an alternative source of electric power, a population of interest might be all geographical locations for which characteristics such as wellhead fluid temperature, pressure, or gas content could be measured. Other examples of populations are:

EXHIBIT 2.1

Population. A statistical population consists of all possible items or units possessing one or more common characteristics under specified experimental or observational conditions.

- all 30-ohm resistors produced by a particular manufacturer under specified manufacturing conditions during a fixed time period;
- all possible fuel-consumption values obtainable with a four-cylinder, 1.7-liter engine using a 10%-methanol, 90%-gasoline fuel blend, tested under controlled conditions on a dynamometer stand;
- all measurements on the fracture strength of one-inch-thick underwater welds on a steel alloy base plate which is located 200 feet deep in a specified salt-water environment; or
- all 1000-lb containers of pelletized, low-density polyethylene produced by a single manufacturing plant under normal operating conditions.

These examples suggest that a population of observations may exist only conceptually, as with the population of fracture-strength measurements. Populations also may represent processes for which the items of interest are not fixed or static; rather, new items are added as the process continues, as in the manufacture of polyethylene.

Populations, as represented by a fixed collection of units or items, are not always germane to an experimental setting. For example, there are no fixed populations in many studies involving chemical mixtures or solutions. Likewise, ongoing production processes do not usually represent fixed populations. The study of physical phenomena such as aging, the effects of drugs, or aircraft engine noise cannot be put in the context of a fixed population of observations. In situations such as these it is a physical process rather than a population that is of interest (see Exhibit 2.2).

EXHIBIT 2.2

Process. A process is a repeatable series of actions that results in an observable characteristic or measurement.

The concepts and analyses discussed in this book relative to samples from populations generally are applicable to processes. For example, one samples both populations and processes in order to draw inferences on models appropriate for each. While one models a fixed population in the former case, one models a "state of nature" in the latter. A simple random sample may be used to provide observations from which to estimate the population model. A suitably conducted experiment may be used to provide observations from which to estimate the process model. In both situations it is the representative collection of observations and the assumptions made about the data that are important to the modeling procedures.

Because of the direct analogies between procedures for populations and for processes, the focus of the discussions in this book could be on either. We shall ordinarily develop concepts and experimental strategies with reference to only one of the two, with the understanding that they should readily be transferrable to the other. In the remainder of this section, we concentrate attention on developing the relationships between samples and populations.

When defining a relevant population (or process) of interest, one must define the exact experimental conditions under which the observations are to be collected. Depending on the experimental conditions, many different populations of observed values could be defined. Thus, while populations may be real or conceptual, they must be explicitly defined with respect to all known sources of variation in order to draw valid statistical inferences.

The items or units that make up a population are usually defined to be the smallest subdivisions of the population for which measurements or observations can take on different values. For the populations defined above, for example, the following definitions represent units of interest. An individual resistor is the natural unit for studying the actual (as opposed to specified) resistance of a brand of resistors. A measurement of fuel consumption from a single test sequence of accelerations and decelerations is the unit for which data are accumulated in a fuel economy study. Individual welds are the appropriate units for investigating fracture strength. A single container of pellets is the unit of interest in the manufacture of polyethylene.

Measurements on a population of units can exhibit many different statistical properties, depending on the characteristic of interest. Thus it is important to define the fundamental qualities or quantities of interest in an experiment. We term these qualities or quantities *variables* (see Exhibit 2.3).

EXHIBIT 2.3

Variable. A property or characteristic on which information is obtained in an experiment.

An *observation*, as indicated in Exhibit 2.4, refers to the collection of information in an experiment, and an *observed value* refers to an actual measurement or attribute that is the result of an individual observation. We often use "observation" in both senses; however, the context of its use should make it clear which meaning is implied.

EXHIBIT 2.4

Observation. The collection of information in an experiment, *or* actual values obtained on variables in an experiment.

A delineation of variables into two categories, response variables (see Exhibit 2.5) and factors, is an important consideration in the modeling of data. In some instances response variables are defined according to some probability model which is only a function of certain (usually unknown) constants. In other instances the model contains one or more factors (see Exhibit 2.6) in addition to (unknown) constants.

EXHIBIT 2.5

Response Variable. Any outcome or result of an experiment.

EXHIBIT 2.6

Factors. Controllable experimental variables that influence the observed values of response variables.

The response variable in a resistor study is the actual resistance measured on an individual resistor. In a study of fuel economy one might choose to model the amount of fuel consumed (response variable) as some function of vehicle type, fuel, driver, ambient temperature, and humidity (factors). In the underwater weld study the response variable is the fracture strength. In the manufacture of polyethylene the response variable of interest might be the actual weight of a container of pellets.

Most of the variables just mentioned are quantitative variables, since each observed value can be expressed numerically. There also exist many qualitative or nonnumerical variables that could be used as factors. Among those variables listed above, ones that could be used as qualitative factors include vehicle type, fuel, and driver.

Populations often are too large to be adequately studied in a specified time period or within designated budgetary constraints. This is particularly true when the populations are conceptual, as in most scientific and engineering experiments, when they represent every possible observation which could be obtained from a manufacturing process, or when the collection of data requires the destruction of the item. If it is not feasible to collect information on every item in a population, inferences on the population can be made by studying a representative subset of the data, a *sample* (see Exhibit 2.7).

EXHIBIT 2.7

Sample. A sample is a group of observations taken from a population or a process.

There are many ways to collect samples in experimental work. A *convenience sample* is one that is chosen simply by taking observations that are easily or inexpensively obtained. The key characteristic of a convenience sample is that all other considerations are secondary to the economic or rapid collection of data. For example, small-scale laboratory studies often are necessary prior to the implementation of a manufacturing process. While this type of pilot study is an important strategy in feasibility studies, the results are generally inadequate for inferring characteristics of the full-scale manufacturing process. Sources of variation on the production line may be entirely different from those in the tightly controlled environment of a laboratory.

Similarly, simply entering a warehouse and conveniently selecting a number of units for inspection may result in a sample of units which exhibits less variation than the population of units in the warehouse. From a statistical viewpoint, convenience samples are of dubious value because the population which they represent may have substantially different characteristics than the population of interest.

Another sampling technique which is frequently used in scientific studies is termed *judgmental sampling*. Here one's experience and professional judgment are used to select representative observations from a population of interest. In the context of a fuel-economy study, an example would be the selection of a particular engine, one fuel blend, and specific laboratory conditions for conducting the study. If the conditions selected for study are not truly representative of typical engine, fuel, and operating conditions, it is difficult to define the relevant population to which the observed fuel-consumption values pertain. A current example of this problem is the E.P.A. fuel-economy ratings posted on automobile stickers. These figures are comparable only under the laboratory conditions under which the estimates are made, not on any typical vehicle, fuel, or operating conditions.

Convenience and judgmental samples are important in exploratory research. The difficulty with these types of samples is that important sources of variation may be held constant or varied over a narrower range than would be the case for the natural occurrence of experimental units from the population of interest. In addition, these sampling schemes may mask the true effects of influential factors. This is an especially acute problem if two or more factors jointly influence a response. Holding one or more of these joint factors constant through convenience or judgmental sampling could lead to erroneous inferences about the effects of the factors on the response.

One of the most important sampling methodologies in experimental work is the *simple random sample*, defined in Exhibit 2.8. In addition to its use in the sampling of observations from a population, simple random sampling has application in the conduct of scientific and engineering experiments. Among the more prominent uses of simple random sampling in experimental work are the selection of experimental units and the randomization of test runs.

EXHIBIT 2.8

Simple Random Sample. In an experimental setting, a simple random sample of size *n* is obtained when items are selected from a fixed population or a process in such a manner that every group of items of size *n* has an equal chance of being selected as the sample.

If one wishes to sample 100 resistors from a warehouse, simple random sampling requires that every possible combination of 100 resistors present in the warehouse have an equal change of being included in the selected sample. Although the requirements of simple random sampling are more stringent than most other sampling techniques, unintentional biases are avoided.

Simple random samples can be obtained in many ways. For example, in the selection of experimental units to be included in an experiment, a common approach is to enumerate or label each item from 1 to N and then use a table of random numbers to select n of the N units. If a test program is to consist of n test runs, the test runs are sequentially numbered from 1 to n and a random-number table is used to select the run order. Equivalently, one can use random-number generators, which are available on computers. Use of such tables or computer algorithms removes personal bias from the selection of units or the test run order.

Simple random samples can be taken *with or without replacement.* Sampling with replacement allows an experimental unit to be selected more than once. One simply obtains n numbers from a random-number table without regard to whether any of the selected numbers occur more than once in the sample. Sampling without replacement prohibits any number from being selected more than once. If a number is sampled more than once, it is discarded after the first selection. In this way n unique numbers are selected. The sequencing of test runs is always performed by sampling without replacement. Ordinarily the selection of experimental units is also performed by sampling without replacement.

Inspection sampling of items from lots in a warehouse is an example for which a complete enumeration of experimental units is possible, at least for those units that are present when the sample is collected. When a population of items is conceptual or an operating production process is being studied, this approach is not feasible. Moreover, while one could conceivably sample at random from a warehouse full of units, the expense suffered through the loss of integrity of bulk lots of product when a single item is selected for inclusion in a sample necessitates alternative sampling schemes.

There are many other types of random sampling schemes besides simple random sampling. *Systematic* random samples are obtained by sampling every kth (e.g., every 5th, 10th, or 100th) unit in the population. This type of sampling often is used to select items from an operating production process or from devices such as filing cabinets that contain individual sample units but for which the task of enumeration would be impossible or uneconomical. To obtain a systematic random sample of every kth unit, one selects a random number between 1 and k to identify the first unit to be selected. One then sequentially selects every kth unit thereafter. If the population size is (approximately) known, one can determine the sampling multiple k by dividing the population size N by the desired sample size n.

Stratified random samples are based on subdividing a heterogeneous population into groups, or *strata*, of similar units and selecting simple random samples from each of the

Table 2.1   Employee Identification Numbers

1	A11401	41	B09087	81	G07704	121	B04256
2	P04181	42	B00073	82	K20760	122	K05170
3	N00004	43	J08742	83	W00124	123	R07790
4	C03253	44	W13972	84	T00141	124	G15084
5	D07159	45	S00856	85	M25374	125	C16254
6	M00079	46	A00166	86	K03911	126	R20675
7	S15552	47	S01187	87	W01718	127	G06144
8	G01039	48	D00022	88	T04877	128	T12150
9	P00202	49	Z01194	89	M22262	129	R07904
10	R22110	50	M32893	90	C00011	130	M24214
11	D00652	51	K00018	91	W23233	131	D00716
12	M06815	52	H16034	92	K10061	132	M27410
13	C09071	53	F08794	93	K11411	133	J07272
14	S01014	54	S71024	94	B05848	134	L02455
15	D05484	55	G00301	95	L06270	135	D06610
16	D00118	56	B00103	96	K08063	136	M31452
17	M28883	57	B29884	97	P07211	137	L25264
18	G12276	58	G12566	98	F28794	138	M10405
19	M06891	59	P03956	99	L00885	139	D00393
20	B26124	60	B00188	100	M26882	140	B52223
21	D17682	61	J21112	101	M49824	141	M16934
22	B42024	62	J08208	102	R05857	142	M27362
23	K06221	63	S11108	103	L30913	143	B38384
24	C35104	64	M65014	104	B46004	144	H08825
25	M00709	65	M07436	105	R03090	145	S14573
26	P00407	66	H06098	106	H09185	146	B23651
27	P14580	67	S18751	107	J18200	147	S27272
28	P13804	68	W00004	108	W14854	148	G12636
29	P23144	69	M11028	109	S01078	149	R04191
30	D00452	70	L00213	110	G09221	150	D13524
31	B06180	71	J06070	111	M17174	151	G00154
32	B69674	72	B14514	112	L04792	152	B19544
33	H11900	73	H04177	113	S23434	153	V01449
34	M78064	74	B26003	114	T02877	154	F09564
35	L04687	75	B26193	115	K06944	155	L09934
36	F06364	76	H28534	116	E14054	156	A10690
37	G24544	77	B04303	117	F00281	157	N02634
38	T20132	78	S07092	118	H07233	158	W17430
39	D05014	79	H11759	119	K06204	159	R02109
40	R00259	80	L00252	120	K06423	160	C18514

strata. This type of sampling is useful when one wishes to ensure that representative observations from each of several groups in the population are included in the sample.

Cluster sampling is based on subdividing the population into groups, or clusters, of units in such a way that it is convenient to randomly sample the clusters and then either randomly sample or completely enumerate all the observations in each of the sampled clusters. Cluster sampling is an important alternative to simple random sampling when, for example, clusters represent geographic locations or lots of product. More details on these and other alternatives to simple random sampling are given in the recommended readings at the end of this chapter.

Regardless of which sampling technique is used, the key idea is that the sample should be representative of the population under study. In experimental settings for which the sampling of populations or processes is not germane, the requirement that the data be representative of the phenomenon or the "state of nature" being studied is still pertinent and necessary. Statistics, as a science, seeks to make inferences about a population, process, or phenomenon based on the information contained in a representative sample or collection of observations.

In order to illustrate the procedures involved in randomly sampling a population, consider the information contained in Table 2.1. The table enumerates a portion of the world-wide sales force of a manufacturer of skin products. The employees are identified in the table by the order of their listing (1–160) and by their employee identification numbers. Such a tabulation might be obtained from a computer printout of personnel records. For the purposes of the study to be described, these 160 individuals form a population that satisfies several criteria set forth in the experimental protocol.

Suppose the purpose of a study involving these employees is to investigate the short-term effects of certain skin products on measurements of skin elasticity. Initial skin measurements are available for the entire population of employees (see Table 2.2). However, the experimental protocol requires that skin measurements be made on a periodic basis, necessitating the transportation of each person in the study to a central measuring laboratory. Because of the expense involved, the researchers would like to limit the participants included in the study to a simple random sample of 25 of the employees listed in Table 2.1.

Since the population of interest has been completely enumerated, one can use a random-number table (e.g., Table A1 of the Appendix) or a computer-generated sequence of random numbers to select 25 numbers between 1 and 160. One such random number sequence is

57, 77, 8, 83, 92, 18, 63, 121, 19, 115, 139, 96, 133,

131, 122, 17, 79, 2, 68, 59, 157, 138, 26, 70, 9.

Corresponding to this sequence of random numbers is the sequence of employee identification numbers which determines which 25 of the 160 employees are to be included in the sample:

B29884, B04303, G01039, W00124, K10061, G12276, S11108,

B04256, M06891, K06944, D00393, K08063, J07272, D00716,

K05170, M28883, H11759, P04181, W00004, P03956, N02634,

M10405, P00407, L00213, P00202.

Table 2.2   Skin Elasticity Measurements

1	31.9	41	36.0	81	36.3	121	33.0
2	33.1	42	28.6	82	36.3	122	37.4
3	33.1	43	38.0	83	41.5	123	33.8
4	38.5	44	39.1	84	33.0	124	35.3
5	39.9	45	39.4	85	36.3	125	37.5
6	36.5	46	30.6	86	36.3	126	31.6
7	34.8	47	34.1	87	30.9	127	33.1
8	38.9	48	40.8	88	32.3	128	38.2
9	40.3	49	35.1	89	39.2	129	31.4
10	33.6	50	34.1	90	35.2	130	35.9
11	36.4	51	36.3	91	35.1	131	37.6
12	34.4	52	35.1	92	33.9	132	35.5
13	35.7	53	35.0	93	42.0	133	34.2
14	33.9	54	39.0	94	35.1	134	34.0
15	36.6	55	34.0	95	34.5	135	31.3
16	36.0	56	35.3	96	35.0	136	32.6
17	30.8	57	36.0	97	35.1	137	34.9
18	31.1	58	34.7	98	35.7	138	35.3
19	37.6	59	39.8	99	36.4	139	35.1
20	35.7	60	35.8	100	39.6	140	35.7
21	29.6	61	35.7	101	35.2	141	32.3
22	37.3	62	39.8	102	37.2	142	38.1
23	31.4	63	36.4	103	33.3	143	36.8
24	31.6	64	36.1	104	33.7	144	38.7
25	34.6	65	37.7	105	37.8	145	40.0
26	34.6	66	32.3	106	34.4	146	35.4
27	33.7	67	35.6	107	36.9	147	34.0
28	30.9	68	38.2	108	31.8	148	34.3
29	34.6	69	39.0	109	35.3	149	32.8
30	37.0	70	34.3	110	38.1	150	30.7
31	35.3	71	40.6	111	34.1	151	34.4
32	36.3	72	37.4	112	35.8	152	34.3
33	31.8	73	37.3	113	33.3	153	35.8
34	38.2	74	36.9	114	33.8	154	37.5
35	34.6	75	29.0	115	36.4	155	34.4
36	36.0	76	39.0	116	36.9	156	35.8
37	40.8	77	33.7	117	35.3	157	31.9
38	39.2	78	32.9	118	37.0	158	36.9
39	33.4	79	33.8	119	33.5	159	34.4
40	34.0	80	36.2	120	40.3	160	30.1

   With periodic measurements taken on only this random sample of employees the researchers wish to draw conclusions about skin elasticity for the population of employees listed in Table 2.1. This statement suggests that a distinction must be made between measured characteristics taken on a population and those taken on a sample. This distinction is made explicit in the next section.

2.2 PARAMETERS AND STATISTICS

Summarization of data can occur in both populations and samples. Parameters, as defined in Exhibit 2.9, are constant population values that summarize the entire collection of observations. Parameters can also be viewed in the context of a stable process or a controlled experiment. In all such settings a parameter is a fixed quantity that represents a characteristic of interest. Some examples are:

- the mean fill level for twelve-ounce cans of a soft drink bottled at one plant,
- the minimum compressive strength of eight-foot-long, residential-grade, oak ceiling supports, and
- the maximum wear on one-half-inch stainless-steel ball bearings subjected to a prescribed wear-testing technique.

EXHIBIT 2.9

Parameters and Statistics. A parameter is a numerical characteristic of a population or a process. A statistic is a numerical characteristic that is computed from a sample of observations.

Parameters often are denoted by Greek letters, such as μ for the mean and σ for the standard deviation (a measure of the variability of the observations in a population), in order to reinforce the notion that they are (generally unknown) constants. Often population parameters are used to define specification limits or tolerances for a manufactured product. Alternatively they may be used to denote hypothetical values for characteristics of measurements that are to be subjected to scientific or engineering investigations.

In many scientific and engineering contexts the term parameter is used as a synonym for variable (as defined in the previous section). The term parameter should be reserved for a constant or fixed numerical characteristic of a population and not used for a measured or observed property of interest in an experiment. To emphasize this distinction we will henceforth use Greek letters to represent population parameters and Latin letters to denote variables. Sample statistics, in particular estimates of population parameters, also will generally be denoted by Latin letters.

The curve in Figure 2.1 often is used as a probability model, the *normal* distribution, to characterize populations and processes for many types of measurements. The *density* or height of the curve above the axis of measurement values, represents the likelihood of obtaining a value. Probabilities for any range of measurement values can be calculated from the probability model once the model parameters are specified. For this distribution, only the mean and the standard deviation are needed to completely specify the probability model.

The peak of the curve in Figure 2.1 is located above the measurement value 50, which is the mean μ of the distribution of data values. Since the probability density is highest around the mean, measurement values around the mean are more likely than measurement values greatly distant from it. The standard deviation σ of the distribution in Figure 2.1 is 3. For normal distributions (Section 5.3), approximately 68% of the measurement values lie between $\mu \pm \sigma$ (47 to 53), approximately 95% between $\mu \pm 2\sigma$ (44 to 56), and approximately 99% between $\mu \pm 3\sigma$ (41 to 59). The mean and the standard deviation are

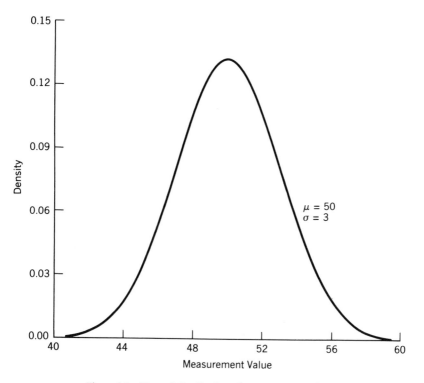

Figure 2.1 Normal distribution of measurement values.

very important parameters for the distribution of measurement values for normal distributions such as that of Figure 2.1.

Statistics are sample values that generally are used to estimate population parameters. For example, the average of a sample of observations can be used to estimate the mean of the population from which the sample was drawn. Similarly, the standard deviation of the sample can be used to estimate the population standard deviation. As we shall see in subsequent chapters, there are often several sample statistics that can be used to estimate a population parameter.

While parameters are fixed constants representing an entire population of data values, statistics are "random" variables and their numerical values depend on which particular observations from the population are included in the sample. One interesting feature about a statistic is that it has its own probability, or *sampling*, distribution: the sample statistic can take on a number of values according to a probability model, which is determined by the probability model for the original population and by the sampling procedure (see Exhibit 2.10). Hence, a statistic has its own probability model as well as

---

EXHIBIT 2.10

Sample Distribution. A sampling distribution is a theoretical model that describes the probability of obtaining the possible values of a sample statistic.

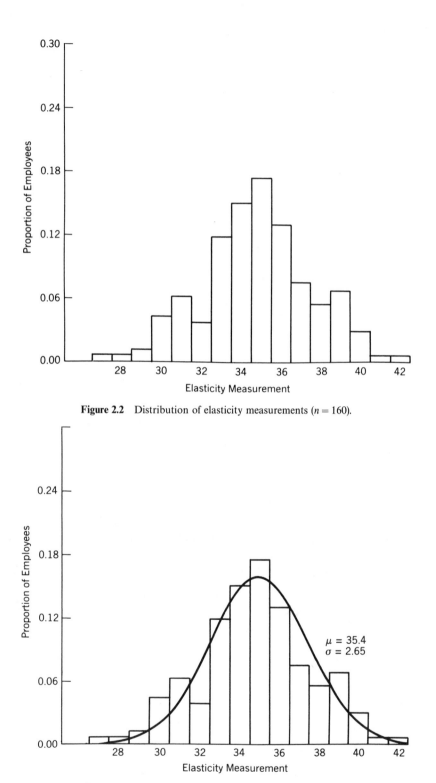

Figure 2.2 Distribution of elasticity measurements ($n = 160$).

$\mu = 35.4$
$\sigma = 2.65$

Figure 2.3 Normal approximation to elasticity distribution.

its own parameter values, which may be quite different from those of the original population.

If the skin elasticity measurements in Table 2.2 are rounded to whole numbers (integers), one obtains a bar chart (histogram; see Section 4.2) of the proportion of measurements having each integer value, as in Figure 2.2. The heights of the bars display a pattern similar in form to the curve in Figure 2.1. On the basis of these data one might postulate a normal probability model for the skin measurements.

Figure 2.3 shows a normal probability model that has the same mean ($\mu = 35.4$) and standard deviation ($\sigma = 2.65$) as the population of values in Table 2.2. Observe that the curve for the theoretical normal model provides a good approximation to the exact distribution of the population of measurements, represented by the vertical bars.

One of the features of a normal model is that averages from simple random samples of size n also follow a normal probability model with the same population mean but with a standard deviation that is reduced by a factor of $\sqrt{n}$ from that of the original population. Thus, averages of random samples of size 4 have standard deviations that are half that of the original population. Figure 2.4 shows the relationship between a normal probability model for individual measurements that have a population mean of $\mu = 35.4$ and a standard deviation of $\sigma = 2.5$ and one for the corresponding population of sample averages of size 4. Note that the latter distribution has $\mu = 35.4$ but $\sigma = 2.5/\sqrt{4} = 1.25$. The distribution of the averages is more concentrated around the population mean

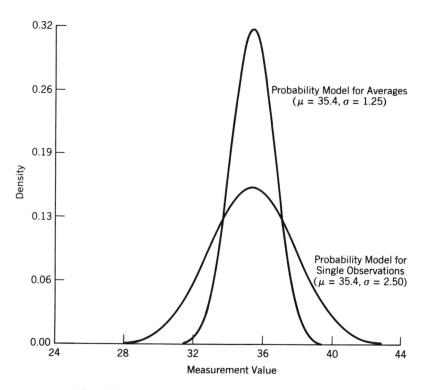

Figure 2.4 Comparison of theoretical normal distributions.

than is the distribution of individual observations. This indicates that it is much more likely to obtain a sample average that is in a fixed interval around the population mean than it is to obtain a single observation in the same fixed interval.

This discussion is intended to highlight the informational content of population parameters and to shed some light on the model-building processes involved in drawing inferences from sample statistics. The final section in this chapter focuses on one additional issue, which helps to distinguish statistical from mathematical problem solving.

2.3 MATHEMATICAL AND STATISTICAL MODELING

Models and model building are commonplace in the engineering and physical sciences. A research engineer or scientist generally has some basic knowledge about the phenomenon under study and seeks to use this information to obtain a plausible model of the data-generating process. Experiments are conducted to characterize, confirm, or reject models—in particular, through hypotheses about those models. Models take many shapes and forms, but in general they all seek to characterize one or more response variables, perhaps through relationship with one or more factors.

Mathematical models, as defined in Exhibit 2.11, have the common trait that the response and predictor variables are assumed to be free of specification error and measurement uncertainty. Mathematical models may be poor descriptors of the physical systems they represent because of this lack of accounting for the various types of errors included in statistical models. Statistical models, as defined in Exhibit 2.12, are approximations to actual physical systems and are subject to specification and measurement errors.

EXHIBIT 2.11

Mathematical Model. A model is termed *mathematical* if it is derived from theoretical or mechanistic considerations that represent exact, error-free assumed relationships among the variables.

EXHIBIT 2.12

Statistical Model. A model is termed *statistical* if it is derived from data that are subject to various types of specification, observation, experimental, and/or measurement errors.

An example of a mathematical model is the well-known fracture mechanics relation:

$$K_{IC} = \gamma S a^{1/2}, \tag{2.1}$$

where K_{IC} is the critical stress intensity factor, S is the fracture strength, a is the size of the flaw that caused the fracture, and γ is a constant relating to the flaw geometry. This formula can be utilized to relate the flaw size of a brittle material to its fracture strength. Its validity is well-accepted by mechanical engineers because it is based on the theoretical foundations of fracture mechanics, which have been confirmed through extensive experimental testing.

Empirical studies generally do not operate under the idealized conditions necessary for a model like equation (2.1) to be valid. In fact, it often is not possible to postulate a mathematical model for the mechanism being studied. Even when it is known that a model like equation (2.1) should be valid, experimental error may become a nontrivial problem. In these situations statistical models are important because they can be used to approximate the response variable over some appropriate range of the other model variables. For example, additive or multiplicative *errors* can be included in the fracture-mechanics model, yielding the statistical models

$$K_{IC} = \gamma S a^{1/2} + e \quad \text{or} \quad K_{IC} = \gamma S a^{1/2} e \tag{2.2}$$

where e is the error. Note that the use of "error" in statistical models is not intended to indicate that the model is incorrect, only that unknown sources of uncontrolled variation, often measurement error, are present.

A mathematical model, in practice, can seldom be proven with data. At best, it can be concluded that the experimental data are consistent with a particular hypothesized model. The chosen model might be completely wrong and yet this fact might go unrecognized because of the nature of the experiment; e.g., data collected over a very narrow range of the variables would be consistent with any of a vast number of models. Hence, it is important that proposed mathematical models be sufficiently "strained" by the experimental design so that any substantial discrepancies from the postulated model can be identified.

In many research studies there are mathematical models to guide the investigation. These investigations usually produce statistical models that may be partially based on theoretical considerations but must be validated across wide ranges of the experimental variables. Experimenters must then seek "lawlike relationships" that hold under a variety of conditions rather than try to build separate statistical models for each new data base. In this type of model generalization, one may eventually evolve a "theoretical" model that adequately describes the phenomenon under study.

REFERENCES

Text References

The distinction between populations and samples and between population parameters and sample statistics is stressed in most elementary statistics textbooks. The references at the end of Chapters 3 and 4 provide good discussions of these concepts. There are many excellent textbooks on statistical sampling techniques, including:

Cochran, W. G. (1977). *Sampling Techniques, Third Edition*, New York: John Wiley and Sons, Inc.

Schaeffer, R. L., Mendenhall, W., and Ott, L. (1979). *Elementary Survey Sampling, Second Edition*, North Scituate, MA: Duxbury Press.
  The first of these texts is a classic in the statistical literature. The second one is more elementary and a good reference for those not familiar with sampling techniques.

Mathematical and statistical modeling is not extensively covered in introductory statistics textbooks. The following text does devote ample space to this important topic:

Box, G. E. P., Hunter, W. G., and Hunter, J. S. (1978). *Statistics for Experimenters*. New York: John Wiley & Sons, Inc., Chapters 9 and 16.

EXERCISES

1 The Department of Transportation (DOT) was interested in evaluating the safety performance of motorcycle helmets manufactured in the United States. A total of 264

helmets were obtained from the major U.S. manufacturers and supplied to an independent research testing firm where impact penetration and chin retention tests were performed on the helmets in accordance with DOT standards.

(a) What is the population of interest?
(b) What is the sample?
(c) Is the population finite or infinite?
(d) What inferences can be made about the population based on the tested samples?

2 List and contrast the characteristics of population parameters and sample statistics.

3 A manufacturer of rubber wishes to evaluate certain characteristics of its product. A sample is made from a warehouse containing bales of synthetic rubber. List some of the possible candidate populations from which this sample can be taken.

4 It is known that the bales of synthetic rubber described in Exercise 3 are stored on pallets with a total of 15 bales per pallet. What type of sampling methodology is being implemented under the following sample scenarios?

(a) Five pallets of bales are randomly chosen; then eight bales of rubber are randomly selected from each pallet.
(b) Forty bales are randomly selected from the 4500 bales in the warehouse.
(c) All bales are sampled on every fifth pallet in the warehouse.
(d) All bales that face the warehouse aisles and can be reached by a forklift truck are selected.

5 Recall the normal distribution discussed in Section 2.2. What is the importance of $\mu \pm 3\sigma$?

6 The population mean and standard deviation of typical cetane numbers measured on fuels used in compression–ignition engines is known to be $\mu = 30$ and $\sigma = 5$. Fifteen random samples of these fuels were taken from the relevant fuel population, and the sample means and standard deviations were calculated. This random sampling procedure was repeated (replicated) nine times.

Replicate No.	n	Sample Mean	Sample Standard Deviation
1	15	32.61	4.64
2	15	28.57	6.49
3	15	29.66	4.68
4	15	30.09	5.35
5	15	30.11	6.39
6	15	28.02	4.05
7	15	30.09	5.35
8	15	29.08	3.56
9	15	28.91	4.88

Consider the population of all sample means of size $n = 15$. What proportion of means from this population should be expected to be between the 30 ± 5 limits? How does this sample of nine averages compare with what should be expected?

7   A research program was directed toward the design and development of self-restoring traffic-barrier systems capable of containing and redirecting large buses and trucks. Twenty-five tests were conducted in which vehicles were driven into the self-restoring traffic barriers. The range of vehicles used in the study included a 1800-lb car to a 40,000-lb intercity bus. Varying impact angles and vehicle speeds were used, and the car damage, barrier damage, and barrier containment were observed.

(a) What is an observation in this study?

(b) Which variables are responses?

(c) Which variables are factors?

8   Space vehicles contain fuel tanks that are subject to liquid sloshing in low-gravity conditions. A theoretical basis for a model of low-gravity sloshing was derived and used to predict the slosh dynamics in a cylindrical tank. Low-gravity simulations were performed in which the experimental results were used to verify a statistical relationship. It was shown in this study that the statistical model closely resembled the theoretical model. What type of errors are associated with the statistical model? Why aren't the statistical and theoretical models exactly the same?

9   Use the table of random numbers (Table A1) in the Appendix to draw 20 simple random samples, each of size $n = 10$, from the population of employees listed in Table 2.1. Calculate the average of the skin elasticity measurements (Table 2.2) for each sample. Graph the distribution of these 20 averages in a form similar to Figure 2.3. Would this graph be sufficient for you to conclude that the sampling distribution of the population of averages is a normal distribution? Why (not)?

10 Use the table of random numbers in the Appendix to choose starting points between 1 and 16 for twenty systematic random samples of the population of employees listed in Table 2.1. Select every 10th employee. Calculate the average of the skin elasticity measurements for each sample. Does a graph of the distribution of these averages have a similar shape to that of Exercise 9?

11 Simple random samples and systematic random samples often result in samples that have very similar characteristics. Give three examples of populations that you would expect to result in similar simple and systematic random samples. Explain why you expect the samples to be similar. Give three examples for which you expect the samples to be different. Explain why you expect them to be different.

PART I

Describing Variability

CHAPTER 3

Descriptive Statistics

In this chapter several statistics that are frequently used to summarize important characteristics of numerical data are introduced. Of particular interest in this chapter are:

- *traditional summary statistics such as the sample mean and the sample standard deviation,*
- *summary statistics that are less sensitive than the traditional ones to the presence of outliers in the data.*

Describing the numerical results of a sequence of tests is a fundamental requirement of most experimental work. This description may utilize a variety of quantitative values, such as averages or medians. In this chapter the calculation and display of summary statistics that describe the center (location) and spread (variation) of a set of data values are introduced.

3.1 TRADITIONAL SUMMARY STATISTICS

The most commonly used numerical measure of the center of a set of data values is the sample mean, or average. The sample mean is easy to calculate and is readily interpretable as a "typical" value of the data set. Although the sample mean is familiar to most readers of this text, we include its definition in Exhibit 3.1 for completeness. Inferences about "typical" values of populations are frequently made by using the sample mean to draw inferences on the population mean.

EXHIBIT 3.1

Sample Mean or Average. The sample mean or average of a set of data values is the total of the data values divided by the number of observations. Symbolically, if $y_1, y_2, \ldots, y_n$ denote n data values, the sample mean, denoted $\bar{y}$, is

$$\bar{y} = n^{-1}\sum y_i = \frac{y_1 + y_2 + \cdots + y_n}{n}.$$

Measures of spread are especially important as measures of typical or expected random variation in data values. It is not sufficient to describe a set of measurements only by reporting the average or some other measure of the center of the data. For example, it is not sufficient to know that bolts which are designed to have thread widths of 1.0 mm have, on the average, a width of exactly 1.0 mm. Half of the bolts could have a thread width of 0.5 mm and the other half a thread width of 1.5 mm. If nuts that are manufactured to fit these bolts all have thread widths of 1.0 ± 0.1 mm, none of the bolts would be usable with the nuts.

Perhaps the simplest measures of the spread of a set of data values are indicated by the extremes of the data set; i.e., the minimum and maximum data values. Although the most frequent use of the extremes of a data set is to indicate limits of the data, there are other important uses. For example, with the use of computers to analyze large numbers of data there is a great likelihood that mistakes in data entry will go unnoticed. Routine examination of the extremes of a data set often aid in the identification of errors in the data entry—e.g., percentages that exceed 100% or negative entries for variables that can only take on positive values.

Perhaps the two most common measures of the spread of a set of data values are the range (Exhibit 3.2) and the sample standard deviation (Exhibit 3.3). The range is quick and easy to compute and is especially valuable where small amounts of data are to be collected and a quick measure of spread is needed, such as by quality-control personnel on a production line. A disadvantage of the range is that it only makes use of two of the data values and can therefore be severely influenced by a single erratic observation.

EXHIBIT 3.2

Range. Range = maximum data value − minimum data value.

The sample standard deviation s is calculated from all the data values using either of the formulas in Exhibit 3.3. The sample standard deviation is based on *deviations*, or differences in the individual observations from the sample mean [see formula (a)]. These deviations, $y_i - \bar{y}$, are squared, the squares are averaged, and the square root is taken of the result. The squaring is performed because a measure of spread should not be affected by whether deviations are positive ($y_i > \bar{y}$) or negative ($y_i < \bar{y}$) but only by the magnitude (the size) of the difference from the mean. Averaging by dividing by $n - 1$ rather than n is done for technical reasons, which need not concern us here; in any event, for large sample sizes the difference in divisors has a negligible effect on the calculation. Because the square root has been taken, the sample standard deviation is measured in units identical to the original observations.

EXHIBIT 3.3

Sample Standard Deviation. The sample standard deviation of a set of data values $y_1, y_2, \ldots, y_n$ can be calculated in either of two equivalent ways:

(a) $s = \left\{ \sum (y_i - \bar{y})^2 / (n - 1) \right\}^{1/2}$

$\quad = \left\{ [(y_1 - \bar{y})^2 + (y_2 - \bar{y})^2 + \cdots + (y_n - \bar{y})^2] / (n - 1) \right\}^{1/2}$

(b) $s = \left\{ [\sum y_i^2 - n^{-1}(\sum y_i)^2] / (n - 1) \right\}^{1/2}$

Formula (b) is easier to use than formula (a) when calculations are made on a desk calculator rather than on a computer. The two expressions for the sample standard deviation are algebraically equivalent, and apart from roundoff error they give the same result.

In view of formula (a) one can interpret the sample standard deviation as a measure of the "typical" variation of data values around the sample mean. The larger the standard deviation, the more the variation in the data values. The smaller the standard deviation, the more concentrated the data values are around the mean. The sample standard deviation often is used as a measure of the precision of a measurement process.

The question of what constitutes excessive spread of a sample of data values is difficult to answer. The interpretation of a measure of spread is enhanced when it can be meaningfully compared either with a standard value such as a specification limit or with values previously obtained from similar measurements. Occasionally several data sets of similar measurements are to be compared and the relative magnitudes of the standard deviations provide valuable information on differences in variability of the processes that generated the data sets.

Figure 3.1 displays the results of a fuel-economy study of four diesel-powered automobiles. All four vehicles were tested under controlled conditions in a laboratory using dynamometers to control the speed and load conditions. For each vehicle, Figure 3.1 graphs the fuel economy (mi/gal) for eight tests using a single test fuel. The eight mileage

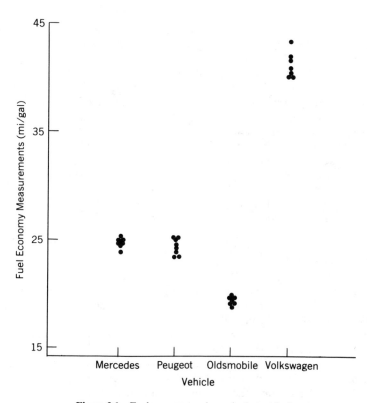

Figure 3.1   Fuel-economy values, single test fuel.

Table 3.1 Fuel Economy for Four Test Vehicles Using a Single Fuel

Obs. No.	Fuel Economy (mi/gal)			
	Mercedes	Peugeot	Oldsmobile	Volkswagen
1	23.8	23.9	18.8	40.1
2	24.7	23.3	18.9	40.1
3	25.1	24.1	19.2	40.5
4	24.7	23.3	19.2	40.4
5	24.5	24.0	19.3	40.9
6	24.8	24.7	19.7	41.8
7	25.0	25.0	19.7	42.0
8	25.0	25.0	19.8	43.3
	Summary Statistics			
Average	24.7	24.2	19.3	41.1
S.D.	0.4	0.7	0.4	1.1
Min	23.8	23.3	18.8	40.1
Max	25.1	25.0	19.8	43.3
Range	1.3	1.7	1.0	3.2

values for the Mercedes and the Peugeot exhibit similar distributions, but the values for the Oldsmobile and the Volkswagen are clearly different from the other two. The visual impressions left by Figure 3.1 are quantified in Table 3.1.

The averages listed in Table 3.1 depict the typical performance of the four vehicles. The Volkswagen is seen to have higher average fuel economy than the other three vehicles for these tests, with the Oldsmobile having a slightly lower average fuel economy than the Mercedes and the Peugeot.

The Oldsmobile test results are most consistent in that the eight fuel-economy values exhibit the least spread as indicated by the ranges and the sample standard deviations.

The statistics discussed in this section are the most commonly used measures of the center and spread of a set of data. They allow one to summarize a set of data with only a few key statistics. They are especially valuable when large numbers of data prevent the display of an entire data set or when the characteristics that are summarized by the sample statistics are themselves the key properties of interest in the data analysis.

3.2 ROBUST SUMMARY STATISTICS

While the summary statistics discussed in the previous section are commonly used to describe the center and spread of a set of data values, there is increasing awareness among data analysts that these statistics can be influenced by *outliers*, or extreme data values. Outliers can occur when equipment temporarily malfunctions, when errors are made in the experimental testing, when data values are erroneously entered in computer files, or when an occasional large or small data value occurs even without any unusual errors in experimentation. The use of the term "outlier" is not intended to imply that an error has been made in the experimental testing, only that an extreme data value has occurred—a value that could have an undue influence on the summary statistics calculated from the data set. In this setting outliers are sometimes referred to as "influential observations" (Chapter 24).

Figure 3.2 displays a second set of eight fuel-economy values for the same four vehicles that are shown in Figure 3.1. The only difference between the two sets of tests is that the values plotted in Figure 3.1 are from a single test fuel, while those plotted in Figure 3.2 are for eight different test fuels. The major difference between the two figures is the low test result for one of the fuels on the Volkswagen. This test is not the result of an error in the experimentation, since the test results were confirmed in subsequent test runs.

Since the low mileage reading for the Volkswagen is a valid experimental measurement, it should be included in any analysis of the test results; nevertheless, this datum does have a major effect on some of the summary statistics discussed in the previous section. Notice from Table 3.2 that the low fuel-economy value drastically increases the range and the sample standard deviation of the Volkswagen data compared with the corresponding values in Table 3.1.

The plot of fuel-economy values in Figure 3.2 clearly indicates that the excessive variability suggested by the summary statistics in Table 3.2 is due to the outlier and not to excessive vehicle variability. This erratic mileage value occurred because one of the test fuels could not be blended according to the experimental design; i.e., it would not produce satisfactory combustion in diesel engines. The test fuel then was altered with additives until it did produce satisfactory combustion. It is this fuel that produces the low mileage

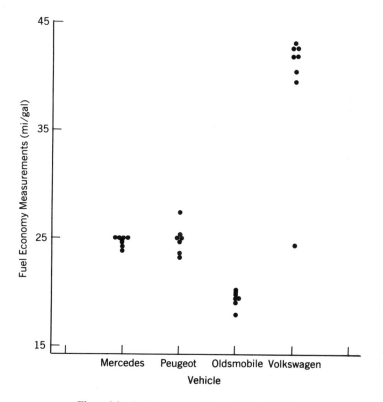

Figure 3.2 Fuel-economy values, eight test fuels.

Table 3.2   Fuel Economy for Four Test Vehicles Using Eight Different Fuels

Fuel No.	Fuel Economy (mi/gal)			
	Mercedes	Peugeot	Oldsmobile	Volkswagen
1	24.8	24.5	19.6	41.7
2	24.7	25.2	20.0	42.2
3	24.9	27.4	19.3	41.7
4	23.9	23.3	19.1	39.3
5	24.9	25.0	19.7	42.8
6	24.9	24.7	20.1	42.4
7	24.9	23.2	17.8	24.4
8	24.5	24.9	19.6	40.6
	Summary Statistics			
Average	24.7	24.8	19.4	39.4
S.D.	0.3	1.3	0.7	6.2
Min	23.9	23.2	17.8	24.4
Max	24.9	27.4	20.1	42.8
Range	1.0	4.2	2.3	18.4

reading on the Volkswagen and also the lowest mileage readings on the Peugeot and the Oldsmobile.

If one were to eliminate this fuel from the Volkswagen data, serious questions would arise when any comparisons were made among vehicles. First, is one biasing the overall fuel economy results for the Volkswagen? Second, should the mileage results for the Peugeot and the Oldsmobile be eliminated for this fuel, since it also produced the lowest readings on those vehicles? Third, although the test fuel was altered to produce combustion, all the data values are valid and reproducible. Should one ignore this information just to make the Volkswagen results more consistent? An alternative to the elimination of these data values is the use of measures of center and spread that are relatively insensitive (robust) to a few extreme observations.

Robust measures of the center and spread of a set of data are less affected by outliers than traditional summary statistics. They also help to identify when the traditional statistics may be unduly influenced by outliers. Two alternatives to the average as a

---

EXHIBIT 3.4

Sample Median. The sample median M is a number that divides ordered data values into two groups of equal size. It is determined as follows:

(i) order the data from the smallest to the largest values, denoting the ordered data values by

$$y_{(1)} \leqslant y_{(2)}, \leqslant \cdots \leqslant y_{(n)};$$

(ii) determine the median as

(a) $M = y_{(q)}$ if n is odd, where $q = (n+1)/2$.

(b) $M = [y_{(q)} + y_{(q+1)}]/2$ if n is even, where $q = n/2$.

---

measure of the center of a set of data values are the sample median (Exhibit 3.4) and the m-estimator (Exhibit 3.5). In the definition in Exhibit 3.4, we distinguish the raw data values $y_1, y_2, \ldots, y_n$ from the ordered data values $y_{(1)}, y_{(2)}, \ldots, y_{(n)}$. The sample median is either the middle data value (if n is odd) or it is the average of the two middle data values (if n is even), after the data have been ordered from smallest to largest. The sample median also is referred to as the 50th percentile of the data, since 50% of the data values are less than or equal to the median and 50% are greater than or equal to it. The sample median is therefore a measure of the center of a set of data values.

The sample mean is generally preferred to the sample median as a summary measure of a set of data values. It is intuitively more reasonable than the median in most data analyses because it is computed using all the data values whereas the median only uses the middle one or two data values. Since the sample median is only based on the middle one or two observations, however, it is less susceptible to the influence of outliers than is the sample mean.

m-estimators (see Exhibit 3.5) are weighted averages of data values. The weights w_i are determined so that typical or nonextreme data values are assigned weights of 1 and extreme data values are given weights less than 1. If all the observations in a data set are sufficiently well behaved, all the weights are equal to 1 and the m-estimate is equal to the sample mean. Calculation of m-estimators serves two important purposes. First, m-estimators are robust alternatives to averages, alternatives that, unlike the median, use all the data values. Second, as is discussed further below, the weights w_i assist in the identification of extreme data values.

EXHIBIT 3.5

m-Estimator. The m-estimator is defined as

$$m = \sum w_i y_i / \sum w_i$$

where

$$w_i = \begin{cases} -tv/(y_i - m) & \text{if} & y_i < m - tv, \\ 1 & \text{if} & m - tv \leqslant y_i \leqslant m + tv, \\ +tv/(y_i - m) & \text{if} & m + tv < y_i. \end{cases}$$

Here $t = 1.345$ or 1.5, and v is a robust measure of the spread in the data.

The "tuning constant" t is usually chosen to be 1.345 or 1.5, depending on how severely one wishes to limit the influence of extreme observations. The value $t = 1.345$ produces a more severe limitation on the influence of outliers than does $t = 1.5$; i.e., the weights w_i tend to be smaller for $t = 1.345$. The robust measure of spread, v, is often chosen to be the median absolute deviation (see Exhibit 3.6). Once t and v are selected, observations corresponding to deviations from the estimate m that are less than tv in absolute value are left unchanged; observations that differ by more than tv units from m are set equal to $w_i y_i$,

---

EXHIBIT 3.6

Median Absolute Deviation. $\text{MAD} = \text{median} \, (|y_i - M|)/0.6745$.

---

where $w_i = \pm tv/(y_i - m)$, depending on the sign of the deviation. In practice the m-estimate is calculated iteratively. The sample median (M) is usually used as an initial estimate of m.

A robust measure of the variation in the data values is the median absolute deviation (MAD), defined in Exhibit 3.6. The constant 0.6745 is used in the definition of the MAD estimator so that with large samples from "well-behaved" populations, such as those associated with normal probability distributions (see Section 2.2), the MAD estimate will provide a good estimate of the population standard deviation. Exhibit 3.7 gives the steps necessary in calculating the sample MAD.

EXHIBIT 3.7   CALCULATION OF MEDIAN ABSOLUTE DEVIATION

1. Determine the sample median M.

2. Calculate the n deviations from the median, $y_i - M$.

3. Take the absolute values of the deviations.

4. Reorder the absolute values of the deviations from smallest to largest.

5. Find the median of the ordered absolute values of the deviations: median($|y_i - M|$).

6. Divide median($|y_i - M|$) by 0.6745.

With this definition of the MAD, an iterative procedure for calculating the m-estimator is described in Exhibit 3.8.

EXHIBIT 3.8 m-ESTIMATION ITERATIVE PROCEDURE

1. Select a tuning constant (e.g., $t = 1.345$), the maximum number of iterations (e.g., $N = 25$), and the relative accuracy desired (e.g., $|m_{k+1} - m_k|/m_k < 0.001$).

2. Calculate the median (M) and the median absolute deviation (MAD), and set $k = 1$, $m_1 = M$, $v = $ MAD.

3. Using the current estimate m_k, update the weights w_i and the m-estimate (denote the new estimate m_{k+1}).

4. If the relative accuracy in the m-estimate is achieved or if the maximum number of iterations occurs, stop the iteration and report $k + 1$, m_{k+1}, and $|m_{k+1} - m_k|/m_k$.

5. Otherwise, replace the previous estimate m_k with the new one m_{k+1} and return to step 3.

---

m-estimates ordinarily are calculated on a computer because of the iterations required. Note that the weights w_i depend on the distance of each observation from the current estimate (m_k) of the center of the data relative to the robust measure of spread (v). The final weights are extremely useful in the identification of influential observations. Weights that are substantially less than 1.0 identify observations that are greatly distant from the center of the data and that therefore could distort the sample mean. m-estimators thus are valuable not only as alternative measures of the center of a set of data values but also as outlier diagnostics.

Quartiles (Exhibit 3.9) are another important set of measures of the spread of data values. They are usually unaffected by a few extreme measurements. The first quartile (25th percentile) is a numerical value that divides the (ordered) data so that 25% of the data values are less than or equal to it, the second quartile (50th percentile) is the sample median, and the third quartile (75th percentile) divides the data so that 75% of the

---

EXHIBIT 3.9 QUARTILES, SEMI-INTERQUARTILE RANGE

Quartiles Q_1, Q_2, and Q_3 are numerical values that divide a sample of observations into groups so that one-fourth (25%) of the data values are less than Q_1, half (50%) of the values are less than Q_2, and three-fourths (75%) of the values are less than Q_3. The second quartile Q_2 is the sample median M. Quartiles are determined as follows:

1. Order the data values:

$$y_{(1)} \leqslant y_{(2)} \leqslant \cdots \leqslant y_{(n)}.$$

2. Let $q = (n + 1)/2$ if n is odd and $q = n/2$ if n is even. Then $Q_2 = M =$

   (a) $y_{(q)}$ if n is odd,

   (b) $(y_{(q)} + y_{(q+1)})/2$ if n is even.

3. (a) If q is odd, let $r = (q + 1)/2$. Then

$$Q_1 = y_{(r)} \quad \text{and} \quad Q_3 = y_{(n+1-r)}.$$

   (b) If q is even, let $r = q/2$. Then

$$Q_1 = \frac{y_{(r)} + y_{(r+1)}}{2}$$

and

$$Q_3 = \frac{y_{(n+1-r)} + y_{(n-r)}}{2}.$$

The semi-interquartile range (SIQR) is

$$\text{SIQR} = \frac{Q_3 - Q_1}{2}.$$

(ordered) data values are less than or equal to its value. Half the difference between the third and the first quartile, referred to as the *semi-interquartile range* (Exhibit 3.9), is a measure of the spread of the data values that is quick to compute and is less affected by extremes in the data than is the sample standard deviation.

Table 3.3 presents robust summary statistics for the two fuel-economy data sets. Note that the estimates of both the center and the spread of the single-fuel data in Table 3.3(a) are very similar to those in Table 3.1. The eight-fuel estimates in Table 3.3(b) for the first three vehicles also are quite similar to the corresponding values in Table 3.2. The major difference between these results and those previously reported occurs for the Volkswagen. The low mileage value has little effect on the sample median or the *m*-estimate. Both of these estimates of the center of the data are similar to the estimates for the single-fuel data. Neither of these estimates is as seriously affected by the extreme mileage reading for the Volkswagen as is the sample mean. In addition, the MAD and semi-interquartile range estimates of the spread are close to the sample standard deviation reported for the single-fuel data.

The final weights for the *m*-estimate indicate how discrepant the individual observations are from the final *m*-estimate of the center of the data. The weights are displayed in Table 3.3(c) for the second data set. Note that the last three vehicles, especially the

Table 3.3   Robust Summary Statistics for Fuel-Economy Comparisons

	Fuel Economy (mi/gal)			
Statistic	Mercedes	Peugeot	Oldsmobile	Volkswagen
		(a) One Common Fuel		
Median	24.8	24.0	19.2	40.7
m-Estimate	24.8	24.2	19.3	41.0
MAD	0.37	1.04	0.59	0.89
Quartiles:				
First	24.5	23.3	18.9	40.1
Third	25.0	24.7	19.7	41.8
SIQR	0.25	0.70	0.40	0.85
		(b) Eight Fuels		
Median	24.8	24.8	19.6	41.7
m-Estimate	24.8	24.7	19.5	41.3
MAD	0.07	0.52	0.52	1.33
Quartiles:				
First	24.5	23.3	19.1	39.3
Third	24.9	25.0	19.7	42.2
SIQR	0.20	0.85	0.30	1.45

(c) m-Estimate Weights

	Weight			
Fuel No.	Mercedes	Peugeot	Oldsmobile	Volkswagen
1	1	1	1	1
2	0.82	1	1	1
3	1	0.26	1	1
4	0.11	0.49	1	0.90
5	1	1	1	1
6	1	1	1	1
7	1	0.46	0.40	0.11
8	0.31	1	1	1

Volkswagen, have been weighted less than 1.0 for fuel 7. There are also small weights on two of the vehicles for fuel 4. An examination of the data presented above in Table 3.2 reveals that fuel 4 produced the lowest fuel economy measurement on the Mercedes and virtually tied with fuel 7 for the lowest measurement on the Peugeot. Thus, examination of these weights has provided interesting insight about one of the fuels. This information might have gone unnoticed if only summary statistics or plots such as Figures 3.1 and 3.2 had been examined.

REFERENCES

Text References

Most textbooks on statistical methods include discussions of summary statistics. Traditional descriptive measures (mean, median, standard deviation, quartiles) receive the most extensive treatment. A selected list of textbooks intended for scientific and engineering audiences are

Guttman, I., Wilks, S. S., and Hunter, J. S. (1982). *Introductory Engineering Statistics*, New York: John Wiley and Sons, Inc., Chapter 5.

Ostle, B. and Malone, L. C. (1988). *Statistics in Research, Fourth Edition*, Ames, IA: Iowa State University Press, Chapter 4.

Snedecor, G. W. and Cochran, W. G. (1980). *Statistical Methods*, Ames, IA: Iowa State University Press, Chapter 2.

Steel, R. G. D. and Torrie, J. H. (1980). *Principles and Procedures of Statistics: A Biometrical Approach*, New York: McGraw-Hill Book Co., Chapter 2.

Robust estimation using procedures such as m-estimation are relatively new techniques and have not yet been incorporated into many elementary statistics textbooks. An informative overview of robust statistical procedures is

Hoaglin, D. C., Mosteller, F., and Tukey, J. W. (1983). *Understanding Robust and Exploratory Data Analysis*, New York: John Wiley and Sons, Inc.

An informative survey of texts on statistics and data analysis is

Hahn, G. J. and Meeker, W. Q. (1984). "An Engineer's Guide to Books on Statistics and Data Analysis," *Journal of Quality Technology*, **16**, 196–218. A condensed version of this survey, without annotated comments, appears in *Chemtech*, **15** (1985), 175–177.

Data References

The fuel-economy data discussed in this chapter are taken from

Hare, C. T. (1985). "Study of the Effects of Fuel Composition, Injection, and Combustion System Type and Adjustment on Exhaust Emissions from Light-Duty Diesels," The Coordinating Research Council, Inc., CRC-APRAC Project No. CAPE-32–80, San Antonio, TX: Southwest Research Institute.

EXERCISES

1 Outliers, or extreme data values, can cause problems in the analysis of a phenomenon or experiment. One way this can occur is if they are not detected during the analysis of the experimental data.

(a) List four ways in which outliers can occur in experimental data.

(b) List three ways in which you can check your data for possible outliers.

2 Algebraically show the equivalence between formulas (a) and (b) for the sample standard deviation.

3 The diameters (mm) of 25 randomly selected piston rings are given below. Are the observed differences in the values of these diameters due to location, dispersion, or both?

76.7   78.7   74.5   78.9   79.6   74.4   79.7   75.4   78.7   79.5   75.1   79.9   75.3

79.8   76.8   79.6   76.0   74.2   79.0   75.9   79.0   73.8   79.2   75.9   79.2

4   A solar energy system was designed and constructed for a North Carolina textile
company. The system was used to preheat boiler feedwater before injection in the
process steam system. The energy delivered to the feedwater was observed for a
48-hour period. The values below represent the energy rate in kbtu/hr:

493   500   507   500   501   489   495   508   490   495   511   498   490   507

488   499   509   499   494   490   490   489   515   493   505   497   490   507

497   492   503   495   513   495   492   492   510   501   530   504   501   491

504   507   496   492   496   511

Calculate the following descriptive statistics for the observed energy rates:

(a) Average.
(b) Sample standard deviation.
(c) Minimum and maximum.
(d) Range.
(e) Median.
(f) MAD.
(g) Quartiles—first, second, and third.
(h) SIQR.

Which is a better measure of the center of these energy rates, the average or the
median? Why? How much does a typical energy rate differ from the average in the
above data set?

5   A study was conducted to examine the explosibility of M-1 propellant dust and to
determine the minimum energy of electrostatic discharge needed to induce an
explosion for dust concentrations of 0.4 g/liter. The ignition energies (joules) were
observed for a series of eighteen experiments:

.23   .30   .35   .33   .64   .36   .27   .20   .23   .31   .22   .21   .16   .24   .22   .27

.20   .25

Calculate the descriptive statistics listed in Exercise 4 for these data. Which of the
measures of center do you prefer for this data set? Why?

6   Comment on the variation of ignition energy data presented in Exercise 5. How do
the range and the sample standard deviation compare in describing the dispersion
of the data? Compare the use of these two measures with that of the median absolute
deviation (MAD). Is the MAD preferable to the other two for these data? Why?

7   A study was conducted to determine the feasibility of using foam sulphur as a

material for shock-isolating large underground structures. A family of rigid sulfur foams was produced, and compressive strengths (psi) were measured on ten random samples of this foam. The calculated m-estimate and the associated weights are given below:

$$m\text{-estimate} = 292.53,$$

$$\text{weights} = 0.07, 1, 1, 0.52, 1, 1, 1, 1, 0.92, 1.$$

What do the weights tell you about the data? What would have been the effect of using a larger tuning constant?

8 Perform the iterations needed to calculate the m-estimate for the Mercedes vehicle evaluated in the fuel-economy study in Section 3.2. What is your interpretation of the m-estimate and weights for these data?

9 Listed below are red-blood-cell counts for patients with various liver diseases. The patients are categorized according to their initial diagnosis upon entering the hospital. If you were to compare location and dispersion measures across diagnosis categories, would you prefer traditional or robust measures? Why?

RED BLOOD CELL COUNTS

Cirrhosis	Hepatitis	Tumor	Other
18	14	3	14
66	17	1	36
18	10	3	4
7	5	6	13
15	27	4	4
6	11	7	5
4	30		3
8	18		6
5	4		11
28	39		6
4	25		33
49			
33			

10 Calculate m-estimates for the red-blood-cell counts in each of the four diagnosis categories in Exercise 9. Examine the weights for each diagnosis category, and make a judgment about whether any of the counts are substantially different from the final value of the m-estimate.

CHAPTER 4

Graphical Displays of Data

In this chapter we introduce several techniques for displaying data, including:

- *raw data displays (point plots, scatterplots, sequence plots),*
- *tabulations and graphical summaries of distributions of data, and*
- *graphical displays used in statistical process control.*

Graphical techniques are extremely useful in a "first look" exploratory phase of experimental data analysis. In this exploratory phase one is seeking to uncover fundamental characteristics of a phenomenon or process. Often one does not know what kind of properties the data will exhibit, perhaps not even the likely magnitudes of the responses. The value of data displays in these situations is that one can frequently gain fundamental insight into models, which can be used to describe the data-generating process, or into relationships between experimental factors and responses. For ongoing industrial processes, data displays provide a visual assessment of the stability (statistical control) of the process.

Graphics not only provide pictorial representations of data; they also assist in the interpretation of formal statistical inference procedures. In this chapter we present a few of the more useful techniques for displaying data. There are many variations on the techniques presented and many more types of data displays that are appropriate for specific uses. The only important limitations on the use of graphics are the creativity of the data analyst and the clarity with which the salient features of the data can be presented.

4.1 RAW DATA DISPLAYS

Table 4.1 reports the partial results of an experiment that was conducted to evaluate allergic reactions of laboratory animals to certain drugs. There are 45 skin thickness measurements taken on one group of animals (the control group) 48 hours after receiving a placebo drug. Another 45 measurements were taken on a second group of animals 48 hours after the administration of an experimental drug. An allergic reaction is indicated by swelling. At the beginning of this experiment skin thickness measurements were believed to be an adequate substitute for a direct measurement of swelling.

The average skin thickness measurement for the 45 observations taken on the control

Table 4.1   Skin Thickness Measurements (mm)

Control Group

.40 .34 .41 .37 .40 .41 .38 .38 .42 .40 .41 .34 .42 .40 .41
.40 .40 .38 .37 .42 .34 .40 .41 .41 .38 .40 .42 .42 .40 .41
.41 .40 .42 .42 .41 .38 .42 .42 .41 .38 .40 .41 .37 .40 .41

Test Group

.40 .40 .38 .42 .40 .43 .38 .38 .43 .40 .41 .40 .43 .38 .40
.38 .40 .38 .49 .41 .44 .40 .40 .41 .38 .38 .41 .40 .47 .38
.40 .42 .38 .41 .40 .44 .38 .42 .40 .38 .43 .40 .50 .38 .40

group is 0.40 mm. Forty-eight hours after the drug was administered, the skin thickness measurements on the second group of animals averaged 0.41 mm. The sample standard deviations were, respectively, 0.02 and 0.03 mm. While there appears to be an indication of swelling, since the 48-hour measurements have a larger average than those for the control group, the difference is not statistically significant; i.e., a swelling of 0.01 mm could have occurred simply by chance.

This conclusion seems reasonable from the numerical summaries provided by the data, but they do not identify a potentially important characteristic, which is clearly suggested by a plot of the data. Figure 4.1 reveals that the two data sets are *skewed*. Both sets of measurements cluster in the interval 0.38–0.42 mm. Several of the control-group measurements fall below this interval and none are larger than its upper limit. In contrast, none of the test-group measurements are less than 0.38 mm, but several are larger than 0.42 mm. These features of the data suggest that the animals in the test group might be suffering from the effects of swelling, perhaps that all the animals in the test group experience varying degrees of swelling. The skewness toward high values in the test-group data also may indicate that some of the animals are suffering from extreme swelling while most of the others are only experiencing minor or moderate amounts of swelling. This is an important finding that is not discernable from a routine examination of the averages and standard deviations.

Just as important from an experimental perspective is the information the plots contain about this experimental design. One cannot determine from either Table 4.1 or Figure 4.1 the amount of swelling experienced by any single animal, since different animals were used in the control and test groups. Comparison of Figures 4.1*a* and *b* strongly suggests that the degree of swelling on an individual animal is the relevant unit of measurement of the allergic effect of this drug, not simply the average skin thickness of groups of animals before and after treatment. A more sensitive statistical evaluation of the allergic effect of the drug can be obtained by designing the experiment so that each animal is its own control and the basic data are differences in skin thickness from two locations on each animal, one location receiving the experimental drug and the other a placebo.

Figures 4.1*a* and *b* are examples of *point plots* (see Exhibit 4.1). Point plots (also called

---

EXHIBIT 4.1   POINT PLOTS

1. Construct a horizontal axis covering the range of data values.
2. Vertically stack repeated data values.

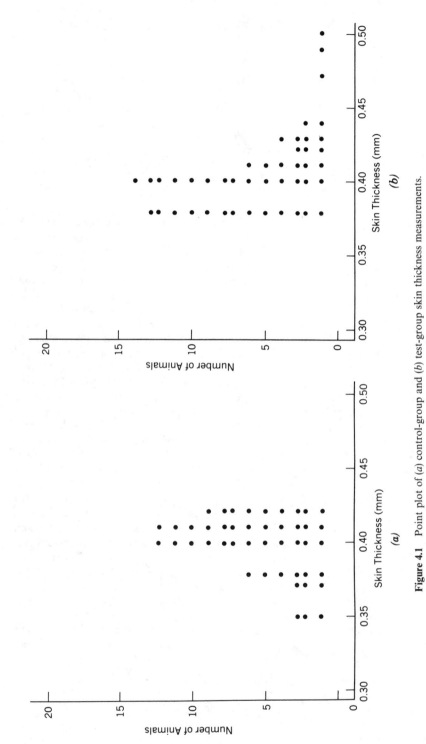

Figure 4.1 Point plot of (*a*) control-group and (*b*) test-group skin thickness measurements.

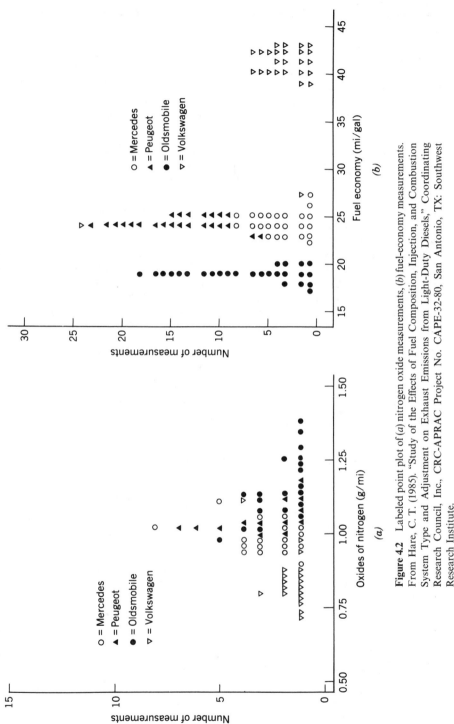

Figure 4.2 Labeled point plot of (*a*) nitrogen oxide measurements, (*b*) fuel-economy measurements. From Hare, C. T. (1985). "Study of the Effects of Fuel Composition, Injection, and Combustion System Type and Adjustment on Exhaust Emissions from Light-Duty Diesels," Coordinating Research Council, Inc., CRC-APRAC Project No. CAPE-32-80, San Antonio, TX: Southwest Research Institute.

"dot diagrams") are graphs of data values along a horizontal measurement scale. Repeated points are stacked vertically.

A second example of a point plot makes use of data from the fuel-economy study in the last chapter. Plotted in Figure 4.2 are 100 measurements collected on nitrogen oxides (nitric oxide, nitrogen dioxide, nitrous oxide) in exhaust emissions and on fuel economy. The plotting symbol identifies the vehicle from which the plotted point was obtained.

As with the previous example, if only summary statistics are reported for these data sets, valuable information will be missing. The averages and the standard deviations are

	Nitrogen Oxides	Fuel Economy
Average	1.00	26.95
Standard deviation	0.13	8.38

While one can interpret these statistics and make some useful inferences from them, the distributions of the two sets of data are quite different and lead to different recommendations about the usefulness of the summary statistics.

The nitrogen oxides measurements tend to cluster in the center of Figure 4.2a, with a symmetric pattern of fewer observations as the measurements get further from the average in either direction. Note that the mean, 1.00, is a good representation of the center of these data values, although there are clear differences in the measurements obtained from the vehicles. Differences in the fuel-economy measurements in Figure 4.2b for the four vehicles result in two or three distinct clusters of data values. Use of the

Table 4.2 Summary Statistics on a Metallic Oxide Chemical Process for 18 Successive Lots of Raw Materials*

Lot Number	Average	Standard Deviation
1	4.12	.07
2	4.09	.12
3	3.52	.04
4	4.18	.07
5	3.34	.27
6	4.11	.09
7	3.84	.09
8	4.00	.11
9	4.02	.35
10	3.90	.11
11	4.18	.05
12	3.20	.00
13	4.10	.28
14	3.84	.23
15	3.91	.09
16	4.04	.27
17	3.02	.88
18	3.89	.41

*Bennett, C. A. (1954). "Effect of Measurement Error on Chemical Process Control." *Industrial Quality Control*, 11, 17–20. Copyright American Society for Quality Control, Inc., Milwaukee, WI. Reprinted by permission.

mean, 26.95, for these data would be far less informative than for the nitrogen oxides data because of the separation of the observed data values into such distinct clusters. The point plot clearly conveys the differences in the distributions of the fuel-economy values for the several vehicles and would alert one to potential problems with the interpretation of summary statistics.

An important two-dimensional plot of individual data values for measurements on pairs of variables is the *scatterplot*. A scatterplot (see Exhibit 4.2) is constructed by plotting pairs of observations (x_i, y_i).

EXHIBIT 4.2 SCATTERPLOTS

1. Construct horizontal and vertical axes that cover the ranges of the two variables.

2. Plot (x_i, y_i) points for each observation in the data set.

Table 4.2 lists summary statistics for a chemical process involving a metallic oxide. The summary statistics are calculated for 18 successive lots of raw materials. It is common practice to plot control charts (Section 4.5) of the summary statistics. One type of control chart used for this purpose is a scatterplot of the summary statistics versus their respective lot numbers.

Figure 4.3 is a control chart of the sample standard deviations versus lot numbers. This

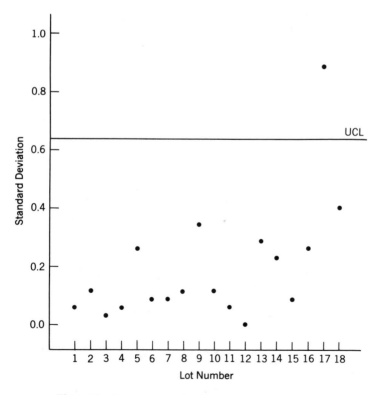

Figure 4.3 Sequence plot of within-lot standard deviations.

control chart allows one to examine the stability of the process through a visual inspection of these estimates of process variability. Superimposed on the plot is an upper control limit (UCL). The UCL provides an upper limit for the variation that should be expected in the sample standard deviations. The large value for the standard deviation calculated from lot 17 is clearly exhibited in Figure 4.3, perhaps more clearly than is apparent from Table 4.2. The procedure for calculating the UCL is explained in Section 4.5.

Control charts are examples of a special type of scatterplot, which is referred to as a *sequence plot* (see Exhibit 4.3). With this type of plot the variable on the horizontal axis relates to a logical or physical sequencing of the data. Frequently this sequencing is chronological: for example, the order in which experimental runs are conducted. Process control charts are often time-ordered plots of averages or sample standard deviations.

EXHIBIT 4.3   SEQUENCE PLOT

A sequence plot is a scatterplot in which there is a natural or chronological ordering to the values on the horizontal axis. Points on a sequence plot are often connected with straight lines for emphasis.

Figure 4.4 is an emphasized sequence plot for the sample standard deviations given in Table 4.2. The line segments connecting the consecutive points are often used as in this

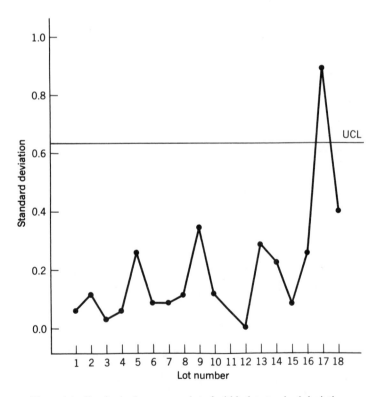

Figure 4.4   Emphasized sequence plot of within-lot standard deviations.

plot to highlight trends occurring among several points and to emphasize outlying observations. The lines connecting the points in Figure 4.4 more dramatically portray the departure of the standard deviation for lot 17 from the other plotted points than does simply plotting the points. There is also a clearer representation of the variability of the other plotted points.

We stress that emphasized sequence plots should be used with caution. The line segments should not be used to describe a continuous trend in successive points (i.e., linear, quadratic, etc.), since the labeling on the horizontal axis is often arbitrary. In this example, the lot numbers are arbitrary; they do not correspond to a time ordering or any other natural sequencing of the standard deviations. The emphasis is only to make characteristics of the plots more pronounced for visual clarity.

A useful variant of point plots and scatterplots involves labeling the plotted points. In a labeled plot the plotting symbols denote the values of an additional variable. Thus the labeling allows the inclusion of additional information without increasing the dimensionality of the graph. In the labeled point plot shown in Figure 4.2 plotting symbols were used to denote data values for each vehicle type ($\bigcirc$ = Mercedes, $\blacktriangle$ = Peugeot, etc.).

4.2 TABULATING AND DISPLAYING DISTRIBUTIONS

The term *distribution* (see Exhibit 4.4) is used throughout this text to refer to the possible values of a variable along with some measure of how frequently the values occur in a sample or a population. Point plots are effective for graphically displaying a small or moderate-sized distribution. Histograms are alternatives to point plots and are among the most common displays for illustrating the distribution of a set of data. They are especially useful when large numbers of data must be processed. Programs for constructing histograms are available in most of the major statistical software libraries on mainframe computers and are rapidly being introduced into microcomputer software libraries.

EXHIBIT 4.4

Distribution. A tabular, graphical, or theoretical description of the values of a variable using some measure of how frequently they occur in a population, a process, or a sample.

Histograms (see Exhibit 4.5) are constructed by dividing the range of the data into several intervals (usually of equal length), counting the number of observations in each interval, and constructing a bar chart of the counts. A by-product of the construction of the histogram is the *frequency distribution*, which is a table of the counts or frequencies for each interval. Frequency distributions and histograms for the nitrogen oxides and fuel-economy data of the previous section are contained in Table 4.3 and Figure 4.5. Note that in Figure 4.5 the heights of the vertical bars equal the data counts in each interval while the width of the bars corresponds to the intervals selected for the measurements on each variable.

Both histograms and the tables of counts that accompany them are sometimes referred to as frequency distributions, since they show how often the data occur in various intervals of the measured variable. The intervals for which counts are made are generally chosen to be equal in width, so that the size (area) of the bar or count is proportional to the number of observations contained in the interval. Selection of the interval width is usually made by

Table 4.3   **Frequency Distribution for Nitrogen Oxides and Fuel Economy Data Sets**

NO$_x$ Emissions* (g/mi)	Frequency	Fuel Economy* (mi/gal)	Frequency
0.70–0.74	2	16–18	1
0.74–0.78	3	18–20	21
0.78–0.82	5	20–22	4
0.82–0.86	8	22–24	8
0.86–0.90	5	24–26	39
0.90–0.94	0	26–28	3
0.94–0.98	11	28–30	0
0.98–1.02	18	30–32	0
1.02–1.06	16	32–34	0
1.06–1.10	10	34–36	0
1.10–1.14	12	36–38	0
1.14–1.18	2	38–40	2
1.18–1.22	0	40–42	11
1.22–1.26	5	42–44	11
1.26–1.30	1		
1.30–1.34	0	Total	100
1.34–1.38	2		
Total	100		

*Intervals include lower limits, exclude upper ones.

simply dividing the range of the data by the number of intervals desired in the histogram or table. Depending on the number of observations, between 8 and 20 intervals are generally selected—the greater the number of observations, the greater the number of intervals.

When the sample size is large, it can be advantageous to construct *relative-frequency* histograms. In these histograms and frequency distributions either the proportions (counts/sample size) or the percentages (proportions $\times$ 100%) of observations in each interval are calculated and graphed, rather than the frequencies themselves. Use of relative frequencies (or percentages) in histograms ensures that the total area under the bars is equal to one (or 100%). This facilitates the comparison of the resultant distribution with that of a theoretical probability distribution, where the total area of the distribution also equals one.

EXHIBIT 4.5   FREQUENCY DISTRIBUTIONS AND HISTOGRAMS

1. Construct intervals, ordinarily equally spaced, which cover the range of the data values.

2. Count the number of observations in each of the intervals. If desirable, form proportions or percentages of counts in each interval.

3. Clearly label all columns in tables and both axes on histograms, including any units of measurement, and indicate the sample or population size.

4. For histograms, plot bars whose

   (a) widths correspond to the measurement intervals, and

   (b) heights are (proportional to) the counts for each interval (e.g., heights can be counts, proportions, or percentages).

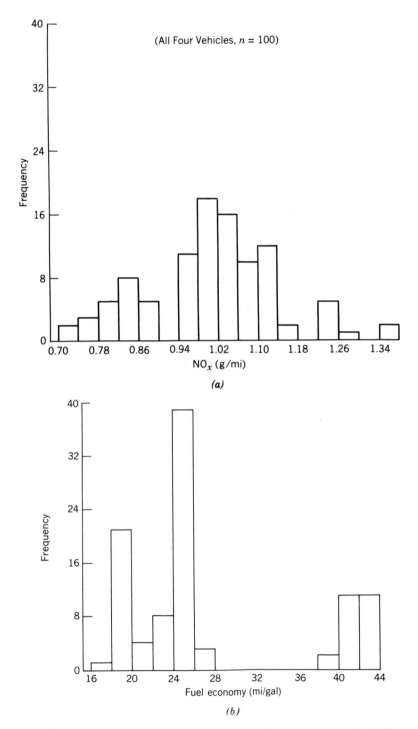

Figure 4.5 (a) Histogram of nitrogen oxide measurements (all four vehicles, $n = 100$). (b) Histogram of fuel-economy measurements.

A plot that closely resembles a histogram is the *stem-and-leaf* display (see Exhibit 4.6). This plot is a rough-and-ready graph that displays features of both a frequency distribution and a histogram. The basic idea is to use the digits of the data to illustrate its range, distributional shape, and density or concentration. Each observation is split into leading digits and trailing digits. All the leading digits are sorted and listed to the left of a vertical line. The trailing digits then are written in the appropriate location to the right of the vertical line. The leading digits are called *stems*, and the trailing digits *leaves*.

For example, consider a value, say 1.12, for NO_x. The stem could be chosen to be 1.1 and its leaf 2. Alternatively, the stem could be 1 and the leaf 0.12 or, rounded to a single digit, 1. For this data set, two-digit stems are needed to provide a sufficient number of stems so relevant characteristics of the data are visually apparent. Treating all the NO_x data in this manner yields the stem-and-leaf diagram shown in Table 4.4.

The *depth* of the stem is a cumulative count of the number of data values from the closer end of the data. The depths increase from each end to the middle of the data. The line containing the middle value (i.e., the median) contains a count of the number of leaves on its line in place of its depth. Depth values are included in stem-and-leaf plots primarily to allow easy identification of medians, quartiles, and other percentiles of a

Table 4.4   Stem-and-Leaf Plot, Nitrogen Oxides Data*

Depth	Stem		Leaves
6	0.7		124679
23	0.8		01112233445567789
41	0.9	Q	455556677778888899
37	1.0	MQ	00011111111222222233444455566677778889
22	1.1		00011111222256
8	1.2		234558
2	1.3		47

*Q: interval includes a quartile; M: interval includes the median.

Table 4.5   Extended Stem-and-Leaf Plot, Nitrogen Oxides Data*

Depth	Stem		Leaves
3	0.7		124
6			679
16	0.8		0111223344
23			5567789
24	0.9		4
41		Q	55556677778888899
24	1.0	M	00011111111222222222334444
35		Q	5556667778889
22	1.1		000111112222
10			56
8	1.2		234
5			558
2	1.3		4
1			7

*Q: interval includes a quartile; M: interval includes the median.

data set. Once the statistics are identified, depth values are usually removed from the display.

By replacing the depth for the median stem with the number of leaves for that stem it is easy to find the exact value for the median. For example, the median for the nitrogen oxides data is the average of the 50th and 51st ordered NO_x values. The depth of the line preceding the median stem in Table 4.4 is 41. Thus the median is the average of the 9th and 10th of the 37 ordered NO_x values in the median stem. Both of these values are 1.01; consequently, the median of the nitrogen oxides emissions is 1.01 g/mi. Quartiles are found similarly, using the stems in which the 25th ordered data values from each end of the distribution are located. The stems containing the median and the quartiles are highlighted by placing an "M" or a "Q", respectively, on the vertical line separating the stem from the leaves.

EXHIBIT 4.6   STEM-AND-LEAF PLOTS

1. Choose a suitable unit to divide the data values into leading and trailing digits.

2. List the sets of all possible leading digits in a column in numerically ascending order.

3. For each observation, list the trailing digit(s) on the same line as the appropriate leading digit(s).

4. Add the depths to each stem by counting the number of data values on each line and all lines closer to the nearest end of the data (for the median line, record the number of data values).

5. Identify the stems containing the median and the quartiles by placing the symbols "M" and "Q", respectively, on the vertical bar separating the stems from the leaves.

The lengths of stem-and-leaf plots can be altered to accommodate samples with narrow or wide ranges. For example, the nitrogen oxides data has a range of only

Table 4.6   Extended Stem-and-Leaf Plot, Fuel-Economy Data*

Depth	Stem		Leaves
0	1		
0			
0			
1			7
22			888999999999999999999
26	2	Q	0000
34			23333333
39		M	444444444444444444444444455555555555555
27		Q	677
24			
24	3		
24			
24			
24			
24			99
22	4		00000001111
11			22222223333

*Q: interval includes a quartile; M: interval include the median.

0.66 g/mi. In order to prevent the data from being concentrated in too few groups, each stem can be split in two: one for leaves 0 to 4 and the other for leaves 5 to 9. This is illustrated in Table 4.5. Alternatively, when the range of the data is too wide, a solution is to truncate the data by rounding some digits on the right. This has been done with the fuel-economy data in Table 4.6. The stems in this table represent the tens digit, and the leaves represent the units digit. The tenths digit has been eliminated by rounding all the fuel-economy values to the nearest integer. There are five stems per tens digit. The five stems represent the following five groupings of the units digits: (0, 1), (2, 3), (4, 5), (6, 7), (8, 9).

The stem-and-leaf plot provides a compact display of a data set. All the numbers are illustrated without necessarily having to list all the digits. Selected summary statistics such as the median and the quartiles can be obtained using the stem depth values. Also, the length of each row provides a visual impression of the number of observations in that row. A stem-and-leaf plot thereby provides a horizontal histogram in which the bars of the histogram are replaced by the values of the leaves. This can be visualized by comparing the shape of the histograms in Figure 4.5 with the stem-and-leaf plots in Table 4.5 and 4.6.

4.3 BOX PLOTS

A simple plot that provides a great amount of information on characteristics of a data set is the *box plot*. Box plots graphically describe the bulk of the data by a rectangle whose lower and upper limits are, respectively, the first and third quartiles. The median, or second quartile, is designated by a horizontal line segment inside the rectangle. The average is denoted by a symbol; e.g., x. In its simplest version, the width of the box is arbitrary and is left to the choice of the data analyst.

Vertical lines, sometimes dashed, extend from the middle of the ends of the box to an upper and a lower *adjacent value*. The upper adjacent value is the largest observation that is less than or equal to the third quartile plus $3 \cdot \text{SIQR}$, where SIQR is the semi-interquartile range. The lower adjacent value is the smallest observation that is greater than or equal to the first quartile minus $3 \cdot \text{SIQR}$. Any sample values beyond the adjacent values are plotted separately on the graph because they identify potential outliers. We summarize the construction of box plots in the instructions in Exhibit 4.7 and in Figure 4.6.

EXHIBIT 4.7   BOX PLOTS

1. Calculate the average and the quartiles of the data set.

2. Calculate the semi-interquartile range, $\text{SIQR} = (Q_3 - Q_1)/2$.

3. Draw a box having

   (a) upper edge at the third quartile.

   (b) lower edge at the first quartile,

   (c) a convenient width,

   (d) a horizontal line identifying the median,

   (e) an x identifying the average.

4. Extend vertical lines from the center of each edge of the box to the most extreme data values that are no farther than $3 \cdot \text{SIQR}$ from each edge.

5. Plot all points that are more extreme than the vertical lines.

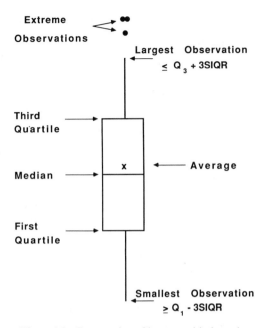

Figure 4.6   Construction of boxes used in box plots.

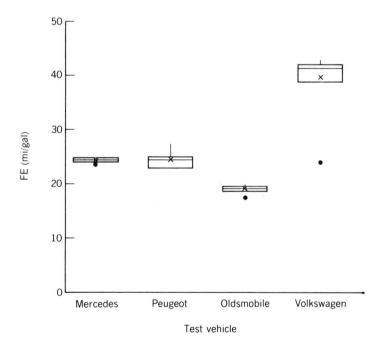

Figure 4.7   Box plot of fuel-economy (FE) measurements.

Figure 4.7 is a comparative box plot for the fuel-economy data from Table 3.2. Ordinarily one would want more than eight observations from which to form a box plot; however, we present this plot so its features can be compared with the summary statistics in Tables 3.2 and 3.3. Another reason we present this comparative box plot is to point out that as long as interpretations are drawn with due recognition of the sampling procedure and the sample size, box plots can be visually compelling even with small sample sizes.

The box plot makes several features of the test results visually apparent, more so than they might be from an examination of the statistics in Table 3.2 or 3.3. First, note the consistency of the test results for the Mercedes and to a lesser extent for the Oldsmobile. These two test vehicles have the lowest measures of spread in Table 3.3(b). Second, the Peugeot fuel-economy values have about the same location values as the Mercedes, but the individual data values are more variable. Third, the reason for the excessive variation in the results for the Volkswagen is clearly indicated by the single extremely low fuel-economy reading. Note too that this low test result causes the average to be pulled toward the lower quartile of the data, whereas the median is much closer to the upper quartile.

4.4 QUANTILE PLOTS

Quantile plots are general-purpose displays that portray many distributional features of a set of data. Quantile plots not only are useful for graphically describing the distribution of a set of data values, but they can also be used to assess the fidelity of a set of data to a hypothesized probability distribution.

Table 4.7   Data Quantiles for Tire-Wear Study

Observation i	Weight Loss (g/mi)	Data Fraction f
1	5	0.05
2	9	0.10
3	9	0.15
4	9	0.20
5	9	0.25
6	9	0.30
7	9	0.35
8	10	0.40
9	10	0.45
10	11	0.50
11	11	0.55
12	11	0.60
13	11	0.65
14	12	0.70
15	12	0.75
16	13	0.80
17	13	0.85
18	14	0.90
19	15	0.95
20	16	1.00

EXHIBIT 4.8

Quantile. A quantile, denoted $Q\{f\}$, is a number that divides a sample or population into two groups so that a specified fraction f of the data values are less than or equal to the value of the quantile.

The computation of quantiles using the definition in Exhibit 4.8 proceeds as follows. Each observation in a data set corresponds to a quantile for the fraction of the observations that are less than or equal to it. Let $y_{(1)}, y_{(2)}, \ldots, y_{(n)}$ denote the ordered observations for a data set; i.e.,

$$y_{(1)} \leqslant y_{(2)} \leqslant \cdots \leqslant y_{(n)}.$$

The ith ordered observation corresponds to $Q\{f\}$ with $f = i/n$.

In order to clarify the notion of quantiles, consider the twenty observations shown in Table 4.7. These observations are weight-loss measurements taken in a study of tire wear. The data are ordered from smallest to largest in the table. Each ordered observation is a data quantile corresponding to a multiple of the fraction $1/n = 0.05$.

Rather than simply reporting numerical values of quantiles, plots of the quantiles versus the corresponding data fraction are constructed. For plotting purposes it is preferable to graph the quantiles versus a slight modification of the data fraction i/n. The modifications that have been recommended result in better agreement between sample quantiles from a theoretical distribution and the actual theoretical quantiles from the distribution. The quantile plots used in this text are constructed using the procedure in Exhibit 4.9. An illustration of a quantile plot for the tire-wear data in Table 4.7 is shown in Figure 4.8.

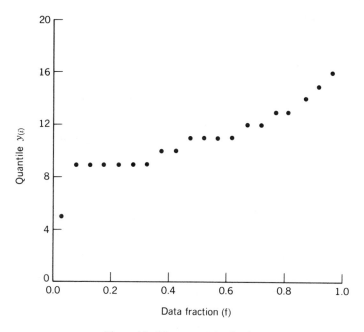

Figure 4.8 Tire-wear quantile plot.

EXHIBIT 4.9   QUANTILE PLOTS

1. Draw a horizontal axis covering the range of the data fractions (0, 1).

2. Draw a vertical axis covering the range of measurements on the response variable.

3. Order the response values from smallest to largest.

4. Plot $y_{(i)}$ versus $f_i = (i - \frac{3}{8})/(n + \frac{1}{4})$.

There are several features of the quantile plot that make it more informative than other types of distributional graphics. First, every observation in the data set is plotted, regardless of whether it is a duplicate. Repeated observations are easily recognized by horizontal bands of points. In general, the more dense the data, the flatter is the trend in the plotted points. Second, the quantile plot displays rather than summarizes the data. Third, no groupings of data (as with frequency distributions and histograms) are required to plot the data. Finally, many properties of the distribution of the data, such as the location of the median and the quantiles, are readily obtained from the plot.

Figures 4.9 and 4.10 are quantile plots for the 100 nitrogen oxide and fuel-economy data values shown in the point plots in Figure 4.2. The flat slope in Figure 4.9 around $Q\{0.5\} = 1.00$ indicates a high concentration of data values around the median. There is a slightly sharper slope for observations above 1.10 than there is for observations below 0.90, indicating that there is less of a concentration of high data values than low ones. These relatively smooth features of the quantile plot in Figure 4.9 can be contrasted with the clusters of points in Figure 4.10. The three sets of concentrated observations (around values of 19, 24, and 41) are clearly indicated in this figure, as they were in Figure 4.2b.

For many analyses the calculation and plotting of quantiles will suffice to enable an

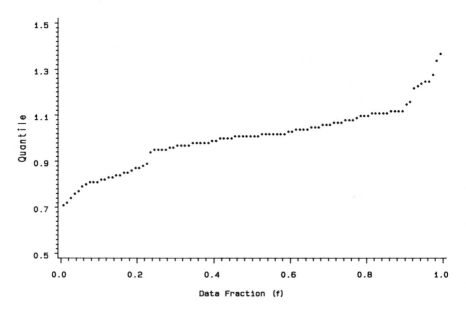

Figure 4.9   Nitrogen-oxide quantile plot.

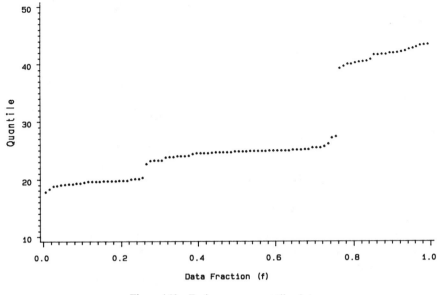

Figure 4.10   Fuel-economy quantile plot.

experimenter to discover most of the salient distributional features of a data set. One can approximate medians, quartiles, or any percentile of the sample from quantile plots such as Figure 4.9 simply by drawing line segments between each adjacent pair of points and interpolating between the points. A vertical line is drawn up from the horizontal axis at the desired data fraction. From the intersection of this vertical line with a plotted point or a line segment, a horizontal line is drawn to the vertical axis. The interpolated quantile is the response value where the horizontal line intersects the vertical axis.

A formula that can be used for quantile interpolation is $Q_I\{f\} = (1 - g)Q\{f_i\} + gQ\{f_{i+1}\}$ for $(n + 1)^{-1} \leqslant f \leqslant n(n + 1)^{-1}$, where $Q_I\{f\}$ denotes an interpolated quantile for the data fraction f, n is the sample size, and i and g are defined by the equation $(n + 1)f = i + g$. The integer portion of the quantity $(n + 1)f$ we denote by i and the fractional portion by g $(0 \leqslant g < 1)$. Then $Q\{f_i\} = y_{(i)}$ and $Q\{f_{i+1}\} = y_{(i+1)}$. One should not use this formula to extrapolate for extremely small or large data fractions outside the interval $(1/(n + 1), n/(n + 1))$.

The above interpolation formula can be used to obtain medians and quartiles that do not correspond to any of the observed data fractions. For example, if $n = 10$, one would approximate the median using the data fraction $f = 0.5$. Then $(n + 1)f = 5.5$, $i = 5$, and $g = 0.5$:

$$\text{Median} = Q_I\{0.5\} = (1 - 0.5)y_{(5)} + (0.5)y_{(6)}$$

$$= \frac{y_{(5)} + y_{(6)}}{2}.$$

Thus the median would be calculated as the average of the fifth and sixth ordered responses, just as was recommended in Section 3.2. Note that the first quartile would be approximated as a weighted average of the second and third ordered response [$f = 0.25$, $(n + 1)f = 2.75$, $i = 2$, $g = 0.75$]:

$$Q_I\{0.25\} = (0.25)y_{(2)} + (0.75)y_{(3)}.$$

This is an alternative to simply using one of the ordered responses for the quartile as was done in the example in Table 3.3 for $n = 8$.

In addition to simply describing the distribution of data values, quantile plots allow one to compare two sets of observations or to compare an observed distribution with a theoretical probability distribution. These topics are covered in the next chapter.

4.5 GRAPHICS FOR PROCESS CONTROL AND IMPROVEMENT

Statistical process-control procedures are integral components of scientific and managerial efforts to improve the quality of goods and services. Statistical methodology is not the sole determinant of success in quality improvement efforts. It may not even be the most important one in many settings. It is clear, however, from the experience of many top-performing companies that statistical design of experiments and statistical analyses of measures of quality are essential to providing product quality information upon which managerial decisions are based.

Measurable characteristics of quality provide the most objective means of assessing the factors influencing customer satisfaction. Data on measures of quality can be obtained through statistically designed experiments (e.g., Chapter 6) or statistically formulated sampling procedures (e.g., Chapter 3). As data are collected, graphical and analytical statistical methods can be used to assess the information obtained. In this section we discuss graphical procedures for displaying relevant features of quality measures. The application of experimental design to the collection of information on quality measures is discussed in Section 6.4.

Many of the monitoring and process-control procedures for measures of product and service quality rely upon simple graphical procedures such as the plotting of measures of location and dispersion. We discuss some of these procedures below. First, however, questions often arise as to what should be monitored. These questions can be answered through the use of frequency distributions, histograms, stem-and-leaf plots, and box plots. A wide variety of product or service characteristics can be summarized with such plots: e.g., counts of customer complaints, the number of items that fail to meet specified quality limits, or the time to first service of a product under warranty.

An especially useful plot for the identification of problems with the quality of a product or service is a bar chart listing the number of quality problems (i.e., yield loss, defectives, complaints, etc.) according to the major causes of these problems. Termed *Pareto diagrams*, these bar charts list the causes of problems according to their frequency of occurrence, from most frequent to least frequent. Usually such bar charts isolate one or a few dominant causes of product quality. Once these dominant causes are identified, sources of the problems are sought, improvement programs are initiated, quality measures are defined, and the monitoring of these measures is begun. Statistically designed experiments often aid in the identification of the sources of these problems.

Figure 4.11 is a Pareto diagram of the number of visual defects of enamel bowls. It is clear that three types of defects account for over 90% of the visual defects. Based on this information, the manufacturer would now seek to determine reasons for these three major

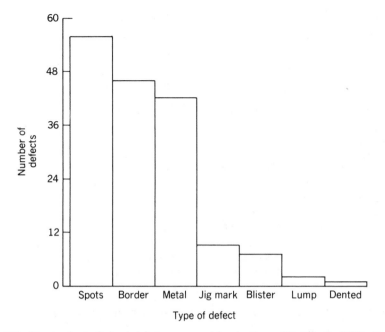

Figure 4.11   Pareto chart of visual defects on enamel bowls. From Ott, Ellis R. (1975). *Process Quality Control*, New York: McGraw–Hill Book Company. Copyright McGraw–Hill, New York: Reprinted with permission.

types of defects, use these counts or other quality measures to monitor the visual defects, and institute process-control procedures to continuously monitor the quality measures.

Control charts are used to monitor and assess the variability of a process thought to be operating under stable conditions. The process characteristics or quality measures plotted on control charts are typically measures of location or of dispersion. The observations from which the location or dispersion measures are derived may be taken from a finished-product variable, or they may be measurements taken during an intermediate stage of the process. Control charts not only aid in monoitoring a process, but they also help minimize overcontrol and undercontrol of the process.

Overcontrol of a process occurs when changes are made without regard to the inherent variability of a monitored characteristic. Overcontrol often is a reaction to the natural variability of the characteristic, and it frequently results in changing the operating conditions of the process. This leads to increased product variability because two sources of variation occur: the natural variation of the product characteristic and the variability associated with the frequent changes in operating conditions.

Undercontrol occurs when there is a delay in changing the operating conditions of a process even though evidence exists that the monitored property has changed in excess of its natural variability. Undercontrol results in periods of off-target operation, again increasing product variability.

A control chart is a sequence plot of a measured characteristic in which the horizontal axis is a time-ordered variable, typically sample number or lot number. In conjunction with the sequence plot, a control chart embodies a decision rule for determining whether the process exhibits excessive variation, i.e., when it is out of statistical contol. Control charts are the traditional means for implementing statistical process control (SPC), defined in Exhibit 4.10.

EXHIBIT 4.10

Statistical Process Control. Statistical process control consists of techniques used to assess statistically the status of a process characteristic relative to a target or aim value.

Before describing specific control-chart techniques, two types of variation of a monitored characteristic need to be distinguished. *Common-cause* variation is due to chance causes or measurement errors, the natural or inherent variation referred to above. This type of variation is often simply designated *random* variation. Variability or bias that exceeds the common-cause variation of a monitored characteristic is termed a *special-cause* or an *assignable-cause* (see also Section 15.1).

When only common-cause variation is present in a process, the distribution of the monitored characteristic is stable and predictable within specified limits. A consistent percentage of the monitored values will fall between lower and upper limits around the target value. These control limits are determined from past process data and are updated

---

EXHIBIT 4.11　　MEASURES OF COMMON CAUSE VARIATION

Average-Range Method

Calculate the range R_i for the $n > 1$ observations sampled at the ith time period ($i = 1, 2, \ldots, k$). The average range is

$$\bar{R} = \frac{R_1 + R_2 + \cdots + R_k}{k}.$$

Average-Standard-Deviation Method

Calculate the sample standard deviation s_i for the $n > 1$ observations sampled at the ith time period. The average standard deviation is

$$\bar{s} = \frac{s_1 + s_2 + \cdots + s_k}{k}.$$

Average-Moving-Range Method

When only $n = 1$ observation is made at each time period, calculate the moving range for the ith sample as

$$MR_i = |y_i - y_{i-1}|, \qquad i = 2, 3, \ldots, k.$$

The average moving range is

$$AMR = \frac{MR_2 + MR_3 + \cdots + MR_k}{k - 1}.$$

The average moving range can be converted to an estimated standard deviation:

$$s_{AMR} = AMR/1.128.$$

---

on a periodic basis (e.g., every six months) or when a significant modification occurs in the process or in the sampling plan used to collect data on the process.

The basic requirements for monitoring a process using control charts are a sampling plan, a target value, and an estimate of the common-cause variability based on ranges or standard deviations. Sampling plans for monitoring stable processes are discussed in the references cited at the end of this chapter. These sampling plans are generally based on simple random samples or systematic random samples (Section 2.1). The appropriate target value for a process or quality characteristic can come from many sources. These include long-term historical process averages, the middle of a specification range, or the nominal value required by the customer. There are several useful measures of common-cause variability. The measures described in Exhibit 4.11 are based on the collection of simple random samples at each of several consecutive time periods.

Adequate data must be obtained to measure common-cause variation satisfactorily. There are two main issues of concern in this regard: (1) the data must cover enough elapsed calendar time to be representative of the variation in the process, and (2) enough data must be obtained to provide statistically stable estimates. At least 30 to 40 samples over approximately a two- to three-month period are recommended.

Two types of control charts typically used for statistical process control are Shewhart charts and cumulative (CUSUM) charts. Shewhart control charts are time sequence plots with control limits superimposed. The upper and lower control limits (UCL and LCL, respectively) for Shewhart control charts are constructed as in Exhibit 4.12. The constants needed to determine the control limits for averages and ranges are given in Table A2 in the Appendix. Those for standard deviations are given in Table A3.

EXHIBIT 4.12   SHEWHART CONTROL CHARTS

1. Randomly sample n observations at each of k time periods (ordinarily equally spaced).
2. Calculate means and ranges (or standard deviations) for each sample. Calculate the average range (or standard deviation) or average moving range, as appropriate.
3. Calculate upper (UCL) and lower (LCR) control limits for averages using the constants in Table A2 of the Appendix:

$$n > 1: \quad \text{LCL} = \text{target} - A_2 \bar{R}, \quad \text{UCL} = \text{target} + A_2 \bar{R};$$
$$n = 1: \quad \text{LCL} = \text{target} - 3s_{AMR}, \quad \text{UCL} = \text{target} + 3s_{AMR}.$$

4. Calculate upper and lower control limits for dispersion when $n > 1$, using the constants in Tables A2 and A3 of the Appendix:

$$\text{Range:} \quad \text{LCL} = D_3 \bar{R}, \quad \text{UCL} = D_4 \bar{R};$$
$$\text{S.D.} \quad \text{LCL} = B_3 \bar{s}, \quad \text{UCL} = B_4 \bar{s}.$$

Either the range chart or the standard deviation chart is used, not both.

5. Plot the averages and the ranges (or standard deviations) on the appropriate chart. If a plotted point falls outside the control limits, the process should be investigated for special causes.

The above steps in constructing control limits for Shewhart charts of averages use the target value as the center line. Shewhart charts are usually constructed with the overall average of the data as the center line. As noted earlier, the overall average is one choice for the target. The choice of the target value will depend on the particulars of the process. If the process is effectively run "on aim," the target value and the long-term average should be essentially the same.

The measures of common-cause variation used to determine control limits for Shewhart charts are calculated from averages of ranges and standard deviations. These quantities have traditionally been used and continue to be used despite the availability of more efficient (precise) estimators. An alternative "pooled" estimator of the standard deviation, which is more efficient than the average range or the average standard deviation, is

$$s_p = \left(\frac{s_1^2 + s_2^2 + \cdots + s_k^2}{k} \right)^{1/2}.$$

The corresponding upper and lower control limits for averages are then calculated as

$$\text{LCL} = \text{target} - \frac{3s_p}{\sqrt{n}}, \qquad \text{UCL} = \text{target} + \frac{3s_p}{\sqrt{n}}.$$

The loss in efficiency due to using average ranges versus s_p usually is small for $n \leqslant 5$ but can be substantial for $n > 5$.

The average-moving-range method for measuring common-cause variation plays an important practical role. It allows statistical process-control procedures to be used when a single observation is collected in time from a process. However, it has limitations. When the observations are positively correlated in time and the bulk of the variability is longer-term, the standard deviation estimate based on the average moving range tends to underestimate the process standard deviation appropriate for constructing control limits. This limitation is more of a concern as data are collected closer and closer together in time i.e., as the sampling frequency increases. What constitutes "too close together in time" is dependent on the dynamics of the physical mechanism generating the data.

Evidence of variation outside the common-cause system is obtained when a plotted value falls outside the control limits. This indicates that the process should be investigated for special causes and, if appropriate, a compensating process change should be made to bring the monitored characteristic back to target.

Figure 4.3 is an example of a Shewhart control chart for $k = 18$ within-lot standard deviations $(n = 2)$ of a metallic oxide chemical process. The average of the 18 standard deviations is 0.196. Using Table A3, the control limits are

$$\text{LCL} = (0)(0.196) = 0$$

and

$$\text{UCL} = (3.27)(0.196) = 0.64.$$

As noted in the discussion of Figure 4.3, the variability of lot 17 exceeds the upper control limit. This anomalous value can be traced to low observations for one sample from the lot. Using this information, an investigation of the cause should be conducted.

Shewhart control charts quickly detect major shifts of the monitored process from a target value. A process-control charting technique that is more sensitive to modest changes in the monitored process characteristic than Shewhart control charts is the cumulative sum (abbreviated CUSUM) charting technique. CUSUM charts (see Exhibit 4.13) can be used when the monitored characteristic is an individual observation, an average, or a measure of dispersion.

EXHIBIT 4.13   CUMULATIVE SUM (CUSUM) CONTROL CHARTS

1. Randomly sample n observations at each of k time periods (ordinarily equally spaced).

2. For CUSUM charts of averages, calculate the sample averages and the average moving range using the $\bar{y}_i$. For charts of dispersion, calculate $\ln s_i$ for each sample and the moving range using the $\ln s_i$.

3. Denote the process characteristic of interest ($\bar{y}$, or $\ln s_i$) by z_i. Calculate the "sum high" (SH) and "sum low" (SL) statistics for the ith sample:

$$SH_i = SH_{i-1} + [z_i - (\text{target} + 0.5s_{AMR})],$$
$$SL_i = SL_{i-1} + [(\text{target} - 0.5s_{AMR}) - z_i],$$

where $SH_0 = SL_0 = 0$. If at any time, SH_i or SL_i is negative, it is set to zero.

4. If either

$$SH_i > 4s_{AMR} \quad \text{or} \quad SL_i > 4s_{AMR},$$

then an investigation for cause is conducted and, if appropriate, corrective action is taken. Both SH_i and SL_i are then set to zero; i.e., the CUSUM is reset, under the assumption that consequences of the corrective action are reflected in the next sample.

5. To implement this technique graphically, plot $\max\{SH_i, SL_i\}$ in a labeled sequence plot, using H and L to designate which sum is being plotted. An upper control limit is drawn at $4s_{AMR}$.

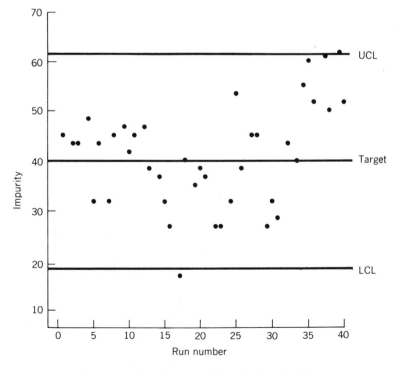

Figure 4.12   Shewhart control chart for impurity data.

Table 4.8 CUSUM Control-Chart Calculations for Impurity Measurements from a Chemical Process

Run No.	Impurities (ppm)	SL	SH	Maximum
1	45	0.0	1.4	High
2	43	0.0	0.8	High
3	44	0.0	1.2	High
4	48	0.0	5.6	High
5	31	5.4	0.0	Low
6	43	0.0	0.0	Low
7	31	5.4	0.0	Low
8	45	0.0	1.4	High
9	47	0.0	4.8	High
10	41	0.0	2.2	High
11	45	0.0	3.6	High
12	47	0.0	7.0	High
13	39	0.0	2.4	High
14	37	0.0	0.0	High
15	32	4.4	0.0	Low
16	26	14.8	0.0	Low
17	16	35.2*	0.0	Low
		CUSUM reset		
18	40	0.0	0.0	Low
19	35	1.4	0.0	Low
20	38	0.0	0.0	Low
21	36	0.4	0.0	Low
22	27	9.8	0.0	Low
23	26	20.2	0.0	Low
24	32	24.6	0.0	Low
25	53	8.0	9.4	High
26	38	6.4	3.8	Low
27	45	0.0	5.2	High
28	45	0.0	6.6	High
29	26	10.4	0.0	Low
30	32	14.8	0.0	Low
31	28	23.2	0.0	Low
32	43	16.6	0.0	Low
33	40	13.0	0.0	Low
34	55	0.0	11.4	High
35	60	0.0	27.8	High
36	52	0.0	36.2*	High
		CUSUM reset		
37	61	0.0	17.4	High
38	50	0.0	23.8	High
39	62	0.0	42.2*	High
		CUSUM reset		
40	51	0.0	7.4	High

*Exceeds control limit of 28.8.

A standard deviation based on the average-moving-range method is used in the CUSUM calculation. For charts of averages, the moving range is calculated with $\bar{y}_i$ inserted in place of y_i in the computation of MR_i. CUSUM control charts for dispersion use $\ln s_i$ in place of y_i in the computation of MR_i.

The CUSUM technique derives its name from the fact that it accumulates successive deviations of the process characteristic from fixed reference points (target $\pm 0.5s_{AMR}$). Evidence of special-cause variation is obtained when a cumulative sum of deviations, on either the high side or the low side of the target, exceeds an action value ($4s_{AMR}$).

Table 4.8 contains impurity data (ppm) from a chemical process. Data from 40 time periods were taken in order to set up a control-chart scheme. These data are single observations taken at roughly equal time intervals. The target value for this process is 40.0 ppm. The standard-deviation estimate based on the average moving range is $s_{AMR} = 7.2$.

Figure 4.12 shows a Shewhart chart for these data. The control limits are $40 \pm 3s_{AMR} = 40 \pm 21.6$. Two of the observations exceed the control limits: those taken at time periods 17 and 39. Based on this information, a process engineer would investigate to determine the causes of the changes in the mean impurity level.

Table 4.8 also lists the SH_i and SL_i values for each time period. Figure 4.13 shows the CUSUM control chart for these data with the UCL at $4s_{AMR} = 28.8$. A difference in the Shewhart and the CUSUM control charts for these data is the identification of a mean-impurity-level change at observation 36 in the CUSUM chart. This is indicative of the greater

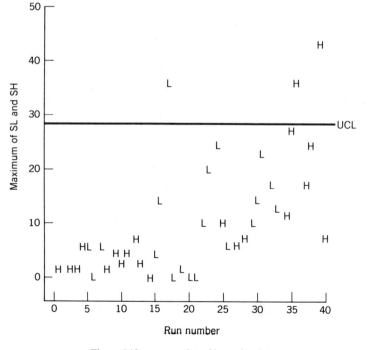

Figure 4.13 CUSUM plot of impurity data.

sensitivity of the CUSUM chart to moderate changes in the mean level of the monitored process characteristic.

The control limits calculated from these 40 observations would now be used to monitor this process as new data are collected. When new data indicate changes in the mean of the monitored characteristic, appropriate action should be quickly taken. Periodically the control limits should be updated to reflect current common-cause process variability.

The properties of the CUSUM control scheme change with the reference points and action values used. The CUSUM calculations are described in this section for typical reference points and action values. Likewise, there are modifications to the Shewhart control-chart procedures described in this section. Further enhancements and details can be found in the references cited at the end of this chapter.

REFERENCES

Text References

A comprehensive discussion of plotting techniques, containing a wide variety of graphical displays for simple and complex data sets, can be found in

Chambers, J. M., Cleveland, W. S., Kleiner, B., and Tukey, P. A. (1983). *Graphical Methods for Data Analysis*, Belmont, CA: Wadsworth International Group.

Explicit instructions for constructing frequency distributions, histograms, and scatterplots can be found in most elementary statistics texts, including:

Freedman, D., Pisani, R., and Purves, R. (1978). *Statistics*, New York: W. W. Norton & Company, Chapters 3 and 7.

Koopmans, L. (1981). *An Introduction to Contemporary Statistics*, Belmont, CA: Duxbury Press, Chapters 1 and 4.

Ott, L. (1977). *An Introduction to Statistical Methods and Data Analysis*, Belmont, CA: Duxbury Press, Chapters 1 and 6.
Stem-and-leaf and box plots are covered in many of the newer elementary texts. The text by Koopmans (Chapters 1 and 2) includes discussions on these techniques. These procedures and many others for initially investigating data are referred to as exploratory data-analysis techniques.

The following references highlight many of the popular exploratory data analysis techniques:

Hoaglin, D. C., Mosteller, F., and Tukey, J. W. (1983). *Understanding Robust and Exploratory Data Analysis*, New York: John Wiley and Sons, Inc.

Tukey, J. (1977). *Exploratory Data Analysis*, Reading, MA: Addison-Wesley Publishing Co.

Velleman, P. F. and Hoaglin, D. C. (1981). *Applications Basics, and Computing of Exploratory Data Analysis*, Reading, MA: Addison-Wesley Publishing Co.
The second text listed above is the seminal reference on exploratory data techniques. Readers wishing to refer to a more elementary exposition should consult the third reference.

Control-chart techniques are available in many sources. The seminal work by Shewhart cited below is still worth reading today.

Burr, I. W. (1976). *Statistical Quality Control Methods*, New York: Marcel Dekker, Inc.

Grant, E. L. and Levenworth, R. S. (1980). *Statistical Quality Control, Fifth Edition*, New York: McGraw-Hill Book Co.

Lucas, J. M. (1976). "The Design and Use of V-Mask Control Schemes," *Journal of Quality Technology*, **8**, 1–12.

Montgomery, D. C. (1985). *Introduction to Statistical Quality Control*, New York: John Wiley and Sons, Inc.

Ott, Ellis R. (1975). *Process Quality Control*, New York: McGraw-Hill, Inc.

Shewhart, W. A. (1931), *Economic Control of Quality of Manufactured Product*, New York: D. Van Nostrand Co.

References for quantile plots are given at the end of the next chapter.

EXERCISES

1 Construct a point plot of the solar-energy data presented in Chapter 3, Exercise 4. Are the sample mean and sample standard deviation adequate representations of location and spread for this data set? Why (not)?

2 A series of valve closure tests were conducted on a 5-inch-diameter speed-control valve. The valve has a spring-loaded poppet mechanism that allows the valve to remain open until the flow drag on the poppet is great enough to overcome the spring force. The poppet then closes, causing flow through the valve to be greatly reduced. Ten tests were run at different spring locking-nut settings, where the flow rate at which the valve poppet closed was measured in gal/min. Produce a scatterplot of these data. What appears to be the effect of nut setting on flow rate?

Nut Setting	Flow Rate	Nut Setting	Flow Rate
0	1250	10	2085
2	1510	12	1503
4	1608	14	2115
6	1650	16	2350
8	1825	18	2411

3 A new manufacturing process is being implemented in a factory that produces automobile spark plugs. A random sample of 50 spark plugs is selected each day over a 15-day period. The spark plugs are examined and the number of defective plugs is recorded each day. Plot the following data in a sequence plot. What does the plot suggest about the new manufacturing process?

Day	No. of Defectives	Day	No. of Defectives
1	3	9	4
2	8	10	3
3	4	11	6
4	5	12	4
5	5	13	4
6	3	14	3
7	4	15	1
8	5		

4 Construct a histogram and a stem-and-leaf plot of the solar-energy data presented in Chapter 3, Exercise 4. Compare the ability of the two types of graphs to highlight general trends and specific details in the data.

5 Construct two histograms from the solar-energy data. Use the following interval

widths and starting lower limits for the first class. What do you conclude about the choices of the interval width for this data set?

	Histogram 1	Histogram 2
Interval width	8	2
Starting lower limit	480	480

6 Construct a box plot from the solar-energy data. Are there any extreme observations depicted? What percentage of the observations lie within the box? Does the sample mean or the sample median (or both) better depict a typical energy rate?

7 Construct a quantile plot from the solar-energy data. Interpolate quantiles for the median, first quartile, and third quartile. How do these interpolated quantiles compare with those calculated in Exercise 4, Chapter 3?

8 The following data were taken from a study of red-blood-cell counts before and after major surgery. Counts were taken on 23 patients, all of whom were of the same sex (female) and who had the same blood type (O +).

Patient	**Count** Pre-op	**Count** Post-op	**Patient**	**Count** Pre-op	**Count** Post-op
1	14	0	13	5	6
2	13	26	14	4	0
3	4	2	15	15	3
4	5	4	16	4	2
5	18	8	17	0	3
6	3	1	18	7	0
7	6	0	19	2	0
8	11	3	20	8	13
9	33	23	21	4	24
10	11	2	22	4	6
11	3	2	23	5	0
12	3	2			

(a) Make stem-and-leaf plots of the pre-op and the post-op blood counts. What distinguishing features, if any, are there in the distributions of the blood counts?

(b) Make a scatter diagram of the two sets of counts. Is there an apparent relationship between the two sets of counts?

9 Use quantiles to approximate the quartiles of the two sets of blood counts in Exercise 8. How do the two sets of data compare?

10 Satellite sensors can be used to provide estimates of the amounts of certain crops that are grown in agricultural regions of the United States. The following data consist of two sets of estimates of the proportions of each of 33 5 × 6-nautical-mile segments of land that are growing corn during one time period during the crop season (the rest of the segment may be growing other crops or consist of roads, lakes, houses, etc.). Use

the descriptive and graphical techniques discussed in this and the preceding chapter to assess whether these two estimation methods are providing similar information on the proportions of these segments that are growing corn.

Segment	Proportion Growing Corn		Segment	Proportion Growing Corn	
	Method 1	Method 2		Method 1	Method 2
1	.49	.24	18	.61	.33
2	.63	.32	19	.50	.20
3	.60	.51	20	.62	.65
4	.63	.36	21	.55	.51
5	.45	.23	22	.27	.31
6	.64	.26	23	.65	.36
7	.67	.36	24	.70	.33
8	.66	.95	25	.52	.27
9	.62	.56	26	.60	.30
10	.59	.37	27	.62	.38
11	.60	.62	28	.26	.22
12	.50	.31	29	.46	.72
13	.60	.56	30	.68	.76
14	.90	.90	31	.42	.36
15	.61	.32	32	.68	.34
16	.32	.33	33	.61	.28
17	.63	.27			

CHAPTER 5

Graphical Comparisons of Distributions

In this chapter box plots and quantile plots are used to provide graphical comparisons of distributions. Specific graphical comparisons discussed in this chapter are:

- *box-plot comparisons of the key statistics that summarize distributions of data,*
- *quantile-plot comparisons of two sample distributions, and*
- *comparison of a sample distribution with a theoretical reference distribution.*

Many statistical analyses require an assessment of the shape of a distribution or an evaluation of whether two or more distributions can be considered to be statistically equivalent. Some analyses involve the comparison of two sample distributions—e.g., in the comparison of treatment and control groups. Other analyses necessitate the comparison of a sample distribution with a theoretical probability distribution such as the normal distribution. Although there are prescribed statistical inference procedures for such comparisons, it often is sufficient to provide a graphical comparison. Even if formal inference procedures are used, graphical displays are informative for illustrating the results of the analysis, especially to a nontechnical audience. In this chapter we use box plots (Section 4.3) and quantile plots (Section 4.4) to analyze experimental results by graphically comparing distributions of data either with one another or with a theoretical reference distribution.

5.1 COMPARISONS USING BOX PLOTS

Experimental programs often are undertaken to diagnose variation in a response that might be attributable to the effects of one or more experimental factors. If several factors are believed to affect a response, their simultaneous effects can only be properly assessed when the factors are jointly varied under similar experimental conditions. One is then faced with the task of reporting the results of such experimentation as a function of the several factors. An important graphical means of accomplishing this is through the use of comparative box plots.

Consider an investigation into the major sources of variation in measurements taken with an infrared scanning instrument. The instrument is used to measure chemical properties of industrial liquids. Factors investigated to determine their effects on the measurement of the chemical response by the scanning instrument are:

- the concentration of a standard chemical in the sample (32% or 36%),
- sample preparations (five samples were prepared for each concentration),
- the number of repeat scans per observation (2, 4, or 16), and
- multiple measurements of the chemical properties for each sample preparation (the scanning instrument measured each sample preparation five times for each number of scans).

In this experiment four potential sources of variability are investigated: the concentration of the chemical, the sample preparations for each concentration, the number of

Table 5.1 Chemical Measurements from an Infrared Scanning Instrument

		Response					
		Conc. = 32%			Conc. = 36%		
Sample Prep.	Deter-mination	No. of Scans			No. of Scans		
		2	4	16	2	4	16
1	1	7.01	7.04	7.03	8.99	9.11	7.59
	2	7.03	7.03	7.05	8.90	9.19	7.39
	3	6.99	7.01	7.03	9.16	9.05	7.45
	4	6.99	7.02	7.00	9.17	8.93	7.50
	5	7.03	7.01	7.02	9.10	8.90	7.56
2	1	7.04	7.05	7.03	9.08	9.35	9.26
	2	7.03	7.05	6.85	9.40	9.20	9.26
	3	7.04	7.04	7.00	9.25	9.23	9.24
	4	7.03	7.07	6.99	9.31	9.18	9.28
	5	7.04	7.04	7.02	9.26	9.33	9.29
3	1	7.02	7.01	6.98	9.05	8.90	8.83
	2	6.99	7.00	7.02	8.33	9.08	8.93
	3	7.04	7.00	7.04	8.22	8.70	8.81
	4	7.03	7.00	7.03	8.82	9.07	8.98
	5	7.04	6.99	7.03	8.35	9.43	8.96
4	1	7.03	7.00	7.03	9.02	8.88	8.99
	2	7.03	7.03	7.02	8.88	8.92	9.07
	3	7.03	7.01	7.02	8.86	9.00	8.99
	4	7.02	7.04	7.03	8.87	9.20	8.90
	5	7.00	7.01	7.01	8.77	9.13	8.92
5	1	7.00	6.98	6.99	8.87	8.90	9.00
	2	7.00	6.99	7.00	9.00	9.15	8.93
	3	6.99	6.99	7.00	8.94	9.17	9.09
	4	6.99	6.99	6.99	9.08	8.96	9.17
	5	7.03	6.98	7.01	8.92	8.99	8.92

repeat scans with the scanning instrument, and the multiple determinations of the chemical response. The concentrations and the numbers of repeat scans are believed to affect the response in a systematic fashion. The sample preparations and the multiple determinations are believed to exert random effects on the response; i.e., their effects on each measurement are not systematic but randomly fluctuate about some constant value. The results of the experimentation are exhibited in Table 5.1. The main conclusion that can be drawn from an examination of the table is that responses for the higher concentration appear to be about 2 units greater than those for the lower concentration.

Figure 5.1 displays a box plot for each combination of concentration and repeat scans. This box plot was made because concentration and the number of repeat scans were believed to have systematic effects on the chemical response of the scanning instrument. The scatter in each box represents the random variation attributable to the other two experimental variables and to other random fluctuations.

The figure clearly indicates that the scanning instrument is sensitive enough to distinguish the two concentrations and that the different numbers of repeat scans do not appear to have any systematic effect on the results. The boxes for the three repeat scans at the lower concentration are consistent with one another in both location and spread. They also exhibit good repeatability; i.e., the 25 measurements (five preparations and five determinations for each preparation) are all tightly clustered.

The three boxes for the higher concentration show features that are markedly different from those of the lower concentration. Although the three boxes have approximately the same medians and interquartile spread, the variation in the 25 measurements for each number of repeat scans is considerably larger than that for the lower concentration. The extremely low measurements for the first and third sets of repeat scans (2 and 16 repeats)

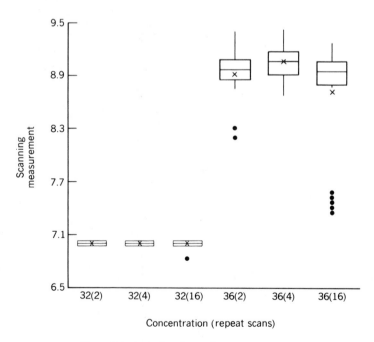

Figure 5.1 Box plot of scanning measurements.

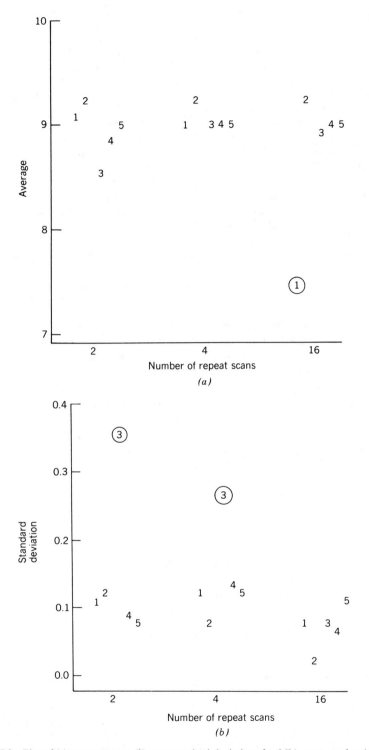

Figure 5.2 Plot of (a) scan averages, (b) scan standard deviations for 36% concentration (plotting symbol designates preparation number).

pulls their respective means well below the medians. The box plot is helpful in focusing attention on the extreme observations for the higher concentration. One can now use other plots and statistics to gain insight into possible reasons for these anomalous observations.

For example, Figures 5.2a and b are plots of the averages and standard deviations, respectively, for the measurements taken on each of the five sample preparations for the higher-concentration samples. Figure 5.2a reveals a very low average for the five measurements on the first sample with 16 repeat scans. The corresponding standard deviation in Figure 5.2b is small, which coupled with the low average measurement indicates a bias in all five determinations for this sample. These five measurements are the five low readings that are indicated below the last box in Figure 5.1.

Observe now the large standard deviation for the five determinations on the third sample with two repeat scans (Figure 5.2b). Note too the large standard deviation for the third sample with four repeat scans. Since the means for these two samples are in good agreement with the other sample means in Figure 5.2a, the large standard deviations indicate that the multiple determinations for these two samples are not as repeatable as those on the other samples.

The box plot in Figure 5.1 has led to the discovery of very valuable information on the effects of the experimental factors on the chemical response of the scanning instrument:

- the laboratory method has the necessary sensitivity to distinguish the 32% concentration from the 36% concentration;
- overall, the average measurements for the sample preparations from the 32% concentration exhibit much closer agreement than those from the 36% concentration;
- the precision of the measurements is affected by the percentage chemical concentration in the sample;
- the five measurements for sample 1 from the 36% concentration with 16 replicate scans are in gross disagreement with the measurements from the other sample preparations for this concentration; and
- the variation in the five measurements from preparation 3 for the 36% concentration with both two and four replicate scans greatly exceeds the variation in all other preparations.

As this example illustrates, comparative box-plot displays help to better understand systematic and random sources of variability and to target problem areas for further technical effort. As with any graphical procedure, the goal is not to arrive at the "ultimate" display. The goal is to be informative so that the graphical displays will result in better knowledge of a data-generating process.

5.2 COMPARING TWO SAMPLE DISTRIBUTIONS USING QUANTILE PLOTS

The distributions of two samples of observations can be readily compared by plotting the quantiles (Section 4.4) of one of the samples versus the corresponding quantiles of the other sample. To make such a plot, denote one set of sample values by $y_i, i = 1, 2, \ldots, n$, and a second set by $x_i, i = 1, 2, \ldots, m$. The two-sample quantile–quantile plot is a graph of $Q_y(f)$ versus $Q_x(f)$ using the same set of f-values. If the two sample sizes (m and n) are equal, the graph is simply a plot of the pairs of ordered observations $(x_{(i)}, y_{(i)})$.

Figure 5.3 is a quantile–quantile plot for measurements of tire wear on two different tires of the same brand. Denote the wear measurements for tire 1 by y and those for tire 2 by x. The line $y = x$ is superimposed on the plot because all the plotted points should lie on or near this line if the two distributions are identical. All points except the one in the upper right-hand corner do lie on or near the line. Thus it appears that the two tires exhibit similar distributions of wear rates. This is expected, since the tires are made by the same manufacturer. The point in the upper right corner of the graph is the result of an unusually high wear rate for tire 2. This occurred in the initial run using this tire and may have been the result of an inadequate break-in period.

A similar quantile–quantile plot for measurements of wear on two tires from different manufacturers is given in Figure 5.4. With this set of data all but one of the paired data points lie above the $y = x$ line. The plotted points appear to lie on a reasonably straight line whose slope is greater than 1. This indicates that the brand-1 tire had a higher wear rate than the brand-2 tire.

The difference in data distributions for the two tire brands is illustrated using a dashed line, $y = 1.3x$, in Figure 5.4. This line fits the wear rates better than the line of equal wear, $y = x$. Estimates of the center and spread of the two distributions differ by approximately the same multiplicative factor, 1.3, but the quantile–quantile plot indicates that the two distributions have the same general shape.

In general, if two sample distributions have the same shape but the values of one

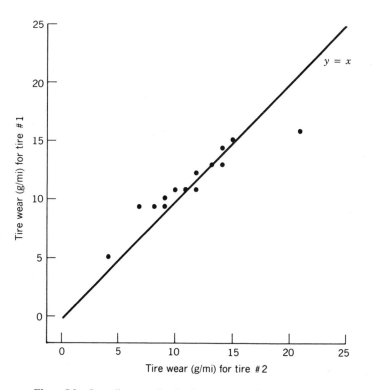

Figure 5.3 Quantile–quantile plot for two tires of the same brand.

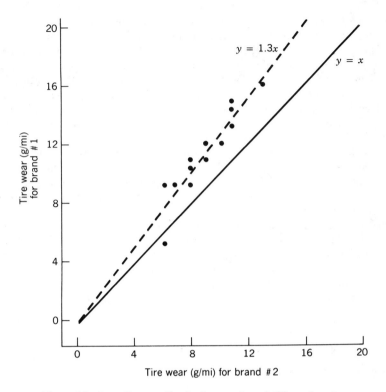

Figure 5.4 Quantile–quantile plot for two tires of different brands.

EXHIBIT 5.1 QUANTILE–QUANTILE PLOTS

1. Denote the smaller sample of ordered responses by

$$y_{(1)} \leqslant y_{(2)} \leqslant \cdots \leqslant y_{(n)}$$

and the larger sample by

$$x_{(1)} \leqslant x_{(2)} \leqslant \cdots \leqslant x_{(m)},$$

2. For each data fraction $f_i = i/n$ in the smaller sample find an interpolated quantile $x'(f_i)$ for the larger sample:

(a) if $n = m$, $x'(f_i) = x_{(i)}$;
(b) if $n < m$, set $h = (m + 1) f_i$ and calculate

$$x'(f_i) = (1 - g)x_{(k)} + gx_{(k+1)},$$

where k is the integer portion of h and $g = h - k$. [If $k \geqslant m$, $x'(f_i) = x(m)$.]

3. Plot $Q_y\{f_i\} = y_{(i)}$ versus $Q_x\{f_i\} = x'(f_i)$, $i = 1, 2, \ldots, n$.

---

response differ from those of another by additive or multiplicative factors, the ordered observations satisfy a straight-line relationship

$$y_{(i)} = a + bx_{(i)},$$

or, in terms of quantiles,

$$Q_y\{f\} = a + bQ_x\{f\}$$

for suitably chosen constants a and b. Relationships of this form lead to quantile–quantile plots similar to Figure 5.4. If the two distributions do not have the same shape, the quantile–quantile plot does not generally result in a straight-line scatter of points.

We now turn to the construction of quantile–quantile plots when the two data sets have an unequal number of observations. One useful procedure is to plot all the points of the smaller sample against the corresponding interpolated values from the larger sample. The general procedure for constructing quantile–quantile plots is given in Exhibit 5.1.

For example, if $n = 10$ and $m = 39$, then $h = 4i$ for $i = 1, 2, \ldots, 10$. Since for this data set each value of h is an integer (i.e., $k = 4i$ and $g = 0$), $y_{(i)}$ is plotted against $x^I(f_i) = x_{(4i)}$. However, if $n = 10$ and $m = 24$, then $h = 2.5i$ and

$$k = 2.5i \qquad \text{and } g = 0 \qquad \text{if } i \text{ is even,}$$
$$k = 2.5i - 0.5 \text{ and } g = 0.5 \qquad \text{if } i \text{ is odd.}$$

An example of the calculation of interpolated quantile values for use in quantile–quantile plots is shown in Table 5.2.

Quantile–quantile plots offer many advantages when the objective is the comparison of the distributions of two data sets. As can be seen from the plots in this section, this technique allows an analyst to see more complex variations in the data than those that are provided by simple summary statistics. The entire range of the distribution can be examined, and subtle shifts in location and spread are easily detectable. One can also use

Table 5.2   Calculation of Interpolated Quantiles for a Quantile–Quantile Plot

Data $(n = 10, m = 12)$

y: 8, 9, 9, 12, 13, 14, 16, 16, 18, 21
x: 6, 8, 9, 11, 12, 14, 16, 17, 19, 22, 23, 25

i	h	k	g	$y_{(i)}$	$x_{(k)}$	$x_{(k+1)}$	$x^I(f_i)$
1	1.3	1	.3	8	6	8	6.6
2	2.6	2	.6	9	8	9	8.6
3	3.9	3	.9	9	9	11	10.8
4	5.2	5	.2	12	12	14	12.4
5	6.5	6	.5	13	14	16	15.0
6	7.8	7	.8	14	16	17	16.8
7	9.1	9	.1	16	19	22	19.3
8	10.4	10	.4	16	22	23	22.4
9	11.7	11	.7	18	23	25	24.4
10	13.0	13	0	21	—	—	25.0

theoretical quantiles in such plots to compare an observed distribution with a theoretical reference distribution, the topic of the next section.

5.3 COMPARISON WITH A REFERENCE DISTRIBUTION

Quantile–quantile plots can be used to compare a sample distribution with a theoretical reference distribution such as the normal probability distribution. The theoretical distribution is called a reference distribution to stress that the true underlying distribution that generates the data is unknown and we are merely assessing whether the observed data values are consistent with one of perhaps many theoretical distributions that could have generated the data. If the plotted points closely approximate a straight line, the sample distribution is similar in shape to the reference distribution. The normal probability distribution is used as a reference distribution for many measurement processes. The normal probability distribution is defined by the density function in Exhibit 5.2. The normal distribution is used to characterize measurements that have relatively high likelihoods of occurrence for values around the mean but lower likelihoods as measurements differ more and more from the mean in either direction. The distribution is "bell-shaped" around the mean μ, with the width or spread of the curve determined by the standard deviation σ. Figure 5.5 shows three normal distributions that have a mean of 50 and standard deviations 1, 3, and 5.

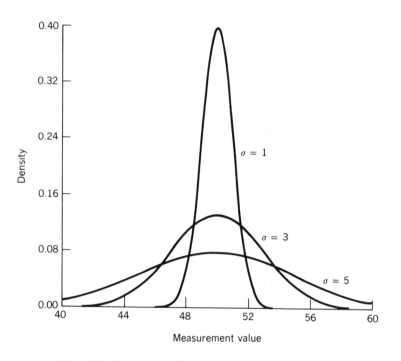

Figure 5.5 Comparison of normal densities with means $= 50$.

EXHIBIT 5.2

Normal Probability Distribution. Measurements whose likelihoods of occurrence are defined by the density function

$$f(y) = (2\pi\sigma^2)^{-1/2}\exp\{-(y-\mu)^2/2\sigma^2\},$$

where y is a measurement value, μ is the population or process mean, and σ is the population or process standard deviation.

In order to construct quantile plots of $Q_N\{f\}$ versus f for the normal probability distribution, one must first compute the quantiles of the distribution. The normal quantile function does not have a closed-form expression, but the following approximation is often used in statistical software programs for a "standard" normal distribution ($\mu = 0$ and $\sigma = 1$):

$$Q_{SN}\{f\} = 4.91\{f^{0.14} - (1-f)^{0.14}\}. \tag{5.1}$$

Quantile plots for a normal distribution with any mean μ and standard deviation σ can then be made from the quantiles for the standard normal distribution using the following formula:

$$Q_N\{f\} = \mu + \sigma Q_{SN}\{f\}.$$

In order to plot quantiles for a sample of observations versus theoretical normal quantiles the procedure in Exhibit 5.3 can be followed.

EXHIBIT 5.3   NORMAL QUANTILE–QUANTILE PLOTS

1. Denote the ordered data values by $y_{(1)} \leqslant y_{(2)} \leqslant \cdots \leqslant y_{(n)}$.

2. For each ordered observation, set $Q_y(f_i) = y_{(i)}$, $i = 1, 2, \ldots, n$.

3. Calculate $Q_{SN}\{f_i\}$ for $f_i = (i - \frac{3}{8})/(n + \frac{1}{4})$ from (5.1):

$$Q_{SN}\{f_i\} = 4.91\,[f_i^{0.14} - (1 - f_i)^{0.14}].$$

4. Plot $Q_y(f_i)$ versus $Q_{SN}\{f_i\}$. Approximate linearity indicates that the sample data are consistent with a normal reference distribution. The sample mean and sample standard deviation estimate the corresponding population or process parameters.

Figure 5.6 is a plot of the nitrogen oxide quantiles (Figure 4.9) versus the quantiles from a standard normal reference distribution. A cursory glance at Figure 5.6 suggests satisfactory agreement with a normal distribution. However, careful examination of the plot, in particular focusing on the two breaks in it, reveals that these data should not be treated as though they come from a single normal distribution.

The two breaks appear to separate distinct normal distributions, since the slopes for the three groups of points are different (check by laying a straightedge through the middle cluster of points). The top and bottom clusters of points each come almost entirely from one vehicle (the Volkswagen and the Oldsmobile, respectively), and the middle cluster

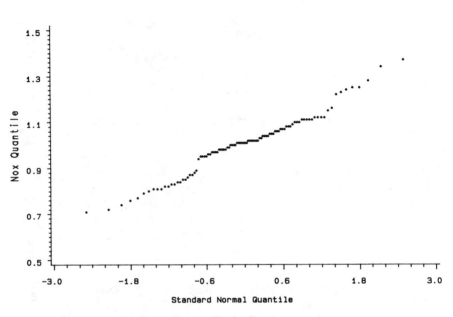

Figure 5.6 Normal quantile–quantile plot for nitrogen-oxide measurements.

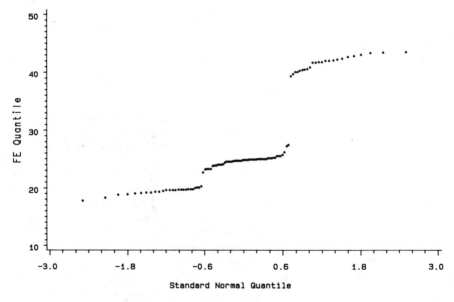

Figure 5.7 Normal quantile–quantile plot for fuel-economy measurements.

84

mainly comes from the other two vehicles (the Mercedes and the Peugeot). Thus there is reason to believe that the data should be considered to have been generated by at least three different normal distributions. This contention is supported by the distribution of the data values in the labeled point plot, Figure 4.2a.

Figure 5.7 is a plot of the quantiles of the fuel-economy data (Figure 4.10) versus standard normal quantiles. The data are obviously not generated from a normal probability distribution. The distinct clusters in Figure 4.2b confirm this conclusion.

In practice one ordinarily would not combine data from four different vehicles as was done in Figures 5.6 and 5.7. One can expect differences in the distributions for data which are subject to different experimental conditions. We combined the data from the four vehicles in order to demonstrate the type of diagnostic information that is available in plots of sample quantiles versus quantiles from a reference distribution.

In this section the normal probability distribution was emphasized in all the examples. Quantile–quantile plots can be made for any reference distribution once the theoretical quantiles are calculated. The same steps that were outlined above for normal quantile–quantile plots can then be followed for any candidate reference distribution.

Probability plots are alternatives to quantile-plot comparisons of a sample distribution with a theoretical reference distribution. Probability plots use graph paper specifically designed for the reference distribution of interest. Normal probability paper, for example, uses a scale for the data fraction f that produces approximate straight-line plots if the observed data quantiles are from a normal probability distribution. One then graphs the ordered data values $y_{(i)}$ versus the corresponding data fraction f_i and examines the plot for approximate linearity. Figure 5.8 is a normal probability plot of the nitrogen oxide data. Note that it provides the same trends as the quantile–quantile plot in Figure 5.6.

An advantage that reference quantile–quantile plots offer over other types of

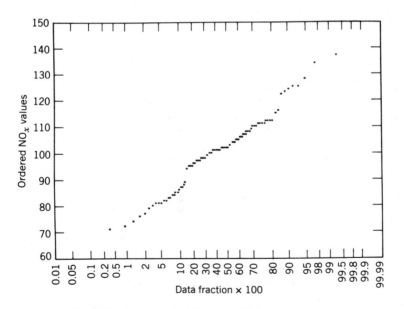

Figure 5.8 Normal probability plot of NO_x measurements.

probability plots is that special graph paper is not needed. The references at the end of this chapter should be consulted for additional information about the relationship between probability plots and reference quantile–quantile plots.

REFERENCES

Text References

Chambers, J. M., Cleveland, W. S., Kleiner, B., and Tukey, P. A. (1983). *Graphical Methods for Data Analysis*, Boston, MA: Duxbury Press.

Nelson, Wayne B. (1979). "How to Analyze Data with Simple Plots," ASQC Technical Conference Transactions, 89–94. Milwaukee, WI: American Society for Quality Control.

Shapiro, S. S. (1980). *How to Test Normality and Other Distributional Assumptions*. Milwaukee, WI: American Society for Quality Control.

Wilk, M. B. and Gnanadesikan, R. (1968). "Probability Plotting Methods for Data Analysis," *Biometrika*, **55**, 1–17.

The first, third, and fourth references contain detailed information on the construction and use of quantile plots. The first and second references detail the comparison of sample distributions with many common reference distributions.

Data References

The tire wear data were abstracted from:

Pierce, R. N., Mason, R. L., Hudson, K. E., and Staph, H. E. (1985). "An Investigation of Low Variability Tire Treadwear Test Procedures and Treadwear Adjustment for Ambient Temperature," Report No. DTNH22-84-C-07106, San Antonio, TX: Southwest Research Institute.

EXERCISES

1 A research program was funded by an oil company to evaluate a drag-reducing polymer additive for a concentrated brine solution. The polymer was tested at concentrations of 5, 20, and 40 parts per million (by weight) in a 95% saturated salt water solution. Ten tests were performed for each polymer concentration. The system flow rates (gal/min) were recorded for each test. Construct a box plot to graphically compare the different levels of polymer concentration.

Concentration	System Flow Rates
5	335, 337, 330, 325, 327, 333, 327, 325, 338, 331
20	348, 381, 335, 372, 355, 377, 361, 382, 335, 354
40	371, 387, 376, 390, 369, 364, 380, 385, 369, 383

2 Two petroleum-product refineries produce batches of fuel using the same refining process. A random sample of batches at each refinery resulted in the following values of cetane number measured on the fuel samples. Use appropriate descriptive statistics and graphics to compare the cetane measurements from the two refineries.

Refinery 1:

      50.3580 47.5414 47.6311 47.6657 47.7793 47.2890 47.5472

      48.2131 46.9531 47.9489 47.3514 48.3738 49.2652 47.2276

      48.6901 47.5654 49.1038 49.8832 48.7042 47.9148

Refinery 2:

      45.8270 45.8957 45.2980 45.4504 46.1336 46.6862 45.6281

      46.1460 46.3159 45.2225 46.1988 45.5000 45.7478 45.5658

3 Through the construction of a box plot, a sample quantile–quantile plot, and any other graphical methods deemed appropriate, assess whether the data from the two refineries in Exercise 2 appear to come from a common theoretical distribution. If they do not, how do the distributions appear to differ?

4 One fuel property that is important in turbine engine combustion is viscosity. Five laboratories were asked to prepare 20 blends of fuels that exhibit viscosities near 3.0 at 40°C. Produce standard normal quantile–quantile plots for the data collected at each of the five laboratories. How do the sample distributions compare with the standard normal distribution? If the sample distribution appears to be nonnormal, suggest reasons related to equipment or personnel that could cause this to happen in a typical laboratory.

Laboratory 1

      2.81 2.47 3.23 3.41 3.25 3.09 2.66 3.27 2.83 3.36

      2.89 4.40 3.64 3.37 3.62 2.90 3.02 3.10 2.61 3.14

Laboratory 2

      3.10 3.45 3.26 3.31 2.20 2.82 3.61 3.30 3.46 3.63

      3.29 3.46 2.46 2.74 2.90 3.67 2.80 2.49 2.89 3.50

Laboratory 3

      2.07 1.69 4.02 2.32 4.10 4.18 4.10 1.91 2.23 2.15

      4.01 4.05 2.12 4.09 3.92 2.15 2.28 3.94 4.08 2.10

Laboratory 4

      2.80 3.67 3.16 2.75 2.75 3.13 2.75 4.05 2.75 3.22

      2.75 2.75 2.77 2.75 3.22 3.58 3.71 2.75 3.75 3.29

Laboratory 5

$$4.00\ 3.86\ 3.92\ 3.71\ 3.88\ 3.94\ 3.90\ 4.00\ 3.88\ 3.93$$

$$3.95\ 3.96\ 3.95\ 3.97\ 3.87\ 3.94\ 3.99\ 3.97\ 3.95\ 3.81$$

5 Make a comparative box plot of the red-blood-cell counts taken pre-op and post-op in Exercise 8 of Chapter 4. Interpret the key comparative features evident in the plots.

6 Tabulated below are summary statistics for a six-week study of the effectiveness of a new product developed to reduce skin dryness. The dryness measurements summarized below were taken on a group of subjects prior to using the product (week 0) and weekly thereafter. A drop of two units in the dryness measure is considered an indication that the product is effective. Construct a comparative box plot using these data, and interpret the results relative to the effectiveness of the product.

Week	Min	Max	Mean	Q_1	Q_3	Median	n
0	1.0	5.0	3.1	2.6	3.6	3.0	64
1	0.2	3.2	2.0	1.7	2.6	2.0	62
2	0.5	4.8	2.0	1.2	2.8	1.6	59
3	0.4	3.0	1.6	1.1	1.8	1.4	63
4	0.2	3.0	1.3	0.8	1.7	1.2	62
5	0.0	2.9	1.4	0.8	1.7	1.2	61
6	0.0	3.5	1.4	0.8	1.9	1.2	62

7 Compare traditional and robust estimates of location and dispersion for the three repeat scans (2, 4, and 16) using the 36% concentration in Table 5.1. Based on these calculations and the box plot in Figure 5.1, which statistics would you prefer to use to summarize location and dispersion for these repeat scans? Why?

8 Calculate the m-estimator weights for the three repeat scans in Exercise 7. Use these weights to identify the outliers displayed in Figure 5.1.

9 The data below are the 25 nitrogen oxide measurements for the Volkswagen, part of the data set used to illustrate several of the graphical procedures introduced in this chapter and the previous one. Make a normal quantile–quantile plot of these data. Do the NO_x measurements for this one vehicle appear to be consistent with an assumption that the measurements are normally distributed? Why (not)?

$$0.71\ 0.72\ 0.74\ 0.76\ 0.84\ 0.94\ 0.83\ 0.86\ 0.81$$

$$0.85\ 0.82\ 0.79\ 0.99\ 1.11\ 0.80\ 0.82\ 0.98$$

$$0.77\ 0.81\ 0.87\ 0.85\ 0.82\ 0.84\ 0.88\ 0.83$$

PART II

Experimental Design

CHAPTER 6

Statistical Principles in Experimental Design

This chapter motivates the use of statistical principles in the design of experiments. Several important facts are stressed:

- *statistically designed experiments are economical,*
- *they allow one to measure the influence of one or several factors on a response,*
- *they allow the estimation of the magnitude of experimental error, and*
- *experiments designed without adhering to statistical principles usually violate one or more of these desirable design goals.*

Test procedures in scientific and engineering experiments are often primarily guided by established laboratory protocol and subjective considerations of practicality. While such experimental procedures may be viewed as economical in terms of the number of test runs that must be conducted, the economy of effort can be deceiving for two reasons. First, economy is often achieved by severely limiting the number of factors whose effects are studied. Second, the sequence of tests may require that only one of the factors of interest be varied at a time, thereby preventing the evaluation of any joint effects of the experimental factors. The effective use of statistical principles in the design of experiments ensures that experiments are designed economically, that they are efficient, and that individual and joint factor effects can be evaluated.

In the next several chapters a variety of statistical experimental designs are presented. In this chapter we discuss general concepts that arise in virtually all experimental settings. The chapter begins with an introduction to the common terminology used in discussing statistical design procedures. Statistical experimental design is then motivated by an examination of problems that frequently arise when statistical principles are not used in the design and conduct of a test program. Special emphasis is placed on the investigation of the joint effects of two or more experimental factors on a response. Finally, a summary of important design considerations is presented.

6.1   EXPERIMENTAL-DESIGN TERMINOLOGY

The terminology of experimental design is not uniform across disciplines or even, in some instances, across textbooks within a discipline. For this reason we begin our discussion of statistical experimental design with a brief definition of terms. Table 6.1 contains definitions of many terms which are in common use. A few of these terms have already been used in previous chapters but are included here for completeness.

The terms "response" and "factor" were defined in Chapter 2 (Section 2.1). A response variable is an outcome of an experiment. It may be a quantitative measurement such as the percentage by volume of mercury in a sample of river water, or it may be a qualitative result such as whether an aircraft engine mounting bolt can withstand a required shearing force. A factor is an experimental variable that is being investigated to determine its effect on a response. It is important to realize that a factor is considered controllable by the experimenter; i.e., the values, or *levels*, of the factor can be determined prior to the beginning of the test program and can be executed as stipulated in the experimental design. While the term "version" is sometimes used to designate categorical or qualitative levels of a factor, we use "level" to refer to the values of both qualitative and quantitative factors. An *experimental region*, or *factor space*, consists of all possible levels of the factors that are candidates for inclusion in the design. For quantitative factors, the factor space often is defined by lower and upper limits for the levels of each factor.

Additional variables that may affect the response but cannot be controlled in an

---

Table 6.1   Experimental-Design Terminology

Block. Group of homogeneous experimental units.

Confounding. One or more effects that cannot unambigously be attributed to a single factor or interaction.

Covariate. An uncontrollable variable that influences the response but is unaffected by any other experimental factors.

Design (layout). Complete specification of experimental test runs, including blocking, randomization, repeat tests, replication, and the assignment of factor–level combinations to experimental units.

Effect. Change in the average response between two factor–level combination or between two experimental conditions.

Experimental region (factor space). All possible factor–level combinations for which experimentation is possible.

Factor. A controllable experimental variable that is thought to influence the response.

Homogeneous experimental units. Units that are as uniform as possible on all characteristics that could affect the response.

Interaction. Existence of joint factor effects in which the effect of each factor depends on the levels of the other factors.

Level. Specific value of a factor.

Repeat tests. Two or more observations that have the same levels for all the factors.

Replication. Repetition of an entire experiment or a portion of an experiment under two or more sets of conditions.

Response. Outcome or result of an experiment.

Test run. Single combination of factor levels that yields an observation on the response.

Unit (item). Entity on which a measurement or an observation is made; sometimes refers to the actual measurement or observation.

---

experiment are called *covariates*. Covariates are not additional responses; i.e., their values are not affected by the factors in the experiment. Rather, covariates and the experimental factors jointly influence the response. For example, in many experiments both temperature and humidity affect a response, but the laboratory equipment can only control temperature; humidity can be measured but not controlled. In such experiments temperature would be regarded as an experimental factor and humidity as a covariate.

A *test run* is a single factor–level combination for which an experimental observation (response) is obtained. *Repeat tests* are two or more experimental observations that are obtained for a specified combination of levels of each of the factors. Repeat tests are conducted under as identical experimental conditions as possible, but they need not be obtained in back-to-back test runs. Repeat tests should not be two or more analytical determinations of the same response; they must be two or more identical but distinct test runs. *Replications* are repetitions of a portion of the experiment (or the entire experiment) under two or more different conditions, e.g., on two or more different days.

Experimental responses are only comparable when they result from observations taken on "homogeneous" experimental units. An experimental unit was described in Chapter 2 (Section 2.1) as either a measurement or material on which a measurement is made. (Note: In keeping with the above discussion, "measurement" as used here can be either quantitative or qualitative.) Unless explicitly stated otherwise, we shall use the term to refer to a physical entity on which a measurement is made. *Homogeneous* experimental units do not differ from one another in any systematic fashion and are as alike as possible on all characteristics that might effect the response. While there is inherent random variation in all experimental units, the ability to detect important factor effects and to estimate these effects with satisfactory precision depends on the degree of homogeneity among the experimental units.

If all the responses for one level of a factor are taken from experimental units that are produced by one manufacturer and all the responses for another level of the factor are taken from experimental units produced by a second manufacturer, any differences noted in the responses could be due to the different levels of the factor, to the different manufacturers, or to both. In this situation the effect of the factor is said to be *confounded* with the effect due to the manufacturers.

When a satisfactory number of homogeneous experimental units cannot be obtained, statistically designed experiments are often *blocked* so that homogeneous experimental units receive each level of the factor(s). Blocking divides the total number of experimental units into two or more groups or blocks (e.g., manufacturers) of homogeneous experimental units so that the units in each block are more homogeneous than the units in different blocks. Factor levels are then assigned to the experimental units in each block. If more than two blocks of homogeneous experimental units can be obtained from each manufacturer, both repeat tests (two or more identical factor–level combinations on units within a block) and replication (repetition of the design for one or more of the blocks from each manufacturer) can be included in the experiment.

The terms *design* and *layout* often are used interchangeably when referring to experimental designs. The layout or design of the experiment includes the choice of the factor–level combinations to be examined, the number of repeat tests or replications (if any), blocking (if any), the assignment of the factor–level combinations to the experimental units, and the sequencing of the test runs.

An *effect* of the design factors on the response is measured by a change in the average response under two or more factor–level combinations. In its simplest form, the effect of a single two-level factor on a response is measured as the difference in the average

Table 6.2 Design for Suspended-Particulate Study*

Run No.	Test Fluid	Pipe Angle (degrees from horizontal)	Flow Rate (ft/sec)
1	1	60	60
2	2	30	60
3	1	60	90
4	2	45	60
5	2	15	90
6	1	15	60
7	2	15	60
8	1	45	60
9	2	45	90
10	1	15	90
11	1	45	90
12	1	30	60
13	2	60	60
14	1	30	90
15	2	30	90
16	2	60	90

*Covariate: temperature (°C).

response for the two levels of the factor; i.e.,

$$\text{factor effect} = \text{average response at one level}$$
$$- \text{average response at a second level.}$$

Factor effects thus measure the influence of different levels of a factor on the value of the response. Individual and joint factor effects are discussed in Section 7.3.

In order to illustrate the usage of these experimental-design terms, two examples are now presented. The design shown in Table 6.2 is for an experiment which is to be

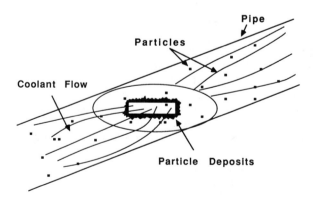

Figure 6.1 Suspended-particulate experiment.

conducted to study the flow of suspended particles in two types of coolants used with industrial equipment. The coolants are to be forced through a slit aperature in the middle of a fixed length of pipe (see Figure 6.1). Using two different flow rates and four different angles of inclination of the pipe, the experimenters wish to study the buildup of particles on the edge of the aperature.

There are three experimental factors in the study: coolant, a qualitative factor at two levels (1, 2); pipe angle, a quantitative factor at four levels (15, 30, 45, 60 degrees from horizontal); and flow rate, a quantitative factor at two levels (60, 90 ft/sec). All sixteen combinations (two coolants × four angles × two rates) of the factor levels are to be included in the design. The sequencing of the tests was determined by randomly assigning a run number to the factor–level combinations. All the test runs are to be conducted during a single day in order to eliminate day-to-day variation. It is believed, however, that the expected 20-degree change in temperature from early morning to late afternoon may have an effect on the test results. For this reason temperature will be measured as a covariate.

The second example, shown in Figure 6.2, represents an agricultural experiment in which five soil treatments (e.g., different types of fertilizer, different methods of plowing) are to be investigated to determine their effects on the yield of soybean plants. The experiment must be conducted on three different fields in order to obtain a sufficient

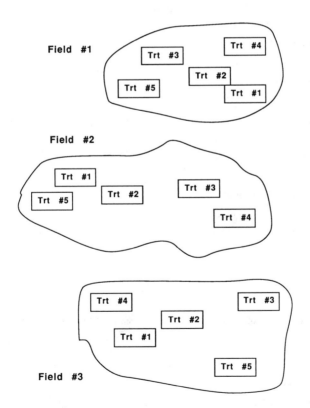

Figure 6.2 Layout for agricultural experiment.

number of homogeneous plots of ground. Each field contains five such homogeneous plots; however, the soil conditions on each field are known to be different from those on the other two fields.

In this experiment the fields are blocks and the plots are the experimental units. One qualitative factor, soil treatment, having five levels is under investigation. The soil treatments are randomly assigned to the plots in each field (block). The response variable is the yield in kilograms per plot.

In Table 6.2 the factor–level combinations are listed in the order in which the test runs will be conducted. There are no physical experimental units per se. Each test run takes the place of an experimental unit. In Figure 6.2 an experimental unit is a plot of ground. Unlike the previous example, there is no test sequence; all the tests are conducted simultaneously. The two examples illustrate two different ways to specify the statistical design of an experiment; in run order or as a physical layout.

6.2   COMMON DESIGN PROBLEMS

When statistical considerations are not incorporated in the design of an experiment, statistical analyses of the results are often inconclusive or, worse yet, misleading. Table 6.3 lists a few of many potential problems that can occur when statistical methodology is not used to design scientific or engineering experiments.

6.2.1   Masking of Factor Effects

Researchers often invest substantial project funds and a great amount of time and effort only to find that the research hypotheses are not supported by the experimental results. Many times the lack of statistical confirmation is due to the inherent variability of the test results.

Consider for example the data listed in Table 6.4. These are test results measuring cylinder pressure in a single-cylinder engine under strictly controlled laboratory conditions. The test results are for 32 consecutive firing cycles of the engine. Note the

Table 6.3   Common Experimental-Design Problems

- Experimental variation masks factor effects.
- Uncontrolled factors compromise experimental conclusions.
- Erroneous principles of efficiency lead to unnecessary waste or inconclusive results.
- Scientific objectives for many-factor experiments may not be achieved with one-factor-at-a-time designs.

Table 6.4   Cylinder-Pressure Measurements under Controlled Laboratory Conditions

229	191	238	231	253	189	224	224
191	201	200	201	197	206	220	214
193	226	237	209	161	187	237	245
181	213	231	217	212	207	242	186

Average = 212.28,   Standard Deviation = 21.63

Table 6.5 Skin Color Measurements

	Color Measurement		
Participant	Week 1	Week 2	Week 3
A	12.1	14.2	13.9
B	19.1	17.6	16.2
C	33.8	34.7	33.2
D	33.0	31.7	30.3
E	35.8	37.7	35.6
F	42.0	38.4	41.5
G	36.8	35.2	35.7

variation in the test results even for this highly controlled experiment. If an experiment is conducted using this engine and the factor effects are of the same order of magnitude as the variation evident in Table 6.4, the effects may go undetected.

Table 6.5 illustrates another kind of variation in test results. In this study of skin color measurements not only is there variation among the participants, but there is also variation for each participant over the three weeks of the study. Experiments that are intended to study factor effects (e.g., suntan products) on skin color must be designed so that the variation in subjects and across time does not mask the effects of the experimental factors.

Figure 6.3 is a schematic representation of the relationship between the detectability of factor effects and the variability of responses. In this figure, test results from two levels of a factor are indicated by squares and circles, respectively. In both cases shown, the

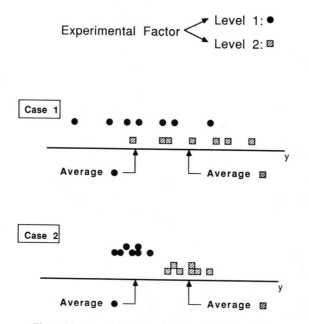

Figure 6.3 Experimental variability and factor effects.

average response for each factor level remains constant; consequently, the factor effect (the difference in the averages of the two levels) does not change. Only the variability of the response changes from case to case.

In case 1, the variability of the test results is so great that one would question whether the factor effect is (a) measuring a true difference in the population or process means corresponding to the two factor levels or (b) simply due to the variation of the responses about a common mean. The data may not provide sufficiently convincing evidence that the two population means are different because of the variability in the responses.

In case 2, the variation of the responses is much less than that in case 1. There is strong evidence, due to the small variation in the responses relative to the large differences in the averages, that the factor effect is measuring a substantial difference between the population or process means corresponding to the two factor levels. Thus, the difference in the means due to the factor levels would be masked by the variability of the responses in case 1 but not in case 2. The implication of this example for the statistical design of experiments is that the variation in case 1 must be compensated for (e.g., by blocking or a large experiment size) in order to ensure that the difference in the means is detectable.

The importance of this discussion is that experimental error variation must be considered in the statistical design of an experiment. Failure to do so could result in true factor effects being hidden by the variation in the observed responses. Blocking and sample size are two key considerations in controlling or compensating for response variability.

6.2.2   Uncontrolled Factors

A second problem listed in Table 6.3 that frequently occurs when statistical considerations are not included in the design phase of a project is the effect of uncontrolled factors on the response. While few researchers would intentionally ignore factors that are known to exert important influences on a response, there are many subtle ways in which failure to carefully consider all factors of importance can compromise the conclusions drawn from experimental results.

Consider, for example, an experiment conducted to compare average carbon monoxide (CO) emissions for a commercial gasoline (base fuel) and five different methanol fuel blends. Suppose further that an adequate experimental design has been selected, including a sufficiently large number of test runs. Figure 6.4 exhibits actual results of such a test program. One of the important conclusions that can be drawn from an analysis of the experimental data is that the last two fuels have significantly lower average CO emissions than the other fuels tested.

Subsequent to a conclusion such as this, researchers often wish to determine which of several fuel properties (e.g., distillation temperatures, specific gravity, oxygen content) contribute to the reduction in the average emission levels. The difficulty with such a determination is that the fuel properties of interest cannot all be specifically controlled in the selection of fuel blends. Because of this, many fuel properties that might subsequently be of interest simultaneously vary across the six fuels, resulting in a confounding of their effects.

This experiment was specifically designed only to investigate the effects of six fuel blends on CO emissions. Studies could be specifically designed to study some of the fuel properties. In such studies the fuel properties would be varied in a systematic fashion and confounding among the properties could be eliminated by the choice of the design. Note that in this example it is the uncontrolled variation of the fuel properties that leads to the confounding of their effects on the response.

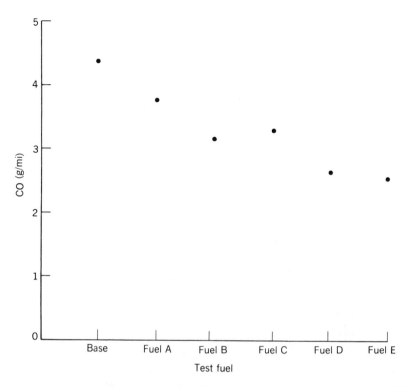

Figure 6.4 Average carbon monoxide emissions. Data from Coordinating Research Council (1984). "Performance Evaluation of Alcohol–Gasoline Blends in 1980 Model Automobiles, Phase II— Methanol–Gasoline Blends," Atlanta, GA: Coordinating Research Council, Table G-8.

Another example in which uncontrolled factors influence a response is depicted in Figure 6.5. In this graph, results of two methods of determining tire wear are plotted: the *groove-depth* and the *weight-loss* method. In studies of this type one sometimes seeks to determine an empirical relationship between the two measurements, perhaps in order to calibrate the two methods of measurement.

The major source of the relationship between the two methods of estimating tread life is not the close agreement between the respective measurements. Some plotted points in Figure 6.5 that represent similar measurements on one method differ by as much as 5000 miles on the other method. The major source of the association among the plotted points is the large variation among the observations due to uncontrolled factors such as road conditions, weather, vehicles, and drivers. Thus it is not so much a close relationship between the methods of estimating tread life as it is the large differences in uncontrolled test conditions (and consequent large differences in tire wear) that contributes to the linear trend observable in Figure 6.5.

6.2.3 Erroneous Principles of Efficiency

The preceding examples demonstrate the need to construct designs in which factors of interest are systematically varied and to consider the likely magnitude of the inherent

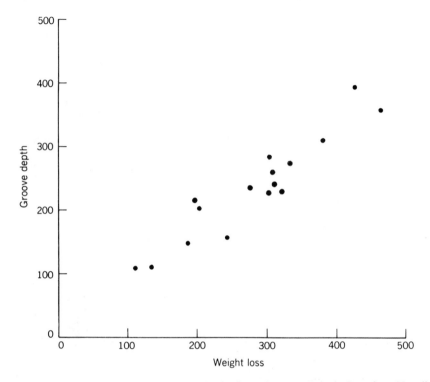

Figure 6.5 Estimates of tread life (in hundreds of miles) using two methods. Data from Natrella (1963, pp. 5–33).

variation in the test results when planning the number of test runs. The third problem listed in Table 6.3 suggests that the desire to run economical experiments can lead to strategies that may in fact be wasteful or even prevent the attainment of the project's goals. The latter difficulty is elaborated on in the next section, where one-factor-at-a-time testing is discussed.

Time and cost efficiencies are always important objectives in experimental work. Occasionally efficiency becomes an overriding consideration and the project goals become secondary. If time or budgetary considerations lead to undue restrictions on the factors and levels that can be investigated, the project goals should be reevaluated relative to the available resources. This may lead to a decision to forgo the experimentation.

The problem of experiment efficiency is most acute when several factors must be investigated in an experiment. When guided only by intuition, many different types of designs could be proposed, each of which might lead to flawed conclusions. Some would choose to hold factors constant that could have important influences on the response. Others would allow many unnecessary changes of factors that are inexpensive to vary and few changes of critical factors that are costly to vary.

Efficiency is achieved in statistically designed experiments because each observation generally provides information on all the factors of interest in the experiment. Table 6.6 shows the number of test runs from Table 6.2 that provide information on each level of the

Table 6.6   Number of Test Runs for Each Factor Level:
Suspended-Particulate Study

Factor	Level	Number of Test Runs
Test fluid	1	8
	2	8
Flow rate	60	8
	90	8
Pipe angle	15	4
	30	4
	45	4
	60	4
Equivalent single-factor experiment size		48

test factors. If each of the test factors had been investigated separately using the same number of test runs shown in Table 6.6, then 48 test runs would have been needed.

Information on individual and joint factor effects can be obtained from a highly efficient experiment such as the one displayed in Table 6.2. It is neither necessary nor desirable to investigate a single factor at a time in order to economically conduct experiments. Because there is a prevalent view that one-factor-at-a-time testing is appropriate when there are several factors to be investigated, we now focus on this type of experimentation.

6.2.4   One-Factor-at-a-Time Testing

Consider an experimental setting in which one wishes to determine which combinations of the levels of several factors optimize a response. The optimization might be to minimize the amount of impurities in the manufacture of a silicon wafer. It might be to maximize the amount of a compound produced from a reaction of two or more chemicals. It might be to minimize the number of defective welds made by a robot on an assembly line.

Table 6.7   One-Factor-at-a Time Test Sequence

Stage 1:	Fix levels of factors $2, 3, \ldots, k$; determine the optimal level of factor 1.
Stage 2:	Use the optimal level of factor 1 from stage 1; fix levels of factors $3, 4, \ldots, k$; determine the optimal level of factor 2.
Stage 3:	Use the optimal levels of factors 1, 2 from stages 1, 2; fix levels of factors $4, 5, \ldots, k$; determine optimal level of factor 3.
$\vdots$	$\vdots$
Stage k:	Use the optimal levels of factors $1, 2, \ldots, k-1$ from stages $1, 2, \ldots, k-1$; determine the optimal level of factor k.

In each of these examples the optimization is a function of several factors, which can be experimentally investigated. Due to the feared complexity of simultaneously investigating the influence of several factors on a response, it is common practice to vary one factor at a time in the search for an optimum combination of levels of the factors. Table 6.7 lists a sequence of steps in the conduct of such an experiment.

The perceived advantages of one-factor-at-a-time testing are primarily two:

(i) the number of test runs is believed to be close to the minimum that can be devised to investigate several factors simultaneously, and

(ii) one can readily assess the factor effects as the experiment progresses, since only a single factor is being studied at any stage.

These attractive features of one-factor-at-a-time testing are more than offset by the potential for failure to achieve the optimization sought in the design of the study.

Figure 6.6 is a typical contour plot of the yield of a chemical reaction. The curves in the figure are curves of constant yield as a function of the two factors of interest, temperature and reaction time. The curve labeled 45, for example, identifies combinations of temperature and reaction time for which the yield of the chemical reaction is 45%.

Many industrial experiments are conducted because contours such as those depicted in Figure 6.6 are unknown. Suppose one does not know the contours in Figure 6.6 and one wishes to conduct an experiment to determine combinations of temperature and reaction time that will maximize the yield. If the range of interest for the temperature variable is

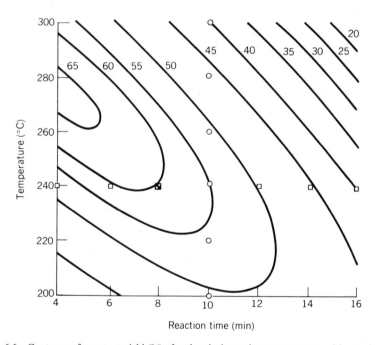

Figure 6.6 Contours of constant yield (%) of a chemical reaction. ○, test runs with reaction time fixed at 10 min; □, test runs with temperature fixed at 240°C; ▣, optimal combination from this experiment.

200–300°C and that of the reaction time is 4–16 min, the region graphed in Figure 6.6 will include the entire region of interest. If one conducts a one-factor-at-a-time experiment by fixing reaction time and varying temperature from 200 to 300°C in increments of 20°, it would be reasonable to set the reaction time to 10 min, the middle of the range of interest.

The points identified by circles in Figure 6.6 represent the six observations that would be taken in this first stage of the experiment. Note that one might stop this stage of the testing after the test run at 260°C, since the yield would be declining noticably thereafter. The optimal level for temperature based on these test runs would be 240°C.

The second stage of the experiment, identified by the squares in Figure 6.6, is a sequence of test runs for different reaction times with temperature set at its "optimal" level of 240°C. The optimal level of reaction time would be determined to be 8 min if observations were taken in 2-min increments as indicated in the figure. Note that the optimal levels of temperature (240) and reaction time (8) determined in this manner produce a yield of 60%.

The true maximum yield for this chemical process exceeds 70%. It is obtainable at a temperature of approximately 270°C and a reaction time of 4 min. Only by experimenting in a suitable grid of points that includes the corners of the experimental region can one locate optimum yields such as the one depicted in Figure 6.6.

This one-factor-at-a-time experiment also does not allow the fitting of a model from which the contours shown in Figure 6.6 can be drawn. This is because the observations along the two paths indicated by the circles and squares cannot characterize the curvature of the region. This is another drawback to one-factor-at-a-time testing, since many times the goal of a study is not only to optimize the response but also to model its behavior over the experimental region (factor space).

A second example of an investigation in which one-factor-at-a-time experimentation may fail to achieve an optimum response is shown in Table 6.8. This investigation is intended to identify optimal combinations of three catalysts for reducing the reaction time of a chemical process. The desired ranges on the catalysts are 3–6% for catalysts A and B, 4–8% for catalyst C. Suppose that in an initial investigation of this process, only combinations of the two extreme levels of these factors are to be examined; i.e., catalyst $A = (3, 6)$, catalyst $B = (3, 6)$, and catalyst $C = (4, 8)$.

Table 6.8 lists the results of two possible one-factor-at-a-time experiments. In

Table 6.8   Reaction Times of a Chemical Process

	Level of Catalyst			Reaction
Run No.	A	B	C	Time (sec)
Experiment 1				
1	3	3	4	> 60
2	6	3	4	> 60
3	3	6	4	> 60
4	3	3	8	> 60
Experiment 2				
5	6	6	4	> 60
6	3	6	8	> 60
7	6	3	8	54.3
8	6	6	8	27.6

experiment 1 the first test run has all three catalysts at their low levels. The reaction time was longer than one minute, an unacceptable duration. The second test run holds catalysts B and C fixed and changes A to its upper level. Again the reaction time is unacceptable. Since the higher amount of catalyst A produced no better results than its lower level, the third test run again sets catalyst A to its lower level and changes catalyst B to its upper level. The reaction time is unacceptable. Finally, the last run of this experiment has catalyst C at its upper level and the other two catalysts at their lower levels. Again the results are unacceptable.

This application of one-factor-at-a-time testing might lead to the conclusion that no combination of catalysts would reduce the reaction time to an acceptable amount. Note, however, that the sequence of test runs for experiment 2 does lead to acceptable reaction times. Experiment 2 systematically tests pairs of factors at their upper levels. This experiment would lead to the optimal choice in four test runs. If experiment 2 were conducted as a continuation of experiment 1, the optimal combination would not be identified until all eight possible combinations of the factor levels have been tested. In this case one-factor-at-a-time testing achieves no economy of effort relative to testing all possible combinations of the factor level.

The drawbacks to this type of testing should now be apparent. Optimal factor combinations may not be obtained when only one factor is varied at a time. Also, the combinations of levels that are tested do not necessarily allow appropriate models to be fitted to the response variable. Additional test runs may have to be added if an estimate of experimental error is to be obtained.

One-factor-at-a-time experimentation is not only used to determine an optimum combination of factors. Often this type of testing is used merely to assess the importance of the factors in influencing the response. This can be an impossible task with one-factor-at-a-time designs if the factors jointly, not just individually, influence the response.

One-factor-at-a-time experimentation does not always lead to incorrect or suboptimal results. The examples used in this section are intended to illustrate the dangers that this type of experimentation pose. As mentioned in the introduction to this chapter, there are economically efficient statistical experimental designs that do permit the fitting of curved response contours, the investigation of joint factor effects, and estimation of experimental-error variation. Many of these designs are discussed in subsequent chapters.

6.3   SELECTING A STATISTICAL DESIGN

In order to avoid many of the potential pitfalls of experimentation that were mentioned in the previous sections of this chapter, several key criteria should be considered in the design of an experiment. Among the more important design considerations are those listed in Table 6.9.

6.3.1   Consideration of Objectives

The first criterion listed in Table 6.9 is the most obvious and necessary for any experiment. Indeed, one always sets goals and determines what variables to measure in any research project. Yet each of the three subheadings in this criterion has special relevance to the statistical design of an experiment.

Consideration of the nature of the anticipated conclusions can prevent unexpected complications when the experiment is finished and the research report is being written. The

Table 6.9   Statistical Design Criteria

Consideration of Objectives

- Nature of anticipated conclusions
- Definition of concepts
- Determination of observable variables

Factor Effects

- Elimination of systematic error
- Measurement of covariates
- Identification of relationships
- Exploration of entire experimental region

Precision

- Estimation of variability (uncertainty)
- Blocking
- Repeat tests, replication
- Adjustment for covariates

Efficiency

- Multiple factors
- Screening designs
- Fractional factorials

Randomization

fuel study discussed in Section 6.2.2 is a good example of the need to clearly resolve the specific objectives of an experiment. If one's goal is to study the effects of various fuel properties on a response, the experimental design might be altogether different than if one's goal is just to compare six fuels.

Concept definition and the determination of observable variables influence both the experimental design and the collection of information on uncontrollable factors. For example, suppose one wishes to study the effects of radiation on human morbidity and mortality. One cannot subject groups of humans to varying levels of radiation, as one would desire to do in a designed experiment, if it is believed that such exposure would lead to increased illness or death. Alternatives include studies with laboratory animals and the collection of historical data on humans. The latter type of study is fraught with the problem of uncontrollable factors similar to that of the tire-wear example in Section 6.2.2. Laboratory studies allow many important factors to be controlled, but the problem of drawing inferences on human populations from animal studies arises.

Apart from these difficulties, statistical issues relating to the measurement of responses and covariates arise. For example, what measure of mortality should be used? Should one use raw death rates (e.g., number of deaths per 100,000 population) or should the death rates be age-, sex-, or race-adjusted? Should covariates such as atmospheric radiation be measured? If so, how will the analysis of the experimental data incorporate such covariate measurements?

6.3.2 Factor Effects

A second criterion that must be considered in the selection of an appropriate statistical design is the likely effects of the factors. Inclusion of all relevant factors, when experimentally feasible, is necessary to ensure that uncontrolled systematic variation of these factors will not bias the experimental results. An accounting for uncontrollable systematic variation through the measurement of covariates is necessary for the same reason.

Anticipated factor effects also influence the choice of a statistical design through their expected relationships with one another. If each factor is believed to affect the response independently of any other factor or if joint effects of the factors are of secondary interest (as in certain pilot studies), screening designs (Chapter 9) can be used to assess the effects of the factors. If the effect of one factor on the response depends on the levels of the other factors, a larger design is needed.

In general, an experimental design should allow the fitting of a general enough model so that the salient features of the response and its relationships with the factors can be identified. For example, the design should permit polynomial terms in the quantitative factors to be included in the fitted model so that curvature in the response function can be assessed. The design should permit an assessment of the adequacy of the fitted model. If the fitted model is judged to be an inadequate representation of the response function, the design should form the basis for an expanded design from which more elaborate models (e.g., higher-order polynomials) can be fitted.

When assessing factor effects it is important to explore the entire experimental region of interest. The combinations of factor levels used in a statistical design should be selected to fill out the experimental region. If a factor is only studied over a narrow portion of the experimental region, important effects may go undetected.

6.3.3 Precision and Efficiency

The next two categories of design criteria listed in Table 6.9, precision and efficiency, will be amply discussed in each of the next several chapters as they relate to specific statistical designs. Since the term "precision" is used frequently throughout this book, we comment briefly on its meaning in this section.

Precision refers to the variability of individual responses or to the variability of effects that measure the influence of the experimental factors (see Exhibit 6.1). Precision is a property of the random variables or statistics and not of the observed values of those variables or statistics. For example, an (observed) effect is said to be sufficiently precise if the standard deviation of the statistic that measures the effect is suitably small. In its simplest form, an effect is simply the difference between two averages. The corresponding statistic is

$$\bar{y}_1 - \bar{y}_2.$$

An observed effect is then said to be sufficiently precise if the standard deviation (or, equivalently, the variance) of this statistic is sufficiently small. In practice, one does not know the value of the standard deviation, but it can be estimated from the data.

EXHIBIT 6.1

Precision. The precision of individual responses refers to the variability inherent in independent observations taken on a response under identical conditions. The precision of a factor effect refers to the variability of the statistic that is used to measure the effect. Satisfactory precision is usually defined in terms of a sufficiently small standard deviation of the response variable or the statistic measuring the effect, respectively.

Blocking, repeat tests, replication, and adjustment for covariates can all increase precision in the estimation of factor effects. Blocking increases precision (decreases variability) by controlling the systematic variation attributable to nonhomogeneous experimental units or test conditions. Adjustment for covariates increases precision by eliminating the effects of uncontrolled factors from the variability that would otherwise be attributed to random error.

Repeat tests and replication increase precision by reducing the standard deviations of the statistics used to estimate effects. For example, as mentioned in Section 2.2 (see also Section 12.1), the standard deviation of a sample mean based on n independent observations from a single population or process is $\sigma/n^{1/2}$, where σ is the population or process standard deviation. Increasing the number of repeat tests or replications increases the sample size n, thereby decreasing the standard deviation.

6.3.4 Randomization

Randomization of the sequence of test runs or the assignment of factor–level combinations to experimental units protects against unknown or unmeasured sources of possible bias. Randomization also helps validate the assumptions needed to apply certain statistical techniques.

The protection that randomization affords against unknown bias is easily appreciated by considering the common problem of instrument drift. If during an experiment instrument drift builds over time as illustrated in Figure 6.7, later tests will be biased because of the drift. If all tests involving one level of a factor are run first and all tests

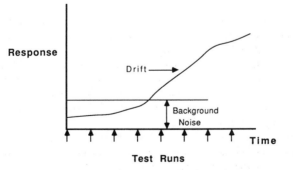

Figure 6.7 Influence of machine drift; test runs taken as indicated by arrows.

involving the second level of a factor are run last, comparisons of the factor levels will be biased by the instrument drift and will not provide a true measure of the effect of the factor.

Randomization of the test runs cannot prevent instrument drift. Randomization can help ensure that all levels of a factor have an equal chance of being affected by the drift. If so, differences in the responses for pairs of factor levels will likely reflect the effects of the factor levels and not the effect of the drift.

The design criteria listed in Table 6.9 are not intended to be comprehensive. They are presented as a guide to some of the more important considerations that must be addressed in the planning stages of most experiments.

6.4 DESIGNING FOR QUALITY

The contributions of statistical methodology to strategies for quality control can be divided into two broad categories: on-line and off-line statistical proceures. The control-chart techniques described in Section 4.5 are an example of on-line procedures. Such procedures assure that the process or system is in statistical control and that it maintains whatever consistency it is capable of achieving.

While in the past it was generally believed that on-line statistical quality-control techniques offer statisfactory assurances of customer satisfaction, today it is recognized that off-line investigations using engineering design techniques and statistical design of experiments provide the greatest opportunity for quality improvement and increased productivity. Off-line experiments are performed in laboratories, pilot plants, and preliminary production runs, ordinarily prior to the complete implementation of production or process operations. As suggested by these considerations, statistical design of experiments is an integral component of off-line quality-improvement studies.

A high-quality product results when key product properties have both the desired average value (target or aim) and small variation around the target (consistency or uniformity). Once a target has been determined, based on customer needs and manufacturing capabilities, quality improvement centers on achieving the target value (on the average) and on reducing variability about the target value.

One widely used Japanese quality-improvement philosophy, the *Taguchi approach*, has statistical design of experiments as its core. This off-line approach to product quality improvement integrates engineering insight with statistically designed experiments. Together these experimental strategies are used to determine process conditions for achieving the target value and to identify variables that can be controlled to reduce variation in key performance characteristics.

It is important to note that these "new" philosophies and experimental strategies are receiving great publicity and are being highly promoted primarily because of the renewed emphasis on product and service quality improvement as a means to attain competitive advantage. Most of the basics of these philosophies have been advocated for many years by quality-control specialists. The fundamentals of the experimental design strategies are also well known. The reason for the new popularity of these philosophies and experimental strategies is the newly competitive environment of today's marketplace in many manufacturing and service industries.

The statistical design techniques discussed in this book can be used to determine desired factor settings so that a process average or a quality characteristic of key product properties are close to the target (on aim) and the variability (standard deviation plus offset from target) is as small as possible. All the procedures recommended, including screening designs, factorial experiments, split-plot designs, and response-surface designs, can be used in suitable quality-improvement settings.

The responses of interest in quality-improvement studies include both measures of product or process location and measures of dispersion. The emphasis in the discussions in this book is on the analysis of location measures. Measures of dispersion can also be used as response variables. Note, however, the distinction between a response that is a measure of dispersion for a specific set of factor levels and a random factor effect (see Section 17.1). In the former case, several observations are made at the same levels of the factors, and a measure of dispersion (e.g., the standard deviation) is calculated and used as the response for that factor-level combination.

When dispersion measures are of interest to an investigator, $\ln s$ is often used as the response, where the standard deviation s is calculated for each of the sets of repeat observations corresponding to each of the factor-level combinations in the design. Ordinarily an equal number r of repeat test runs is required at each factor-level combination, with $r > 4$.

If both location and dispersion measures are of interest to an investigator, the two response functions (one for level and one for dispersion) can be overlaid graphically with contour plots to explore the tradeoffs in achieving optimal levels of the design factors. Many times the goal will not be to find optimal points (maxima or minima) in the response surfaces that coincide, but rather to locate flat regions (mesas or plains) that give stability of the responses. This is particularly true in product design (robustness) and process design (process control).

Some of the major benefits associated with off-line quality improvement procedures can be extended to production or process facilities in full operation. Evolutionary operation (EVOP) is used as an experimental strategy in such environments when only two or three factors can be varied at a time, and only small changes in the factor levels can be tolerated. As such, EVOP is a hybrid of on-line and off-line quality improvement techniques.

EVOP implements statistical designs on operating production or process facilities as part of the normal operations of these facilities. In this manner information for process improvement is collected by operating personnel, and normal production can continue undisturbed. Two-level factorial experiments (Chapter 7) around a center point are typically used. As operating conditions that lead to improved process characteristics are identified, the experimental region is moved to explore around this new set of conditions. This procedure is repeated until no further improvement is obtained.

The EVOP strategy has many of the desirable statistical design criteria listed in Table 6.9. However, certain risks are inherent due to exploring a limited experimental region and a small number of factors.

A consequence of the EVOP approach for process improvement is that many repeat test runs are needed at each set of factor-level combinations. This large number of repeat tests is needed because factor levels can only be changed by small amounts so that existing quality levels will not be seriously degraded at some of the factor-level settings. Because of this requirement, there is a weak "signal" (change in the response) relative to the "noise" (experimental error or process variation). This usually results in the need to collect many observations so that the standard deviations of the statistics used to measure the effects are sufficiently small and statistically significant effects can be detected.

From this discussion, it is apparent that implementation of EVOP procedures requires a commitment by management to an extended experiment before major improvements in quality can be realized. The extended investigation is necessitated by the large number of observations that are required and the limitation to studying only two or three factors at a time. The dangers of having confounded effects and of not detecting important joint factor effects adds to the necessity for a commitment by management to an ongoing, long-lasting (preferably permanent) program.

Thus, experimentation as part of off-line quality improvement investigations or on-line ones using EVOP methodology, as well as on-line quality monitoring using control charts, can utilize many of the design and analysis techniques discussed in this book. The benefits in terms of customer satisfaction and retention, reduction of scrap or rework, and increased productivity for these quality-improvement programs well outweigh the costs and other requirements for implementation of any of these on-line or off-line procedures.

REFERENCES

Text References

Box, G. E. P., Hunter, W. G., and Hunter, J. S. (1978). *Statistics for Experiments*, New York: John Wiley and Sons, Inc.
  Chapter 1 of this text contains an excellent introduction to the philosophy of scientific experimentation, including the role of statistical experimental design in such experimentation.

Cox, D. R. (1958). *Planning of Experiments*, New York: John Wiley and Sons, Inc.

Davies, O. L. (Ed.) (1971). *Design and Analysis of Industrial Experiments*, New York: John Wiley and Sons, Inc.
  These texts contain detailed discussions on the principles of statistical experimental design. Comparative experiments, randomization, sample-size considerations, blocking, the use of covariates, and factorial experiments all receive extensive treatment. Much of the discussion centers around examples.

Natrella, M. G. (1963). *Experimental Statistics*, National Bureau of Standards Handbook 91, Washington, D.C.: U.S. Government Printing Office. (Reprinted by John Wiley and Sons, Inc.)
  Chapter 11 contains short discussions on statistical considerations in the design of industrial experiments.

  Designing for quality is specifically discussed in the following references. The above reference by Box, Hunter, and Hunter also addresses issues relating to this topic.

Box, G. E. P. and Draper, N. R. (1969). *Evolutionary Operation*, New York: John Wiley and Sons, Inc.

Gunter, B. (1987). "A Perspective on the Taguchi Method," *Quality Progress*, **20**, 44–52.

Hahn, G. J. (1976). "Process Improvement Using Evolutionary Operation," *Chemtech*, **6**, 204–206.

Hunter, J. S. (1985) "Statistical Design Applied to Product Design," *Journal of Quality Technology*, **17**, 210–221.

Scherkenbach, W. W. (1986). *The Deming Route to Quality and Productivity*, Rockville, MD: Mercury Press.

EXERCISES

1 A test program was conducted to evaluate the quality of epoxy–glass-fiber pipes. The test program required sixteen pipes, half of which were manufactured at each of two manufacturing plants. Each pipe was produced under one of two operating conditions at one of two water temperatures. The following test conditions constituted the experimental protocol:

Run No.	Plant	Operating Conditions	Water Temp. (°F)
1	1	Normal	175
2	1	Normal	150
3	1	Severe	150
4	2	Severe	175
5	1	Normal	175
6	1	Normal	150
7	2	Normal	150
8	1	Severe	175
9	2	Normal	175
10	2	Severe	150
11	2	Normal	150
12	1	Severe	175
13	2	Severe	175
14	2	Severe	150
15	1	Severe	150
16	2	Normal	175

Identify which of the following statistical design features are included in the test program:

(a) Factor(s)   **(e)** Repeat tests

(b) Factor levels   **(f)** Replications

(c) Blocks   **(g)** Test runs

(d) Experimental units   **(h)** Covariate(s)

2   An accelerator for the hydration and hardening of cement is being developed to aid in the structural construction and integrity of bridge trusses. Two mixes of fast-setting cement were tested by measuring the compressive strength (psi) of five 2-inch cubes of mortar, 24 hours after setting. This experiment was conducted twice and the following measurements of compressive strength were determined:

	Compressive Strength (psi)	
Experiment	Mix A	Mix B
1	2345	2285
	2119	2143
	2150	2362
	2386	2419
	2297	2320
2	2329	2387
	2310	2345
	2297	2352
	2311	2360
	2314	2349

Identify which of the statistical design features (a) to (h) of Exercise 1 occur in this experiment.

3   An experiment is to be designed to study the effects of several factors on the quality of electronic signals transmitted over telephone lines. The factors of interest are the baud rate (speed of transmission), signal intensity, and type of transmission device. The project leader elects to investigate baud rates from 300 to 9600 bps in increments of 100 bps, three different signal intensities, and two of three transmission devices. The use of only two transmission devices is necessitated by the cost of the project. The projected costs for the experimentation are estimated to be $500 for each baud rate, $2000 for each signal intensity, and $10,000 for each transmission device. These costs are primarily for equipment, personnel, and data processing and are largely unaffected by the number of test runs in the experiment. Identify the advantages and disadvantages of the proposed experiment. Suggest remedies for the disadvantages.

4   Automobile emission-control systems are known to deteriorate as a vehicle is used. A concern in many experiments is that emission results may be affected by how far test vehicles have previously been driven prior to being tested. Below are carbon monoxide (CO) emissions for several test runs (under identical conditions) for each of three automobiles. Also recorded are the odometer mileages (MILES) at the start of each test run. Use a labeled scatterplot to assess whether initial mileage should be used as a covariate in a test program involving these three vehicles.

Vehicle 1		Vehicle 2		Vehicle 3	
MILES	CO	MILES	CO	MILES	CO
6814	1.36	7155	2.12	6842	1.33
6843	1.39	7184	2.01	6879	1.45
6942	1.68	7061	1.83	6952	1.54
6970	1.61	7091	1.81	6991	1.70
6885	1.52			7027	1.98
6913	1.29			7056	1.92
7059	1.80				

5   Water flows through tubes in a fluidized-bed combustor, where it is converted to steam. The bed of the combustor consists of sorbent particles, which are used to absorb pollutants given off during combustion. The particles are suspended in the combustor by a gas flowing up from the bottom of the particle bed. The problem under investigation concerns the corrosion effects of the exterior surface of the tubes used to carry the water through the pipes. The following factors may influence the corrosion effects:

Factor	Levels
Bed temperature (°C)	700, 1000
Tube temperature below the bed temperature (°C)	0, 200, 400
Particle size (μm)	1000, 3000, 4000
Environment	Oxidizing, sulfidizing
Particle material	Limestone, dolomite
Tube material	Stainless steel, iron

Calculate the number of test runs needed for all possible combinations of the factor levels to be included in the experiment. How many of these test runs are used in calculating the average corrosion measurements for each of the above factor levels? Suppose each factor was investigated in a separate experiment using the same number of test runs just computed for each level. What would the total experiment size be for the separate investigations of each factor?

6   For the experiment in Exercise 2, calculate the averages and the sample standard deviations of the compressive strengths for each mix in each of the two experiments. From the averages just calculated, calculate the effect of the two mixes, separately for each experiment. What tentative conclusions can be drawn from the calculated effect of the mixes for each of the experiments? Do the sample standard deviations affect the conclusions? Why (not)?

7   Spectral properties of a compound mixed in a chemical solution are such that measurements of the refraction of light passing through the solution are about 5% greater than for the solution without the presence of the compound. Below are 10 measurements of light refraction (ratio of the angle of incidence to the angle of refraction) for the chemical solution with and without the compound. What do these data suggest about the ability of this procedure to detect the compound?

Solution without Compound	Solution with Compound
1.41	1.55
1.39	1.62
1.50	1.88
1.47	1.92
1.42	1.59
1.48	1.91
1.43	1.30
1.45	1.71
1.45	1.26
1.48	1.41

8   Sketch two geometrically different types of contours plots for which one-factor-at-a-time testing will lead to a result near the vicinity of the optimal response value within an appropriate experimental region. How could these contours be modified so that the same test sequence leads to a value that is not optimal in the experimental region?

9   A new prototype diesel engine is being tested by its manufacturer to determine at what level speed (rpm) and load (lb-ft) the brake-specific fuel consumption (BFSC, lb/bph-hr) is minimized. Suppose that the BSFC response surface for this engine can be modeled as follows:

$$\text{BSFC} = 0.9208 - 0.0016 \times \text{load} - 0.0003 \times \text{speed} + 3.1164 \times 10^{-6} \times (\text{load})^2$$
$$+ 8.3849 \times 10^{-8} \times (\text{speed})^2 - 2.0324 \times 10^{-7} \times \text{load} \times \text{speed}.$$

In this model, the engine operates at speeds from 1500 to 2800 rpm and loads from 100 to 500 lb-ft. Using this model, calculate responses for a one-factor-at-a-time experiment to find the levels of load and speed that minimize BSFC. Plot rough contours of this surface (e.g., plot response values for a grid of speed and load values, and interpolate between the plotted points). Does the one-factor-at-a-time experiment lead to the vicinity of the optimal speed–load combination?

10 Refer to the engine experiment of Exercise 9. Another response of interest in this experiment is the "knock" produced by pressure changes in the engine. Suppose the response surface for this response is

$$knock = -525.5132 + 0.2642 \times speed + 3.6783 \times load$$
$$-0.0015 \times 10^{-2} \times (speed)^2 - 0.2278 \times 10^{-2} \times (load)^2$$
$$-0.0975 \times 10^{-2} \times load \times speed.$$

Repeat the procedures requested in Exercise 9 for the minimization of knock. Note from the contour plots that the minimization of BSFC and the minimization of knock do not lead to the same optimal speed–load combination. Overlay the two sets of contour plots to determine a reasonable compromise for speed and load that will come close to minimizing both responses.

CHAPTER 7

Factorial Experiments in Completely Randomized Designs

The most straightforward statistical designs to implement are those for which the sequencing of test runs or the assignment of factor combinations to experimental units can be entirely randomized. In this chapter we introduce completely randomized designs for factorial experiments. Included in this discussion are the following topics:

- *completely randomized designs, including the steps needed to randomize a test sequence or assign factor-level combinations to experimental units;*
- *factorial experiments, the inclusion of all possible factor-level combinations in a design; and*
- *calculation of factor effects as measures of the individual and joint influences of factor levels on a response.*

Experiments are conducted in order to investigate the effects of one or more factors on a response. When an experiment consists of two or more factors, the factors can influence the response individually or jointly. Often, as in the case of one-factor-at-a-time experimentation, an experimental design does not allow one to properly assess the joint effects of the factors.

Factorial experiments conducted in completely randomized designs are especially useful for evaluating joint factor effects. Factorial experiments include all possible factor-level combinations in the experimental design. Completely randomized designs are appropriate when there are no restrictions on the order of testing, or when all the experimental units to be used in the experiment can be regarded as homogeneous.

In the first two sections of this chapter factorial experiments, completely randomized designs, and joint factor effects are characterized. The construction of completely randomized designs, including the randomization procedures, are detailed. The last two sections of this chapter detail the calculation of factor effects. The calculation of factor effects is shown to be a valuable aid in interpreting the influence of factor levels on a response.

7.1   FACTORIAL EXPERIMENTS

A (complete) factorial experiment includes all possible factor-level combinations in the experimental design. Factorial experiments can be conducted in a wide variety of experimental designs. One of the most straightforward designs to implement is the *completely randomized design.*

In a completely randomized design all the factor-level combinations in the experiment are randomly assigned to experimental units, if appropriate, or to the sequence of test runs. Randomization is important in any experimental design because an experimenter cannot always be certain that every major influence on a response has been included in the experiment. Even if one can identify and control all the major influences on a response, unplanned complications are common. Instrument drift, power surges, equipment malfunctions, technician or operator errors, or a myriad of other undetectable influences can bias the results of an experiment.

Randomization does not prevent any of the above experimental complications from occurring. As mentioned in the last chapter, randomization affords protection from bias by tending to average the bias effects over all levels of the factors in the experiment. When comparisons are made among levels of a factor, the bias effects will tend to cancel and the true factor effects will remain. Randomization is not a guarantee of bias-free comparisons, but it is certainly inexpensive insurance.

There are numerous ways to achieve randomization in an experimental design. Any acceptable randomization procedure must, however, adhere to procedures that satisfy the definition given in Exhibit 7.1. With this definition of randomization, the steps used to construct a completely randomized design are given in Exhibit 7.2. Note that the randomization procedure described in Exhibit 7.2 is equivalent to obtaining a simple random sample without replacement (Section 2.1) of the integers 1 to N.

EXHIBIT 7.1

Randomization. Randomization is a procedure whereby factor-level combinations are (a) assigned to experimental units or (b) assigned to a test sequence in such a way that every factor-level combination has an equal chance of being assigned to any experimental unit or position in the test sequence.

EXHIBIT 7.2   CONSTRUCTION OF FACTORIAL EXPERIMENTS IN COMPLETELY RANDOMIZED DESIGNS

1. Enumerate all factor-level combinations. Include repeat tests.
2. Number the factor-level combinations, including any repeat tests, sequentially from 1 to N.
3. From a random-number table (e.g., Table A1 of the Appendix) or from a computer-generated sequence of random numbers, obtain a random sequence of the integers 1 to N.
4. Assign the factor-level combinations to the experimental units or conduct the test runs in the order specified by the random number sequence. If both a randomized assignment to experimental units and a randomized test sequence are to be included in the design, use two random number sequences.

Figure 7.1 lists factors which are of interest in a laboratory investigation of torque forces experienced by rotating shafts. In this experiment a rotating shaft is to be braced by

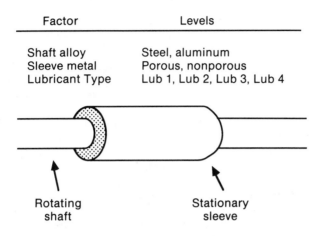

Factor	Levels
Shaft alloy	Steel, aluminum
Sleeve metal	Porous, nonporous
Lubricant Type	Lub 1, Lub 2, Lub 3, Lub 4

Rotating
shaft

Stationary
sleeve

Figure 7.1 Factors for torque experiment.

a stationary cylindrical sleeve and lubricants are to be applied to the inner wall of the sleeve in order to reduce the amount of friction between the shaft and the sleeve. The investigators wish to study shafts made of steel and aluminum alloys. The sleeves are to be made of two types of metal, one porous and one nonporous, in order to determine whether the lubricants are more effective when they adhere to the sleeve. Four lubricants of widely differing properties are to be selected from among many commercial brands available.

This experiment is to be conducted in a laboratory using a small-scale version of the shaft–sleeve apparatus found in large industrial equipment. It is feasible to consider experimental designs that include all possible combinations of the factors. Complete randomization of the run order is also feasible for this laboratory experiment. Thus a completely randomized factorial experiment is appropriate.

Table 7.1   Factor-Level Combinations for Torque Study

Combination Number	Shaft Alloy	Sleeve Metal	Lubricant Type
1	Steel	Porous	Lub 1
2	Steel	Porous	Lub 2
3	Steel	Porous	Lub 3
4	Steel	Porous	Lub 4
5	Steel	Nonporous	Lub 1
6	Steel	Nonporous	Lub 2
7	Steel	Nonporous	Lub 3
8	Steel	Nonporous	Lub 4
9	Aluminum	Porous	Lub 1
10	Aluminum	Porous	Lub 2
11	Aluminum	Porous	Lub 3
12	Aluminum	Porous	Lub 4
13	Aluminum	Nonporous	Lub 1
14	Aluminum	Nonporous	Lub 2
15	Aluminum	Nonporous	Lub 3
16	Aluminum	Nonporous	Lub 4

Table 7.1 lists all of the sixteen possible factor-level combinations for the factors from Figure 7.1. Suppose that row 20 and column 3 is selected as a starting point in the random-number table, Table A1 in the Appendix. Numbers between 1 and 16 can then be selected as consecutive two-digit numbers (proceeding left to right across each row), ignoring all numbers greater than 16 (this is only one of many ways to use the table). The following random test sequence is then obtained:

$$8, \quad 13, \quad 4, \quad 7, \quad 5, \quad 1, \quad 11, \quad 15, \quad 9, \quad 3, \quad 12, \quad 10, \quad 6, \quad 14, \quad 16, \quad 2.$$

Assigning this test order to the factor-level combinations in Table 7.1 results in the completely randomized design shown in Table 7.2.

Experimental designs should, whenever possible, include repeat test runs. Repeat test runs, as discussed in Section 6.1, are two or more experimental tests or observations in which the factor-level combinations are identical. Responses from repeat tests exhibit variability only from uncontrolled sources of variation such as variation due to changes in test conditions and to measurement error. The sample standard deviations from repeat tests can be used to estimate the uncontrolled experimental variability.

An alternative to the inclusion of repeat tests in an experiment occurs when some factor interactions (Section 7.2) can be assumed to be zero. It is often true that experiments involving large numbers of factors have interactions that can be assumed to be zero or negligible relative to the uncontrolled error variation. In such circumstances the statistics that would ordinarily measure the interactions of these factors can be used to measure experimental variability. Fractional factorial and screening experiments (Chapter 9) also exploit this assumption in order to reduce the number of test runs needed to investigate all the design factors.

As mentioned above, when possible, repeat tests should be included in the design of an experiment. Even if fractional factorial or screening designs are used, the inclusion of several repeat tests allows one to (a) estimate uncontrolled experimental-error variation

Table 7.2   Randomized Test Sequence for Torque Study

Run Number	Combination Number	Shaft Alloy	Sleeve Metal	Lubricant Type
1	8	Steel	Nonporous	Lub 4
2	13	Aluminum	Nonporous	Lub 1
3	4	Steel	Porous	Lub 4
4	7	Steel	Nonporous	Lub 3
5	5	Steel	Nonporous	Lub 1
6	1	Steel	Porous	Lub 1
7	11	Aluminum	Porous	Lub 3
8	15	Aluminum	Nonporous	Lub 3
9	9	Aluminum	Porous	Lub 1
10	3	Steel	Porous	Lub 3
11	12	Aluminum	Porous	Lub 4
12	10	Aluminum	Porous	Lub 2
13	6	Steel	Nonporous	Lub 2
14	14	Aluminum	Nonporous	Lub 2
15	16	Aluminum	Nonporous	Lub 4
16	2	Steel	Porous	Lub 2

and (b) investigate the adequacy of the fitted model (Section 22.3) without having to make the assumption that some of the interactions are zero. It is especially important to include repeat tests if one is unsure about the validity of an assumption that interactions are zero.

The inclusion of repeat tests in an experimental design does not necessarily require that each factor-level combination be repeated or that each be repeated an equal number of times. When possible, *balance*—the inclusion of all factor-level combinations an equal number of times—should be the goal in a complete factorial experiment. The only requirement is that a sufficient number of repeat tests be included so that a satisfactory estimate of experimental error can be obtained. In general, it is unwise to select a single factor-level combination and repeat it several times. A better approach when all

Table 7.3   Randomized Test Sequence for Torque Study Including Repeat Tests

Run Number	Combination Number	Shaft Alloy	Sleeve Metal	Lubricant Type
1	4	Steel	Porous	Lub 4
2	2	Steel	Porous	Lub 2
3	7	Steel	Nonporous	Lub 3
4	16	Aluminum	Nonporous	Lub 4
5	10	Aluminum	Porous	Lub 2
6	4	Steel	Porous	Lub 4
7	11	Aluminum	Porous	Lub 3
8	12	Aluminum	Porous	Lub 4
9	8	Steel	Nonporous	Lub 4
10	16	Aluminum	Nonporous	Lub 4
11	7	Steel	Nonporous	Lub 3
12	8	Steel	Nonporous	Lub 4
13	12	Aluminum	Porous	Lub 4
14	3	Steel	Porous	Lub 3
15	5	Steel	Nonporous	Lub 1
16	11	Aluminum	Porous	Lub 3
17	5	Steel	Nonporous	Lub 1
18	1	Steel	Porous	Lub 1
19	10	Aluminum	Porous	Lub 2
20	9	Aluminum	Porous	Lub 1
21	1	Steel	Porous	Lub 1
22	9	Aluminum	Porous	Lub 1
23	15	Aluminum	Nonporous	Lub 3
24	15	Aluminum	Nonporous	Lub 3
25	6	Steel	Nonporous	Lub 2
26	13	Aluminum	Nonporous	Lub 1
27	13	Aluminum	Nonporous	Lub 1
28	6	Steel	Nonporous	Lub 2
29	3	Steel	Porous	Lub 3
30	14	Aluminum	Nonporous	Lub 2
31	2	Steel	Porous	Lub 2
32	14	Aluminum	Nonporous	Lub 2

combinations cannot be repeated in an experiment is to randomly select, without replacement, the factor-level combinations to be repeated.

If one wishes to have two repeat tests for each factor level in the torque study, one will list each factor-level combination twice. Then random numbers will be obtained until two of each combination are included. This random number sequence then dictates the run order. One such sequence produced the run order shown in Table 7.3.

Several benefits accompany the factorial experiments shown in Tables 7.2 and 7.3. First, the inclusion of each factor level with a variety of levels of other factors means that the effects of each factor on the response are investigated under a variety of different conditions. This allows more general conclusions to be drawn about the factor effects than if each factor effect were studied for a fixed set of levels of the other factors.

A second benefit of these designs is that the randomization protects against unknown biases, including any unanticipated or unobservable "break-in" effects due to greater or lesser care in conducting the experiment as it progresses. Note too that the randomization of the repeat tests in Table 7.3 ensures that responses from repeat tests give a valid estimate of the experimental error of the test runs. If "back-to-back" repeat tests are conducted, the estimate of experimental error can be too small because any variability associated with setting up and tearing down the equipment would not be present.

A third major benefit of factorial experiments conducted in completely randomized designs is the ability to investigate joint factor effects. There are joint factor effects among the factors in the torque study (see Section 16.3). The importance of planning experiments so that joint factor effects can be measured is the topic of discussion in the next section.

7.2 INTERACTIONS

Interaction means the presence of joint factor effects (see Exhibit 7.3). The definition in Exhibit 7.3 implies that the presence of interactions precludes an assessment of the effects of one factor without simultaneously assessing the effects of other factors. This is the essence of an interaction effect: when interactions occur, the factors involved cannot be evaluated individually.

EXHIBIT 7.3

Interaction. An interaction exists among two or more factors if the effect of one factor on a response depends on the levels of other factors.

In order to better understand the implications of interaction effects, three examples will now be presented. The first example is based on the data shown in Table 7.4. This table presents data on the life (in hours) of certain cutting tools used with lathes. There are two types of cutting tools of interest in this example, labeled type A and type B. Data on the lifetimes of these cutting tools are collected at many different operating speeds of the lathe. Figure 7.2 is a plot of tool life versus lathe speed for each tool type.

A statistical analysis of these data using covariance analysis (Section 19.2) reveals that the data for each tool type can be well fitted using straight lines with equal slope. The intercepts for the two lines differ by about 15 hours. These fits are superimposed on the plotted points in Figure 7.2.

For these data, the two variables lathe speed and tool type do not interact. The effect of

Table 7.4   Cutting-Tool Life Data*

Tool Life (hr)	Lathe Speed (rpm)	Tool Type
18.73	610	A
14.52	950	A
17.43	720	A
14.54	840	A
13.44	980	A
24.39	530	A
13.34	680	A
22.71	540	A
12.68	890	A
19.32	730	A
30.16	670	B
27.09	770	B
25.40	880	B
26.05	1000	B
33.49	760	B
35.62	590	B
26.07	910	B
36.78	650	B
34.95	810	B
43.67	500	B

*Data from Montgomery, D. C., and Peck, E. C. (1982), *Introduction to Linear Regression Analysis*, New York: John Wiley & Sons, Inc. Copyright 1982 by John Wiley & Sons, Inc. Used by permission.

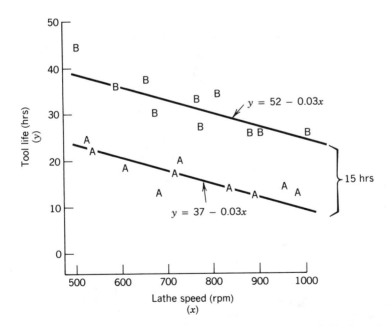

Figure 7.2   Effects of tool type and lathe speed on tool life. (Plotting symbol is the tool type.)

tool type is the same at all lathe speeds: Tool B lasts 15 hours longer, on the average, than tool A at all lathe speeds. Likewise, the effect of lathe speed is the same for both tool types: increasing lathe speed decreases the tool life by approximately 0.03 hours per rpm for both tool types. The lack of interaction is graphically indicated by the parallel lines in Figure 7.2.

Now consider the data shown in Figure 7.3. These data are average chemical yields of a process in an investigation conducted on a pilot plant. The factors of interest in this study are the operating temperature of the plant, the type of catalyst used in the process, and the concentration of the catalyst.

Figure 7.4 contains alternative plots of the chemical-yield averages. In these plots the averages for each concentration and each temperature are plotted separately for each catalyst. We join the averages for each catalyst in Figure 7.4 by dashed lines for visual emphasis. The reader should not infer that there is a straight-line relationship between yield and concentration (Figure 7.4a) or temperature (Figure 7.4b) for each catalyst. The lines simply highlight the change in average yield for each catalyst as a function of concentration or temperature levels. Dashed rather than solid lines are used as a reminder that the relationships are not necessarily linear.

The visual impression left by Figure 7.4a is similar to that of Figure 7.2 and is the reason for connecting the averages by dashed lines. The effect of concentration on yield, as suggested by the parallel lines, is the same for each catalyst. An increase from 20 to 40% in concentration produces a large, approximately equal decrease for each catalyst. Thus, concentration has approximately the same effect on yield for each catalyst, and vice versa. There is no two-factor interaction between catalyst and concentration.

Contrast these results with the trends indicated in Figure 7.4b. The average chemical yield increases slightly for catalyst C_1 as the temperature is increased from 160 to 180°C; however, the chemical yield increases much more dramatically for catalyst C_2 as

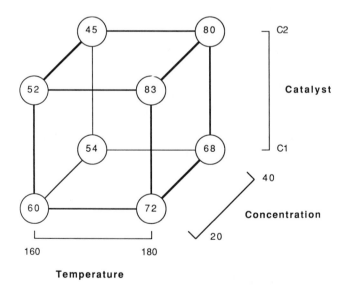

Figure 7.3 Average chemical yields for pilot-plant experiment. Data from Box, G. E. P., Hunter, W. G., and Hunter, J. S. (1978). *Statistics for Experimenters*, New York: John Wiley & Sons, Inc. Used by permission.

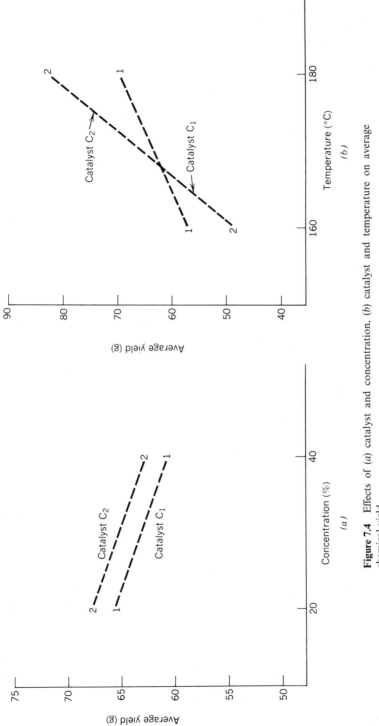

Figure 7.4 Effects of (a) catalyst and concentration, (b) catalyst and temperature on average chemical yield.

temperature is increased. In fact, the higher yield at 160°C occurs with catalyst C_1, but at 180°C the higher yield occurs with catalyst C_2.

Figure 7.4b is a graphical representation of an interaction. As in Figure 7.4a, the dashed lines connecting the plotted points are used for visual emphasis only, not to suggest a linear effect of temperature. With this understanding, the visual impression of Figure 7.4b is that the effect on yield of changing temperature from 160 to 180°C depends on which catalyst is used. In addition, the optimum (higher) yield depends on both the temperature and the catalyst. If the plant is operated at 160°C, catalyst C_1 produces a higher average yield. If the plant is operated at 180°C, catalyst C_2 produces a higher average yield. Consequently, one cannot draw conclusions about either of these factors without specifying the level of the other factor. Even though catalyst C_2 has a higher average yield (65) than catalyst C_1 (63.5), it would be misleading to conclude that catalyst C_2 is always preferable to catalyst C_1. The preference for one of the catalysts depends on the operating temperature of the plant.

In general, interactions are indicated graphically by nonparallel trends in plots of average responses. The dashed line segments connecting the average responses for each level of the factors need not intersect as they do in Figure 7.4b. Any nonparallel changes in the average responses is an indication of an interaction. As with all statistical analyses, one confirms the existence (or absence) of interaction effects implied by graphical displays with formal statistical inference procedures that account for experimental variation. Thus, the lack of parallelism in the changes of the average responses must be sufficiently large that they represent a real factor effect, not just uncontrolled experimental variability.

The final example of interaction effects is taken from the data generated from the torque study of the last section. Torque measurements were obtained using the test sequence of the completely randomized design in Table 7.3. The torque measurements were averaged over the repeat runs, and the averages (see Table 16.3) are plotted in Figure 7.5.

In Figure 7.5a the average torque measurements for the aluminum shaft are plotted as a function of the sleeve and lubricant used. The dashed line segments connecting the points are not parallel, suggesting the presence of an interaction between sleeve and lubricant on the average torque. Both the porous and the nonporous sleeves produce average torque measurements of approximately the same magnitude when lubricant 2 is used. The average torque measurements differ greatly between sleeve types for each of the other lubricants.

In Figure 7.5b the average torque measurements for the steel shaft are plotted. There is a strong indication of an interaction between the sleeve and lubricant factors; however, the trends in this plot are not similar to those in Figure 7.5a. As can be confirmed by a statistical analysis of the averages (see Section 16.3), there is a three-factor interaction among the three design factors. With the aluminum shaft, low average torque measurements are achieved with all the lubricants when the nonporous sleeve is used. Contrast this result with the substantially lower average torque measurement for lubricant 2 and the nonporous sleeve than for the other lubricant–sleeve combinations with the steel shaft. Moreover, if lubricant 4 is used, the porous sleeve yields lower average torque measurements than does the nonporous sleeve with the steel shaft.

From these examples it should be clear that the presence of interactions requires that factors be evaluated jointly rather than individually. It should also be clear that one must design experiments to measure interactions. Failure to do so can lead to misleading, even incorrect, conclusions. Factorial experiments enable all joint factor effects to be estimated. If one does not know that interaction effects are absent, factorial experiments should be seriously considered.

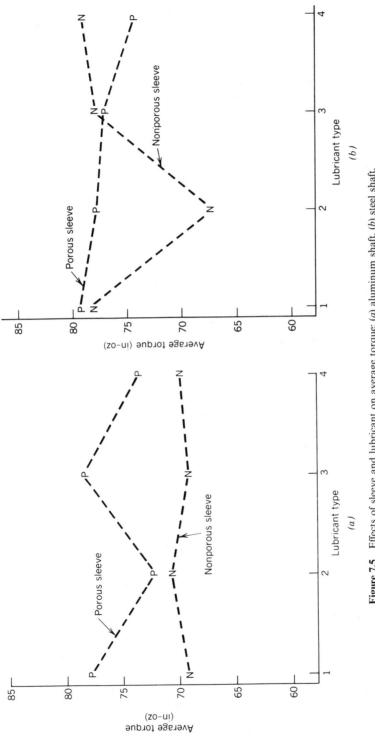

Figure 7.5 Effects of sleeve and lubricant on average torque: (*a*) aluminum shaft. (*b*) steel shaft.

The graphical procedures illustrated in this section allow a visual inspection of factor effects. Statistical analyses, especially interval estimation and hypothesis testing, utilize numerical estimates of the factor effects. In the next section, the calculation of factor effects is detailed. These calculations supplement a graphical assessment of interactions and are useful in quantifying the influence that changing factor levels have on average response values.

Because of the possible existence of interactions, complete factorial experiments should be conducted whenever possible. In experiments with a large number of factors, however, complete factorial experiments may not be economically feasible even if it is believed that some interactions among the factors may exist. If it is not possible to conduct a complete factorial experiment, fractional factorial and screening experiments are important alternatives. Designs for these experiments are discussed in Chapter 9.

7.3   CALCULATION OF FACTOR EFFECTS

An effect was defined in Table 6.1 as a change in the average response corresponding to a change in factor-level combinations or to a change in experimental conditions. We now wish to specify more completely several types of effects for factors that have only two levels (see Exhibit 7.4). In the appendix to this chapter we generalize these effects to factors whose numbers of levels are greater than two, with special emphasis on factors whose numbers of levels are powers of two. The calculation of effects is facilitated by the introduction of algebraic notation for individual responses and averages of responses. This notation is also helpful in clarifying the concepts of confounding and design resolution presented in Chapter 9.

We denote factors in designed experiments by uppercase Latin letters: $A, B, C, \ldots, K$. An individual response is represented by a lowercase Latin letter, usually y, having one or more subscripts. The subscripts, one for each factor, designate the specific factor-level combination from which the response is obtained. There may also be one or more subscripts designating repeat observations for fixed factor levels. For example, in the torque study described in Figure 7.1 the factors and their levels could be designated as follows:

Factor	Factor Symbol	Level	Subscript Symbol
Alloy	A	Steel	$i = 1$
		Aluminum	$i = 2$
Sleeve	B	Porous	$j = 1$
		Nonporous	$j = 2$
Lubricant	C	Lub 1	$k = 1$
		Lub 2	$k = 2$
		Lub 3	$k = 3$
		Lub 4	$k = 4$

A fourth subscript l could be used to denote the repeat-test number, provided there is at least one repeat test; otherwise, the fourth subscript is superfluous and not included.

With this convention in notation, responses from a three-factor experiment with r

---

EXHIBIT 7.4 EFFECTS FOR TWO-LEVEL FACTORS

Main effect. The difference between the average responses at the two levels of a factor.

Two-factor interaction. Half the difference between the main effects of one factor at the two levels of a second factor.

Three-factor interaction. Half the difference between the two-factor interaction effects at the two levels of a third factor.

---

repeat tests per factor–level combination would be denoted y_{ijkl} with $i = 1, \ldots, a$, $j = 1, \ldots, b$, $k = 1, \ldots, c$, and $l = 1, \ldots, r$. The upper limits a, b, c, and r on the subscripts denote the numbers of levels of the three factors A, B, C and the number of repeat observations, respectively.

Average responses are represented by replacing one or more of the subscripts by a dot and placing a bar over the symbol for the response. Using dot notation,

$$\bar{y}_{\dots} = n^{-1} \sum_{ijkl} y_{ijkl}$$

denotes the overall average response for a three-factor experiment with repeat tests, where $n = abcr$ and the summation is over all the observations (all possible values of the subscripts). Similarly,

$$\bar{y}_{i\dots} = (bcr)^{-1} \sum_{jkl} y_{ijkl}$$

is the average response for all responses having the ith level of factor A. The symbol $\bar{y}_{2 \cdot 1}$ denotes the average response for all test runs having the second level of the first factor and the first level of the third factor. We denote by $\bar{y}_{ijk}$, a typical average response, across repeat observations, for one of the factor-level combinations.

Consider now a factorial experiment with each factor having two levels and r repeat observations for each combination of levels of the factors. The factor-level combinations in the experiment can be represented in any of several equivalent ways. An especially useful way to represent the factor-level combinations is using the *effects representation*, which we now describe.

Let one of the levels of a factor be coded -1 and the other level be coded $+1$. It is arbitrary which level receives each code, although it is customary for quantitative factors to let the lower level be -1. A straightforward way to list all the unique combinations of the factor levels (ignoring repeats) using this coding is given in Exhibit 7.5.

EXHIBIT 7.5 EFFECTS CODING OF FACTOR LEVELS

1. Designate one level of each factor as -1 and the other level as $+1$.

2. Lay out a table with column headings for each of the factors A, B, C, ..., K.

3. Let $n = 2^k$, where k is the number of factors.

4. Set the first $n/2$ of the levels for factor A equal to -1 and the last $n/2$ equal to $+1$. Set the first $n/4$ levels of factor B equal to -1, the next $n/4$ equal to $+1$, the next $n/4$ equal to -1, and the last $n/4$ equal to $+1$. Set the first $n/8$ of the levels for factor C equal to -1, the next $n/8$ equal to $+1$, etc. Continue in this fashion until the last column (for factor K) has alternating -1 and $+1$ signs.

Table 7.5 shows the effects representation for a two-factor factorial experiment. Observe that each time one of the factor levels is coded -1 its actual value is its "lower" level, while each time it is coded a $+1$ its actual value is the factor's "upper" level. The designation of lower and upper is completely arbitrary for a qualitative variable. The effects representation in Table 7.5 can be related to the calculation of main effects. Before doing so, we discuss the effects coding for interactions.

The effects coding for interactions is similar to that of individual factors and is derivable directly from the effects coding for the individual factors. Table 7.6 shows the effects coding for a two-factor, two-level factorial experiment, including the interaction column. The elements in the interaction column AB are the products of the individual elements in the columns labeled A and B. For example, the second element in the AB column is the product of the second elements in the A and B columns: $(-1)(+1) = -1$.

Also listed in Table 7.6 are the symbolic average responses (these would be individual responses if $r = 1$) for each of the factor-level combinations designated by the effects coding for the factors. A subscript 1 denotes the lower level of a factor, and a subscript 2 denotes the upper level. The first average response shown is $\bar{y}_{11.}$, since $A = -1$ and $B = -1$ denotes the combination with both factors at their lower levels. The other average responses shown in the table are identified in a similar manner.

Now consider the main effect for factor A, designated $M(A)$, for a two-factor, two-level experiment with repeat tests. From the definition of a main effect as the difference in average responses at each level of a factor, the main effect for A can be calculated as

$$M(A) = \bar{y}_{2..} - \bar{y}_{1..}$$
$$= \frac{-\bar{y}_{11.} - \bar{y}_{12.} + \bar{y}_{21.} + \bar{y}_{22.}}{2}.$$

Table 7.5 Equivalent Representations for Factor Levels in a Two-Factor Factorial Experiment

Factor Levels*		Effects Representation	
Factor A	Factor B	A	B
Lower	Lower	-1	-1
Lower	Upper	-1	$+1$
Upper	Lower	$+1$	-1
Upper	Upper	$+1$	$+1$

*Either factor level can be designated "lower" or "upper" if the levels are qualitative.

Table 7.6 Effects Representation for Main Effects and Interaction in a Two-Factor Factorial Experiment

Effects			Average
A	B	AB	Response
-1	-1	$+1$	$\bar{y}_{11.}$
-1	$+1$	-1	$\bar{y}_{12.}$
$+1$	-1	-1	$\bar{y}_{21.}$
$+1$	$+1$	$+1$	$\bar{y}_{22.}$

The latter equality holds because, for example, $\bar{y}_{1..} = (\bar{y}_{11.} + \bar{y}_{12.})/2$. Observe that the signs on the response averages for this main effect are the same as those in the effects representation for A in Tables 7.5 and 7.6. In the same way one can readily verify that

$$M(B) = \frac{-\bar{y}_{11.} + \bar{y}_{12.} - \bar{y}_{21.} + \bar{y}_{22.}}{2}.$$

Again, the signs on the response averages in the main effect for B are the same as those in the B columns of Tables 7.5 and 7.6.

The definition of a two-factor interaction effect is that it is half the difference between the main effects for one of the factors at the two levels of the other factor. The main effect for factor A at each level of factor B is calculated as follows:

$$M(A)_{j=1} = \bar{y}_{21.} - \bar{y}_{11.}$$

and

$$M(A)_{j=2} = \bar{y}_{22.} - \bar{y}_{12..}$$

Note that the main effect for factor A at the first (lower) level of B is again the difference of two averages, those for all responses at each of the levels of A using only those observations that have B fixed at its lower level ($j = 1$). Similarly, the main effect for A at the second (upper) level of B is the difference of the average responses for each level of A using only those observations that have the second level of B.

From these effects the interaction between factors A and B, denoted $I(AB)$, can be calculated as

$$I(AB) = \frac{M(A)_{j=2} - M(A)_{j=1}}{2}$$

$$= \frac{(\bar{y}_{22.} - \bar{y}_{12.}) - (\bar{y}_{21.} - \bar{y}_{11.})}{2}$$

$$= \frac{+\bar{y}_{11.} - \bar{y}_{12.} - \bar{y}_{21.} + \bar{y}_{22.}}{2}.$$

The signs on the average responses for this interaction effect are the same as those in the AB column of Table 7.6.

This example illustrates a general procedure (see Exhibit 7.6) for the calculation of main effects and interaction effects for two-level factorial experiments. This procedure can be used with any number of factors so long as there are an equal number of repeat observations for each of the factor-level combinations included in the experiment (i.e., the design is balanced).

EXHIBIT 7.6   CALCULATION OF EFFECTS FOR TWO-LEVEL FACTORS

1. Construct the effects representation for each main effect and each interaction.
2. Calculate linear combinations of the average responses (or individual responses if $r = 1$), using the signs in the effects column for each main effect and for each interaction.
3. Divide the respective linear combinations of average responses by 2^{k-1}, where k is the number of factors in the experiment.

Table 7.7 Effects Representation and Calculated Effects for the Pilot-Plant Chemical-Yield Study*

Factors

Symbol	Designation
A	Temperature
B	Concentration
C	Catalyst

Results

			Effect Representation				Average Yield
A	B	C	AB	AC	BC	ABC	$(r = 2)$
-1	-1	-1	$+1$	$+1$	$+1$	-1	60
-1	-1	$+1$	$+1$	-1	-1	$+1$	52
-1	$+1$	-1	-1	$+1$	-1	$+1$	54
-1	$+1$	$+1$	-1	-1	$+1$	-1	45
$+1$	-1	-1	-1	-1	$+1$	$+1$	72
$+1$	-1	$+1$	-1	$+1$	-1	-1	83
$+1$	$+1$	-1	$+1$	-1	-1	-1	68
$+1$	$+1$	$+1$	$+1$	$+1$	$+1$	$+1$	80

Calculated effects

23.0	-5.0	1.5	1.5	10.0	0.0	0.5
				$s_e = 2.828$		

*Data from Box, G. E. P., Hunter, W. G., and Hunter, J. S. (1978). *Statistics for Experimenters*, New York: John Wiley & Sons, Inc. Copyright 1978 by John Wiley & Sons, Inc. Used by permission.

Table 7.7 shows the effects representation of the main effects and interactions for the chemical-yield study. The averages listed are obtained from Figure 7.3. Also shown in the table are the calculated main effects and interactions.

The estimated standard error (see Section 15.3) of any one of these effects is $2s_e/n^{1/2}$, where $n = 16$ is the total number of test runs ($k = 3$ and $r = 2$ for this experiment) and s_e is the estimated standard deviation of the uncontrolled experimental error. For this experiment, $s_e = 2.828$ (see Section 15.3). Hence, the estimated standard error of any one of the effects shown in Table 7.7 is 1.414. Comparing the calculated effects with this estimated standard error, it is apparent that the only interaction that is substantially larger than the standard error is that between temperature and catalyst. These calculations confirm the graphical conclusions drawn from Figure 7.4. In addition, although none of the interactions involving concentration are substantially larger than the standard error, the main effect for concentration is over 3.5 times larger than the effects standard error.

The effects representation of main effects and interactions is used in Chapter 9 to show how the intentional confusing or confounding of effects known to be zero or small can be used to control for extraneous variability in the designing of experiments and to reduce the number of test runs required for a complete factorial experiment. Statistical analyses of effects for complete factorial experiments are detailed in Chapters 15 to 18, depending on the design used and the effects being analyzed. Completely randomized designs are discussed in Chapter 15.

7.4   YATES'S METHOD FOR CALCULATING EFFECTS

For balanced two-level complete factorial experiments, an alternative computational procedure to that given in the last section is available for the calculation of factor effects. This procedure is referred to as *Yates's method*.

EXHIBIT 7.7   STANDARD ORDER

1. Arrange the factors in a convenient order; number the factors from 1 to k.

2. Code the levels of each factor so that -1 represents one level and $+1$ represents the other level.

3. Arrange the factor levels in a table so that the first factor has successive -1 values followed by $+1$ values, the second factor has pairs of -1 values followed by pairs of $+1$ values, ..., the kth column has 2^{k-1} values of -1 followed by 2^{k-1} values of $+1$.

4. Write the average response (or individual response if $r = 1$) for the factor-level combinations in a column to the right of the last factor.

Yates's method will be illustrated using the three-factor, two-level pilot-plant factorial experiment. To use this method for calculating factor effects, the coded factor levels must first be arranged in *standard order* (see Exhibit 7.7). Standard order of the factor levels is the reverse of the order used in the effects representation described in the last section. Yates's method yields the factor effects through patterned addition, subtraction, and finally division. It proceeds as indicated in Exhibit 7.8.

EXHIBIT 7.8   YATES'S METHOD

1. Arrange the factors in standard order.

2. Successively pair consecutive response averages; i.e., pair the first two responses, the next two, etc.

3. Generate a new column whose first $k/2$ entries are the sums of the successive pairs of responses averages and whose next $k/2$ entries are the differences

second average $-$ first average

for the two response averages in each pair.

4. Generate a second column from the new column in step 3 by performing the same additions and subtractions on successive pairs of entries in the new column.

5. Working each newly generated column, continue the patterned addition and substraction of successive pairs of entries until a total of k new columns have been generated.

6. Divide the first entry in the kth column by 2^k; divide the remaining entries in this column by 2^{k-1}. This column of values contains the constant effect, main effects, and interaction effects of the model factors.

7. Identify the factor effects by noting the coded factor levels in the first k columns of the table. The row of minus signs in the first k columns (the first row of the table) denotes the overall constant effect (the overall average) of the responses. Each of the next k rows contains a single $+1$, indicating the main effect for that factor. The next $k(k-1)/2$ rows have two factors with $+1$ values, indicating the two-factor interaction effects. Continue in this fashion until all main effects and interactions have been identified.

Table 7.8 Calculations According to Yates's Method for Pilot-Plant Chemical-Yield Data*

Coded Factor Levels			Average Yield	Generated Columns			Divisor	Effect	Effect Name
A	B	C		1	2	3			
−1	−1	−1	60	132	254	514	8	64.25	Average
+1	−1	−1	72	122	260	92	4	23.00	A
−1	+1	−1	54	135	26	−20	4	−5.00	B
+1	+1	−1	68	125	66	6	4	1.50	A × B
−1	−1	+1	52	12	−10	6	4	1.50	C
+1	−1	+1	83	14	−10	40	4	10.00	A × C
−1	+1	+1	45	31	2	0	4	0.00	B × C
+1	+1	+1	80	35	4	2	4	0.50	A × B × C

* A = Temperature, B = Concentration, C = Catalyst.

Table 7.8 shows the calculations for Yates's method for the pilot-plant data. The calculated factor effects are the same as those shown in Table 7.7. The "constant effect" calculated with Yates's method is simply the overall response average.

7.5 PLOTTING EFFECTS

A visual examination of calculated effects provides important information about the influence of factors on the response variable. Plotting the effects is especially important when the experimental design does not permit satisfactory estimation of the uncontrolled error variation.

Highly fractionated designs used for screening experiments (see Chapter 9) often are used to obtain information about only the major factor effects on a response. These designs are generally conducted so that only main effects are estimated and few degrees of freedom are available for estimating the uncontrolled error standard deviation. With such designs, the imprecise estimate of the error standard deviation may limit the usefulness of the interval estimation and the testing procedures discussed in Chapter 15.

Factor effects can be calculated for any complete experiment using the techniques discussed in the previous two sections (7.3 and 7.4). These techniques can also be used to calculate effects for fractional factorial experiments designed using the procedures described in Chapter 9. Once they are calculated, point plots (Section 4.1) and normal quantile-quantile plots (Section 5.3) of the effects highlight those that have unusually large magnitudes.

To illustrate the information obtainable from quantile plotting of effects, we include Figure 7.6, a normal quantile–quantile plot from the pilot plant chemical yield study. The calculated effects are shown in Tables 7.7 and 7.8. As is characteristic of this type of normal quantile–quantile plot, most of the effects are small and appear to approximate a straight line; however, three of them deviate markedly from the others.

A straight line has been superimposed on Figure 7.6 over the four small effects. The main effects for temperature (A) and concentration (B), and the interaction between temperature and catalyst (AC) appear to deviate substantially from the straight line. As is confirmed in Chapter 15 (see Table 15.6), these three effects are statistically significant and are the dominant ones.

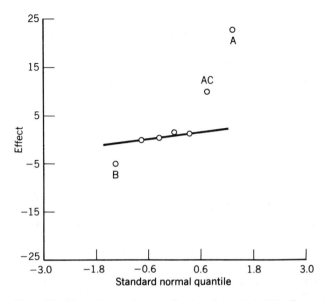

Figure 7.6   Normal quantile–quantile plot of chemical yield effects.

Plotting the effects is useful even if statistical tests can be performed on them. A visual comparison of the effects aids in the assessment of whether they are statistically significant because they are measured with great precision, resulting in very high power (Section 12.5) for the corresponding statistical tests, or because the effects truly are the dominant ones.

As with all visual comparisons, one should whenever possible confirm conclusions drawn from an analysis of plotted effects with the calculation of confidence intervals or tests of appropriate statistical hypotheses. This confirmation is desirable because of the need to compare apparent large effects with an appropriate estimate of uncontrolled experimental error variation. Such comparisons reduce the chance of drawing inappropriate conclusions about the existence or dominance of effects. In the absence of estimates of experimental error variation, however, the plotting of effects is still a vital tool for the analysis of factor effects.

APPENDIX: CALCULATION OF EFFECTS FOR FACTORS WITH MORE THAN TWO LEVELS

The general procedures detailed in Section 7.3 for representing effects of two-level factors as differences of averages can be extended to any number of factor levels, but the representation is not unique. In this appendix we briefly outline an extension to factors whose numbers of levels are greater than two, with special emphasis on factors whose numbers of levels are powers of two.

A main effect for a two-level factor can be uniquely defined as the difference between the average responses at the two levels of the factor. When a factor has more than two levels, say k, there is more than one main effect and there are many ways of defining the main effects. One way is to calculate differences between averages for levels 1 to $k - 1$ from the average for level k:

$$\bar{y}_1. - \bar{y}_k., \bar{y}_2. - \bar{y}_k., \ldots, \bar{y}_{k-1}. - \bar{y}_k..$$

An alternative way is to calculate the differences of the averages at the various levels from the overall average:

$$\bar{y}_1. - \bar{y}..., \bar{y}_2. - \bar{y}..., \ldots, \bar{y}_k. - \bar{y}...$$

Note that main effects calculated from one of these definitions can be determined from the other by taking linear combinations of the effects (see the exercises).

When factor levels are quantitative, an alternative to the two above definitions of main effects is to determine linear combinations of the factor-level averages that measure the linear, quadratic, cubic, etc. effects of the quantitative levels. This is accomplished using *contrasts* of the averages. We discuss this application in Sections 16.2 and 16.3.

In the remainder of this appendix, attention is directed to factors whose numbers of levels is a power of two. The generalization of the effects representation of two-level factors to this situation allows many of the design features of fractional factorial experiments (Chapter 9) to be extended to factors having more than two levels.

One way to specify main effects for a factor whose number of levels, say k, is a power of two, $k = 2^m$, is to define m two-level factors to represent the k-level factor. One can show that any main effect for the original factor that is defined as a difference in averages can be expressed as a linear combination of the columns of main effects and interactions for the m two-level factors.

For example, suppose A is a four-level factor ($k = 4$). Define two two-level factors A_1 and A_2 ($m = 2$) as follows:

Level of Factor A	Level of A_1	Level of A_2
1	-1	-1
2	-1	$+1$
3	$+1$	-1
4	$+1$	$+1$

Represent the response averages in the equivalent forms $\bar{y}_1., \bar{y}_2., \bar{y}_3., \bar{y}_4.$ and $\bar{y}_{11}., \bar{y}_{12}., \bar{y}_{21}., \bar{y}_{22}.$ corresponding to the two representations of the four factor levels. One can show that any definition of a main effect for A that can be written in the form

$$c_1\bar{y}_1. + c_2\bar{y}_2. + c_3\bar{y}_3. + c_4\bar{y}_4.,$$

where the sum of the coefficients $c_1, \ldots, c_4$ is zero (contrasts of the averages), can be written as a linear combination of the main effects A_1, A_2 and the "interaction" A_1A_2, calculated from the $\bar{y}_{ij}$.

It is important to note that all the "main effects" and "interactions" of the m constructed two-level factors are needed to represent just the main effects of the original factors. Continuing the above example, the effects represented by A_1, A_2, and A_1A_2 each represent a main effect for the original factor A. This is because one cannot express all the differences in the factor-level averages

$$\bar{y}_{1\cdot} - \bar{y}_{k\cdot}, \qquad i = 1, 2, \ldots, k-1,$$

or

$$\bar{y}_{1\cdot} - \bar{y}_{\cdot\cdot}, \qquad i = 1, 2, \ldots, k,$$

just in terms of A_1 and A_2. One can, however, express these differences as linear combinations of A_1, A_2, and $A_1 A_2$. Similarly, $A_1 B$, $A_2 B$, and $A_1 A_2 B$ each represent a two-factor interaction of the four-level factor A with a second factor B. In particular, the interaction $A_1 A_2 B$ does not represent a three-factor interaction, since $A_1 A_2$ is a main effect for factor A.

This approach can be extended to factors having eight ($m = 3$), sixteen ($m = 4$), or any number of levels that is a power of two. In each case, the complete effects representation of the m constructed two-level factors, including all $2^m - 1$ main effects and interactions, is needed to completely specify the main effects of the original factors.

REFERENCES

Text References

Extensive coverage of complete factorial experiments appears in:

Box, G. E. P., Hunter, W. G., and Hunter, J. S. (1978). *Statistics for Experimenters*, New York: John Wiley and Sons, Inc.

Davies, O. L. (Ed.) (1971). *The Design and Analysis of Industrial Experiments*, New York: Macmillan Co.

Diamond, W. J. (1981). *Practical Experimental Designs*, Belmont, CA: Wadsworth, Inc.
    Each of these texts discusses the construction and use of factorial experiments.

Data References

The average chemical yields for the pilot-plant experiment in Figure 7.3 are taken from

Box, G. E. P., Hunter, W. G., and Hunter, J. S. (1978). *Statistics for Experimenters*, New York: John Wiley and Sons, Inc., Table 10.1, p. 308.

The cutting tool life data are taken from

Montgomery, D. C. and Peck, E. C. (1982). *Introduction to Linear Regression Analysis*, New York: John Wiley and Sons, Inc.

EXERCISES

1 A test program is to be conducted to evaluate the performance of artillery projectiles against a variety of targets. The six target types to be included in the experiment are aluminum, steel, titanium, topsoil, sand, and simulated fuel tanks. The projectiles will be fired from two different angles (measured from horizontal to line of flight): 60 and 45°. Construct a completely randomized design for this factorial experiment.

2 The operating characteristics of a truck diesel engine that operates on unleaded

gasoline are to be investigated. The time (sec) needed to reach stable operation under varying conditions is the response of primary interest. Tests are to be run at four engine speeds (1000, 1200, 1400, and 1600 rpm) and four temperatures (70, 40, 20, and 0°F). Since a large budget is available for this project, repeat tests can be conducted at each factor-level combination. Construct a completely randomized design for this experiment.

3   An experiment is to be conducted to study heat transfer in article molds used in the manufacture of drinking glasses. Three molten-glass temperatures are to be varied in the experiment: 1100, 1200 (the industry standard), and 1300°C. Two cooling times are also to be varied in the experiment: 15 and 20 sec. In addition, two types of glass, type A and type B, are to be studied. Due to budget constraints, repeat tests can only be run for the 1200°C test runs. Construct a completely randomized design for this factorial experiment.

4   The fuel economy of two lubricating oils for locomotive engines is to be investigated. Fuel economy is measured by determining the brake-specific fuel consumption (BSFC) after the oil has operated in the engine for 10 minutes. Each oil is to be tested five times. Suppose that the run order and the resulting data are as follows:

Run	BFSC	Run	BFSC
1	.536	6	.550
2	.535	7	.552
3	.538	8	.559
4	.537	9	.563
5	.542	10	.571

Use graphical and numerical methods to discuss how conclusions about the effectiveness of the two oils might be compromised if the order that the oils were tested was

A, A, A, A, A, B, B, B, B, B.

Would the same difficulties occur if a randomization of the order of testing resulted in the following test sequence

A, B, A, A, B, B, A, B, B, A?

How would conclusions about the effectiveness of the two oils change, based on which test sequence was used?

5   Suppose that in the experiment described in Exercise 3 the variation in mold temperature (a measure of heat transfer) is the primary response of interest. Further, suppose budget constraints dictate that only the 1200° test runs (with repeats) can be conducted. The data below show test results for two plants that manufacture drinking glasses. Use graphical techniques to assess whether an interaction effect exists between glass type and cooling time. Assess the presence of interactions separately for the two plants.

PLANT 1

Glass Type	Cooling Time (sec)	Mold Temperature (°C)
A	15	467
B	15	462
A	20	469
B	20	467
A	15	470
B	15	463
A	20	472
B	20	469

PLANT 2

Glass Type	Cooling Time (sec)	Mold Temperature (°C)
A	15	473
B	15	469
A	20	466
B	20	465
A	15	462
B	15	462
A	20	463
B	20	467

6 Suppose that the experiment described in Exercise 3 can be conducted as stated except that no funds are available for repeat tests. The data below represent test results from such an experiment. The response of interest is, as in the previous exercise, the mold temperature. Use graphical techniques to assess whether interactions exist among the design factors.

Molten-Glass Temperature (°C)	Glass Type	Cooling Time	Mold Temperature
1300	A	15	482
1200	A	20	469
1300	B	15	471
1100	A	15	459
1200	B	20	470
1300	A	20	480
1200	B	15	475
1200	A	15	460
1100	B	20	479
1100	A	20	462
1300	B	20	469
1100	B	15	471

7 Calculate the main effects and the two-factor interaction between glass type and

cooling time for each plant in Exercise 5. Do these calculated effects reinforce the conclusions drawn from the graphs?

8 In Exercise 6, consider only the data for the 15-second cooling time. Calculate the main effects and two-factor interaction effect for glass type and molten-glass temperature, using only

(a) 1100 and 1200°C,

(b) 1100 and 1300°C as the factor levels.

Interpret these calculated effects along with suitable graphics displaying the effects.

9 In Exercise 6, suppose only two molten glass temperatures had been investigated, 1100 and 1300°C (i.e., remove the data for 1200°C). Calculate the main effects, two-factor interaction effects, and the three-factor interaction effect. Interpret these effects with the help of suitable graphs displaying them.

10 Construct a table of the effects representation for the main effects and two-factor interaction of glass type and cooling time

(a) for only the portion of the experiment involving plant 1 described in Exercise 5, and

(b) the entire experiment.

Using the effects representation, calculate the numerical values of the main effects and interactions. Are these calculated effects consistent with the visual interpretations of the interaction plots?

11 An experiment is to be conducted in which the effects of the angle of incidence, incident light intensity, wavelength, concentration of a compound, acidity of a chemical solution, and opaqueness of a glass flask affect the intensity of light passing through a mixture of the solution and the compound. Each of the above factors is to be investigated at two levels. Construct a table of the effects representation for all main effects and interactions for a complete factorial experiment.

12 An experiment consists of a single factor that has four levels. Suppose each level of the factor is repeated r times during the experiment. Consider the following two ways of defining the effects of the four levels of the factor:

(a) $\bar{y}_1. - \bar{y}_4., \bar{y}_2. - \bar{y}_4., \bar{y}_3. - \bar{y}_4.$:

(b) $\bar{y}_1. - \bar{y}.., \bar{y}_2. - \bar{y}.., \bar{y}_3. - \bar{y}.., \bar{y}_4. - \bar{y}..$

Show that each of these sets of main effects can be determined from the other.

13 Construct two two-level factors, say A_1 and A_2, to represent the four-level factor in Exercise 12. Make a symbolic effects representation table similar to Table 7.6 for these new factors and their interaction. Show that each of the sets of main effects in Exercise 12 can be expressed as a linear function of the main effects and interaction of the two constructed variables. (Hint: Denote the three columns of effects by $\mathbf{a}_1, \mathbf{a}_2,$

and $\mathbf{a}_{12}$. Express any of the main effects in (a) or (b) as $b_1y_1. + b_2y_2. + b_3y_3. + b_4y_4.$. Put the coefficients $b_1, \ldots, b_4$ in a column vector $\mathbf{b}$, and show that $\mathbf{b} = c_1\mathbf{a}_1 + c_2\mathbf{a}_2 + c_3\mathbf{a}_{12}$ for suitably chosen constants c_1, and c_2, and c_3.)

14 Construct a main-effects table for a factor that has eight levels by defining three two-level factors to represent the eight levels of the original factor. Then write out seven main effects for the eight-level factor similar to one of the two definitions in Exercise 12. Show that one (or more) of these main effects can be written as a linear function of the seven effects for the three constructed factors.

15 An experiment is to have three factors each at two levels, and one factor at four levels. Create two two-level factors from the four-level factor. List the factor-level combinations for a complete factorial experiment in these five factors. Show by comparison that this is the same as a complete factorial experiment in the original four factors.

16 Repeat Exercise 15 for an experiment that is to have two two-level factors and one eight-level factor.

CHAPTER 8

Blocking Designs

In this chapter we introduce blocking designs as an experimental technique for controlling variability, thereby allowing more sensitive estimation of factor effects. Of special relevance to this discussion are the following topics:

- *the use of replication and local control in reducing the variability of estimators of factor effects,*
- *the construction and use of randomized complete block designs when there are sufficiently many homogeneous experimental units so that all factor combinations of interest can be run in each block,*
- *the use of balanced incomplete block designs when the number of homogeneous experimental units in each block is less than the number of factor–level combinations of interest, and*
- *the use of latin-square and crossover designs to control variability when special restrictions are placed on the experimentation.*

In Section 6.2 the presence of excessive experimental variability was shown to contribute to the imprecision of response variables and of estimators of factor effects. This imprecision affects statistical analyses in that real factor effects can be hidden because of the variation of individual responses. Excessive variability can occur from the measurement process itself, the test conditions at the time observations are made, or a lack of homogeneity of the experimental units on which measurements are made.

Chief among the statistical design techniques for coping with variability in an experiment are repeat tests, replication, and local control. Evolving from these techniques will be the usefulness of blocking designs, particularly randomized complete block designs and balanced incomplete block designs. When restrictions are placed on the experimentation in addition to the number of test runs that can be conducted under homogeneous test conditions or on homogeneous experimental units, alternative blocking designs such as latin squares and crossover designs can be effective in controlling variability. Each of these topics is discussed in the following sections of this chapter.

8.1   CONTROLLING EXPERIMENTAL VARIABILITY

There are several ways that an experimenter can, through the choice of a statistical design, control or compensate for excessive variability in experimental results. Three of the most important are the use of repeat tests, replication, and local control. Although these techniques neither reduce nor eliminate the inherent variability of test results, their proper application can greatly increase the precision of statistics used in the calculation of factor effects.

As mentioned earlier in Table 6.1, we distinguish between repeat tests and replications (see Exhibit 8.1). Repeat tests enable one to estimate experimental measurement-error variability, whereas replications generally provide estimates of error variability for the factors or conditions that are varied with the replication. Not only do repeat tests and replication provide estimates of each type of variability, but increasing the number of repeat tests or replications also increases the precision of estimators of factor effects that are subject to each type of error.

As an illustration of the distinction between the use of repeat tests and replications, consider an experiment designed to investigate the cutoff times of automatic safety switches on lawnmowers. One version of this type of safety mechanism requires that a lever

Table 8.1   Experimental Design for Lawnmower Automatic Cutoff Times (Nonrandomized)

Test Run	Manufacturer	Lawnmower	Speed
1	A	1	Low
2	A	1	Low
3	A	1	High
4	A	1	High
5	A	2	Low
6	A	2	Low
7	A	2	High
8	A	2	High
9	A	3	Low
10	A	3	Low
11	A	3	High
12	A	3	High
13	B	1	Low
14	B	1	Low
15	B	1	High
16	B	1	High
17	B	2	Low
18	B	2	Low
19	B	2	High
20	B	2	High
21	B	3	Low
22	B	3	Low
23	B	3	High
24	B	3	High

be pressed by the operator while the mower is being used. Release of this lever by the operator results in the mower engine automatically stopping. The cutoff time of interest is the length of time between the release of the lever and the activation of the engine cutoff.

Table 8.1 displays a nonrandomized test sequence for an experiment that could be used to evaluate the cutoff times of lawnmowers produced by two manufacturers. Three lawnmowers from each manufacturer are included in the design. The cutoff times are to be evaluated at two engine operating conditions, low and high speed.

In this design there are two repeat tests for each manufacturer, lawnmower, and speed combination. Note too that the four test runs for each manufacturer (the two high-speed and the two low-speed test runs) are replicated by using three lawnmowers from each manufacturer. There are then two sources of experimental-error variability: error due to using different lawnmowers and error due to uncontrolled variation in the repeat tests.

Figure 8.1 depicts the combining of these two sources of error variation into the overall variability for an observation. In this figure the repeat tests have an uncontrolled error which is depicted as normally distributed around zero. The variability of the lawnmowers is also depicted as normally distributed around zero, but with a variance that is three times larger than that of the repeat test errors. The uppermost distribution shows the combining of these errors in the distribution of responses from one of the manufacturers. Assuming that these errors are statistically independent (see Section 12.1), the distribution of the total error is normal about zero with variance equal to the sum of the component variances.

Any comparisons between the cutoff times for the two manufacturers must take account of the variability due to both sources of error. This is because the effect of the manufacturers is calculated as the difference of the average cutoff times for the two manufacturers (Section 7.3), and each average is based on observations from all three lawnmowers made by that manufacturer and on the two repeat tests for each of the speeds. In particular, any comparison of manufacturer A with manufacturer B requires averages over different lawnmowers (and repeat tests); no comparison can be made on a single lawnmower.

Comparisons between the two operating speeds are ordinarily more precise than those between the manufacturers. This is because comparisons between the two speeds are only affected by the repeat-test variability. To appreciate this fact note that speed effects can be calculated for each individual lawnmower by taking differences of averages across repeat tests. Each of these six individual speed effects is subject to only the repeat-test variability, because each is based on averages for one lawnmower from one of the manufacturers. The overall effect of the two speeds (the difference of the overall averages for the high and the low speeds) can be calculated by taking the average of the six individual speed effects. Consequently, the overall effect of the speeds on the cutoff times, like the speed effects for the individual lawnmowers, is subject to only the uncontrolled variability due to the repeat tests.

As will be seen in Section 17.6, estimates of variability suitable for comparing

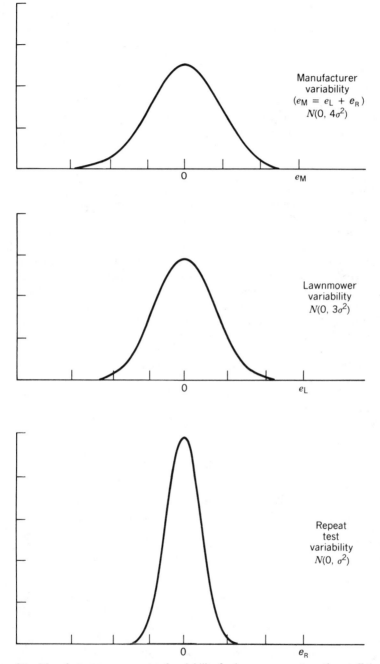

Manufacturer
variability
$(e_M = e_L + e_R)$
$N(0, 4\sigma^2)$

e_M

Lawnmower
variability
$N(0, 3\sigma^2)$

e_L

Repeat
test
variability
$N(0, \sigma^2)$

e_R

Figure 8.1 Manufacturer components of variability for lawnmower automatic-cutoff times.

manufacturers and for comparing speeds can be obtained from this experiment. The variability estimate needed for a comparison of the speeds is available from the repeat tests. The variability estimate needed for a comparison of the manufacturers is available from the replications. The greater precision in the comparisons of the speeds occurs because the estimate of error variability due to the repeat tests is ordinarily less than that of the replications.

The importance of this example is that the use of replication and repeat tests in an experiment design may be necessary to adequately account for different sources of response variability. Had no repeat tests been conducted, any estimate proposed for use as an estimate of the uncontrolled experimental error might be inflated by effects due to the lawnmowers or to one or more of the design factors (manufacturers, speed conditions). If the replicate observations on the six lawnmowers were not available, comparisons between the manufacturers would ordinarily be made relative to the estimate of the uncontrolled measurement error. If so, an erroneous conclusion that real differences exist between the manufacturers could be drawn when the actual reason was large variation in cutoff times of the lawnmowers made by each manufacturer. With no replication (only one lawnmower from each manufacturer), the effect due to the manufacturers would be confounded with the variation due to lawnmowers.

An important application of the use of repeat tests and replication is in the design of interlaboratory testing studies. In these studies experiments are conducted in order to evaluate a measurement process, or *test method*. The primary goal of such experiments is to estimate components of variation of the test method. Two key components of the total variation of the test method are variability due to repeat testing at a single laboratory and that due to replicate testing at different laboratories. The variability associated with the difference of two measurements at a single laboratory is referred to as the *repeatability* of the measurement process. The variability associated with the difference of two measurements taken at two different laboratories is referred to as the *reproducibility* of the measurement process.

Another major technique for controlling or compensating for variability in an experiment is termed *local control* (see Exhibit 8.2). It consists of the planned grouping of experimental units.

EXHIBIT 8.2   LOCAL CONTROL

Local control involves the grouping, blocking, and balancing of the experimental units.

Grouping simply refers to the placement of experimental units or of test runs into groups.

Blocking refers to the grouping of experimental units so that units within each block are as homogeneous as possible.

Balancing refers to the allocation of factor combinations to experimental units so that every factor–level combination occurs an equal number of times in a block or in the experiment.

The grouping of experimental units or of test runs often is undertaken because it is believed that observations obtained from tests within a group will be more uniform than those between groups. For example, one might choose to conduct certain test runs during one day because of known day-to-day variation. Repeat tests are sometimes conducted back to back in order to remove all sources of variability other than those associated with measurement error. While this type of grouping of test runs can result in unrealistically low estimates of run-to-run uncertainty, it is sometimes used to provide an estimate of the minimum uncertainty that can be expected in repeat testing. The grouping of a sequence of tests with one or more factors fixed is sometimes necessitated because of economic or experimental constraints.

Blocking is a special type of grouping whose origin is found in agricultural experiments. Agricultural land often is separated into contiguous blocks of homogeneous plots. Within each block the plots of land are experimental units which receive levels of the experimental factors (fertilizers, watering regimens, etc.). Thus, the experimental errors among plots of ground within a block are relatively small because the plots are in close physical proximity. Errors between plots in two different blocks are usually much larger than those between plots in the same block, because of varying soil or environmental conditions.

Blocking or grouping can occur in a wide variety of experimental settings. Since the experimental design concepts relating to blocking and grouping are ordinarily identical, we shall henceforth use the term "blocking" to refer to any planned grouping of either test runs or experimental units. In some experiments blocks may be batches of raw materials. In others, blocks may be different laboratories or groups of subjects. In still others, blocking may be the grouping of test runs under similar test conditions or according to the levels of one or more of the design factors. Blocking is thus an experimental design technique which removes excess variation by grouping experimental units or test runs so that those within a block are more homogeneous than those in different blocks. Comparisons of factor effects are then achieved by comparing responses or averages of observations within each block, thereby removing variability due to units or test conditions in different blocks.

The lawnmowers in the cutoff-time example are blocks, since the four observations for each lawnmower are subject only to repeat-test variability whereas observations on different lawnmowers are also subject to variations among the lawnmowers. The agricultural experiment depicted in Figure 6.2 also illustrates a block design with the three fields constituting blocks.

Balancing is a property of a design that assures that each factor–level combination occurs in the experiment an equal number of times. One often seeks to have balance in a design to ensure that the effects of all factor levels are measured with equal precision. Designs that are not balanced necessarily have certain factor effects that are measured with less precision than others. For ease of interpretation and analysis, we generally desire that designs be balanced.

8.2 RANDOMIZED COMPLETE BLOCK DESIGNS

One of the most often used designs is the randomized complete block (RCB) design. The steps in constructing a RCB design are given in Exhibit 8.3. It is assumed that no

---

EXHIBIT 8.3 RANDOMIZED COMPLETE BLOCK DESIGN

1. Group the experimental units or test runs into blocks so that those within a block are more homogeneous than those in different blocks.

2. Randomly assign all factor combinations of interest in the experiment to the units or test sequence in a block; use a separate randomization for each block. Complete or fractional factorial experiments can be included in each block.

---

interaction effects exist between the blocks and the experimental factors. If this is not the case, designs that allow for interaction effects must be used (see Chapter 7).

A common application of the blocking concept occurs in experiments in which responses are collected on pairs of experimental units. Pairing may occur because only two experimental units can be produced from each shipment of raw materials, because experiments are performed on sets of nearly identical experimental units such as twins, or because experimental units in close geographical proximity (such as plots of ground) are more uniform than widely separated ones.

In each of the above illustrations, the blocks represented physical units. Closely related to observations taken on pairs of experimental units are pairs of observations that are collected under similar experimental conditions. Repeat observations taken at several different locations, pre- and post-testing of the same unit or individual, and experimental conditions that limit the experimenter to two test runs per day are examples of pairs of observations that are blocked.

The construction of an RCB design can be illustrated by considering an experiment to develop a laboratory protocol for the study of allergic reactions to environmental pollutants. The laboratory procedure envisioned would utilize ten mice from a single strain. The responses of interest are skin thickness measurements at two locations on each animal. Skin thickness measurements are to be taken immediately prior to and 45 minutes following an injection of a solution containing a known pollutant. Repeat skin thickness measurements are to be taken at each location.

The design factors of interest in this experiment are the location (back, shoulder) and the time (0, 45 minutes) of the measurement. It is well known that substantial variation can be expected from the animals. It is not expected that interactions between animals and the design factors will exist in this experiment. Each animal is to be used as a block in this design, so that comparisons of the locations and the times will not be affected by animal-to-animal variation.

Table 8.2 displays a design layout for this experiment. Notice the features of this design. It has replicates (animals), repeat tests, and balance. Two levels of randomization afford protection against unknown sources of bias. The first level of randomization occurs with the random order of the testing of the animals at each measurement time. The second level of randomization occurs with the random order of the measurements on each of the animals.

RCB designs are particularly useful in the physical and engineering sciences. This is because experimentation often is fragmented across several time periods, different laboratories, separate groups of materials, or different test subjects. Blocking (grouping) on the sources of uncontrolled variability allows more sensitive comparisons of factor effects.

Despite the many advantages in using a RCB design, there are experimental conditions that necessitate the use of alternative blocking designs. For instance, the RCB designs described above only control for one extraneous source of variability. If two or more sources of extraneous variability exist, one can sometimes block on all the sources simultaneously using latin-square (Section 8.4) or other alternative blocking designs.

Even when there is only one source of extraneous variability that is of concern, there may not be a sufficient number of homogeneous experimental units available to include all factor-level combinations of interest in each block. In such situations one can often utilize designs that do not include all factor–level combinations in each block, but that are balanced. One such class of designs is the balanced incomplete block (BIB) design, the topic of the next section.

Table 8.2   Design Layout for Allergic-Reaction Experiment

Animal Number	Measurement Location
	Initial Measurements
6	Back, Shoulder, Shoulder, Back
2	Shoulder, Back, Shoulder, Back
10	Back, Shoulder, Shoulder, Back
3	Shoulder, Shoulder, Back, Back
1	Back, Back, Shoulder, Shoulder
9	Back, Shoulder, Shoulder, Back
7	Back, Shoulder, Back, Shoulder
5	Back, Shoulder, Shoulder, Back
8	Shoulder, Back, Back, Shoulder
4	Shoulder, Shoulder, Back, Back
	45 Minutes after Injection
10	Back, Shoulder, Back, Shoulder
3	Back, Back, Shoulder, Shoulder
1	Back, Shoulder, Shoulder, Back
4	Shoulder, Back, Back, Shoulder
7	Back, Shoulder, Shoulder, Back
9	Shoulder, Back, Shoulder, Back
5	Back, Shoulder, Shoulder, Back
6	Shoulder, Shoulder, Back, Back
2	Shoulder, Back, Shoulder, Back
8	Shoulder, Back, Back, Shoulder

8.3   BALANCED INCOMPLETE BLOCK DESIGNS

Situations can occur in constructing a blocking design where it is not possible to include all factor combinations in every block. The result is an incomplete block design. There are several different types of incomplete block designs available, but one of the most popular is the balanced incomplete block (BIB) design.

BIB designs can be found in Table 8A.1 in the appendix to this chapter. Table 8A.1 lists 17 BIB designs according to several design parameters. The design parameters are given in Exhibit 8.4. Although Table 8A.1 only lists designs for eight or fewer factor-level combinations, BIB designs are available for experiments having more than eight

---

EXHIBIT 8.4   KEY PARAMETERS IN A BIB DESIGN

f = number of factor-level combinations

r = number of times each factor-level combination occurs in the design

b = number of blocks

k = number of experimental units in a block

p = number of blocks in which each pair of factor-level combinations occurs together

---

combinations. However, BIB designs do not exist for all possible choices of the design parameters. This is why only select designs are presented in Table 8A.1.

Balance is achieved in BIB designs because every factor-level combination occurs an equal number of times (r) in the design and each pair of factors occur together in an equal number (p) of blocks in the design. Because of the desire for balance in these designs, certain restrictions are placed on them. These restrictions prevent BIB designs from existing for every possible choice of the above design parameters. The nature of some of the restrictions is indicated by the properties of BIB designs shown in Exhibit 8.5. Note in particular that p must be an integer and that b must be at least as great as f.

EXHIBIT 8.5   PROPERTIES OF A BALANCED INCOMPLETE BLOCK DESIGN

- Total number of observations:

$$N = fr = bk.$$

- Number of blocks in which a pair of factor–level combinations occur:

$$p = r\frac{k-1}{f-1}.$$

- Relation between the number of blocks and the number of factor–level combinations:

$$b \geqslant f.$$

The factor-level combinations referred to in BIB designs can be all the levels of a single factor or combinations of levels of two or more factors. Both complete and fractional experiments can be conducted in BIB designs, subject to the existence of a design for the chosen design parameters. A BIB design is chosen following the procedure in Exhibit 8.6.

EXHIBIT 8.6   BALANCED INCOMPLETE BLOCK DESIGN

1. Group the experimental units or test runs into blocks so that those within a block are more homogeneous than those in different blocks.

2. Check to be certain that the number of factor-level combinations to be used in the experiment is greater than the number of units or test runs available in each block.

3. Refer to Table 8A.1 to select a design for the chosen values of the design parameters.

4. Assign the factor-level combinations to the blocks as dictated by the BIB design selected.

5. Randomly assign the factor-level combinations to the experimental units or to the test sequence in a block; use a separate randomization for each block.

To illustrate the use of BIB designs, consider an experiment intended to measure the fuel consumption (gal/mi) of four engine oils. A single test engine is to be used in the experiment in order to eliminate as many extraneous sources of variation as possible. The engine is run on a dynamometer test stand, which is recalibrated after every second test run because each test run is the equivalent of several thousand road miles. Three replicates with each oil are considered necessary.

Table 8.3 Balanced Incomplete Design for Oil-Consumption Experiment

Block (Recalibration)	Oils Tested
1	A, B
2	C, D
3	A, C
4	B, D
5	A, D
6	B, C

In this example a block consists of the two measurements taken between calibrations. In terms of the design parameters, f is the number of engine oils (4), r is the number of replicate tests on each oil (3), and k is the number of tests between calibrations (2). Using the first of the above BIB design equations, the testing must be conducted using $b = 4(3/2) = 6$ blocks. Using the second of the three design equations, each pair of oils will appear together in exactly $p = 3(1/3) = 1$ block. These quantities dictate that the dynamometer will need to be recalibrated six times during the course of the experiment, and that between calibrations any pair of oils being tested will appear together exactly once. Using BIB design 1 from Table 8A.1, the layout of the experiment is as shown in Table 8.3. Prior to actually running this experiment one would randomize both the order of the blocks and the order of the testing of the two oils in each block.

The layout in Table 8.3 illustrates the key features of a BIB design. There are fewer units (2) available in each block than there are factor levels (4). Every oil appears in exactly three blocks and appears with every other oil in exactly one block. Finally, it is reasonable to assume that there is no interaction effect between oil and calibration.

The analysis of experiments conducted in BIB designs is presented in Section 18.2. Because of the restrictions introduced in the construction of BIB designs, it is not appropriate to calculate factor effects using differences of ordinary averages. The averages must be adjusted to take account of the fact that each factor level (or factor–level combination) does not occur in the same blocks as each of the other levels (combinations). Details of this adjustment are presented in Section 18.2.

8.4 LATIN-SQUARE AND CROSSOVER DESIGNS

The only type of restriction considered thus far in the discussion of blocking designs is a restriction on the number of experimental units or test runs that are available in each block. In this section we discuss latin-square designs and crossover designs to illustrate the rich variety of blocking designs that are available for implementation in experimental work. Latin-square designs are appropriate when two sources of extraneous variation are to be controlled. Crossover designs are useful when two or more factor levels are to be applied in a specified sequence in each block.

8.4.1 Latin Square Designs

Latin-square designs are ordinarily used when there is one factor of interest in the experiment and the experimenter desires to control two sources of variation. These designs can also be used to control two sources of variability with two or more factors of interest. In the latter setting, the factor levels referred to in the following discussion would be replaced by the factor-level combinations of interest.

Latin-square designs can be used when the number of levels of the factor of interest and the numbers of levels of the two blocking factors are all equal, each number being at least three. It is assumed that no interactions exist between the factor(s) of interest and the two blocking factors. The assumption of no interactions between the three factors is critical to the successful implementation of latin-square designs.

Although either of the blocking factors in a latin-square design could be replaced by another design factor, this implementation is strongly discouraged because of the possibility of interactions between the design factors. If two or all three of the factors in a latin-square design are design factors, the design is actually a form of a fractional factorial experiment. In such experiments it is preferable to conduct fractional factorial experiments in completely randomized, randomized block, or balanced incomplete block designs, as appropriate.

The basic construction of latin-square designs is straightforward. The design is laid out in rows and columns; the number of each equals the number of levels of the factor. Denote the levels of the factor of interest by capital letters: A, B, C,..., K. The letters, in the order indicated, constitute the first row of the design:

	Col. 1	Col. 2	Col. 3	$\cdots$	Col. k
Row 1	A	B	C	$\cdots$	K

Each succeeding row of the design is obtained by taking the first letter in the previous row and placing it last, shifting all other letters forward one position. Examples of latin-square designs for factors having from three to seven levels are given in Table 8A.2 of the appendix to this chapter. The procedure for constructing and using latin-square designs, including the randomization procedure, is given in Exhibit 8.7.

EXHIBIT 8.7   CONSTRUCTION OF LATIN SQUARE DESIGNS

1. Set up a table having k rows and k columns.
2. Assign the letters, $A, B, C, ..., K$ to the cells in the first row of the table.
3. For the second through kth rows of the table, place the first letter of the previous row in the last position and shift all other letters forward one position.
4. Randomly assign one of the blocking factors to the rows, the other one to the columns, and assign the design factor to the body of the table.
5. Randomly assign the levels of the three factors to the row positions, column positions, and letters, respectively.

A latin-square experimental design was used in the road testing of four tire brands for a tire wholesaler. The purpose of the study was to assess whether tires of different brands would result in demonstrably different fuel efficiencies on heavy-duty commercial trucks. In order to obtain reasonably reliable measures of fuel efficiency, a single test run would have to be of several hundred miles. It was therefore decided to use several test trucks and conduct the test so that each type of tire was tested on each truck. It was also necessary to conduct the experiment over several days. The experimenters then required that each tire be tested on each day of the test program.

Table 8.4   Latin-Square Design for Tire-Test Study

		Day			
		4	3	1	2
	3	Tire 4	Tire 2	Tire 1	Tire 3
Truck	4	Tire 2	Tire 1	Tire 3	Tire 4
	2	Tire 1	Tire 3	Tire 4	Tire 2
	1	Tire 3	Tire 4	Tire 2	Tire 1

A latin-square design was chosen in order to meet the requirements of the test program. Four trucks of one make were selected, and a four-day test period was identified. Randomly assigning the two blocking factors (trucks, days) to the design resulted in the rows of the design corresponding to the trucks, the columns to the days, and the letters to the tires. The layout of the design is given in Table 8.4. Note that each tire is tested on each truck and on each day of the test program.

In the above experiment the factor of interest was the tire brands. The truck and day factors were included because they add extraneous variation to the response. By including these factors, one is blocking (or grouping) on sources of variation and thereby allowing for more precise measurement of the effects due to the tire brands. Failure to include these factors in the design would result in an inflated experimental-error estimate. True differences due to the tire brands might then be masked by the variation due to trucks and days.

As mentioned above, the levels of factors used in latin-square designs can be factor combinations from two or more factors. For example, the four brands could represent four combinations of two tire types, each with two tread designs. Effects on the response due to both tire type and tread design can be analyzed. One can also assess whether an interaction due to tires and treads exists. One cannot, however, assess any interactions between tires or treads and the two other factors.

8.4.2   Crossover Designs

Crossover designs are used in experiments in which experimental units receive more than one factor-level combination in a preassigned time sequence. Many biological and medical experiments use crossover designs so that each subject or patient can act as its own control. This type of design is a blocking design in which each experimental unit (subject, patient) is a block.

Crossover designs are especially useful when there is considerable variation among experimental units. If excessive variation exists among experimental units, comparisons of factor effects could be confounded by the effects of the groups of experimental units that receive the respective factor levels.

There is an extensive literature on crossover designs. We discuss only very simple crossover designs in this section and refer the interested reader to the references for more detailed information. Part of the reason for the extensive literature is the problem of *carryover*, or *residual*, effects. This problem occurs when a previous factor-level combination influences the response for the next factor-level combination applied to the experimental unit. These carryover effects are common in drug testing, where the influence of a particular drug cannot always be completely removed prior to the administering of the

Table 8.5 Crossover Design Layout for a Clinical Study of Three Experimental Drugs

Sequence Number	Time Period			Subject Numbers
	1	2	3	
1	Drug 1	Drug 2	Drug 3	20, 4, 2
2	Drug 1	Drug 3	Drug 2	17, 9, 16, 3
3	Drug 2	Drug 1	Drug 3	13, 19, 5
4	Drug 2	Drug 3	Drug 1	12, 7, 1
5	Drug 3	Drug 1	Drug 2	14, 11, 6
6	Drug 3	Drug 2	Drug 1	8, 18, 10, 15

next drug. We use the simplifying assumptions in Exhibit 8.8 in the remaining discussion of crossover designs.

EXHIBIT 8.8 CROSSOVER DESIGN ASSUMPTIONS

- Excessive uncontrollable variation exists among the experimental units to be included in the experiment.
- Time periods are of sufficient duration for the factor effects, if any, to be imparted to the response.
- No residual or carryover effects remain from one time period to the next.
- Factor effects, if any, do not change over time (equivalently, no time–factor interactions).

For very small numbers of factor–level combinations, latin-square designs are frequently used as crossover designs. The rows of the latin square represent experimental units, the columns represent time periods, and the symbols represent the factor levels or factor–level combinations. For example, the three-level latin square in Table 8A.2 could be used for a single factor that has three levels. The design can be replicated (with a separate randomization for each) to accommodate more than three experimental units. The four-level design can be used for a single factor that has four levels or for all four factor–level combinations of a 2^2 factorial experiment.

Larger numbers of factor–level combinations can be accommodated by listing all possible time-order sequences for the factor–level combinations and randomly assigning the sequences to the experimental units. For illustration, Table 8.5 lists all six time orderings for the testing of three experimental drugs. There are 20 subjects who are available for the experimental program. They have been randomly assigned to the six time orderings. Similar designs can be constructed when there are many more than the three factor levels used in this illustration.

APPENDIX: BALANCED INCOMPLETE BLOCK AND LATIN SQUARE DESIGNS

Table 8A.1 Selected Balanced Incomplete Block Designs for Eight or Fewer Factor–Level Combinations

f	k	r	b	p	Design No.
4	2	3	6	1	1
	3	3	4	2	2
5	2	4	10	1	3
	3	6	10	3	4
	4	4	5	3	5
6	2	5	15	1	6
	3	5	10	2	7
	3	10	20	4	8
	4	10	15	6	9
	5	5	6	4	10
7	2	6	21	1	11
	3	3	7	1	12
	4	4	7	2	13
	6	6	7	5	14
8	2	7	28	1	15
	4	7	14	3	16
	7	7	8	6	17

DESIGN 1: $f = 4, k = 2, r = 3, b = 6, p = 1$

Block	Rep. I		Block	Rep. II		Block	Rep. III	
(1)	1	2	(3)	1	3	(5)	1	4
(2)	3	4	(4)	2	4	(6)	2	3

DESIGN 2: $f = 4, k = 3, r = 3, b = 4, p = 2$

Block				Block			
(1)	1	2	3	(3)	1	3	4
(2)	1	2	4	(4)	2	3	4

DESIGN: 3: $f = 5, k = 2, r = 4, b = 10, p = 1$

Block	Reps. I, II		Block	Reps. III, IV	
(1)	1	2	(6)	1	4
(2)	3	4	(7)	2	3
(3)	2	5	(8)	3	5
(4)	1	3	(9)	1	5
(5)	4	5	(10)	2	4

DESIGN 4: $f = 5, k = 3, r = 6, b = 10, p = 3$

Block	Reps. I, II, III			Block	Reps. IV, V, VI		
(1)	1	2	3	(6)	1	2	4
(2)	1	2	5	(7)	1	3	4
(3)	1	4	5	(8)	1	3	5
(4)	2	3	4	(9)	2	3	5
(5)	3	4	5	(10)	2	4	5

DESIGN 5: $f = 5, k = 4, r = 4, b = 5, p = 3$

Block	Reps. I–IV			
(1)	1	2	3	4
(2)	1	2	3	5
(3)	1	2	4	5
(4)	1	3	4	5
(5)	2	3	4	5

DESIGN 6: $f = 6, k = 2, r = 5, b = 15, p = 1$

Block	Rep. I		Block	Rep. II		Block	Rep. III	
(1)	1	2	(4)	1	3	(7)	1	4
(2)	3	4	(5)	2	5	(8)	2	6
(3)	5	6	(6)	4	6	(9)	3	5

Block	Rep. IV		Block	Rep. V	
(10)	1	5	(13)	1	6
(11)	2	4	(14)	2	3
(12)	3	6	(15)	4	5

DESIGN 7: $f = 6, k = 3, r = 5, b = 10, p = 2$

Block				Block			
(1)	1	2	5	(6)	2	3	4
(2)	1	2	6	(7)	2	3	5
(3)	1	3	4	(8)	2	4	6
(4)	1	3	6	(9)	3	5	6
(5)	1	4	5	(10)	4	5	6

DESIGN 8: $f = 6, k = 3, r = 10, b = 20, p = 4$

Block	Rep. I			Block	Rep. II			Block	Rep. III		
(1)	1	2	3	(3)	1	2	4	(5)	1	2	5
(2)	4	5	6	(4)	3	5	6	(6)	3	4	6

Block	Rep. IV			Block	Rep. V			Block	Rep. VI		
(7)	1	2	6	(9)	1	3	4	(11)	1	3	5
(8)	3	4	5	(10)	2	5	6	(12)	2	4	6

Block	Rep. VII			Block	Rep. VIII			Block	Rep. IX		
(13)	1	3	6	(15)	1	4	5	(17)	1	4	6
(14)	2	4	5	(16)	2	3	6	(18)	2	3	5

Block	Rep. X		
(19)	1	5	6
(20)	2	3	4

DESIGN 9: $f = 6, k = 4, r = 10, b = 15, p = 6$

Block	Reps. I, II				Block	Reps. III, IV			
(1)	1	2	3	4	(4)	1	2	3	5
(2)	1	4	5	6	(5)	1	2	4	6
(3)	2	3	5	6	(6)	3	4	5	6

Block	Reps. V, VI				Block	Reps. VII, VIII			
(7)	1	2	3	6	(10)	1	2	4	5
(8)	1	3	4	5	(11)	1	3	5	6
(9)	2	4	5	6	(12)	2	3	4	6

Block	Reps. IX, X			
(13)	1	2	5	6
(14)	1	3	4	6
(15)	2	3	4	5

DESIGN 10: $f = 6, k = 5, r = 5, b = 6, p = 4$

Block						Block					
(1)	1	2	3	4	5	(4)	1	2	4	5	6
(2)	1	2	3	4	6	(5)	1	3	4	5	6
(3)	1	2	3	5	6	(6)	2	3	4	5	6

DESIGN 11: $f = 7, k = 2, r = 6, b = 21, p = 1$

Block	Reps. I, II		Block	Reps. III, IV		Block	Reps. V, VI	
(1)	1	2	(8)	1	3	(15)	1	4
(2)	2	6	(9)	2	4	(16)	2	3
(3)	3	4	(10)	3	5	(17)	3	6
(4)	4	7	(11)	4	6	(18)	4	5
(5)	1	5	(12)	5	7	(19)	2	5
(6)	5	6	(13)	1	6	(20)	6	7
(7)	3	7	(14)	2	7	(21)	1	7

DESIGN 12: $f = 7, k = 3, r = 3, b = 7, p = 1$

Block				Block			
(1)	1	2	4	(5)	5	6	1
(2)	2	3	5	(6)	6	7	2
(3)	3	4	6	(7)	7	1	3
(4)	4	5	7				

DESIGN 13: $f = 7, k = 4, r = 4, b = 7, p = 2$

Block					Block					Block				
(1)	3	5	6	7	(4)	1	2	3	6	(7)	2	4	5	6
(2)	1	4	6	7	(5)	2	3	4	7					
(3)	1	2	5	7	(6)	1	3	4	5					

DESIGN 14: $f = 7, k = 6, r = 6, b = 7, p = 5$

Block							Block						
(1)	1	2	3	4	5	6	(5)	1	2	4	5	6	7
(2)	1	2	3	4	5	7	(6)	1	3	4	5	6	7
(3)	1	2	3	4	6	7	(7)	2	3	4	5	6	7
(4)	1	2	3	5	6	7							

DESIGN 15: $f = 8, k = 2, r = 7, b = 28, p = 1$

Block	Rep. I		Block	Rep. II		Block	Rep. III	
(1)	1	2	(5)	1	3	(9)	1	4
(2)	3	4	(6)	2	8	(10)	2	7
(3)	5	6	(7)	4	5	(11)	3	6
(4)	7	8	(8)	6	7	(12)	5	8

Block	Rep. IV		Block	Rep. V		Block	Rep. VI	
(13)	1	5	(17)	1	6	(21)	1	7
(14)	2	3	(18)	2	4	(22)	2	6
(15)	4	7	(19)	3	8	(23)	3	5
(16)	6	8	(20)	5	7	(24)	4	8

Block	Rep. VII	
(25)	1	8
(26)	2	5
(27)	3	7
(28)	4	6

DESIGN 16: $f = 8, k = 4, r = 7, b = 14, p = 3$

Block	Rep. I				Block	Rep. II				Block	Rep. III			
(1)	1	2	3	4	(3)	1	2	7	8	(5)	1	3	6	8
(2)	5	6	7	8	(4)	3	4	5	6	(6)	2	4	5	7

Block	Rep. IV				Block	Rep. V				Block	Rep. VI			
(7)	1	4	6	7	(9)	1	2	5	6	(11)	1	3	5	7
(8)	2	3	5	8	(10)	3	4	7	8	(12)	2	4	6	8

Block	Rep. VII			
(13)	1	4	5	8
(14)	2	3	6	7

DESIGN 17: $f = 8, k = 7, r = 7, b = 8, p = 6$

Block							
(1)	1	2	3	4	5	6	7
(2)	1	2	3	4	5	6	8
(3)	1	2	3	4	5	7	8
(4)	1	2	3	4	6	7	8
(5)	1	2	3	5	6	7	8
(6)	1	2	4	5	6	7	8
(7)	1	3	4	5	6	7	8
(8)	2	3	4	5	6	7	8

Table 8A.2   Latin-Square Designs for Three Through Eight Levels of Each Factor

(3)		
A	B	C
B	C	A
C	A	B

(4)			
A	B	C	D
B	C	D	A
C	D	A	B
D	A	B	C

(5)				
A	B	C	D	E
B	C	D	E	A
C	D	E	A	B
D	E	A	B	C
E	A	B	C	D

(6)					
A	B	C	D	E	F
B	C	D	E	F	A
C	D	E	F	A	B
D	E	F	A	B	C
E	F	A	B	C	D
F	A	B	C	D	E

(7)						
A	B	C	D	E	F	G
B	C	D	E	F	G	A
C	D	E	F	G	A	B
D	E	F	G	A	B	C
E	F	G	A	B	C	D
F	G	A	B	C	D	E
G	A	B	C	D	E	F

(8)							
A	B	C	D	E	F	G	H
B	C	D	E	F	G	H	A
C	D	E	F	G	H	A	B
D	E	F	G	H	A	B	C
E	F	G	H	A	B	C	D
F	G	H	A	B	C	D	E
G	H	A	B	C	D	E	F
H	A	B	C	D	E	F	G

REFERENCES

Text References

Cox, D. R. (1958). *Planning of Experiments*, New York: John Wiley and Sons, Inc.
    This text introduces many of the design concepts discussed in this chapter in a nontechnical exposition. It avoids mathematical formulas and stresses intuitive understanding of basic design concepts. Much of the discussion centers around examples.

The following references can be consulted for additional information on complete and incomplete block designs. The references by Cochran and Cox, Davies, and Natrella contain numerous layouts for balanced incomplete block designs. The books by Cochran and Cox; Cox: Federer: John: and Milliken and Johnson discuss various aspects of crossover designs.

Anderson, R. L. and Bancroft, T. A. (1952). *Statistical Theory in Research*, New York: McGraw-Hill.

Anderson, V. L. and McLean, R. A. (1974). *Design of Experiments: A Realistic Approach*, New York: Marcel Dekker, Inc.

Box, G. E. P., Hunter, W.G., and Hunter, J. S. (1978). *Statistics for Experimenters*, New York: John Wiley and Sons, Inc.

Cochran, W.G. and Cox, D. R. (1957). *Experimental Designs*, New York: John Wiley and Sons, Inc.

Cox, D. R. (1958). *Planning of Experiments*, New York: John Wiley and Sons, Inc.

Daniel, C. (1976). *Design and Analysis of Industrial Experiments*, New York: The Macmillan Co.

Federer, W. T. (1955). *Experimental Design: Theory and Application*, New York: The Macmillan Co.

Hicks, C. R. (1982). *Fundamental Concepts in the Design of Experiments*, New York: Holt, Rinehart, and Winston.

John, P. W. M. (1980). *Incomplete Block Designs*, New York: Marcel Dekker, Inc.

John, P. W. M. (1971). *Statistical Design and Analysis of Experiments*, New York: The Macmillan Co.

Johnson, N. L. and Leone, F. (1977). *Statistics and Experimental Design in Engineering and the Physical Sciences, Volume II*, New York: John Wiley and Sons, Inc.

Milliken, G. A. and Johnson, D. A. (1984). *Analysis of Messy Data: Design of Experiments*, New York: Van Nostrand Reinhold.

Natrella, M. G. (1963). *Experimental Statistics*, National Breau of Standards Handbook 91, Washington, D. C.: U.S. Government Printing Office. (Reprinted by John Wiley and Sons, Inc.)

EXERCISES

1   An industrial hygienist is concerned with the exposure of workers to chemical vapors during bulk liquid transfer operations on a chemical tanker. The downwind concentration of vapors displaced from loading tanks is being investigated. The industrial hygienist intends to measure the vapor concentrations (ppm) of three different chemicals (benzene, vinyl acetate, and xylene) being loaded into a tanker on five consecutive days. Design an experiment to control for the likely variation in days.

2   Three technicians in a chemical plant are given the task of investigating the chemical yield of a raw material by applying one of five experimental treatments. Three repeat tests are to be made with each of the five treatments. Design an experiment appropriate for this study. Is there a blocking factor in the design? If so, what is it?

3   Design an experiment to investigate deposits on a filter of a cooling system. The factors of interest are:

$$\text{Flow rate:} 5, 10 \, \text{gps}$$
$$\text{Filter diameter:} 0.5, 1, 2 \, \text{cm}$$
$$\text{Fluid temperature:} 75, 100, 125°\text{F}$$

A maximum of 20 test runs can be made with each batch of cooling fluid because of impurities that enter the cooling system. It is desired that several batches of cooling fluid be used, subject to a maximum of 75 total test runs.

4   Design an experiment involving three factors, each at two levels, in which no more than ten test runs can be conducted on each of several test days. It is desired that all main effects and interactions be estimated when the experiment is concluded. A maximum of ten test days are available for the experiment.

5   Design an experiment in which two factors are to be investigated, one at two levels and one at three levels. The experiment must be run in blocks, with no more than four test runs per block. Up to 20 blocks can be used in the experiment.

6   A study is to be conducted to assess the effects of kiln firing of clay sculptures on the color of the fired clay. Spectral measurements are to be taken on sculptures made from four different clay compositions. Location variations within the kiln and variations in temperature are also believed to influence the final color. Four locations in the kiln are to be investigated, as are four temperatures. The clay must be made in batches. Each batch can produce a maximum of 20 clay sculptures. No more than 50

clay sculptures can be included in the experiment. Construct a suitable design for this experiment.

7   A study is to be conducted to investigate seven computer algorithms for handling the scheduling of deliveries of parcels by a private delivery company. Because of the wide variety of scheduling scenarios that can arise, a collection of 50 commonly encountered problems have been carefully constructed on which to test the algorithms. A good algorithm is one that can solve the problems rapidly; consequently, the time needed to solve the problems is the key response variable. All the problems are complex and require intensive computing effort to solve. Because of this, it is prohibitively expensive for all the algorithms to solve all the problems. It is decided that approximately 40–50 problems should be used in the experiment, that no more than four of the algorithms should be given any one problem, and that each algorithm must be give at least ten of the problems. Design an experiment to satisfy all these requirements.

8   An experiment is to be designed to study the lift on airplane wings. Two wing models from each of five wing designs are to be included in the study. The wing models are to be placed in a wind tunnel chamber, in which two takeoff angles (5, 10° from horizontal) can be simulated. The lift (lb) is to be measured twice on each wing at each angle. Design a suitable experiment to study the lift on these wing designs.

9   The U.S. National Weather Service wants to commission an experiment to study the performance of weather balloons made of different materials. Nylon, a polyester blend, and a rayon–nylon blend are the materials to be studied. Six balloons made from each of the materials are to be tested. Two balloons of each material will be loaded with instrumentation having one of three weights (15, 20, and 25 lb). Each balloon will be released at the same geographic location. The response of interest is the height the balloons rise to before exploding due to atmospheric pressure. Design an appropriate experiment for this study.

10 An experiment is to be designed to determine the amount of corrosion for five different tube materials (stainless steel, iron, copper, aluminum, and brass) in a study similar to the fluidized-bed experiment in Exercise 5 of Chapter 6. Six repeat tests are to be run with each type of tube material. Only three tests can be run per month with the accelerated-aging machine before the equipment that pumps the water through the pipes must be overhauled. Design an experiment for this study.

11 The U.S. Army wishes to study the effects of two methanol fuel blends on the wear rate of iron in jeep engines. Nine jeeps are to be provided for this study. Both fuels are to be used in each jeep. Each fuel is to be used in a jeep for one month. Only three months are available for the study. Thus, three wear measurements will be available from each jeep. Design a study that will control for the sequencing of the fuels in the jeeps.

12 A clinical study is to be conducted on the effectiveness of three new drugs in reducing the buildup of fluid in the ears of infants during winter months. Twenty-five infants are to be included in the four-month study. Each infant is to receive each of the drugs at least once. Lay out an experimental design appropriate for this study.

CHAPTER 9

Fractional Factorial Experiments

Complete factorial experiments cannot always be conducted, due to economic, time, or other constraints. Fractional factorial experiments are widely used in such circumstances. In this chapter fractional factorial experiments for two-level factors are detailed. Extensions to more than two levels are discussed in the chapter appendix. The following topics are emphasized:

- *confounding and design resolution, properties of fractional factorial experiments that determine which factor-level combinations are to be included in the experiment,*
- *construction of fractional factorial experiments in completely randomized designs,*
- *construction of complete and fractional factorial experiments in incomplete block designs,*
- *the use of fractional factorial experiments in screening experiments, and*
- *sequentially building experiments using fractional factorials.*

Fractional factorial experiments are important alternatives to complete factorial experiments when budgetary, time, or experimental constraints preclude the execution of complete factorial experiments. In addition, experiments that involve many factors are routinely conducted as fractional factorials because it is not necessary to test all possible factor–level combinations in order to estimate the important factor effects, generally the main effects and low-order interactions.

A study in which a fractional factorial experiment was necessitated is summarized in Table 9.1. The experiment was designed to study the corrosion rate of a reactor at a chemical acid plant. There were six factors of interest, each having two levels. Initially it was unknown whether changing any of these six process variables (factors) would have an effect on the reactor corrosion rate. In order to minimize maintenance and downtime, it was desirable to operate this acid plant under process conditions which produce a low corrosion rate. If one were to test all possible combinations of levels of the six process variables, 64 test runs would be required.

A practical difficulty with this investigation was that the plant needed to cease commercial production for the duration of the study. An experimental design that required 64 test runs would have been prohibitively expensive. A small screening experiment was designed and executed in which only gross factor effects were identified. The cost of experimentation was greatly reduced. In addition, substantial production

Table 9.1　Factor Levels for Acid-Plant Corrosion-Rate Study

Factor	Levels
Raw-material feed rate	3000, 6000 pph
Gas temperature	100°, 220°C
Scrubber water	5, 20%
Reactor-bed acid	20, 30%
Exit temperature	300, 360°C
Reactant distribution point	East, west

savings resulted from the identification and subsequent adjustment of factors that have a large effect on the corrosion rate.

In this chapter we discuss fractional factorial experiments. In order to provide a systematic mechanism for constructing such experiments, we begin the chapter by discussing confounding and design resolution for them. In Sections 9.2 and 9.3, fractional factorial experiments conducted in completely randomized and blocking designs are discussed, respectively. The final section of this chapter outlines the use of fractional factorial experiments in screening designs and in sequential experimentation.

9.1　CONFOUNDING AND DESIGN RESOLUTION

Whenever fractional factorial experiments are conducted, some effects are confounded with one another. The goal in the design of fractional factorial experiments is to ensure that the effects of primary interest are either unconfounded with other effects or, if that is not possible, confounded with effects that are not likely to have appreciable magnitudes. In this section the confounding of effects in fractional factorial experiments is discussed. The following three sections utilize the planned confounding of factor effects in the construction of fractional factorial experiments.

In order to make some of the concepts related to confounding and design resolution clear, reconsider the pilot-plant example introduced in Section 7.2. In this study the yield of a chemical process was investigated as a function of three design factors: operating temperature of the plant (factor A), concentration of the catalyst used in the process (factor B), and type of catalyst (factor C). The graphical (Figure 7.4) and numerical (Tables 7.7 and 7.8) analyses of these data suggest that temperature and type of catalyst have a strong joint effect on the chemical yield and that concentration of the catalyst has a strong main effect.

Suppose now that the experimenters wish to design an experiment in order to confirm these results on a full-scale manufacturing plant. Because of the time required to make changes in the process, suppose that only four test runs can be conducted during each eight-hour operating shift. The ability to run only four of the eight possible factor–level combinations during a single shift introduces the possibility that some of the factor effects could be confounded with the effects of different operators or operating conditions on the shifts.

Confounding was defined in Table 6.1 as the situation where an effect cannot unambiguously be attributed to a single main effect or interaction. In Section 7.3 and the appendix to Chapter 7, factor effects were defined in terms of differences or *contrasts* (linear combinations whose coefficients sum to zero) of response averages. The expanded definition in Exhibit 9.1 relates the confounding of effects explicitly to their computation. According to this definition, one effect is confounded with another effect if the effects representations of the two effects are identical, apart from a possible sign reversal. Effects that are confounded in this way are called *aliases*. The examples presented in this and the next section will make this point clear.

EXHIBIT 9.1

Confounded effects. Two or more experimental effects are confounded if calculated effects can only be attributed to their combined influence on the response, not to their individual ones. Two or more effects are confounded if the calculation of one effect uses the same (apart from sign) difference or contrast of the response averages as the calculation of the other effects.

In some experimental designs factor effects are confounded with other factor effects (see Section 9.2). In other designs factor effects are confounded with block effects (see Section 9.3). It is imperative that an experimenter know which effects are confounded, and with what other effects they are confounded, when an experiment is designed. This knowledge is necessary in order to ensure that conclusions about the effects of interest will not be compromised because of possible influences of other effects.

In the manufacturing-plant example there are three factors, each at two levels. Table 9.2 lists the effects representation (Section 7.3) for each of the main effects and interactions. Table 9.2 also contains a column for the *constant* effect, represented by the letter I. The constant effect is simply the overall average response, a measure of location for the response in the absence of any effects due to the design factors.

For simplicity in this discussion, suppose only eight test runs are to be made ($r = 1$) at the manufacturing plant, so that a typical response for the ith level of temperature (factor A), the jth level of concentration (factor B), and the kth level of catalyst (factor C) is denoted y_{ijk}. The last column of Table 9.2 algebraically lists the responses for each factor–level combination.

Note that the signs on the constant effect are all positive. The constant effect is simply the overall response average, and it is obtained by summing the individual factor–level responses (since $r = 1$) and dividing by the total number of factor–level combinations, $2^k = 8$ for $k = 3$ factors. All other effects are obtained by taking the contrasts

Table 9.2 Effects Representation for Manufacturing-Plant Factorial Experiment*

Factor–Level Combination	Constant (I)	A	B	C	AB	AC	BC	ABC	Response
1	+1	−1	−1	−1	+1	+1	+1	−1	y_{111}
2	+1	−1	−1	+1	+1	−1	−1	+1	y_{112}
3	+1	−1	+1	−1	−1	+1	−1	+1	y_{121}
4	+1	−1	+1	+1	−1	−1	+1	−1	y_{122}
5	+1	+1	−1	−1	−1	−1	+1	+1	y_{211}
6	+1	+1	−1	+1	−1	+1	−1	−1	y_{212}
7	+1	+1	+1	−1	+1	−1	−1	−1	y_{221}
8	+1	+1	+1	+1	+1	+1	+1	+1	y_{222}

*A = temperature ($-1 = 160°C$, $+1 = 180°C$); B = concentration ($-1 = 20\%$, $+1 = 40\%$); C = catalyst ($-1 = C_1$, $+1 = C_2$).

of the responses indicated in the respective columns and dividing by half the number of factor–level combinations, $2^{k-1} = 4$ (see Sections 7.3 and 7.4).

Suppose the response values in Table 9.2 are the responses that would be observed in the absence of any shift effect. Suppose further that the first four factor–level combinations listed in Table 9.2 are taken during the first shift and the last four are taken during the second shift. Observe that the first four combinations have temperature at its lower level (160°C) and the second four all have temperature at its upper (180°C) level. The influence of the operating conditions and operators on each shift is simulated in Table 9.3 by adding a constant amount, s_1 for the first shift and s_2 for the second, to each observation taken during the respective shifts. The average responses for each shift are thereby increased by the same amounts.

One way to confirm the confounding of the main effect for temperature and the effect due to the shifts is to observe that these two effects have the same effect representation. The main effect for temperature is the difference between the first four and the last four factor–level combinations in Table 9.3. But this is the same difference in averages that one would take to calculate the effect of the shifts. Thus, the effects representation for factor A in Table 9.3 is, apart from changing all of the signs, also the effects representation for the shift effect.

A second way of confirming the confounding of the main effect for temperature and the shift effect is to symbolically compute the main effect using the effects representation and the responses shown in Table 9.3. Doing so yields the following result:

$$M(A) = (\bar{y}_{2..} + s_2) - (\bar{y}_{1..} + s_1)$$
$$= \bar{y}_{2..} - \bar{y}_{1..} + s_2 - s_1.$$

While $\bar{y}_{2..} - \bar{y}_{1..}$ is the actual effect of temperature, the calculated effect is increased by an amount equal to $s_2 - s_1$, the shift effect.

The other two main effects and all the interaction effects have an equal number of positive and negative signs associated with the responses in each shift. When these effects

Table 9.3 Influence of Shifts on Responses from a Factorial Experiment for the Manufacturing-Plant Chemical-Yield Study

Factor–Level Combination	Effects Representation*			Shift	Observed Response
	A	B	C		
1	-1	-1	-1	1	$y_{111} + s_1$
2	-1	-1	1	1	$y_{112} + s_1$
3	-1	1	-1	1	$y_{121} + s_1$
4	-1	1	1	1	$y_{122} + s_1$
				Average	$\bar{y}_{1..} + s_1$
5	1	-1	-1	2	$y_{211} + s_2$
6	1	-1	1	2	$y_{212} + s_2$
7	1	1	-1	2	$y_{221} + s_2$
8	1	1	1	2	$y_{222} + s_2$
				Average	$\bar{y}_{2..} + s_2$

*A = Temperature $(-1 = 160°C, 1 = 180°C)$; B = Concentration $(-1 = 20\%, 1 = 40\%)$; C = Catalyst $(-1 = C_1, 1 = C_2)$.

are calculated, any shift effect will cancel, since half the responses in each shift will add the shift effect and the other half of the responses will subtract the shift effect. For example,

$$M(B) = \tfrac{1}{4}[(y_{121} + s_1) + (y_{122} + s_1) + (y_{221} + s_2)$$
$$+ (y_{222} + s_2) - (y_{111} + s_1) - (y_{112} + s_1)$$
$$- (y_{211} + s_2) - (y_{212} + s_2)]$$
$$= \bar{y}_{.2.} - \bar{y}_{.1.}.$$

In this example, the complete factorial experiment is partitioned so that half of the complete factorial is conducted during each of the two shifts. This is a blocking design with half the complete factorial in each of two blocks. The fractions chosen for testing during each shift are not the best ones that could be selected given the information known from the pilot study. It would be preferable to design the experiment so that one of the interactions that showed little influence on the response in the pilot study would be confounded with the shift effect. In this way, all the effects believed to appreciably influence the response could be estimated even though the responses are split between the two shifts. It is this type of planned confounding that is the goal of the statistical design of fractional factorial experiments.

In general, when designing fractional factorial experiments one seeks to confound either effects known to be negligible relative to the uncontrolled experimental error variation, or in the absence of such knowledge, high-order interactions, usually those involving three or more factors. Confounding of high-order interactions is recommended because frequently these interactions either do not exist or are negligible relative to main effects and low-order interactions. Thus, in the absence of the information provided by the pilot-plant study, it would be preferable to confound the three-factor interaction in the manufacturing-plant experiment rather than the main effect for temperature. To do so one merely needs to assign all factor–level combinations that have one sign (e.g., 1, 4, 6, 7) in the three-factor interaction ABC in Table 9.2 to the first shift, and those with the other sign (2, 3, 5, 8) to the second shift.

Not only does confounding occur when a complete factorial experiment is conducted in blocks; it also occurs when only a portion of all the possible factor–level combinations are included in the design. Confounding occurs because two or more effects representations (apart from a change in all the signs) are the same. A calculated effect then represents the combined influence of the effects. In some instances (e.g., one-factor-at-a-time experiments) the confounding pattern may be so complex that one cannot state with assurance that any of the calculated effects measure the desired factor effects. This again is why planned confounding, confounding in which important effects either are unconfounded or are only confounded with effects that are believed to be negligible, is the basis for the statistical constructions of fractional factorial experiments.

Consider again the manufacturing-plant experiment. Suppose that because of the expense involved, only four test runs could be conducted. While this is a highly undesirable experimental restriction for many reasons, it does occur. Suppose further that only the first four test runs are conducted. Then the effects representations for the various experimental effects are those shown in the first four rows of Table 9.2. These representations are redisplayed in Table 9.4, grouped in pairs.

The effects representations are grouped in pairs because the effects in each pair are calculated, apart from a change in all the signs, from the same linear combination of

Table 9.4   Effects Representation for a Fractional Factorial Experiment for the Manufacturing-Plant Chemical-Yield Study

Factor–Level Combination	Factor Effects							
	I	A	B	AB	C	AC	BC	ABC
1	1	-1	-1	1	-1	1	1	-1
2	1	-1	-1	1	1	-1	-1	1
3	1	-1	1	-1	-1	1	-1	1
4	1	-1	1	-1	1	-1	1	-1

the responses. For example, the main effect for B and the interaction AB are, respectively,

$$M(B) = \tfrac{1}{2}(-y_{111} - y_{112} + y_{121} + y_{122})$$

$$= \bar{y}_{.2.} - \bar{y}_{.1.}.$$

and

$$I(AB) = \tfrac{1}{2}(+y_{111} + y_{112} - y_{121} - y_{122})$$

$$= -M(B).$$

The negative sign in the equation $I(AB) = -M(B)$ indicates that the two effects representations are identical except that one has all its signs opposite those of the other.

Similarly, $M(C) = -I(AC)$ and $I(BC) = -I(ABC)$. Calculation of the constant effect, denoted CONST, differs from the calculation of the other effects only in that its divisor is twice the others. Nevertheless, it too is confounded, since CONST $= -M(A)/2$. In terms of the effects representation for these factor effects, one would express this confounding pattern as

$$I = -A, \qquad B = -AB, \qquad C = -AC, \qquad BC = -ABC.$$

Thus, this fractional factorial experiment results in each experimental effect being confounded with one other experimental effect. Moreover, no information is available on the main effect of temperature (factor A), because only one of its levels was included in the design. This is why temperature is confounded with the constant effect.

By constructing a fractional factorial experiment with a different set of four factor–level combinations, a different confounding pattern would result. Every choice of four factor–level combinations will result in each factor effect being confounded with one other effect. One can, however, choose the combinations so that no main effect is confounded with the constant effect. For example, by choosing combinations 2, 3, 5, and 8, the confounding pattern is

$$I = ABC, \qquad A = BC, \qquad B = AC, \qquad C = AB.$$

With this design, all information is lost on the three-factor interaction, since all the factor–level combinations in the design have the same sign on the effects representation for the interaction ABC.

This latter design would not be advisable for the manufacturing-plant example because both the main effect of concentration, $M(B)$, and the joint effect of temperature and catalyst type, $I(AC)$, are known to have appreciable effects on the chemical yield, at least for the pilot plant. The above design would confound these two effects, since $B = AC$. This design would, however, be suitable if prior experimentation had shown that all interactions were zero or at least negligible relative to the uncontrolled experimental error. If this were the experimental situation, any large calculated main effects could be attributed to individual effects of the respective factors.

In the next two sections of this chapter the construction of fractional factorial experiments in completely randomized designs and in blocking designs will be detailed. The focus in these discussions is on the planned confounding of effects so that the effects of most interest are either unconfounded or only confounded with effects that are believed to be negligible relative to the uncontrolled experimental error. When prior knowledge about the magnitude of effects is not available, attention is concentrated on the confounding of main effects and low-order interactions only with high-order interactions.

An important guide that will be use in selecting fractional factorial experiments in the next two sections is the concept of *design resolution*. (see Exhibit 9.2). Design resolution identifies for a specific design the order of confounding of main effects and interactions.

EXHIBIT 9.2

Design Resolution. An experimental design is of resolution R if all effects containing s or fewer factors are unconfounded with any effects containing fewer than $R - s$ factors.

Design resolution is defined in terms of which effects are unconfounded with which other effects. One ordinarily seeks fractional factorial experiments that have main effects ($s = 1$) and low-order interactions (say those involving $s = 2$ or 3 factors) unconfounded with other main effects and low-order interactions—equivalently, in which they are confounded only with high-order interactions.

The resolution of a design is usually denoted by capital Roman letters; e.g., III, IV, V. The symbol R in the above definition denotes the Arabic equivalent of the Roman numeral; e.g., $R = 3, 4, 5$.

A design in which main effects are not confounded with other main effects but are confounded with two-factor and higher-order interactions is a resolution-III (also denoted R_{III}) design. Such a design is of resolution III because effects containing $s = 1$ factor are unconfounded with effects containing fewer than $R - s = 3 - 1 = 2$ factors. The only effects containing fewer than $s = 2$ factors are other main effects. As should be clear from the confounding pattern described above, the four-combination (2, 3, 5, 8) fractional factorial experiment with $I = ABC$ for the chemical yield experiment is R_{III} design.

At least one main effect is confounded with a two-factor interaction in a R_{III} design; otherwise the design would be R_{IV}. In a resolution-IV design, main effects are unconfounded with effects involving fewer than $R - s = 4 - 1 = 3$ factors, i.e., main effects and two-factor interactions. Two-factor interactions ($s = 2$) are unconfounded with effects involving fewer than $R - s = 2$ factors, i.e., main effects ($s = 1$). There are some two-factor interactions confounded with other two-factor interactions in R_{IV} designs.

A design of resolution V has main effects and two-factor interactions ($s = 1, 2$) unconfounded with one another. Some main effects are confounded with four-factor and higher-order interactions; some two-factor interactions are confounded with three-factor and higher-order interactions. In resolution-VII designs, main effects, two-factor interactions, and three-factor interactions are mutually unconfounded with one another.

This discussion of design resolution is intended to provide a quick means of assessing whether an experimental design allows for the estimation of all important effects (assumed to be main effects and low-order interactions) without the confounding of these effects with one another. Techniques for determining the resolution of a design are presented in the next section.

9.2 COMPLETELY RANDOMIZED DESIGNS

The number of test runs needed for complete factorial experiments increases quickly as the number of factors increases, even if each factor is tested at only two levels. The intentional confounding of factor effects using fractional factorials can reduce the number of test runs substantially for large designs if some interaction effects are known to be nonexistent or negligible relative to the uncontrolled experimental error variation.

The next two subsections of this section detail the construction of fractional factorial experiments in completely randomized designs. The basic procedures are introduced in Section 9.2.1 for half fractions. These procedures are extended to smaller fractions $(\frac{1}{4}, \frac{1}{8},$ etc.) in Section 9.2.2.

9.2.1 Half Fractions

Half fractions of factorial experiments are specified by first stating the *defining contrast* for the fraction. The defining contrast is the effect that is confounded with the constant effect. It can be expressed symbolically as an equation, the *defining equation*, by setting the confounded effect equal to I, which represents the constant effect. For the manufacturing-plant example in the last section, two half fractions of the complete 2^3 experiment were discussed. The defining equations for the two fractional factorial experiments were, respectively, $I = -A$ and $I = ABC$.

Ordinarily the effect chosen for the defining contrast is the highest-order interaction among the factors. Once the defining contrast is chosen, the half fraction of the factor-level combinations that are to be included in the experiment are those that have either the positive or the negative signs in the effects representation of the defining contrast (see Exhibit 9.3). The choice of the positive or the negative signs is usually made randomly (e.g., by flipping a coin).

As an illustration of this procedure, consider the acid-plant corrosion-rate study discussed in the introduction to this chapter. Suppose that a half fraction of the 2^6

EXHIBIT 9.3   DESIGNING HALF FRACTIONS OF TWO-LEVEL FACTORIAL EXPERIMENTS IN COMPLETELY RANDOMIZED DESIGNS

1. Choose the defining contrast: the effect that is to be confounded with the constant effect.

2. Randomly decide whether the experiment will contain the factor–level combinations with the positive or the negative signs in the effects representation of the defining contrast.

3. Form a table containing the effects representations of the main effects for each of the factors. Add a column containing the effects representation of the defining contrast.

4. Select the factor–level combinations that have the chosen sign in the effects representation of the defining contrast. Randomly select any repeat tests that are to be included in the design.

5. Randomize the assignment of the factor–level combinations to the experimental units or to the test sequence, as appropriate.

Table 9.5 Half Fraction of the Complete Factorial Experiment for the Acid-Plant Corrosion-Rate Study

Factor–Level Combination	Effects Representation						
	A	B	C	D	E	F	ABCDEF
1	−1	−1	−1	−1	−1	1	−1
2	−1	−1	−1	−1	1	−1	−1
3	−1	−1	−1	1	−1	−1	−1
4	−1	−1	−1	1	1	1	−1
5	−1	−1	1	−1	−1	−1	−1
6	−1	−1	1	−1	1	1	−1
7	−1	−1	1	1	−1	1	−1
8	−1	−1	1	1	1	−1	−1
9	−1	1	−1	−1	−1	−1	−1
10	−1	1	−1	−1	1	1	−1
11	−1	1	−1	1	−1	1	−1
12	−1	1	−1	1	1	−1	−1
13	−1	1	1	−1	−1	1	−1
14	−1	1	1	−1	1	−1	−1
15	−1	1	1	1	−1	−1	−1
16	−1	1	1	1	1	1	−1
17	1	−1	−1	−1	−1	−1	−1
18	1	−1	−1	−1	1	1	−1
19	1	−1	−1	1	−1	1	−1
20	1	−1	−1	1	1	−1	−1
21	1	−1	1	−1	−1	1	−1
22	1	−1	1	−1	1	−1	−1
23	1	−1	1	1	−1	−1	−1
24	1	−1	1	1	1	1	−1
25	1	1	−1	−1	−1	1	−1
26	1	1	−1	−1	1	−1	−1
27	1	1	−1	1	−1	−1	−1
28	1	1	−1	1	1	1	−1
29	1	1	1	−1	−1	−1	−1
30	1	1	1	−1	1	1	−1
31	1	1	1	1	−1	1	−1
32	1	1	1	1	1	−1	−1

	Levels	
Factor	−1	+1
A = Raw-material feed rate (pph)	3000	6000
B = Gas temperature (°C)	100	220
C = Scrubber water (%)	5	20
D = Reactor-bed acid (%)	20	30
E = Exit temperature (°C)	300	360
F = Reactant distribution point	East	West

complete factorial experiment is economically feasible. If no prior information is available about the existence of interaction effects, the defining contrast is the six-factor interaction, $ABCDEF$. The effects representation for this defining contrast is determined by multiplying the respective elements from the effects representations of the six main effects. The sign ($+$ or $-$) of the elements in the effects representation is randomly chosen and the corresponding 32 factor–level combinations to be included in the experiment are thereby identified.

Table 9.5 lists the 32 combinations for the half fraction of the acid-plant corrosion-rate experiment using the defining equation $I = -ABCDEF$. If a few repeat tests can be included in the experiment, they will be randomly selected from those shown in Table 9.5. The test sequence of these factor–level combinations should be randomized prior to the execution of the experiment.

The resolution of a half fraction of a complete factorial experiment equals the number of factors included in the defining contrast. Half fractions of highest resolution, therefore, are those that confound the highest-order interaction with the constant effect. The highest resolution a half fraction can attain is equal to the number of factors in the experiment.

The confounding pattern of a half fraction of a complete factorial is determined by symbolically multiplying each side of the defining equation by each of the factor effects. The procedure given in Exhibit 9.4 is used to determine the confounding pattern of effects.

EXHIBIT 9.4   CONFOUNDING PATTERN FOR EFFECTS IN HALF FRACTIONS OF TWO-LEVEL COMPLETE FACTORIAL EXPERIMENTS

1. Write the defining equation of the fractional factorial experiment.

2. Symbolically multiply both sides of the defining equation by one of the factor effects.

3. Reduce the right side of the equation using the following algebraic convention: For any factor, say X,

$$X \cdot I = X \quad \text{and} \quad X \cdot X = X^2 = I.$$

4. Repeat steps 2 and 3 until each factor effect is listed in either the left or the right side of one of the equations.

The symbolic multiplication of two effects is equivalent to multiplying the individual elements in the effects representations of the two effects. Note that the constant effect has all its elements equal to 1. Consequently, any effect multiplying the constant effect remains unchanged. This is the reason for the convention $X \cdot I = X$. Similarly, any effect multiplying itself will result in a column of ones because each element is the square of either $+1$ or -1. Consequently, an effect multiplying itself is a column of $+1$ values: $X \cdot X = I$.

As an illustration of this procedure for determining the confounding pattern of effects, recall the first of the two half fractions of the manufacturing-plant example discussed in the last section. Since the first four factor–level combinations were used, those with the negative signs on the main effect for A, the defining equation is $I = -A$. Multiplying this equation by each of the other effects, using the algebraic convention defined in step 4, results in the following confounding pattern for the factor effects:

$$I = -A, \qquad B = -AB, \qquad C = -AC, \qquad BC = -ABC.$$

This is the same confounding pattern that was found in the previous section by calculating the individual effects. One can use this same procedure to confirm the confounding pattern reported in the last section for the half fraction with defining equation $I = ABC$.

The latter half fraction for the manufacturing-plant example has three factors in the defining contrast. Because of this, the completely randomized design is of resolution III. Main effects are unconfounded with other main effects, but main effects are confounded with two factor interactions. The confounding pattern makes this explicit; however, the confounding pattern did not have to be derived in order to make this determination. Simply knowing the defining contrast allows one to ascertain the resolution of the design and therefore the general pattern of confounding.

As a second illustration, consider again the acid-plant corrosion-rate study. The highest-resolution half fraction has as its defining contrast $ABCDEF$, the six-factor interaction. This is a resolution-VI design, since six factors occur in the defining contrast. Because the design is R_{VI}, all main effects and two-factor interactions are unconfounded with one another, but some three-factor interactions are confounded with other three-factor interactions.

Suppose one chooses the negative signs in the six-factor interaction to select the factor–level combinations as in Table 9.5. Then the defining equation is $I = -ABCDEF$. If one wishes to write out the explicit confounding pattern of the effects, the defining equation will be multiplied by all of the effects until each effect appears on one or the other side of one of the equations. For example, the confounding pattern of the main effects is as follows:

$$A = -AABCDEF = -BCDEF,$$

$$B = -ABBCDEF = -ACDEF,$$

$$C = -ABCCDEF = -ABDEF,$$

$$D = -ABCDDEF = -ABCEF.$$

$$E = -ABCDEEF = -ABCDF,$$

$$F = -ABCDEFF = -ABCDE.$$

Note that one does not now have to evaluate the confounding pattern of the five-factor interactions, since each of them already appears in one of the above equations. Confounding patterns for two-factor and higher-order interactions are determined similarly. By determining the confounding pattern of all the factor effects, one can confirm that the design is R_{VI}.

9.2.2. Quarter and Smaller Fractions

Quarter and smaller fractions of two-level complete factorial experiments are constructed similarly to half fractions. The major distinction is that more than one defining contrast is needed to partition the factor–level combinations.

Consider designing an experiment that is to include a quarter fraction of a six-factor complete factorial experiment. For example, suppose only 16 of the 64 test runs can be conducted for the acid-plant corrosion-rate study. By using two defining contrasts, $\frac{1}{4}$ of the factor–level combinations can be selected. Suppose the two defining contrasts chosen are $ABCDEF$ and ABC. Suppose further that the negative sign is randomly assigned to the first defining contrast and a positive sign is randomly selected for the second defining contrast. This means that a factor–level combination is only included in the experiment if it has both a negative sign on the effects representation for the six-factor interaction $ABCDEF$ and a positive sign on the effects representation for the three-factor interaction ABC. Table 9.6 lists the 16 factor–level combinations that satisfy both requirements.

Although only two defining equations need to be chosen to select the factor–level combinations for a quarter fraction of a complete factorial experiment, a third equation is also satisfied. Note in the last column of Table 9.6 that all the factor–level combinations that satisfy the first two defining equations, $I = -ABCDEF$ and $I = ABC$, also satisfy $I = -DEF$. A third defining equation is always satisfied when two defining contrasts are chosen. This third implicit defining contrast can be identified by symbolically multiplying the other two contrasts. In this example,

$$(-ABCDEF)(ABC) = -AABBCCDEF = -DEF.$$

A concise way of expressing the defining contrasts is $I = -ABCDEF = ABC (= -DEF)$, with the implied contrast in parentheses being optional, since it can be easily be determined from the other two.

Table 9.6 Quarter Fraction of the Complete Factorial Experiment for the Acid-Plant Corrosion-Rate Study

Factor–Level Combination	Effects Representation								
	A	B	C	D	E	F	ABCDEF	ABC	DEF
1	−1	−1	1	−1	−1	−1	−1	1	−1
2	−1	−1	1	−1	1	1	−1	1	−1
3	−1	−1	1	1	−1	1	−1	1	−1
4	−1	−1	1	1	1	−1	−1	1	−1
5	−1	1	−1	−1	−1	−1	−1	1	−1
6	−1	1	−1	−1	1	1	−1	1	−1
7	−1	1	−1	1	−1	1	−1	1	−1
8	−1	1	−1	1	1	−1	−1	1	−1
9	1	−1	−1	−1	−1	−1	−1	1	−1
10	1	−1	−1	−1	1	1	−1	1	−1
11	1	−1	−1	1	−1	1	−1	1	−1
12	1	−1	−1	1	1	−1	−1	1	−1
13	1	1	1	−1	−1	−1	−1	1	−1
14	1	1	1	−1	1	1	−1	1	−1
15	1	1	1	1	−1	1	−1	1	−1
16	1	1	1	1	1	−1	−1	1	−1

Factors	Levels	
	−1	+1
A = Raw material feed rate (pph)	3000	6000
B = Gas temperature (°C)	100	220
C = Scrubber water (%)	5	20
D = Reactor bed acid (%)	20	30
E = Exit temperature (°C)	300	360
F = Reactant distribution point	East	West

The resolution of quarter fractions of complete factorials constructed in this fashion equals the number of factors in the smallest of the defining contrasts, including the one implied by the two chosen for the design. Thus in the above example, the resulting quarter fraction is a resolution-III design because the smallest number of factors in the defining contrasts is three (ABC or DEF). A higher-resolution fractional factorial design can be constructed by using, apart from sign, the defining equations $I = ABCD = CDEF$ ($= ABEF$). This would be a R_{IV} design, since the smallest number of factors in the defining contrast is four.

Just as there are three defining contrast in a quarter fraction of a complete factorial experiment, each of the factor effects in the experiment is confounded with three additional effects. The confounding pattern is determined by symbolically multiplying the defining equations, including the implied one, by each of the factor effects. For example, if the defining equations for a quarter fraction of a six-factor factorial are

$$I = ABCD = CDEF = ABEF,$$

then the main effects are confounded as follows:

$$A = BCD \quad = ACDEF = BEF,$$
$$B = ACD \quad = BCDEF = AEF,$$
$$C = ABD \quad = DEF \quad = ABCEF,$$
$$D = ABC \quad = CEF \quad = ABDEF,$$
$$E = ABCDE = CDF \quad = ABF,$$
$$F = ABCDF = CDE \quad = ABE.$$

As required by the fact that this is a resolution-IV design, the main effects are not confounded with one another. One can show in the same manner that the two-factor interactions are confounded with one another.

The general procedure for constructing fractional factorial experiments for two-level factors is a straightforward generalization of the procedure for quarter fractions. A $(\frac{1}{2})^p$ fraction requires that p defining contrasts be chosen, none of which is obtainable by algebraic multiplication of the other contrasts. An additional $p(p-1)/2$ implicit defining contrasts are then determined by algebraically multiplying the p chosen contrasts. The resolution of the resulting completely randomized fractional factorial experiment equals the number of factors in the smallest defining contrast, including the implicit ones. The confounding pattern is obtained by multiplying the defining equations by the factor effects.

Table 9A.1 in the appendix to this chapter is a list of defining equations, apart from sign, for fractional factorial experiments involving from 5 to 11 factors. Included in the table are the defining equations and the resolutions of the designs. Resolution-IV and resolution-V designs are sought because the primary interest in multifactor experiments is generally in the analysis of main effects and two-factor interactions. The last column of the table is an adaptation of the defining equations to the construction of fractional factorial experiments using added factors. This type of construction is explained in the appendix to this chapter.

9.3 BLOCKING DESIGNS

The planned confounding of factor effects in blocking designs is an effective design strategy for experiments in which all factor–level combinations of interest cannot be tested under uniform conditions. The lack of uniform conditions can occur because either there are not a sufficient number of homogeneous experimental units or environmental or physical changes during the course of the experiment affect the responses.

Randomized complete block, balanced incomplete block, latin-square, and crossover designs were introduced in the last chapter as alternatives to completely randomized designs when factor–level combinations must be blocked or grouped. In each of these designs there is no confounding of factor effects with block effects. Each of these designs has added requirements, however, which may render their use impractical for a specific experiment. For example, randomized complete block designs require that the number of experimental units be at least as large as the total number of factor–level combinations of interest. Balanced incomplete block designs do not exist for every combination of the design parameters. Latin-square designs require no interactions between the experimental and blocking factors. Crossover designs have restrictions on the order of testing.

There are many situations in which the intentional confounding of factor effects with block effects can satisfy the requirements of the experiment and still allow the evaluation of all the factor effects of interest. Unlike fractional factorial experiments, only some of the factor effects are confounded with block effects when complete factorial experiments are partitioned into blocks. The goal of conducting factorial experiments in blocks is to plan the confounding so that only those effects that are known to be zero or that are of little interest are confounded with the block effects.

The manufacturing-plant example in Section 9.1 illustrates the key concepts in constructing block designs for complete factorial experiments. Care must be taken to ensure that effects of interest are not confounded with block effects. Table 9A.2 in the appendix to this chapter lists the defining equations for some block designs involving three to seven factors. In addition to the specified confounded effects, other effects are implicitly confounded as with fractional factorial experiments. The additional confounded effects are identified by taking the algebraic products of the defining contrasts, as described in the last section.

The designing of a randomized block design in which one or more effects from a

---

EXHIBIT 9.5 CONSTRUCTION OF RANDOMIZED INCOMPLETE BLOCK DESIGNS FOR TWO-LEVEL COMPLETE FACTORIAL EXPERIMENTS

1. Determine a block size that can be used in the experiment. If there are to be no repeats, the block size should be a power of 2.

2. Choose the effects that are to be confounded with the block effects. Table 9A.2 can be used to choose defining contrasts for experiments with three to seven factors.

3. List the effects representation of the defining contrasts, and randomly select the sign to be used with each.

4. Randomly select a block, and·place in that block all factor-level combinations that have a specific sign combination in the defining contrasts.

5. Repeat step 4 for each of the groups of factor–level combinations.

6. Randomize the test sequence or the assignment to experimental units separately for each block.

complete factorial experiment are intentionally confounded with blocks proceeds as described in Exhibit 9.5. The construction of randomized incomplete block designs requires 2^p blocks of experimental units if p effects are to be confounded. If there are k factors in the experiment, each block consists of at least 2^{k-p} experimental units. If more than 2^{k-p} experimental units are available in one or more of the blocks, repeat observations can be taken. This will result in the design being unbalanced, but it will enable estimation of the uncontrolled experimental-error variation without the possible influences of factor or block effects.

In some experiments there are an ample number of blocks. In such circumstances the construction of incomplete block designs as described in Exhibit 9.5 can be modified to include replicates of the complete factorial experiment. In each replicate different effects can be confounded with blocks so that every effect can be estimated from at least one of the replicates. This type of design is referred to as a *partially* confounded design, and the effects are referred to as *partially confounded*.

To illustrate the construction of complete factorial experiments in randomized incomplete block designs, consider the experiment illustrated in Figure 9.1. This experiment is to be conducted to investigate the wear on drill bits used for oil exploration. A scale model of a drilling platform is to be built in order to study five factors believed to affect drill-bit wear: the rotational (factor A) and longitudinal (factor B) velocities of the drill pipe, the length of the drill pipe (factor C), the drilling angle (factor D), and the geometry (factor E) of the tool joint used to connect sections of the pipe.

A factorial experiment is feasible for this study, but only eight test runs, one per hour, can be run each day. Daily changes in the equipment setup may influence the drill-bit wear

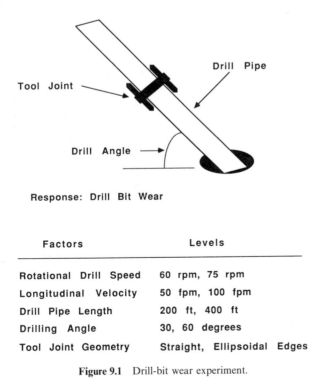

Response: Drill Bit Wear

Factors	Levels
Rotational Drill Speed	60 rpm, 75 rpm
Longitudinal Velocity	50 fpm, 100 fpm
Drill Pipe Length	200 ft, 400 ft
Drilling Angle	30, 60 degrees
Tool Joint Geometry	Straight, Ellipsoidal Edges

Figure 9.1 Drill-bit wear experiment.

Table 9.7 Effects Representation for Blocking in the Drill-Bit Wear Experiment

Factor–Level Combination	Effects Representation						
	A	B	C	D	E	ABC	CDE
1	−1	−1	−1	−1	−1	−1	−1
2	−1	−1	−1	−1	1	−1	1
3	−1	−1	−1	1	−1	−1	1
4	−1	−1	−1	1	1	−1	−1
5	−1	−1	1	−1	−1	1	1
6	−1	−1	1	−1	1	1	−1
7	−1	−1	1	1	−1	1	−1
8	−1	−1	1	1	1	1	1
9	−1	1	−1	−1	−1	1	−1
10	−1	1	−1	−1	1	1	1
11	−1	1	−1	1	−1	1	1
12	−1	1	−1	1	1	1	−1
13	−1	1	1	−1	−1	−1	1
14	−1	1	1	−1	1	−1	−1
15	−1	1	1	1	−1	−1	−1
16	−1	1	1	1	1	−1	1
17	1	−1	−1	−1	−1	1	−1
18	1	−1	−1	−1	1	1	1
19	1	−1	−1	1	−1	1	1
20	1	−1	−1	1	1	1	−1
21	1	−1	1	−1	−1	−1	1
22	1	−1	1	−1	1	−1	−1
23	1	−1	1	1	−1	−1	−1
24	1	−1	1	1	1	−1	1
25	1	1	−1	−1	−1	−1	−1
26	1	1	−1	−1	1	−1	1
27	1	1	−1	1	−1	−1	1
28	1	1	−1	1	1	−1	−1
29	1	1	1	−1	−1	1	1
30	1	1	1	−1	1	1	−1
31	1	1	1	1	−1	1	−1
32	1	1	1	1	1	1	1

Factors	Levels	
	−1	+1
A = Rotational drill speed (rpm)	60	75
B = Longitudinal drill speed (fpm)	50	100
C = Drill pipe length (ft)	200	400
D = Drilling angle (deg)	30	60
E = Tool joint geometry	Straight	Ellipse

so the test runs will be blocked with eight test runs per block. Two weeks are set aside for the testing. A complete factorial experiment can be conducted during each week by running a quarter fraction on each of four days. The other day can be reserved, if needed, for setting up and testing equipment. Since two replicates of the complete factorial experiment can be conducted, different defining contrasts will be used each week so that all factor effects can be estimated when the experiment is concluded.

In Table 9A.2, the defining equations for blocking are $I = ABC = DCE \ (= ABDE)$. The effects representations for these interactions are shown in Table 9.7. To conduct the experiment, the factor–level combinations that have negative signs on the effects representations for both defining contrasts would be randomly assigned to one of the four test days in the first week. Those with a positive sign on ABC and a negative sign on CDE would be randomly assigned to one of the remaining three test days in the first week. The other two blocks of eight factor–level combinations would similarly be randomly assigned to the remaining two test days in the first week. Prior to conducting the test runs, each of the eight runs assigned to a particular test day would be randomly assigned a run number.

For the second week of testing, a similar procedure would be followed except that two different three-factor interactions would be confounded with blocks. One might choose as the defining equations $I = BCD = ADF \ (= ABCF)$.

A further adaptation of randomized incomplete block designs is their use with fractional factorial experiments. With such designs there is the confounding of factor effects as with any fractional factorial experiment, and there is confounding with blocks because all of the factor–level combinations in the fractional factorial cannot be run in each block.

Table 9A.3 lists sets of defining contrasts for fractional factorial experiments conducted in randomized incomplete block designs. All of the designs resulting from the use of Table 9A.3 are at least of Resolution V. To conduct these experiments, one first selects the factor–level combinations to be included in the experiments from the defining contrasts for the factors in Table 9A.3. Once these factor–level combinations are identified, they are blocked according to their signs on the defining contrasts for the blocks in Table 9A.3.

For example, to conduct a half fraction of the acid-plant corrosion-rate study in two blocks of 16 test runs each, one would first select the factor–level combinations to be tested using the defining equation $I = ABCDEF$, with the sign randomly chosen. Suppose the negative signs are selected so that the combinations are as listed in Table 9.5. Next these 32 factor–level combinations would be blocked according to whether the effects representation for the ABC interaction is positive or negative. Thus factor–level combinations 1–4, 13–16, and 21–29 would be assigned to one of the blocks, and the remaining combinations to the other block.

9.4 SCREENING DESIGNS AND SEQUENTIAL EXPERIMENTATION

Screening experiments are conducted when a large number of factors are to be investigated but limited resources mandate that only a few test runs be conducted. Screening experiments are conducted in order to identify a small number of dominant factors, often with the intent to conduct a more extensive investigation involving only these dominant factors.

An important application of screening experiments is with *ruggedness tests* for laboratory test methods. The purpose of a ruggedness test is to determine environmental factors or test conditions (technicians, equipment fluctuations, maintenance cycles, etc.) that influence measurements obtained from the test method. Once these factors are identified, either they are tightly controlled or the test method is changed to reduce or eliminate their effects. The resulting test method then retains sensitivity, accuracy, and precision in spite of normal changes in operating conditions. One major goal of ruggedness testing is to permit the test method to be reliably used in different laboratories.

Ruggedness testing and the application of screening designs can be extended to product and manufacturing process design. A rugged product or process possesses the ability to withstand a wide variety of uses and conditions. Two examples of ruggedness tests are provided in Section 18.4.

Fractional factorial experiments are most often used as screening designs. Resolution-III designs allow the estimation of main effects that are unconfounded with one another. Often R_{III} designs result in experiments that clearly identify the dominant factors without the major time and expense that would be required for higher-resolution designs.

A special class of fractional factorial experiments that is widely used in screening experiments was proposed by Plackett and Burman (see the references). These experiments have resolution III when conducted in completely randomized designs and are often referred to as *Plackett–Burman* designs.

The designs discussed by Plackett and Burman are available for experiments that have the number of test runs equal to a multiple of four. Table 9A.4 lists design generators for experiments having 12, 16, 20, 24, and 32 test runs. The rows of the table denote the design factors, and the elements in each column are coded factor levels: a minus sign denotes one level of a factor and a plus sign denotes the other factor level. To allow for an estimate of experimental error, it is recommended that the design have at least six more test runs than the number of factors included in the experiment.

The particular design selected depends on both the number of factors in the experiment and the number of test runs. Each design generator can be used to construct a screening design having up to one fewer factor than the number of test runs. For example, the 16-run design generator can be used to construct screening designs for 1–15 factors; however, we again recommend that no more than 10 factors be used with a 16-run design.

To illustrate the construction of screening experiments using the design generators

Table 9.8   Twelve-Run Screening Designs

Run No.	Factor No.										
	1	2	3	4	5	6	7	8	9	10	11
1	+	+	−	+	+	+	−	−	−	+	−
2	+	−	+	+	+	−	−	−	+	−	+
3	−	+	+	+	−	−	−	+	−	+	+
4	+	+	+	−	−	−	+	−	+	+	−
5	+	+	−	−	−	+	−	+	+	−	+
6	+	−	−	−	+	−	+	+	−	+	+
7	−	−	−	+	−	+	+	−	+	+	+
8	−	−	+	−	+	+	−	+	+	+	−
9	−	+	−	+	+	−	+	+	+	−	−
10	+	−	+	+	−	+	+	+	−	−	−
11	−	+	+	−	+	+	+	−	−	−	+
12	−	−	−	−	−	−	−	−	−	−	−

in Table 9A.4, consider the construction of a twelve-run design. The first column of Table 9A.4 is used as the first row of a (nonrandomized) twelve-run screening design. Succeeding rows are obtained, as with the latin-square designs discussed in Section 8.4, by taking the sign in the first position of the previous row and placing it last in the current row, and then shifting all other signs forward one position. After all cyclical arrangements have been included in the design (a total of one less than the number of test runs), a final row of minus signs is added. The complete twelve-run design is shown in Table 9.8.

If, as recommended, fewer factors than test runs are included in the design, several columns of the generated design are discarded. Any of the columns can be eliminated. Elimination can be based on experimental constraints or be random. All rows (test runs) for the retained columns are included in the design.

The acid-plant corrosion-rate study was actually conducted as a screening experiment. The twelve-run screening experiment shown in Table 9.8 satisfied the recommended requirement of six more test runs than factors. For ease of discussion, the first six columns of Table 9.8 are used; the remaining columns are discarded. The combination of factor levels to be used in the experimental program are shown in Table 9.9. To minimize the possibility of bias effects due to the run order, the experimental test sequence should be randomize as should the assignment of the factors to the columns.

Screening designs often lead to further experimentation once dominant factors are identified. More generally, one often can plan experiments so that preliminary results determine the course of the experimentation.

Sequential experimentation is highly advisable when one embarks on a new course of experimental work in which little is known about the effects of any of the factors on the response. It would be unwise, for example, to design a complete factorial experiment in seven or more factors in such circumstances. The result of such a large test program might be that a small number of main effects and two-factor interactions are dominant. If so, a great amount of time and effort would have been wasted when a much smaller effort could have achieved the same results.

If an experiment is contemplated in which a large number of factors are to be

Table 9.9 Actual Factor Levels for a Twelve-Run Screening Design for the Acid-Plant Corrosion-Rate Study

Factor–Level Combination	Feed Rate (pph)	Gas Temp. (°C)	Scrubber Water (%)	Bed Acid (%)	Exit Temp. (°C)	Distribution Point
1	6000	220	5	30	360	West
2	6000	100	20	30	360	East
3	3000	220	20	30	300	East
4	6000	220	20	20	300	East
5	6000	220	5	20	300	West
6	6000	100	5	20	360	East
7	3000	100	5	30	300	West
8	3000	100	20	20	360	West
9	3000	220	5	30	360	East
10	6000	100	20	30	300	West
11	3000	220	20	20	360	West
12	3000	100	5	20	300	East

investigated, a key question that should be asked is whether the project goals will be met if only the dominant factor effects are identified. If so, a screening experiment should be performed, followed by a complete or fractional factorial experiment involving only the dominant factors. If not, perhaps a resolution-V fractional factorial experiment would suffice.

If a large experiment is required to satisfactorily investigate all the factor effects of interest, consideration should be given to blocking the test runs so that information can be gained about the factors as the experiment progresses. In the previous section, blocking designs for complete and fractional experiments were discussed. For the latter type of experiments, Table 9A.3 showed blocking patterns that ensured that no main effects or two-factor interactions were confounded. These blocking patterns can be exploited to allow sequential testing in which information about key factors can be evaluated as the experiment proceeds, not simply when it is completed.

An interesting feature of fractional factorial experiments is that a design of resolution R is a complete factorial experiment in any $R - 1$ factors. Thus a half fraction of a three factor factorial experiment is a complete factorial in any two of the three factors if the defining contrast involves the three-factor interaction (a resolution-III design). This can be seen graphically in Figure 9.2, where the four points on the cube are projected into

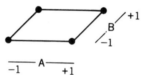

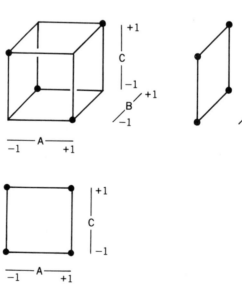

Figure 9.2   Projections of a half-fraction of a three-factor complete factorial experiment ($I = ABC$).

each of the three-two-dimensional planes of pairs of the factors. Examination of the columns for A, B and C in Table 9.2 for combinations with $+1$ values for the ABC interaction also reveals the complete factorial arrangement of each pair of factors.

Because of this property, one can often block the experiment so that preliminary information on main effects and two-factor interactions can be obtained from each of the blocks. This information is considered preliminary because information on all the main effects and two-factor interactions may not be obtainable from each of the blocks and because each of the effects will be estimated with greater precision at the completion of the experiment.

Recall the discussion in the last section about conducting a half fraction of the complete factorial for the acid-plant corrosion-rate study in two blocks of 16 test runs. The defining contrast for the half fraction is $I = -ABCDEF$. The defining contrast for the blocking is $I = ABC$. After the first block of test runs is completed, a quarter fraction of the complete factorial experiment will have been conducted. This quarter fraction has as its defining equations either $I = -ABCDEF = ABC$ or $I = -ABCDEF = -ABC$, depending on which block of factor–level combinations was run first. Note that this quarter fraction is itself a resolution-III design. Thus, from the first block of test runs one can estimate main effects that are unconfounded with one another. Based on this information, decisions may be made about continuing the experimentation or terminating it, depending on whether the main effects show that one or more of the factors is having a strong effect on the corrosion rates.

This example illustrates how blocking designs can be used to exploit the planned confounding of factor effects in sequential experimentation. Care must be exercised in the interpretation of results from highly fractionated designs so that one does not draw erroneous conclusions because of the confounding pattern of the effects. Nevertheless, conducting experiments in blocks can alert one early to dominant effects, which in turn can influence one's decision to continue the experimentation.

APPENDIX: ADDED FACTORS AND FACTORS HAVING MORE THAN TWO LEVELS

1. Added Factors

For a large number of factors, a time-saving device for constructing fractional factorial experiments is the technique of *added factors*. Rather than examining the effects representation of all the factor–level combinations and then selecting a portion of them for the design, new factors are added to complete factorial experiments involving fewer than the desired number of factors in order to construct fractional factorial experiments involving a large number of factors. For example, consider setting a sixth factor equal to a five-factor interaction. By setting the levels of this sixth factor equal to those in the effects representation of the five-factor interaction, one would construct a half-fraction of a complete factorial in six factors from the complete factorial in five factors. Symbolically, this substitution can be expressed (apart from the sign) as

$$F = ABCDE.$$

Algebraically multiplying both sides of this equation by F produces the defining equation for a half-fraction of a six-factor factorial: $I = ABCDEF$.

This approach can be effectively utilized to construct fractional factorial experiments from complete factorials of fewer factors. In Table 9A.1, the last factor listed in each defining equation can be equated to the product of the previous factors without changing the contrasts, simply by multiplying both sides of each defining equation by the factor. This process can be implemented on a computer by the following procedure.

Let $x_1, x_2, \ldots, x_k$ denote the k factors in a fractional factorial experiment. For each factor level in the complete factorial experiment, let $x_j = 0$ if the jth factor is at its lower level (-1), and let $x_j = 1$ if the factor is at its upper level $(+1)$. Replace the defining contrasts for the fractional factorial experiment desired by an equal number of sums of

Table 9A.1 Selected Fractional Factorial Experiments

Number of Factors	Number of Test Runs	Fraction	Resolution	Defining Equations	Added Factors
5	8	$\frac{1}{4}$	III	$I = ABD$	$4 = 12$
				$I = ACE$	$5 = 13$
6	8	$\frac{1}{8}$	III	$I = ABD$	$4 = 12$
				$I = ACE$	$5 = 13$
				$I = BCF$	$6 = 23$
	16	$\frac{1}{4}$	IV	$I = ABCE$	$5 = 123$
				$I = BCDF$	$6 = 234$
7	8	$\frac{1}{16}$	III	$I = ABE$	$4 = 12$
				$I = ACE$	$5 = 13$
				$I = BCF$	$6 = 23$
				$I = ABCG$	$7 = 123$
	16	$\frac{1}{8}$	IV	$I = ABCE$	$5 = 123$
				$I = BCDF$	$6 = 234$
				$I = ACDG$	$7 = 134$
	32	$\frac{1}{4}$	IV	$I = ABCDF$	$6 = 1234$
				$I = ABDEG$	$7 = 1245$
8	16	$\frac{1}{16}$	IV	$I = BCDE$	$5 = 234$
				$I = ACDF$	$6 = 134$
				$I = ABCG$	$7 = 123$
				$I = ABDH$	$8 = 124$
	32	$\frac{1}{8}$	IV	$I = ABCF$	$6 = 123$
				$I = ABDG$	$7 = 124$
				$I = BCDEH$	$8 = 2345$
	64	$\frac{1}{4}$	V	$I = ABCDG$	$7 = 1234$
				$I = ABEFH$	$8 = 1256$
9	16	$\frac{1}{32}$	III	$I = ABCE$	$5 = 123$
				$I = BCDF$	$6 = 234$
				$I = ACDG$	$7 = 134$
				$I = ABDH$	$8 = 124$
				$I = ABCDJ$	$9 = 1234$
	32	$\frac{1}{16}$	IV	$I = BCDEF$	$6 = 2345$
				$I = ACDEG$	$7 = 1345$
				$I = ABDEH$	$8 = 1245$
				$I = ABCEJ$	$9 = 1235$

Table 9A.1 (continued)

Number of Factors	Number of Test Runs	Fraction	Resolution	Defining Equations	Added Factors
	64	$\frac{1}{8}$	IV	$I = ABCDG$	$7 = 1234$
				$I = ACEFH$	$8 = 1356$
				$I = CDEFJ$	$9 = 3456$
	128	$\frac{1}{4}$	VI	$I = ACDFGH$	$8 = 13467$
				$I = BCEFGJ$	$9 = 23567$
10	16	$\frac{1}{64}$	III	$I = ABCE$	$5 = 123$
				$I = BCDF$	$6 = 234$
				$I = ACDG$	$7 = 134$
				$I = ABDH$	$8 = 124$
				$I = ABCDEJ$	$9 = 1234$
				$I = ABK$	$10 = 12$
	32	$\frac{1}{32}$	IV	$I = ABCDF$	$6 = 1234$
				$I = ABCEG$	$7 = 1235$
				$I = ABDEH$	$8 = 1245$
				$I = ACDEJ$	$9 = 1345$
				$I = BCDEK$	$10 = 2345$
	64	$\frac{1}{16}$	IV	$I = BCDFG$	$7 = 2346$
				$I = ACDFH$	$8 = 1346$
				$I = ABDEJ$	$9 = 1245$
				$I = ABCEK$	$10 = 1235$
	128	$\frac{1}{8}$	V	$I = ABCGH$	$8 = 1237$
				$I = BCDEJ$	$9 = 2345$
				$I = ACDFK$	$10 = 1346$
11	16	$\frac{1}{128}$	III	$I = ABCE$	$5 = 123$
				$I = BCDF$	$6 = 234$
				$I = ACDG$	$7 = 134$
				$I = ABDH$	$8 = 124$
				$I = ABCDJ$	$9 = 1234$
				$I = ABK$	$10 = 12$
				$I = ACL$	$11 = 13$
	32	$\frac{1}{64}$	IV	$I = ABCF$	$6 = 123$
				$I = BCDG$	$7 = 234$
				$I = CDEH$	$8 = 345$
				$I = ACDJ$	$9 = 134$
				$I = ADEK$	$10 = 145$
				$I = BDEL$	$11 = 245$
	64	$\frac{1}{32}$	IV	$I = CDEG$	$7 = 345$
				$I = ABCDH$	$8 = 1234$
				$I = ABFJ$	$9 = 126$
				$I = BDEFK$	$10 = 2456$
				$I = ADEFL$	$11 = 1456$
	128	$\frac{1}{16}$	V	$I = ABCGH$	$8 = 1237$
				$I = BCDEJ$	$9 = 2345$
				$I = ACDFK$	$10 = 1346$
				$I = ABCDEFGL$	$11 = 1234567$

the form

$$x_w = x_u + \cdots + x_v,$$

where x_w represents the added factor and $x_u, \ldots, x_v$ represent the other factors involved in the defining contrast. These sums are calculated for each factor–level combination in the complete factorial for the factors on the right side of the added-factor equation.

For example, to construct a half fraction of a six-factor factorial experiment, let $x_1, \ldots, x_6$ represent, respectively, factors $A, \ldots, F$. An effects-representation table for the complete factorial experiment in the first five factors can be constructed by setting the elements in the columns for $x_1, \ldots, x_5$ equal to 0 or 1 rather than -1 or 1, respectively. The added factor is then

$$x_6 = x_1 + x_2 + x_3 + x_4 + x_5.$$

For each factor–level combination in the complete five-factor factorial experiment, compute the value of x_6 from the above equation. If x_6 is even (0 mod 2), assign the sixth factor its upper level. If x_6 is odd (1 mod 2), assign the sixth factor its lower level. The 32 factor–level combinations in the half fraction of this 2^6 factorial are identified by replacing the zeros in the table with the lower levels of each factor and the ones with the upper levels of each.

This process is valid for smaller fractions. If one desires to construct a $\frac{1}{16}$ fraction of an eleven-factor factorial experiment, a resolution-V design can be obtained using the added factors listed in Table 9A.1:

$$x_8 = x_1 + x_2 + x_3 + x_7, \qquad x_9 = x_2 + x_3 + x_4 + x_5, \qquad \text{etc.}$$

For each factor–level combination in a complete seven-factor factorial experiment, one would compute x_8, x_9, x_{10}, and x_{11}. As these values for x_j are computed, if the sign on the corresponding defining contrast is $+$, the value of $x_j (\text{mod } 2)$ is used. If the sign is $-$, $x_j + 1 \ (\text{mod } 2)$ is used. Once the values of $x_8, \ldots, x_{11}$ are determined, all the zeros and ones for the eleven factors are replaced by the lower and upper levels of the respective factors.

2. Factors with More Than Two Levels

Construction of fractional factorial experiments when factors have more than two levels is far more complicated than for two-level factors. A considerable amount of information is available for three-level factors, but its presentation is beyond the scope of this book. Key references on this material are given at the end of this chapter.

Beyond two- or three-level factors very little can be said in general about the properties of fractional factorial experiments. One ordinarily must evaluate each design to determine its properties. The properties of two-level factorial experiments can, however, be extended to factors whose number of levels is a power of two. In doing so, it is imperative that one understand the nature of effects for factors having more than two levels (see the appendix to Chapter 7).

Consider as an illustration an experiment in which three factors are to be investigated, factor A having four levels and factors B and C each having two levels. As recommended in the appendix to Chapter 7, define two new factors A_1 and A_2, each having two levels,

Table 9A.2 Selected Randomized Incomplete Block Designs For Complete Factorial Experiments

No. of Factors	Block Size	Fraction	Defining Equations*
3	2	$\frac{1}{4}$	$I = AB = AC$
	4	$\frac{1}{2}$	$I = ABC$
4	2	$\frac{1}{8}$	$I = AB = BC = CD$
	4	$\frac{1}{4}$	$I = ABD = ACD$
	8	$\frac{1}{2}$	$I = ABCD$
5	2	$\frac{1}{16}$	$I = AB = AC = CD = DE$
	4	$\frac{1}{8}$	$I = ABE = BCE = CDE$
	8	$\frac{1}{4}$	$I = ABC = CDE$
	16	$\frac{1}{2}$	$I = ABCDE$
6	2	$\frac{1}{32}$	$I = AB = BC = CD = DE = EF$
	4	$\frac{1}{16}$	$I = ABF = ACF = CDF = DEF$
	8	$\frac{1}{8}$	$I = ABCD = ABEF = ACE$
	16	$\frac{1}{4}$	$I = ABCF = CDEF$
	32	$\frac{1}{2}$	$I = ABCDEF$
7	2	$\frac{1}{64}$	$I = AB = BC = CD = DE = EF$ $= FG$
	4	$\frac{1}{32}$	$I = ABG = BCG = CDG = DEG$ $= EFG$
	8	$\frac{1}{16}$	$I = ABCD = EFG = CDE = ADG$
	16	$\frac{1}{8}$	$I = ABC = DEF = AFG$
	32	$\frac{1}{4}$	$I = ABCFG = CDEFG$
	64	$\frac{1}{2}$	$I = ABCDEFG$

*Each of the defining contrasts is confounded with a block effect.

Table 9A.3 Selected Randomized Incomplete Block Designs For Fractional Factorial Experiments

Number of Factors	Block Size	Resolution	Defining Equations* Factors	Blocks
5	None	V	$I = ABCDE$	
6	16	VI	$I = ABCDEF$	$I = ABC$
7	8	VII	$I = ABCDEFG$	$I = ABCD = ABEF$ $= ACEG$
8	16	V	$I = ABCDG$ $= ABEFH$	$I = ACE = CDH$
9	16	VI	$I = ACDFGH$ $= BCEFGJ$	$I = ACH = ABJ$ $= GHJ$
10	16	V	$I = ABCGH$ $= BCDEJ$ $= ACDFK$	$I = ADJ = ABK$ $= HJK$
11	16	V	$I = ABCGH$ $= BCDEJ$ $= ACDFK$ $= ABCDEFGL$	$I = ADJ = ABK$ $= HJK$

*No main effects or two-factor interactions are confounded with one another or with block effects.

to represent the four levels of factor A. One can construct a half fraction of the complete factorial experiment in these factors using the defining equation (apart from sign)

$$I = A_1 A_2 BC.$$

In interpreting the properties of this design, one might conclude that the design is of resolution IV, since four factors are present in the defining equation. One must recall, however, that $A_1 A_2$ represents one of the main effects for the four-level factor A. Thus, this design is not R_{IV}; it is R_{III}, since the confounding pattern for the main effects is

$$A_1 = A_2 BC, \qquad A_2 = A_1 BC, \qquad A_1 A_2 = BC, \qquad B = A_1 A_2 C, \qquad C = A_1 A_2 B.$$

Table 9A.4 Screening Design Generators

Factor No.	No. of Test Runs					Factor No.
	12	16	20	24	32	
1	+	+	+	+	−	1
2	+	−	+	+	−	2
3	−	−	−	+	−	3
4	+	−	−	+	−	4
5	+	+	+	+	+	5
6	+	−	+	−	−	6
7	−	−	+	+	+	7
8	−	+	+	−	−	8
9	−	+	−	+	+	9
10	+	−	+	+	+	10
11	−	+	−	−	+	11
12		−	+	−	−	12
13		+	−	+	+	13
14		+	−	+	+	14
15		+	−	−	−	15
16			−	−	−	16
17			+	+	−	17
18			+	−	+	18
19			−	+	+	19
20			−		+	20
21				−	+	21
22				−	+	22
23				−	−	23
24					−	24
25					+	25
26					+	26
27					−	27
28					+	28
29					−	29
30					−	30
31					+	31

This design is not R_{IV} because, for example, the main effect $A_1 A_2$ is confounded with the two-factor interaction BC.

REFERENCES

Test References

Extensive discussion of complete and fractional factorial experiments appears in:

Box, G. E. P., Hunter, W. G., and Hunter, J. S. (1978). *Statistics for Experimenters*, New York: John Wiley and Sons, Inc. Perhaps the best single reference for the topics covered in this chapter.

Davies, O. L. (Ed.) (1971). *The Design and Analysis of Industrial Experiments*, New York: Macmillan Co. This text explores confounding and fractional factorial experiments extensively. Examples are the primary vehicle for exposition of these topics.

Diamond, W. J. (1981). *Practical Experimental Designs*, Belmont, CA: Wadsworth, Inc. Many fractional designs not ordinarily presented in texts are discussed. Three-quarter fractions and alternative resolution V designs are presented.

An extensive list of fractional factorial experiments in completely randomized and blocking designs is contained in Box, Hunter, and Hunter (1978) and in the following references:

Anderson, V. L. and McLean, R. A. (1984). *Applied Factorial and Fractional Designs*, New York: Marcel Dekker, Inc.

Cochran, W. G. and Cox, D. R. (1957). *Experimental Designs*, New York: John Wiley and Sons, Inc. *Included in Anderson and McLean and Cochran and Cox are fractional experiments for three-level factors and combinations of two- and three-level factors.*

The material in Tables 9A.1 to 9A.3 is compiled from [see also Box, Hunter, and Hunter (1978)]

Box, G. E. P and Hunter, J. S. (1961). "The 2^{k-p} Fractional Factorial Designs, Part I, II," *Technometrics*, **3**, 311–351, 449–458.

The fractional factorial screening designs proposed by Plackett and Burman are derived in the following article:

Plackett, R. L. and Burman, J. P. (1946). "The Design of Optimum Multifactorial Experiments," *Biometrika*, **34**, 255–272.

These screening designs are also discussed in the books by Cochran and Cox and by Diamond cited above.

EXERCISES

1 Construct a half fraction of the complete factorial experiment for the experiment on the intensity of light described in Exercise 11 of Chapter 7. What is the defining contrast for the half fraction? What is the resolution of the design? What is the confounding pattern for the effects?

2 Construct a quarter fraction of the complete factorial experiment on light intensity. What are the defining contrasts? What is the resolution of the design? What is the confounding pattern for the effects?

3 Construct a half fraction (of highest resolution) of a four-factor, two-level complete factorial experiment. Show that this design is a complete factorial in any three of the factors. (Optional: Construct this design using added factors.)

4   Construct a resolution-V design for an eight-factor, two-level experiment in which
the number of test runs cannot exceed 70. Verify that this a complete factorial
experiment in any one set of five factors. (Optional: Construct this design using added
factors.)

5   An experiment is to have three factors each at two levels and one factor at four levels.
Suppose a maximum of 20 test runs can be made. Construct a half fraction of the
five-factor experiment. Specify the resolution of the design, and write out the
confounding pattern.

6   Design an experiment involving three factors, each at two levels, in which no more
than six test runs can be conducted on each of several test days. It is desired that all
main effects and interactions be estimated when the experiment is concluded. A
maximum of five test days are available for the experiment.

7   Consider redesigning the torque study using the factors and levels displayed in
Figure 7.1 of Chapter 7. Suppose that only eight test runs can be made on each day
of testing. Design an experiment to be conducted on five test-days. List the test runs
to be made each day, the confounding, if any, and the resolution of the design.

8   Design an experiment in which a 2^8 complete factorial is conducted in blocks of no
more than 20 test runs. The design must be at least of resolution V.

9   A ruggedness test is to be conducted on laboratory procedures for the determination
of acid concentration in a solution. The factors of interest are the temperature and
the humidity of the laboratory, the speed of the blender used to mix the solution,
the mixing time, the concentration of acid in the solution, volume of the solution,
the size of the flask containing the solution, the presence or absence of a catalyst,
and the laboratory technicians. Each of these factors is to be investigated at two
levels. Design a screening experiment consisting of no more than 20 test runs to
investigate the main effects of these factors.

10 Can a fractional factorial experiment of greater than resolution III be designed for
the experiment in Exercise 9? If so, state the defining contrasts and the resolution
of the design.

11 An experiment is to be conducted to determine factors that would increase the
displacement efficiency of material used in cementing operations in wellbore drilling
holes in oil fields. It is suspected that any of a number of factors might affect the
displacement efficiency, including the following:

Factor	Levels
Hole angle	45, 60°
Cement viscosity	30, 40 centipoise
Flow rate	50, 175 gal/min
Production-string rotation speed	0, 60 rpm
Type of wash material	Plain, caustic water
Wash volume	5, 10 barrels/annular volume
Production string eccentricity	Concentric, 30° offset

Design a screening experiment to investigate the effects of these factors. What properties does this screening design possess (e.g., resolution, confounding).

12   The endurance strength (psi) of several different stainless steels is being studied under varying fatigue cycles and temperatures. It is suspected that the temperature and fatigue cycles to which each specimen is subjected are jointly contributing to the strength of the material. Four stainless steels (302 SS, 355 SS, 410 SS, 430 SS) are to be tested under four temperatures and two fatigue cycles. Design an experiment to study the effects of these factors on the strength of stainless steel. Design for three alternatives:

   (a) any reasonable number of test runs can be conducted,

   (b) at most 25 test runs can be conducted,

   (c) at most 15 test runs can be conducted.

   Discuss the properties of these three designs. Are there any that you would recommend against? Why (not)?

13   Refer to the experiment described in Exercise 5 of Chapter 7. Suppose only one plant is to be included in the experiment and that operating restrictions dictate that only four test runs can be made on each of two days. If the experiment is conducted as indicated below, which effect is confounded with days?

DAY 1

Molten-Glass Temperature	Cooling Time	Glass Type
1300	15	A
1100	20	B
1300	20	B
1100	15	A

DAY 2

Molten-Glass Temperature	Cooling Time	Glass Type
1100	15	B
1100	20	A
1300	20	A
1300	15	B

CHAPTER 10

Nested Designs

Frequently experimental situations require that unique levels of one factor occur within each level of a second factor. This nesting of factors can also arise when an experimental procedure restricts the randomization of factor–level combinations. In this chapter we discuss nested experimental designs, with special emphasis on:

- *the distinction between crossed and nested factors,*
- *designs for hierarchically nested factors,*
- *staggered nested designs in which the number of levels of a factor can vary within a stage of the nesting, and*
- *split-plot designs, which include both nested and crossed factors.*

Consider an experiment that is to be conducted using three samples of a raw material from each of two vendors. In such an experiment there is no physical or fundamental relationship between the samples labeled 1, 2, and 3 from each vendor. In experimental design terminology, the factor "samples" would be said to be nested within the factor "vendor." A design that includes nested factors is referred to as a nested design.

Experiments conducted to diagnose sources of variability in manufacturing processes or in a laboratory method often use nested designs. Experiments in which subjects, human or animal, are given different experimental treatments are nested if only one of the treatments is given to each subject. Nested designs also occur when restrictions of cost or experimental procedure require that some factor–level combinations be held fixed while others are varied.

The first section of this chapter clarifies the main differences between nested and crossed factors. The second section is devoted to a discussion of hierarchically nested designs in which each factor is progressively nested within all preceding factors. The third section discusses split-plot designs in which some of the factors are nested and others are crossed. The final section shows how certain types of restricted randomization result in split-plot designs.

10.1 CROSSED AND NESTED FACTORS

Crossed factors occur in an experiment when each level of the factor(s) has a physical or fundamental property that is the same for every level of the other factors in the experiment (see Exhibit 10.1). Complete factorial experiments, by definition, involve crossed factors because each level of each factors occurs in the experiment with each level of every other factor. Crossed factors are often used in an experiment when the levels of the factors are the specific values of the factor for which inferences are desired. In preceding chapters the levels of factors such as pipe angle (Table 6.2), soil treatment (Figure 6.2), catalyst (Table 6.8), alloy (Figure 7.1), manufacturer (Table 8.1), measurement location (Table 8.2), oil (Table 8.3), reactor feed rate (Table 9.1), and many others are crossed factors.

EXHIBIT 10.1

Crossed Factors. A crossed factor contains levels that have a physical or fundamental property that is the same for all levels of the other factors included in the experiment.

In contrast to crossed factors, nested factors (see Exhibit 10.2) have levels that differ within one or more of the other factors in the experiment. Nested factors often are included in experiments when it is desired to study components of response variation that can be attributable to the nested factors. Variation in chemical analyses (Table 1.1) and in lawnmowers (Table 8.1) are two examples of nested factors used in previous chapters. In each of these examples the factor level is simply an identifying label.

EXHIBIT 10.2

Nested Factor. A nested factor has unique levels within each level of one or more other factors.

As an example of an experiment where factors are nested, consider an investigation to determine the major sources of variability in a laboratory test for measuring the density of polyethylene. The test method requires that polyethlyene pellets be heated, melted, and pressed into a plaque. A small disk is then cut from the plaque, and the density of the disk is measured. The test is nondestructive, so that more than one measurement can be made on each disk. Data are to be collected for two shifts of technicians working in a laboratory. Each shift is to prepare two plaques and four disks from each plaque.

There are three factors of interest in this experiment: shift, plaque, and disk. Since the plaques made on one shift have no physical relationship with those made on another

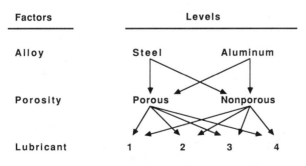

Figure 10.1 Crossed factors.

shift, plaques are nested within shifts. Similarly, since the disks cut from one plaque have no physical relationship with those cut from another plaque, disks are nested within plaques. Note that disk 1, 2, etc. and plaque 1, 2, etc. are only labels. It is the variation in the density measurements attributable to these plaques and disks that is of concern, not any inferences on the individual plaques and disks used in the experiment.

Figures 10.1 and 10.2 depict the difference between crossed and nested factors. In Figure 10.1 the factor levels for the torque study (Figure 7.1) are listed. The arrows from one factor to the next indicate that each level of each factor can be used in combination with any level of the other factors. One can obtain all factor–level combinations shown in Table 7.2 by following a particular sequence of arrows in Figure 10.1. For example, one arrow goes from steel to nonporous, another from nonporous to lubricant 4: this is factor–level combination 8 in Table 7.2. Note the crossing pattern of these arrows.

The arrows in Figure 10.2 for the density study do not cross one another. One cannot connect shift 2 with disk 1 cut from plaque 1 by following any of the arrows in Figure 10.2. This is because disk 1 of plaque 1 was produced during the first shift. It is a unique entity, different from disk 9 cut from plaque 3, even though each of these disks is the first one cut from the first plaque produced on a shift.

Both nested and crossed factors can be included in an experimental design. The experiment conducted to investigate automatic cutoff times of lawnmowers (Table 8.1) is an example of a design with both types of factors: manufacturer (A, B) and speed (high,

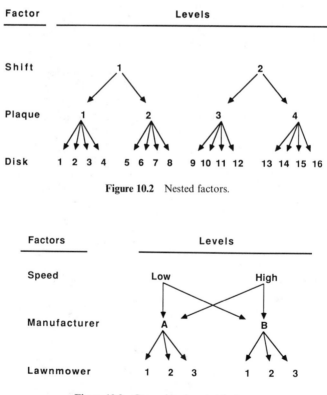

Figure 10.2   Nested factors.

Figure 10.3   Crossed and nested factors.

low) are crossed factors, and lawnmower is a nested factor. Figure 10.3 illustrates the crossing of the speed and manufacturer factors and the nesting of lawnmowers within manufacturers.

10.2 HIERARCHICALLY NESTED DESIGNS

Hierarchically nested designs are appropriate when each of the factors in an experiment is progressively nested within the preceding factor. The factors in the polyethylene density study in Section 10.1 are hierarchically nested: disks within plaques, plaques within shifts. The layout for a hierarchically nested experiment (see Exhibit 10.3) is similar to that of Figure 10.2. The design is usually balanced. Balance is achieved by including an equal number of levels of one factor within each of the levels of the preceding factor. Wherever possible, randomization of testing the factor levels is included in the specification of the design.

EXHIBIT 10.3   CONSTRUCTION OF HIERARCHICALLY NESTED DESIGNS

1. List the factors to be included in the experiment.
2. Determine the hierarchy of the factors.
3. Select, randomly if possible, an equal number of levels for each factor within the levels of the preceding factor.
4. Randomize the run order or the assignment of factor–level combinations to experimental units.

The allocation of experimental resources in hierarchically nested designs yields more information on factors that are lower in the hierarchy than on those that are higher. For example, only two shifts are studied in the design of Figure 10.2, whereas sixteen disks are included in the design. In some circumstances it may not be desirable to have

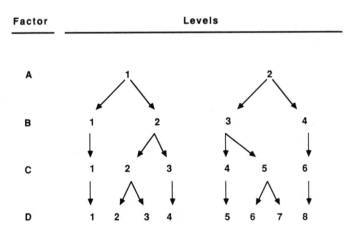

Figure 10.4   Staggered nested design.

many factor levels at lower stages of the hierarchy. In such circumstances, "staggered" nested designs can be used to reduce the number of factor levels.

Figure 10.4 shows a typical layout for a staggered nested design for four factors. In this figure, factor A occurs at two levels. Two levels of factor B are nested within each level of factor A. Rather than selecting two levels of factor C to be nested within each level of factor B, two levels of factor C occur within only one of the levels of factor B. The other level of factor B has only one level of factor C. This process continues at each remaining stage of the construction of staggered nested designs.

Note the pattern in the figure. Whenever a factor level has only a single level of the next factor nested within it, all subsequent factors also have only a single level nested within them. The unbalanced pattern in Figure 10.4 can easily be extended to more factors or more levels within each factor.

An illustration of the use of this type of design is given in Figure 10.5. This experiment is designed to investigate sources of variability in a continuous polymerization process that produces polyethelyne pellets. In this process, lots (100 thousand-pound boxes of pellets) are produced. From each lot, two boxes are selected. Two sample preparations are made from pellets from one of the boxes; one sample preparation is made from pellets from the other box. Two strength tests are made on one of the sample preparations from the first box. One strength test is performed on each of the other preparations. The completed design consists of strength tests from 30 lots.

The advantage of the design just described over the traditional hierarchical design is

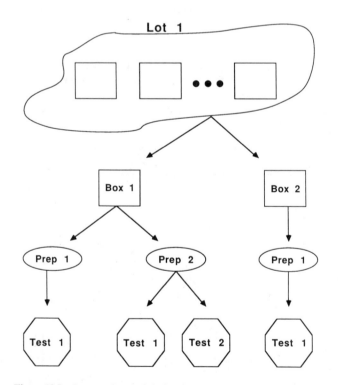

Figure 10.5   Staggered nested design for polymerization-process study.

Table 10.1 Number of Levels for Hierarchical and Staggered Nested Designs: Polymerization Study

	Number of Levels	
Factor	Hierarchical Design	Staggered Nested Design
Lots	30	30
Boxes	60	60
Preparations	120	90
Strength tests	240	120

that many fewer boxes, preparations, and tests need to be included in the design. If one chose to test two boxes, two preparations from each box, and two strength tests on each preparation, the number of factor levels for each type of nested design would be as shown in Table 10.1. Half the number of strength tests are required with the staggered nested design.

10.3 SPLIT-PLOT DESIGNS

Many experiments require that crossed factors occur within each level of a nested factor. This may be due to the physical nature of the experimental units or to restrictions on the randomization that can be allowed in the experiment. A design that is often used in this type of experimentation is the *split-plot design* (see Exhibit 10.4).

EXHIBIT 10.4   CONSTRUCTION OF SPLIT-PLOT DESIGNS

1. Identify whole plots and the factors to be assigned to them. List all combinations of the whole-plot factors that are to be tested.

2. Identify split plots and the factors to be assigned to them. List all combinations of the split-plot factors that are to be tested.

3. Randomly select a whole plot. Randomly assign a whole-plot factor–level combination to the whole plot.

4. Randomly select split plots from the whole plot chosen in step 3. Randomly assign combinations of the split-plot factors to the split plots.

5. Repeat steps 3 and 4 for all the whole plots.

6. Steps 1–5 can be extended to include additional nesting of experimental units (e.g., split-split-plots) as needed.

The term "split-plot" derives from agricultural experimentation, where it is common to experiment on plots of ground. Levels of one or more factors are assigned to geographical areas referred to as *whole plots* (e.g., individual farms), while levels of other factors are applied to smaller *split plots* (e.g., fields within each farm). For example, one might need to fertilize or control insects by spraying entire farms from airplanes. Thus, different fertilizers or insecticides would be applied to entire farms. Different varieties of crops or different crop spacings could be used in individual fields within a farm. Such an experimental layout, with appropriate randomization, would constitute a split-plot design.

In this agricultural example, the split-plot nature of the design occurs because there are two types of experimental units in the study. The whole plots and the split plots are different experimental units to which one or more of the factor levels are applied. Insecticides are applied to the farms, and varieties or spacings are assigned to the fields. Thus, there are five factors in this study: insecticides, farms (whole plots), varieties, spacings, fields (split plots).

Figure 10.6 illustrates a design for an experiment similar to the one just described. Two insecticides, A and B, are each applied to two farms. Two types of seed, seed 1 and seed 2, are planted in each of three fields on each of the farms. In this example, insecticide is the whole-plot factor and the farms are the whole plots. Seed is the subplot factor and the fields are the subplots.

The replicate farms in this example are blocks. Because of this, the advantages associated with blocking (see Section 8.1) are obtainable with this and with many other split-plot designs. For example, the effect of the seeds is calculated from differences of averages over the fields within each farm (block). These differences eliminate the farm effect, and therefore the seed effect is measured with greater precision than the effect of the insecticides. The latter effect is calculated from the difference between the averages

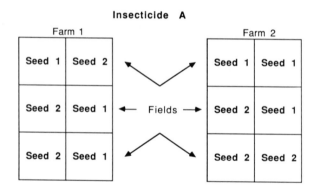

Figure 10.6 A split-plot design.

Table 10.2 Experimental Factors for Connector-Pin Study

Factor	Levels
Vendor	Vendor 1, vendor 2
Environment	Ambient, high humidity
Heat treatment	Yes, no

over all the observations in the replicate farms. These averages are subject to variability due to both fields and farms.

This example illustrates one of the main applications of split-plot designs: the estimation of components of variability due to two or more sources. Frequently responses are subject to variability from a variety of sources, some of which can be controlled through the use of appropriate blocking or nested designs. Hierarchical designs and split-plot designs, as well as the other blocking designs discussed in Chapter 8, can be used to allow estimation of the components of variability associated with the sources.

Although split-plot designs are extremely useful in agricultural experimentation, they are also beneficial in other scientific and in engineering research. As an illustration, consider an investigation of one of the components of an electronic connector. The pins on the connector are made from small-diameter bronze alloy wire. The factors of interest in this investigation are the vendors who supply the wire, the operational environment in which the pins are used, and the heat treatment applied to the pins. The factors and their levels are shown in Table 10.2.

This experiment is to be conducted using a split-plot design because there are two types of experimental units to which the levels of the factors are applied. Each vendor (whole-plot factor) supplies the wire on reels. The whole plots are reels of wire. Six-inch samples of wire are cut from the reels. The environment and heat-treatment factors (split-plot factors) are applied to the wire samples. The split plots are the wire samples taken from each reel.

A second example of the use of split-plot designs in industrial experimentation is the lawnmower automatic-cutoff-time example presented in Table 8.1. The whole-plot factor in this experiment is the manufacturer. The whole plots (blocks) are the lawnmowers. The split-plot factor is the operating speed. In this example there are replicates and repeat tests. There are no split plots; the individual test runs take the place of the split plots.

When designing split-plot experiments one should remember that experimental error is contributed by both the whole-plot and the split-plot experimental units. In order to estimate both types of experimental error it is necessary to replicate the whole plots and conduct repeat tests on the split plots. Replication and/or repeat tests may not be necessary if one can assume that certain interactions involving the whole-plot and split-plot factors are zero or at least negligible. This consideration will be discussed in the next section and in Section 17.6.

10.4 RESTRICTED RANDOMIZATION

There are occasions when the randomization of factor–level combinations to experimental units or to the run-order sequence must be restricted. The restrictions may be necessary because some of the factor levels are difficult or expensive to change. This restriction of the randomization of factor–level combinations is similar to the restrictions placed on the assignment of factor–level combinations to whole plots and to split plots in split-plot designs.

In split-plot designs the split-plot factors can be randomized over any of the experimental units (plots) in the design. The whole-plot factors must be assigned only to the whole plots and in that respect are similar to factors that are difficult or costly to change.

As an illustration of this type of experiment, consider an investigation of wear on the cylinder walls of automobile engines. The factors of interest in such an investigation might include those shown in Figure 10.7. One might select a half fraction of a complete factorial experiment using the defining equation $I = ABCDEF$ for this investigation. A randomized test sequence for the 32 factor–level combinations for this factorial experiment is shown in Table 10.3.

Of concern to the research engineer who is conducting this investigation is the number of times the piston rings must be changed. There are 20 changes of the piston rings in the design shown in Table 10.3. Each change of the piston ring requires that the engine be partially disassembled and reassembled, and that various pieces of test equipment be disconnected and reconnected. The design was considered very burdensome because of these requirements and because it extended the duration of the test program.

Factors	Levels
Piston Rings (A)	Type 1, Type 2
Engine Oil (B)	Oil A, Oil B
Engine Speed (C)	1500, 3000 rpm
Intake Temperature (D)	30, 90 degrees F
Air-Fuel Mixture (E)	Lean, Rich
Fuel Oxygen Content (F)	2.5, 7.5 %

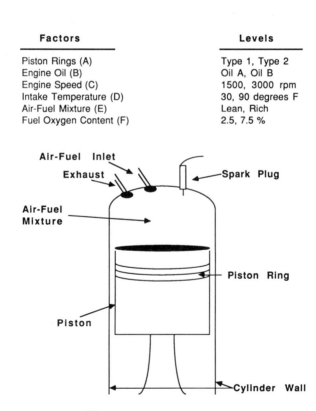

Figure 10.7 Cylinder wear study.

Table 10.3   Cylinder-Wear Fractional Factorial Experiment: Completely Randomized Design

Run	Piston-Ring Type	Engine Oil	Engine Speed (rpm)	Intake Temperature (°F)	Air–Fuel Mixture	Oxygen Content (%)
1	2	A	3000	90	Lean	7.5
2	1	B	1500	90	Lean	2.5
3	2	A	1500	90	Rich	7.5
4	1	A	3000	90	Lean	2.5
5	1	A	1500	90	Lean	7.5
6	1	B	3000	90	Lean	7.5
7	2	A	3000	30	Lean	2.5
8	2	A	1500	30	Rich	2.5
9	1	B	3000	30	Rich	7.5
10	2	A	3000	90	Rich	2.5
11	1	A	3000	30	Rich	2.5
12	2	B	3000	90	Rich	7.5
13	1	B	1500	30	Lean	7.5
14	1	B	1500	30	Rich	2.5
15	2	A	1500	30	Lean	7.5
16	1	B	3000	90	Rich	2.5
17	1	A	3000	30	Lean	7.5
18	2	B	1500	90	Lean	7.5
19	2	A	1500	90	Lean	2.5
20	2	B	1500	30	Rich	7.5
21	2	B	3000	30	Rich	2.5
22	1	A	1500	90	Rich	2.5
23	1	A	3000	90	Rich	7.5
24	2	A	3000	30	Rich	7.5
25	2	B	1500	90	Rich	2.5
26	1	A	1500	30	Lean	2.5
27	2	B	3000	90	Lean	2.5
28	2	B	1500	30	Lean	2.5
29	1	A	1500	30	Rich	7.5
30	1	B	3000	30	Lean	2.5
31	2	B	3000	30	Lean	7.5
32	1	B	1500	90	Rich	7.5

One solution to this problem is to change the piston rings only once, as indicated in Table 10.4. Within each set of 16 test runs for a given piston ring, the sequence of factor levels is completely randomized. Note the similarity with split-plot designs: piston-ring levels act as whole plots, and the run order within a piston ring constitutes the split-plot portion of the design.

The design in Table 10.4 consists of two groups of 16 test runs, one group corresponding to each piston ring. If a group effect results from some change in experimental conditions, it will be confounded with the piston-ring effect. This group effect is similar to the whole-plot error effect in a split-plot design. If each half of the design in Table 10.4 can be replicated, estimates of the variation due to the whole-plot (group) testing can be obtained. Likewise, if some of the test runs can be repeated, the

Table 10.4 Cylinder-Wear Fractional Factorial Experiment: Split-Plot Design

Run	Piston-Ring Type	Engine Oil	Engine Speed (rpm)	Intake Temperature (°F)	Air–Fuel Mixture	Oxygen Content (%)
1	1	A	3000	30	Rich	2.5
2	1	B	1500	90	Lean	2.5
3	1	A	1500	90	Rich	2.5
4	1	A	1500	30	Rich	7.5
5	1	B	1500	30	Lean	7.5
6	1	B	1500	90	Rich	7.5
7	1	B	1500	30	Rich	2.5
8	1	A	3000	90	Lean	2.5
9	1	B	3000	30	Rich	7.5
10	1	A	1500	90	Lean	7.5
11	1	B	3000	90	Rich	2.5
12	1	A	3000	30	Lean	7.5
13	1	A	3000	90	Rich	7.5
14	1	B	3000	30	Lean	2.5
15	1	A	1500	30	Lean	2.5
16	1	B	3000	90	Lean	7.5
17	2	B	1500	90	Rich	2.5
18	2	A	3000	90	Rich	2.5
19	2	B	1500	90	Lean	7.5
20	2	B	3000	30	Rich	2.5
21	2	A	1500	90	Lean	2.5
22	2	A	3000	30	Lean	2.5
23	2	A	3000	90	Lean	7.5
24	2	B	1500	30	Rich	7.5
25	2	A	1500	30	Lean	7.5
26	2	B	3000	90	Rich	7.5
27	2	B	3000	30	Lean	7.5
28	2	A	1500	30	Rich	2.5
29	2	B	1500	30	Lean	2.5
30	2	A	3000	30	Rich	7.5
31	2	A	1500	90	Rich	7.5
32	2	B	3000	90	Lean	2.5

split-plot error variation can be estimated. This is similar to the inclusion of replicate farms and several fields on which seed treatments were repeated in Figure 10.6.

The use of restricted randomization without replication and/or repeat tests can only be advocated when experiments are so well controlled that the whole-plot error can be ignored; i.e., it is nonexistent or negligible. As will be discussed in Section 17.6, if certain interactions involving split-plot factors are zero, the split-plot error variation can be estimated from the quantities that would ordinarily be used to measure these interaction effects. So too, if whole-plot error is not negligible but interactions involving only the whole-plot factors are negligible, the quantities used to measure the interaction effects can be used to estimate the whole-plot error variation. If interactions involving both whole-plot factors and split-plot factors can be assumed to be negligible, replication and/or repeat tests need not be conducted. This is an important consideration in many types of scientific and engineering studies for which fractional factorial experiments are conducted.

REFERENCES

Text References

Nested designs, in particular split-plot designs, are not covered in many elementary engineering statistics texts. The following books include adequate coverage of nested designs:

Anderson, V. L. and McLean, R. A. (1974). *Design of Experiments: A Realistic Approach*, New York: Marcel Dekker, Inc.

Cochran, W. G. and Cox, D. R. (1957). *Experimental Designs*, New York: John Wiley and Sons, Inc.

Daniel, C. (1976). *Applications of Statistics to Industrial Experimentation*, New York: John Wiley and Sons, Inc.

Federer, W. T. (1955). *Experimental Design: Theory and Application*, New York: The Macmillian Co.

Johnson, N. L. and Leone, F. C. (1977). *Statistics and Experimental Design in Engineering and the Physical Sciences*, New York: John Wiley and Sons, Inc.

Kempthorne, O. (1952). *The Design of Experiments*, New York: John Wiley and Sons, Inc.

Steel, R. G. D. and Torrie, J. H. (1980). *Principles and Procedures of Statistics: A Biometrical Approach*, New York: McGraw-Hill Book Company.

Staggered nested designs are discussed in the following articles:

Bainbridge, T. R. (1965). "Staggered, Nested Designs for Estimating Variance Components," *Industrial Quality Control*, **22**, 12–20.

Smith, J. R. and Beverly, J. M. (1981). "The Use and Analysis of Staggered Nested Factorial Designs," *Journal of Quality Technology*, **13**, 166–173.

> *The latter article not only discusses the type of staggered nested designs introduced in Section 10.2, it also includes designs in which some factors have a factorial relationship and others are nested.*

The use of split-plot designs with factorial or fractional factorial experiments without including replicates and/or repeat tests is discussed in the following article and in Daniel (1976).

Addleman, S. (1964). "Some Two-Level Factorial Plans with Split-Plot Confounding," *Technometrics*, **6**, 253–258.

> *In addition to discussing the use of interaction effects to estimate whole-plot and split-plot error variation, the article provides a list of defining contrasts for constructing fractional factorial experiments in split-plot designs with high-order confounding of interaction effects.*

EXERCISES

1   The plume dispersion of liquid chemicals from a ruptured hole in a chemical tanker is to be studied in a simulated river channel. Two factors of interest are the speed of the river (10, 100, 300 ft/min) and the spill rate of the chemical (50, 150, 400 gal/min). The experimenter is interested in comparing the plume dispersion of sodium silicate and ethyl alcohol. Design an experiment for this study. Which factors, if any, are crossed? Which, if any, are nested? Sketch a layout of the experiment similar to the figures in this chapter.

2   A ballistics expert wishes to investigate the penetration depth of projectiles fired

from varying distances into an armor plate. Three different gun types that fire the same-size projectiles are to be studied. Two guns of each type are to fire four projectiles at ranges of 5, 50, and 100 feet. Design an experiment for this study. Which factors, if any, are crossed? Which, if any, are nested? Sketch a layout of the experimental design.

3   Refer to the heat-transfer study of Exercise 5, Chapter 7. For a specific glass type and cooling temperature, suppose an experiment is to be conducted to assess the variability of the mold temperatures that is inherent in the production process. In this experiment, three disks used to make the drinking glasses are to be studied. Each disk contains molds to produce three drinking glasses. Five measurements of mold temperature are to be taken on each mold from each disk at each of the two plants. Design an experiment for this study. Which factors, if any, are crossed? Which, if any, are nested? Sketch a layout of the experimental design.

4   Suppose the number of measurements to be taken in the experiment described in Exercise 3 is considered excessive. Design two alternative hierarchically nested designs, one a staggered nested design, which would reduce the number of measurements yet retain the ability to estimate measurement error from repeat measurements on the mold temperature.

5   Sketch a layout of the experimental design for the wing design study in Exercise 8 of Chapter 8. What type of design is it? Identify whether the factors are crossed or nested and whether there is replication or repeat tests.

6   An interlaboratory study of the stress (ksi) of titanium is to be designed, involving three laboratories in the United States and three in Germany. The laboratories in each of the two countries can be considered representative of the many laboratories making stress measurements in their countries. Two temperatures (100, 200°F), and four strain rates (1, 10, 100, 1000 sec^{-1}) are to be investigated. Two titanium specimens are available for each lab–temperature–strain combination. Design an experiment for this study. Identify crossed and nested factors, if any, and sketch a layout of the design.

7   The quality of dye in a synthetic fiber being manufactured by a textile factory is to be investigated. The response of interest is a quantitative measurement of the dye in a specific fiber. Four factors are to be studied: shifts (three shifts of eight hours each on a given work day), operators (two on each shift), packages of bobbins, and individual bobbins. Each operator produces three packages of bobbins on each shift; each package contains two bobbins. Design an experiment to study these four factors. The manufacturer restricts the experiment to one work day and no more than 20 bobbins. Sketch a layout of the design.

8   A chemical laboratory is experimenting with factors that may contribute to the success of plating gold on a base metal. The factors being studied in the current investigation are plating-bath type (two levels), anode type (two levels), and current density (three levels). It is noted that current density is easy to change but that the other two factors are much more difficult. Design an experiment for this study. Include replicates and/or repeat tests where desirable.

9 A study of alloying processes leads to an experiment to compare the effects of heat treatment on the strength of three alloys. Two ingots are to be made of each of the alloys. These ingots are to be cut into four pieces, each of which will receive one of three heat treatments. Design an experiment for this investigation. Sketch a layout of the design.

10 The technical training office of a large international manufacturer is experimenting with three methods of teaching the concepts of statistical process control to its work force. Five large manufacturing plants are chosen to be part of the study. In each plant, employees on a particular shift are to be given one of the three methods of instruction. The shift to receive each type of instruction is to be randomly selected at each plant. A random sample of the employees in each shift are to be further divided into two groups. One of the groups on each shift is to be trained in a workshop to implement the process-control techniques on microcomputers, with appropriate software for calculations and graphics. The other group is to utilize calculators and graph paper. Design an experiment for this study involving no more than 10 workers from each shift at each plant.

CHAPTER 11

Response-Surface Designs

This chapter contains a discussion of designs that are useful when an experimenter wants to explore an unknown response surface. Topics to be addressed include:

- *uses of response-surface methodology,*
- *methods for locating an appropriate experimental region,*
- *description of designs for fitting response surfaces, and*
- *designs that can be used when a response is a function of only the relative proportion of the components (factors in a mixture or formulation).*

A common goal in many types of experimentation is to characterize the relationship between a response and a set of factors of interest to the researcher. This is accomplished by constructing a model that describes the response over the applicable ranges of the factors of interest. In many industrial applications the fitted model is referred to as a *response surface* because the response can then be graphed as a curve in one dimension (one factor of interest) or a surface in two dimensions (two factors of interest). The response surface can be explored to determine important characteristics such as optimum operating conditions (i.e., factor levels that produce the maximum or minimum estimated response), or relevant tradeoffs when there are multiple responses. The objective of this chapter is to present the design strategies for developing and analyzing these types of response functions.

Two special classes of designs that are discussed in Section 11.3 are central composite designs and Box–Behnken designs. These serve as alternative designs to the factorial layouts given in Chapter 7 and are particularly useful for conducting experiments in a sequential manner. Efficiency is achieved by reducing the number of factor–level combinations from what would be required using complete or fractional factorial experiments.

A third special class of designs that is useful for exploring response surfaces is referred to as *mixture* designs. Mixture designs are used when the response of interest is influenced not by the actual amounts of the factors (components of the mixture) but only by their relative amounts. These designs are introduced in Section 11.4.

11.1 USES OF RESPONSE-SURFACE METHODOLOGY

Certain types of scientific problems involve expressing a response variable, such as the viscosity of a fluid, as an empirical function of one or more quantitative factors, such as reaction time and reaction temperature. This is accomplished using a *response function* to model the relationship:

$$\text{Viscosity} = f(\text{reaction time, reaction temperature}).$$

A general form of this type of response function is

$$y = f(x_1, x_2, \ldots, x_k), \tag{11.1}$$

where y is the response and $x_1, x_2, \ldots, x_k$ are quantitative levels of the factors of interest. Knowledge of the form of the function, f, often found by fitting models to data obtained from designed experiments, allows one to both summarize the results of the experiment and predict the response for values of the quantitative factors. The function f defines the response surface (see Exhibit 11.1).

EXHIBIT 11.1

Response Surface. A response surface is the geometric representation obtained when a response variable is plotted as a function of one or more quantitative factors.

Designing experiments to study or fit response surfaces is important for several reasons, including the following:

- the response function is characterized in a region of interest to the experimenter,
- statistical inferences can be made on the sensitivity of the response to the factors of interest,
- factor levels can be determined for which the response variable is optimum (e.g., maximum or minimum), and
- factor levels can be determined that simultaneously optimize several responses; if simultaneous optimization is not possible, tradeoffs are readily apparent.

Each of these uses will now be discussed.

A response surface can have various shapes, depending on the form of the response function in equation (11.1). For example, if it is a second-order (i.e., quadratic) function (see Chapter 23) of only one factor, the surface might look similar to the curve shown in Figure 11.1. The plotted points represent pairs of observed responses (y) for each of three quantitative values of the factor (x). The fitted model, represented by the smooth curve, characterizes the response surface and identifies where the maximum response is obtained.

One experimental strategy to use when a response is believed to be a quadratic function of a single factor is to select a three-level, completely randomized design with repeat tests. Note that using a two-level design would only allow a straight line to be fitted. Repeat tests are included in order to provide an estimate of the uncontrolled experimental-error variation. The plotted points in Figure 11.1 depict the location of the design points and the corresponding responses for a typical experiment.

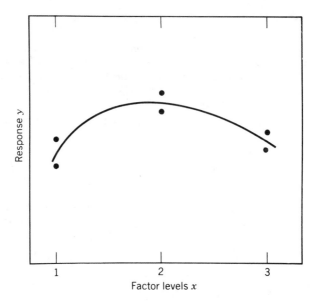

Figure 11.1   Quadratic response surface in one factor.

When the response is a function of two factors, the experimental situation can still be depicted graphically. As before, assume the response function is a second-order polynomial. In this situation the response surface might be graphed similarly to the one shown in Figure 11.2. An experimental strategy to use in characterizing this surface might consist of a completely randomized design in two factors, each at three levels. Typical design points are illustrated on the graph (without repeats).

An alternative approach to plotting the two-factor response surface is to plot contours (see Exhibit 11.2) of constant response as a function of the two factors. These are similar to the contours of equal elevation on a topographical map.

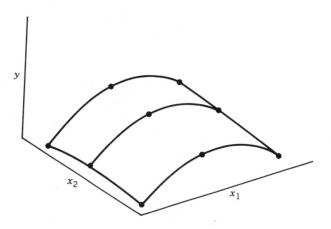

Figure 11.2   Quadratic response surface in two factors.

EXHIBIT 11.2

Contour Plot. A contour plot is a series of lines or curves that identify values of the factors for which the response is constant. Curves for several (usually equally spaced) values of the response are plotted.

Contour plots can be constructed in several ways. If the response function is a simple enough function of the two factors, it can be solved for one of the factors as a function of the response and the other factor. Fixing the response at a value then enables one to plot the contour having that response value by plotting one of the factors in terms of the other one.

When the response function is complicated, a more direct solution is to calculate the values of the response variable for a grid of values of the two factors. Instead of plotting points, one can then plot the numerical values of the response on a graph as a function of the two factors; i.e., the two axes represent the factor values and the numerical value of a calculated response is placed on the grid at the intersection of the two factor values used to

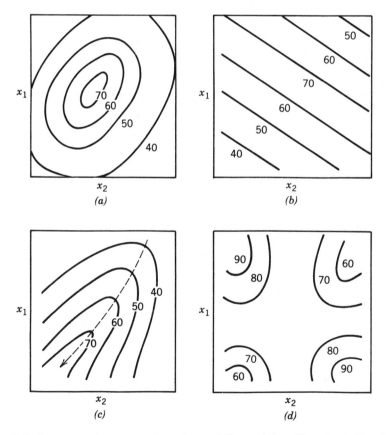

Figure 11.3 Response-surface contours in two factors: (a) "mound-shaped" maximum; (b) stationary ridge; (c) rising ridge; (d) saddle.

calculate it. Contours can then be approximated by interpolating between values of the response variable.

Figure 11.3 illustrates several common forms of contour plots from two-factor response surfaces. Each of the plots shows curves for values of the response in increments of 10 units. Figure 11.3a illustrates the shape of contours when a conical or symmetrical "mound-shaped" surface is located in the center of the experimental region. Figure 11.3b depicts the identification of a *stationary ridge*, a sequence of flat contours with a maximum along one of the contours. With response surfaces such as this, there is a wide range of values of the two factors that produce an optimal response. Sometimes fitted response surfaces look like Figure 11.3b, but the response values are all increasing in one direction; i.e., there is no contour with a maximum response. This indicates that the fitted surface is a plane and that the region of optimum response is distant from the experimental region.

Figure 11.3c shows a *rising ridge*, such that the location of a minimum response is just outside the experimental region. A path up the ridge line (dashed in Figure 11.3c) leads to the optimum response. Figure 11.3d depicts a *saddle*, on which one can either decrease or increase the response by selecting factor levels along a 45° or a 135° line, respectively, from the center of the region.

The variety of response-surface contours shown in Figure 11.3 demonstrate why

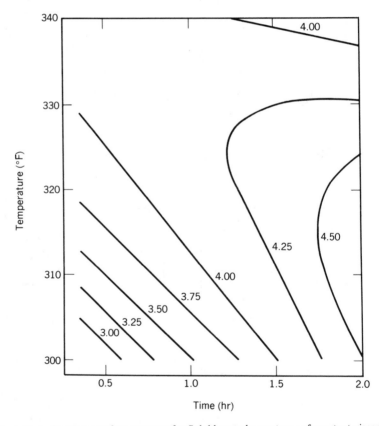

Figure 11.4 Response-surface contours for Sulphlex study: contours of constant viscosity.

response-surface characterization is one primary reason for many studies of responses and the factors that influence them. An equally important reason for fitting and studying response surfaces is the determination of the sensitivity of the response to the various factors. If the contours in Figure 11.3b were more horizontal it would indicate graphically that the response was very sensitive to the first factor (x_1) and relatively insensitive to the second one (x_2). This is because contours that are almost horizontal indicate that the horizontal factor could change greatly and the response would remain fairly constant.

Polynomials models of first and second order (i.e., linear and quadratic equations with interactions) are routinely used to model the response surface in equation (11.1). Experimental designs are selected that ensure that information about the response is gathered in an efficient manner in the regions of interest (Sections 11.2 to 11.4). Fitted response surfaces or their contours are graphed to illustrate the changes in the response as a function of the various factors in the study. Statistical inference techniques (Chapter 23) can be used to assess the importance of the individual factors, the appropriateness of their functional form, and the sensitivity of the response to each factor.

A third use of response-surface experimentation is to find the factor levels that provide

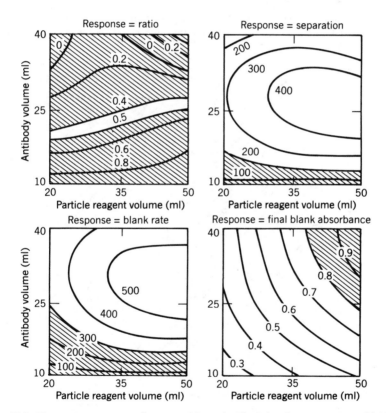

Figure 11.5 Four response contours for gentamicin study. (Shaded regions are unacceptable.) From Myers, G. C., Jr. (1985). "Use of response surface methodology in clinical chemistry," in Snee, R. D., Hare, L. B., and Trout, J. R. (Eds.), *Experiments in Industry: Design, Analysis and Interpretation of Results*, Milwaukee, WI: American Society for Quality Control. Copyright American Society for Quality Control, Inc., Milwaukee, WI. Used by permission.

the optimum response. This could be either a maximum or a minimum response. For example, in a study on Sulphlex binders used in developing synthetic asphalt, a response-surface approach was used to determine the reaction temperature and reaction time that would yield a binder with the highest possible viscosity. Using a 3^2 factorial experiment with several repeat tests, a second-order response surface was fitted to the observed viscosities, and the contour plot given in Figure 11.4 was obtained. It can be seen that the highest viscosity values occurred at the higher reaction times and at the lower temperatures. The contours suggest that even higher viscosities could be obtained for moderate temperatures with longer reaction times. This was the direction taken in further experimentation.

A fourth use of response-surface design is to find factor regions that produce the best combinations of several different responses. To illustrate this use, consider an experiment that was run to study the sensitivity of an analytical method used to assay blood serum for the presence of the therapeutic antibiotic drug gentamicin. The purpose was to determine the volumes of two reagents, a particle reagent and an antibody, used in an analytical assay test pack that were necessary to maintain uniform performance.

Four responses were measured as a function of the volumes of these two reagents: (a) separation, (b) ratio, (c) blank, and (d) final blank absorbance. The experimental region of interest included volume combinations of the reagents for which separation was greater

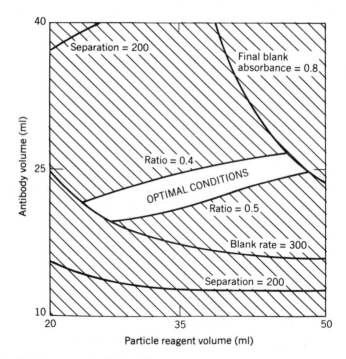

Figure 11.6 Simultaneous optimization of four responses: gentamicin study. From Myers, G. C., Jr. (1985). "Use of response surface methodology in clinical chemistry," in Snee, R. D., Hare, L. B., and Trout, J. R. (Eds.), *Experiments in Industry: Design, Analysis and Interpretation of Results*, Milwaukee, WI: American Society for Quality Control. Copyright American Society for Quality Control, Inc., Milwaukee, WI. Used by permission.

than 200 milliabsorbance units per minute (ma/min), ratio was between 0.4 and 0.5, blank rate exceeded 300 ma/min, and final blank absorbance was less than 0.8 absorbance units (800 ma).

The purpose of the response-surface investigation was to find the particle reagent and antibody volumes that maximized the separation and blank rates subject to the above restrictions. A 3^2 factorial experiment with three repeat tests was used to develop the response-surface contours shown in Figure 11.5. These contours are overlaid in Figure 11.6. The region of optimal conditions for the two factors is indicated in Figure 11.6 by the unshaded region.

In this example there exists a portion of the experimental region for which all four responses are optimized. There is no guarantee that this will always occur. Often there will be regions in which some responses are optimized but others are not. Constructing contour plots similar to Figure 11.5 will allow reasonable compromises to be made.

11.2 LOCATING AN APPROPRIATE EXPERIMENTAL REGION

As noted in Section 11.1, one of the uses of a response-surface design is to identify values of the factors that produce optimum or near optimum values of the response. This often is accomplished through a series of experiments whereby one searches for the region of the optimum response. One method of experimentation that is popular in this process is the one-factor-at-a-time strategy. Each factor is individually increased or decreased in an effort to find the maximum response. The combination of these optimum factor levels is then chosen as the conditions for obtaining the overall maximum. Unfortunately, as discussed in Section 6.2, this method frequently fails to locate the region of the optimum response, since the procedure does not take account of any joint effects of the factors on the response.

A preferred alternative procedure involves up to three stages, the second of which is a sequence of test runs conducted along a path of *steepest ascent*. In the initial stage of this procedure a simple experiment is conducted over some small area of the region of interest. Usually this initial experiment does not include the extremes of the experimental region, but it may include the extremes of a desirable subregion. A first-order model is then fitted to the data, and the resultant equation is used to determine the direction to move toward the surface optimum. Observations are sequentially taken along this path of steepest ascent until the vicinity of the optimum of the surface is located. A more comprehensive experiment is then run in this new region, and usually a higher-order model is fitted in order to characterize the surface. The procedure is outlined in Exhibit 11.3. The fitting of response-surface models is discussed in Section 23.2. In the remainder of this section and in the next two sections of this chapter we consider the three design-related steps in the steepest-ascent procedure: steps 1, 4, and 6.

Complete or fractional factorial experiments are the most common initial experiments in the study of response surfaces. We shall use data from the Sulphlex study to illustrate the method of steepest ascent. Figure 11.7 shows four design points (a 2^2 factorial) taken from the Sulphlex study introduced in the previous section. Linear contours for response values from 3.25 to 4.25 in increments of 0.25 are shown.

The path of steepest ascent can be determined graphically as indicated in Figure 11.7. The direction indicated by the arrow, perpendicular to the contour lines, is the direction in which additional test runs will be taken if the goal is to maximize viscosity. Ordinarily, test runs are equally spaced along the path of steepest ascent.

EXHIBIT 11.3   LOCATING AN OPTIMUM RESPONSE

1. Conduct a small-scale experiment (often a two-level fractional factorial) in a region of the factor space that is believed to include the optimum response.

2. Fit a model (usually first-order) to the data collected in step 1.

3. Use the fitted model (equation) from step 2 to find the direction of the steepest ascent or descent in the response. The path is perpendicular to the contour lines. If the model is nonlinear and is a function of three or more factors, a canonical analysis (Chapter 23) may be needed to determine the direction of steepest ascent.

4. Conduct a series of test runs along the path determined in step 3 until the increase or decrease in the response becomes small or reverses.

5. Repeat steps 1–4 in the new region of the factors, if needed, until the region of the optimum response is located.

6. When the factor region containing the optimum response is located, conduct a more extensive experiment that will permit the fitting of higher-order models so the curvature of the response surface can be adequately approximated.

The path of steepest ascent is usually determined from the center of the experimental region. The determination can be facilitated by coding the region so that the lower and upper levels of each factor are -1 and $+1$, respectively. To do so for each factor, use the coding technique given in Exhibit 11.4. Using this technique, the coded levels of the two factors, $x_1 =$ time and $x_2 =$ temperature, in the Sulphlex study are

$$t = \frac{x_1 - 1.25}{0.75}, \qquad T = \frac{x_2 - 320}{20}.$$

As indicated on the axes in Figure 11.7, the contours can be plotted using the coded or the raw factor levels.

Since the contours in Figure 11.7 are straight lines, the path of steepest ascent is a line from the center of the region that is perpendicular to the contours (a general technique for finding the path of steepest ascent from the fitted model is given in Chapter 23). Perpendicular lines to these contours have slopes opposite in sign and inverse in

---

EXHIBIT 11.4   TWO-LEVEL FACTOR CODING

For each factor:

1. Determine the average of the two factor-level values. Denote it by AVG.

2. Determine the midrange of the factor levels:

$$MID = \frac{\text{upper level} - \text{lower level}}{2}.$$

3. Code the factor levels:

$$\text{coded level} = \frac{\text{level} - AVG}{MID}.$$

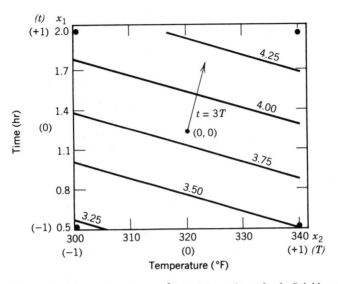

Figure 11.7   Response-surface contours from a 2^2 factorial experiment for the Sulphlex study: coded (t, T) and original (x_1, x_2) factor levels.

magnitude to those in the figure; i.e., the lines

$$t = bT \quad \text{and} \quad t = cT$$

are perpendicular when $c = -1/b$.

All the contour lines in Figure 11.7 have slopes equal to -0.33; consequently, the perpendiculars to these lines have slopes equal to $+3.00$. Thus the path of steepest ascent goes from the center of the region along the line $t = 3.00T$, as indicated in Figure 11.7. Table 11.1 shows the path of steepest ascent in both the coded and the original scales of the factors. These values were determined (arbitrarily) by incrementing the coded temperature

Table 11.1   Path of Steepest Ascent for Sulphlex Data*

Coded Factors		Original Factors	
Time t	Temperature T	Time x_1 (hr)	Temperature x_2 (°C)
0.0	0.0	1.25	320
1.5	0.5	2.38	330
3.0	1.0	3.50	340
4.5	1.5	4.62	350
6.0	2.0	5.75	360
7.5	2.5	6.88	370
.	.	.	.
.	.	.	.
.	.	.	.

*Increments of 0.5 Units in Coded Temperature.

factor by 0.5. Other increments can be used if they are deemed more reasonable by the experimenter. Note that one need not start the test runs at the center of the current design; rather, one could begin at the first combination (e.g., $t = 1.5$, $T = 0.5$) that is outside the region shown in Figure 11.7.

The objective now is to advance along the path $t = 3.00T$ until the maximum of the response is obtained. As experimentation continues along the path of steepest ascent, the increases in the response will become smaller; eventually the responses will decrease. One should then conduct a new set of experimental test runs to ensure that the optimum experimental region has been located. The new test runs should be selected so that if the optimum region has been identified, additional runs can be added to these to allow the fitting of higher-order models. The inclusion of repeat tests with these new test runs will allow an assessment of the adequacy of the fit.

A good design for this latter purpose is again a two-level factorial with the levels chosen so that a third level for each factor can be added if it is determined that the optimum experimental region has been located. Repeat tests can be made at the center of this new experimental region.

For a large number of factors, the initial experiment in the steepest-ascent procedure may be a two-level fractional factorial. When using two-level complete or fractional factorial experiments, caution should be exercised in moving along the path of steepest ascent. A biasing effect may lead one along an incorrect path if higher-order effects are substantial. This is especially true when nearing the region of the optimum response, where curvature and joint factor effects are likely to be most pronounced.

Figure 11.8 shows response contours for the coded Sulphlex factors after fitting interaction and quadratic terms from a complete 3^2 design. Since short reaction times were of interest in this experiment, the middle level of time was chosen closer to the lower level than to the upper one. If a two-level factorial experiment had been run in this region, the response contours would have been those in Figure 11.7. By using a three-level factorial experiment and fitting a model including terms for curvature, the path of steepest ascent is as indicated in the figure, quite different from that shown in Figure 11.7.

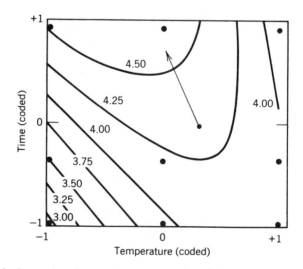

Figure 11.8   Response-surface contours from a 3^2 factorial experiment: Sulphlex study.

Table 11.2   Typical Factor Levels for a 3^2 Factorial

Coded Factors		Original Factors	
Factor A	Factor B	Factor A	Factor B
−1	−1	Low	Low
−1	0	Low	Middle
−1	+1	Low	High
0	−1	Middle	Low
0	0	Middle	Middle
0	+1	Middle	High
+1	−1	High	Low
+1	0	High	Middle
+1	+1	High	High

11.3   DESIGNS FOR FITTING RESPONSE SURFACES

Several different types of designs are useful when one is attempting to fit a response surface. Complete and fractional factorial experiments in completely randomized designs are extremely useful when one is exploring the factor space in order to identify the region where the optimum response is located. As stressed at the end of the last section, two-level factorials are highly efficient but must be used with some caution. They allow the fitting of only first-order models with or without interaction terms and cannot detect curvature.

When one has located the region of the optimum response, curvature can be pronounced. Three-level factorial experiments are often conducted in order to fit such response surfaces. The nine equally spaced factor–level combinations for a 3^2 experiment are shown in Table 11.2. The coded factor levels are designated as low (−1), middle (0), and high (+1).

The gentamicin study described in Section 11.1 uses a 3^2 factorial. The antibody volume (factor A) has three levels (10, 25, and 40 ml), and the particle-reagent volume (factor B) has three levels (20, 35, and 50 ml). The factor space and design points are illustrated in Figure 11.9. Three repeat tests were taken at each of the nine design points.

As the number of factors increases, the 3^k factorials become inefficient and impractical. These experiments need large numbers of observations; e.g., $3^5 = 243$ and $3^{10} = 59,049$. Further, these designs do not give equal precision for fitted responses at points (factor–level combinations) that are at equal distances from the center of the factor space. A design that has this property is termed a *rotatable* design (see Exhibit 11.5). Rotatability is a desirable property for response-surface models because prior to the collection of data and the fitting of the response surface, the orientation of the design with respect to the surface is unknown. Thus, the exploration of the response surface is dependent on the orientation of the design. In particular, procedures such as the method of steepest ascent, which utilize

---

EXHIBIT 11.5

Rotatable Design. When fitting specified response-surface models, a design is rotatable if fitted models estimate the response with equal precision at all points in the factor space that are equidistant from the center of the design.

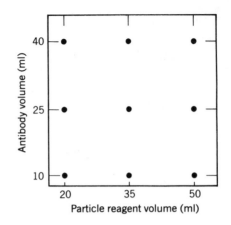

Figure 11.9 3^2 factorial layout used in gentamicin study.

the fitted response surface, can be jeopardized if some estimated responses are less precise than others.

In general, rotatable designs can be constructed from equally spaced points on circles or spheres. If one is interested in constructing rotatable designs for use with models in which only linear terms, with or without interactions, are to be included, points in regular polygons about the center of the experimental region can be used. For example, in a two-factor experiment the vertices of a square (two-level factorial), pentagon, hexagon, or octagon could be used. These are two-dimensional polygons centered at the origin with vertices on a circle. Coordinates of such designs are multiples of those shown in Table 11.3. Ordinarily, in such designs, repeat observations are taken at the origin. If there are to be repeat tests conducted at the vertices of the design, there must be an equal number of repeat tests at all the design points for the design to be rotatable.

For more than two factors, design points should lie on a sphere, or a hypersphere in four or more dimensions. The design points must also form a regular geometric figure such as a cube for the design to be rotatable. All 2^k complete factorials are rotatable, but 3^k factorials are not. Fortunately, there are classes of designs for two or more factors

Table 11.3 Some Rotatable Designs for Two Factors*

Factor–Level Combination	Square		Pentagon		Hexagon		Octagon	
	x_1	x_2	x_1	x_2	x_1	x_2	x_1	x_2
1	-1	-1	1	0	1	0	1	0
2	-1	1	.309	.951	.500	.866	.707	.707
3	1	-1	$-.809$	.588	$-.500$	.866	0	1
4	1	1	$-.809$	$-.588$	-1	0	$-.707$	.707
5			.309	$-.951$	$-.500$	$-.866$	-1	0
6					.500	.866	.707	$-.707$
7							0	-1
8							$-.707$	$-.707$

*Coordinates are cosines and sines of multiples of $360/k$ degrees, where k is the number of points in the design.

that can be used in place of 3^k factorials for fitting second-order polynomials to response surfaces.

Two designs that make more efficient use of the experimental units or test runs than the 3^k factorial experiments are the central composite design and the Box–Behnken design. Both of these designs are fractions of the 3^k factorials, but they only require enough observations to estimate the second-order effects of the response surface. Both of these designs can be made rotatable, or approximately so.

11.3.1 Central Composite Designs

The central composite design is constructed following the steps given in Exhibit 11.6.

EXHIBIT 11.6 CENTRAL COMPOSITE DESIGN

1. Construct a complete or fractional 2^k factorial layout, depending on the need for efficiency and the ability to ignore interaction effects.

2. Add $2k$ *axial*, or *star*, points along the coordinate axes. Each pair of star points is denoted, using coded levels, as follows:

$$(\pm a, \quad 0, \quad 0, \dots, \quad 0),$$
$$(0, \quad \pm a, \quad 0, \dots, \quad 0),$$
$$(0, \quad 0, \quad 0, \dots, \quad \pm a),$$

where a is a constant, which can be chosen to make the design rotatable or to satisfy some other desirable property.

3. Add m repeat observations at the design center:

$$(0, \quad 0, \quad 0, \dots, \quad 0).$$

4. Randomize the assignment of factor–level combinations to the experimental units or to the run sequence, whichever is appropriate.

The total number of test runs in a central composite design based on a complete 2^k factorial is $n = 2^k + 2k + m$. This count usually is less than 3^k, so that fewer observations are required than in a 3^k factorial. The central composite design can be made to be rotatable by choosing $a = F^{1/4}$, where F is the number of factorial points (e.g., $F = 2^k$ when a complete factorial is used). An illustration of the three-factor central composite design based on a complete 2^k factorial is given in Figure 11.10.

We now present an example of a central composite design that was used to evaluate the performance of a method for determining the level of the hormone thyroxine in blood serum. After initial experimentation, it was found that the key factors in this study were two reagent concentrations and the serum sample volume. The two reagents included an enzyme and a thyroxine conjugate, which acts as an enzyme inhibitor.

A rotatable central composite design could be used for this experiment. The design would consist of the eight corner points of the 2^3 cube, the six star points, and m center points. The star points would have $a = 8^{1/4} = 1.68$. If three center points were selected, the design (prior to randomization) would be as given in Table 11.4.

The actual design used is illustrated in Figure 11.11. It is a special form of the central

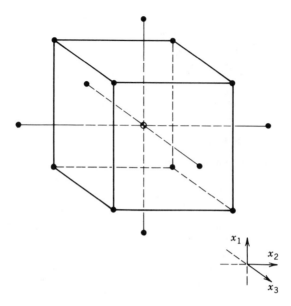

x_1
x_2
x_3

Figure 11.10 Central composite design in three factors.

Table 11.4 Rotatable Central Composite Design for Thyroxine Study*

| Run No. | Coded Factor Levels | | |
	Enzyme	Conjugate	Sample Volume
1	−1	−1	−1
2	−1	−1	+1
3	−1	+1	−1
4	−1	+1	+1
5	+1	−1	−1
6	+1	−1	+1
7	+1	+1	−1
8	+1	+1	+1
9	−1.68	0	0
10	+1.68	0	0
11	0	−1.68	0
12	0	+1.68	0
13	0	0	−1.68
14	0	0	+1.68
15	0	0	0
16	0	0	0
17	0	0	0

*Nonrandomized.

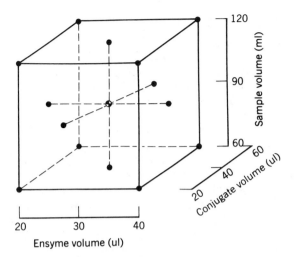

Figure 11.11 Face-centered central composite design for thyroxine study.

composite design in that the star points are on the face of the cube formed from the 2^3 factorial. Thus, $a = 1$ and the design is called a *face-centered cube design*. The face-centered cubic design was chosen over alternatives such as a rotatable design because the face-centered design only uses three levels of each factor, whereas other central composite designs would require five levels of each $(0, \pm 1, \pm a)$. Having three levels instead of five was cited as desirable because it reduced the preparation time and lessened the potential for mistakes in preparing the test serum. Three replicates were taken at the design center, so that the total number of observations was $n = 8 + 6 + 3 = 17$. This is slightly over half the number of observations that would be required for a three-level factorial without repeats $(3^3 = 27)$.

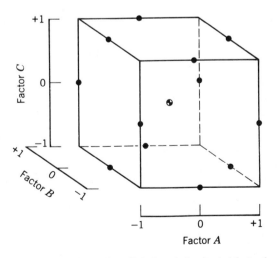

Figure 11.12 Three-factor Box–Behnken design (coded factor levels).

11.3.2 Box–Behnken Designs

The Box–Behnken design is another alternative to the 3^k factorial. The designs are formed by combining 2^k factorials with incomplete block designs. The result is a design that makes efficient use of the experimental units and is also rotatable or nearly so. Useful Box–Behnken designs are listed in Table 11A.1 in the appendix to this chapter. Sequences of ± 1 signs in the rows of Table 11A.1 mean that each level of the factors is to be run with each level of all the other factors in that row of the design that have ± 1.

The three-factor Box–Behnken design is shown in Figure 11.12. The equivalent design for the thyroxine study is given in Table 11.5 (without randomization). The total number of test runs needed for this design is 15, fewer than the 17 required for a central composite design with the same number of repeats and the 27 required for a 3^3 factorial without repeats. Three repeat tests are included at the center of the design, as was done with the central composite design. The Box–Behnken design in Table 11.5 only requires three levels of each factor. It is preferable to the face-centered central composite design not only because it requires fewer test runs but also because it is rotatable.

As indicated in Figure 11.12, Box–Behnken designs do not contain any points at the extremes of the cubic region created by the two-level factorial. All of the design points are either on a sphere or at the center of a sphere. This design is advantageous when the points on one or more corners of the cube represent factor–level combinations that are prohibitively expensive or impossible to test due to physical constraints on the experimentation.

A careful examination of the three-level face-centered central composite design (Figure 11.11) and the three-level Box–Behnken design (Figure 11.12) reveals an interesting geometric pattern. When the two designs are overlaid, one obtains a complete 3^3 factorial as shown in Figure 11.13. Thus, these alternative designs are simply fractional parts of

Table 11.5 Box–Behnken Design for Thyroxine Study*

	Coded Factor Levels		
Run No.	Enzyme	Conjugate	Sample Volume
1	−1	−1	0
2	−1	+1	0
3	+1	−1	0
4	+1	+1	0
5	−1	0	−1
6	−1	0	+1
7	+1	0	−1
8	+1	0	+1
9	0	−1	−1
10	0	−1	+1
11	0	+1	−1
12	0	+1	+1
13	0	0	0
14	0	0	0
15	0	0	0

*Nonrandomized.

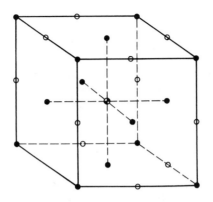

Figure 11.13 3^3 factorial layout: combination of Box–Behnken (O) and face-centered central composite (●) designs.

the three-level factorial. As such, both provide good results for a wide range of practical problems.

11.4 MIXTURE DESIGNS

The use of a mixture design is necessary when the response of interest is a function of the relative proportions of the factors (components) $x_1, x_2, \ldots, x_p$ that appear in the mixture or formulation. This situation differs from the response-function setting of the previous section, in which the response is a function of the actual numerical values of the factors. In a mixture experiment the factor space of allowable component values is constrained (see Exhibit 11.7). A particular combination of the components in a formulation is often referred to as a *recipe*.

EXHIBIT 11.7   CONSTRAINED FACTOR SPACE

Let $x_1, x_2, \ldots, x_p$ denote the proportions of p components that are to be used in a mixture design. The factor space of permissible component values is

$$0 \leqslant x_i \leqslant 1, \qquad i = 1, 2, \ldots, p, \tag{11.2a}$$

and

$$x_1 + x_2 + \cdots x_p = 1 \qquad \text{(i.e., 100\%)}. \tag{11.2b}$$

The constraints in (11.2) are not always realistic, because one or more of the components might not be allowed to be zero. As an example, suppose a civil engineer wishes to study the properties of concrete. The key response of interest might be the hardness, which depends on the relative amounts of cement, water, and aggregate in the formulation. This is a three-component mixture problem with x_1 = cement, x_2 = water, and x_3 = aggregate. The lower bounds of zero in the constraints (11.2) must be altered in order to produce concrete with acceptable properties. A formulation of 0% cement, 100% water, and 0% aggregate is not acceptable. The general form of a lower-bound constraint is

$$l_i \leqslant x_i \leqslant 1, \qquad i = 1, 2, \ldots, p, \tag{11.3}$$

where l_i is the lower limit for the ith mixture component.

The constraints mentioned thus far are restrictions on the factor space where experimentation is possible. This type of restriction is illustrated graphically in Figures 11.14 and 11.15. Figure 11.14a depicts the constrained factor space of possible factor–level

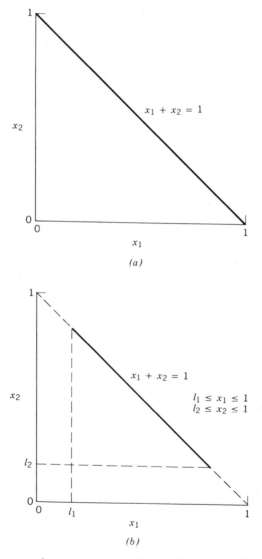

Figure 11.14 Factor space for two-component mixtures: (a) constrained factor space; (b) lower-bound constraints.

combinations for two mixture components. The factor space includes all values of the two components that lie on the line segment $x_1 + x_2 = 1.0$, with each component bounded by zero and one. Lower-bound constraints such as those indicated in the inequalities (11.3) restrict experimentation to a portion of the line segment between $x_1 = l_1$ and $x_2 = l_2$, as indicated in Figure 11.14b.

When there are three mixture components, Figure 11.15 shows two views of the

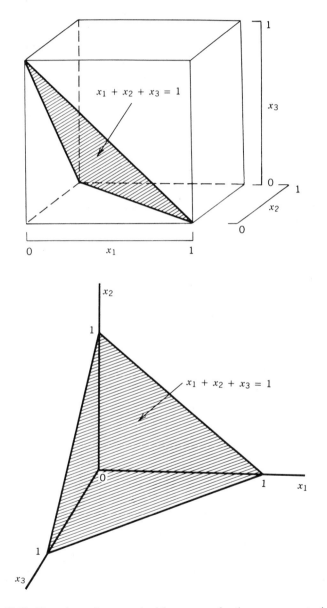

Figure 11.15 Two views of a constrained factor space for three-component mixtures.

constrained factor space. The space is a triangle with vertices corresponding to formulations consisting of *pure* mixtures, i.e., mixtures that are 100% of a single component. When lower-bound constraints (11.3) are part of the experimental environment, experimentation is confined to a possibly irregular-shaped portion of the constrained factor space.

When there are three mixture components, the constrained experimental region can be conveniently represented on trilinear coordinate paper. Each of the three sides of the graph paper in Figure 11.16 represents a mixture that has none of one of the components, the component labeled on the opposite vertex of the graph. The vertices again correspond to pure mixtures. The nine grid lines in each direction represent 10% increments in the respective components. Constraints of the form (11.3) are indicated on the trilinear graph paper by lines drawn parallel to the respective base lines at a height determined by the l_i.

Figure 11.17 shows a constrained simplex for the concrete example. The shaded area in Figure 11.17 denotes formulations that satisfy the following lower-bound constraints:

$$\text{cement,} \quad x_1 > 0.50 \quad (50\%),$$
$$\text{water,} \quad x_2 > 0.20 \quad (20\%),$$
$$\text{aggregate,} \quad x_3 > 0.10 \quad (10\%).$$

Simplex designs are used to study the effects of mixture components on a response. A commonly used class of simplex designs is given in Table 11A.2 for five or fewer components and in Table 11A.3 for six or more. The designs given for six or more components are screening designs because of the large number of components. Based on an analysis of the results using this type of screening design, a smaller number of components identified as important may be further studied using the experimental design

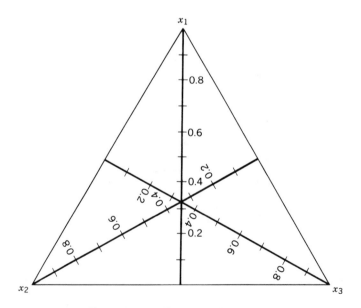

Figure 11.16 Trilinear coordinate system.

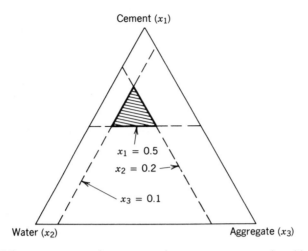

Figure 11.17 Mixture-component factor space for a concrete example with lower-bound constraints.

shown in Table 11A.2. Figure 11.18 displays the simplex design from Table 11A.2 for $p = 3$ components.

To illustrate the selection of a simplex mixture design, consider an experiment that is to be conducted to evaluate certain properties of a liquid feed additive. The feed additive is a formulation of a monomer (a chemical compound that can undergo polymerization), an oligomèr (a polymer containing a few structural units), and water. In order to study the effects of the $p = 3$ mixture components on the feed additive, a mixture experiment was designed using the simplex design formulations shown in Table 11.6. Note that each formulation is repeated, resulting in a total of twenty experimental test runs, with the randomized run order indicated in the table.

In order for the animal-feed additive to be efficacious, a minimum amount of each of the mixture components must be present. These minimum amounts are

$$\text{monomer} > 0.30 \quad (30\%),$$

$$\text{oligomer} > 0.30 \quad (30\%),$$

$$\text{water} > 0.10 \quad (10\%).$$

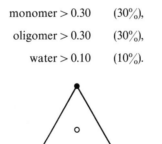

Figure 11.18 Simplex design for three components. (Optional points = ○.)

Table 11.6 Simplex Design Formulations for Animal-Feed Additive Example

Component			
Monomer	Oligomer	Water	Run Order
1	0	0	14, 2
0	1	0	3, 9
0	0	1	19, 5
$\frac{1}{2}$	$\frac{1}{2}$	0	18, 10
$\frac{1}{2}$	0	$\frac{1}{2}$	12, 6
0	$\frac{1}{2}$	$\frac{1}{2}$	4, 1
$\frac{1}{3}$	$\frac{1}{3}$	$\frac{1}{3}$	13, 7
$\frac{2}{3}$	$\frac{1}{6}$	$\frac{1}{6}$	16, 11
$\frac{1}{6}$	$\frac{2}{3}$	$\frac{1}{6}$	15, 8
$\frac{1}{6}$	$\frac{1}{6}$	$\frac{2}{3}$	17, 20

The design laid out in Table 11.6 cannot be run as stipulated, since it includes recipes (factor–level combinations) that violate the lower-bound constraints.

In order to circumvent this problem and not have to construct new designs for each constrained mixture problem, one can use *pseudocomponents* rather than the original mixture components. A pseudocomponent is a reexpression of the mixture components (see Exhibit 11.8).

EXHIBIT 11.8   MIXTURE PSEUDOCOMPONENTS

Let $x_1, x_2, \ldots, x_p$ denote the original p mixture components, at least one of which is subject to a lower-bound constraint (11.3) in addition to the usual mixture constraints (11.2). Define the jth pseudocomponent as

$$s_j = \frac{x_j - l_j}{(1 - \sum_{i=1}^{p} l_i)}, \qquad \sum_{i=1}^{p} l_i < 1. \tag{11.4}$$

The permissible values of the pseudocomponents s_j form a constrained factor space as in (11.2).

The use of pseudocomponents allows the simplex design to be applicable when lower-bound constraints are a part of the experimental environment. The formulations specified by the simplex design for the pseudocomponents are transformed to formulations for the actual components by reversing the transformation. If s_j is the value assigned to the jth pseudocomponent on one of the test runs in the simplex design, the value for the original jth mixture component is

$$x_j = l_j + \left(1 - \sum_{i=1}^{p} l_i\right) s_j. \tag{11.5}$$

Table 11.7 shows the simplex design, in run order, for the animal-feed additive example. Both the pseudocomponents and the original components are displayed. The calculation

Table 11.7 Simplex Design Layout for Animal-Feed Additive Example

Run No.	Pseudocomponents			Original Components		
	Monomer	Oligomer	Water	Monomer	Oligomer	Water
1	0	$\frac{1}{2}$	$\frac{1}{2}$	.30	.45	.25
2	1	0	0	.60	.30	.10
3	0	1	0	.30	.60	.10
4	0	$\frac{1}{2}$	$\frac{1}{2}$	.30	.45	.25
5	0	0	1	.30	.30	.40
6	$\frac{1}{2}$	0	$\frac{1}{2}$	.45	.30	.25
7	$\frac{1}{3}$	$\frac{1}{3}$	$\frac{1}{3}$	.40	.40	.20
8	$\frac{1}{6}$	$\frac{2}{3}$	$\frac{1}{6}$	.35	.50	.15
9	0	1	0	.30	.60	.10
10	$\frac{1}{2}$	$\frac{1}{2}$	0	.45	.45	.10
11	$\frac{2}{3}$	$\frac{1}{6}$	$\frac{1}{6}$	.50	.35	.15
12	$\frac{1}{2}$	0	$\frac{1}{2}$	.45	.30	.25
13	$\frac{1}{3}$	$\frac{1}{3}$	$\frac{1}{3}$	.40	.40	.20
14	1	0	0	.60	.30	.10
15	$\frac{1}{6}$	$\frac{2}{3}$	$\frac{1}{6}$	.35	.50	.15
16	$\frac{2}{3}$	$\frac{1}{6}$	$\frac{2}{3}$	.50	.35	.15
17	$\frac{1}{6}$	$\frac{1}{6}$	$\frac{1}{6}$	.35	.35	.30
18	$\frac{1}{2}$	$\frac{1}{2}$	0	.45	.45	.10
19	0	0	1	.30	.30	.40
20	$\frac{1}{6}$	$\frac{1}{6}$	$\frac{2}{3}$	.35	.35	.30

of the original mixture-component value for oligomer from run No. 1 is as follows:

$$\text{oligomer} = 0.30 + (1 - 0.70)(\tfrac{1}{2}) = 0.45.$$

The above procedures for constructing simplex designs can be altered for use when upper-bound constraints occur on one or more of the mixture components. One simply replaces $x_j - l_j$ and $1 - \sum l_j$ in equation (11.4) with $u_j - x_j$ and $\sum u_j - 1$, respectively, where the u_j are the upper-bound constraints on the mixture components subject to the restriction $\sum u_j - \min(u_j) < 1$. These techniques are not appropriate if the components have both lower and upper constraints.

APPENDIX: BOX–BEHNKEN AND SIMPLEX DESIGNS

Table 11A.1   Box–Behnken Designs

No. of Factors	Coded Factor Levels									No. of Points
	1	2	3	4	5	6	7	8	9	
3	±1	±1	0							4
	±1	0	±1							4
	0	±1	±1							4
	0	0	0							3
										15
4	±1	±1	0	0						4
	±1	0	±1	0						4
	±1	0	0	±1						4
	0	±1	±1	0						4
	0	±1	0	±1						4
	0	0	±1	±1						4
	0	0	0	0						3
										27
5	±1	±1	0	0	0					4
	±1	0	±1	0	0					4
	±1	0	0	±1	0					4
	±1	0	0	0	±1					4
	0	±1	±1	0	0					4
	0	±1	0	±1	0					4
	0	±1	0	0	±1					4
	0	0	±1	±1	0					4
	0	0	±1	0	±1					4
	0	0	0	±1	±1					4
	0	0	0	0	0					6
										46
6	±1	±1	0	±1	0	0				8
	0	±1	±1	0	±1	0				8
	0	0	±1	±1	0	±1				8
	±1	0	0	±1	±1	0				8
	0	±1	0	0	±1	±1				8
	±1	0	±1	0	0	±1				8
	0	0	0	0	0	0				6
										54
7	0	0	0	±1	±1	±1	0			8
	±1	0	0	0	0	±1	±1			8
	0	±1	0	0	±1	0	±1			8
	±1	±1	0	±1	0	0	0			8
	0	0	±1	±1	0	0	±1			8
	±1	0	±1	0	±1	0	0			8
	0	±1	±1	0	0	±1	0			8
	0	0	0	0	0	0	0			6
										62

No. of Factors	Coded Factor Levels									No. of Points
	1	2	3	4	5	6	7	8	9	
9	±1	0	0	±1	0	0	±1	0	0	8
	0	±1	0	0	±1	0	0	±1	0	8
	0	0	±1	0	0	±1	0	0	±1	8
	±1	±1	±1	0	0	0	0	0	0	8
	0	0	0	±1	±1	±1	0	0	0	8
	0	0	0	0	0	0	±1	±1	±1	8
	±1	0	0	0	±1	0	0	0	±1	8
	0	0	±1	±1	0	0	0	±1	0	8
	0	±1	0	0	0	±1	±1	0	0	8
	±1	0	0	0	0	±1	0	±1	0	8
	0	±1	0	±1	0	0	0	0	±1	8
	0	0	±1	0	±1	0	±1	0	0	8
	±1	0	0	±1	0	0	±1	0	0	8
	0	±1	0	0	±1	0	0	±1	0	8
	0	0	±1	0	0	±1	0	0	±1	8
	0	0	0	0	0	0	0	0	0	10
										130

Table 11A.2 Simplex Designs for Five or Fewer Mixture Components

Location	No.	Component				
		x_1	x_2	x_3	$\cdots$	x_p
Vertices	1	1	0	0	$\cdots$	0
	2	0	1	0	$\cdots$	0
	3	0	0	1	$\cdots$	0
	$\vdots$	$\vdots$	$\vdots$	$\vdots$		
	p	0	0	0	$\cdots$	1
Edges	1	$\frac{1}{2}$	$\frac{1}{2}$	0	$\cdots$	0
	2	$\frac{1}{2}$	0	$\frac{1}{2}$	$\cdots$	0
	3	$\frac{1}{2}$	0	0	$\cdots$	0
	$\vdots$	$\vdots$	$\vdots$	$\vdots$		$\vdots$
	$p(p-1)/2$	0	0	0	$\cdots$	$\frac{1}{2}$
Centroid	1	$1/p$	$1/p$	$1/p$	$\cdots$	$1/p$
Optional*	1	$(p+1)/2p$	$1/2p$	$1/2p$	$\cdots$	$1/2p$
	2	$1/2p$	$(p+1)/2p$	$1/2p$	$\cdots$	$1/2p$
	3	$1/2p$	$1/2p$	$(p+1)/2p$	$\cdots$	$1/2p$
	$\vdots$	$\vdots$	$\vdots$	$\vdots$		$\vdots$
	p	$1/2p$	$1/2p$	$1/2p$	$\cdots$	$(p+1)/2p$

*Optional formulations are interior points in the experimental region; their inclusion is recommended.

Table 11A.3 Simplex Designs for Six or More Mixture Components

Location	No.	Component x_1	x_2	x_3	$\cdots$	x_p
Vertices	1	1	0	0	$\cdots$	0
	2	0	1	0	$\cdots$	0
	3	0	0	1	$\cdots$	0
	$\vdots$	$\vdots$	$\vdots$	$\vdots$		
	p	0	0	0	$\cdots$	1
Interior	1	$(p+1)/2p$	$1/2p$	$1/2p$	$\cdots$	$1/2p$
	2	$1/2p$	$(p+1)/2p$	$1/2p$	$\cdots$	$1/2p$
	3	$1/2p$	$1/2p$	$(p+1)/2p$	$\cdots$	$1/2p$
	$\vdots$	$\vdots$	$\vdots$	$\vdots$		$\vdots$
	p	$1/2p$	$1/2p$	$1/2p$	$\cdots$	$(p+1)/2p$
Centroid	1	$1/p$	$1/p$	$1/p$	$\cdots$	$1/p$
Optional*	1	0	$1/(p-1)$	$1/(p-1)$	$\cdots$	$1/(p-1)$
	2	$1/(p-1)$	0	$1/(p-1)$	$\cdots$	$1/(p-1)$
	3	$1/(p-1)$	$1/(p-1)$	0	$\cdots$	$1/(p-1)$
	$\vdots$	$\vdots$	$\vdots$	$\vdots$		$\vdots$
	p	$1/(p-1)$	$1/(p-1)$	$1/(p-1)$	$\cdots$	0

*Optional formulations are used to check for the effect on the response of complete elimination of one component.

REFERENCES

Text References

Extensive coverage of response surface designs appears in the following texts:

Box, G. E. P., Hunter, W. G., and Hunter, J. S. (1978). *Statistics for Experimenters*, New York: John Wiley and Sons, Inc.

Davis, O. L. (Ed.) (1971). *The Design and Analysis of Industrial Experiments*, New York: Macmillan Co.

Myers, R. H. (1971). *Response Surface Methodology*, Boston: Allyn and Bacon, Inc.

The following journal articles contain additional information on a variety of designs and models that are useful for fitting response surfaces. Some of these articles are highly technical.

Box, G. E. P. (1954). "The Exploration and Exploitation of Response Surfaces: Some General Considerations and Examples," *Biometrics*, **10**, 16–60.

Box, G. E. P. (1959). "A Basis for the Selection of a Response Surface Design," *Journal of the American Statistical Association*, **54**, 622–654.

Box, G. E. P. and Draper, N. R. (1963). "The Choice of a Second Order Rotatable Design," *Biometrika*, **50**, 335–352.

Box, G. E. P. and Youle, P. V. (1955). "The Exploration and Exploitation of Response Surfaces: An Example of the Link between the Fitted Surface and the Basic Mechanism of the System," *Biometrics*, **11**, 287–323.

Hunter, J. S. (1958–1959). "Determination of Optimum Operating Conditions by Experimental Methods," *Industrial Quality Control*, **15**, Part II-1 (Dec. 1958), Part II-2 (Jan. 1959), **16**, Part II-3 (Feb. 1959).

An extensive list of Box–Behnken designs for 3–12 factors is contained in:

Box, G. E. P. and Behnken, D. W. (1960). "Some New Three Level Designs for the Study of Quantitative Variables," *Technometrics*, **2**, 455–476.

An important reference on mixture designs is:

Cornell, J. A. (1981). *Experiments with Mixtures: Designs, Models, and the Analysis of Mixture Data*, New York: John Wiley & Sons, Inc. *The designs in Table 11A.2 are adaptations of the simplex-centroid designs discussed in Section 2.11 of this text.*

The designs in Table 11A.3 are the simplex-screening designs discussed in the following article:

Snee, R. D. and Marquardt, D. W. (1976). "Screening Concepts and Designs for Experiments with Mixtures," *Technometrics*, **18**, 19–30.

Data References

Designs for the gentamicin method study and the thyroxin experiment were taken from the following article:

Myers, G. C., Jr. (1985). "Use of Response Surface Methodology in Clinical Chemistry," in Snee, R. D., Hare, L. B., and Trout, J. R. (Eds.), *Experiments in Industry: Design, Analysis and Interpretation of Results*, Milwaukee, WI: American Society for Quality Control. Copyright American Society for Quality Control, Inc., Milwaukee, WI. Reprinted by permission.

The design for the Sulphlex experiment was taken from:

Dale, J. M. (1984). "Design for Sulphlex® Binders," Federal Highway Administration, Contract No. DTFH-61-82-C-00049. San Antonio, TX: Southwest Research Institute.

EXERCISES

1 Sketch three-dimensional surfaces for the contours graphed in Figure 11.3.

2 A fitted second-order model for a response surface has the following equation:

$$y = 300.0 + 70.0x_1 + 70.5x_2 + 3.0x_1x_2 - 10.5x_1^2 - 10.0x_2^2.$$

Sketch a contour plot of this three-dimensional surface by calculating the responses for a grid of values of the two factors and interpolating between adjacent responses. The ranges of interest on each of the two factors are 0 to 10. Plot contours for response values of 100 to 500 in increments of 100.

3 Use the Sulphlex-study contour plot in Figure 11.4 to plan a sequence of test runs along the path of steepest ascent to find the maximum viscosity (see also Figure 11.8). Specify ten factor–level combinations.

4 A ceramic diesel engine is being tested in buses to measure the amount of combustion soot being deposited in the engines. It is known that engine temperature plays an important role in the depositing of soot in engines. A study is being conducted to determine the maximum temperature (°F) in the engine during a combustion test. Two factors are of interest in one phase of the study: the time (sec) during which combustion takes place, and the engine operating speed (rpm). Initially, a 2^2 factorial experiment was conducted with the results shown below. Design a sequence of tests using the path of steepest ascent. List at least five test runs.

Time (sec)	Speed (rpm)	Temperature (°F)
10	1000	350
100	1000	625
10	1600	200
100	1600	400

5 A disk-type test rig was designed and fabricated to measure the wear of graphite material under specified test conditions. Two central composite designs were constructed using the following ranges on the two factors

    temperature: 500–1000°F,
    contact pressure: 15–21 psi.

The designs below are in coded factor levels. Replace the coded levels with the actual factor levels. Is either design rotatable? Why (not)? If not, construct a rotatable central composite design for this project.

	Design 1		Design 2	
Run No.	Temperature	Pressure	Temperature	Pressure
1	− 1	− 1	− 1	− 1
2	− 1	1	− 1	1
3	1	− 1	1	− 1
4	1	1	1	1
5	0	− 1	0	− 1.5
6	0	1	0	1 5
7	− 1	0	− 1.5	0
8	1	0	1.5	0
9	0	0	0	0
10	0	0	0	0
11	0	0	0	0

6 Construct a Box–Behnken design for the previous exercise. Under which of the following conditions would the various central composite designs or the Box–Behnken design be preferable? Why?

(a) The limits in Exercise 5 are not constraints, only suggested starting ranges.

(b) No design points can exceed the ranges specified in Exercise 5, but test runs are desired as close to the limits as possible.

(c) No design points can exceed the ranges specified in Exercise 5, and test runs are not desired in the extreme corners of the region.

7 Suppose that in addition to temperature and contact pressure, sliding speed and disk hardness are important factors in the experiment described in Exercise 5. These latter

two factors have the following ranges:

$$\begin{array}{ll} \text{Sliding speed} & 54\text{--}60\,\text{ft/sec} \\ \text{Disk hardness} & 58\text{--}60\,R_c \end{array}$$

Construct a rotatable central composite design for this experiment. Include six repeat tests. List the factor–level combinations in both coded and noncoded form.

8 Construct a Box–Behnken design for the experiment in Exercise 7.

9 A diesel fuel is being blended to meet a required cetane number. Three blending components are used to make the diesel fuels: light cycle oil, straight-run diesel, and hydrocracked distillate. It is known from previous experimentation that the following upper-bound constraints are needed for the three components:

$$\begin{array}{ll} \text{Light cycle oil} & < 0.45 \\ \text{Straight-run diesel} & < 0.30 \\ \text{Hydrocracked distillate} & < 0.50 \end{array}$$

Sketch the mixture factor space for this experiment, including the constraints using trilinear coordinate axes. Construct a simplex design for this experiment. Plot the design points on the trilinear graph.

10 A chef is experimenting with four ingredients for a new salad dressing: vinegar, water, oil, special seasonings. From his own tastes, the chef imposes the following constraints on the ingredients:

$$\begin{array}{ll} \text{Vinegar} & > 0.25 \\ \text{Water} & > 0.15 \\ \text{Oil} & > 0.35 \\ \text{Special seasonings} & > 0.10 \end{array}$$

Design a simplex experiment to investigate the components of this salad dressing. List the design points both as pseudocomponents and as original components.

11 A soft-drink manufacturer is planning to market a new beverage consisting of a mixture of several fruit juices. Current production facilities can use a maximum of four fruit juices. An experiment is planned to screen eight potential juices: orange, pineapple, apple, apricot, grapefruit, lemon, lime, and papaya. Design a simplex screening design for this experiment, and provide a specific sequence of test runs.

PART III

Analysis of Designed Experiments

CHAPTER 12

Statistical Principles in Data Analysis

The material covered in the first two parts of this text has concentrated on an exposition of statistical modes of experimentation and on graphical and quantitative summarization of data. The main focus in the next two parts of the text is on using the data collected from an experiment to draw inferences on the phenomena or processes under investigation. Specifically, this chapter:

- *discusses sampling distributions and their role in statistical inference,*
- *introduces the central limit property and shows its relevance to the statistical analysis of data,*
- *develops the fundamental concepts relating to interval estimation and hypothesis testing for drawing statistical inferences, and*
- *illustrates techniques that are useful for the computation of sample sizes in certain types of experiments.*

Most statistically designed experiments are conducted with the intent of better understanding populations, observable phenomena, or stable processes. In some instances, sample statistics are computed in order to draw conclusions about the corresponding population parameters. In others, factor effects are calculated in order to compare the influences of different factor levels. In both of these examples and in many other applications of statistical methodology, statistical models are postulated and inferences are drawn relative to the specification of the model. Any such inference or conclusion involves some degree of uncertainty because of the uncontrolled experimental variability of the observed responses.

The data displayed in Table 12.1 are tensile-strength measurements on wire that is used in industrial equipment. The wire is manufactured by drawing the base metal through a die that has a specified diameter. The data in Table 12.1 are tensile strengths of eighteen samples taken from large spools of wire made with each of three dies. It is of interest to determine whether the use of different dies will affect the tensile strength of the wire.

Table 12.2 contains a variety of summary information (see Chapter 3) compiled from each of the sets of tensile-strength measurements. The comparative stem-and-leaf plot

Table 12.1   Tensile Strengths of Wire from Three Different Dies

Tensile Strength (10^3 psi)					
Die 1		Die 2		Die 3	
85.769	86.725	79.424	82.912	82.423	81.768
86.725	84.292	81.628	83.185	81.941	83.078
87.168	84.513	82.692	86.725	81.331	83.515
84.513	86.725	86.946	84.070	82.205	82.423
84.513	84.513	86.725	86.460	81.986	83.078
83.628	82.692	85.619	83.628	82.860	82.641
82.912	83.407	84.070	84.513	82.641	82.860
84.734	84.070	85.398	84.292	82.592	82.592
82.964	85.337	86.946	85.619	83.026	81.507

Table 12.2   Summary Information on Tensile-Strength Data

(a) *Stem-and-Leaf Plots*

Tensile Strength	Data		
	Die 1	Die 2	Die 3
79		4	
80			
81		6	9385
82	97	79	4209664696
83	604	26	0151
84	5735155	1153	
85	83	646	
86	777	97975	
87	2		

(b) *Summary Statistics*

Statistic	Value		
	Die 1	Die 2	Die 3
Average	84.733	84.492	82.470
Median	84.513	84.403	82.592
m-estimate	84.692	84.648	82.488
Standard deviation	1.408	2.054	0.586
MAD estimate	1.476	2.007	0.609
n	18	18	18

suggests that the measurements from wire produced by die 3 are less variable then those for the other two dies. The traditional and robust location statistics for each individual die are quite similar, and there does not appear to be any need for concern about influential effects due to outliers. The tensile strengths for the wires from die 3 average about 2000 psi lower than those of the other dies. The stem-and-leaf plot suggests that this difference in average tensile strengths is due to generally smaller tensile strengths for wire from die 3. Several interesting questions emerge from the information provided in Table 12.2. One question is whether this information is sufficient to conclude that the third die produces wire with a smaller mean tensile strength than the other two dies.

This question is difficult to answer if one only examines the information shown in Table 12.2. While the average tensile strength for the wires from die 3 is lower than those for the other two dies, the wires from the other two dies exhibit greater variability, as measured by the sample standard deviations and the MAD estimates, than those from die 3. In addition, the wire sample with the lowest tensile strength measurement, 79.424, is made from die 2 not die 3. Under some circumstances comparisons among dies are fairly straightforward; for instance, when all the measurements for one die are lower than the minimum tensile strengths for the other dies. When measurements overlap, as they do in the stem-and-leaf plots in Table 12.2, decisions on the relative performance of the dies are more difficult. It is in these circumstances that statistical analyses that incorporate measures of uncontrolled variability are needed.

Most of the statistical procedures used in this book to analyze data from designed experiments are based on modeling the response variable. The statistical models used include various parameters (constants) that, depending on the experimental situation, characterize either location or the influence of factor levels. These models also include variables that represent either random assignable causes or uncontrolled experimental variability. It is assumed that these variables follow certain probability distributions, most frequently a normal probability distribution.

In Section 12.1, probability distributions and the calculation of probabilities are discussed. Also discussed are sampling distributions for sample means and variances. Sections 12.2 and 12.3 are devoted to an exposition of interval estimation in which measures of variability are explicitly included in estimation procedures. Section 12.4 details the use of statistical hypothesis-testing principles. The determination of sample-size requirements for certain types of experimental settings is presented in Section 12.5.

12.1 PROBABILITY CONCEPTS

A probability is a number between zero and one that expresses how likely an event, an action, or a response is to occur. A probability close to one indicates that an event is very likely to occur, whereas a probability close to zero indicates that an event is very unlikely to occur. Probabilities sometimes are expressed as percentages, in which case we shall refer to them as chances; i.e., chance = probability $\times$ 100%.

Probabilities are calculated in a variety of ways. One of the simplest methods of calculating probabilities occurs when there are N equally likely outcomes (responses) that could occur. The probability of observing any specified outcome is then $1/N$. If m of these outcomes constitute an event of interest, the probability p of observing the event of interest is then $p = m/N$. Games of chance (e.g., drawing cards from a well-shuffled deck) often satisfy the requirements for calculating probabilities this way.

Probabilities can be calculated for any population or process if a probability distribution can be assumed for the response variables. Probability distributions specify the possible values of a response variable and either the probability of observing each of the response values or a density function for the values.

Probabilities for many probability distributions are obtainable, as in the above illustration, from formulas that relate an individual response to its probability. These distributions are for discrete response variables, those whose permissible values form a finite or countable infinite set. The probability distribution for N equally likely outcomes is expressed as a listing of each of the outcomes and a tabular or graphical display of the probabilities, $1/N$, for each. The binomial and the Poisson distributions discussed in Chapter 20 are additional examples of discrete probability distributions for which formulas are available to calculate probabilities for individual responses.

A *density* is a function that defines the likelihood of obtaining various response values for *continuous* response variables (those whose permissible values are any real number in a finite or infinite interval). Probabilities for continuous response variables are obtained as areas under the curve defined by the density function.

The normal probability distribution was introduced in Section 5.3 as an appropriate reference distribution for many types of response variables. The form of the normal density function is

$$f(y) = (2\pi\sigma^2)^{-1/2} \exp\left(-\frac{(y-\mu)^2}{2\sigma^2} \right) \tag{12.1}$$

for

$$-\infty < y < \infty, \qquad -\infty < \mu < \infty, \qquad \sigma > 0.$$

In the density function (12.1), y is a response variable, μ is the population mean, and σ is the population standard deviation.

The degree to which the probability distribution of the responses assumed by a model (the reference distribution) must agree with the actual distribution of the responses in a population or from a process depends on the statistical methodology used. Some statistical procedures are very robust to specific differences between the assumed and the actual distribution of the responses. Others are very sensitive to such differences. It is incumbent on the researcher to identify any assumptions needed to implement statistical procedures and to assess whether the needed assumptions are reasonable for the data being analyzed. The assessment of model assumptions is dealt with in detail in Chapters 24 to 26 for linear regression models. These techniques can be applied to experimental-design models if effects representations or orthogonal polynomials are used to represent the factors (see Section 22.1).

The normal probability distributions is a possible reference distribution for the tensile-strength measurements of the wires from any one of the dies. In this setting, μ is the mean tensile strength for all wire produced by a particular die. The standard deviation σ controls the spread in the distribution of the responses (see Figure 5.5). This spread represents the cumulative effects of a variety of sources contributing to the uncontrolled variability of the measurements.

In order to determine whether a normal probability distribution for the tensile-strength measurements is a reasonable assumption, measurements on 35 samples of wire from a die similar to those in this experiment were plotted versus a standard normal ($\mu = 0$, $\sigma = 1$) reference distribution in a quantile–quantile plot (see Section 5.3). The straight-line trend in the plot, Figure 12.1, indicates that the normal probability distribution is a reasonable assumption for these tensile-strength measurements. It is now of interest to consider how one would compute probabilities for a normal distribution.

Probabilities for continuous response variables are obtained as areas under the plotted density function. In some cases these areas can be obtained analytically through integration of the density function. In other cases, the integral has no closed-form solution, but numerical solutions for the intergrals can be obtained using appropriate computer software. Many of the density functions used in this text have been conveniently tabulated

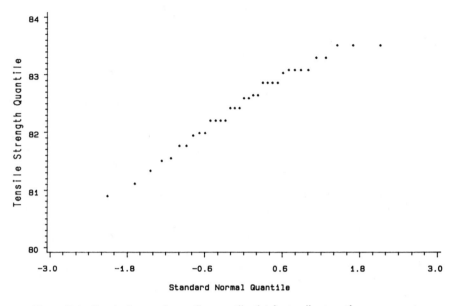

Figure 12.1 Standard normal quantile–quantile plot for tensile-strength measurements.

so that no integration is needed. Such is the case for the standard normal distribution (see Table A4 in the appendix). Use of probability tables for the standard normal probability distribution is explained in the appendix to this chapter. Probabilities for any normal distribution can be calculated from those of a standard normal distribution, if the mean μ and standard deviation σ are known, by transforming the original normal variable to a standard normal variable (see Exhibit 12.1).

EXHIBIT 12.1   STANDARD NORMAL VARIABLES

If y is a response variable following a normal distribution with population or process mean μ and standard deviation σ, the variate

$$z = \frac{y - \mu}{\sigma} \tag{12.2}$$

follows a standard normal distribution; i.e., z is normally distributed with mean 0 and standard deviation 1.

In Chapter 3 the distinctions between populations and samples and between parameters and statistics are stressed. Experiments are performed to collect data (samples) from which to draw inferences on populations or processes. Frequently these inferences are conclusions drawn about parameters of models or probability distributions, conclusions that are based on the calculation of statistics from the individual responses. Many of the statistics used and the resulting inferences require an assumption that the individual responses are independent (see Exhibit 12.2).

EXHIBIT 12.2

Statistical Independence. Observations are statistically independent if the value of one of the observations does not influence the value of any other observation. Simple random sampling produces independent observations.

Sample statistics follow sampling distributions (see Section 2.2) just as individual response variables follow probability distributions. This is because there are many samples (perhaps an infinite number) of a fixed size n that could be drawn from a population or a process. Each sample that could be drawn results in a value for a sample statistic. The collection of these possible values forms the sampling distribution.

It is important to the proper understanding of a sampling distribution to distinguish between a statistic and a realization of a statistic. Prior to collecting data, a statistic has a sampling distribution from which a realization will be taken. The sampling distribution is determined by how the data are to be collected, i.e., the experimental design or the sampling procedure. The realization is simply the observed value of the statistic once it is calculated from the sample of n responses. This distinction is often stressed in the terminology used in the estimation of parameters. An *estimator* is the statistic used to estimate the parameter, and an *estimate* is the realized or calculated value from the sample responses.

If individual response variables are statistically independent and follow a normal probability distribution, the population of all possible sample means of a fixed sample size also follows a normal probability distribution (see Exhibit 12.3).

EXHIBIT 12.3   SAMPLING DISTRIBUTION OF SAMPLE MEANS

If independent observations

$$y_1, y_2, y_3, \ldots$$

follow a normal probability distribution with mean μ and standard deviation σ, then the distribution of all possible sample means $\bar{y}_1, \bar{y}_2, \ldots$ of size n from this population is also normal with population mean μ but with a standard error (standard deviation) of $\sigma/n^{1/2}$.

The term *standard error* (see Exhibit 12.4) is used to distinguish the standard deviation of a sample statistic from that of an individual observation, just as the term *sampling distribution* is used to distinguish the distribution of the statistic from that of the individual observation. The standard error is the most frequently used measure of the precision of a parameter estimator. The standard error measures the precision of an estimator as a function of both the standard deviation of the original population of measurements and the sample size. For example, the standard error of a sample mean of n independent observation is

$$\sigma_{\bar{y}} = \sigma/n^{1/2}. \tag{12.3}$$

Note that this standard error can be small if either the population standard deviation σ is small or the sample size n is large. Thus, both the standard deviation of the measurements and the sample size contribute to the standard error, and hence to the precision, of the sample mean.

EXHIBIT 12.4

Standard Error. The standard error of a statistic is the standard deviation of its sampling distribution. The standard error is usually obtained by taking the positive square root of the variance of the statistic.

The distribution of sample means of independent observations from a normal distribution can be standardized as in (12.2):

$$z = \frac{\bar{y} - \mu}{\sigma/n^{1/2}} = \frac{n^{1/2}(\bar{y} - \mu)}{\sigma} \tag{12.4}$$

follows a standard normal distribution. If σ is known in (12.4), inferences on an unknown mean μ can be made using the standard normal distribution and the inference procedures described in Sections 12.2 and 12.4 (see also Chapter 13).

When the population standard deviation is unknown, especially if the sample size is not large, the standard normal variate (12.4) cannot be used to draw inferences on the population mean. On replacing the standard deviation σ in (12.4) with the sample standard deviation s (Section 3.1), the sampling distribution of the resulting variate is not a standard normal distribution. Rather, the sampling distribution of the variate

$$t = \frac{n^{1/2}(\bar{y} - \mu)}{s} \tag{12.5}$$

is a *Student's t-distribution*. Probabilities for Student's t-distribution depend on the sample size n through its *degrees of freedom*. If n independent observations are taken from a normal distribution with mean μ and standard deviation σ, the degrees of freedom are $n - 1$. Probability calculations for Student's t-distribution are described in the appendix to this chapter. Probability tables for this distribution are given in Table A5 of the appendix to this book.

Frequently interest in an industrial process centers on measuring variation rather than on estimates of a population mean. Inferences on population or process standard deviations for independent observations from normal probability distributions can be made using the variate

$$X^2 = (n - 1)s^2/\sigma^2. \tag{12.6}$$

The sampling distribution of the variate (12.6) is a *chi-square* distribution. This distribution, like Student's t, depends on the sample size through the degrees of freedom. The form of the statistic (12.6) is appropriate for inferences on σ based on independent observations from a normal distribution, in which case the degrees of freedom again are $n - 1$. Probability calculations for this distribution are explained in the appendix to this chapter and probability tables are given in Table A6 in the appendix to this book.

Another frequently used sampling distribution is needed when one wishes to compare the variation in two populations or processes. The ratio of two sample variances (s^2), each divided by its population variance (σ^2), is termed an *F-variate*:

$$F = \frac{s_1^2/\sigma_1^2}{s_2^2/\sigma_2^2} = \frac{s_1^2\sigma_2^2}{s_2^2\sigma_1^2}. \tag{12.7}$$

Note that this form of the F-variate is the ratio of two independent chi-square variates, $(n_i - 1)s_i^2/\sigma_i^2$, each divided by its number of degrees of freedom, $n_i - 1$. The sampling distribution of this F-variate depends on the numbers of degrees of freedom, $n_1 - 1$ and $n_2 - 1$, of both of the samples. Probability calculations for the F-distribution are illustrated in the appendix to this chapter. Probability tables for the F-distribution are given in Table A7 of the appendix to this book.

The sampling distributions for the variates (12.5)–(12.7) have density functions from which probabilities are obtained as areas under the respective curves. These density functions are used to tabulate the probabilities in the various tables in the appendix. There are many different forms for normal, t, chi-square, and F statistics, depending on how the response variables are obtained (including the statistical design used, if appropriate) and the model used to define the relationship between the response variable and the assignable causes and uncontrolled random variation. While the forms of these statistics may change (and the degrees of freedom for the last three), the tables in the appendix can be used for all the forms discussed in this book.

Each of the sampling distributions for the variates in (12.4)–(12.7) requires that the individual response variables in a sample be independent and that they come from a single normal probability distribution. In practice there are many situations where the assumption of a normal probability distribution is not reasonable. Fortunately there is a powerful theoretical result, the central limit property (see Exhibit 12.5), which enables the use of these statistics in certain situations when the individual response variables are not normally distributed. The implications of the central limit property are important for inferences on population means. Regardless of the true (usually unknown) probability distribution of individual observations, the standard normal distribution and Student's t-distribution are sampling distributions that can be used with the sample mean if the conditions listed in Exhibit 12.5 are met. As a practical guide, if underlying distributions are not badly skewed, samples of size 5 to 10 are usually adequate to invoke the central limit property. The more radical the departure from the form of a normal distribution (e.g., highly skewed distributions), the larger the sample size that is required before the normal approximation will be adequate.

The central limit property can be applied to variables that are discrete (e.g., frequency counts) or continuous (e.g., measurements). In the next several chapters, guidelines for the application of the central limit property will be provided for many important inference procedures.

EXHIBIT 12.5   CENTRAL LIMIT PROPERTY

If a sample of n observations $y_1, y_2, \ldots, y_n$ are

(a) independent, and

(b) from a single probability distribution having mean μ and standard deviation σ,

then the sampling distribution of the sample mean is well approximated by a normal distribution with mean μ and standard deviation $\sigma/n^{1/2}$ if the sample size is sufficiently large.

One note of caution about the central limit property is needed. It is stated in terms of the sample mean. It can be extended to include the sample variance and thereby permit the use of the chi-square distribution (12.6); however, larger sample sizes are needed than for the sample mean. It is also applicable to any statistic that is a linear function of the response variables—e.g., factor effects and least-squares regression estimators (see Chapters 21–22). There are many functions of sample statistics, such as the ratio of sample variances in the F-statistic (12.7), to which the central limit property does not apply.

Randomization affords a second justification for the use of the above sampling distributions when the assumption of normality is not met. The mere process of randomization induces sampling distributions for the statistics in (12.4)–(12.7) that are quite similar to the sampling distributions of these statistics derived under the assumption that the responses are statistically independent and normally distributed. Thus, randomization not only affords protection against possible bias effects due to unexpected causes during the course of an experiment, it also provides a justification for the sampling distributions of the statistics (12.4)–(12.7) when the assumption of normality is not satisfied. Further discussion of the robustness properties of these statistics is provided in Section 25.4.

12.2 INTERVAL ESTIMATION

Parameter estimates do not by themselves convey any information about the adequacy of the estimation procedure. In particular, individual estimates do not indicate the precision with which the parameter is estimated. For example, the sample mean of the tensile-strength measurements for die 1 in Table 12.1 is 84.733. Can one state with any assurance that this estimate is close to the population mean tensile strength for die 1? By itself, this average provides no such information.

Standard errors provide information on the precision of estimators. Estimates of standard errors should therefore be reported along with the parameter estimates. A convenient and informative way to do so is with the use of *confidence intervals*. Confidence intervals are intervals formed around parameter estimates. The length of a confidence interval provides a direct measure of the precision of the estimator: the shorter the confidence interval, the more precise the estimator. Confidence intervals also allow one to be more specific about plausible values of the parameter of interest than just reporting a single value.

Consider now the construction of a confidence interval for the population mean of a normal probability distribution when the standard deviation is known. Let $z_{\alpha/2}$ denote the standard normal deviate (value) corresponding to an upper-tail probability of $\alpha/2$. This value is obtained from Table A4 in the appendix. Since the variate in (12.4) has a standard normal probability distribution,

$$\Pr\left\{-z_{\alpha/2} < \frac{n^{1/2}(\bar{y} - \mu)}{\sigma} < z_{\alpha/2}\right\} = 1 - \alpha. \tag{12.8}$$

If one solves for μ in the inequality, the probability statement can be rewritten as

$$\Pr\{\bar{y} - z_{\alpha/2}\sigma_{\bar{y}} < \mu < \bar{y} + z_{\alpha/2}\sigma_{\bar{y}}\} = 1 - \alpha, \tag{12.9}$$

where $\sigma_{\bar{y}} = \sigma/n^{1/2}$ is the standard error of $\bar{y}$, equation (12.3). The limits in the probability statement (12.9) are said to form a $100(1 - \alpha)\%$ confidence interval on the unknown population mean. A general definition of a confidence interval is given in Exhibit 12.6. The lower and upper limits, $L(\hat{\theta})$ and $U(\hat{\theta})$, in this definition are functions of the estimator $\hat{\theta}$ of the parameter θ. Thus, it is the limits that are random, not the parameter enclosed by the limits. These limits can conceivably be calculated for a large number of samples. In the long run, $100(1 - \alpha)\%$ of the intervals will include the unknown parameter.

EXHIBIT 12.6

Confidence Interval. A $100(1 - \alpha)\%$ confidence interval for a parameter θ consists of limits $L(\hat{\theta})$ and $U(\hat{\theta})$ that will bound the parameter with probability $1 - \alpha$.

The $100(1 - \alpha)\%$ confidence interval for the mean of a normal distribution with known standard deviation is, from (12.9),

$$\bar{y} - z_{\alpha/2}\sigma_{\bar{y}} < \mu < \bar{y} + z_{\alpha/2}\sigma_{\bar{y}} \quad \text{or} \quad \bar{y} \pm z_{\alpha/2}\sigma_{\bar{y}}. \tag{12.10}$$

Note that the limits are a function of the sample mean. The midpoint of the interval in (12.10) is the parameter estimator $\bar{y}$. The length of the confidence interval depends on the size of the standard error $\sigma_{\bar{y}}$ and on the confidence level $100(1 - \alpha)\%$. Confidence levels are ordinarily desired to be high, e.g., 90–99%. A narrow confidence interval, one that indicates a precise estimate of the population mean, ordinarily requires a small standard error.

Interval estimates for means and standard deviations of normal probability models can be constructed using the technique described above and the appropriate sampling distribution. The construction of confidence intervals for means and standard deviations will be considered more fully in Chapters 13 and 14, including illustrations with the analysis of data.

It is important to stress that confidence intervals provide bounds on a population parameter: it is the bounds that are random, not the parameter value. Confidence intervals can be interpreted in several useful ways. In Exhibit 12.7, we present some statements about confidence intervals that are valuable when interpreting the numerical bounds obtained when sample values are inserted into confidence interval formulas. These interpretations will be reinforced in the next several chapters. The first of the interpretations in Exhibit 12.7 is commonly used to state the results of the confidence interval procedure. The second and third interpretations follow from the probability statement on which the confidence interval is based, e.g., equation (12.9). Ordinarily a single sample is collected and a single interval is calculated using (12.10). Based on the probability statement (12.9), if a large number of samples could be taken, $100(1 - \alpha)\%$ of them would contain the unknown population parameter, in this case the population mean. Because of this, we state that we are $100(1 - \alpha)\%$ confident that the one interval we have computed using the inequality (12.10) does indeed contain the unknown population mean.

EXHIBIT 12.7   INTERPRETATION OF CONFIDENCE INTERVALS

- With $100(1 - \alpha)\%$ confidence, the confidence interval includes the parameter.
- The procedures used (including the sampling technique) provide bounds that include the parameter with probability $1 - \alpha$.
- In repeated sampling, the confidence limits will include the parameter $100(1 - \alpha)\%$ of the time.

12.3 STATISTICAL TOLERANCE INTERVALS

Tolerance intervals are extremely important in evaluating production, quality, and service characteristics in many manufacturing and service industries. Two types of tolerance intervals must be distinguished in this context: engineering tolerance intervals and statistical tolerance intervals.

Engineering tolerance intervals define an acceptable range of performance for a response. Often expressed as specification limits, engineering tolerance intervals dictate a range of allowable variation for the response variable within which the product or service will meet stated requirements.

Statistical tolerance intervals (see Exhibit 12.8) reflect the actual variability of the product or service. This actual variability may or may not be within the desired engineering tolerances. Statistical tolerance intervals are not intervals around parameter values; they are intervals that include a specified portion of the observations from a population or a process.

EXHIBIT 12.8

Statistical Tolerance Intervals. A statistical tolerance interval establishes limits that include a specified proportion of the responses in a population or a process with a prescribed degree of confidence.

Statistical tolerance intervals are derived from probability statements similar to the derivation of confidence intervals. For example, if a response variable y can be assumed to follow a normal probability distribution with known mean and standard deviation, one can write the following probability statement using the standard normal response variable (12.2):

$$\Pr(\mu - z_\alpha \sigma < y < \mu + z_\alpha \sigma) = p.$$

From this probability statement, it immediately follows that $100p\%$ of the distribution of responses are contained within the interval

$$(\mu - z_\alpha \sigma, \mu + z_\alpha \sigma), \tag{12.11}$$

where z_α is the critical value from Table A4 in the appendix corresponding to an upper-tail probability of $\alpha = (1 - p)/2$. The limits in (12.11) are referred to as *natural process limits* because they are limits that include $100p\%$ of the responses from a process that is in statistical control. Intervals similar to (12.11) form the basis for many control-charting techniques (see Section 4.5).

No "confidence" is attached to the interval (12.11), because all the quantities are known and fixed. This tolerance interval is special in that regard. Most applications of tolerance intervals, especially for new products and services, require the estimation of means and standard deviations.

When responses follow a normal probability distribution but the mean and standard deviation are unknown, it is natural to consider intervals of the form

$$(\bar{y} - ks, \bar{y} + ks), \tag{12.12}$$

where k is a critical value from some appropriate sampling distribution corresponding to an upper-tail area of $\alpha = (1 - p)/2$. This interval, unlike (12.11), is not an exact $100p\%$ tolerance interval for the responses. One reason this interval is not exact is that the sample mean and standard deviation are only estimates of the corresponding parameters. Because of this, the interval itself is random, whereas the true interval (12.11) is constant.

Procedures have been devised to incorporate not only the randomness of the response variable y but also the randomness of the sample statistics $\bar{y}$ and s in the construction of statistical tolerance intervals. In doing so, one can only state with a prescribed degree of confidence that the calculated interval contains the specified proportion of the responses.

Tables A8 and A9 in the appendix provide factors for determining one- and two-sided tolerance intervals, respectively. Factors for upper tolerance limits, $\bar{y} + ks$, are obtained from Table A8. Lower tolerance limits, $\bar{y} - ks$, use the same factors. Two-sided tolerance intervals have the form (12.12), where k is obtained from Table A9. These tolerance factors can be used when the sample mean and standard deviation are calculated from n independent normally distributed observations $y_1, y_2, \ldots, y_n$. The tolerance factor k is determined by selecting the confidence coefficient γ, the proportion p of the population of observations around which the tolerance interval is desired, and the sample size n.

12.4　TEST OF STATISTICAL HYPOTHESES

Statistical hypotheses are statements about theoretical models or about probability or sampling distributions. In this section we introduce the concepts needed to effectively conduct tests of statistical hypotheses. To clarify the basic principles involved in the testing of statistical hypotheses, we focus attention on tests involving the mean of a normal probability distribution when the population or process standard deviation is known.

There are two hypotheses that must be specified in any statistical testing procedure: the *null* hypothesis and the *alternative* hypothesis (see Exhibit 12.9). In the context of a statistically designed experiment, the null hypothesis, denoted H_0, defines parameter values or other distributional properties that indicate no experimental effect. The alternative hypothesis, denoted H_a, specifies values that indicate change or experimental effect for the parameter or distributional property of interest.

EXHIBIT 12.9　TYPES OF STATISTICAL HYPOTHESES

Null Hypothesis (H_0). Hypothesis of no change or experimental effect.

Alternative hypothesis (H_a). Hypothesis of change or experimental effect.

In the tensile-strength example discussed in the introduction to this chapter, the key parameters of interest might be the mean tensile-strength measurements for all wire samples drawn from each die. Suppose die 1 is a standard die, and dies 2 and 3 are new designs that are being examined. One hypothesis of interest is whether the new dies have mean tensile-strength measurements that differ from the mean for the standard die. Thus one set of hypotheses of interest in a comparison of the first two dies is

$$H_0: \mu_1 = \mu_2 \quad \text{vs} \quad H_a: \mu_1 \neq \mu_2,$$

where μ_1 denotes the average tensile strength measurement for die 1, and μ_2 is the average for die 2. Note that H_a is a hypothesis of experimental effect, i.e., that the mean tensile-strength measurement for wire drawn from die 2 differs from that for the standard die. The hypothesis of no effect (no difference) is the null hypothesis.

Hypotheses can be one-sided if the experimental effect is believed to be one-sided or if interest is only in one-sided effects. For example, one might only be interested in whether the two experimental dies increase the mean of the tensile-strength measurements. If so, the only comparison of interest between the first two dies might be posed as follows:

$$H_0 : \mu_1 \geqslant \mu_2 \quad \text{vs} \quad H_a : \mu_1 < \mu_2.$$

Again, it is the alternative hypothesis that designates the hypothesis of experimental effect. The null hypothesis in this case could be written as $H_0 : \mu_1 = \mu_2$, but the above statement makes it explicit that only an increase in the mean tensile-strength measurement for die 2 over die 1 constitutes an experimental effect of interest.

The procedures used in statistical hypothesis testing have an interesting analogy with those used in the U.S. judicial system. In a U.S. court of law the defendant is presumed innocent (H_0). The prosecution must prove the defendant guilty (H_a). To do so, the evidence must be such that the innocence of the defendant can be rejected "beyond a reasonable doubt." Thus the burden of proof is on the prosecution to prove the defendant guilty. Failure to do so will result in a verdict of "not guilty." The defendant does not have to prove innocence; the failure to prove guilt suffices.

Table 12.3 shows the possible true status of the defendant and the possible decisions a judge or jury could make. If the defendant is innocent and the decision is "not guilty," a correct decision is made. Similarly, if the defendant is guilty and the decision is "guilty," a correct decision is made. The other two possibilities result in incorrect decisions.

Table 12.4 displays the corresponding situation for a test of a statistical hypothesis. The

Table 12.3 Relation of Statistical Hypothesis Testing to U.S. Judicial System

		Defendant's True Status	
		Innocent	Guilty
Decision	Not Guilty	Correct Decision	Error
	Guilty	Error	Correct decision

Table 12.4 Statistical Hypothesis Testing

		True Hypothesis	
		H_0 True	H_a True
Decision	Do Not Reject H_0	Correct decision	Type II error (probability $= \beta$)
	Reject H_0	Type I error (probability $= \alpha$)	Correct decision

decision, based on data collection and analysis, corresponds to the hearing of evidence and judgement by a judge or jury. Again, there are two possible correct decisions and two possible incorrect ones. Note that the two decisions are labeled "do not reject H_0" and "reject H_0." There is no decision labeled "accept H_0," just as there is no judicial decision labeled "innocent."

Consider testing the hypotheses

$$H_0 : \mu \leqslant 10 \quad \text{vs} \quad H_a : \mu > 10,$$

where μ is the mean of a normal probability distribution with known standard deviation. A sample mean sufficiently greater than 10 would lead one to reject the null hypothesis and accept the alternative hypothesis. How large must the sample mean be in order to reject the null hypothesis?

Assuming that the null hypothesis is true, we can use the transformation (12.4) to convert the sample mean to a standard normal variate, $z = n^{1/2} (\bar{y} - 10)/\sigma$. We could then decide on the following decision rule:

$$\text{Decision:} \quad \text{Reject } H_0 \text{ if } z > z_\alpha, \tag{12.13}$$

where z_α is a *critical value* from the standard normal table corresponding to an upper-tail probability of α. Note that if the null hypothesis is true, there is only a probability of α that the standard normal variate z will be in the rejection region defined by (12.13). This is the probability of a Type I error in Table 12.4. The Type I error probability is termed the *significance level* of the test. The term "confidence level" is also used. The confidence level, $100(1 - \alpha)\%$, actually denotes the chance of failing to reject the null hypothesis when it is true.

Table 12.5 lists several of the terms in common usage in statistical hypothesis-testing procedures. The significance level of a test is controlled by an experimenter. The significance level is set sufficiently small so that if the test statistic falls in the rejection region, the experimenter is willing to conclude that the null hypothesis is false. Once the significance level is set, the test procedure is objective: the test statistic is compared with the

Table 12.5   Terms Used in Statistical Hypothesis Testing

Alternative hypothesis. Hypothesis of experimental effect or change.

Confidence level. Probability of failing to reject H_0 when H_0 is true $(1 - \alpha)$.

Critical value(s). Cutoff value(s) for a test statistic used as limits for the rejection region; determined by the alternative hypothesis and the significance level.

Null hypothesis. Hypothesis of no experimental effect or change.

Operating characteristic curve. Graph of the probability of a Type II error (β) as a function of the hypothetical values of the parameter being tested.

Power. Probability of correctly rejecting H_0 when H_0 is false $(1 - \beta)$; usually unknown.

Rejection (critical) region. Large and/or small values of a test statistic that lead to rejection of H_0.

Significance level. Probability of a Type I error (α); controlled by the experimenter.

Significance probability (p-value). Probability of obtaining a value for a test statistic that is as extreme as or more extreme than the observed value, assuming the null hypothesis is true.

Type I error. Erroneous rejection of H_0; probability $= \alpha$.

Type II error. Erroneous failure to reject H_0; probability $= \beta$.

critical value, and the null hypothesis is either rejected or not, depending on whether the test statistic is in the rejection region.

The significance probability of a test statistic quantifies the degree of discrepancy between the estimated parameter value and its hypothesized value. The significance probability is the probability of obtaining a value of the test statistic that is as extreme as or more extreme than the observed value, assuming the null hypothesis is true. In the above test on the mean of a normal distribution, suppose the standard normal variate had a value of 3.10. Since the alternative hypothesis specifies that large values of the sample mean (and therefore the standard normal variate) lead to rejection of the null hypothesis, the significance probability for this test statistic would be

$$p = \Pr\{Z > 3.10\} = 0.0001.$$

Note that if the p-value is smaller than the significance level, the null hypothesis is rejected. Comparison of the p-value with the significance level is equivalent to comparing the test statistic to the critical value.

Common values of the significance level are 0.05 and 0.01, although any small value can be selected for α. The more serious the consequences of a Type I error, the smaller one will choose the significance level. The choice of a significance level will be discussed further in the examples presented in the next several chapters.

In the analogy with the U.S. judicial system, the determination of a rejection region is equivalent to the judge's charge to the jury, especially regarding points of law and requirements needed to find the defendant guilty. The significance level is equivalent to a definition of what constitutes "reasonable doubt." If after an evaluation of the evidence and comparison with the points of law the jurors are confident "beyond a reasonable doubt" that the defendant committed the crime, they are required to find the defendant guilty; otherwise, the defendant is found not guilty.

The significance probability is equivalent to the "strength of conviction" with which the jury finds the defendant guilty. If the p-value is greater than the significance level, the jury does not have sufficient evidence to convict the defendant. If the p-value is much smaller than the significance level, the evidence is overwhelmingly in favor of conviction. If the p-value is only marginally smaller than the significance level, the jury has sufficient evidence for conviction but the strength of their conviction is less than if the significance probability were very small.

As mentioned above, the term "confidence" is often used in place of significance level or significance probability. It is not the preferred usage when testing hypotheses but it has become common in some disciplines. The confidence generally quoted for a statistical test is $100(1 - p)\%$. Thus, one might quote statistical significance for the above test "with 99.99% confidence." This use of "confidence" requires caution. It is incorrect to report "with 95% confidence" that the null hypothesis is true. This is equivalent to stating that a jury is "95% confident" that a defendant is innocent. Recall that both in the courtroom and when conducting statistical tests, evidence is collected to show guilt (reject the null hypothesis). Failure to prove guilt (failure to reject the null hypothesis) does not prove innocence (null hypothesis is true) at any "confidence level."

Unlike the significance level, the probability of a Type II error is almost never known to the experimenter. This is because the exact value of the parameter of interest, if it differs from that specified in the null hypothesis, is generally unknown. The probability of a Type II error can only be calculated for specific values of the parameter of interest. The probability of a Type II error, operating characteristic curves, and power are extremely important for assessing the adequacy of an experimental design, especially for determining sample sizes. These considerations are the subject of the next section.

12.5 SAMPLE SIZE AND POWER

Although the probability of a Type II error is almost never known to an experimenter, calculation of these probabilities can be made for hypothetical values of the parameter of interest. Suppose, for example, one wishes to test the hypotheses

$$H_0 : \mu \leqslant 10 \quad \text{vs} \quad H_a : \mu > 10$$

when the population standard deviation is known to equal 2. If a significance level of 0.05 is chosen for this test, the null hypothesis will be rejected if

$$z = \frac{n^{1/2}(\bar{y} - 10)}{2} > z_{0.05} = 1.645.$$

Next suppose that it is critically important to the experimenter that the null hypothesis be rejected if the true population mean is 11 rather than the hypothesized value of 10. If the experimenter wishes to be confident that the null hypothesis will be rejected when the true population mean is 11, the power $(1 - \beta)$ of the test must be acceptably large; equivalently, the probability β of a Type II error must be acceptably small. Figure 12.2 is a typical graph

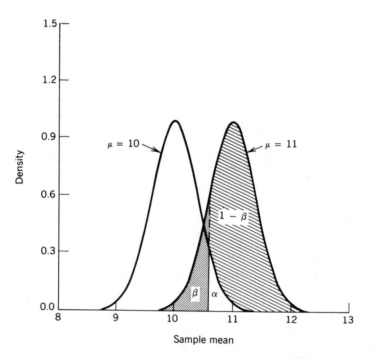

Figure 12.2 Sampling distributions and error probabilities.

of the sampling distributions of the sample mean for the two cases $\mu = 10$ and $\mu = 11$. The significance level, the probability of a Type II error, and power of the test are indicated.

If $\mu = 11$, then $z = n^{1/2}(\bar{y} - 11)/2$ has a standard normal distribution, while $n^{1/2}(\bar{y} - 10)/2$ is not a standard normal variate. Using the sampling distribution of the mean under the alternative hypothesis, the power of the test can be calculated as follows:

$$
\begin{aligned}
1 - \beta &= \Pr\left\{\frac{n^{1/2}(\bar{y} - 10)}{2} > 1.645\right\} \\
&= \Pr\left\{\frac{n^{1/2}(\bar{y} - 11)}{2} > 1.645 - n^{1/2}\frac{11 - 10}{2}\right\} \\
&= \Pr\{z > 1.645 - 0.5n^{1/2}\}.
\end{aligned}
\tag{12.14}
$$

Once the sample size is determined, the power (12.14) of the above test can be determined. If the experimenter decides to use a sample size of $n = 25$, the power of the test is

$$
1 - \beta = \Pr\{z > -0.855\} = 0.804.
$$

Thus the experiment has a probability of 0.804 of correctly rejecting the null hypothesis when the true population mean is $\mu = 11$. If this power is not acceptably large to the experimenter, the sample size must be increased. A sample size of $n = 50$ would produce a power of

$$
1 - \beta = \Pr\{z > -1.891\} = 0.971.
$$

Figure 12.3 illustrates an analogy that may assist in understanding the connections between the three quantities that determine the properties of a statistical test: sample size, significance level, and power. This schematic drawing shows the reciprocal of the sample

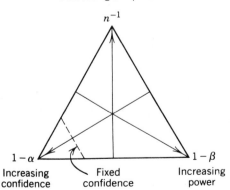

Figure 12.3 Simplex analogy between sample size (n), significance level (α), and power $(1 - \beta)$.

size, the confidence level $(1 - \alpha)$ of the test, and the power of the test plotted on trilinear coordinate paper. (Trilinear plotting was discussed in Chapter 11; see Figures 11.16 and 11.17.) The scales of the axes in Figure 12.3 are nonlinear and have been omitted for clarity.

Note that if one fixes any two of these quantities, the third one is necessarily determined. So too, if one fixes any one of the components, the other two components are restricted to a curve in the trilinear coordinate system. Again for simplicity, the curve for fixed confidence is shown in Figure 12.3 as a straight line. The important point to remember is that for any statistical test procedure one can only increase the power for a fixed significance level by increasing the sample size.

A graph of power probabilities can be made as a function of the sample size for a fixed value of the alternative hypothesis or as a function of the values of the parameter of interest for a fixed sample size. Equivalently one can plot the probability of Type II error. The latter type of graph is called an *operating-characteristic (OC) curve.* Power curves and OC curves are extremely valuable for evaluating the utility of an experimental design and for assessing the size of a sample needed to reject the null hypothesis for specified values of the alternative hypothesis.

Returning to the analogy with the judicial system, power refers to the probability of obtaining a conviction when the defendant is guilty. The resources needed (sample size) to assure conviction with a high probability depend on the magnitude of the crime relative to the other factors that might prevent discovery. For example, if a defendant embezzles a very large sum of money from a company, there might not need to be as much effort put into obtaining evidence definitively linking the defendant to the crime than if a much smaller sum of money were embezzled. For a very large sum of money, perhaps only a few people at the company would have access to the funds, and therefore only a few people would need to be investigated.

The foregoing discussion of sample size as related to significance levels and power serves two useful purposes. First, it establishes relationships among the three quantities that are useful for designing experiments. Second, it alerts the experimenter to the need to consider questions relating to the attainable power of a statistical test when sample sizes are fixed and cannot be changed. Many experiments are of that kind. Often the experiment size is predetermined by budget or other constraints. In such settings sample-size considerations are moot. Nevertheless, if great expenditures are being invested in a project, it may be critical to the success of the project to use the procedures described in this section, in Chapters 13 and 14, or in Section 15.5 to evaluate the potential for determining the existence of experimental effects of importance.

APPENDIX: PROBABILITY CALCULATIONS

1. Normal Distribution

Table A4 in the appendix to this book lists probabilities for a standard normal probability distribution. The probabilities tabulated are *cumulative* probabilities: for any standard normal variate z, the probability in the body of the table is

$$\Pr\{z \leqslant z_c\},$$

where z_c is a value of that variable. For example,

$$\Pr\{z \leqslant 1.65\} = 0.9505.$$

The entire area under the standard normal distribution is 1.00. Consequently, "upper-tail" probabilities are obtained by subtraction:

$$\Pr\{z > 1.65\} = 1 - 0.9505 = 0.0495.$$

Probabilities for negative standard normal values can be obtained from the corresponding probabilities for positive values. This can be done because the normal distribution is symmetric around the mean; i.e., the curve to the left of $z = 0$ is the mirror image of the curve to the right of $z = 0$. Thus,

$$\Pr\{z \geqslant -1.65\} = \Pr\{z \leqslant 1.65\} = 0.9505$$

and

$$\Pr\{z < -1.65\} = \Pr\{z > 1.65\} = 0.0495.$$

Also because of the symmetry of the normal distribution, the area under the distribution to the left of $z = 0$ is 0.5000 and the area to the right of $z = 0$ is 0.5000. With this information one can compute probabilities for any range of a standard normal variable. For example,

$$\Pr\{0 \leqslant z \leqslant 2.00\} = \Pr\{z \leqslant 2.00\} - 0.5000$$
$$= 0.9772 - 0.5000 = 0.4772,$$
$$\Pr\{-1.00 \leqslant z \leqslant 0\} = \Pr\{0 \leqslant z \leqslant 1.00\}$$
$$= 0.8413 - 0.5000 = 0.3413,$$
$$\Pr\{0.79 \leqslant z \leqslant 1.47\} = \Pr\{z \leqslant 1.47\} - \Pr\{z \leqslant 0.79\}$$
$$= 0.9292 - 0.7852 = 0.1440.$$

2. Student's *t*-Distribution

Probabilities for Student's t-distribution depend on the degrees of freedom of the statistic much like probabilities for a normal distribution depend on the mean and standard deviation of the normal variable. Unlike the standard normal distribution, there is no transformation of the t-variable that will allow one t-distribution to be used for all t-variables.

The degrees of freedom for a t-variable are essentially an adjustment to the sample size based on the number of distributional parameters that must be estimated. Consider a t statistic of the form

$$t = \frac{n^{1/2}(\bar{y} - \mu)}{s},$$

where the sample mean and the sample standard deviation are estimated from n independent observations from a normal probability distribution. If the standard deviation were known, a standard normal variable, equation (12.4), could be used instead of this t-statistic by inserting σ for s. The standard normal distribution can be shown to be a t-distribution with an infinite number of degrees of freedom. The above t-statistic has $v = n - 1$ degrees of freedom, based on the stated assumptions. So the number of degrees of freedom, $v = n - 1$ instead of $v = \infty$, represents an adjustment to the sample size for having to estimate the standard deviation.

Table A5 in the appendix lists cumulative t-probabilities for several t-distributions. The values in the table are t-values that give the cumulative probabilities listed at the top of each column. Thus, for a t-variable with 10 degrees of freedom,

$$\Pr\{t \leqslant 1.812\} = 0.95.$$

If, as is generally the case, a computed t-value does not exactly equal one of the table values, the exact probability is reported as being in an interval bounded by the probabilities in the table. For example, for a t-variable with eight degrees of freedom $\Pr\{t \leqslant 2.11\}$ is not available from the table. Note that 2.11 lies between the table values 1.860 and 2.306. Denoting the desired probability by p, one would report

$$0.950 < p < 0.975.$$

Similarly, if one wishes to compute the "upper-tail" probability $\Pr\{t > 3.02\}$ for a t-variable having 18 degrees of freedom, one would report

$$p < 0.005.$$

Probabilities for negative t-values are obtained from Table A5 much the same way as are normal probabilities. The t-distribution is symmetric around $t = 0$. Thus, for a t-variable with five degrees of freedom

$$\Pr\{t \geqslant -2.015\} = \Pr\{t \leqslant 2.015\} = 0.95$$

and

$$\Pr\{t < -3.365\} = \Pr\{t > 3.365\} = 0.01.$$

Intervals for nontabulated t-values would be reported similarly to the above illustrations.

3. Chi-Square Distribution

Chi-square distributions depend on degrees of freedom in much the same way as t-distributions. Table A6 in the appendix lists cumulative and upper-tail probabilities for several chi-square distributions, each having a different number of degrees of freedom. Chi-square variates can only take nonnegative values, so there is no need to compute probabilities for negative chi-square values. Use of Table A6 for chi-square values follows the same procedures as was outlined above for nonnegative t-values.

4. F-Distribution

F-statistics are ratios of two independent chi-squares, each divided by the number of its degrees of freedom. There are two sets of degrees of freedom for an F variable, one corresponding to the numerator statistic (v_1) and one corresponding to the denominator statistic (v_2).

Table A7 in the appendix contains F-values and their corresponding cumulative and upper-tail probabilities. Due to the need to present tables for a large number of possible numerator and denominator degrees of freedom, Table A7 only contains F-values corresponding to cumulative probabilities of 0.75, 0.90, 0.95, 0.975, 0.99, and 0.995. The F-statistic is nonnegative, so the use of Table A7 is similar to the use of Table A6 for the chi-square distribution. Critical F-values for lower-tail cumulative probabilities of 0.25, 0.10, 0.05, 0.025, 0.01, and 0.005 can be obtained from these tables by using the following relationship:

$$F_\alpha(v_1, v_2) = 1/F_{1-\alpha}(v_2, v_1).$$

REFERENCES

Text References

Theoretical derivations of sampling distributions and discussions of results such as the central limit property are presented in most mathematical-statistics texts. The following graduate-level texts provide comprehensive coverage of these topics:

Hogg, R. V. and Craig, A. T. (1978). *Introduction to Mathematical Statistics*, New York: The Macmillan Co.

Mood, A. M., Graybill, F. A., and Boes, D. C. (1974). *Introduction to the Theory of Statistics*, New York: McGraw-Hill Book Co.

The following mathematical-statistics texts are less rigorous than the above ones. They are also directed toward engineers and scientists:

Bethea, R. M., Duran, B. S., and Boullion, T. L. (1985). *Statistical Methods for Engineers and Scientists*, New York: Marcel Dekker, Inc.

Bowker, A. H. and Lieberman, G. J. (1972). *Engineering Statistics*, Englewood Cliffs, NJ: Prentice-Hall, Inc.

Devore, J. L., (1982). *Probability and Statistics for Engineering and Sciences*, Monterey, CA: Brooks Cole Publishing Co.

Guttman, I., Wilks, S. S., and Hunter, J. S. (1982). *Introductory Engineering Statistics*, New York: John Wiley and Sons, Inc.

Hines, W. W. and Montgomery, D. C. (1980). *Probability and Statistics in Engineering and Management*, New York: John Wiley and Sons, Inc.

Mendenhall, W. and Sincich, T. (1984). *Statistics for the Engineering and Computer Sciences*, Santa Clara, CA: Dellen Publishing Co.

Walpole, R. E. and Myers, R. H. (1972). *Probability and Statistics for Engineers and Scientists*, New York: The Macmillan Co.

Each of the above texts discusses confidence intervals and tests of statistical hypotheses. The treatment is more theoretical than in this book, and most of the texts provide detailed derivations. As specific confidence-interval and hypothesis-testing procedures are covered in the following chapters, references to more applied texts will be provided. The text by Bowker and Lieberman presents material on tolerance intervals.

The following two books also discuss tolerance intervals. The first of these books contains extensive tables of tolerance factors.

Odeh, R. E. and Owen, D. B. (1980). *Tables for Normal Tolerance Limits, Sampling Plans, and Screening*. New York: Marcel Dekker, Inc.

Ostle, B. and Malone, L. C. (1988). *Statistics in Research*, Fourth Edition. Ames, Iowa: Iowa State University Press.

Some useful books of tables of probabilities, sample sizes, and power calculations are:

Beyer, W. H. (1968). *CRC Handbooks of Tables for Probability and Statistics*, Cleveland, OH: The Chemical Rubber Co.

Odeh, R. E. and Fox, M. (1975). *Sample Size Choice: Charts for Experiments with Linear Models,* New
York: Marcel Dekker, Inc.

Odeh, R. E., Owen, D. B., Birnaum, Z. W., and Fisher, L. (1976). *Pocket Book of Statistical Tables,*
New York: Marcel Dekker, Inc.

Pearson, E. S. and Hartley, H. O. (1969). *Biometrika Tables for Statisticians,* Cambridge: Cambridge
University Press.

EXERCISES

1   Use the normal probability tables to determine the following probabilities for a
standard normal response variable z:

(a) $\Pr\{z > 1.98\}$
(b) $\Pr\{z < 3\}$
(c) $\Pr\{-2.5 < z < -1.2\}$
(d) $\Pr\{1.6 < z < 3.0\}$
(e) $\Pr\{z < 0\}$
(f) $\Pr\{z > 2\}$
(g) $\Pr\{z < 2\}$
(h) $\Pr\{z < 1.64\}$
(i) $\Pr\{z > 1.96\}$
(j) $\Pr\{z < 2.58\}$

2   For the following t and chi-square (X) variates, determine the probabilities indicated.
In each case use $n-1$ as the number of degrees of freedom.

(a) $\Pr\{t < 2.467\},\ n = 29$
(b) $\Pr\{t > 2.074\},\ n = 23$
(c) $\Pr\{t < -1.65\},\ n = 12$
(d) $\Pr\{t > 3.011\},\ n = 30$
(e) $\Pr\{t > 1.5\},\ n = 7$
(f) $\Pr\{X < 53.5\},\ n = 79$
(g) $\Pr\{X > 85.7\},\ n = 56$
(h) $\Pr\{X < 16.2\},\ n = 32$
(i) $\Pr\{X > 27.1\},\ n = 15$
(j) $\Pr\{X > 42.8\},\ n = 6$

3   For the following F-variates, determine the probabilities indicated. Use the numbers
of degrees of freedom shown.

(a) $\Pr\{F < 4.19\},\ v_1 = 3,\ v_2 = 4.$
(b) $\Pr\{F > 4.39\},\ v_1 = 9,\ v_2 = 12.$
(c) $\Pr\{F < 3.50\},\ v_1 = 3,\ v_2 = 7.$

4 A manufacturer of thin polyester film used to fabricate scientific balloons reports the standard deviation of a run of film to be 0.01 mm. A sample of nine strips of the film was taken, and the thickness of each strip was measured. The average thickness was found to be 0.5 mm. Construct a 98% confidence interval for the mean thickness of this type of film. Interpret the confidence interval in the context of this exercise. What assumptions did you make in order to construct the interval?

5 Suppose the mean film thickness in the previous exercise is known to be 0.46 mm and the standard deviation is known to be 0.01 mm. Determine natural process limits for the film thickness that will include 99% of all film-thickness measurements. Interpret the process limits in the context of this exercise.

6 A container manufacturer wishes to ensure that cartons manufactured for a furniture-moving company have sufficient strength to protect the contents during shipment. Twenty-five cartons are randomly selected from the manufacturer's large inventory, and the crushing strength of each is measured. The average crushing strength is 126.2 psi and the standard deviation is calculated to be 5.0 psi. Construct a 95% tolerance interval for the crushing strengths of these cartons that will include 99% of the individual carton measurements. What assumptions are you making in order to construct this interval?

7 Suppose in the previous exercise that the desired minimum crushing strength is 100 psi for an individual carton. What can you conclude from the statistical tolerance interval calculated in Exercise 6 about the ability of the manufacturer's cartons to meet this engineering specification limit?

8 It has been determined that the distribution of lengths of nails produced by a particular machine can be well represented by a normal probability distribution. A random sample of ten nails produced the following lengths (in inches):

$$1.14, 1.15, 1.11, 1.16, 1.13, 1.15, 1.18, 1.12, 1.15, 1.12.$$

Calculate a 90% tolerance interval that includes 95% of the nail lengths produced by this machine.

9 A rocket was designed by an aerospace company to carry an expensive payload of scientific equipment on a satellite to be launched in orbit around the earth. On the launch date, scientists debated whether to launch due to possible adverse weather conditions. In the context of a test of a statistical hypothesis, the relevant hypotheses might be posed as follows:

H_0: the satellite will launch successfully,

H_a: the satellite will not launch successfully.

Describe the Type I and the Type II errors for these hypotheses. Which error appears to be the more serious? How, in the context of this exercise, would you decide whether to launch in order to make the Type I error acceptably small? What about the Type II error?

10 In Exercise 4, suppose interest was in testing whether the mean film thickness is less than 0.48 mm versus the alternative that it is greater than 0.48. Using a significance level of 0.01, conduct an appropriate test of this hypothesis. Draw a conclusion in the context of this exercise.

11 Construct OC curves and power curves for the test in the previous exercise. In each case, draw separate curves for samples of size $n = 10, 15, 20$, and 25. Use 0.48 mm as the hypothesized mean and 0.01 as the standard deviation. Use a significance level of 0.01. Evaluate the Type II error probability and the power for hypothesized mean values in an interval from 0.48 to 0.55. Suppose one wishes to ensure that a mean of 0.53 will be detected with a probability of at least 0.90. What is the minimum sample size that will suffice?

12 A manufacturer of ball bearings collects a random sample of 11 ball bearings made during a day's production shift. The engineering specifications on the bearings are that they should have diameters that are 3.575 ± 0.001 mm. Do the data below indicate that the diameters of the ball bearings are meeting this specification?

$$3.573 \quad 3.571 \quad 3.575 \quad 3.576 \quad 3.570 \quad 3.580$$
$$3.577 \quad 3.572 \quad 3.571 \quad 3.578 \quad 3.579$$

CHAPTER 13

Inferences on Means

In this chapter we present statistical techniques for drawing inferences on location parameters, primarily means. The construction of confidence intervals and procedures for performing tests of statistical hypotheses involving parameters from one, two, or several populations or processes are discussed. Sample-size determination is also detailed. Specific topics covered include:

- *single-sample inferences on a mean,*
- *comparisons of two means using either paired or independent samples,*
- *comparisons of several means when independent samples are available to estimate each, and*
- *nonparametric location tests for one or two samples.*

The basic concepts associated with confidence intervals and tests of statistical hypotheses were introduced in Chapter 12. The purpose of this chapter is to expand this framework to include specific inference procedures for common experimental situations that require inferences involving location parameters from one or more populations or processes.

Most of the techniques used in this chapter are derived under the assumption that the response variables are statistically independent and follow a normal probability distribution. The normality assumption is not, however, critical for most of these inference procedures. As discussed in Section 12.1, the central limit property can be used to justify the use of the normal distribution as an approximate sampling distribution for the sample mean. The t-distribution is known to be especially robust to departures from normality when inferences are desired on a mean. The randomization of test runs can also be used to justify these inference procedures. Further discussion on this topic is presented in Section 13.5.

Sections 13.1–13.3 provide inferential techniques for means when a normal distribution for the responses can be either assumed or justified by one of the above arguments. When the normality cannot be assumed or satisfactorily justified by one of these arguments, nonparametric statistical inference techniques may be applicable. Nonparametric procedures for inferences on one or two population or process means are presented in Section 13.4. Nonparametric inferences for more than two means are deferred until Chapter 15.

13.1 SINGLE SAMPLE

When independent observations are obtained from one process or population in an experiment, one can often model the responses as

$$y_i = \mu + e_i, \qquad i = 1, 2, \ldots, n. \tag{13.1}$$

In this representation y_i is a measurement on a continuous variate, μ is the unknown mean of the process or population, and e_i is a random error component associated with the variation in the observations. These errors are assumed to be statistically independent and to have a common probability distribution possessing a mean of zero and a constant but unknown standard deviation of σ. In the tensile-strength experiment of Chapter 12, y_i represents a tensile-strength measurement on one of the dies, μ is the population average tensile strength for all wire produced from that die, and e_i is the difference between the observed tensile strength and the true (unknown) average tensile strength μ.

The sample mean or average $\bar{y}$ is an estimator of the unknown constant μ in equation (13.1). Under an assumption that the errors are statistically independent and normally distributed, the response variables are independent and normally distributed with mean μ and standard deviation σ. The sampling distribution of the sample mean is then normal with mean μ and standard error $\sigma/n^{1/2}$ (see Section 12.1). If σ is known, interval-estimation and hypothesis-testing procedures using standard normal variates can be applied. These procedures were discussed in Sections 12.2 and 12.4, respectively. In the remainder of this section, we discuss the relevant procedures for the more common situation in which the standard deviation is unknown.

13.1.1 Confidence Intervals

Suppose it is of interest to construct a confidence interval for the model mean in equation (13.1) when the standard deviation σ is unknown and the response variables y_i can be considered normally distributed. The average of the sample values $y_i, y_2, \ldots, y_n$ is denoted by $\bar{y}$, and the sample standard deviation by s.

From the distributional results presented in Section 12.1,

$$t = \frac{\bar{y} - \mu}{s/n^{1/2}} = \frac{n^{1/2}(\bar{y} - \mu)}{s} \tag{13.2}$$

follows a t-distribution with $n - 1$ degrees of freedom. In this expression, $s_{\bar{y}} = s/n^{1/2}$ is the estimated standard error of the sample mean $\bar{y}$. The degrees of freedom, $n - 1$, for this t-statistic are associated with the estimation of the standard deviation (or variance).

A t-statistic is the ratio of two quantities. The numerator is a standard normal variate. The denominator is the square root of a chi-square variate divided by its number of degrees of freedom; i.e., the general form of a t-statistic is

$$t = \frac{z}{(X^2/v)^{1/2}}, \tag{13.3}$$

where z is a standard normal variate and X^2 is a chi-square variate having v degrees of freedom. The degrees of freedom of a t statistic equal those of the chi-square statistic from which it is formed.

As mentioned in Section 12.1, $z = n^{1/2}(\bar{y} - \mu)/\sigma$ is a standard normal variate. The sample mean, and hence z, is statistically independent of the sample variance s^2. The variate $X^2 = (n-1)s^2/\sigma^2$ follows a chi-square distribution with $v = n - 1$ degrees of freedom. Inserting these expressions for z and X^2 into (13.3) results in the t-variate (13.2).

Under the normality and independence assumptions stated above, the degrees of freedom for the chi-square variate $X^2 = (n-1)s^2/\sigma^2$ are $n - 1$. The degrees of freedom indicate how many statistically independent responses, or in some instances functions of the responses, are used to calculate the variate. In the calculation of the sample variance not all n of the differences $y_i - \bar{y}$ are statistically independent, since they sum to zero. Knowing any $n - 1$ of them and the constraint that all n sum to zero enables one to determine the last one: the last one has the same magnitude as the sum of the $n - 1$ known differences, but opposite sign.

In many instances, as is the case for the sample variance, the number of degrees of freedom equals the number of independent observations less the number of additional parameters that must be estimated to calculate the statistic. The mean μ must be estimated (by $\bar{y}$) to calculate the sample variance (and hence) the sample standard deviation. If μ were known, the sample variance could be estimated as

$$s^2 = \sum \frac{(y_i - \mu)^2}{n}$$

and $X^2 = ns^2/\sigma^2$ would follow a chi-square distribution with $v = n$ degrees of freedom. The t-statistic obtained by inserting this chi-square variate in (13.3) would then also have n degrees of freedom.

Following the procedure outlined in Section 12.2, a $100(1 - \alpha)\%$ confidence interval for μ can be derived by starting with the probability statement

$$\Pr\left\{ -t_{\alpha/2} < \frac{n^{1/2}(\bar{y} - \mu)}{s} < t_{\alpha/2} \right\} = 1 - \alpha. \tag{13.4}$$

where $t_{\alpha/2}$ is the critical value from Table A5 in the appendix for a t-variate having $v = n - 1$ degrees of freedom and an upper-tail probability of $\alpha/2$. The confidence interval is formed by algebraically manipulating the above probability statement to isolate the mean between the inequalities. The resulting confidence interval is

$$\bar{y} - t_{\alpha/2}s_{\bar{y}} < \mu < \bar{y} + t_{\alpha/2}s_{\bar{y}},$$

or

$$\bar{y} \pm t_{\alpha/2}s_{\bar{y}}.$$

The probability statement (13.4) states that the bounds computed from the sample will cover the true value of the mean with probability $1 - \alpha$. Once the sample is collected and the bounds are computed, we state that we are $100(1 - \alpha)\%$ confident that the confidence limits do cover the unknown mean. A 95% confidence interval for the average tensile strength of wire produced from the first die using the 18 tensile-strength measurements in Table 12.1 is

$$84.733 - 2.110\frac{1.408}{(18)^{1/2}} < \mu < 84.733 + 2.110\frac{1.408}{(18)^{1/2}},$$

or

$$84.033 < \mu < 85.433.$$

As stressed in Chapter 12, the confidence limits 84.033 and 85.433 should not be used to imply that there is a 0.95 probability that μ is in the interval defined by these limits. The probability statement (13.4) is valid only prior to the collection and analysis of the data. Since μ is a fixed constant, it either is or is not in the computed confidence interval. The confidence coefficient $100(1 - \alpha)\%$ refers to the statistical methodology; i.e., the methodology used will result in confidence limits that include the true parameter value with a probability of $1 - \alpha$.

To emphasize this point further, Figure 13.1 schematically depicts confidence intervals from several samples laid out on the same scale as the measurement of interest and its associated normal probability distribution. Suppose the true mean is 85. Each time a sample is drawn and a confidence interval is constructed, the calculated interval either includes $\mu = 85$ or the interval fails to include it. If one repeatedly draws samples of size n from this normal distribution and constructs a confidence interval with each sample, then approximately $100(1 - \alpha)\%$ of the intervals will contain $\mu = 85$ and $100\alpha\%$ will not.

In practice we do not take many samples; usually only one is drawn. The interval estimate obtained from this sample either brackets μ or it does not. We do not know if

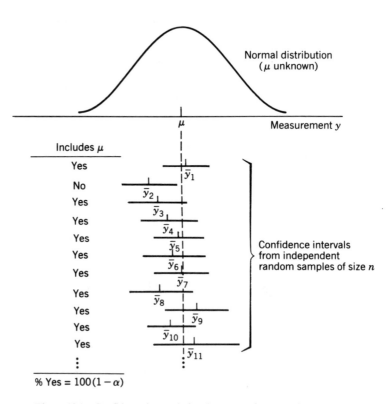

Figure 13.1 Confidence intervals for the mean of a normal population.

Table 13.1 Confidence Intervals for Means of Normal Probability Distributions*

σ Known	σ Unknown

(a) *Two-Sided*

$$\bar{y} - z_{\alpha/2}\sigma_{\bar{y}} < \mu < \bar{y} + z_{\alpha/2}\sigma_{\bar{y}} \qquad\qquad \bar{y} - t_{\alpha/2}s_{\bar{y}} < \mu < \bar{y} + t_{\alpha/2}s_{\bar{y}}$$

(b) *One-Sided Upper*

$$\mu < \bar{y} + z_{\alpha}\sigma_{\bar{y}} \qquad\qquad \mu < \bar{y} + t_{\alpha}s_{\bar{y}}$$

(c) *One-Sided Lower*

$$\bar{y} - z_{\alpha}\sigma_{\bar{y}} < \mu \qquad\qquad \bar{y} - t_{\alpha}s_{\bar{y}} < \mu$$

*$\sigma_{\bar{y}} = \sigma/n^{1/2}$, $s_{\bar{y}} = s/n^{1/2}$, $z_{\alpha} =$ standard normal critical value, $t = t$ critical value, $v = n - 1$.

the interval does include μ, but our confidence rests with the procedure used: $100(1 - \alpha)\%$ of the time the mean will be included in an interval constructed in this manner. Thus we state that we are $100(1 - \alpha)\%$ confident that the one interval we have computed does indeed include the mean.

It occasionally is of interest to an experimenter to construct a one-sided rather than a two-sided confidence interval. For example, in the tensile-strength experiment, one may only be concerned with a lower bound on the average tensile strength. One could then begin with a one-sided probability statement [using the upper limit in (13.4)] and derive the following one-sided lower confidence interval for μ:

$$\bar{y} - t_{\alpha}s_{\bar{y}} < \mu,$$

where t_{α} is used in place of $t_{\alpha/2}$ for one-sided intervals. Formulas for these and other one-sided and two-sided confidence intervals are given in Table 13.1.

13.1.2 Hypothesis Tests

Hypotheses testing, as introduced in Chapter 12, consists of the four basic steps given in Exhibit 13.1.

EXHIBIT 13.1 STATISTICAL HYPOTHESIS TESTING

1. State the null and alternative hypotheses.

2. Draw a sample and calculate the appropriate test statistic.

3. Compare the calculated value of the test statistic to the critical value(s) corresponding to the significance level selected for the test; equivalently, compare the significance probability of the test statistic to the significance level selected for the test.

4. Draw the appropriate conclusion and interpret the results.

One of the most straightforward statistical tests involves determining if a mean μ differs from some hypothesized value, say μ_0. This is symbolized by

$$H_0:\mu = \mu_0 \quad \text{vs} \quad H_a:\mu \neq \mu_0.$$

The appropriate test statistic for a single sample of independent observations is

$$t = \frac{\bar{y} - \mu_0}{s_{\bar{y}}}.$$

The decision rule used for these hypotheses is:

Decision: Reject H_0 if $t < -t_{\alpha/2}$

or if $t > t_{\alpha/2}$,

where α is the chosen significance level of the test.

Consider again the tensile-strength experiment. Suppose one wishes to test

$$H_0:\mu = 84 \quad \text{vs} \quad H_a:\mu \neq 84.$$

From the summary statistics in Table 12.2, the test statistic is

$$t = \frac{84.733 - 84}{1.408/(18)^{1/2}} = 2.209.$$

If one chooses a significance level of $\alpha = 0.05$, the decision would be to reject H_0, since $t = 2.209$ exceeds $t_{0.025}(17) = 2.110$. Equivalently, since the two-tailed significance probability is $0.02 < p < 0.05$ (twice the upper-tail probability limits), which is less than the significance level selected for the test, the null hypothesis is rejected.

In stating the conclusion to a hypothesis test, some experimenters elect to state the significance level of the test, while others present the p-value associated with the test statistic. A third approach, which is less commonly used, involves relating either the significance level or the p-value to a confidence level. Hence, in the above example, one could state 95% confidence in the decision to reject the null hypothesis, since $1 - \alpha = 0.95$. One might also report that one's confidence is between 95% and 98%, since $0.95 < 1 - p < 0.98$. The dangers of having such statements misinterpreted were explained in Section 12.4. We recommend that this terminology be used with caution and only when there is a clear explanation of the meaning.

Since hypothesis testing and the construction of confidence intervals involve the same statistics and the same probability statements, there is a unique correspondence between the two. If a computed confidence interval excludes a hypothesized value of the parameter of interest, a test of the hypothesis that the parameter equals the hypothesized value will be rejected. Similarly, if the interval includes the hypothesized parameter value, the hypothesis would not be rejected by the corresponding statistical test. For example, a two-sided 95% confidence interval for the mean of the tensile-strength measurements from die 1 was given above as (84.033, 85.433). Since this interval does contain $\mu = 85$, a test of the hypothesis $H_0:\mu = 85$ vs $H_a:\mu \neq 85$ would not be rejected. On the other hand, a test of $H_0:\mu = 84$ vs $H_a:\mu \neq 84$ would be rejected at the $\alpha = 0.05$ significance level.

Rules for testing hypotheses about the mean of a normal population using a single

**Table 13.2 Decision Rules and p-Value Calculations for Tests
on the Mean of a Normal Distribution***

	Small Sample	Known σ or Large Sample

(a) $H_0: \mu = \mu_0$ vs $H_a: \mu \neq \mu_0$

	Small Sample	Known σ or Large Sample
Reject H_0 if:	$\|t_c\| > t_{\alpha/2}$	$\|z_c\| > z_{\alpha/2}$
p-Value:	$p = 2 \Pr\{t > \|t_c\|\}$	$p = 2 \Pr\{z > \|z_c\|\}$

(b) $H_0: \mu \geq \mu_0$ vs $H_a: \mu < \mu_0$

Reject H_0 if:	$t_c < -t_\alpha$	$z_c < -z_\alpha$
p-Value:	$p = \Pr\{t < t_c\}$	$p = \Pr\{z < z_c\}$

(c) $H_0: \mu \leq \mu_0$ vs $H_a: \mu > \mu_0$

Reject H_0 if:	$t_c > t_\alpha$	$z_c > z_\alpha$
p-Value:	$p = \Pr\{t > t_c\}$	$p = \Pr\{z > z_c\}$

*$t = n^{1/2}(\bar{y} - \mu_0)/s$; $z = n^{1/2}(\bar{y} - \mu_0)/\sigma$; t_c, z_c = calculated values of test statistics; z_α = standard normal critical value; $t_\alpha = t$ critical value; $v = n - 1$.

sample of independent observations are summarized in Table 13.2. The procedures outlined for large samples are identical to those described in Section 12.4 for models in which the standard deviations are known. The equivalence of these procedures is based on the near-equality of critical values for the normal distribution and Student's t distribution when the degrees of freedom of the latter are sufficiently many, say more than 30 (cf. Tables A4 and A5).

13.1.3 Choice of a Confidence Interval or a Test

In the previous subsection, the equivalence between confidence-interval procedures and procedures for testing statistical hypotheses was illustrated. The equivalence between confidence-interval and statistical-testing procedures is a general one that can be shown to hold for the parameters of many probability distributions. Because of this equivalence, either could be used to perform a test on the stated parameter. There are circumstances, however, when one may be preferable to the other.

In general, a statistical test simply allows one to conclude whether a null hypothesis should be rejected. The p-value often provides a degree of assurance that the decision is the correct one given the observed data and the assumptions, but fundamentally a statistical test is simply a reject–no-reject decision.

When both can be used to draw inferences on a parameter of interest, confidence intervals provide more information than that afforded by a statistical test. Not only does a confidence interval allow one to assess whether a hypothesis about a parameter should be rejected, but it also provides information on plausible values of the parameter. In particular, a tight confidence interval, one whose upper and lower bounds are sufficiently close to the estimate of the parameter, implies that the parameter has been estimated with a high degree of precision. If the confidence interval is wide, the experimenter knows that the current data do not provide a precise estimate of the parameter. Either of these conclusions might cast doubt on the result of a statistical test. As input to a decision-making process, either could be more important than the result of a statistical test in the determination of a course of action to be taken.

Of what value, then, is a statistical test? Statistical testing procedures allow an assessment of the risks of making incorrect decisions. Specification of a significance level requires consideration of the consequences of a Type I error. Power curves and OC curves describe the risk of a Type II error in terms of possible values of the parameter of interest, the significance level of the test, and the sample size.

Confidence intervals and statistical testing procedures supplement one another. As with any statistical methodology, either of these procedures can be used inappropriately. In some settings it would be inappropriate, for example, to test a hypothesis about a mean and ignore whether it is estimated with satisfactory precision. In others, confidence-interval estimation may be inadequate without the knowledge of the risk of a Type II error. In general, it is recommended that statistical tests be accompanied by interval estimates of parameters. This philosophy is stressed throughout the remainder of this book, although for clarity of presentation many examples include only interval estimation or testing results.

13.1.4 Sample Size

An important consideration in the design of many experiments is the determination of the number of sample observations that need to be obtained. In the estimation of parameters, sample sizes are selected to ensure satisfactory precision. In hypothesis testing, the sample size is chosen in order to control the value of β, the probability of a Type II error (see Section 12.5).

The sample size can be chosen to provide satisfactory precision in the estimation of parameters if some knowledge is available about the variability of responses. For example, in the estimation of the mean when responses are normally distributed and statistically independent, the length of the confidence interval is [see (12.10)]:

$$2z_{\alpha/2}\sigma/n^{1/2}. \tag{13.5}$$

If one wishes to have the length of this confidence interval be no larger than L, say, the required sample size is obtainable by setting (13.5) equal to L and solving for n:

$$n = \left(\frac{2z_{\alpha/2}\sigma}{L}\right)^2. \tag{13.6}$$

When the standard deviation σ is not known, several approximate techniques can be used. In some instances, a worst-case sample size can be determined by inserting a reasonable upper bound on σ into (13.6). If data are available from which to estimate the standard deviation, the sample estimate can be inserted into (13.6) if it is deemed to be sufficiently precise. If an estimate is available but it is not considered sufficiently precise, an upper limit from a one-sided confidence interval on σ (see Section 14.1) can be used to provide a conservative sample-size determination.

In the last section a 95% confidence interval for the mean tensile strength for wire produced by die 1 was found to be (84.033, 85.433). The length of this confidence interval is 1.400. Suppose this is not considered precise enough and in future investigations of the tensile strength of wire from dies such as this it is desired that the confidence intervals have a length of approximately 1 mm; i.e., the confidence interval is desired to be approximately $\bar{y} \pm 0.5$ mm.

From an analysis of the summary statistics in Table 13.2 and perhaps analyses of other tensile-strength data, suppose the experimenters are willing to use $\sigma = 2$ mm as a conservative value for the standard deviation of wire samples made from dies similar to these three. If a 95% confidence interval is to be formed, the required sample size is, from (13.6),

$$n = \left(\frac{(2)(1.96)(2)}{(1)} \right)^2 = 61.47,$$

or approximately 62 observations.

To select sample sizes for hypothesis-testing procedures, use is made of operating-characteristic (OC) curves. OC curves are plots of β against values of the parameter of interest for several different sample sizes. Equivalently, power curves (plots of $1 - \beta$) can be graphed versus the sample size. Separate sets of curves usually are provided for different values of α, the probability of a Type I error.

Suppose interest is in testing the hypothesis

$$H_0 : \mu = \mu_0 \quad \text{vs} \quad H_a : \mu \neq \mu_0,$$

where μ_0 is known and the response values are normally distributed. Suppose further that the true value of the population mean is $\mu = \mu_0 + \delta$, where $\delta \neq 0$. The test statistic is a standard normal:

$$z = \frac{n^{1/2}(\bar{y} - \mu_0)}{\sigma}. \tag{13.7}$$

The power or the Type II error probability for this standard normal test statistic can be computed as detailed in Section 12.5. Alternatively, the Type II error probabilities can be read from the Tables A10(a)–(h) in the appendix.

Tables A10(a)–(h) are graphs of the probabilities of not rejecting H_0 for the test statistic (13.7) for one- and two-sided tests. Charts are presented for four significance levels: 0.10, 0.05, 0.025, and 0.01. The Type II error probabilities are determined by the proposed sample size n and the scaled difference between the hypothesized mean μ_0 and the true mean $\mu = \mu_0 + \delta$:

$$\phi = \frac{|\mu - \mu_0|}{\sigma} = \frac{|\delta|}{\sigma}. \tag{13.8}$$

To demonstrate the use of these charts, consider again the tensile-strength example and suppose it is known from past experimentation that $\sigma = 2.0$. If one wishes to test $H_0 : \mu = 84$ vs $H_a : \mu \neq 84$, the test statistic (13.7) is used. Now suppose that an experimenter is concerned about detecting a one-unit difference between the hypothesized mean, 84, and the true mean. Suppose further that the experimenter is willing to tolerate no larger than a 10% chance of failing to detect this large a difference. Thus, $\beta = 0.10$ when $|\delta| = 1$; equivalently, the power of the test must be $1 - \beta = 0.90$ when $\mu = 84 \pm 1$.

Using these parameter values, $\phi = 0.5$. If a significance level of $\alpha = 0.05$ is to be used in

Table 13.3 Sample size Determination for Tests on the Mean of a Normal Distribution

	σ Known	σ Unknown
(a) $H_0:\mu = \mu_0$ vs $H_a:\mu \neq \mu_0$		
Value of ϕ:	$|\delta|/\sigma$	$|\delta|/\sigma$
Table:	A10(a), A10(d)	11
(b) $H_0:\mu \geqslant \mu_0$ vs $H_a:\mu < \mu_0$		
Value of ϕ:	$-\delta/\sigma$	$-\delta/\sigma$
Table:	A10(e), A10(h)	11
(c) $H_0:\mu \leqslant \mu_0$ vs $H_a:\mu > \mu_0$		
Value of ϕ:	δ/σ	δ/σ
Table:	A10(e), A10(h)	11

the test, Table A10(b) can be used to determine an appropriate sample size. Drawing a vertical line up from the horizontal axis at $\phi = 0.5$ and a horizontal line across the chart at $\beta = 0.1$, the two lines interesect at the curve for $n = 40$. Thus, 40 observations must be taken in order to ensure a Type II error probability of no more than 0.1 if the true mean tensile strength differs from the hypothesized one by at least one unit.

The remaining Tables A10(a)–(h) of the appendix are used in a similar fashion. Similar charts can be prepared for t-tests on means when the standard deviation is unknown. Alternatively, Table A11 in the appendix can be used. This table shows the minimum sample size needed to achieve the stated Type II error probability as a function of significance level and the scaled difference in the means, equation (13.8). As in the determination of interval estimates of the mean that have satisfactory precision, known or conservative values for $|\delta|$ and σ, or their ratio ϕ, must be available. One could also determine sample sizes for a range of choices of these parameters and select a sample size that affords acceptable protection from Type II errors over a reasonable range of the parameter values. For the same set of parameter values used above ($\alpha = 0.05$, $\beta = 0.10$, $\phi = 0.5$), Table A11 indicates that $n = 44$ observations are required.

Table 13.3 summarizes the various one- and two-sided tests on the mean of a normal distribution. For each of these situations, the value of δ and the appropriate sample-size table are indicated. Note that the tables in the appendix involve three parameters: δ, σ, and n. Specifying any two of these parameters allows one to use Tables A10(a)–(h) to select the third one. While interest usually centers on selection of the sample size, an experimenter might in some instances wish to know how sensitive is an experiment of a fixed size to differences in the means or to the size of the standard deviation.

13.2 PAIRED SAMPLES

Comparing population or process means using data from two samples requires consideration of whether the responses are collected on pairs of experimental units (or under similar experimental conditions), or whether the responses from the two samples are mutually independent. This section contains a discussion of the analysis of paired samples,

Table 13.4 Percentage Solids by Volume at Two Pipeline Locations

Sample Number	Solids (%)		Difference $d = B - A$
	Port A	Port B	
1	1.03	1.98	0.95
2	0.82	1.71	0.89
3	0.77	1.63	0.86
4	0.74	1.39	0.65
5	0.75	1.33	0.58
6	0.74	1.45	0.71
7	0.73	1.39	0.66
8	0.66	1.24	0.58
9	0.65	1.30	0.65
Average	0.77	1.49	0.73
S.D.	0.11	0.24	0.14

which is a special case of a randomized complete block design (Section 8.2). Analyses of data from block designs are discussed in Chapter 18. Analyses of means for data from two unpaired samples are given in Section 13.3.

The data in Table 13.4 contain pairs of measurements on the percentage of solids in a fixed volume of fluid flowing through a pipeline. The pairs of measurements are taken at two port locations along the pipeline. It is believed that measurements in different samples are statistically independent but the two measurements in each sample are not, because of the proximity of the ports. The investigators collected these data in order to compare the mean solids measurements from the two ports.

When two populations (or processes) are being sampled in an experiment, one can often model the responses as

$$y_{ij} = \mu_i + e_{ij}, \qquad i = 1, 2, \quad j = 1, 2, \ldots, n, \tag{13.9}$$

where y_{ij} is the jth measurement taken from the ith population, μ_i is the unknown mean of the ith population, and e_{ij} is a random error component associated with the variation in the measurements. If the observations are paired, differences in the respective pairs of observations in model (13.9) can be expressed as

$$d_j = y_{1j} - y_{2j} = (\mu_1 - \mu_2) + (e_{1j} - e_{2j})$$
$$= \mu_d + e_j, \tag{13.10}$$

where $\mu_d = \mu_1 - \mu_2$ and $e_j = e_{1j} - e_{2j}$.

The model (13.10) for the differences in the pairs of observations looks very similar to the model (13.1) for a sample of independent observations from a single population. In fact, the two models are identical. The differences d_j are statistically independent because the observations on different pairs $(j = 1, 2, \ldots, n)$ are assumed to be independent. If the original responses are normally distributed, the differences d_j can be shown to be independently normally distributed with common mean $\mu_d = \mu_1 - \mu_2$ and a common standard deviation σ_d.

Because the differences d_j can be modeled as a sample of independent observations from a normal population, the confidence-interval, hypothesis-testing, and sample-size determination procedures presented in the last section can be directly applied to drawing inferences on the mean difference μ_d. One simply replaces $\bar{y}$ and $s_{\bar{y}}$ with $\bar{d}$ and $s_{\bar{d}}$, where

$$\bar{d} = \frac{1}{n}\sum d_i = \bar{y}_1 - \bar{y}_2, \qquad s_{\bar{d}} = \frac{s_d}{n},$$

and

$$s_d = \left(\frac{1}{n-1}\sum (d_i - \bar{d})^2\right)^{1/2}. \tag{13.11}$$

Using the summary statistics in Table 13.4 on the differences, a 95% confidence interval on the mean difference $\mu_d = \mu_B - \mu_A$ is

$$0.635 < \mu_d < 0.817.$$

Since this interval does not include the value zero, one can infer from the confidence interval that a statistical test of the equality of the means would be rejected. Indeed, using the t-statistic (13.2) with the appropriate substitutions for d and $s_{\bar{d}}$, a test of $H_0: \mu_d = 0$ vs $H_a: \mu_d \neq 0$ is rejected, since $t = 15.71$ ($p < 0.001$).

Determining sample sizes needed for inferential procedures on the mean difference μ_d proceeds just as in Section 13.1. The tables used are still applicable when σ or s is replaced by σ_d or s_d, respectively, and $\delta = \mu - \mu_0$ is replaced by $\delta = \mu_1 - \mu_2$.

13.3 TWO SAMPLES

The solids measurements shown in Table 13.4 could be statistically independent if the experiment were conducted differently from the description in the last section. For example, measurements at each port could be taken at different times, the measurement sequence randomized as in a completely randomized design. Alternatively, measurements could be taken with two instruments, a different one at each port. In such an experiment the instruments would have to be calibrated relative to one another to ensure that no bias would be incurred; however, random measurement errors would still occur.

If the measurements in Table 13.4 for the two port locations are statistically independent, the procedures of the previous section are not appropriate. This is because there is no physical or logical way to form pairs of observations. The sample numbers are simply arbitrary labels. One can still model the responses as in equation (13.9), but the errors, and hence the responses, are now mutually independent. Inferences on means based on independent samples from two populations are a special case of inferences using one-factor linear models from completely randomized experimental designs (see Section 13.4 and Chapter 15).

Denote observations from two samples as $y_{11}, y_{12}, \ldots, y_{1n_1}$ and $y_{21}, y_{22}, \ldots, y_{2n_2}$, respectively. Denote the respective sample means by $\bar{y}_1$ and $\bar{y}_2$ and the sample standard deviations by s_1 and s_2. Our discussion of inferences on the mean difference from two independent samples distinguishes two cases, equal and unequal population standard deviations.

Consider first the case where the population (or process) standard deviations are assumed to be equal. Inference techniques for this assumption are discussed in

Section 14.2. When the population standard deviations are equal, inferences on the difference in the population means are made using the following statistic:

$$t = \frac{(\bar{y}_1 - \bar{y}_2) - (\mu_1 - \mu_2)}{[s_p^2(1/n_1 + 1/n_2)]^{1/2}}.$$ (13.12)

The t statistic in (13.12) uses a *pooled* estimate of the common standard deviation:

$$s_p = \left(\frac{(n_1 - 1)s_1^2 + (n_2 - 1)s_2^2}{n_1 + n_2 - 2} \right)^{1/2}$$ (13.13)

$$= \left(\frac{s_1^2 + s_2^2}{2} \right)^{1/2} \quad \text{if} \quad n_1 = n_2.$$

The denominator of the two-sample t-statistic (13.12) is the estimated standard error of the difference of the sample means:

$$SE(\bar{y}_1 - \bar{y}_2) = s_p \left(\frac{1}{n_1} + \frac{1}{n_2} \right)^{1/2}.$$

The t-statistic (13.12) follows a Student's t-distribution with $v = n_1 + n_2 - 2$ degrees of freedom. Note that the number of degrees of freedom equals the sum of the numbers for the two sample variances. Using this t-statistic, the confidence interval and hypothesis testing procedures discussed in Section 13.1 can be applied to inferences on the difference of the two means.

If one wishes to obtain a confidence interval for the difference in the mean tensile strengths of wires produced from the first two dies using the sample measurements in Table 12.1, the two-sample t-statistic can be used if one can reasonably make the assumption that the variability of tensile-strength measurements is about equal for wires from the two dies. If so, the pooled estimate of the standard deviation, using the summary statistics in Table 12.2 and equation (13.13), is

$$s_p = \left(\frac{1.408 + 2.054}{2} \right)^{1/2} = 1.761.$$

A 95% confidence interval for the difference between the mean tensile strengths for dies 1 and 2 is then

$$(84.733 - 84.492) - (2.038)(1.761)(\tfrac{1}{18} + \tfrac{1}{18})^{1/2} < \mu_1 - \mu_2$$
$$< (84.733 - 84.492) + (2.038)(1.761)(\tfrac{1}{18} + \tfrac{1}{18})^{1/2},$$

or

$$-0.955 < \mu_1 - \mu_2 < 1.437,$$

where 2.038 is the critical value for a Student t-variate having 34 degrees of freedom (linearly interpolated between $v = 30$ and $v = 60$ in Table A5 of the appendix) corresponding to an upper-tail probability of $\alpha/2 = 0.025$. Since this confidence interval includes zero, the means would not be judged significantly different by a statistical test of $H_0: \mu_1 = \mu_2$ vs $H_a: \mu_1 \neq \mu_2$.

Suppose now that $\sigma_1 \neq \sigma_2$. In this situation one should not use the pooled estimate of the variance, s_p^2. Rather than using the t-statistic (13.12), we use the following statistic which approximately follows a Student's t-distribution for normally distributed response variables:

$$t = \frac{(\bar{y}_1 - \bar{y}_2) - (\mu_1 - \mu_2)}{(s_1^2/n_1 + s_2^2/n_2)^{1/2}}. \tag{13.14a}$$

The number of degrees of freedom for this statistic is taken to be the (rounded) value of

$$v = \frac{(w_1 + w_2)^2}{w_1^2/(n_1 - 1) + w_2^2/(n_2 - 1)}, \tag{13.14b}$$

where $w_1 = s_1^2/n_1$ and $w_2 = s_2^2/n_2$.

Tables 13.5 and 13.6 summarize confidence interval and hypothesis-testing procedures for the difference of two means when independent (unpaired) observations from both populations or processes are available. Table 12 of the appendix, using the values of ϕ shown in Table 13.7, provide the necessary sample sizes when the two population standard deviations are equal. Note that the sample size read from this table applies to each sample; i.e., samples of size n are taken from each population.

Table 13.5   Confidence Intervals for the Difference in Means of Two Normal Populations: Independent (Unpaired) Samples

---

(a) *Two-Sided*

$$(\bar{y}_1 - \bar{y}_2) - t_{\alpha/2}\,\mathrm{SE} < \mu_1 - \mu_2 < (\bar{y}_1 - \bar{y}_2) + t_{\alpha/2}\,\mathrm{SE}$$

(b) *One-Sided Upper*

$$\mu_1 - \mu_2 < (\bar{y}_1 - \bar{y}_2) + t_\alpha\,\mathrm{SE}$$

(c) *One-Sided Lower*

$$(\bar{y}_1 - \bar{y}_2) - t_\alpha\,\mathrm{SE} < \mu_1 - \mu_2$$

$$\sigma_1 = \sigma_2$$

$$v = n_1 + n_2 - 2, \qquad \mathrm{SE} = s_p\left(\frac{1}{n_1} + \frac{1}{n_2}\right)^{1/2}, \qquad s_p^2 = \frac{(n_1 - 1)s_1^2 + (n_2 - 1)s_2^2}{v}$$

$$\sigma_1 \neq \sigma_2$$

$$v = \frac{(w_1 + w_2)^2}{w_1^2/(n_1 - 1) + w_2^2/(n_2 - 1)}, \qquad \mathrm{SE} = \left(\frac{s_1^2}{n_1} + \frac{s_2^2}{n_2}\right)^{1/2}, \qquad w_j = \frac{s_j^2}{n_j}$$

Table 13.6 Decision Rules and p-Value Calculations for Tests on the Difference of Means of Two Normal Populations: Independent (Unpaired) Samples*

(a) $H_0: \mu_d = \mu_0$ vs $H_a: \mu_d \neq \mu_0$

Reject H_0 if: $|t_c| > t_{\alpha/2}$

p-Value: $p = 2 \Pr\{t > |t_c|\}$

(b) $H_0: \mu_d \geq \mu_0$ vs $H_a: \mu_d < \mu_0$

Reject H_0 if: $t_c < -t_\alpha(v)$

p-Value: $p = \Pr\{t < t_c\}$

(c) $H_0: \mu_d \leq \mu_0$ vs $H_a: \mu_d > \mu_0$

Reject H_0 if: $t_c > t_\alpha$

p-Value: $p = \Pr\{t > t_c\}$

*$\mu_d = \mu_1 - \mu_2$, $t = [(\bar{y}_1 - \bar{y}_2) - \mu_0]/\text{SE}$, $t_c =$ calculated value of t; v and SE are defined in Table 13.5.

Table 13.7 Sample-Size Requirements for Testing the Difference of Means from Two Normal Populations ($\sigma_1 = \sigma_2 = \sigma$): Independent (Unpaired) Samples

(a) $H_0: \mu_1 - \mu_2 = \mu_0$ vs $H_a: \mu_1 - \mu_2 \neq \mu_0$

$$\phi = \frac{|\mu_d - \mu_0|}{\sigma} = \frac{|\delta|}{\sigma}$$

(b) $H_0: \mu_1 - \mu_2 \geq \mu_0$ vs $H_a: \mu_1 - \mu_2 < \mu_0$

$$\phi = -\frac{(\mu_d - \mu_0)}{\sigma} = -\frac{\delta}{\sigma}$$

(c) $H_0: \mu_1 - \mu_2 \leq \mu_0$ vs $H_a: \mu_1 - \mu_2 > \mu_0$

$$\phi = \frac{\mu_d - \mu_0}{\sigma} = \frac{\delta}{\sigma}$$

13.4 SEVERAL SAMPLES

There are many procedures available for drawing inferences on means of several populations or processes. The technique discussed in this section compares each of several means with an overall population or process mean.

The models (13.1) and (13.9) for one and two samples are readily generalizable to more than two samples. For experiments or samples involving k means, a model of the following form can be used:

$$y_{ij} = \mu_i + e_{ij}, \qquad i = 1, 2, \ldots, k, \quad j = 1, 2, \ldots, n. \qquad (13.15)$$

This model is equivalent to a one-factor model used to analyze data resulting from a completely randomized experimental design in Chapter 15. In the context of comparing means using data from designed experiments, general procedures are presented in Chapters 15–19 that can be used for experiments involving one or more factors using models that can be far more complex than equation (13.15). In this section we present a generalization of the techniques presented in the two previous sections of this chapter.

A technique referred to as the *analysis of means* (ANOM) is especially suited to comparisons of means from several populations or processes that have equal standard deviations. This technique can be extended to more general situations than the one considered in this section. The analysis is similar to the use of control charts (Section 4.5), but it differs in one major respect. Control-charting techniques are appropriate for the comparison of a single response or a single average response with a (known or estimated) target value. ANOM is suitable for simultaneous inferences on the comparison of several means μ_i with a target value, in this case the overall mean $\mu = \sum \mu_i / k$.

Assume that the model (13.15) is appropriate for mutually independent observations from k populations or processes. The sample sizes (n_i for the ith sample) need not be equal, but we shall only present the analysis for equal sample sizes. Assume further that the errors in the model (13.15) are normally distributed with mean zero and common standard deviation σ. Techniques for assessing the common-standard-deviation assumption are given in Section 14.3. The purpose of an ANOM is to assess whether the means in the model (13.15) are significantly different from the overall mean; i.e., whether a common-mean model such as (13.1) is appropriate. The analysis-of-means technique is summarized in Exhibit 13.2.

An example of an experiment in which these assumptions are reasonable is one that was conducted to compare the effects of filtration on deposit-forming characteristics of eight lubricants in a turbine engine. The response of interest is a deposit rating, selected to quantify different types and thicknesses of deposits on a hot-wall specimen in a test rig. In one portion of the project, the effects of different filters were ignored and deposit ratings were obtained on test runs in which no filter was present in the engine. Summary statistics for this portion of the experiment are displayed in Table 13.8.

Table 13.8   Summary Statistics for Lubricant-Deposit Study

Lubricant	Sample Size	Average	Standard Deviation
1	9	24.89	20.11
2	9	78.11	37.96
3	9	64.94	23.95
4	9	72.50	29.15
5	9	88.61	29.46
6	9	78.94	26.48
7	9	61.94	12.95
8	9	51.50	33.52

Overall average = 65.18
Pooled standard deviation = 27.68
$n = 72$

EXHIBIT 13.2 ANALYSIS-OF-MEANS PROCEDURE

1. Calculate the average response $\bar{y}_i$ for each of the k samples. Calculate the overall average $\bar{y}$ of all the data.

2. Calculate the sample standard deviation s_i for each of the k samples. Calculate a pooled standard deviation s_p by taking the positive square root of the pooled sample variance s_p^2:

$$s_p^2 = \sum s_i^2 / k.$$

3. Calculate upper (UDL) and lower (LDL) decision lines:

$$\text{UDL} = \bar{y} + h_\alpha(k, v) s_p \left(\frac{k-1}{N} \right)^{1/2}$$

$$\text{LDL} = \bar{y} - h_\alpha(k, v) s_p \left(\frac{k-1}{N} \right)^{1/2}$$

where $h_\alpha(k, v)$ is a critical value corresponding to a significance level α from Table A13 in the Appendix, $v = k(n-1)$ is the total number of degrees of freedom in the sample variances s_i^2 (and the degrees of freedom for the pooled variance estimate), and $N = nk$.

4. Plot the k sample averages versus their sample number, and superimpose the decision lines,

5. Conclude that the means are significantly different if at least one average is outside the decision lines. The mean corresponding to any average falling outside the decision lines is judged to be significantly different from the overall mean.

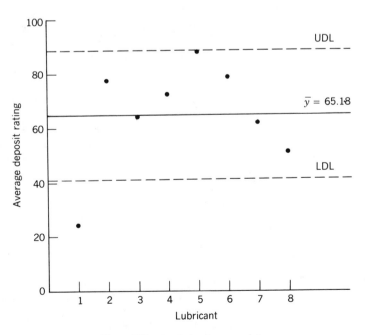

Figure 13.2 Analysis-of-means plot.

The summary statistics provided in Table 13.8 can be used to evaluate whether there are significant differences in the mean deposit ratings of the six lubricants. Using a 5% significance level with $k = 8$ and $v = 64$, the critical value from Table A13 is $h_{0.05}(8, 64)$, approximately 2.80. The respective decision lines are then

$$\text{UDL} = 65.18 + (2.80)(27.68)(\tfrac{7}{72})^{1/2} = 89.35,$$
$$\text{LDL} = 65.18 - (2.80)(27.68)(\tfrac{7}{72})^{1/2} = 41.01.$$

The lubricant averages are plotted in Figure 13.2 along with the decision lines. Based on a 5% significance level, the mean deposit ratings for only lubricant 1 would be judged to be significantly different from the overall mean deposit rating. Note that this mean is not necessarily different from those of each of the other lubricants in this study. Techniques appropriate for comparing the individual lubricant means with one another are detailed in Chapter 16.

13.5 NONPARAMETRIC LOCATION TESTS

The inference procedures for means presented in the previous sections of this chapter assume that the responses are normally distributed. As has been stressed several times, these techniques are highly robust to the effects of nonnormal responses. Two justifications have been given for applying these procedures when responses are not normally distributed: the central limit property and randomization. In general, these two justifications can be used to support the use of the normal and t sampling distributions for the sample mean even with small samples for responses from many nonnormal probability distributions.

Concern does sometimes arise over invoking the central limit property or randomization as a justification for applying the techniques of the previous sections. These concerns generally arise when data are from distributions that are highly skewed or evidence other radical departures from normality, and few observations are available. An alternative to the techniques of the previous section is a class of inference procedures referred to as *nonparametric* or *distribution-free*.

The nonparametric procedures discussed in this section are location tests, primarily on medians of distributions. They require few assumptions about the distribution of the response variable. As discussed in the next subsection, these median tests can be used as tests for means if the distribution of the responses is symmetric. Because these location tests are appropriate for many different distributions, they have less power than those of the previous sections when an assumption of normally distributed responses, the central limit property, or randomization can be used to justify the normal or t sampling distributions. They also do not readily provide interval estimates for parameters, although some nonparametric interval estimation procedures are available.

13.5.1 Location Tests for Single Samples

The term "location" is used in a general sense to denote the center of a distribution. The location parameter referred to in most nonparametric tests is the median of the distribution. These procedures may or may not be appropriate for testing hypotheses about means. If the distribution is symmetric (see Exhibit 13.3), the median and the mean coincide; consequently, a test for medians is simultaneously a test for means. The normal distribution is symmetric around its mean μ, and Student's t-distribution is symmetric around zero.

EXHIBIT 3.3

Symmetric Distribution. The distribution of a random variable y is symmetric around the value $y = m$ if the distribution to the left of $y = m$ is the mirror image of the distribution to the right of $y = m$.

One of the simplest location tests for single samples of observations is the *sign test* (see Exhibit 13.4). The sign test is used to test hypotheses on the median of a population (or a process). Responses can be continuous or discrete, so long as they are statistically independent. Let M denote the median of a population (see Section 3.2). The sign test is appropriate for testing the hypotheses

$$H_0:M = c \quad \text{vs} \quad H_a:M \neq c,$$

where c is a hypothesized value for the median.

EXHIBIT 13.4   SIGN TEST

1. Calculate the differences $d_i = y_i - c$ for each observation in a data set. Discard zero differences, if any, and reduce the sample size by the number discarded.
2. Count the number of positive values of d_i. Denote the number of positive differences by x.
3. Using Table A14 in the appendix with $\theta = \frac{1}{2}$, reject the hypothesis $H_0:M = c$ if x is in the upper or lower $100(\alpha/2)\%$ tail of the distribution.
4. For one-sided tests $H_0:M \leqslant c$ vs $H_a:M > c$, reject H_0 if x is in the $100(1-\alpha)\%$ upper tail of the distribution. To test $H_0:M \geqslant c$ vs $H_a:M < c$, reject H_0 if x is in the $100(1-\alpha)\%$ lower tail of the distribution.

The binomial distribution in Table A14 of the appendix to this book is described in the appendix to Chapter 20. The sign test is a specific application of a test of hypothesis for a binomial distribution. This specific application does not require a comprehensive discussion of the binomial probability distribution; consequently, we defer a general discussion of this distribution until Section 20.1.

To illustrate the sign test, consider again the tensile-strength data presented in Table 12.1. Suppose that one wishes to test the hypothesis $H_0:M = 84$ vs $H_a:M \neq 84$ for the tensile strengths of wire produced from die 1. Note that this is a test on the median of the distribution. The null hypothesis stipulates that die 1 produces wire for which half the tensile strengths are greater than 84,000 psi and half are less. This hypothesis is equivalent to stating that the mean tensile strength for wire produced by die 1 equals 84,000 psi if the distribution of tensile strengths is symmetric.

The second column of Table 13.9 lists the differences $y_i - 84$. There are $x = 13$ positive differences. From Table A14 in the appendix, the probability of obtaining 13 or more positive differences out of $n = 18$ comparisons when $\theta = \frac{1}{2}$ is obtained by adding the probabilities for $x = 13, 14, \ldots, 18$. The upper-tail probability equals 0.049. Since this is a two-tailed test, the significance probability is $p = 2(0.049) = 0.098$.

Table 13.9 Differences and Rankings for Tensile-Strength
Responses, Die 1

Tensile Strength y	Difference $y - 84$	Sign	Rank of $\|y - 84\|$
85.769	1.769	+	14
86.725	2.725	+	16
87.168	3.168	+	18
84.513	0.513	+	5.5
84.513	0.513	+	5.5
83.628	− 0.372	−	3
82.912	− 1.088	−	11
84.734	0.734	+	9
82.964	− 1.036	−	10
86.725	2.725	+	16
84.292	0.292	+	2
84.513	0.513	+	5.5
86.725	2.725	+	16
84.513	0.513	+	5.5
82.692	− 1.308	−	12
83.407	− 0.593	−	8
84.070	0.070	+	1
85.337	1.337	+	13

Inferences on the mean tensile strength should be made using the t-statistic (13.2) and not the sign test. Use of the t-statistic for the responses from die 1 can be recommended for any of the three reasons mentioned earlier in this section. We use this data set to illustrate nonparametric location tests in this section primarily to allow comparisons with the previous results. Recall that in Section 13.1 a test of $H_0: \mu = 84$ vs $\mu \neq 84$ was conducted using the t-statistic (13.2). The significance probability of that test was $0.02 < p < 0.05$. The t-test is significant at a 5% significance level, but the sign test is not. This is due to the inability of the sign test to incorporate information on the magnitudes of the differences d_i.

Once again, when normal or t sampling distributions for the mean can be justified, they are the preferred method of drawing inferences on location parameters. With some data sets, the sign test and the t-test yield similar conclusions. This will not always occur.

A nonparametric test that utilizes information on both the signs and the magnitudes of differences is the *Wilcoxon signed-rank test* (see Exhibit 13.5). If the assumed distribution of the responses is symmetric, the signed-rank test is simultaneously a test for the median or the mean of the distribution. The assumptions needed to apply this test are that the distribution is continuous and symmetric (if testing the mean) and that the responses are statistically independent. The continuity assumption is made primarily to ensure that there will be few, if any, differences equal to zero. The test can be used without this assumption so long as there are not a large number of tied ranks. When both the sign test and the signed-rank test are appropriate tests for location, the signed-rank test is preferred, because it uses both the signs and the ranks of the differences, whereas the sign test only uses their signs.

Since the tensile strength measurements for die 1 are continuous, the signed-rank test can be used to test $H_0: \mu = 84$ vs $H_a: \mu \neq 84$ if one can also assume the distribution is

---

EXHIBIT 13.5 SIGNED-RANK TEST

1. Calculate the differences $d_i = y_i - c$, where c is a hypothesized value for the median (mean) of the distribution of the responses. Omit differences that are zero and reduce the sample size accordingly.

2. Let n denote the sample size, reduced for zero d_i if necessary. Order the absolute values of the differences, $|d_i|$, from smallest to largest. Assign ranks 1 to n to the ordered absolute values. If several $|d_i|$ have the same value, assign to each the average of the ranks that would have been assigned to these observations had there been no ties.

3. Determine the sum T_+ of the ranks assigned to the positive d_i. Also determine the sum T_- of the ranks assigned to negative d_i. Let

$$T = \min(T_+, T_-). \tag{13.16}$$

4. Reject $H_0: M = c$ in favor of the alternative $H_a: M \neq c$ if T is in the lower $100(\alpha/2)\%$ of the distribution in Table A15 of the appendix. For a one-tailed test, use either T_- or T_+, according as the null hypothesis is $M \leq c$ or $M \geq c$. In either case the null hypothesis is rejected if the test statistic is in the lower $100\alpha\%$ of the distribution in Table A15.

5. If $n > 50$, use the approximate normal test statistic

$$z = \frac{T - n(n+1)/4}{[n(n+1)(2n+1)/24]^{1/2}}. \tag{13.17}$$

Reject $H_0: M = c$ for a two-tailed test if z is in the lower $100(\alpha/2)\%$ tail of the standard normal distribution. One-sided tests are conducted by replacing T with T_- or T_+, whichever is expected to be smaller under the alternative hypothesis, and rejecting if z is in the lower $100\alpha\%$ tail of the normal distribution.

---

symmetric. This is a reasonable assumption for this example. In the fourth column of Table 13.9 the rankings assigned to the absolute values of the differences from the second column are displayed. Note the average ranking for the four values of 0.513. The average ranking, 5.5, is the average of the rankings (4, 5, 6, 7) that these differences would receive were they all slightly different from one another.

The sum of the ranks for the positive differences is $T_+ = 127$; the sum for the negative differences is $T_- = 44$. Thus $T = 44$. With $n = 18$, this value is not smaller than the lower cutoff value from Table A15 (40 if $\alpha = 0.05$, 28 if $\alpha = 0.01$).

This example points out the need for careful consideration of assumptions appropriate for drawing inferences on parameters. The sign test does not reject the hypothesis that the median of the distribution of tensile strengths is equal to 84,000 psi, since $p = 0.098$. If one is willing to add an assumption of symmetry, the signed-rank test also does not reject that the mean (median) equals 84,000 psi ($0.05 < p < 0.10$). If one is further willing to assume that the responses are normally distributed, the t-test performed in Section 13.1 rejects the hypothesis that the mean equals 84,000 psi, since $0.02 < p < 0.05$.

While these conclusions may appear contradictory, they have much in common. None of the tests reject the respective null hypotheses at a significance level of 0.01. Both the signed-rank test and the t-test have significance probabilities around 0.05. The conclusion one ultimately draws depends on the strength of one's conviction about the model assumptions and on the consequences of making a Type I error, mistakenly concluding that the mean or median tensile strength of wire from die 1 is not 84,000 psi.

If the consequences of making a Type I error are substantial, one will probably choose a small significance level and all three procedures will lead to nonrejection of the null hypothesis. A larger sample size could be effective in resolving the seeming inconsistencies in the above conclusions if one is unsure of which of the above sets of assumptions is most appropriate for these data. A larger sample size would also increase the power of the test for all the above procedures.

13.5.2 Location Tests for Paired Samples

When observations are paired, differences in the pairs of responses can be analyzed using the sign test or the signed-rank test. The sign test is appropriate for testing

$$H_0: M_d = c \quad \text{vs} \quad H_a: M_d \neq c,$$

where M_d is the median of the distribution of the differences $d_i = y_{1i} - y_{2i}$. The signed-rank test is appropriate if the differences d_i are continuous or if there are few tied ranks. If the distribution is also symmetric, the signed-rank test can be used to test that the mean difference $\mu_d = \mu_1 - \mu_2 = c$. In particular, it can be used to test that the two means are equal, i.e., $\mu_d = 0$ or $\mu_1 = \mu_2$.

The procedures for these tests are the same as those described for single samples. One simply replaces the $d_i = y_i - c$ used in the single-sample procedures with $d_i = (y_{1i} - y_{2i}) - c$.

13.5.3 Location Tests for Two Samples

The two-sample t-statistic (13.12) requires equal population (process) standard deviations. The approximate t-statistic (13.14) can be used when this assumption does not hold. We now introduce a nonparametric location test for responses from two continuous distributions when independent (unpaired) samples are available. The distributions are assumed to be identical except possibly in location. A nonparametric test for difference in location is the *Wilcoxon rank-sum test* (see Exhibit 13.6).

EXHIBIT 13.6   RANK-SUM TEST

1. Order the combined sample of $n = n_1 + n_2$ responses. Assign the ranks 1 to n to the ordered responses. Tied responses are assigned the average of the ranks that would have been assigned had the responses not been tied.

2. Suppose $n_1 \leqslant n_2$. Let T be the sum of the ranks for the smaller sample, the sample having n_1 observations.

3. For a two-tailed test of $H_0: \mu_1 = \mu_2$ vs. $H_a: \mu_1 \neq \mu_2$, reject H_0 if T is either less than the lower or greater than the upper critical value in Table A16 in the appendix. For a one-tail test, reject H_0 if T exceeds the critical value that is appropriate for the alternative hypothesis: less than the lower value if T is expected to be small under the alternative: greater than the upper value if T is expected to be large.

4. For sample sizes larger than those in Table A16, use the following normal approximation and either one- or two-tailed critical values, as appropriate:

$$z = \frac{T - n_1(n+1)/2}{[n_1 n_2 (n+1)/12]^{1/2}}. \tag{13.18}$$

Table 13.10 Rankings for Combined Tensile-Strength Responses, Dies 1 and 2

Die 1		Die 2	
Tensile Strength	Rank	Tensile Strength	Rank
85.769	27	79.424	1
86.725	31	81.628	2
87.168	36	82.692	3.5
84.513	19	86.946	34.5
84.513	19	86.725	31
83.628	10.5	85.619	25.5
82.912	5.5	84.070	13
84.734	22	85.398	24
82.964	7	86.946	34.5
86.725	31	82.912	5.5
84.292	15.5	83.185	8
84.513	19	86.725	31
86.725	31	84.070	13
84.513	19	86.460	28
82.692	3.5	83.628	10.5
83.407	9	84.513	19
84.070	13	84.292	15.5
85.337	23	85.619	25.5

The assumption that the two distributions differ, if at all, only in location implies that the two distributions have the same shape; if overlaid on one another, the distributions would be identical. With the assumption that the two populations are identical except for a possible location shift, the rank-sum test can be used to test for the equality of means or of medians.

To illustrate the rank-sum test, consider a comparison of the tensile strengths in Table 12.1 of wire produced by the first two dies. The sum of the ranks for the wires from die 1 in the combined sample (Table 13.10) is $R_1 = 341$. For the wires from die 2, $R_2 = 325$. Since $n_1 = n_2$, T can be either sum. Let $T = 325$. From Table A16 with $n_1 = n_2 = 18$, this is not statistically significant. This conclusion is in agreement with the two-sample t-test performed in Section 13.2.

REFERENCES

Text References

Some useful texts for engineers and scientists that cover much of the material in this chapter are (see also Chapter 12):

Dixon, W. J. and Massey, F. J. (1983). *Introduction to Statistical Analysis*, New York: McGraw-Hill, Inc.

Hines, W. W. and Montgomery, D. C. (1980). *Probability and Statistics in Engineering and Management*, New York: John Wiley and Sons, Inc.

Neter, J., Wasserman, W., and Whitmore, G. A. (1982). *Applied Statistics*, Boston: Allyn and Bacon, Inc.

Ostle, B. and Malone, L. C. (1988). *Statistics in Research*, Fourth Edition, Ames, IA: Iowa State University Press.

Snedecor, G. W. and Cochran, W. G. (1980). *Statistical Methods*, Ames, IA: Iowa State University Press.

Walpole, R. E. and Myers, R. H. (1978). *Probability and Statistics for Engineers and Scientists*, New York: Macmillan Co.

The analysis-of-means procedure is detailed in the following book:

Ott, E. R. (1975). *Process Quality Control*, New York: McGraw-Hill Book Co.
   Several informative review articles also appear in the Journal of Quality Technology, Volume 15 (January 1983). Nonparametric statistical procedures can be found in most of the books listed above.

In addition, the following texts provide extensive coverage of these techniques:

Conover, W. J. (1980). *Practical Nonparametric Statistics, Second Edition.* New York: John Wiley and Sons, Inc.

Gibbons, J. D. (1976). *Nonparametric Methods for Quantitative Analysis*, New York: Holt, Reinhart, and Winston.

Hollander, M. and Wolfe, D. A. (1973). *Nonparametric Statistical Methods*, New York: John Wiley and Sons, Inc.

Data Reference

The lubrication data in Section 13.4 were taken from:

Tyler, J. C., Cuellar, J. P. Jr., and Mason, R. L. (1983). "Effects of Wear Metal on Lubrication Deposition," Aero Propulsion Laboratory, Wright-Patterson Air Force Base, Contract No. F33615-81-C-2021, San Antonio, TX: Southwest Research Institute.

EXERCISES

1   Halon 1301 is an extinguishing agent for fires resulting from flammable vapors and liquids. A chemical company producing fire-extinguishing mixtures involving Halon 1301 sampled 20 batches and measured the concentrations by volume of Halon 1301 to be:

$$6.11 \quad 5.47 \quad 5.76 \quad 5.31 \quad 5.37 \quad 5.74 \quad 5.54 \quad 5.43 \quad 6.00 \quad 6.03$$

$$5.70 \quad 5.34 \quad 5.98 \quad 5.22 \quad 5.57 \quad 5.35 \quad 5.82 \quad 5.68 \quad 6.12 \quad 6.09$$

Find a 95% confidence interval for the mean concentration, and interpret the results in the context of this exercise.

2   Using the data in Exercise 1, test the following hypotheses:

$$H_0: \mu = 5.7,$$

$$H_a: \mu \neq 5.7.$$

Use a significance level of $\alpha = 0.10$. Calculate the significance probability for the observed value of the test statistic. Interpret the test results in the context of this exercise.

3 Two research laboratories were asked to shoot a 0.05-in. steel cube into a 0.125-in.-thick aluminum target and measure the resulting hole area in the target. Assuming the standard deviations of the measurements of hole areas are equal for both labs, compute a 95% confidence interval on the difference of the two mean hole area measurements. Using the calculated confidence interval, test the hypotheses

$$H_0: \mu_1 = \mu_2,$$

$$H_a: \mu_1 \neq \mu_2.$$

The data on the hole areas (in.^{2}) are as follows:

Laboratory 1:

> .304 .305 .302 .310 .294 .293 .300 .296 .300 .298
>
> .316 .304 .303 .309 .305 .298 .292 .294 .301 .307

Laboratory 2:

> .315 .342 .323 .229 .410 .334 .247 .299 .227 .322
>
> .259 .278 .361 .349 .250 .321 .298 .329 .315 .294

4 Assuming the standard deviations of the measurements of hole size for the two research laboratories in Exercise 3 are not equal, compute an approximate 95% confidence interval on the difference of hole-area means. Intepret the confidence interval in the context of this exercise. Use the confidence interval to test the hypothesis

$$H_0: \mu_1 = \mu_2,$$

$$H_a: \mu_1 \neq \mu_2.$$

5 Using the data from Table 12.1 for all three dies, use an analysis-of-means procedure to assess whether the mean tensile strengths for wire from all three dies are equal. What assumptions are you making in order to conduct the test? Do these assumptions appear to be reasonable? Why (not)? Interpret the test results in the context of this exercise.

6 A new alloy is proposed for use in protecting inner walls of freight containers from rupture. The engineering design specification requires a mean penetration of no greater than 2.250 mm when a beam of specified weight and size is rammed into the wall at a specified velocity. Sixty test runs resulted in an average penetration of 2.147 mm and a standard deviation of 0.041 mm. Does it appear from these data that the design specifications are being met? Why (not)? Cite any assumptions needed to perform your analysis.

7 An investigation of technician differences was conducted in the hematology laboratory of a large medical facility. In one portion of the investigation, blood specimens from seven donors were given to two laboratory technicians for analysis.

Are any substantive differences detectable from the following analyses of the seven blood samples? Cite any assumptions needed to perform your analyses.

Specimen	Measurement	
	Technician 1	Technician 2
1	1.27	1.33
2	1.36	1.82
3	1.45	1.77
4	1.21	1.41
5	1.19	1.48
6	1.41	1.52
7	1.38	1.66

8   In a study to determine the effectiveness of a fortified feed in producing weight gain in laboratory animals, two groups of animals were administered a standard feed and the fortified feed over a six-week period. The weight gain (g) of each animal in the two groups is shown below. Do these data indicate that the fortified feed produces a greater mean weight gain than the standard feed? What assumptions are you making in order to perform your analysis?

Standard: 8.9 3.0 8.2 5.0 3.9 2.2 5.7 3.2 9.6 3.1 8.8

Fortified: 5.7 12.0 10.1 13.7 6.8 11.9 11.7 10.4 7.3 5.3 11.8

9   Perform a sign test and a signed-rank test on the data in Exercise 7. How do the results compare with an analysis in which the data are assumed to be normally distributed? Which analysis is most appropriate for these data? Why?

10   Perform a rank-sum test on the data in Exercise 8. How do the results compare with an analysis in which the data are assumed to be normally distributed? Which analysis is more appropriate for these data? Why?

11   The safety mechanism of a prototype of a new type of table saw was tested to determine whether the safety switch activated within 0.05 seconds after a safety cover over the saw blade was lifted. On 30 test runs, 10 of the activation times were less 0.05 seconds. Is this sufficient evidence to conclude that the median activation time is greater than 0.05 seconds? Why (not)?

12   The minor traffic accidents at the fifteen most dangerous locations in a metropolitan area were counted for ten randomly selected days during a six-month period. During the next six months, an intensive program of education and enforcement was instituted in the metropolian area. Then the accidents were again counted for ten randomly selected days during the next six-month period. Do the data below indicate a change in the mean number of accidents? Justify any assumptions needed to perform your analyses.

Prior to Program	Following Program
18, 7, 24, 11, 28,	11, 10, 3, 6, 15,
9, 15, 20, 16, 15	9, 12, 8, 6, 13

CHAPTER 14

Inferences on Standard Deviations

This chapter presents inferential techniques for measures of variability. The construction of confidence intervals and tests of hypotheses on population or process standard deviations is the primary focus. Inferential procedures are presented for parameters from one, two, or several populations. The specific topics covered in this chapter include:

- *single-sample inferences on a standard deviation,*
- *comparisons of two standard deviations using independent samples from the corresponding populations or processes, and*
- *comparisons of several standard deviations when independent samples are available to estimate each.*

The variability of response variables is characterized in statistical models by the inclusion of one or more model terms denoting uncontrollable variation. This variation may be due to numerous small influences that are either unknown or unmeasurable. For example, the observable variability in measurements of the output of an electronic circuit may be due to small voltage fluctuations, minor electrical interference from other power sources, or a myriad of other possible influences.

The error components of statistical models are not fixed constants. They are assumed to be random variables that follow a probability distribution, frequently the normal distribution. The assumption that model errors follow a probability distribution is an explicit recognition of the uncontrolled nature of response variability. This assumption enables the response variability to be characterized (e.g., through the shape of the distribution) and its magnitude to be specified through one or more of the model parameters (e.g., the standard deviation). A goal of many experiments is to draw inferences on the variability of the model errors.

When the normal probability distribution is a reasonable assumption for model errors, the variability in the distribution is entirely determined by the standard deviation σ or its square, the variance σ^2. This chapter presents inference procedures for population or process standard deviations. These procedures are based on sampling distributions for the sample variance. Although these techniques can be used to make inferences on either variances or standard deviations, our emphasis is on the standard deviation because it is a more natural measure of variation than the variance, in part because it is measured in the same units as the response variable.

14.1 SINGLE SAMPLE

Inferences on a population standard deviation from a single sample of statistically independent observations are made using the sample variance:

$$s^2 = \frac{1}{n-1}\sum(y_i - \bar{y})^2.$$

The sample standard deviation, first discussed in Section 3.1, is the positive square root of the sample variance. Note that since both the sample variance and the sample standard deviation are always nonnegative, inferences on the sample standard deviation can be made by making inferences on the sample variance. For example, a test of the hypothesis that $\sigma = 2$, say, can be made by testing that $\sigma^2 = 4$.

In Section 12.1 we noted that the variate

$$X^2 = (n-1)s^2/\sigma^2 \tag{14.1}$$

has a chi-square probability distribution with $n-1$ degrees of freedom. The chi-square distribution for the statistic (14.1) is critically dependent on the independence and normality assumptions. To obtain an interval estimate of σ we begin with the probability statement

$$\Pr\{X^2_{1-\alpha/2} < (n-1)s^2/\sigma^2 < X^2_{\alpha/2}\} = 1-\alpha,$$

or, after some algebraic rearrangement,

$$\Pr\{(n-1)s^2/X^2_{\alpha/2} < \sigma^2 < (n-1)s^2/X^2_{1-\alpha/2}\} = 1-\alpha, \tag{14.2}$$

where $X^2_{1-\alpha/2}$ and $X^2_{\alpha/2}$ are, respectively, lower and upper critical points for a chi-square variate having $v = n-1$ degrees of freedom and a probability of $\alpha/2$ in each tail of the distribution. The latter probability statement provides the basis for the following $100(1-\alpha)\%$ two-sided confidence interval for σ^2:

$$(n-1)s^2/X^2_{\alpha/2} < \sigma^2 < (n-1)s^2/X^2_{1-\alpha/2}. \tag{14.3}$$

Taking the square root of each side of the inequality (14.3) provides a $100(1-\alpha)\%$ confidence interval for the standard deviation:

$$\left(\frac{(n-1)s^2}{X^2_{\alpha/2}}\right)^{1/2} < \sigma < \left(\frac{(n-1)s^2}{X^2_{1-\alpha/2}}\right)^{1/2}. \tag{14.4}$$

The confidence interval (14.3) for σ and the corresponding one-sided intervals are exhibited in Table 14.1.

Using the tensile-strength data for die 1 given in Table 12.1, the sample variance is $s^2 = 1.982$ and the sample standard deviation is $s = (1.982)^{1/2} = 1.408$. A 95% confidence interval for the variance is

$$\frac{(17)(1.982)}{30.19} < \sigma^2 < \frac{(17)(1.982)}{7.56},$$

Table 14.1 Confidence Intervals for Standard Deviations: Normal Probability Distribution*

(a) *Two-Sided*

$$\left(\frac{(n-1)s^2}{X_{\alpha/2}^2}\right)^{1/2} < \sigma < \left(\frac{(n-1)s^2}{X_{1-\alpha/2}^2}\right)^{1/2}$$

(b) *One-Sided Upper*

$$\sigma < \left(\frac{(n-1)s^2}{X_{1-\alpha}^2}\right)^{1/2}$$

(c) *One-Sided Lower*

$$\left(\frac{(n-1)s^2}{X_{\alpha}^2}\right)^{1/2} < \sigma$$

*X_{α}^2 = chi-square critical value, $v = n - 1$.

or

$$1.116 < \sigma^2 < 4.457,$$

where 7.56 and 30.19 are the $X_{0.975}^2$ and $X_{0.025}^2$ values, respectively, based on $v = 17$ degrees of freedom. From these confidence limits, we obtain the limits for the standard deviation:

$$1.056 < \sigma < 2.111$$

In order to test a hypothesis of the form

$$H_0 : \sigma^2 = \sigma_0^2 \quad \text{vs} \quad H_a : \sigma^2 \neq \sigma_0^2$$

where σ_0^2 is a hypothesized value of σ^2, use is once again made of the variate (14.1) but with the hypothesized value of the population variance inserted:

$$X^2 = (n-1)s^2/\sigma_0^2.$$

Using this statistic, the decision rule can be formulated as follows:

Decision: Reject H_0 if $X^2 < X_{1-\alpha/2}^2$

or if $X^2 > X_{\alpha/2}^2$,

where α is the chosen significance level of the test.

This test procedure is also used to test the corresponding hypotheses on the population standard deviation. The test procedure provides exact Type I and Type II error probabilities because $\sigma = \sigma_0$ if and only if $\sigma^2 = \sigma_0^2$.

Consider again the tensile-strength experiment. Suppose we wish to test the hypothesis

$$H_0: \sigma = 2 \quad \text{vs} \quad H_a: \sigma \neq 2.$$

The equivalent test for the population variance is

$$H_0: \sigma^2 = 4 \quad \text{vs} \quad H_a: \sigma^2 \neq 4.$$

The test statistic is

$$X^2 = (17)(1.982)/4 = 8.424 \qquad (0.05 < p < 0.10).$$

If we select a significance level of $\alpha = 0.05$, the null hypothesis is not rejected, because X^2 is neither greater than $X^2_{0.025} = 30.19$ nor less than $X^2_{0.975} = 7.56$.

As discussed in Section 13.1.3, there is an equivalence between many interval estimation techniques and tests of hypotheses. The test just conducted could have been performed by examining whether the hypothesized value of the standard deviation, $\sigma = 2$, is within the limits of the confidence interval for σ. If it is, the hypothesis is not rejected; otherwise, it is rejected. Since $\sigma = 2$ is in the computed interval (1.056, 2.111), the hypothesis that $\sigma = 2$ is not rejected.

Table 14.2 contains various one-sided and two-sided tests for standard deviations. Included are both decision rules and methods for calculating the appropriate significance probabilities.

Operating-characteristic curves for determining the sample sizes to be used in testing

Table 14.2   Decision Rules and p-Value Calculations for Tests on the Standard Deviation of a Normal Distribution*

(a) $H_0: \sigma = \sigma_0 \quad \text{vs} \quad H_a: \sigma \neq \sigma_0$

Reject H_0 if: $\quad X_c^2 < X_{1-\alpha/2}^2 \quad \text{or} \quad X_c^2 > X_{\alpha/2}^2$

p-Value: $\quad\quad p = 2\min(p_l, p_u)$

$$p_l = \Pr\{X^2 < X_c^2\}$$
$$p_u = \Pr\{X^2 > X_c^2\}$$

(b) $H_0: \sigma \geq \sigma_0 \quad \text{vs} \quad H_a: \sigma < \sigma_0$

Reject H_0 if: $\quad X_c^2 < X_\alpha^2$

p-Value: $\quad\quad p = \Pr\{X^2 < X_c^2\}$

(c) $H_0: \sigma \leq \sigma_0 \quad \text{vs} \quad H_a: \sigma > \sigma_0$

Reject H_0 if: $\quad X_c^2 > X_\alpha^2$

p-Value: $\quad\quad p = \Pr\{X^2 > X_c^2\}$

---

*$X^2 = (n-1)s^2/\sigma_0^2$, $X_c^2 = $ calculated value of test statistic, $X_\alpha^2 = $ chi-square critical value, $v = n - 1$.

hypotheses on the standard deviation (or the variance) of a normal distribution can be constructed in a fashion similar to those for the mean of a normal distribution. Table A17 in the appendix contains sample-size requirements for one-sided tests on standard deviations as a function of $v = n - 1$. One enters this table by specifying the significance level α, the Type II error probability β, and the ratio $\lambda = \sigma^2/\sigma_0^2$. In this variance ratio λ, the numerator is a value for the variance under the alternative hypothesis.

Table A17 is constructed for a one-tailed test of the hypothesis

$$H_0: \sigma \leqslant \sigma_0 \quad \text{vs} \quad H_a: \sigma > \sigma_0.$$

To use the table, the experimenter must specify a minimum value for the standard deviation σ that is important to detect. The value of the variance ratio is then set equal to $\lambda = \sigma^2/\sigma_0^2$. Consideration of the Type I and Type II errors and their consequences leads to a selection of the two error probabilities α and β. The significance levels included in Table A17 are 0.05 and 0.1. The Type II error probabilities included are 0.01, 0.05, 0.1 (power 0.99, 0.95, 0.90, respectively). One enters Table A17 with the values of α, β, and λ and reads the number of degrees of freedom, $v = n - 1$, from the first column of the table.

Sample sizes for lower-tail tests on standard deviations can also be obtained from this table. To do so, reverse the roles of the two error probabilities and use the reciprocal of the variance ratio. Thus, if a significance level of 0.01 and a Type II error probability of 0.05 are desired for a lower-tailed test of the hypotheses

$$H_0: \sigma \geqslant \sigma_0 \quad \text{vs} \quad H_a: \sigma < \sigma_0,$$

enter the table with $\alpha = 0.05$, $\beta = 0.01$, and $\lambda = \sigma_0^2/\sigma^2$. The statistical test is, of course, performed with the stated significance level of 0.01.

To illustrate sample-size determination using this table, suppose in the tensile-strength example the experimenters designed a new die that is intended to reduce the variability of the measurements. A test that may be of interest is $H_0: \sigma \geqslant 2$ versus $H_a: \sigma < 2$. Rejection of this hypothesis would suggest that the new die has smaller variability than, for example, die 1. Suppose further that the experimenters wish to detect a drop of 0.5 unit in the standard deviation using a significance level of $\alpha = 0.01$ and a Type II error probability of $\beta = 0.05$.

Entering Table A17 with $\alpha = 0.05$, $\beta = 0.01$, and $\lambda = (2/1.5)^2 = 1.78$, the degrees of freedom are between 60 and 120. Thus a large number of observations are needed to detect a half-unit shift in the standard deviation with the significance level and Type II error probability selected. This example illustrates that inferences on even moderate departures from hypothesized values of standard deviations can require large sample sizes.

14.2 TWO SAMPLES

The analysis of the standard deviations from two independent samples often preceeds a two-sample t-test on the equality of two means. An inference that the two population or process standard deviations are not significantly different from each other would lead one to use the two-sample t-statistic (13.12). A conclusion that the standard deviations are significantly different would lead one to use the two-sample approximate t-statistic (13.14a). A comparison of two standard deviations is also often performed when one is interested in comparing the variability between two processes. For example, one may desire to compare the variability of the results received from two different laboratories, or from two different types of machinery.

When testing the equality of two standard deviations from normal populations the following F-variate is used:

$$F = \frac{s_1^2/\sigma_1^2}{s_2^2/\sigma_2^2}, \tag{14.5}$$

where s_1^2 is the sample variance of a random sample of size n_1 from a normal distribution with variance σ_1^2, and s_2^2 is the sample variance of a random sample of size n_2 from a normal distribution with variance σ_2^2. This F-statistic follows an F-distribution with $v_1 = n_1 - 1$ and $v_2 = n_2 - 1$ degrees of freedom. Both the assumption of independent observations between and within the samples and the assumption of normal distributions are critical assumptions for the validity of the F-distribution for this variate. It is important, therefore, that the normality assumption be examined using the graphical techniques of Chapter 5 and the procedures recommended in Section 25.1 before this variate is used to compare standard deviations.

Comparisons of standard deviations are often made through the use of the corresponding variances, for the same reasons as cited in the previous section. Using the F-variate (14.5) and its corresponding probability distribution, the following two-sided confidence interval for the ratio of population variances is obtained:

$$\frac{s_1^2}{s_2^2 F_{\alpha/2}} < \frac{\sigma_1^2}{\sigma_2^2} < \frac{s_1^2}{s_2^2 F_{1-\alpha/2}}, \tag{14.6}$$

where $F_{1-\alpha/2}$ and $F_{\alpha/2}$ are lower and upper critical points from an F-distribution having $v_1 = n_1 - 1$ and $v_2 = n_2 - 1$ degrees of freedom and a probability of $\alpha/2$ in each tail of the distribution. The upper critical value $F_{\alpha/2}$ can be read directly from Table A7 in the appendix. The lower critical value can also be obtained from this table through the following calculation:

$$F_{1-\alpha/2}\{v_1, v_2\} = 1/F_{\alpha/2}\{v_2, v_1\}, \tag{14.7}$$

where, for example, $F_{1-\alpha/2}\{v_1, v_2\}$ is an F critical value with v_1 and v_2 degrees of freedom and upper-tail probability $1 - \alpha/2$.

The sample variances for dies 1 and 2 in the tensile-strength data (see Table 12.2) are, respectively, $s_1^2 = 1.982$ and $s_2^2 = 4.219$. Thus the sample standard deviations are $s_1 = 1.408$ and $s_2 = 2.054$. A 95% confidence interval for the ratio of the population variances is

$$\frac{1.982}{(4.219)(2.68)} < \frac{\sigma_1^2}{\sigma_2^2} < \frac{1.982}{(4.219)(0.37)},$$

or

$$0.175 < \frac{\sigma_1^2}{\sigma_2^2} < 1.259,$$

where $F_{0.025}\{17, 17\} = 2.68$ and $F_{0.975}\{17, 17\} = 1/F_{0.025}\{17, 17\} = 0.37$. On the basis of

Table 14.3 Confidence Intervals for the Ratio of Standard Deviations of Two Normal Populations*

(a) *Two-Sided*

$$\left(\frac{s_1^2}{s_2^2 F_{\alpha/2}}\right)^{1/2} < \frac{\sigma_1}{\sigma_2} < \left(\frac{s_1^2}{s_2^2 F_{1-\alpha/2}}\right)^{1/2}$$

(b) *One-Sided Upper*

$$\frac{\sigma_1}{\sigma_2} < \left(\frac{s_1^2}{s_2^2 F_{1-\alpha}}\right)^{1/2}$$

(c) *One-Sided Lower*

$$\left(\frac{s_1^2}{s_2^2 F_{\alpha}}\right)^{1/2} < \frac{\sigma_1}{\sigma_2}$$

*F_a = critical value of F; $v_1 = n_1 - 1$, $v_2 = n_2 - 1$. Subscripts denote upper-tail probabilities.

this procedure, we can state with 95% confidence that the true tensile-strength variance ratio for the two dies is between 0.175 and 1.259.

One-sided confidence intervals for the ratio of two variances can be obtained from the two-sided limits in equation (14.6) by using only the upper or the lower bound, as appropriate, and replacing either $\alpha/2$ with α or $1 - \alpha/2$ with $1 - \alpha$. Confidence intervals for the ratio of the standard deviations can be obtained by taking the square roots of the limits for the variance ratio. Table 14.3 displays one- and two-sided confidence intervals for the ratio of two standard deviations.

An important hypothesis that often is of interest in an experiment is

$$H_0: \sigma_1^2 = c^2 \sigma_2^2 \quad \text{vs} \quad H_a: \sigma_1^2 \neq c^2 \sigma_2^2,$$

where c is the constant specified by the experimenter. Frequently, $c = 1$ and the test is on the equality of the two variances. Note that the equivalent test for standard deviations is

$$H_0: \sigma_1 = c\sigma_2 \quad \text{vs} \quad H_a: \sigma_1 \neq c\sigma_2.$$

The test statistic for use in testing either of these hypotheses is

$$F = s_1^2 / c^2 s_2^2. \tag{14.8}$$

If one is testing the equality of the population variances, the above F-statistic is simply the ratio of the sample variances. The decision rule for this test is

Decision: Reject H_0 if $F > F_{\alpha/2}$

or if $F < F_{1-\alpha/2}$.

Table 14.4   Decision Rules and p-Value Calculations for Tests on the Standard Deviations of Two Normal Distributions*

(a) $H_0: \sigma_1 = c\sigma_2$ vs $H_a: \sigma_1 \neq c\sigma_2$

Reject H_0 if: $F_c < F_{1-\alpha/2}$ or $F_c > F_{\alpha/2}$
p-Value: $p = 2 \min(p_l, p_u)$

$$p_l = \Pr\{F < F_c\}$$
$$p_u = \Pr\{F > F_c\}$$

(b) $H_0: \sigma_1 \geqslant c\sigma_2$ vs $H_a: \sigma_1 < c\sigma_2$

Reject H_0 if: $F_c < F_\alpha$
p-Value: $p = \Pr\{F < F_c\}$

(c) $H_0: \sigma_1 \leqslant c\sigma_2$ vs $H_a: \sigma_1 > c\sigma_2$

Reject H_0 if: $F_c > F_\alpha$
p-Value: $p = \Pr\{F > F_c\}$

*$F = s_1^2/cs_2^2$, $v_1 = n_1 - 1$, $v_2 = n_2 - 1$; F_c = calculated F-value, F_α = critical value of F.

Table 14.4 contains various one-sided and two-sided tests for the population variances. Included are both decision rules and methods for calculating the appropriate significance probabilities. Note again that these tests are identical to those that would be used to test the corresponding hypotheses on the standard deviations.

Table A18 in the appendix specifies the sample size required in order to have a specified Type II error probability when comparing two variances using independent sample variances from two normal populations. The sample sizes are stated in terms of the number of degrees of freedom, $v = n - 1$, assuming that the two sample sizes are equal. The test for which this table is appropriate is the one-sided test

$$H_0: \sigma_1 \geqslant \sigma_2 \quad \text{vs} \quad H_a: \sigma_1 < \sigma_2.$$

The table is entered by specifying values of α, β, and $\lambda = \sigma_2^2/\sigma_1^2$.

14.3   SEVERAL SAMPLES

Several useful statistical tests exist for testing the hypothesis of equal error standard deviations for samples from more than two populations or processes. Two of these are described below. Both are highly sensitive to departures from the assumption of normality; consequently, they should be used only after verification that the assumption of normally distributed errors is reasonable.

When using ANOVA models with data from designed experiments, a valuable assessment of the assumption of constant standard deviations across k factor–level combinations is given by the F-max test (see Exhibit 14.1). The F-max test is used to test the hypotheses

$$H_0: \sigma_1 = \sigma_2 = \cdots = \sigma_k \quad \text{vs} \quad H_a: \text{at least two } \sigma_i \text{ differ.}$$

The F-max test requires that repeat observations be available so that sample standard deviations can be calculated for the observations at each factor–level combination. This test can also be used when the variabilities of k populations or processes are to be compared and random samples are available from each.

EXHIBIT 14.1   *F*-MAX TEST

1. Calculate the sample standard deviation of the responses from each population or process, or each factor–level combination, as appropriate. Denote these standard deviations by s_i and the corresponding sample sizes by n_i.

2. Calculate the ratio

$$F_{max} = \left(\frac{\max(s_i)}{\min(s_i)} \right)^2. \tag{14.9}$$

3. If $n_1 = n_2 = \cdots = n_k = n$, use the critical values in Table A19 of the appendix with $v = n - 1$ to determine whether to reject the hypothesis of equal standard deviations. If the n_i are unequal but not too different, use the harmonic mean of the n_i in place of n; i.e., use $v = n - 1$ with

$$n = k \left(\sum n_i^{-1} \right)^{-1}.$$

The F-max test can be conducted for any set of k samples of observations. Small p-values imply that the assumption of equal error standard deviations is not reasonable. Otherwise, the assumption is accepted as tenable, indicating that there is no strong evidence that one or more standard deviations are much larger than the others.

To illustrate the F-max test, we return to the tensile-strength example introduced in Chapter 12. In a manufacturing process, attention is often focused on variability. The variability of the tensile strengths may be of just as great importance as the average strengths of wires produced by the dies. Using the summary statistics given in Table 12.2, $F_{max} = (2.054/0.586)^2 = 12.29$. This large a value of the F_{max} statistic causes rejection ($p < 0.001$) of the hypothesis of equal error standard deviations.

An examination of the stem-and-leaf plots in Table 12.2 reveals the reason for the rejection of this statistical test. Wire produced from die 3 is much less variable than wire from the other two dies. Even elimination of the lowest observation for die 2 (see the stem-and-leaf plot in Table 12.2) does not sufficiently reduce the variability of the remaining observations for die 2 to result in a nonsignificant F-max statistic ($F_{max} = 8.10$, $p < 0.001$).

A second commonly used test for equal error standard deviations is Bartlett's test (see Exhibit 14.2). Bartlett's test does not require equal sample sizes and is generally regarded as the best test available for the hypothesis of equal standard deviations. Its main drawbacks are its sensitivity to nonnormal errors and the somewhat more difficult calculations than the F-max test.

For the tensile-strength data, $B = 32.40$ and $v = 2$, resulting in a significance probability of $p < 0.001$. As with the F-max test, the hypothesis of equal standard deviations is rejected.

Table 14.5 illustrates the calculations for Bartlett's test for the data on the lubricant-

Table 14.5   Calculations for Bartlett's Test, Lubricant-Deposit Study

Lubricant	n_i	s_i	$(n_i - 1)\log_{10}(s_i^2)$
1	9	20.11	20.85
2	9	37.96	25.27
3	9	23.95	22.07
4	9	29.15	23.43
5	9	29.46	23.51
6	9	26.48	22.77
7	9	12.95	17.80
8	9	33.52	24.40
		Total	180.10

$$s_p^2 = 766.19 \qquad \sum(n_i - 1) = 64$$
$$B = 2.3026[\log_{10}(766.19)(64) - 180.10] = 10.36$$

EXHIBIT 14.2   BARTLETT'S TEST

1. For each of the k samples, denote the standard deviations by s_i and the corresponding sample sizes by n_i.

2. Calculate the pooled sample variance

$$s_p^2 = \frac{\sum(n_i - 1)s_i^2}{\sum(n_i - 1)}.$$

3. Compute the test statistic

$$B = \ln(s_p^2)\sum(n_i - 1) - \sum(n_i - 1)\ln(s_i^2)$$
$$= 2.3026\left\{\log_{10}(s_p^2)\sum(n_i - 1) - \sum(n_i - 1)\log_{10}(s_i^2)\right\} \qquad (14.10)$$

4. Reject the hypothesis of equal standard deviations if B exceeds an upper-tail chi-square critical value with significance level α and degrees of freedom $v = k - 1$.

---

deposit study presented in Section 13.4. The calculated value of Bartlett's statistic is 10.36, which is not statistically significant at the 1% or the 5% significance level ($v = 7$, $0.10 < p < 0.25$).

REFERENCES

Text References

Most of the references at the end of Chapters 12 and 13 discuss the topics covered in this chapter.

EXERCISES

1   A brand of electrical wire is being studied to assess its resistivity characteristics. It is known from information furnished by the manufacturer that resistivity measure-

ments can be considered normally distributed. Construct 95% confidence intervals on the mean and the standard deviation of the resistivity measurements using the data provided below. If one were testing hypotheses about the respective parameters, what values of the parameters would lead to nonrejection of a two-sided test for each?

0.141 0.138 0.144 0.142 0.139 0.146 0.143 0.142

2 Suppose the electrical wire in the previous exercise is to be compared with samples of wire from a second manufacturer using the data provided below. In particular, the means and standard deviations are to be compared to determine whether the two manufacturers' processes result in wire that have similar resistivity properties. Conduct appropriate analyses and draw conclusions about the resistivity characteristics of the two types of wire.

0.133 0.142 0.151 0.148 0.140 0.141 0.150 0.148 0.135

3 Calculate a 99% confidence interval for the standard deviation of Halon 1301 concentrations using the data from Exercise 1 in Chapter 13. Interpret the confidence interval in the context of that exercise.

4 Use the Halon 1301 concentrations in Exercise 1 of Chapter 13 to test the hypotheses

$$H_0: \mu = 0.5 \quad \text{vs} \quad H_a: \mu \neq 0.5.$$

Choose an appropriate significance level, calculate the p-value of the test, and interpret the results in the context of the exercise.

5 In Exercise 7 of Chapter 13, an examination of the laboratory techniques of two technicians was described. Investigate the variability of the blood-sample measurements taken by the technicians. Construct an appropriate confidence interval and interpret the results.

6 Investigate the variability of the weight gains of the laboratory animals in Exercise 8 of Chapter 13. Do the two feeds appear to induce equal variation in weight gains? Why (not)?

7 Construct individual 95% confidence intervals for the standard deviations of skin thickness measurements for animals receiving palcebo and experimental drugs, using the data in Table 4.1 of Chapter 4. Construct a 95% confidence interval on the ratio of the standard deviations. Compare the results of the two confidence-interval procedures. What information does each provide that the other does not? How does each of these procedures compare with a statistical test of the hypothesis that the standard deviations are equal?

8 Compare the variability of the fuel-economy measurements for the four vehicles using a single fuel in Table 3.1 of Chapter 3. Interpret the results in the context of that experiment.

9 Repeat Exercise 8 on the data for eight different fuels in Table 3.2 of Chapter 3.

Interpret the result of this comparison relative to the discussion in Section 3.1 (note Figure 3.2).

10 Four different processes can be used to manufacture circular aluminum tops for food containers. A critically important measure in the determination of whether the tops can be adequately sealed on the food containers is a "sphericity index." The data below represent sphericity index measurements for several tops from one of the four processes. The target values for the process mean and the process standard deviation are, respectively, 0.75 and 0.02. Assess whether the sphericity index measurements from this process are consistent with the target values.

Process A Sphericity Index Measurements

0.709	0.731	0.706	0.722	0.713
0.734	0.720	0.711	0.729	0.722

CHAPTER 15

Analysis of Completely Randomized Designs

In this chapter techniques for the analysis of data obtained from a completely randomized design are presented. Analysis-of-variance procedures are introduced as a methodology for the simultaneous assessment of factor effects. The analysis of complete factorial experiments in which factor effects are fixed (constant effects) is stressed. Adaptations of these procedures to designs with nested factors, to blocking designs, and to alternative models are detailed in subsequent chapters. The major topics included in this chapter are:

- *analysis-of-variance decomposition of the variation of the observed responses into components due to assignable causes and to uncontrolled experimental error,*
- *estimation of model parameters for both qualitative and quantitative factors,*
- *statistical tests of factor effects, and*
- *sample-size considerations.*

The previous two chapters of this book detail inference procedures for means and standard deviations from one or more samples. The analysis of data from many diverse types of experiments can be placed in this context. Many others, notably the multifactor experiments described in Chapters 6 to 10, require alternative inferential procedures. In this chapter we introduce general procedures for analyzing multifactor experiments.

In order to more easily focus on the concepts involved in the analysis of multifactor experiments, we restrict attention in this chapter to complete factorial experiments involving factors whose effects on a response are, apart from uncontrollable experimental error, constant. Analytic techniques for data obtained from completely randomized designs are stressed. Thus, the topics covered in this chapter are especially germane to the types of experiments discussed in Chapter 7.

Analysis-of-variance procedures are introduced in Section 15.1 for experiments involving a single factor. These procedures are generalized in Section 15.2 to multifactor experiments. Estimation of analysis-of-variance model parameters in Section 15.3 includes interval estimation and the estimation of polynomial effects for quantitative factors. Statistical tests for factor effects are presented in Section 15.4. Sample-size considerations are discussed in Section 15.5.

15.1   SINGLE-FACTOR EXPERIMENTS

Single-factor experiments with two or more fixed levels of the design factor constitute the simplest version of the fixed-effects analyses discussed in this chapter. In the following subsections, we introduce the analysis-of-variance procedures for single-factor experiments. The basic concepts and computations introduced with this simple model extend in later sections and chapters to more complex experiments.

15.1.1   Fixed Factor Effects

Factor effects are called *fixed* if the levels of the factor are specifically selected because they are the only ones for which inferences are desired (see Exhibit 15.1). These effects are modeled by unknown parameters in statistical models that relate the response variable to the factor levels (see Section 15.1.3). Fixed factor levels are not randomly selected from some population; they are intentionally chosen for investigation. This is the primary distinction between fixed and random factor levels (see Section 17.1).

EXHIBIT 15.1

Fixed Factor Effects.   Factors have fixed effects if the levels chosen for inclusion in the experiment are the only ones for which inferences are desired.

Figure 15.1 is a schematic diagram of an experiment that was conducted to study flow rates (cc/min) of a particle slurry that was pumped under constant pressure for a fixed time period through a cylindrical housing. The factor of interest in this experiment is the type of filter placed in the housing. The test sequence was randomized so that the experiment was conducted using a completely randomized design. The purpose of the experiment was to assess the effects of the filter types, if any, on the average flow rates.

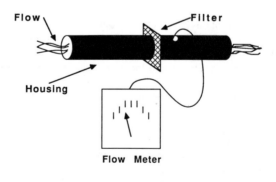

Filter	Flow Rates			
A	0.233	0.197	0.259	0.244
B	0.259	0.258	0.343	0.305
C	0.183	0.284	0.264	0.258
D	0.233	0.328	0.267	0.269

Figure 15.1   Flow-rate experiment.

The only factor in this experiment is the filter type chosen for each test run. This factor has four levels. Since inferences are desired on these specific filter types, the factor is said to have fixed effects. If only two filter types were included in the experiment, the two-sample inference procedures detailed in Chapter 13 could be used to assess the filter effects on the flow rates. In order to assess the influence of all four filter types, the inferential techniques discussed in Chapter 13 must be extended to accommodate more than two factor levels (equivalently, more than two population or process means). This is accomplished through an analysis-of-variance partitioning of the response variability.

15.1.2 Analysis of Variance

Analysis-of-variance (ANOVA) procedures separate or partition the variation observable in a response variable into two basic components: variation due to assignable causes and to uncontrolled or random variation (see Exhibit 15.2). Assignable causes refer to known or suspected sources of variation from variates that are controlled (experimental factors) or measured (covariates) during the conduct of an experiment. Random variation includes the effects of all other sources not controlled or measured during the experiment. While random sources of response variability are usually described as including only chance variation or measurement error, any factor effects that have been neither controlled nor measured are also included in the measurement of the component labeled "random" variation.

EXHIBIT 15.2   SOURCES OF VARIATION

Assignable Causes. Due to changes in experimental factors or measured covariates.

Random Variation. Due to uncontrolled effects, including chance causes and measurement errors.

Throughout earlier chapters of this book the sample variance or its square root, the standard deviation, was used to measure variability. The first step in an analysis-of-variance procedure is to define a suitable measure of variation for which a partitioning into components due to assignable causes and to random variation can be accomplished. While there are many measures that could be used, the numerator of the sample variance is used for a variety of computational and theoretical reasons. This measure of variability is referred to as the *total sum of squares* (TSS):

$$\text{TSS} = \sum_{i=1}^{n} (y_i - \bar{y})^2.$$

The total sum of squares adjusts for the overall average $\bar{y}$ by subtracting it from each individual response; consequently, it is sometimes referred to as the total *adjusted* sum of squares, or TSS(adj). Since we shall always adjust for the sample mean, we drop the word "adjusted" in the name.

For a single-factor experiment, let A denote the experimental factor. Let y_{ij} denote the observed response for the jth repeat observation ($j = 1, 2, \ldots, n_i$) corresponding to the ith level ($i = 1, 2, \ldots, a$) of the experimental factor. The total sum of squares in this type of experiment consists of the sum of the squared differences of the observations y_{ij} from the overall average response $\bar{y}_{..}$:

$$\text{TSS} = \sum_i \sum_j (y_{ij} - \bar{y}_{..})^2. \tag{15.1}$$

In order to partition the total sum of squares into components for the assignable cause (factor A) and for random (uncontrolled) variation, add and subtract the factor–level averages $\bar{y}_{i.}$ from each term in the summation for the total sum of squares. Then

$$\text{TSS} = \sum_i \sum_j (y_{ij} - \bar{y}_{i.} + \bar{y}_{i.} - \bar{y}_{..})^2$$

$$= \sum_i n_i(\bar{y}_{i.} - \bar{y}_{..})^2 + \sum_i \sum_j (y_{ij} - \bar{y}_{i.})^2$$

$$= \text{SS}_A + \text{SS}_E. \tag{15.2}$$

The first component in (15.2) is the sum of squares attributable to variation in the effects of the levels of factor A. As was mentioned in the appendix to Chapter 7 and will be discussed further in Sections 15.2 and 15.3.2, one measure of the effect of a factor–level on the response is $\bar{y}_{i.} - \bar{y}_{..}$, the difference between the average response for the ith factor–level and the overall response average. These measures are termed *main effects* (see Section 7.3 for an alternative definition for two-level factors). The sum of squares for factor A is thus a weighted sum of squares of the factor effects, the weights being the respective sample sizes n_i. This sum of squares is zero only when all the main effects are zero.

The second component in (15.2) is the sum of squares associated with the uncontrolled variability in the responses. This sum of squares is actually the numerator of a pooled sample variance [cf. equation (13.13) for $k = 2$]; i.e., $s_p^2 = \text{SS}_E/v$, where SS_E is as expressed in (15.2) or, equivalently,

$$\text{SS}_E = \sum_i (n_i - 1)s_i^2, \qquad v = \sum_i (n_i - 1), \tag{15.3}$$

and $s_i^2 = \sum_j(y_{ij} - \bar{y}_{i.})^2/(n_i - 1)$ is the sample variance of the repeat observations for the ith level of the experimental factor. Thus SS_E provides an overall measure of uncontrolled variability using the variance estimates from the repeat tests for each of the factor levels.

The sum of squares for the experimental factor is a function of the averages for the factor levels. In the context of sampling from different populations, these averages are sometimes referred to as *group* averages and SS_A as the sum of squares *between groups*. The error sum of squares is then referred to as the sum of squares *within groups*.

Associated with the total sum of squares and each of its partitioned components is a quantity called the number of its *degrees of freedom*. Degrees of freedom indicate how many statistically independent response variables or functions of the response variables comprise a sum of squares. For example, if $n = \sum n_i$ response variables y_{ij} are statistically independent, the statistic

$$\sum_i \sum_j y_{ij}^2$$

has n degrees of freedom, because each term in the summation is statistically independent of each of the other terms. The total sum of squares (15.1) does not have $n = \sum_i n_i$ degrees of freedom, because each of the terms is not statistically independent of all the others. The n terms $y_{ij} - \bar{y}_{..}$ sum to zero; consequently, any one of the terms equals the negative of the sum of the remaining $n - 1$ terms. For this reason, TSS has only $n - 1$ degrees of freedom.

Each of the terms $\sum_j(y_{ij} - \bar{y}_{i.})^2$ $(i = 1, 2, \ldots, a)$ in SS_E has the same zero constraint as the total sum of squares. The individual terms therefore have $n_i - 1$ degrees of freedom, and SS_E has $v = \sum(n_i - 1) = n - a$ degrees of freedom. The sum of squares for factor A has a similar zero constraint, viz., $\sum n_i(\bar{y}_i - \bar{y}..) = 0$. Thus, SS_A has $a - 1$ degrees of freedom.

The partitioning of the total sum of squares is summarized in an analysis-of-variance (ANOVA) table. Table 15.1 is a symbolic ANOVA table. Along with the sums of squares and the numbers of degrees of freedom, additional quantities are included in ANOVA tables. The mean squares are the respective sums of squares divided by their numbers of degrees of

Table 15.1 Symbolic Analysis-of-Variance Table for Single-Factor Experiments

ANOVA

Source of Variation	Degrees of Freedom (df)	Sum of Squares	Mean Square	F-Value
A	$a - 1$	SS_A	$MS_A = SS_A/(a - 1)$	$F_A = MS_A/MS_E$
Error	$n - a$	SS_E	$MS_E = SS_E/(n - a)$	
Total	$n - 1$	TSS		

Calculations

$$TSS = \sum_i \sum_j y_{ij}^2 - SS_M \qquad SS_M = n^{-1}(y..^2)$$

$$SS_A = \sum_i \frac{y_{i.}^2}{n_i} - SS_M \qquad SS_E = TSS - SS_A$$

$$y.. = \sum_i \sum_j y_{ij} \qquad y_{i.} = \sum_j y_{ij} \qquad n = \sum_i n_i$$

Table 15.2 Analysis of Variance Table for Flow-Rate Data

ANOVA

Source of Variation	Degrees of Freedom (df)	Sum of Squares	Mean Square	F-Value
Filter	3	0.00820	0.00273	1.86
Error	12	0.01765	0.00147	
Total	15	0.02585		

Data

Filter	Flow Rates				Total
A	0.233	0.197	0.259	0.244	0.933
B	0.259	0.258	0.343	0.305	1.165
C	0.183	0.284	0.264	0.258	0.989
D	0.233	0.328	0.267	0.269	1.097
Total					4.184

freedom. The F-statistic is the mean square due to the experimental factor divided by the mean square due to error. This F-statistic generalizes the two-sample t-statistic (13.12) to comparisons of more than two population or process means. Also shown in Table 15.1 are computational formulas for the calculation of the sums of squares. These formulas are recommended when computations are performed on a calculator, and they are applicable to the calculation of sums of squares for any single-factor experiment.

Table 15.2 displays summary statistics and the ANOVA table for the flow rate data of Figure 15.1. It is interesting to note that the F-statistic is not appreciably larger than 1. In Section 15.4 we shall see how to interpret this finding.

Quantities in an ANOVA table are functions of parameter estimates. The parameters being estimated belong to statistical models that are used to characterize the relationship between the response variable, the assignable causes, and random error. In Chapters 13 and 14, statistical models were introduced to describe the relationship between a response variable and key parameters of distributions: the means and standard deviations. We now wish to extend those models to responses from single-factor experiments and to show the relationship between the ANOVA sums of squares and the parameter estimates of the statistical models.

15.1.3 ANOVA Models

One can model the responses in a single-factor fixed-effects experiment as

$$y_{ij} = \mu_i + e_{ij}, \qquad i = 1, 2, \ldots, a, \quad j = 1, 2, \ldots, n_i. \tag{15.4}$$

The assignable-cause portion of the model can be expressed as

$$\mu_i = \mu + \alpha_i. \tag{15.5}$$

The expression of the model as in equation (15.4) stresses the connection between single-factor experiments and sampling from a number of populations or processes that differ only in location. The expression (15.5) of the model means as an overall mean μ plus parameters α_i facilitates an understanding of the analysis of these effects using the ANOVA procedures discussed in this chapter.

The α_i can be reexpressed using (15.5) as $\alpha_i = \mu_i - \mu$. In this form, these parameters are similar to the expressions for main effects, $\bar{y}_{i.} - \bar{y}_{...}$ Since the α_i in the model (15.4) are fixed constants, the model is termed a fixed-effects ANOVA model. The analysis of experiments containing a single fixed-effects factor can be accomplished through the fitting of model (15.4), reexpressed using (15.5):

$$y_{ij} = \mu + \alpha_i + e_{ij}. \tag{15.6}$$

The model (15.6) does not uniquely define the fixed-effects parameters μ and α_i. Since $\mu_i = \mu + \alpha_i$, one might think that a unique value of μ_i would result in unique values for μ and α_i. This is not so, because for any choices of μ and α_i for which $\mu_i = \mu + \alpha_i$, it is also true that $\mu_i = \mu^* + \alpha_i^*$, where $\mu^* = \mu + c$, $\alpha_i^* = \alpha_i - c$, and c is any constant. In order to uniquely specify μ and α_i a constraint must be imposed on the values of these parameters. The most commonly imposed constraint is $\sum n_i \alpha_i = 0$ or, for balanced designs, the equivalent constraint $\sum \alpha_i = 0$. Multiplying both sides of (15.5) by n_i, summing, and imposing this constraint results in the subsequent constraint that μ is a weighted average of the μ_i:

$$\mu = n^{-1} \sum n_i \mu_i.$$

If the design is balanced, μ becomes the ordinary average of the μ_i.

The error terms in the model (15.6) are generally assumed to follow a common distribution, usually a normal distribution. Estimation of the common error standard deviation permits the construction of interval estimates and the testing of hypotheses about model parameters. We discuss the estimation of the parameters of the models (15.4) and (15.6) in Section 15.3. We now wish to generalize the developments of this section to multifactor experiments.

15.2 MULTIFACTOR EXPERIMENTS

The pilot-plant chemical-yield study introduced in Section 7.2 is an experiment in which all the factors have fixed effects. Table 15.3 lists the factor–level combinations and responses for this experiment. The two temperatures, the two concentrations, and the two catalysts were specifically chosen and are the only ones for which inferences can be made. One cannot infer effects on the chemical yield for factor levels that are not included in the experiment—e.g., that the effect of a temperature of 170°C would be between those of 160 and 180°C.

At times one is willing to assume, apart from experimental error, that there is a smooth functional relationship between the quantitative levels of a factor and the response. For example, one might be willing to assume that temperature has a linear effect on the chemical yield. If so, the fitting of a model using two levels of temperature would allow one to fit a straight line between the average responses for the two levels and infer the effect of a temperature of, say, 170°C. In this setting the levels of temperature are still fixed and inferences can only be drawn on the temperature effects at these two

Table 15.3 Test Results from Pilot-Plant Chemical-Yield Experiment*

Temperature (°C)	Concentration (%)	Catalyst	Yield (g)
160	20	C_1	59, 61
		C_2	50, 54
	40	C_1	50, 58
		C_2	46, 44
180	20	C_1	74, 70
		C_2	81, 85
	40	C_1	69, 67
		C_2	79, 81

*Two repeat test runs for each factor–level combination.
Data from Box, G. E. P., Hunter, W. G., and Hunter, J. S. (1978), *Statistics for Experimenters*, New York: John Wiley & Sons, Inc. Copyright 1978, John Wiley & Sons, Inc. Used with permission.

levels. It is the added assumption of a linear effect of temperature that allows inferences for temperatures other than the two actually included in the experiment.

In the previous section, the assumptions for a single-factor, fixed-effects ANOVA model were described. The assumptions that accompany fixed-effects ANOVA models are listed in Exhibit 15.3.

EXHIBIT 15.3   FIXED-EFFECTS MODEL ASSUMPTIONS

1. The levels of all factors in the experiment represent the only levels of which inferences are desired.

2. The analysis-of-variance model contains parameters (unknown constants) for all main effects and interactions of interest in the experiment.

3. The experimental errors are statistically independent.

4. The experimental errors are satisfactorily modeled by the normal probability distribution with mean zero and (unknown) constant standard deviation.

A statistical model for the pilot-plant yield example is

$$y_{ijkl} = \mu_{ijk} + e_{ijkl}, \qquad i = 1, 2, \quad j = 1, 2, \quad k = 1, 2, \quad l = 1, 2. \tag{15.7}$$

In this model, μ_{ijk} represents the effects of the assignable causes and e_{ijkl} represents the random error effects. The assignable cause portion of the model can be further decomposed into terms representing the main effects and the interactions among the three model factors:

$$\mu_{ijk} = \mu + \alpha_i + \beta_j + \gamma_k + (\alpha\beta)_{ij} + (\alpha\gamma)_{ik} + (\beta\gamma)_{jk} + (\alpha\beta\gamma)_{ijk}. \tag{15.8}$$

The subscript i in the model (15.7) refers to one of the levels of temperature, the subscript j to one of the levels of concentration, the subscript k to one of the catalysts, and the subscript l to one of the two repeat tests. The form of the model (15.7) is a generalization of the model (15.4) for single-factor experiments, and (15.8) generalizes (15.5).

The parameters in (15.8) having a single subscript represent main effects for the factor identified by the subscript. Two-factor interactions are modeled by the terms having two subscripts, and the three-factor interaction by terms having three subscripts. The first term in equation (15.8) represents the overall response mean, i.e., a term representing an overall population or process mean from which the various individual and joint factor effects are measured.

The interaction components of (15.8) model joint effects that cannot be suitably accounted for by the main effects. Each interaction effect measures the incremental joint contribution of factors to the response variability over what can be measured by the main effects and lower-order interactions. Thus, a two-factor interaction is present in a model only if the effects of two factors on the response cannot be adequately modeled by the main effects. Similarly, a three-factor interaction is included only if the three main effects and the three two-factor interactions do not satisfactorily model the effects of the factors on the response.

The parameters representing the assignable causes in (15.8) can be expressed in terms of the model means μ_{ijk}. In order for this representation to be unique, we must again impose constraints on the values of the parameters. The constraints we impose are similar to the constraints imposed on the parameters of the model (15.6) for single-factor models:

$$\sum_i \alpha_i = 0, \qquad \sum_j \beta_j = 0, \qquad \sum_k \gamma_k = 0, \qquad \sum_i (\alpha\beta)_{ij} = 0, \qquad \sum_j (\alpha\beta)_{ij} = 0, \qquad \text{etc.}$$

Thus, the constraints require that each of the parameters in (15.8) must sum to zero over any of its subscripts.

With these constraints, the effects parameters can be expressed in terms of averages of the model means. For example,

$$\mu = \bar{\mu}_{\cdots} \quad \alpha_i = \bar{\mu}_{i\cdots} - \bar{\mu}_{\cdots} \quad \beta_j = \bar{\mu}_{\cdot j\cdot} - \bar{\mu}_{\cdots} \quad \gamma_k = \bar{\mu}_{\cdot\cdot k} - \bar{\mu}_{\cdots},$$

$$(\alpha\beta)_{ij} = \bar{\mu}_{ij\cdot} - \bar{\mu}_{i\cdots} - \bar{\mu}_{\cdot j\cdot} + \bar{\mu}_{\cdots}, \qquad \text{etc.},$$

$$(\alpha\beta\gamma)_{ijk} = \mu_{ijk} - \bar{\mu}_{ij\cdot} - \bar{\mu}_{i\cdot k} - \bar{\mu}_{\cdot jk}$$

$$+ \bar{\mu}_{i\cdots} + \bar{\mu}_{\cdot j\cdot} + \bar{\mu}_{\cdot\cdot k} - \bar{\mu}_{\cdots}.$$

$$(15.9)$$

Note that main-effect parameters are simply the differences between the model means for one level of a factor and the overall model mean. The two-factor interaction parameters are the differences in the main effects for one level of one of the factors at a fixed level of a second and the main effect of the first factor:

$$(\alpha\beta)_{ij} = (\bar{\mu}_{ij\cdot} - \bar{\mu}_{\cdot j\cdot}) - (\bar{\mu}_{i\cdots} - \bar{\mu}_{\cdots})$$

$$= (\text{main effect for } A \text{ at level } j \text{ of } B)$$

$$- (\text{main effect for } A).$$

The three-factor interaction can be similarly viewed as the difference in the two-factor interaction between any two of the factors, say A and B, at a fixed level of the third factor, say C, and the two-factor interaction between A and B:

$$(\alpha\beta\gamma)_{ijk} = (\mu_{ijk} - \bar{\mu}_{i\cdot k} - \bar{\mu}_{\cdot jk} + \bar{\mu}_{\cdot\cdot k})$$

$$- (\bar{\mu}_{ij\cdot} - \bar{\mu}_{i\cdots} - \bar{\mu}_{\cdot j\cdot} + \bar{\mu}_{\cdots})$$

$$= (A \times B \text{ interaction at level } k \text{ of } C)$$

$$- (A \times B \text{ interaction}).$$

It is important to note that while the means in the model (15.7) are uniquely defined, main-effect and interaction parameters can be introduced in many ways; i.e., they need not be defined as in (15.8). The form (15.8) is used in this book explicitly to facilitate an understanding of inference procedures for multifactor experiments.

Even though the main-effect and interaction parameters can be defined in many ways, certain functions of them are uniquely defined in terms of the model means. The functions that are uniquely defined are precisely those that allow comparison among factor–level effects. We shall return to this topic in Section 15.3.

When considering whether to include interaction terms in a statistical model, one convention that is adopted is to use only hierarchical factor effects. A model is hierarchical (see Exhibit 15.4) if any interaction term involving k factors is included in the specification of the model only if the main effects and all lower-order interactions involving two, three,..., $k-1$ of the factors are also included in the model. In this way, a high-order interaction term is only added to the model if the main effects and lower-order interactions are not able to satisfactorily model the response.

EXHIBIT 15.4

Hierarchical Model. A statistical model is hierarchical if an interaction term involving k factors is included only when the main effects and all lower-order interaction terms involving the k factors are also included in the model.

We return to analysis-of-variance models in the next section, where the estimation of the model parameters is discussed. We now wish to examine the analysis-of-variance partitioning of the total sum of squares for multifactor experiments.

The response variability can be partitioned into components for each of the main effects and interactions and for random variability. This partitioning is accomplished as a straightforward generalization of the decomposition for single-factor models when the design is balanced. When the design is unbalanced, the partitioning can still be accomplished, but it requires a slightly different technique. We consider each of these design settings in the following two subsections.

15.2.1 ANOVA Tables for Balanced Designs

Balanced complete factorial experiments have an equal number of repeat tests for all factor–level combinations. The pilot-plant experiment shown in Table 15.3 is balanced, because each of the factor–level combinations appears twice in the experiment. We shall use this example and the three-factor model (15.7) to illustrate the construction of analysis-of-variance tables for complete factorial experiments from balanced designs. The three-factor model is sufficiently complex to demonstrate the general construction of main effects and interaction sums of squares.

Generalize the model (15.7) to three factors, of which factor A has a levels (subscript i), factor B has b levels (subscript j), and factor C has c levels (subscript k). Let there be r repeat tests for each combination of the factors. Thus the experiment consists of a complete factorial experiment in three factors with r repeat tests per factor–level combination.

Designate the contribution of all the assignable causes as the *model* sum of squares, MSS:

$$\text{MSS} = r \sum_i \sum_j \sum_k (\bar{y}_{ijk\bullet} - \bar{y}_{\dots\dots})^2.$$

The model sum of squares can be partitioned into the following components:

$$\text{MSS} = \text{SS}_A + \text{SS}_B + \text{SS}_C + \text{SS}_{AB} + \text{SS}_{AC} + \text{SS}_{BC} + \text{SS}_{ABC}, \tag{15.10}$$

where SS_A, SS_B, and SS_C measure the individual contributions (main effects of the three experimental factors) to the total sum of squares, SS_{AB}, SS_{AC}, and SS_{BC} measure the two-factor joint effects of the factors on the response, and SS_{ABC} measures the joint effects of all three experimental factors above the contributions measured by the main effects and the two-factor interactions.

The sums of squares for the factor main effects are multiples of the sum of the squared differences between the averages for each level of the factor and the overall average; e.g.,

$$SS_A = bcr \sum_{i=1}^{a} (\bar{y}_{i...} - \bar{y}_{....})^2.$$ (15.11)

The differences between the averages for each level of a factor and the overall average yields a direct measure of the individual factor–level effects, the main effects of the factor. The sum of squares (15.11) is a composite measure of the combined effects of the factor levels on the response. Note that the multiplier in front of the summation sign is the total number of observations that are used in the calculation of the averages $\bar{y}_{i...}$ for each level of factor A. Sums of squares for each of the main effects are calculated as in (15.11). (Note: computational formulas are shown below in Table 15.4.)

Just as the main effect for a factor can be calculated by taking differences between individual averages and the overall average, interaction effects can be calculated by taking differences between (a) the main effects of one factor at a particular level of a second factor and (b) the overall main effect of the factor. For example, the interaction effects for factors A and B can be obtained by calculating:

(Main effect for A at level j of B) $-$ (main effect for A)

$$= (\bar{y}_{ij..} - \bar{y}_{.j..}) - (\bar{y}_{i...} - \bar{y}_{....})$$

$$= \bar{y}_{ij..} - \bar{y}_{i...} - \bar{y}_{.j...} + \bar{y}_{.....}.$$

This is the same expression one would obtain by reversing the roles of the two factors. The interaction effects can be combined in an overall sum of squares for the interaction effects of factors A and B:

$$SS_{AB} = cr \sum_{i=1}^{a} \sum_{j=1}^{b} (\bar{y}_{ij..} - \bar{y}_{i...} - \bar{y}_{.j..} + \bar{y}_{....})^2.$$ (15.12)

The sum of squares for the three-factor interaction, SS_{ABC}, can be obtained in a similar fashion by examining the differences between the interaction effects for two of the factors at fixed levels of the third factor and the overall interaction effect of the two factors. The result of combining these three-factor interaction effects is

$$SS_{ABC} = r \sum_{i=1}^{a} \sum_{j=1}^{b} \sum_{k=1}^{c} (\bar{y}_{ijk.} - \bar{y}_{ij..} - \bar{y}_{i.k.} - \bar{y}_{.jk.}$$

$$+ \bar{y}_{i...} + \bar{y}_{.j..} + \bar{y}_{..k.} - \bar{y}_{....})^2.$$ (15.13)

The final term needed to complete the partitioning of the total sum of squares is the error sum of squares, SS_E. One can show algebraically that the error sum of squares is simply the sum of the squared differences between the individual repeat-test responses and the averages for the factor–level combinations:

$$SS_E = \sum_{i=1}^{a} \sum_{j=1}^{b} \sum_{k=1}^{c} \sum_{l=1}^{r} (y_{ijkl} - \bar{y}_{ijk.})^2.$$ (15.14)

The information about factor effects that is obtained from the partitioning of the total sum of squares is summarized in an analysis-of-variance table. A symbolic ANOVA table for the three-factor interaction model we have been considering is shown in Table 15.4. In this table, the first column, "source of variation," designates the partitioning of the response variability into the various components included in the statistical model.

The second column, "degrees of freedom," partitions the sample size into similar components that relate the amount of information obtained on each factor, the interaction, and the error term of the model. Note that each main effect has one less degree of freedom than the number of levels of the factor. Interaction degrees of freedom are conveniently calculated as products of the degrees of freedom of the respective main effects. The number of error degrees of freedom is the number of factor–level combinations multiplied by one less than the number of repeat tests for each combination. These degrees of freedom can all be found by identifying constraints similar to those described in Section 15.1.2.

The third column of Table 15.4 contains the sums of squares for various main effects and interactions, and the error components of the model. The fourth column, "mean squares," contains the respective sums of squares divided by their numbers of degrees of freedom. These statistics are used for forming the F-ratios in the next column, each main effect and interaction mean square being divided by the error mean square.

Table 15.5 lists summary statistics used in the calculation of the sums of squares for the pilot-plant chemical-yield experiment. The sums of squares can be calculated using these

Table 15.4 Symbolic Analysis-of-Variance Table for Three Factors

Source of Variation	Degrees of Freedom (df)	Sum of Squares	Mean Square	F-Value
A	$a-1$	SS_A	$MS_A = SS_A/df(A)$	$F_A = MS_A/MS_E$
B	$b-1$	SS_B	$MS_B = SS_B/df(B)$	$F_B = MS_B/MS_E$
C	$c-1$	SS_C	$MS_C = SS_C/df(C)$	$F_C = MS_C/MS_E$
AB	$(a-1)(b-1)$	SS_{AB}	$MS_{AB} = SS_{AB}/df(AB)$	$F_{AB} = MS_{AB}/MS_E$
AC	$(a-1)(c-1)$	SS_{AC}	$MS_{AC} = SS_{AC}/df(AC)$	$F_{AC} = MS_{AC}/MS_E$
BC	$(b-1)(c-1)$	SS_{BC}	$MS_{BC} = SS_{BC}/df(BC)$	$F_{BC} = MS_{BC}/MS_E$
ABC	$(a-1)(b-1)(c-1)$	SS_{ABC}	$MS_{ABC} = SS_{ABC}/df(ABC)$	$F_{ABC} = MS_{ABC}/MS_E$
Error	$abc(r-1)$	SS_E	$MS_E = SS_E/df(Error)$	
Total	$abcr-1$	TSS		

ANOVA (table header)

Calculations

$$\text{TSS} = \sum_i \sum_j \sum_k \sum_l y_{ijkl}^2 - SS_M \qquad SS_M = y_{....}^2/n$$

$$SS_A = \sum_i y_{i...}^2/bcr - SS_M \qquad SS_{AB} = \sum_i \sum_j y_{ij..}^2/cr - SS_M - SS_A - SS_B$$

$$SS_B = \sum_j y_{.j..}^2/acr - SS_M \qquad SS_{AC} = \sum_i \sum_k y_{i.k.}^2/br - SS_M - SS_A - SS_C$$

$$SS_C = \sum_k y_{..k.}^2/abr - SS_M \qquad SS_{BC} = \sum_j \sum_k y_{.jk.}^2/ar - SS_M - SS_B - SS_C$$

$$SS_{ABC} = \sum_i \sum_j \sum_k y_{ijk.}^2/r - SS_M - SS_A - SS_B - SS_C - SS_{AB} - SS_{AC} - SS_{BC}$$

$$SS_E = \text{TSS} - SS_A - SS_B - SS_C - SS_{AB} - SS_{AC} - SS_{BC} - SS_{ABC}$$

$$y_{....} = \sum_i \sum_j \sum_k \sum_l y_{ijkl} \qquad y_{i...} = \sum_j \sum_k \sum_l y_{ijkl} \qquad n = abcr$$

$$y_{ij..} = \sum_k \sum_l y_{ijkl} \qquad y_{ijk.} = \sum_l y_{ijkl} \qquad df(A) = a-1, \quad \text{etc.}$$

Table 15.5 Summary Statistics for Pilot-Plant Chemical-Yield Experiment

	Factor	
Symbol	Subscript	Label
A	i	Temperature
B	j	Concentration
C	k	Catalyst

$$\sum_i \sum_j \sum_k \sum_l y_{ijkl}^2 = 68{,}748 \qquad y_{....}^2 = (1028)^2 = 1{,}056{,}784$$

$$\sum_i y_{i...}^2 = (422)^2 + (606)^2 = 545{,}320$$

$$\sum_j y_{.j..}^2 = (534)^2 + (494)^2 = 529{,}192$$

$$\sum_k y_{..k.}^2 = (508)^2 + (520)^2 = 528{,}464$$

$$\sum_i \sum_j y_{ij..}^2 = (224)^2 + (198)^2 + (310)^2 + (296)^2 = 273{,}096$$

$$\sum_i \sum_k y_{i.k.}^2 = (228)^2 + (194)^2 + (280)^2 + (326)^2 = 274{,}296$$

$$\sum_j \sum_k y_{.jk.}^2 = (264)^2 + (270)^2 + (244)^2 + (250)^2 = 264{,}632$$

$$\sum_i \sum_j \sum_k y_{ijk.}^2 = (120)^2 + (104)^2 + (108)^2 + (90)^2$$
$$+ (144)^2 + (166)^2 + (136)^2 + (160)^2 = 137{,}368$$

Table 15.6 ANOVA Table for Pilot-Plant Chemical-Yield Study

Source of Variation	df	Sum of Squares	Mean Square	F-Value
Temperature T	1	2116.00	2116.00	264.50
Concentration Co	1	100.00	100.00	12.50
Catalyst Ca	1	9.00	9.00	1.13
$T \times$ Co	1	9.00	9.00	1.13
$T \times$ Ca	1	400.00	400.00	50.00
Co $\times$ Ca	1	0.00	0.00	0.00
$T \times$ Co $\times$ Ca	1	1.00	1.00	0.13
Error	8	64.00	8.00	
Total	15	2699.00		

statistics and the computational formulas shown in Table 15.4. The complete ANOVA table is displayed in Table 15.6. Several of the F-statistics in the table are much larger than 1. The interpretation of these large F-values will be discussed in Section 15.4.

15.2.2 ANOVA Tables for Unbalanced Designs

Use of the formulas shown in Table 15.4 to measure main effects and interactions is only appropriate for factorial experiments conducted in certain types of balanced completely randomized, randomized complete block, and nested designs. They can be used with all balanced completely randomized designs involving complete factorial experiments.

Fractional factorial experiments, screening experiments, experiments involving covariates, and a wide variety of experiments that require unbalanced designs may utilize models similar to (15.7), but the determination of the sums of squares and degrees of freedom for the ANOVA table for such experiments cannot always be done as indicated above. We now present a general technique for investigating factor and covariate effects, which can be used with any experimental design. These procedures can also be used with nonexperimental data such as often accompany the fitting of regression models (Chapters 21–22).

Two statistical models can be readily compared with respect to their ability to adequately account for the variation in the response if one of the models has terms that are a subset of the terms in the other model. Such *subset hierarchical models* (see Exhibit 15.5) allow one to assess main effects and interactions by comparing error sums of squares from different model fits. Error sums of squares can be calculated for any model, whether the design is balanced or not, and regardless of whether the model terms represent design factors or covariates.

EXHIBIT 15.5

Subset Hierarchical Models. Two statistical models are referred to as *subset hierarchical models* when the following two conditions are statisfied:

    (1) each model contains only hierarchical terms, and
    (2) one model contains terms that are a subset of those in the other model.

With two subset hierarchical models, one can calculate the reduction in the error sum of squares (see Exhibit 15.6) between the simpler model and the more complex model as a measure of the effect of adding the specific terms in the larger model. With complete factorial experiments in balanced designs these reductions in error sums of squares are exactly equal to the sums of squares for the main effects and interactions (e.g., Table 15.4). When the designs are unbalanced or covariates are included, the reduction in error sums of squares provides the appropriate measure of the effects of the various model factors and covariates on the response.

EXHIBIT 15.6   REDUCTION IN ERROR SUMS OF SQUARES

Symbolically denote two statistical models by M_1 and M_2. Assume that the models are subset hierarchical models with the terms in M_2 a subset of those in M_1. Denote the respective error sums of squares by SSE_1 and SSE_2. The reduction in error sums of squares, $R(M_1|M_2)$, is defined as

$$R(M_1|M_2) = SSE_2 - SSE_1.$$

The number of degrees of freedom for this reduction equals the difference in the numbers of degrees of freedom for the two error sums of squares.

The reason for introducing the concept of reduction in error sums of squares is that even for very complicated models and/or experimental designs this procedure can be used to assess the effects of model factors and covariates. Statistical analyses of such models will ordinarily be performed with the aid of suitable computer software, not computing formulas. Thus, the ability to correctly formulate models and identify the appropriate sums of squares on computer printout can enable one to investigate many of these complicated designs and models. Examples of the use of reductions in error sums of squares are given in Chapter 18 for blocking designs and in Chapter 19 for models with covariates. The procedures are general, however, and are not restricted to any specific design or model.

To illustrate the principle of reduction in error sums of squares, the pilot-plant chemical-yield data of Table 15.3 were altered as shown in Table 15.7. The formulas in Table 15.4 cannot be now used to obtain the sums of squares, because the design is unbalanced. One factor–level combination does not have any response values. The formulas required to calculate the sums of squares depend on the model being fitted and the number of observations for the various factor–level combinations. Rather than discuss formulas that would be appropriate for only this example, we rely on computer software to fit models to these data.

Table 15.7 Reduction in Error Sums of Squares for Unbalanced Pilot-Plant Chemical-Yield Experiment

Temperature (°C)	Concentration (%)	Catalyst	Yield (g)
160	20	C_1	59
		C_2	50,54
	40	C_1	
		C_2	46,44
180	20	C_1	74,70
		C_2	81
	40	C_1	69
		C_2	79

Reduction in Error Sums of Squares

Source of Variation	df	Sum of Squares	Mean Square	F-Value
(a) With Temperature × Catalyst				
Model	6	1682.40	280.40	46.73
Error	3	18.00	6.00	
Total	9	1700.40		
(b) Without Temperature × Catalyst				
Model	5	1597.07	319.41	12.36
Error	4	103.33	25.83	
Total	9	1700.40		
(c) Temperature × Catalyst F-Statistic				
Temperature × catalyst	1	85.33	85.33	14.22
Error	3	18.00	6.00	

Suppose the only models of interest are those that do not include the three-factor interaction. If one wishes to draw inferences on, say, the two-factor interaction between temperature and catalyst, one fits two models: one with temperature-by-catalyst interaction terms and one without them, both models also excluding the three-factor interaction terms. Table 15.7 lists the model and error sums of squares for these two fits, as well as the reduction in error sums of squares attributable to the effect of the temperature-by-catalyst interaction. The F-statistic for the temperature-by-catalyst interaction is formed by dividing the mean square due to this interaction by the error mean square from the larger model, the one that includes the interaction. Note that, as in Table 15.6, the temperature-by-catalyst interaction has a large F-ratio.

15.3 PARAMETER ESTIMATION

Inferences on specific factor effects requires the estimation of the parameters of ANOVA models. In Section 15.3.1 estimation of the error standard deviation is discussed. In Section 15.3.2 the estimation of both the means of fixed-effects models and parameters representing main effects and interactions are discussed. Relationships are established between estimates of the means of the fixed-effects models and estimates of the parameters representing the main effects and interactions. Alternative estimates of factor effects for quantitative factor levels are presented in Section 15.3.3.

15.3.1 Estimation of the Error Standard Deviation

A key assumption in the modeling of responses from designed experiments is that the experimental conditions are sufficiently stable that the model errors can be considered to follow a common probability distribution. Ordinarily the distribution assumed for the errors is the normal probability distribution. There are several justifications for this assumption.

First, the model diagnostics discussed in Chapter 25 can be used to assess whether this assumption is reasonable. Quantile plots, similar to those discussed in Section 5.3, provided an important visual assessment of the reasonableness of the assumption of normally distributed errors. Second, fixed-effects models are extensions of the one- or two-sample experiments discussed in Chapter 13. The central limit property (Section 12.1) thus provides a justification for the use of normal distributions for averages and functions of averages that constitute factor effects even if the errors are not normally distributed. Finally, randomization provides a third justification: the process of randomization induces a sampling distribution on factor effects that is closely approximated by the theoretical distributions of the t and F statistics used to make inferences on the parameters of ANOVA models.

Assuming that the errors follow a common distribution, the error mean square from the ANOVA table is a pooled estimator of the common error variance. If the model consists of all main effect and interaction parameters (a *saturated* model) and the experiment is a complete factorial with repeat tests, the error sum of squares is similar in form to equation (15.14) for three factors. With the balance in the design, the error mean square is simply the average of the sample variances calculated from the repeat tests for each factor–level combination:

$$s_e^2 = \mathrm{MS}_E = \frac{1}{abc}\sum_i\sum_j\sum_k s_{ijk}^2. \tag{15.15}$$

When the design is unbalanced and the model is saturated, estimation of the error variance still requires that repeat tests be available. If there are repeat tests, the error mean square is still a pooled estimator of the common variance. The sample variances are now based on different numbers of degrees of freedom for each factor–level combination, and some factor–level combinations that do not have repeat tests will not contribute to the estimate of the variance. The general form of the error mean square can be shown to equal

$$s_e^2 = \mathrm{MS}_E = \frac{1}{v}\sum_i\sum_j\sum_k (n_{ijk}-1)s_{ijk}^2, \qquad v = \sum_i\sum_j\sum_k (n_{ijk}-1), \qquad (15.16)$$

where the summations are only over those factor–level combinations for which there are at least two repeat tests ($n_{ijk} > 1$). Note the similarity between (15.15) and (15.3) for single-factor models.

If there are no repeat tests and the model is saturated, there is no estimate of experimental error available. Equations (15.15) and (15.16) are not defined if none of the n_{ijk} are greater than 1, since none of the sample variances in either expression can be calculated. In this situation an estimate of experimental-error variation is only obtainable if some of the parameters in the saturated model can be assumed to be zero.

To illustrate why estimates of experimental-error variation are available when certain model parameters are assumed to be zero, consider again the single-factor model (15.6). Suppose that the design for the single-factor experiment is balanced. If the factor levels all have the same constant effect on the response, all the α_i in the model (15.6) are zero. Then $\bar{y}_{i.} = \mu + \bar{e}_{i.}$ and $\bar{y}_{..} = \mu + \bar{e}_{..}$, so that the main effect for factor level i is

$$\bar{y}_{i.} - \bar{y}_{..} = \bar{e}_{i.} - \bar{e}_{..}.$$

Thus SS_A [equation (15.2)] does not estimate any function of the main effect parameters. It measures variability in the averages of the errors. One can show that both MS_A and MS_E in this situation are estimators of the error variance σ^2.

As mentioned in Section 15.1, any effects not included in the specification of the ANOVA model are measured in the random error component of the partitioning of the total sum of squares. Thus, if one can assume that certain model parameters (equivalently, the corresponding factor effects, apart from random error) are zero, estimation of the error variance can be accomplished even if there is no replication in the design.

Ordinarily, it is reasonable to assume that high-order interactions are zero. This principle was used in Chapter 9 to construct fractional experiments. It can also be used to justify the estimation of experimental-error variation in the absence of repeat tests. We stress here, as in Chapters 6–11, that repeat tests provide the best estimate of experimental-error variation; however, there are many experimental constraints that dictate that repeat tests cannot be included in the design of an experiment. When such constraints occur, an examination of proposed ANOVA models should be made to see whether experimental error can be estimated from high-order interactions that can reasonably be assumed to be zero.

The confidence-interval techniques presented in Section 14.1 can be used with analysis-of-variance models to estimate the error variance or the error standard deviation. Under an assumption of statistically independent, normally distributed errors, $v \cdot \mathrm{MS}_E/\sigma^2$ follows a chi-square sampling distribution with v degrees of freedom, where v is the number of error degrees of freedom from the ANOVA table. Confidence intervals for the error standard deviation can be constructed as shown in Table 14.1 with the replacement of s^2 by MS_E and the number of degrees of freedom $(n-1)$ by v.

15.3.2 Estimation of Effects Parameters

The parameters associated with the main effects and interactions in analysis-of-variance models are estimated by functions of the factor–level response averages. For any fixed-effects ANOVA model, including the one- and two-sample models (13.1) and (13.9), the model means are estimated by the corresponding response averages. For example, in the single-factor model (15.4), the factor–level averages $\bar{y}_{i\cdot}$ estimate the model means μ_i. In the three-factor model (15.7), the averages for the factor–level combinations, $\bar{y}_{ijk\cdot}$, estimate the model means μ_{ijk}. .

Confidence intervals can be placed on any of the model means using the procedures discussed in Section 13.1. A $100(1-\alpha)\%$ confidence interval for μ_{ijk} from a balanced three-factor factorial experiment is

$$\bar{y}_{ijk\cdot} - \frac{t_{\alpha/2} s_e}{r^{1/2}} < \mu_{ijk} < \bar{y}_{ijk\cdot} + \frac{t_{\alpha/2} s_e}{r^{1/2}}, \tag{15.17}$$

where $t_{\alpha/2}$ is a Student's t critical value with v degrees of freedom corresponding to an upper-tail probability of $\alpha/2$, $s_e = (MS_E)^{1/2}$, and the error mean square has $v = abc(r-1)$ degrees of freedom. Confidence intervals for other model means can be obtained in a similar fashion. For example, an interval estimate of $\bar{\mu}_{i\cdot\cdot}$ can be obtained by replacing μ_{ijk} with $\bar{\mu}_{i\cdot\cdot}$, $\bar{y}_{ijk\cdot}$ with $\bar{y}_{i\cdot\cdot\cdot}$, and r with bcr in (15.17). Note that while the confidence interval (15.17) is based on a balanced three-factor factorial experiment, means from models for unbalanced designs can be estimated similarly by replacing r with n_{ijk}, bcr with $n_{i\cdot\cdot}$, etc. depending on the mean being estimated.

Comparisons among factor–level means are usually of great interest in experiments involving factors at two or more levels. Confidence intervals for mean differences such as $\bar{\mu}_{i\cdot\cdot} - \bar{\mu}_{j\cdot\cdot}$ can be constructed using the general expression

$$\hat{\theta} - t_{\alpha/2} s_e m^{1/2} < \theta < \hat{\theta} + t_{\alpha/2} s_e m^{1/2}, \tag{15.18}$$

where

$$\theta = \sum_i \sum_j \sum_k a_{ijk} \mu_{ijk}$$

is some linear combination of the model means,

$$\hat{\theta} = \sum_i \sum_j \sum_k a_{ijk} \bar{y}_{ijk\cdot}$$

is the corresponding linear combination of the response averages, and

$$m = \sum_i \sum_j \sum_k \frac{a_{ijk}^2}{n_{ijk}}.$$

This formula simplifies greatly for certain comparisons. For example, if a confidence interval is desired for some linear combination of the factor–level means for a single factor, the quantities in (15.18) can be rewritten as

$$\theta = \sum a_i \mu_i, \qquad \hat{\theta} = \sum a_i \bar{y}_{i\cdot}, \qquad m = \sum a_i^2 / n_i. \tag{15.19}$$

If a confidence interval on $\theta = \mu_i - \mu_j$ is desired, then $\hat{\theta} = \bar{y}_{i\cdot} - \bar{y}_{j\cdot}$, and $m = n_i^{-1} + n_j^{-1}$ are inserted in (15.18) because $a_i = 1$, $a_j = -1$, and the other $a_k = 0$ for $k \neq i, j$ in (15.19).

It was noted in Section 15.2 that the main-effect and interaction parameters are not uniquely defined in terms of the model means. As an illustration of this, consider the single-factor model (15.4). Rather than using (15.5) to define the main-effect parameters, define them as follows:

$$\mu_1 = \mu, \qquad \mu_i = \mu + \alpha_i, \quad i = 2, 3, \ldots, a.$$

This is an equally meaningful way to define main-effect parameters. The main-effect parameters α_2 to α_a in this representation measure differences in factor effects from the mean of the first level rather than from an overall mean as in (15.5).

Because of the nonuniqueness of the definitions of main-effect and interaction parameters, the estimation of these parameters depends on how the parameters are defined. For example, estimation of the individual parameters α_i for single-factor models depends on whether (15.5) or the above definition is used to relate the parameters to the model means. In (15.5) $\alpha_2 = \mu_2 - \mu$, while in the above definition $\alpha_2 = \mu_2 - \mu_1$.

Since model means are unique, estimates of functions of them are also unique, even though these functions can be represented in different ways using the main-effect and interaction model parameters. Estimation of model means was stressed above so the estimation procedures could be discussed without special conventions that depend on how the main effects and interactions relate to the model means.

Main-effect and interaction parameters for ANOVA models defined as in (15.8) are used only to focus more clearly on estimation and testing procedures. Relationships between model means and these main-effect and interaction parameters are established in (15.9). Insertion of the averages for their corresponding model means in (15.9) yields the following estimators of the parameters:

$$\hat{\mu} = \bar{y}_{\ldots}, \quad \hat{\alpha}_i = \bar{y}_{i\ldots} - \bar{y}_{\ldots}, \quad \hat{\beta}_j = \bar{y}_{\cdot j\ldots} - \bar{y}_{\ldots}, \quad \hat{\gamma}_k = \bar{y}_{\cdot\cdot k\cdot} - \bar{y}_{\ldots},$$

$$\widehat{(\alpha\beta)}_{ij} = \bar{y}_{ij\cdot\cdot} - \bar{y}_{i\ldots} - \bar{y}_{\cdot j\ldots} + \bar{y}_{\ldots}, \quad \text{etc.,}$$

$$\widehat{(\alpha\beta\gamma)}_{ijk} = \bar{y}_{ijk\cdot} - \bar{y}_{ij\cdot\cdot} - \bar{y}_{i\cdot k\cdot} - \bar{y}_{\cdot jk\cdot} \qquad (15.20)$$

$$+ \bar{y}_{i\ldots} + \bar{y}_{\cdot j\ldots} + \bar{y}_{\cdot\cdot k\cdot} - \bar{y}_{\ldots}.$$

These quantities are the calculated main effects and interactions for the experimental factors.

In Section 7.3, main effects and interactions for two-level factors are defined as differences in factor–level averages; e.g., $\bar{y}_{2\ldots} - \bar{y}_{i\ldots}$. The expressions in (15.20) are alternative to those in Section 7.3 and can be directly related to them; e.g., $\bar{y}_{2\ldots} - \bar{y}_{i\ldots} = \hat{\alpha}_2 - \hat{\alpha}_1$. The advantage to the representation (15.20) is that it is immediately extendable to any number of factors having any number of levels. These effects and the parameter estimates in (15.20) are also directly related to the sums of squares in an ANOVA tables, as is demonstrated in Section 15.4.2.

While (15.17) or (15.18) can be used to construct individual confidence intervals for any of the model means, the overall confidence that all the intervals simultaneously contain the means is not $100(1 - \alpha)\%$ when two or more confidence intervals are computed. This is because the same data are being used to form many confidence intervals separately. The derivation of the confidence interval in Section 13.1 is based on a single population or process mean and uses a single sample of independent observations from that population

Table 15.8 Interval Estimates for Flow-Rate Model Parameters

Filter	Average	95% Confidence Interval on Model Mean μ_i
A	0.233	(0.191, 0.275)
B	0.291	(0.249, 0.333)
C	0.247	(0.205, 0.289)
D	0.274	(0.232, 0.316)

Filter Pair	Difference in Averages	95% Confidence Interval on Difference in Main Effects $(\alpha_i - \alpha_j \text{ or } \mu_i - \mu_j)$
A–B	−0.058	(−0.117, 0.001)
A–C	−0.014	(−0.073, 0.045)
A–D	−0.041	(−0.100, 0.018)
B–C	0.044	(−0.015, 0.103)
B–D	0.017	(−0.042, 0.076)
C–D	−0.027	(−0.086, 0.032)

or process. For a set of confidence intervals for which one desires a confidence coefficient of $100(1 - \alpha)\%$, the simultaneous inference procedure discussed in Chapter 16 should be used.

Individual confidence-interval estimates of the means and the differences in the factor–level effects for the flow-rate data of Figure 15.1 are displayed in Table 15.8. The filter averages and the estimated error variance can be obtained from the summary information in Table 15.2. The confidence intervals for the means is calculated using (15.17) with the appropriate substitutions for the factor–level means μ_i, averages $\bar{y}_{i.}$, and sample sizes ($r = 4$). The confidence intervals for the differences in the main effects is calculated using (15.18) and (15.19), which are equivalent to the two-sided interval in Table 13.5 for pairwise differences in means.

We again stress that these individual intervals are presented for illustration purposes and that each interval does not have a confidence coefficient of 95%, since several intervals were formed from the same set of data. It is interesting to note that all of the intervals for the differences in factor effects include zero. This suggests that the factor effects are not significantly different from one another. A test statistic appropriate for testing this hypothesis is presented in Section 15.4.

15.3.3 Quantitative Factor Levels

The estimation techniques for main effects and interactions that were presented in the last section are applicable to the levels of any fixed-effects factor. When factor levels are quantitative, however, it is of interest to assess whether factor effects are linear, quadratic, cubic, or possibly of higher order. To make such an assessment, one must define linear combinations of the response averages that measure these effects.

Although polynomial effects can be calculated for most balanced and unbalanced designs, we restrict attention in this section to balanced experiments in which factor levels are equally spaced. Quantitative levels of a factor X are equally spaced if consecutive levels differ by the same constant amount—i.e., if, algebraically, the value of the ith level x_i can be expressed as $x_i = c_0 + c_1 i$ for suitably chosen constants c_0 and c_1. When designs are balanced, the coefficients of the linear combinations of the response averages that measure the linear, quadratic, etc. effects of the factor levels on the response can be conveniently tabulated. Table A20 in the Appendix contains coefficients for such polynomials.

To illustrate the calculation of polynomial factor effects, consider an experiment conducted to evaluate the warping of copper plates. The response variable is a measurement of the amount of warping, and the factors of interest are temperature (50, 75, 100, and 125°C) and the copper content of the plates (40, 60, 80, and 100%). Both of the factors are equally spaced. Both are fixed effects, since the levels of each were specifically chosen for inclusion in the design. It is of interest, assuming a smooth functional relationship between the factor levels and the mean of the response variable, to

Table 15.9 Analysis of Polynomial Effects of Copper Content on Warping*

Warping Measurements

		Copper Content (%)				
		40	60	80	100	Total
	50	17, 20	16, 21	24, 22	28, 27	175
Temperature	75	12, 9	18, 13	17, 12	27, 31	139
(°C)	100	16, 12	18, 21	25, 23	30, 23	168
	125	21, 17	23, 21	23, 22	29, 31	187
	Total	124	151	168	226	669

ANOVA

Source of Variation	Degrees of Freedom (df)	Sum of Squares	Mean Square	F-Value
Copper Content	3	698.34	232.78	34.33
Temperature	3	156.09	52.03	7.67
Content × temp.	9	113.78	12.64	1.86
Error	16	108.50	6.78	
Total	31	1076.71		

Polynomial Coefficients

Copper Content (%)	Coefficient			
	Linear	Quadratic	Cubic	Average
40	−3	1	−1	15.50
60	−1	−1	3	18.88
80	1	−1	−3	21.00
100	3	1	1	28.25
Scaled effect	25.51	5.47	4.04	
95% Confidence Interval	(19.99, 31.03)	(−0.05, 10.99)	(−1.48, 9.56)	

*Data from Johnson, N. L., and Leone, F. C. (1977), *Statistics and Experimental Design in Engineering and the Physical Sciences*, New York: John Wiley & Sons, Inc. Copyright 1977 John Wiley & Sons, Inc. Used by permission.

characterize quantitatively the relationship between the factor levels and the mean amount of warping.

Table 15.9 lists the data from the experiment. Observe from the analysis-of-variance table that the interaction between copper content and temperature is not statistically significant (see Section 15.4). The two main effects are statistically significantly ($p < 0.001$ for copper content, $p = 0.002$ for temperature).

Also included are scaled linear, quadratic, and cubic factor effects for copper content (those for temperature are left as an exercise). Because the coefficients in Table A20 of the linear, quadratic, and cubic effects are not all of the same magnitude, some effects can appear to be larger simply because they have larger coefficients. The scaling is performed to make the magnitudes of the effects comparable. Using the notation of (15.19), the scaled effects are

$$\hat{\theta}_s = \frac{1}{(D/n)^{1/2}} \sum_i a_i \bar{y}_{i\cdot}, \qquad D = \sum_i a_i^2, \tag{15.21}$$

where the a_i are the coefficients for one of the polynomial effects taken from Table A20 and n is the number of responses used in the calculation of each of the averages. The sum of squares D of the coefficients is also included for each polynomial effect in Table A20.

Confidence intervals, comparable to (15.18) and (15.19), for the scaled polynomial effects are

$$\hat{\theta}_s - t_{\alpha/2} s_e < \theta_s < \hat{\theta}_s + t_{\alpha/2} s_e, \tag{15.22}$$

when n is the number of responses entering into each average. The calculated confidence

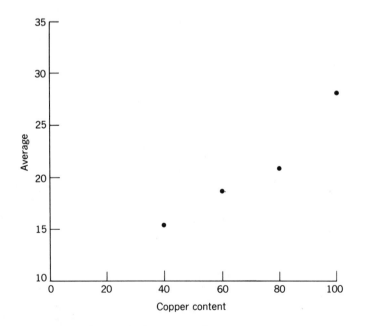

Figure 15.2   Average warping measurement.

interval for each effect is shown in Table 15.9. Note from Table 15.9 that the scaled linear effect for copper content is quite large relative to the estimated model standard deviation $s = 2.60$. Note too that zero is included in the confidence intervals for the quadratic and the cubic parameter effects, but not for the linear effect. These two findings suggest that the dominant quantitative effect of copper content is linear. This suggestion is confirmed in Figure 15.2.

Polynomial effects can also be calculated for interactions between two equally spaced quantitative factors. The total number of such polynomial effects equals the product of the individual degrees of freedom for the main effects if the design is a balanced complete factorial. Each such polynomial effect is calculated by forming a table of coefficients, the elements of which are products of the main-effect coefficients.

In the construction of tables of coefficients for interaction effects, the rows of a table correspond to the levels of one of the factors and the columns correspond to the levels of the other factor. The body of the table contains the products, element by element, of the coefficients for each of the orthogonal polynomials.

For example, consider the copper-plate warping example. Linear, quadratic, and cubic main effects can be calculated for each factor. The nine degrees of freedom shown for the interaction sum of squares in Table 15.9 can be accounted for by nine joint polynomial effects. Letting copper content be factor A and temperature factor B, the nine quadratic effects are: linear $A \times$ linear B, linear $A \times$ quadratic B, linear $A \times$ cubic B, quadratic $A \times$ linear B, quadratic $A \times$ quadratic B, quadratic $A \times$ cubic B, cubic $A \times$ linear B, cubic $A \times$ quadratic B, and cubic $A \times$ cubic B.

Table 15.10 lists the coefficients for each of the nine joint effects. To calculate scaled joint effects, multiply the coefficients in the table by the corresponding response averages, sum the products, and divide by the square root of the ratio of the sum of the squares of the coefficients to the number of repeat responses in each average, similarly to (15.21). The sum of the squares of the coefficients equals the product of the two D-values in Table A20 for the corresponding two main effects. This scaling of interaction effects is a straightforward extension of (15.21).

Table 15.10   Interaction Coefficients for Four-Level Factors

		Linear				Quadratic				Cubic			
		−3	−1	1	3	1	−1	−1	1	−1	3	−3	1
Linear	−3	9	3	−3	−9	−3	3	3	−3	3	−9	9	−3
	−1	3	1	−1	−3	−1	1	1	−1	1	−3	3	−1
	1	−3	−1	1	3	1	−1	−1	1	−1	3	−3	1
	3	−9	−3	3	9	3	−3	−3	3	−3	9	−9	3
Quadratic	1	−3	−1	1	3	1	−1	−1	1	−1	3	−3	1
	−1	3	1	−1	−3	−1	1	1	−1	1	−3	3	−1
	−1	3	1	−1	−3	−1	1	1	−1	1	−3	3	−1
	1	−3	−1	1	3	1	−1	−1	1	−1	3	−3	1
Cubic	−1	3	1	−1	−3	−1	1	1	−1	1	−3	3	−1
	3	−9	−3	3	9	3	−3	−3	3	−3	9	−9	3
	−3	9	3	−3	−9	−3	3	3	−3	3	−9	9	−3
	1	−3	−1	1	3	1	−1	−1	1	−1	3	−3	1

The scaled linear copper content by linear temperature interaction effect is

$$(A_L \times B_L) = \frac{9(18.5) + 3(18.5) - 3(23.0) - 9(27.5) + 3(10.5) + \cdots + 9(30.0)}{[(20)(20)/2]^{1/2}}$$

$$= -\frac{6.5}{(200)^{1/2}} = -0.46.$$

Note that this scaled effect is small relative to the linear main effect of copper content and to the estimated error standard deviation.

15.4 STATISTICAL TESTS

Tests of statistical hypothesis can be performed on the parameters of ANOVA models. These tests provide an alternative inferential methodology to the interval-estimation procedures discussed in the last section. In this section several commonly used statistical tests are discussed. We separate this discussion according to whether tests are desired for (a) a single parameter or a single function of the model parameters or (b) groups of parameters or parametric functions.

15.4.1 Tests on Individual Parameters

Tests of hypotheses on individual model means are straightforward extensions of the single-sample t-tests of Section 13.1. For example, in the ANOVA model for a three-factor balanced complete factorial experiment a test of $H_0: \mu_{ijk} = c$ versus $H_a: \mu_{ijk} \neq c$, where c is a specified constant (often zero), is based on the t-statistic

$$t = n_{ijk}^{1/2} \frac{\bar{y}_{ijk\cdot} - c}{s_e}, \tag{15.23}$$

where $s_e = (\mathrm{MS_E})^{1/2}$. Hypotheses about other model means can be tested by making the appropriate substitutions in (15.23).

Testing hypotheses about a single linear combination of model means is accomplished in a similar manner. The hypothesis

$$H_0: \theta = c \quad \text{vs} \quad H_a: \theta \neq c,$$

where $\theta = \sum_i \sum_j \sum_k a_{ijk} \mu_{ijk}$, is tested using the single-sample t-statistic

$$t = \frac{\hat{\theta} - c}{s_e m^{1/2}}, \tag{15.24}$$

where $\hat{\theta} = \sum_i \sum_j \sum_k a_{ijk} \bar{y}_{ijk\cdot}$, $m = \sum_i \sum_j \sum_k a_{ijk}^2 / r$, and r is the number of repeat tests for each factor–level combination. Note the equivalence of this test procedure with the confidence-interval approach using (15.18).

Tests of the statistical significance of polynomial effects can also be performed using (15.24) and the coefficients in Table A20. If the scaled form (15.21) of the polynomial effect is used, the t-statistic (15.24) is simply $\hat{\theta}_s / s_e$.

15.4.2 F-Tests for Factor Effects

One of the difficulties with performing separate t-tests for each main effect or each interaction effect is that the overall chance of committing one or more Type I errors can greatly exceed the stated significance level for each of the individual tests. In part this is due to multiple testing using the same data set. There are a number of test procedures that can be used to simultaneously test the equality of all the main effects for a factor or for all the interaction effects of two or more factors. In this section we discuss F-tests based on the mean squares from an ANOVA table.

The numerators of the F-statistics in fixed effects ANOVA tables are the main effects and interaction mean squares. The sums of squares for these main effects and interactions in a balanced three-factor complete factorial experiment can be written as in equations (15.11)–(15.13). Comparison of these sums of squares with the factor effects in (15.20) reveals that the sums of squares are functions of the estimated effects parameters when the parametrization (15.8) is used to relate the effects parameters to the model means; e.g.,

$$ SS_A = bcr \sum_i \hat{\alpha}_i^2, \qquad SS_B = acr \sum_j \hat{\beta}_j^2, \qquad SS_{AB} = cr \sum \sum (\widehat{\alpha\beta})_{ij}^2, \qquad \text{etc.} $$

The sums of squares in an ANOVA table test the hypothesis that the parameters corresponding to the main effects and interactions are zero. For example, SS_A tests the hypothesis $H_0 : \alpha_1 = \alpha_2 = \cdots = \alpha_a = 0$ versus $H_a : \alpha_i \neq 0$ for at least one factor level i. The equivalent hypothesis in terms of the model means is $H_0 : \bar{\mu}_{1..} = \bar{\mu}_{2..} = \cdots = \bar{\mu}_{a..}$ versus $H_a : \bar{\mu}_{i..} \neq \bar{\mu}_{j..}$ for at least one pair of factor levels.

Under the hypothesis that a particular main effect or interaction is zero, the corresponding F-ratio should be around 1, since both the numerator and the denominator of the F-statistic are estimating the same quantity, the error variance. On the other hand, if the stated null hypothesis is false, the numerator mean square will tend to be larger than the error mean square (see Section 17.2). Thus, large F-ratios lead to rejection of the hypotheses of no factor effects.

The analysis-of-variance table for the flow-rate data is shown in Table 15.2. The main effect for the three filters has an F-ratio of 1.86. Comparison of this F-ratio with F critical values in Table A7 reveals that the filter effects are not statistically significant ($0.10 < p < 0.25$). Thus the mean flow rates attributable to the filters do not significantly differ for the three filters. Stated another way, the response variability attributable to the filter means is not significantly greater than the variability attributable to uncontrolled experimental error.

The ANOVA table for the pilot-plant study is shown in Table 15.6. Using a significance level of $\alpha = 0.05$, the temperature-by-catalyst interaction and the main effect of concentration are statistically significant. In keeping with the hierarchical modeling of the response, we ignore whether the main effects of temperature and catalyst are statistically insignificant. The reason we ignore the test for these main effects is that the significant interaction of these two factors indicates that they have a joint influence on the response; consequently, there is no need to examine the main effects.

The next step in the analysis of these data would be to examine which of the factor levels are affecting the response. For example, one would now like to know which combinations of temperature and catalyst produced significantly higher or lower yields than other combinations. The multiple-comparison procedures discussed in the next chapter should be used for this purpose. We defer detailed consideration of the effects of individual factor–level combinations until we discuss some of these procedures.

15.5 SAMPLE-SIZE CONSIDERATIONS

Formulas are available for determining the sample size necessary to assure the power desired for detecting a minimum difference in parameter effects that is important to an experimenter. The equations presented below provide values of the sample size when the desired significance level of a test is 0.05 and the desired power is 0.90. We provide sample-size formulas for both main effects and for interactions. Sample-size requirements for other significance levels and other power values are available in the references.

The equation for determining the sample size for detecting a main effect is

$$n = (4as_p/\delta)^2, \tag{15.25}$$

where

  $a =$ the number of levels of the factor,
  $s_p =$ a planning estimate of the standard deviation of individual response variable, and
  $\delta =$ the minimum difference of importance between any two main effects; equivalently, between any two factor–level means.

The sample size necessary for detecting an interaction effect is given by

$$n = \frac{9s_p^2(v+1)c}{\delta^2 2^{m-2}}, \tag{15.26}$$

where s_p is as for main effects and

  $v =$ number of interaction degrees of freedom,
  $c =$ number of factor–level combinations for the factors that are involved in the interaction,
  $m =$ number of factors involved in the interaction, and
  $\delta =$ minimum difference of interest among the interaction effects.

Both equations (15.25) and (15.26) are appropriate for few (say 10 or fewer) degrees of freedom for the error. A correction for more error degrees of freedom is to multiply each of the equations by 0.5.

Suppose in the flow-rate experiment discussed in Section 15.12 that for a future experiment it is desired that a true difference in filter mean flow rates of 0.1 cc/min be detectable with a power of 0.90. Using a significance level of 0.05 and a standard-deviation estimate from the ANOVA table (Table 15.2), the required sample size is

$$n = (4 \times 4 \times 0.0383/0.1)^2 = 37.63.$$

Thus each filter should be tested approximately 9–10 times in the experiment.

One can also use equations (15.25) and (15.26) in reverse. That is, n can be specified and the equations solved for δ. If one is able to include, say, eight repeat tests in future filter flow-rate experiments, the minimum difference in mean flow rates that can be detected among the four filters with $\alpha = 0.05$ and $\beta = 0.10$ is

$$\delta = 4 \times 4 \times 0.0383/(32)^{1/2}$$

$$= 0.11 \text{ mm.}$$

For illustration purposes, suppose now that the experimenters wished to study flow rates for the four filters using two mesh sizes. In this setting both the filters and the mesh sizes are fixed factors. If the experimenters wish to design the experiment so that a predetermined minimum difference in any two interaction means μ_{ij} can be detected, equation (15.26) should be used to determine the sample size. Suppose again that the experimenters wish to detect any difference of at least 0.1 mm. The required sample size is at least

$$n = \frac{9(0.00147)(3 + 1)8}{(0.1)^2 2^0}$$

$$= 42.34.$$

Thus, approximately five repeat tests are to be made with each of the eight filter–mesh-size combinations.

REFERENCES

Text References

Analysis-of-variance assumptions, models, and calculations are covered in a traditional manner in the following two texts. The first is at a more elementary level.

Guenther, W. C. (1964). *Analysis of Variance*, Englewood Cliffs, NJ: Prentice-Hall, Inc.

Ostle, B. and Malone, L. C. (1988). *Statistics in Research*, Fourth Edition, Ames, IA: The Iowa State University Press.

The following texts are on a more advanced level. The first two describe analysis-of-variance calculations in detail, the second text using the Statistical Analysis Software (SAS) library, including designs in which the data are unbalanced. The last two texts have wealth of information on the principle of reduction in error sums of squares.

Johnson, N. L. and Leone, F. C. (1964). *Statistical and Experimental Design for Engineers and Scientists*, New York: John Wiley and Sons, Inc.

Milliken, G. A. and Johnson, D. E. (1984). *Analysis of Messy Data—Volume I: Designed Experiments*, Belmont, NJ: Wadsworth Publishing Co.

Searle, S. R. (1971). *Linear Models*, New York: John Wiley and Sons, Inc.

The following journal article contains technical details for the sample-size results given in Section 15.5.

Wheeler, Robert E. (1974). "Portable Power," *Technometrics* **16**, 193–201.

   Another set of useful articles on sample-size tables for analysis of variance is given in

Bratcher, T. L., Mason, M. A., and Zimmer, W. J. (1970). "Tables of Samples Sizes in the Analysis of Variance," *Journal of Quality Technology*, **2**, 156–164.

Nelson, L. S. (1985). "Sample Size Tables for Analysis of Variance," *Journal of Quality Technology*, **17**, 167–169.

Data Reference

The copper-plate warping example is extracted from the above text by Johnson and Leone (p. 98). The chemical pilot-plant data are taken from

Box, G. E. P., Hunter, W. G., and Hunter, J. S. (1978). *Statistics for Experimenters*, New York: John Wiley and Sons, Inc.

EXERCISES

1 In a nutrition study, three groups of month-old laboratory mice were fed supplements to their usual feed. The percentage weight gains over the next month were recorded and are shown below. Assess whether the supplements differ in their ability to produce weight gain.

Group 1: 7.1 7.0 7.0 8.6 8.2 6.6 6.8 6.7 7.7

Group 2: 7.6 7.3 7.3 8.3 7.6 6.6 6.7 6.8 7.0

Group 3: 9.2 8.3 9.1 9.0 8.9 9.0 9.2 7.6 8.1

2 The following data are elasticity measurements on skin that has been exposed to varying intensities of light for a fixed time period. The light intensities are equally spaced but have been coded for convenience. Evaluate whether there are significant differences in the elasticity measurements for the various light intensities.

Light Intensity (Coded)	Elasticity Measurements						
1	0.54	1.98	0.65	0.52	1.92	1.48	0.97
2	1.76	1.24	1.82	1.47	1.39	1.25	1.29
3	2.05	2.18	1.94	2.50	1.98	2.17	1.83
4	7.92	4.88	9.23	6.51	6.77	4.25	3.72

3 Using the data in Exercise 2, construct a confidence interval on the mean elasticity measurement for the highest light intensity.

4 Treating the light intensities in Exercise 2 as equally spaced factor levels, estimate the linear, quadratic, and cubic effects of light intensity. Which, if any, of these effects are statistically significant? Which, if any, of these effects appears to be the most dominant? Make a plot of the averages to confirm your conclusions visually.

5 A consumer testing agency tested the lifetimes of five brands of dot-matrix computer printer ribbons using a single printer. The results are tabulated below. Assess the performance of these printer ribbons.

Brand	Lifetime (hr)			
A	20.5	17.7	20.0	19.2
B	24.0	26.2	21.2	26.1
C	27.4	35.2	31.2	28.2
D	17.1	18.1	18.5	16.7
E	36.5	33.9	26.9	27.0

6 The ballistic limit velocity for 10-mm rolled homogeneous armor was measured for projectiles with blunt, hemispherical, and conical nose shapes fired at 0° and 45° obliquity to the target. Three test runs were made with each type of projectile, with the firing order of the 18 test runs randomized. Analyze the effects of nose shape and firing angle on the ballistic velocities.

| | **Ballistic Velocities (m/sec)** | |
Nose Shape	Angle: 0°	45°
Blunt	938	1162
	942	1167
	943	1163
Conical	889	1151
	890	1145
	892	1152
Hemispherical	876	1124
	877	1125
	881	1128

7 Determine the linear, quadratic, and cubic effects of temperature for the copper-plate warping experiment in Table 15.9. Place individual confidence intervals on the mean for each of these polynomial effects. What inferences appear reasonable from these results?

8 Calculate the polynomial interaction effects between copper content and temperature for the copper-plate warping example. Which, if any, of these effects are statistically significant?

9 The table below contains elasticity measurements similar to those reported in Exercise 2. These measurements are taken on four subjects who were specifically chosen to represent four types of skin classes. Monthly measurements were taken on the last day of five consecutive months. Assess whether there are significant effects due to either skin group or time.

| | **Elasticity** | | | | |
Skin Group	Month 1	Month 2	Month 3	Month 4	Month 5
A	1.21	1.32	1.44	1.38	1.26
B	0.89	1.11	1.26	1.05	0.82
C	3.67	4.69	4.88	4.33	4.02
D	2.22	2.36	2.58	2.46	2.13

10 Calculate polynomial effects for the main effect for time in the previous exercise. Which polynomial effects, if any, are statistically significant?

11 Refer to Exercise 10, Chapter 14. A second question of interest to process engineers is whether the four processes for manufacturing aluminum tops for food containers are

equivalent to one another. These processes are considered equivalent if they possess the same process means and process standard deviations. Examine whether the processes can be considered equivalent, using the data provided below.

Process	Sphericity Index Measurements			
A	0.723	0.721	0.707	0.723
B	0.734	0.789	0.796	0.761
C	0.811	0.752	0.902	0.864
D	0.721	0.786	0.742	0.777

CHAPTER 16

Multiple Comparisons

In this chapter we discuss various techniques for analyzing data in greater detail than by simply identifying statistically significant effects in an analysis-of-variance table. Some of the techniques discussed can be applied without reference to an ANOVA *table; others ordinarily are used subsequent to the determination of significant factor effects in an analysis of variance. Specifically, we examine techniques for comparing means or groups of means. These techniques have application to the comparison of individual and joint factor–level effects. Major topics covered in this chapter include*

- *the investigation of contrasts of factor–level means,*
- *the evaluation of polynomial effects for equally spaced qualitative factor levels, and*
- *a discussion of preferred mean-comparison techniques.*

Detailed exploration of data often stems from a desire to know what effects are accounting for the results obtained in an experiment. One's interest thus is directed to the comparison of specific factor–level means or of groups of means rather than strictly to the detection of a statistically significant main effect or interaction. In these situations, procedures utilizing multiple comparisons of means are appropriate.

Multiple comparisons of means frequently involve preselected comparisons that address specific questions of interest to a researcher. In such circumstances one sometimes has little interest in the existence of overall experimental effects; consequently, the summaries provided by an ANOVA table are of secondary interest, perhaps only to provide an estimate of the experimental-error variance. In contrast, researchers in many experimental settings do not know which factor effects may turn out to be statistically significant. If so, the *F*-statistics in an ANOVA table provide the primary source of information on statistically significant factor effects. However, after an *F*-test in an ANOVA table has shown significance, an experimenter may desire to conduct further analyses to determine which pairs or groups of means are significantly different from one another.

Both of the above types of comparisons are examined in the following sections. Specific attention is given to comparisons involving quantitative factor levels, comparisons based on *t*-statistics, and multiple-range tests. The final section of this chapter contains a discussion of the preferred procedures to use when making multiple comparisons and the experimental conditions where each technique is recommended.

329

16.1 PHILOSOPHY OF MEAN-COMPARISON PROCEDURES

The estimation of linear combinations of means is detailed in this section to aid in the understanding of the philosophy of multiple-comparison procedures. There is a close connection between the material in this section and the discussions in Sections 15.3 and 15.4.

Analysis of linear combinations of means is a major objective of many experiments. For example, one may be interested in comparing the average fuel economy, $\bar{y}_1$, achieved by a test oil in a laboratory experiment with the averages, $\bar{y}_2$ and $\bar{y}_3$, of the fuel economies of two different reference oils. A linear combination of the sample means that would be used for the comparison is $\bar{y}_1 - (\bar{y}_2 + \bar{y}_3)/2$. Linear combinations of means such as these are termed *contrasts* (see Exhibit 16.1).

EXHIBIT 16.1

Contrast. A linear combination of k averages, denoted by

$$a_1\bar{y}_1 + a_2\bar{y}_2 + \cdots + a_k\bar{y}_k,$$

where $\bar{y}_i$ is the ith average and the a_i are constants, at least two of which are nonzero, is termed a contrast if the coefficients (the a's) sum to zero, i.e., if

$$a_1 + a_2 + \cdots + a_k = 0.$$

In Chapter 7 we defined main effects and interactions as contrasts of factor–level averages (see Section 7.3 and the appendix to Chapter 7). In Section 15.3 we related main effects, interactions, and other factor effects to linear combinations of the model means and to linear combinations of the main-effect and the interaction model parameters. One generally wishes to draw inferences on linear combinations of means or model parameters using the corresponding linear combinations of factor–level averages. Contrasts are of special importance because only contrasts of the main-effect and interaction parameters can be estimated in an ANOVA model.

Consider the single-factor ANOVA model introduced in Section 15.1, equation (15.6):

$$y_{ij} = \mu + \alpha_i + e_{ij}, \qquad i = 1, 2, \ldots, a, \quad j = 1, 2, \ldots, r, \tag{16.1}$$

where the model parameters are related to the factor–level means through equation (15.5), $\mu_i = \mu + \alpha_i$. The equal number of repeat tests (r) is imposed in (16.1) only to simplify the following discussion; the essence of the following arguments holds for unequal numbers of repeat tests.

Suppose that one is interested in estimating some linear combination of the factor–level means $\theta = \sum a_i\mu_i$. Estimation of this linear combination of means is accomplished by inserting the factor–level averages for the means as in (15.19). Observe that in terms of the model parameter in (16.1),

$$\theta = \sum a_i\mu_i = \mu\sum a_i + \sum a_i\alpha_i = \sum a_i\alpha_i \qquad \text{if} \quad \sum a_i = 0.$$

The reason that contrasts are so important in the comparison of model means is that the constant term μ appears in all linear combinations of model means except for contrasts, in which case $\sum a_i = 0$. Thus, comparisons of the effects of individual factor levels must be made using contrasts.

The use of contrasts is not a limitation on comparisons of factor–level means. Note that a direct comparison of two means is accomplished by estimating the difference in the means. Such differences are contrasts. Comparisons of three or more means can be made by pairwise comparisons of differences of the means or any other contrasts that are deemed informative. We examine several such informative contrasts in this chapter.

The above comparison between the test oil and the two reference oils using the linear combination $\bar{y}_1 - (\bar{y}_2 + \bar{y}_3)/2$ is a contrast, since the sum of the coefficients is zero. To avoid fractional coefficients, it is common to rewrite the comparison as $2\bar{y}_1 - (\bar{y}_2 + \bar{y}_3)$. Since averages are estimators of fixed-effects portions of ANOVA models, these linear combinations of averages are estimators of the same linear combinations of model means.

For a fixed-effects model of the form (16.1) containing only oils as a factor, α_i denotes the fixed-effect model parameter corresponding to the ith oil. The above linear combination of averages is an estimator of

$$2(\mu + \alpha_1) - [(\mu + \alpha_2) + (\mu + \alpha_3)] = 2\alpha_1 - (\alpha_2 + \alpha_3).$$

Note that this linear combination of model parameters does not contain the constant term μ, and it is zero when all three factor levels have an equal effect ($\alpha_1 = \alpha_2 = \alpha_3$) on the response. These two characteristics are present in all contrasts of factor effects.

When making multiple comparisons of factor effects, an additional property of contrasts is needed. In order to make these comparisons statistically independent of one another, the contrasts involved must be mutually orthogonal (see Exhibit 16.2).

EXHIBIT 16.2

Orthogonal Contrasts. Two contrasts,

$$C_1 = \sum a_i \bar{y}_i \quad \text{and} \quad C_2 = \sum b_i \bar{y}_i$$

are said to be orthogonal if the sum of the products of the corresponding coefficients in the two contrasts is zero, i.e., if

$$a_1 b_1 + a_2 b_2 + \cdots + a_k b_k = 0.$$

Three or more contrasts are said to be mutually orthogonal if all pairs of contrasts are orthogonal.

To illustrate the use of orthogonal contrasts, consider the data in Table 15.3 (see also Figure 7.3) on the chemical yield of a manufacturing process in a pilot plant. For simplicity, we will examine only two factors, catalyst (A) and concentration (B), each of which has two levels. If a 2^2 factorial experiment (see Chapter 7) had been conducted, three comparisons of interest among the four factor–level means μ_{ij} might be:

(i) the effect of the first catalyst compared with that of the second catalyst,

(ii) the effect of the high level of concentration compared with that of the low level of concentration, and

(iii) the difference in the effects of the two catalysts at the high concentration compared with their difference at the low concentration.

Denote the average responses for the four factor–level combinations as $\bar{y}_{11}, \bar{y}_{12}, \bar{y}_{21}$, and $\bar{y}_{22}$. In this representation, the first subscript refers to the catalyst (1 = catalyst 1, 2 = catalyst 2) and the second refers to the concentration (1 = 20%, 2 = 40%). The above three comparisons can now be made using these factor–level averages to estimate the corresponding model means. The comparisons of interest can then be expressed as:

(i) $(\bar{y}_{21} + \bar{y}_{22})/2 - (\bar{y}_{11} + \bar{y}_{12})/2$,

(ii) $(\bar{y}_{12} + \bar{y}_{22})/2 - (\bar{y}_{11} + \bar{y}_{21})/2$,

(iii) $(\bar{y}_{22} - \bar{y}_{12}) - (\bar{y}_{21} - \bar{y}_{11})$.

The coefficients in these three comparisons, apart from the divisor of 2 in the first two, are:

Contrast	Factor–Level Response Average			
	$\bar{y}_{11}$	$\bar{y}_{12}$	$\bar{y}_{21}$	$\bar{y}_{22}$
(i)	−1	−1	+1	+1
(ii)	−1	+1	−1	+1
(iii)	+1	−1	−1	+1

Notice that the sum of the coefficients of each of these comparisons is zero, indicating that each is a contrast. Further, the products of the corresponding coefficients of any two contrasts sum to zero, indicating that the three contrasts are mutually orthogonal. Finally, note that the above contrasts are the effects representations of the main effects and interaction for two two-level factors shown in Table 7.6.

The sums of squares corresponding to fixed effects in any ANOVA table can be partitioned into component sums of squares, each component of which corresponds to one of the degrees of freedom. This partitioning corresponds to the orthogonal contrasts that can be formed from the means that go into the sum of squares. For main effects, there are $k - 1$ mutually orthogonal contrasts that can be formed from the means for the k levels of a factor. The degrees of freedom for interactions are design dependent, but frequently (e.g., for complete factorial experiments) the number of degrees of freedom and the number of mutually orthogonal contrasts equal the product of the numbers of degrees of freedom for the main effects corresponding to the factors in the interaction.

For a set of averages each of which consists of n observations, the formula for the sum of squares corresponding to a contrast $C = \sum a_i \bar{y}_i$ is given by

$$SS(C) = \frac{n(\sum a_i \bar{y}_i)^2}{\sum a_i^2}. \tag{16.2}$$

This sum of squares, since it has only one degree of freedom, is a mean square. Divided by the error mean square from the ANOVA table for these response variables, the ratio $SS(C)/MS_E$ is an F-statistic, which can be used to test the hypothesis

$$H_0: \sum a_i \mu_i = 0 \quad vs \quad H_a: \sum a_i \mu_i \neq 0, \tag{16.3}$$

where μ_i is the mean of the ANOVA model corresponding to the average $\bar{y}_1$. The degrees of freedom of this F statistic are $v_1 = 1$ and $v_2 = v$, where v is the number of degrees of freedom of the error mean square.

This F-statistic is exactly equivalent to the use of a t-statistic with the effects calculated as in (15.19) or (15.20). For example, replacing n_i with n in (15.19) and squaring the resulting t-statistic yields

$$t^2 = \left(\frac{\hat{\theta}}{s_e m^{1/2}} \right)^2 = \frac{\text{SS}(C)}{\text{MS}_E}.$$

This equivalence allows the estimation procedures detailed in Section 15.3 or the testing procedure outlined in Section 15.4 to be applied to the general discussions in this chapter.

It is possible to obtain a nonsignificant overall F-statistic for a main effect or interaction in an ANOVA table, yet find that one or more of the component contrasts are statistically significant. This result generally occurs when one or two of the orthogonal contrasts are significant but the remaining ones are not. When totaled to give the sum of squares for the effect, the nonsignificant contrasts dilute the significant one(s), yielding a nonsignificant main effect or interaction. Similarly, it is possible to have a significant overall F-statistic for a factor yet find that no pairwise comparison between any two averages is significant.

Mutual orthogonality is not a prerequisite for analyzing a set of contrasts. An experimenter may choose to examine whichever set of linear combinations of means is of interest or value. Orthogonal contrasts are usually selected because (a) contrasts eliminate the constant term from mean comparisons, (b) comparisons of interest among factor–level means can usually be expressed as contrasts, and (c) the sums of squares calculated from orthogonal contrasts are statistically independent for balanced experimental designs.

A critical question in making multiple comparisons concerns the choice of the significance level of the tests or the confidence level for interval estimates. Since several statistical inference procedures (several interval estimates or several tests) are to be made on the same data set, a distinction must be made between two types of error rates. We describe these Type I error rates (see Exhibit 16.3) in terms of significance levels for tests of hypotheses, but they are equally germane to interval estimation because of the equivalence between statistical tests and interval estimation that was established in the preceding chapters (e.g. Chapter 12). We return to this point below.

EXHIBIT 16.3   TYPE I ERROR RATES

Comparisonwise Error Rate. The probability of erroneously rejecting a null hypothesis when making a single statistical test.

Experimentwise Error Rate. The probability of erroneously rejecting at least one null hypothesis when making statistical tests of two or more null hypotheses using the data from a single experiment.

Comparisonwise Type I error rates essentially indicate the significance level associated with a single statistical test. All previous discussions of significance levels have been comparisonwise error rates. Experimentwise Type I error rates measure the significance level associated with multiple tests using the same data set. When making several statistical tests using the same data set, the experimentwise Type I error rate can be much larger than the significance level stated for each test.

An experimentwise error rate of 0.05 is much more stringent than a comparisonwise error rate of 0.05. Consider the comparison of k population means. If statistically independent contrasts are used to make $k - 1$ comparisons among the sample averages and the error standard deviation is assumed known, the experimentwise error rate E is related to the comparisonwise error rate (significance level of each comparison) through the formulas

$$E = 1 - (1 - \alpha)^{k-1} \quad \text{or} \quad \alpha = 1 - (1 - E)^{1/(k-1)}. \tag{16.4}$$

For example, suppose the comparisonwise significance level is selected to be $\alpha = 0.05$ and $k = 6$ means are to be compared. If all six population means are equal, the probability of incorrectly rejecting one or more of the five orthogonal comparisons is

$$E = 1 - (1 - 0.05)^5 = 0.23,$$

a value almost five times larger than the stated significance level for each test. If one desires an experimentwise significance level of $E = 0.05$, the individual comparisonwise significance levels should be

$$\alpha = 1 - (1 - 0.05)^{0.2} = 0.01.$$

This example illustrates that an experimentwise error rate yields fewer Type I errors than the same comparisonwise error rate. This advantage is counterbalanced by the fact that more Type II errors are likely to occur when controlling the experimentwise error rate. The formulas in (16.4) are not valid when several nonorthogonal comparisons among means are made with the same set of data. In such cases, the experimentwise error rate can be larger than that indicated by equation (16.4). Likewise, these formulas are not strictly valid when the same error mean square is used in the comparisons; however, these formulas are still used as approximations to the true error rates in such situations.

In practice, an experimenter must choose whether to control the comparisonwise or the experimentwise error rate. The more appropriate error rate to control depends on the degree to which one wants to control the Type I error rate. In situations involving sets of orthogonal contrasts, the comparisonwise error rate is often selected as the preferred choice when there are not a large number of comparisons to be made. When there are many comparisons to make or the comparisons are not based on orthogonal contrasts, the experimentwise error rate usually is controlled. This is especially true when using nonorthogonal contrasts because then the outcome of one comparison may affect the outcomes of subsequent comparisons.

The connection of this discussion with interval estimation is readily established. A Type I error occurs for a statistical test when the null hypothesis is erroneously rejected. The equivalent error for confidence intervals occurs when the confidence interval does not include the true value of the parameter. Thus comparisonwise error rates α and experimentwise error rates E are related to individual confidence levels $100(1 - \alpha)\%$ for one confidence interval and overall confidence levels $100(1 - E)\%$ for two or more intervals formed from one data set.

While an experimenter should be concerned about controlling overall error rates and confidence levels, it is important to stress that the main objective of multiple comparison procedures is to learn as much about populations, processes, or phenomena as is possible from an experiment. A low error rate is not the sole purpose of an experiment; it is a guide to careful consideration of objectives and to an awareness of the proper use of statistical methodology. In a particular experiment an acceptably low error rate may be 20% or it may be 0.1%. Further remarks on the choice of error rates are provided in Section 16.5.

16.2 GENERAL COMPARISONS

There are many ways to select specific comparisons of model means, although the choice of an experimental design effectively determines which contrasts are available for analysis. If one intends to conduct an analysis of variance of the data, one might choose to partition the degrees of freedom for one or more of the main effects or interactions into a specific set of mutually orthogonal, single-degree-of-freedom contrasts. In an analysis involving a quantitative factor one might decide to use a partitioning of the sum of squares into a linear effect, a quadratic effect, a cubic effect, and so forth. In studies involving a control group, one might choose to make select comparisons of the treatment groups with the control group, as well as comparing the treatment groups with one another. Finally, suggestions for comparisons might result from plots of the experimental results.

To illustrate how specific comparisons can be made, consider again the wire-die example given in Table 12.1. The purpose of the experiment was to compare the tensile strengths of wire samples taken from three different dies. Suppose that die 3 is a control die and the other two dies are new ones that have different geometric designs that could potentially allow higher production rates. Two comparisons which might be of interest are:

(1) The mean tensile strength of wire made from die 3 versus the mean tensile strength of wire made from dies 1 and 2, and

(2) the mean tensile strength of wire made from die 1 vs. that for wire made from die 2.

Note that the first comparison examines whether the experimental dies differ from the standard [i.e., whether $\mu_3 = (\mu_1 + \mu_2)/2$], while the second one determines whether the two experimental dies differ [i.e., whether $\mu_1 = \mu_2$].

The comparisons stated above can be made using contrasts among the three average tensile strengths:

(1) $C_1 = (\bar{y}_1 + \bar{y}_2)/2 - \bar{y}_3$, and
(2) $C_2 = \bar{y}_1 - \bar{y}_2$.

Writing these contrasts as

$$C_1 = \sum a_i \bar{y}_i \quad \text{with} \quad (a_1, a_2, a_3) = (0.5, 0.5, -1),$$
$$C_2 = \sum b_i \bar{y}_i \quad \text{with} \quad (b_1, b_2, b_3) = (1, -1, 0),$$

one can see that these contrasts are orthogonal. The sums of squares corresponding to these two contrasts constitute an additive partition of the sum of squares for the die factor.

Using equation (16.2) and the summary statistics given in Table 12.2, we find that

Table 16.1   Analysis-of-Variance Table for Tensile-Strength

Experiment Source	df	SS	MS	F
Dies	2	55.592	27.796	12.742
Error	51	111.254	2.181	
Total	53	116.846		

the sums of squares for contrasts (1) and (2) are given by

$$SS(C_1) = 18 \frac{\{(84.733 + 84.492)/2 - 82.470\}^2}{1 + 0.25 + 0.25}$$

$$= 55.067,$$

$$SS(C_2) = 18 \frac{(84.733 - 84.492)^2}{1 + 1}$$

$$= 0.525.$$

These sums of squares add up to the sum of squares due to the die factor shown in Table 16.1:

$$SS_D = SS(C_1) + SS(C_2) = 55.067 + 0.525 = 55.592.$$

We can now test the significance of these two contrasts by dividing each by the error mean square in Table 16.1. The two F-statistics are

$$F(C_1) = \frac{MS(C_1)}{MS_E} = \frac{55.067}{2.181} = 25.249 \qquad (p < 0.001),$$

$$F(C_2) = \frac{MS(C_2)}{MS_E} = \frac{0.525}{2.181} = 0.241 \qquad (p > 0.50).$$

Since there are only two comparisons to be made, comparisons that are orthogonal contrasts, a comparisonwise error rate can be used to test these hypotheses. Any reasonable choice of a significance level will result in the rejection of the first hypothesis and the nonrejection of the second one. Thus, there is not sufficient evidence from these data to conclude that the mean tensile strengths of wire made from dies 1 and 2 differ from one another. There is, however, sufficient evidence to conclude that the mean tensile strength of wire made from die 1 is different from the average of the means for the other two dies.

As mentioned above, multiple comparisons need not be orthogonal. They can be any comparisons that are of interest to an experimenter, provided that the data and the model allow them to be analyzed. In the tensile-strength example we could have chosen to compare each of the test dies with the standard die. The two contrasts then would have the form

(3) (control versus die 1) $C_3 = \bar{y}_1 - \bar{y}_3$, and

(4) (control versus die 2) $C_4 = \bar{y}_2 - \bar{y}_3$.

These are not orthogonal contrasts, since the coefficients for the two linear combinations of the three average tensile strengths, when multiplied together, are not zero:

$$(+1)(0) + (0)(+1) + (-1)(-1) = 1.$$

Nevertheless, F-statistics appropriate for testing that the corresponding contrasts of the factor–level means are zero would be calculated in the same manner as those for orthogonal contrasts. Using equation (16.2) and the data in Table 12.1,

$$SS(C_3) = 18 \frac{(84.733 - 82.470)^2}{1+1} = 46.088,$$

$$SS(C_4) = 18 \frac{(84.492 - 82.470)^2}{1+1} = 36.796.$$

Using the MS_E from Table 16.1, the F-statistics are

$$F(C_3) = \frac{MS(C_3)}{MS_E} = \frac{46.088}{2.182} = 21.127 \qquad (p < 0.001),$$

$$F(C_4) = \frac{MS(C_4)}{MS_E} = \frac{36.796}{2.182} = 16.858 \qquad (p < 0.001).$$

Since these two contrasts are not orthogonal, it is appropriate to use an experimentwise error rate rather than a comparisonwise error rate. Letting $E = 0.05$, we use equation (16.4) to obtain a comparisonwise significance level of 0.025 for testing these two contrasts. Since both significance probabilities are less than 0.001, we reject both of the hypotheses; i.e., we conclude that each of the test dies produces wire that has an average tensile strength different from that of the standard die. An examination of the averages in Table 12.1 indicates that the experimental dies produce wire that has significantly higher average tensile strengths than the standard die.

The steps involved in multiple comparisons of means are summarized in Exhibit 16.4 in a general way.

As mentioned in the previous section, there is an equivalence between the multiple-comparison testing procedures described in this chapter and the use of multiple-confidence-

---

EXHIBIT 16.4   MULTIPLE COMPARISONS

1. State the comparisons (hypotheses) of interest in a form similar to (16.3).
2. Calculate the individual sums of squares for the comparisons using equation (16.2). (Equivalently, the t-statistics discussed in Sections 15.3 and 15.4 can be used.)
3. Divide the comparison sums of squares by the mean squared error from the ANOVA table to form F-statistics for testing the respective hypotheses.
4. Choose an appropriate comparisonwise or experiment wise error rate based on
   (a) whether the comparisons are orthogonal contrasts, and
   (b) the number of comparisons to be made.
5. Compare the significance probabilities with the selected significance level and draw the resulting conclusions.

---

interval procedures. Multiple-confidence-interval procedures provide important inform-ation on the precision of calculated effects and are often more informative than a series of statistical tests. The philosophy of multiple comparisons described in this section is applicable to multiple confidence intervals if the individual confidence intervals are calculated using a confidence coefficient of $100(1 - \alpha)\%$, where α is determined from (16.4). The overall confidence coefficient of the multiple interval is then $100(1 - E)\%$.

In the next two sections of this chapter we introduce alternative multiple-comparison techniques to the general ones described in this section. Prior to doing so, we wish to return to the investigation of quantitative factor levels that was first discussed in Section 15.3. We return to this topic because of the importance of modeling the effects of quantitative factor levels on a response when the response can be assumed to be a smooth function of the factor levels. Testing of the significance of polynomial effects is a frequent application of multiple-comparison procedures.

The procedures that have been illustrated above are applicable whether the experi-mental factor is qualitative or quantitative. When a quantitative factor is included in an experiment, interest often centers on determining whether the quantitative factor has a linear, quadratic, cubic, etc. effect on the response. To make such a determination, one must define linear combinations of the response averages that measure these effects.

The coefficients a_i used in equation (16.2) to measure an rth-degree polynomial effect on a response are obtained from a linear combination of the first r powers of the quantitative factor:

$$a_i = b_1 x_i + b_2 x_i^2 + \cdots + b_r x^r \qquad (16.5)$$

where a_i is the coefficient on $\bar{y}_i$ for the rth-degree polynomial effect and x_i^k is the ith level of the factor raised to the kth power. When these coefficients a_i are selected so that the rth polynomial (e.g., cubic, $r = 3$) is orthogonal to all lower-order polynomials (e.g., linear, $r = 1$, and quadratic, $r = 2$), the resulting coefficients are referred to as *orthogonal polynomials*.

When a factor has equally spaced levels in a balanced design, the fitting of orthogonal polynomial terms can be accomplished similarly to the fitting of orthogonal contrasts. Tables of coefficients a_i for orthogonal polynomials are available in Table A20 of the Appendix. These tables provide the a_i needed in equation (16.2) to compute the effects (linear, quadratic, cubic, etc.) of the factor on the response (see also Section 15.3.3). The tables are constructed so that the coefficients are orthogonal contrasts; consequently, all the procedures described above for contrasts involving qualitative factors are applicable.

Consider an experiment conducted to examine the force (lb) required to separate a set of electrical connectors at various angles of pull. The force exerted is the response variable, and the factor of interest is the pull angle. This factor has four equally spaced levels: $0, 2, 4,$ and $6°$. The experiment was conducted using five connectors and five repeat tests for each angle with each connector. The factor–level averages, based on 25 observations per angle, were as follows:

Angle (deg)	Average Force (lb)
0	41.94
2	42.36
4	43.82
6	46.30

Just as there are only $k-1$ orthogonal contrasts for a qualitative factor having k levels, one only can fit a polynomial expression up to degree $k-1$ for a quantitative factor having k equally spaced levels. Thus for the pull angle, we can fit up to a cubic polynomial. The coefficients for the linear, quadratic, and cubic terms in the polynomial are, from Table A20 of the appendix,

$$\text{Linear } (L): \quad (-3, -1, 1, 3),$$
$$\text{Quadratic } (Q): \quad (1, -1, -1, 1),$$
$$\text{Cubic } (C): \quad (-1, 3, -3, 1).$$

Using these coefficient with the above averages in equation (16.2) yields the following sums of squares for the orthogonal polynomials:

$$SS(L) = 264.26, \quad SS(Q) = 26.52, \quad SS(C) = 0.01.$$

The appropriate mean square for testing the significance of these polynomial terms is the interaction mean square between connectors (B) and the pull angles (A), $MS_{AB} = 37.92$ with 12 degrees of freedom (see Section 17.4). Dividing each of the above sums of squares by this mean square yields the following F-statistics:

$$F(L) = 6.97, \quad F(Q) = 0.70, \quad F(C) = 0.0003.$$

Since this is a small number of mutually orthogonal contrasts, we select a comparisonwise error rate of 0.05 and use an F-table with 1 and 12 degrees of freedom. Only the linear

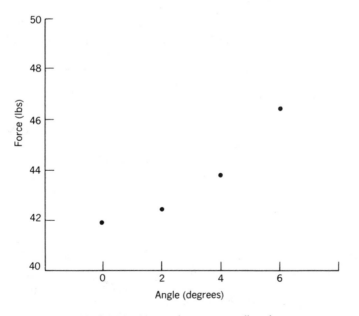

Figure 16.1 Average force versus pull angle.

effect of pull angle is statistically significant ($p < 0.001$). Hence, as the pull angle increases, the separation force increases at approximately a linear rate.

It should be noted that comparisons can be composed of groups of orthogonal contrasts rather than merely of individual orthogonal contrasts. When using orthogonal contrasts the sum of squares for the group comparison is obtained by adding the sums of squares for the individual contrasts. The mean square for the group comparison is obtained by dividing this sum of squares by the number of contrasts (i.e., of degrees of freedom). An F-statistic for testing the significance of this group comparison can be calculated by dividing this mean square by the error mean square.

Figure 16.1 is a plot of the average force measurements as a function of pull angle. This figure well illustrates the desirability of accompanying statistical tests with estimation and plotting procedures. The above tests indicated that the only statistically significant polynomial effect of pull angle is linear. However, the plot clearly suggests a quadratic effect. The reason the quadratic effect is not statistically significant is that orthogonal polynomials adjust each effect for lower-order effects in order to make the contrasts orthogonal. Consequently, once the force averages are adjusted for the linear effect of pull angle, the quadratic effect is not statistically significant.

As an indication that there is a statistically significant quadratic effect due to pull angle, we can calculate the combined effect of the linear and the quadratic effects. The sum of squares for the combined effects is the sum of the component sums of squares (since the effects are based on orthogonal contrasts), and the mean square is this combined sum of squares divided by 2. The resulting F-statistic, with 2 and 12 degrees of freedom, is 3.83. This F-statistic has a p-value of about 0.05. As an alternative to this analysis, one can fit polynomial regression models (Chapter 23) to the force values and assess the desirability of including a quadratic effect due to pull angle.

16.3 COMPARISONS BASED ON t-STATISTICS

The most commonly used multiple comparison tests are those based on t-statistics. Chief among these are the tests for contrasts presented in the last section. As mentioned above, rather than using an F-statistic, a contrast can be tested using a t-statistic. The tests are exactly equivalent. For testing the hypotheses shown in equation (16.3), the test statistic formed from (15.2) can be written as

$$t = \frac{\sum a_i \bar{y}_i}{\text{SE}_c},\qquad(16.6)$$

where SE_c, the estimated standard error of the contrast, is given by

$$\text{SE}_c = \left[\text{MSE}\left(\sum \frac{a_i^2}{r} \right) \right]^{1/2}.\qquad(16.7)$$

Contrasts tested using (16.6) and (16.7) often involve the pairwise comparisons of means. One of the oldest and most popular techniques for making multiple pairwise comparisons of means is Fisher's least-significant-difference (LSD) procedure (see Exhibit 16.5). It is termed a *protected* LSD test if the procedure is applied only after a significant F-test is obtained using an analysis of variance. An *unprotected* LSD procedure occurs when pairwise comparisons of means are made regardless of whether a main effect or interaction F-test is statistically significant. The technique consists of applying the two-sample Student's t-test (i.e., see Section 13.3) to all $k(k-1)/2$ possible pairs of the k factor–level means.

EXHIBIT 16.5 FISHER'S LSD

Two averages, $\bar{y}_i$ and $\bar{y}_{j'}$, in an analysis of variance are declared to be significantly different if

$$|\bar{y}_i - \bar{y}_j| > \text{LSD}, \qquad (16.8)$$

where

$$\text{LSD} = t_{\alpha/2}(v)\,[\text{MS}_E(n_i^{-1} + n_j^{-1})]^{1/2}.$$

In this expression n_i and n_j denote the numbers of observations used in the calculation of the respective averages, v denotes the number of degrees of freedom for error, and $t_{\alpha/2}(v)$ denotes a critical value for the t-distribution having v degrees of freedom and an upper-tail probability of $\alpha/2$.

Fisher's LSD procedure consists of pairwise t-tests, each with a significance level of α. Individual factor–level means or interaction means can be compared using the appropriate response averages for each. Fisher's protected LSD procedure is usually applied with a comparisonwise significance level that equals that for the F-statistic used to make the overall comparison of factor or interaction levels. The unprotected LSD procedure should be used with a significance level that is chosen to control the experimentwise error rate. Bonferroni comparisonwise error rates (see below) are often used: the comparisonwise error rate is α/m, where m is the number of pairwise comparisons to be made.

Fisher's LSD procedure is simple and convenient to use. It can be applied with unequally repeated individual or joint factor levels. It can be used to provide confidence intervals on mean differences; e.g.,

$$(\bar{y}_i - \bar{y}_j) - \text{LSD} \leqslant \mu_i - \mu_j \leqslant (\bar{y}_i - \bar{y}_j) + \text{LSD} \qquad (16.9)$$

EXHIBIT 16.6 LEAST SIGNIFICANT INTERVAL PLOT

1. Plot each of the k averages (vertical axis) versus its level (horizontal axis).

2. If all averages have an equal number of observations (r), calculate LSD using (16.8):

$$\text{LSD} = t_{\alpha/2}(v)\,[\text{MS}_E(2/r)]^{1/2}$$

3. If the k averages are based on different numbers of observations (r_i), calculate LSD as above with r replaced by the harmonic mean ($\bar{r}$) of the r_i:

$$\bar{r} = k\left[\sum r_i^{-1}\right]^{-1}.$$

4. Extend vertical bars a distance equal to LSI = LSD/2 above and below each average to visually display the uncertainty intervals.

5. Any two averages for which the uncertainty intervals indicated by the vertical bars do not overlap are declared statistically significant.

---

is a $100(1 - \alpha)\%$ confidence interval for $\mu_i - \mu_j$. All of these features make its use attractive.

Displays such as graphs of the average responses versus the levels of a factor are valuable for visually presenting the results involving the comparison of means. For example, the LSD procedure can be performed graphically by using a *least-significant-interval* (LSI) plot (see Exhibit 16.6).

In Section 13.4 a study was described involving the effects of eight lubricants on engine deposits. Figure 16.2 is an LSI plot based on the summary statistics provided in Table 13.8. For this example, $t_{0.025}(64) = 2.00$ (approximately), $s_p = 27.68$, and $r = 9$, so that LSD $= 26.10$ and LSI $= 13.05$. Based on Figure 16.2, one could conclude that the mean deposit rating for lubricant 1 is lower than the mean deposit ratings of all the other lubricants, the mean deposit rating for lubricant 8 is lower than those for lubricants 2, 5, and 6, and the mean deposit rating for lubricant 7 is lower than that for lubricant 5.

Plots of averages and LSI plots also can be made when interactions are judged to be statistically significant. Interaction plots of averages for two-factor interactions are given in Exhibit 16.7. Interaction plots for three-factor or higher interactions can be

---

EXHIBIT 16.7   INTERACTION PLOTS FOR TWO FACTORS

1. Calculate the average response for each combination of the levels of the two factors.
2. Place the levels of one of the two factors on the horizontal axis. Construct a suitable scale for the response averages on the vertical axis.
3. For each level of the second factor, plot the two-factor response averages $\bar{y}_{ij}$ versus the levels of the factor on the horizontal axis, i.e., for each j plot $\bar{y}_{ij}$ versus level i of the first factor.
4. Connect the plotted averages for a common level of the second factor by a dashed line.

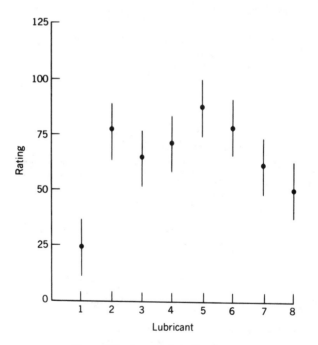

Figure 16.2 Average deposit ratings.

made by constructing two-factor interaction plots for each level or combination of levels of the other factors.

In Section 7.1, a complete factorial experiment was constructed to study the effects of three factors on the torque of a rotating shaft. The randomized test sequence, including two repeat test runs per factor–level combination, is shown in Table 7.3. Table 16.2 is an ANOVA table for the resulting data. The three-factor interaction is judged to be statistically significant ($p = 0.002$). Therefore all three factors are exerting an effect on the torque measurements. Hence, there is no need for further testing of two-factor interactions or of main effects.

At this juncture in the analysis, it is recommended that one plot the averages for the factor–level combinations of the three factors. The three-factor interaction plot is shown in Figure 7.5 (Section 7.2). If the experimenter wishes to determine the factor–level combination that produced minimum average torque, the interaction plot reveals that it was the steel shaft with a porous sleeve and lubricant 2. It is now of interest to determine whether the average torque measurement for this combination of factor levels is significantly lower than for all the other combinations of the factor levels.

Table 16.3 lists the average torque measurements for all combinations of the factor levels. Each average in the table is calculated from two repeat measurements. Using Fisher's protected LSD and a significance level of 0.05, any two averages in this table are significantly different if their difference exceeds

$$\text{LSD} = 2.12[5.25(\tfrac{1}{2} + \tfrac{1}{2})]^{1/2} = 4.86.$$

Table 16.2 Analysis-of-Variance Table for the Torque Study

Source	df	SS	MS	F	p-Value
Alloy	1	105.12	105.12	20.02	0.000
Sleeve	1	105.12	105.12	20.02	0.000
Lubricant	3	93.38	31.12	5.93	0.006
$A \times S$	1	36.12	36.12	6.88	0.018
$A \times L$	3	21.38	7.13	1.36	0.292
$S \times L$	3	48.38	16.13	3.07	0.058
$A \times S \times L$	3	118.38	39.46	7.52	0.002
Error	16	84.00	5.25		
Total	31	611.88			

Table 16.3 Factor–Level Averages for the Torque Study

		Torque			
Alloy	Sleeve	Lubricant 1	2	3	4
Aluminum	Nonporous	69.5	70.5	69.0	70.0
	Porous	78.0	72.0	78.5	73.5
Steel	Nonporous	78.5	67.0	77.5	79.0
	Porous	79.0	77.5	77.0	74.5
		LSD = 4.86			

The corresponding LSI value is 2.43. An LSI plot for the three-factor interaction is shown in Figure 16.3.

The LSD criterion indicates that the lowest torque value in Table 16.3, that for the factor–level combination of steel shaft, nonporous sleeve, and lubricant 2, is significantly different from all the other averages for the steel alloy and also from the averages for the aluminum alloy with the porous sleeve. The combination of steel, nonporous sleeve, and lubricant 2 does not produce a significantly lower average torque measurement than the aluminum alloy with the nonporous sleeve for any choice of lubricant. Thus there are four factor–level combinations that are not significantly different from the combination that produces the lowest average torque measurement. This conclusion also can be observed by visually comparing the averages in the LSI plot.

The major disadvantage of the LSD technique is that its error rate is not satisfactory for testing all possible pairs of mean differences when there are a moderate or large number to be compared. In repeated testing of this type, the experimentwise error rate can be much greater than the desired significance level. The experimentwise error rate can be better controlled by using a Bonferroni procedure (see Exhibit 16.8) to select the comparisonwise significance level. Bonferroni comparisons control the experimentwise error rate by adjusting the t critical value to compensate for many tests on the same set of data. The inequality (16.8) can again be used to make pairwise comparisons of means, but the comparisonwise (two-tailed) significance level is α/m, where α is the desired experimentwise error rate and m is the number of comparisons to be made. Bonferroni pairwise comparisons should not be used when there are a very large number of comparisons to be made, because the comparisonwise significance level can become too

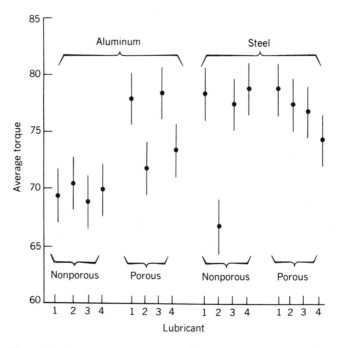

Figure 16.3   Least-significant-interval plot for torque data. LSD = 4.86.

small to be of value. In these situations multiple-range tests (Section 16.4) offer a compromise between the desired experimentwise error rate and an unacceptably small comparisonwise error rate.

EXHIBIT 16.8 BONFERRONI COMPARISONS

1. Let $\theta = \sum a_i \mu_i$ represent one of m linear combinations of the means μ_i for which one is interested in testing $H_0 : \theta = 0$ vs $H_a : \theta \neq 0$.

2. Reject H_0 if $|\hat{\theta}| = |\sum a_i \bar{y}_i|$ exceeds

$$BSD = t_{\alpha/2m}[MS_E \sum a_i^2/n_i]^{1/2},$$

where n_i is the number of observations used in the calculation of $\bar{y}_i$, and $t_{\alpha/2m}$ is an upper-tail t critical value based on v degrees of freedom, the number of degrees of freedom for MS_E, and an upper-tail probability of $\alpha/2m$.

Bonferroni techniques can be used to test any contrasts or linear combinations of interest using the t-statistic from equation (16.6) or those discussed in Sections 15.3 and 15.4. Confidence intervals can also be constructed using these t-statistics.

Using the wire-die example, we can apply Bonferroni's proceudre to the comparisons of the three dies in the tensile-strength study. Using an experimentwise significance level of $\alpha = 0.05$, the two-tailed comparisonwise significance level is $\alpha/m = 0.05/3 = 0.017$. We can choose to be slightly more conservative and use a two-tailed significance level of $\alpha = 0.01$ in order to obtain a critical value from Table A5 in the appendix. Doing so produces the following results:

Die:	3	1	2
Average:	82.470	84.492	84.733

In this analysis, each average is based on 18 observations (see Table 13.2), the mean squared error from the ANOVA table (Table 16.1) is $MSE = 2.181$, and the t critical value is 2.68; consequently, the LSD cutoff value for the mean differences is $LSD = 1.32$. The line underscoring the two test dies is used to indicate that dies 1 and 2 do not have significantly different averages. The absence of a line underscoring the average for die 3 and for either of the other indicates that dies 1 and 2 are significantly different from the standard die, die 3.

16.4 MULTIPLE-RANGE TESTS

There are a variety of multiple range tests; however, in this section we shall only consider Tukey's significant-difference (TSD) test and Duncan's multiple-range test. Both of these procedures utilize the "studentized range" statistic. Critical points for this statistic are contained in Table A21 in the appendix.

Tukey's procedure controls the experimentwise error rate for multiple comparisons when all averages are based on the same number of observations (see Exhibit 16.9). The stated experimentwise error rate is very close to the correct value even when the sample sizes are not equal. The technique is similar to Fisher's LSD procedure. It differs in that the critical value used in the TSD formula is the upper $100\alpha\%$ point for the difference between the largest and smallest of k averages. This difference is the range of the k averages, and the critical point is obtained from the distribution of the range statistic, not from the t-distribution.

EXHIBIT 16.9 TUKEY'S TSD

Two averages, $\bar{y}_i$ and $\bar{y}_j$, based on n_i and n_j observations respectively, are significantly different if

$$|\bar{y}_i - \bar{y}_j| > \text{TSD},$$

where

$$\text{TSD} = q(\alpha; k, v)\left(\text{MS}_E \frac{n_i^{-1} + n_j^{-1}}{2}\right)^{1/2}, \qquad (16.10)$$

in which $q(\alpha; k, v)$ is the studentized range statistic, k is the number of averages being compared, MS_E is the mean squared error from an ANOVA fit to the data based on v degrees of freedom, and α is the experimentwise error rate.

Tukey's TSD is stated in terms of an analysis of variance, but it can be applied in more general situations, as can all of the multiple-comparison procedures discussed in this chapter. All that is needed is a collection of k statistically independent averages and an independent estimate of the common population variance based on v degrees of freedom. The estimate of σ^2 would replace MS_E in these applications.

Tukey's TSD can provide simultaneous confidence intervals on all pairs of mean differences $\mu_i - \mu_j$. These confidence intervals collectively have a confidence coefficient of $100(1 - \alpha)\%$; i.e., in the long run, the confidence intervals will simultaneously cover all the mean differences with a probability of $1 - \alpha$. The simultaneous confidence intervals are defined as

$$(\bar{y}_i - \bar{y}_j) - \text{TSD} \le \mu_i - \mu_j \le (\bar{y}_i - \bar{y}_j) + \text{TSD}, \qquad (16.11)$$

where TSD is given in (16.10).

As an example of a multiple-range test using Tukey's TSD reconsider the wire-tensile-strength study. Using an experimentwise error rate of 0.05, the TSD factor for this example is

$$\text{TSD} = 3.418(2.181/18)^{1/2} = 1.19,$$

where 3.418 is the studentized-range critical value obtained from Table A21 in the appendix (linearly interpolating to approximate the critical point for $v = 51$ degrees of freedom). Testing the equality of every pair of means, we arrive at the same conclusion as with the Bonferroni procedure in the last section: there is not enough evidence to conclude that the mean tensile strengths of wire produced from dies 1 and 2 are different, but there is sufficient evidence to conclude that the mean for die 3 is different from those for each of the other two.

The second multiple-comparison procedure to be considered in this section is Duncan's multiple-range test (see Exhibit 16.10). It is similar to Tukey's TSD procedure except that the protection level varies with the number of means in the comparison. This modification provides greater protection than Fisher's unprotected LSD against Type

I errors, and it has greater power than Tukey's TSD; however, unlike Tukey's TSD, it does not control the experimentwise error rate. It is a valuable compromise between Fisher's LSD and Tukey's TSD for a moderate number of comparisons.

EXHIBIT 16.10 DUNCAN'S MULTIPLE RANGE TEST

The largest and smallest of p averages $\bar{y}_i$ and $\bar{y}_j$, each based on a sample of size n, are significantly different if

$$|\bar{y}_i - \bar{y}_j| > R_p,$$

where

$$R_p = q(\alpha_p; p, v)(MS_E/n)^{1/2} \qquad (16.12)$$

and $q(\alpha_p; p, v)$ is the studentized-range critical point based on comparing the largest and smallest of p averages, MS_E is the mean squared error based on v degrees of freedom, and the experimentwise significance level is α_p. The experimentwise significance level is related to a comparisonwise level α through the equation

$$\alpha_p = 1 - (1 - \alpha)^{p-1}. \qquad (16.13)$$

The critical values $q(\alpha_p; p, v)$ in Table A22 of the appendix are presented for two comparisonwise significance levels: $\alpha = 0.01$ and 0.05. The experimentwise significance level α_p can be calculated for any p from equation (16.13).

Application of Duncan's multiple-range test requires that the averages be ordered from smallest to largest. The two most extreme averages are compared first. The difference between the largest and the smallest of the $p = k$ factor–level (or interaction) averages is compared using R_k in equation (16.12) with an experimentwise significance level of α_k. The critical values $q(\alpha_p; p, v)$ can be obtained from Table A22 in the appendix for $\alpha = 0.01$ or 0.05. If these two averages are not judged significantly different with $k = p$, testing stops and all the averages are declared not significantly different at the $100\alpha_k\%$ significance level. This is equivalent to nonrejection of $H_0: \mu_1 = \mu_2 = \cdots = \mu_k$. If the two extreme averages are significantly different, testing continues.

The next step is to compare the largest average with the second smallest and the smallest average with the second largest, each test using (16.12) and (16.13) with $p = k - 1$. If neither of these tests is statistically significant, testing ceases and only the two extreme averages are judged significantly different. If one or both of the tests are statistically significant, testing continues with the group(s) of averages for which the two extremes have been declared significantly different. Testing continues in this fashion until no further significant differences are obtained.

In order to illustrate Duncan's multiple-range test, consider the four averages for the steel alloy using the nonporous sleeve in Table 16.3. In order to apply the above procedure, first reorder the averages from smallest to largest:

Lubricant:	2	3	1	4
Average:	67.0	77.5	78.5	79.0

The first comparison is between the smallest and the largest of these $p = 4$ averages. The critical value for Duncan's multiple range test for a comparisonwise significance level of

$\alpha = 0.05$ ($\alpha_4 = 0.143$) is, from Table A22 in the appendix, 3.235. Thus, $R_4 = 3.71$. Since the averages for lubricant 2 and lubricant 4 differ by more than 3.71, they are significantly different from one another.

The next step in the analysis is to compare the averages for lubricants 2 and 1 and for lubricants 3 and 4. Proceeding as above on each of these two comparisons of the extremes of $p = 3$ averages, one will find that averages 1 and 2 are significantly different but averages 3 and 4 are not. Since the averages for lubricants 3 and 4 are not significantly different, no testing is performed on the difference between averages 1 and 3 or between averages 1 and 4. One does, however, continue the process and compare averages 2 and 3 and averages 1 and 3 using the Duncan procedure for the extremes of $p = 2$ averages. Only the test between averages 2 and 3 is statistically significant. Thus, Ducan's multiple-range test determines that the mean for the second lubricant is lower than the means for each of the other lubricants, but there is not sufficient evidence to conclude that the means for any of the other lubricants are different from one another.

16.5 PREFERRED MULTIPLE-COMPARISON PROCEDURES

The procedures discussed in the sections of this chapter provide techniques for making detailed inferences on means beyond the overall conclusions arrived at in an analysis of variance. Two general approaches have been discussed. When a researcher seeks answers to particular questions about the factors in the experiment that can be posed in terms of orthogonal contrasts, t or F statistics can be used to make the comparisons. It is common to simply perform t or F tests using a comparisonwise significance level. If a number of tests are to be made, an adjusted comparisonwise level (e.g., Bonferroni) should be used. Confidence-interval procedures using comparisonwise confidence levels are equally effective and often more informative.

When comparisons involving nonorthogonal contrasts or when many comparisons are to be made, one of the techniques that offer experimentwise error-rate protection should be used. There are numerous techniques that can be used in this situation. Some of the more more popular procedures have been described in Sections 16.3 and 16.4. Selecting the most useful procedure generally depends on the choice of controlling the experimentwise or the comparisonwise error rate.

If one wishes to control the comparisonwise error rate, the perferred methods discussed in this chapter are t-tests, Fisher's protected and unprotected LSD procedures, or Duncan's technique. Fisher's protected LSD tests and Duncan's multiple-range tests offer more protection from Type I errors than the other two and are less conservative than a procedure based on controlling the experimentwise error rate.

If one desires to control the experimentwise error rate, the useful methods include Bonferroni t-tests and Tukey's procedure. Both of these techniques have strong advocates. Bonferroni tests have the advantage of using an ordinary t-statistic. Their main disadvantage is the small size of the comparisonwise significance level when making a large number of comparisons.

Our experiences lead us to recommend Fisher's protected LSD and Tukey's TSD for multiple pairwise comparisons, depending on whether one wishes to control the comparisonwise or the experimentwise error rates. Each offers sufficient protection against Type I errors, and they are relatively easy to implement. Since they can both be written as t-statistics (they only differ in the choice of a critical value), they use a statistic that is familiar to most users of statistical methodology.

An important class of alternatives to the above multiple comparison techniques is ranking and selection procedures. The goal of these techniques is to choose the best population from a set of populations that are, in fact, different. "Best" is defined in this setting as the population with the largest (or smallest) mean. These procedures are based on nonparametric statistical methods that do not require that the data be normally distributed or that an ANOVA model be used to relate responses to factor effects. These techniques are beyond the scope of this text. References to some of them are given below.

REFERENCES

Text References

Most advanced statistical textbooks contain sections devoted to a discussion of multiple-comparison tests. Two excellent resources that provide comprehensive discussions of the multiple-comparison procedures discussed in this chapter, and other multiple-comparison techniques, are:

Chew, V. (1977). "Comparisons among Treatment Means in an Analysis of Variance," Washington: United States Department of Agriculture, Agricultural Research Services (ARS-H-6).

Miller, R. G. (1981). *Simultaneous Statistical Inference,* New York: Springer-Verlag.

Other useful textbooks that include helpful descriptions of multiple comparison procedures include:

Dixon, W. J. and Massey, Jr., F. J. (1983). *Introduction to Statistical Analysis, Fourth Edition,* New York: McGraw-Hill Book Co.

Milliken, G. A. and Johnson, D. E. (1984). *Analysis of Messey Data, Volume 1: Designed Experiments,* New York: Van Nostrand Reinhold Co.

Ostle, B. and Malone, L. C. (1988). *Statistics in Research, Fourth Edition,* Ames, IA: Iowa State University Press.

Snedecor, G. W. and Cochran, W. G. (1980). *Statistical Methods, Seventh Edition,* Ames, IA: Iowa State University Press.

Steel, R. G. D. and Torrie, J. H. (1980). *Principles and Procedures of Statistics: A Biometrical Approach, Second Edition,* New York: McGraw-Hill Book Co.

Walpole, R. E. and Myers, R. H. (1979). *Probability and Statistics for Engineers and Scientists, Second Edition,* New York: Macmillan Co.

Further details on the graphical least significant interval procedure can be found in:

Andrews, H. P., Snee, R. D., and Sarner, M. H. (1980). "Graphical Display of Means," *The American Statistician,* **34**, 195–199.

Useful texts that provide information on nonparametric multiple-comparison texts are:

Hollander, M. and Wolfe, D. A. (1973). *Nonparametric Statistical Methods,* New York: John Wiley and Sons, Inc.

Gibbons, J. D. (1976). *Nonparametric Methods for Quantitative Analysis,* New York: Holt, Reinhart and Winston.

Ranking and selection methods are detailed in the following text:

Gibbons, J. D., Olkin, I., and Sobel, M. (1977). *Selecting and Ordering Populations: A New Statistical Methodology,* New York: John Wiley and Sons, Inc.

Data References

The example describing the force needed to separate electrical connectors in Section 16.2 is from Chew (1977).

EXERCISES

1 Construct an effects representation for the main effects of a single four-level factor using two two-level factors (see the appendix to Chapter 7). Show that the three main effects so constructed are orthogonal contrasts. Show that the sum of squares for the main effect of pull angle in the electric-connector study (Section 16.2) equals the sum of the individual sums of squares for these three contrasts.

2 Show that the orthogonal polynomials for a four-level factor can be expressed as linear combinations of the three main-effects constructed in Exercise 1, algebraically confirming that effects representations and orthogonal polynomials are two ways to partition a sum of squares into single-degree-of-freedom components.

3 Five brands of automobile tires are being tested to evaluate their stopping distances (ft) on wet concrete surfaces. Four tires of each brand were mounted on a mid-sized sedan. The vehicle then accelerated to a speed of 60 mph and the brakes were applied. The data below show ten test runs for each tire brand:

MEASURED STOPPING DISTANCES (FT)

Tire Brand

Run	A	B	C	D	E
1	194.1	188.7	185.0	183.0	194.6
2	184.4	203.6	183.2	193.1	196.6
3	189.0	190.2	186.0	183.6	193.6
4	188.8	190.3	182.8	186.3	201.6
5	188.2	189.4	179.5	194.4	200.2
6	186.7	206.5	191.2	198.7	211.3
7	194.7	203.1	188.1	196.1	203.7
8	185.8	193.4	195.7	187.9	205.5
9	182.8	180.7	189.1	193.1	201.6
10	187.8	206.4	193.6	195.9	194.8

Perform a single-factor analysis of variance on these data. Assess the statistical significance of these tire brands using the F-statistic for the main effect of tire brands. What assumptions are necessary in order to conduct this test?

4 Construct individual 95% confidence intervals for the mean stopping distances for each of the tire brands in Exercise 3. Use the error standard deviation estimate from the analysis of variance in each of the intervals. Construct simultaneous 95% confidence intervals for the means using a Bonferroni procedure. How do the intervals compare?

5 Construct a set of orthogonal contrasts that can be used to compare the mean stopping times in Exercise 3. Calculate the F or t statistic for testing that each of the contrasts in the means is zero. Interpret the results of these tests in the context of the exercise.

6 Perform Fisher's LSD, Tukey's TSD, and Duncan's multiple-range test on all pairwise comparisons of the average stopping distances. Interpret the results of each of these procedures.

7 Construct an LSI plot for the mean comparisons in Exercise 6. Visually compare the conclusions drawn using Fisher's LSD with the LSI plot.

8 A study was conducted to investigate the effects of three factors on the fragmentation of an explosive device. The factors of interest are pipe diameter, a restraint material, and the air gap between the device and the restraint material. Three repeat tests were conducted on each combination of the factor levels. The data are shown below. Construct an ANOVA table for these data, and perform appropriate tests on the main effects and interactions.

<table>
<tr><td colspan="6" align="center">Pipe Diameter</td></tr>
<tr><td></td><td></td><td colspan="2" align="center">0.75 in.</td><td colspan="2" align="center">1.75 in.</td></tr>
<tr><td></td><td>Material:</td><td>Sand</td><td>Earth</td><td>Sand</td><td>Earth</td></tr>
<tr><td rowspan="4">Air Gap</td><td rowspan="3">0.50 in.</td><td>.0698</td><td>.0659</td><td>.0625</td><td>.0699</td></tr>
<tr><td>.0698</td><td>.0651</td><td>.0615</td><td>.0620</td></tr>
<tr><td>.0686</td><td>.0676</td><td>.0619</td><td>.0602</td></tr>
<tr><td rowspan="3">0.75 in.</td><td>.0618</td><td>.0658</td><td>.0589</td><td>.0612</td></tr>
<tr><td>.0613</td><td>.0635</td><td>.0601</td><td>.0598</td></tr>
<tr><td>.0620</td><td>.0633</td><td>.0621</td><td>.0594</td></tr>
</table>

9 Use Yates's method (Section 7.4) to calculate the main effects and interactions for the factors in Exercise 8. Show that the sums of squares for each of these effects equals the square of the calculated effect multiplied by $r2^{k-2}$, where r is the number of repeat tests ($r = 3$ here) and k is the number of factors ($k = 3$ here).

10 Perform suitable multiple-comparison procedures to determine which combinations of the factor levels in Exercise 8 have significantly different average fragment thicknesses.

11 Two companies were contacted by a major automobile manufacturer to design a new type of water pump for their off-the-road vehicles. Each company designed two water-pump prototypes. Ten pumps of each design were tested in four-wheel-drive vehicles, and the number of miles driven before failure of each water pump was recorded. Investigate the following specific comparisons among the response means:

(a) mean failure mileage for the two companies (ignoring design types),
(b) mean failure mileage for the two designs, separately for each company, and
(c) mean failure mileage for the four designs.

| Company A | | Company B | |
Design 1	Design 2	Design 3	Design 4
31, 189	24, 944	24, 356	27, 077
31, 416	24, 712	24, 036	26, 030
30, 643	24, 576	24, 544	26, 573
30, 321	25, 488	26, 233	25, 804
30, 661	24, 403	23, 075	25, 906
30, 756	24, 625	25, 264	27, 190
31, 316	24, 953	25, 667	26, 539
30, 826	25, 930	21, 613	27, 724
30, 924	24, 215	21, 752	26, 384
31, 168	24, 858	26, 135	26, 712

12 High-pressure and high-stress tests were used in a project for an aerospace research firm to identify materials suitable for use in syngas coolers. This particular study focused on the effects of temperature on the corrosion of a copper material. Specimens from the copper material were tested for 1000 hours of exposure at 11,000 psi. Eight specimens were tested at each temperature level. The data collected are shown below. Plot the average corrosion percentages versus the temperature levels. Use orthogonal polynomials to assess the polynomial effects of temperature on corrosion.

Temp. (°C)	Copper Corrosion (%)							
	Specimen 1	2	3	4	5	6	7	8
300	5.1	5.2	5.8	4.0	5.5	4.7	5.5	4.3
350	6.5	8.2	2.0	6.1	8.0	4.3	7.2	10.6
400	5.9	4.1	6.4	5.4	3.7	5.5	7.6	7.5
450	11.2	9.8	13.6	12.7	15.1	8.8	13.0	12.7
500	13.8	16.1	18.9	15.7	17.0	15.1	17.3	17.0

CHAPTER 17

Analysis of Designs with Random Factor Levels

In this chapter the analysis of statistically designed experiments in which one or more of the experimental factors have random levels is discussed. There are important differences between the analysis of experiments in which all the factors have fixed effects and those in which some of the factors have random levels. The following topics are emphasized in this chapter:

- *the distinction between fixed and random factor effects,*
- *estimation of variance components for random factor effects, and*
- *modifications of ANOVA test procedures needed to accommodate random factor levels.*

In many experiments the levels of one or more of the factors under study are not selected because they are the only ones for which inferences are desired. Random factor levels are intended to be representative of a much larger population of levels. For example, specific vehicles used in an automobile mileage study are representative of a much larger population of vehicles having the same body style, power train, etc. We term factors that have random factor levels *random-effects* factors.

Experiments in which one or more of the factors have random effects are ordinarily conducted with the intent to draw inferences not on the specific factor levels included in the experiment, but on the variability of the population or process from which the levels are obtained. Inferences on fixed factor effects are made on means associated with the specific factor levels included in the experiment (see Chapter 15). Inferences on random factor effects are made on the variances of the factor–level effects.

In this chapter we introduce random effects and the techniques needed to accommodate inferences on factors having random effects. Section 17.1 distinguishes random from fixed factor effects. Section 17.2 describes the calculation of *expected mean squares*, quantities that determine the appropriate mean squares to use in testing hypotheses about random factor effects. In Section 17.3, interval estimation and statistical tests involving random factor effects are discussed. Section 17.4 extends the discussion of random effects to include experiments that have both random and fixed factors. The analysis of hierarchically nested designs is discussed in Section 17.4, and split-plot designs in Section 17.5. These two types of designs are included in this chapter because they ordinarily involve one or more factors that have random effects.

17.1 RANDOM FACTOR EFFECTS

In many experimental programs, the levels of the factors under study are only a sample of all possible levels of interest. For example, in a process capability study, production lots are often one of the factors. The actual lots (levels) chosen for the study are of no particular interest to the engineer; rather, characteristics of all possible lots that might be produced by the process are of importance. In these situations the factors are said to have *random effects* (see Exhibit 17.1).

EXHIBIT 17.1

Random Effects. A factor effect is random if the levels included in the experiment are only a random subset of a much larger population of the possible factor levels. The model terms that represent random factor levels are themselves random variables.

Designs that include factors that have random effects can be identical to those with factors that have fixed effects. Blocking designs and nested designs frequently include one or more random-effects factors. The ANOVA models for random factor effects can be written in a form that appears to be identical to those for fixed factor effects. There are, however, important differences in the assumptions that accompany models having random factor effects (see Exhibit 17.2). We first consider models for which all factors have random effects. Models having both fixed and random effects are discussed in Section 17.4.

EXHIBIT 17.2 RANDOM-EFFECTS MODEL ASSUMPTIONS

1. The levels of all factors in the experiment represent only a random subset of the possible factor levels of interest.
2. The ANOVA model contains random variables representing main effects and interactions. These random variables are assumed to be statistically independent.
3. The experimental errors are statistically independent of each other and of the random-effects model terms.
4. The model terms for each main effect, each interaction, and the experimental errors are satisfactorily modeled by normal probability distributions, which have zero means but possibly different standard deviations.

Consider a two-factor model in which both factors are random and r repeat tests are conducted on each combination of the factor levels. The model, using Latin letters to denote random effects, would be expressed as follows:

$$y_{ijk} = \mu + a_i + b_j + (ab)_{ij} + e_{ijk}, \qquad i = 1, 2, \ldots, a, \quad j = 1, 2, \ldots, b, \quad k = 1, 2, \ldots, r \quad (17.1)$$

Note that this model is similar in form to a two-factor version of the model (15.7)–(15.8). We now add the following assumptions to complete the definition of the model (17.1):

Model Terms	Distribution	Mean	Standard Deviation	
a_i	Normal	0	σ_a	
b_j	Normal	0	σ_b	(17.2)
$(ab)_{ij}$	Normal	0	σ_{ab}	
e_{ijk}	Normal	0	σ	

In addition, we assume that all of the above random variables are statistically independent.

The responses in random-effects models do not have the same distributional properties as those in fixed-effects models. If the factors in the above model were fixed, the responses y_{ijk} would be normally distributed with means

$$\mu_{ij} = \mu + \alpha_i + \beta_j + (\alpha\beta)_{ij}$$

and common standard deviation σ. In contrast, when the factor effects are all random, the responses y_{ijk} have a common normal distribution with mean μ and standard deviation

$$\sigma_y = (\sigma^2 + \sigma_a^2 + \sigma_b^2 + \sigma_{ab}^2)^{1/2}. \tag{17.3}$$

It is the four *variance components* in equation (17.3) that contain the information about the populations of factor effects, since the main effects and interaction factor levels are all assumed to be normally distributed with zero means.

The variance terms in (17.3) are called variance components because of the variance of the responses, σ_y^2, is the sum of the individual variances for the main effects (σ_a^2, σ_b^2), interaction effects (σ_{ab}^2), and uncontrolled error (σ^2). Thus, in contrast to fixed factor levels, the random factor levels contribute to the overall variation observable in the response variable, not to differences in factor–level means. Random effects that have large standard deviations can be expected to influence the response more than effects that have small standard deviations.

To focus more clearly on these concepts, reconsider the data presented in Table 6.5 on skin color measurements. Measurements in this table were taken on seven participants during each of three weeks of the study. Note that it is not the specific participants who are of interest to the manufacturer of the skin-care product, but the population of possible purchasers of this product. By modeling the effects of the participants as a normal random variable, the standard deviation of the distribution measures the variability in the measurements that is attributable to different skin types among intended purchasers.

The week-to-week effect can also be modeled as a random effect. Because environmental conditions and other influences—such as daily hygiene, exposure to the sun, etc.—that can affect skin color measurements are not controllable, each individual does not have the same skin color measurement each week of the study. By modeling the weekly effects on skin color as random effects, the standard deviation of its assumed normal distribution measures the contribution of week-to-week variation to the overall variability of the skin color measurements.

One can also consider the inclusion of a random interaction effect between participants and weeks of the study. An interaction component is included if it is believed that weekly effects differ randomly from participant to participant. Thus, random main effects and interaction effects lead to the model (17.1) for the skin color measurements.

17.2 EXPECTED MEAN SQUARES

The functions of the model parameters that are estimated by the mean squares in an ANOVA table are called *expected mean squares*. For factorial experiments in which all the factor effects are fixed, consideration of expected mean squares is not of great importance, because all the main effects and interactions in the ANOVA table are tested against the error mean square; i.e., as is shown below, all the F-ratios have MS_E in the denominator. This is always not true for experiments in which all the factors are random or for experiments in which some factor effects are fixed and some are random.

Expected mean squares for fixed-effects ANOVA models all have the same form. Using a three-factor balanced complete factorial experiment for illustration purposes, the general form is expressible as indicated in Table 17.1. Each expected mean square equals the error variance plus a sum of squares of the corresponding effects parameters. These expressions are valid when the model is parametrized as in (15.8) and the zero constraints are imposed.

The sums of squares in ANOVA tables for balanced complete factorial experiments are statistically independent regardless of whether the factors have fixed or random effects. For fixed-effects models, a hypothesis that main-effect or interaction model parameters are zero results in the corresponding expected mean squares simply equaling the error variance. One can show that under such hypotheses the sums of squares divided by the error variance follow chi-square distributions. Consequently, the ratio of the mean square for any of the main effects or interactions to the error mean square follows an F-distribution (see Section 12.1). This is the justification for the use of F-statistics in Section 15.4 to evaluate the statistical significance of main effects and interactions for fixed-effects models.

Expected mean squares play a fundamental role in the determination of appropriate test statistics for testing hypotheses about main effects and interactions. Expected mean squares determine which hypotheses are tested by each mean square. An F-statistic can only be formed when, under appropriate hypotheses, two expected mean squares have the same value. For fixed-effects experiments, therefore, the expected mean squares identify the hypotheses being tested with each mean square and reveal that under each hypothesis the effects mean square should be tested using the error mean square in the denominator.

Table 17.1 Expected Mean Squares for Typical Main Effects and Interactions: Fixed Factor Effects

Factor Effect	Source of Variation	Typical Mean Square	Expected Mean Square
Main effect	A	MS_A	$\sigma^2 + bcrQ_\alpha$
Interaction	AB	MS_{AB}	$\sigma^2 + crQ_{\alpha\beta}$
Interaction	ABC	MS_{ABC}	$\sigma^2 + rQ_{\alpha\beta\gamma}$
Error	Error	MS_E	σ^2

$$Q_\alpha = \sum_i \frac{\alpha_i^2}{(a-1)}, \qquad Q_{\alpha\beta} = \sum_i \sum_j \frac{(\alpha\beta)_{ij}^2}{(a-1)(b-1)}$$

$$Q_{\alpha\beta\gamma} = \sum_i \sum_j \sum_k \frac{(\alpha\beta\gamma)_{ijk}^2}{(a-1)(b-1)(c-1)}$$

Experiments do not have to be balanced for the above general statements to hold for fixed-effects models. If sums of squares are formed using the principle of reduction in error sums of squares (Section 15.2), the form of the expected mean squares is similar to that of Table 17.1. Unbalanced designs do introduce complications, because the hypotheses being tested are not simply sums of squares of the effects parameters. They are weighted functions of the model means, with the weights determined by the number of observations for each of the factor–level combinations.

Nevertheless, F-tests are still very useful for assessing the overall significance of factor effects. One should accompany such tests with a careful examination of the factor–level averages. Multiple comparisons of means (Chapter 16) can be used either as an alternative to the F-tests or to investigate any comparisons of interest subsequent to the identification of a statistically significant F-statistic.

Expected mean squares for random factor effects differ from those of fixed effects in one important respect: they involve the variances of the effects rather than the sums of squares of the effects parameters. This follows because of the distributional assumptions [e.g., (17.2)] that accompany the specification of the random-effects model. Table 17.2 displays the expected mean squares for typical mean squares from a three-factor balanced complete factorial experiment. There are two major differences between Tables 17.1 and 17.2: (a) the replacement of the Q's in Table 17.1 with the variance components in Table 17.2, and (b) the hierarchical nature of the expected mean squares for the random effects in Table 17.2.

The pattern in Table 17.2 should be clear enough to adapt to models that have either fewer or more factors than three. Any effect has components in its expected mean square for its own variance component, all interactions included in the model involving the factor(s) contained in the mean square, and the error variance. The coefficients of the variance components equal the numbers of observations used in the calculation of an average for the effects represented by the variance components. General rules for determining these expected mean squares are provided in the appendix to this chapter and are illustrated in the next section.

The difference in expected mean squares between random and fixed-effects models suggests one important difference in the analysis of the two types of experiments. When analyzing experiments having fixed effects, main effects and low-order interactions are not ordinarily evaluated if they are included in a statistically significant higher-order interaction. This is because any factor–level combinations of interest involving the main effects or low-order interactions can be evaluated from the averages for the higher-order interaction. There is no prohibition from analyzing the main effects or low-order interactions, but it is usually superfluous to do so.

With random-effects experiments, all the main effects and interactions are investigated.

Table 17.2 Expected Mean Squares for Typical Main Effects and Interactions, Random Factor Effects

Factor Effect	Source of Variation	Typical Mean Square	Expected Mean Square
Main effect	A	MS_A	$\sigma^2 + r\sigma_{abc}^2 + cr\sigma_{ab}^2$ $+ br\sigma_{ac}^2 + bcr\sigma_a^2$
Interaction	AB	MS_{AB}	$\sigma^2 + r\sigma_{abc}^2 + cr\sigma_{ab}^2$
Interaction	ABC	MS_{ABC}	$\sigma^2 + r\sigma_{abc}^2$
Error	Error	MS_E	σ^2

The expected mean square for each main effect and each interaction contains a unique variance component. Confidence interval estimation or hypothesis testing procedures should be used to assess whether each of the variance components differs significantly from zero.

The expected mean squares in Table 17.2 indicate that for three-factor random-effects models there are no exact F-tests for the main effects. This is because under the null hypothesis that a main-effect variance component is zero, the expected mean square for that main effect does not equal any of the other expected mean squares in the table. Each two-factor interaction mean square is tested against the three-factor interaction mean square because under a hypothesis that one of the two-factor interaction variance components is zero, the expected mean square for the corresponding interaction effect equals that of the three-factor interaction mean square. The three-factor interaction mean square is again tested against the error mean square.

Because the main effects in three-factor random-effects models and many other models of interest do not possess exact F-tests, approximate procedures have been developed that are similar to the two-sample t-test for means when the two variances are unequal and unknown. A direct extension of the approximate t-statistic (13.14a) is Satterthwaite's approximate F-statistic (see Exhibit 17.3).

To apply this procedure to the testing of a main effect for a three-factor random-effects model, either of the following two approximate F-statistics can be used:

(i) $F = L_1/M_1$, with

$$L_1 = MS_A \quad \text{and} \quad M_1 = MS_{AB} + MS_{AC} - MS_{ABC},$$

EXHIBIT 17.3   SATTERTHWAITE'S APPROXIMATE TEST PROCEDURE

1. Examine the expected mean squares to ensure that an exact test of the hypothesis of interest does not exist.
2. Denote the mean squares in the ANOVA table by $MS_1, MS_2, \ldots, MS_g$.
3. Find a linear combination of mean squares

$$L = a_1 MS_1 + a_2 MS_2 + \cdots + a_g MS_g$$

for which the same linear combination of their expected mean squares under the null hypothesis of interest equals the expected mean square for some mean square M from the ANOVA table. Note that

(a) L may involve as few as two mean squares;

(b) the second mean square M may also be a linear combination of mean squares, but its expected mean square should be valid regardless of whether the null hypothesis is true.

4. Use the following approximation for the number degrees of freedom associated with L:

$$df(L) = \frac{L^2}{(a_1 MS_1)^2/v_1 + \cdots + (a_g MS_g)^2/v_g},$$

where v_i is the number of degrees of freedom associated with MS_i.

5. Test the hypothesis of interest using $F = L/M$, with F following an approximate F-distribution having $v_1 = df(L)$ and $v_2 = df(M)$ degrees of freedom, the numbers of degrees of freedom associated with M.

---

or

(ii) $F = L_2/M_2$, with

$$L_2 = MS_A + MS_{ABC} \quad \text{and} \quad M_2 = MS_{AB} + MS_{AC}.$$

Note that with either of these statistics, the numerator mean square only has the same expected mean square as the denominator under the hypothesis that the main-effect variance component for factor A is zero. The first of these approximate F-statistics is often preferred because only the denominator must be approximated; however, the denominator can be negative. The second approximation is always nonnegative.

With unbalanced designs the reduction in error sums of squares results in mean squares that have the same general form as in Table 17.2, but the multipliers of the variance components are complicated functions of the numbers of repeat tests for each of the factor–level combinations. F-tests can still be appropriate for testing hypotheses about the variance components, but the expected mean squares must be known to confirm this.

Expected mean squares are also important for the estimation of the variance components of random effects. In the next section we discuss estimation of the variance components.

Main-effect and interaction sums of squares for balanced complete factorial experiments having random effects are the same as those for fixed-effects models [e.g., see equations (15.11)–(15.14) and Table 15.4]. As in the fixed-effects model, the mean squares in a random-effects ANOVA table estimate combinations of model parameters. For the random-effects model, the mean squares estimate linear combinations of the variance components rather than the squares of linear combinations of model means. Also as in the fixed-effects model, ratios of appropriate mean squares have an F-distribution with numerator and denominator degrees of freedom equal to those of the corresponding mean squares. We justify these last two properties in the discussions of expected mean squares and inference procedures in the next two sections.

17.3 VARIANCE-COMPONENT ESTIMATION

The expected mean squares for random factor effects provide an important source of information on the estimation of variance components for the factor effects. Estimates of the variance components are obtained by setting expected mean squares for an ANOVA table equal to their respective mean squares and solving for the variance component of interest (see Exhibit 17.4). Thus, estimates of the error variance and the three-factor interaction variance component in a three-factor balanced complete factorial experiment are, from Table 17.2,

$$s_e^2 = MS_E,$$

$$s_{abc}^2 = \frac{MS_{ABC} - MS_E}{r},$$

where we denote the variance component estimate for σ^2 and σ_{abc}^2 by s_e^2 and s_{abc}^2, respectively. Other expected mean squares can be obtained in a similar fashion, by taking appropriate linear combinations of the mean squares.

EXHIBIT 17.4　ESTIMATION OF VARIANCE COMPONENTS

1. Calculate the mean squares for the main effects, interactions, and error terms in the ANOVA model.

2. If the design is a complete factorial with an equal number of repeat tests per factor–level combination, determine the expected mean squares as in Table 17.2 for random-effects models. If the design is nested or if some factors are fixed and others are random, refer to the rules presented in Section 17.4.

3. Equate the mean squares to their expected mean squares and solve the resulting system of simultaneous equations.

Interval estimates for the ratio of two expected mean squares can be found using the methodology presented in Section 14.2. Denote one of the mean squares by s_1^2 and its corresponding expected mean square by σ_1^2. Denote the second mean square by s_2^2 and its expected mean square by σ_2^2. Then the inequality (14.6) provides an interval estimate for the ratio of the expected mean squares.

Interval estimates for ratios of expected mean squares are most often of interest as an intermediate step in the estimation of the ratio of one of the random-effects variance components to the error variance. Once an appropriate interval estimate for the ratio of the main effect or interaction expected mean square to the error variance is obtained, the interval can often be solved for the ratio of the variance component of interest to the error variance.

In Table 17.3, the ANOVA table for the skin color measurements from Table 6.5 is displayed. The F-statistics clearly indicate that the effect of the variability of the participants is statistically significant. The week-to-week variability is not significantly different from the uncontrolled variability of the measurements. Based on the significance of the main effect due to participants, it is now of interest to compare the variability due to the participants with that due to uncontrolled variability.

The estimates of the error and the participant standard deviations are

$$s_e = (1.71)^{1/2} = 1.31,$$

$$s_a = \left(\frac{317.43 - 1.71}{3} \right)^{1/2} = 10.26.$$

The participant standard deviation is almost 8 times larger than the uncontrolled error standard deviation. A 95% confidence interval for the ratio of the standard deviations is

Table 17.3　Analysis-of-Variance Table for the Skin Color Measurements

Source	df	SS	MS	F	E(MS)
Subjects	6	1904.58	317.43	186.17	$\sigma^2 + 3\sigma_a^2$
Weeks	2	2.75	1.37	0.81	$\sigma^2 + 7\sigma_b^2$
Error	12	20.46	1.71		σ^2
Total	20	1927.78			

obtainable from the interval estimate for the ratio of the two expected mean squares:

$$\frac{317.43}{(1.71)(3.73)} < \frac{\sigma^2 + 3\sigma_a^2}{\sigma^2} < \frac{(317.43)(5.37)}{1.71},$$

or

$$49.92 < \frac{\sigma^2 + 3\sigma_a^2}{\sigma^2} < 966.84.$$

Solving this inequality for the ratio of the estimated standard deviations yields

$$4.03 < \sigma_a/\sigma < 17.94.$$

Estimated variance components can be negative or very small in magnitude. The estimated variance component for the week-to-week variation in skin color measurements is negative. From Table 17.3, $s_b^2 = (1.37 - 1.71)/7 = -0.05$. When estimated variance components are negative, one should inspect the data to ensure that outliers (Chapter 24) are not present. A single outlier in a data set can cause estimates of variance components to be negative. Ordinarily when negative estimates of variance are obtained and the data have been examined to ensure that the negative estimate is not caused by an outlier, the estimated variance is set equal to zero.

Confidence intervals for individual variance components can be obtained, but the procedures for doing so are complicated and beyond the scope of this book. The above interval estimation procedures usually suffice for drawing inferences on variance components for most experimental settings.

17.4 COMBINATIONS OF FIXED AND RANDOM EFFECTS

Experiments often are conducted in which some of the factors are fixed and others are random. The corresponding ANOVA models are termed *mixed* models. Sometimes the random factors are the primary focus of interest. Other times the fixed factor effects are, but the random factors are included in order to generalize the range of applicability of the results. When this occurs, the variance components are termed *nuisance parameters*. That is, they must be accounted for in order to correctly test for fixed factor effects, but they are not the primary focus of the investigation. Variances of block effects are often nuisance parameters.

Sums of squares and mean squares for mixed models are calculated using the same formulas as with models having either all fixed effects or all random effects. As with random-effects models, the F-statistics used to assess main effects and interactions are not all formed with the error mean square in the denominator. Expected mean squares are critically important in the correct use of ANOVA tables for mixed models.

Expected mean squares for mixed models depend not only on the type of design but also on the structure of the ANOVA model that is used to relate the response variable to the design factors. For this reason, the formulation of expected mean squares for many unbalanced designs can only be determined once the design is specified. There are, however, general rules for determining expected mean squares when designs are balanced.

In the appendix to this chapter, rules are listed for the determination of expected mean squares for balanced experimental designs. These rules can be used for either crossed or nested designs. They can be used for fixed, random, or mixed models. In the following discussions, we examine these rules for crossed factors only. In the next two sections, designs involving nested factors are discussed. For balanced complete factorial experiments involving only fixed factor effects, application of these rules results in expected mean squares similar to those in Table 17.1. If all factor effects are random, application of these rules results in expected mean squares similar to those in Table 17.2.

A situation in which mixed-effects models often are used is in toxicology studies with laboratory animals. Table 17.4 contains data from such a study. The measurements shown in the table are on the amount of swelling that occurred after the skin of laboratory mice was subjected to a chemical intended to produce an allergic reaction. The purpose of the study was to determine whether the area around a mouse's ear or on its back was more sensitive to the chemical. The allergic reaction is localized to small portions of a mouse's skin, so the amount of swelling is believed to be independent for the two locations. Two measurements were taken at each location of each mouse, with sufficient time for the reaction to wear off prior to the second application of the chemical.

Since two specific locations were selected for comparison, the location factor is a fixed effect. The mice chosen for the study, however, represent random levels. Interest does not focus on the six specific mice that were included in the study, but on the conceptually infinite population of mice that could be subjected to the chemical. The experimenters did not believe that location effects would change with different mice; consequently, the ANOVA model does not contain an interaction term.

The ANOVA table for this experiment is shown in Table 17.5. Both the fixed effects of location and the random effects of animals are highly significant. Since the average swelling for the ear location, 0.72 mm, is larger than that for the back location, 0.30 mm, we conclude that the ear location has a significantly higher mean swelling than the back location. Since the mice are assumed to be a random effect, we estimate the standard deviations for measurement error and the mouse effect as indicated in the

Table 17.4   Skin Swelling Measurements

| Animal | Repeat Measurement | Swelling (mm) | |
		Ear	Back
1	1	0.53	0.09
	2	0.33	0.01
2	1	0.67	0.38
	2	0.65	0.31
3	1	0.72	0.31
	2	0.61	0.30
4	1	0.71	0.36
	2	0.83	0.38
5	1	0.85	0.36
	2	1.05	0.49
6	1	0.92	0.28
	2	0.74	0.29

Table 17.5 ANOVA **Table for Skin Sensitivity Study**

Source of Variation	df	Sum of Squares	Mean Square	F-Value
Location	1	1.0626	1.0626	141.88
Animals	5	0.4436	0.0887	11.85
Error	17	0.1273	0.0075	
Total	23	1.6335		

last section:

$$s_e = (MS_E)^{1/2} = 0.0865, \qquad s_a = \left(\frac{MS_A - MS_E}{5}\right)^{1/2} = 0.2853.$$

Thus, as measured by these standard deviations, the variability attributable to the laboratory mice is over three times larger than that of the measurement error.

A special application of the estimation of variance components in mixed ANOVA models is in the estimation of *repeatability* and *reproducibility* for interlaboratory studies of test procedures. In its simplest form, several materials or specimens are sent to each of several laboratories. Each laboratory is instructed to measure a characteristic of the materials with at least two independent repeat measurements. The data are then analyzed to obtain estimates of the between-laboratory variability (reproducibility) and the within-laboratory variability (repeatability).

In the experimental setting just described, the materials are specifically chosen to cover a broad range of measurement values of the characteristic of interest. Materials are therefore a fixed effect. The laboratories often are viewed as representative of a large number of laboratories that could perform these tests and are therefore considered a random effect.

Reproducibility is defined in terms of the limits of variability that can be expected on the difference of two independent test results from two different laboratories. A test result is defined to be the average of a specified number m of measurements of the characteristic of interest. This number of repeat tests need not equal the number of repeat tests in the interlaboratory study, but it often does. One can show that the variance of an average defined in this manner is $\sigma_L^2 + \sigma^2/m$, where σ_L^2 is the variance of the main effect for laboratories and σ^2 is the variance of the uncontrolled experimental error. The variance of the difference in two averages is then twice this value.

Reproducibility limits are obtained by analogy with confidence-interval estimation of the difference of two means. Inserting $z_{0.05} = 1.96$ in place of a t-statistic in the two-sided confidence interval formula in Table 13.5 yields the approximate 95% confidence limits for reproducibility:

$$\pm 1.96 \left[2\left(s_L^2 + \frac{s_e^2}{m} \right) \right]^{1/2} = \pm 2.77 \left(s_L^2 + \frac{s_e^2}{m} \right)^{1/2},$$

where s_L^2 and s_e^2 are respectively the variance component estimates for the laboratory and error variances.

Approximate 95% confidence limits for repeatability are obtained in a similar fashion.

Repeatability limits are based on the expected difference in two test results for independent measurements at one laboratory. The variance of a difference of two averages, each based on m repeat tests, at one laboratory is $2\sigma^2/m$. Thus, repeatability limits are

$$\pm 1.96 \left(2\frac{s_e^2}{m} \right)^{1/2} = \pm 2.77 \frac{s_e}{m^{1/2}}.$$

17.5 HIERARCHICALLY NESTED DESIGNS

Nested designs (Chapter 10) are designs in which the levels of one or more factors differ with each level of one or more other factors. With hierarchically nested designs, each level of nesting is imbedded within a previous level of nesting. The polyethelene density experiment discussed in Sections 10.1 and 10.2 (see Figure 10.2) is an example of a hierarchically nested design.

Hierarchically nested designs are most frequently used with random factor levels. The analysis we present in this section requires that all but the first factor be a random factor. There are situations in which one or more of the nested factors correspond to fixed factor levels. The analysis of such designs depends on which factors are fixed and which are random. Computation of expected mean squares using the rules in the appendix to this chapter is necessary to determine appropriate F-statistics for testing hypotheses and for estimating variance components for random effects.

Consider an experiment having three hierarchically nested factors A, B, and C. Suppose that factor C is nested within factor B and that factor B is nested within factor A. Assume that there are an equal number of replications at each level of nesting. The model for such an experiment can be written as follows:

$$y_{ijkl} = \mu + a_i + b_{j(i)} + c_{k(ij)} + e_{ijkl}$$
$$i = 1, 2, \ldots, a, \quad j = 1, 2, \ldots, b, \quad k = 1, 2, \ldots, c, \quad l = 1, 2, \ldots, r, \qquad (17.4)$$

where the subscripts in parentheses denote the factor levels within which the leading factor level is nested; for example, the kth level of factor C is nested within the ith level of factor A and the jth level of factor B. Since Latin letters have been used to denote the factor levels, we are assuming that all factors are random and denote the variance components of the nested factors by σ_a^2, σ_b^2, and σ_c^2, respectively.

Sums of squares for nested factors are computed differently from those for crossed factors. Each nested sum of squares is computed as a sum of squared differences between the average response for a factor–level combination and the average of the responses at the previous level of nesting. For example, the sums of squares for the model (17.4) are

$$SS_A = bcr \sum_i (\bar{y}_{i\ldots} - \bar{y}_{\ldots})^2, \quad SS_{B(A)} = cr \sum_i \sum_j (\bar{y}_{ij\ldots} - \bar{y}_{i\ldots})^2,$$
$$SS_{C(AB)} = r \sum_i \sum_j \sum_k (\bar{y}_{ijk\cdot} - \bar{y}_{ij\cdot\cdot})^2. \qquad (17.5)$$

These nested sums of squares are related to the sums of squares for main effects and interactions, treating all the factors as crossed factors, as follows:

$$SS_{B(A)} = SS_B + SS_{AB}, \quad SS_{C(AB)} = SS_C + SS_{AC} + SS_{BC} + SS_{ABC}.$$

Table 17.6 displays a symbolic ANOVA table, along with the expected mean squares, for this hierarchically nested model. Hierarchically nested models having more than three factors have ANOVA tables with the same general pattern. Note that each expected mean square contains all the terms of the expected mean square that follows it in the table. Because of this pattern, to test the statistical significance of any factor, the appropriate F-statistic is the ratio of the mean square for the factor of interest divided by the mean square for the factor that follows it. For example, to test the hypotheses

$$H_0:\sigma_b^2 = 0 \quad \text{vs} \quad H_a:\sigma_b^2 \neq 0,$$

the appropriate F-ratio is $F = \mathrm{MS}_{B(A)}/\mathrm{MS}_{C(AB)}$.

Estimation of the variance components for the various model factors follows the general procedures outlined in Section 17.3. Estimation of these variance components is especially easy for hierarchically nested factors. For three-factor random-effects models one simply uses differences between the mean square for the factor of interest and the

Table 17.6 Symbolic ANOVA Table for a Three-Factor, Hierarchically Nested Random-Effect Model

ANOVA

Source	df	SS	MS	Expected Mean Square
A	$a-1$	SS_A	MS_A	$\sigma^2 + r\sigma_c^2 + cr\sigma_b^2 + bcr\sigma_a^2$
$B(A)$	$a(b-1)$	$\mathrm{SS}_{B(A)}$	$\mathrm{MS}_{B(A)}$	$\sigma^2 + r\sigma_c^2 + cr\sigma_b^2$
$C(AB)$	$ab(c-1)$	$\mathrm{SS}_{C(AB)}$	$\mathrm{MS}_{C(AB)}$	$\sigma^2 + r\sigma_c^2$
Error	$abc(r-1)$	SS_E	MS_E	σ^2
Total	$abcr-1$	TSS		

Computations

$$\mathrm{TSS} = \sum_i \sum_j \sum_k \sum_l y_{ijkl}^2 - \mathrm{SS}_M, \quad \mathrm{SS}_M = n^{-1} y_{....}^2$$

$$\mathrm{SS}_A = \sum_i \frac{y_{i...}^2}{bcr} - \mathrm{SS}_M$$

$$\mathrm{SS}_{B(A)} = \sum_i \sum_j \frac{y_{ij..}^2}{cr} - \sum_i \frac{y_{i...}^2}{bcr}$$

$$\mathrm{SS}_{C(AB)} = \sum_i \sum_j \sum_k \frac{y_{ijk.}^2}{r} - \sum_i \sum_j \frac{y_{ij..}^2}{cr}$$

$$\mathrm{SS}_E = \mathrm{TSS} - \mathrm{SS}_A - \mathrm{SS}_{B(A)} - \mathrm{SS}_{C(AB)}$$

$$y_{....} = \sum_i \sum_j \sum_k \sum_l y_{ijkl}, \quad y_{i...} = \sum_j \sum_k \sum_l y_{ijkl}, \quad \text{etc.}$$

one following it in Table 17.6. For example,

$$s_b^2 = \frac{MS_{B(A)} - MS_{C(AB)}}{rc}.$$

Staggered nested designs are an alternative to hierarchically nested designs when it is desired to obtain more evenly distributed information on each of the factors. A staggered nested design for a polymerization process study was described in Section 10.2 (see Figure 10.5). The data from this study are shown in Table 17.7. Note that there are two boxes selected from each of 30 lots, but only one preparation is made from the second box of each lot, whereas two preparations are made from the first box. Likewise, only one strength test was made on preparation 1 from each box, but two strength tests were made on preparation 2 from the first box. By reducing the number of preparations and the

Table 17.7   Strength Data for Polymerization Process Study

Strength	Lot	Box	Prep.	Strength	Lot	Box	Prep.
11.91	1	1	1	4.08	29	1	1
9.76	1	1	2	4.88	29	1	2
9.24	1	1	2	4.96	29	1	2
9.02	1	2	1	4.76	29	2	1
10.00	7	1	1	6.73	30	1	1
10.65	7	1	2	9.38	30	1	2
7.77	7	1	2	8.02	30	1	2
13.69	7	2	1	6.99	30	2	1
8.02	8	1	1	6.59	32	1	1
6.50	8	1	2	5.91	32	1	2
6.26	8	1	2	5.79	32	1	2
7.95	8	2	1	6.55	32	2	1
9.15	9	1	1	5.77	47	1	1
8.08	9	1	2	7.19	47	1	2
5.28	9	1	2	7.22	47	1	2
7.46	9	2	1	8.33	47	2	1
7.43	10	1	1	8.12	48	1	1
7.84	10	1	2	7.93	48	1	2
5.91	10	1	2	6.48	48	1	2
6.11	10	2	1	7.43	48	2	1
7.01	11	1	1	3.95	49	1	1
9.00	11	1	2	3.70	49	1	2
8.38	11	1	2	2.86	49	1	2
8.58	11	2	1	5.92	49	2	1
11.13	12	1	1	5.96	51	1	1
12.81	12	1	2	4.64	51	1	2
13.58	12	1	2	5.70	51	1	2
10.00	12	2	1	5.88	51	2	1
14.07	13	1	1	4.18	52	1	1
10.62	13	1	2	5.94	52	1	2
11.71	13	1	2	6.28	52	1	2
14.56	13	2	1	5.24	52	2	1

Table 17.7 Strength Data for Polymerization Process Study (continued)

Strength	Lot	Box	Prep.	Strength	Lot	Box	Prep.
11.25	14	1	1	4.35	53	1	1
9.50	14	1	2	5.44	53	1	2
8.00	14	1	2	5.38	53	1	2
11.14	14	2	1	7.04	53	2	1
9.51	15	1	1	2.57	54	1	1
10.93	15	1	2	3.50	54	1	2
12.16	15	1	2	3.88	54	1	2
12.71	15	2	1	3.76	54	2	1
16.79	16	1	1	3.48	55	1	1
11.95	16	1	2	4.80	55	1	2
10.58	16	1	2	4.46	55	1	2
13.08	16	2	1	3.18	55	2	1
7.51	19	1	1	4.38	56	1	1
4.34	19	1	2	5.35	56	1	2
5.45	19	1	2	6.39	56	1	2
5.21	19	2	1	5.50	56	2	1
6.51	21	1	1	3.79	57	1	1
7.60	21	1	2	3.09	57	1	2
6.72	21	1	2	3.19	57	1	2
6.35	21	2	1	2.59	57	2	1
6.31	23	1	1	4.39	58	1	1
5.12	23	1	2	5.30	58	1	2
5.85	23	1	2	4.72	58	1	2
8.74	23	2	1	6.13	58	2	1
4.53	24	1	1	5.96	59	1	1
5.28	24	1	2	7.09	59	1	2
5.73	24	1	2	7.82	59	1	2
5.07	24	2	1	7.14	59	2	1

number of strength tests (i.e., not having two preparations from each box and two strength tests for each preparation), the total number of tests was reduced from 240 to 120 (see Table 10.1). In spite of the reduction in the number of tests, there are still ample observations that can be used to assess the effects of the various experimental factors.

The analysis of staggered nested designs is similar to the analysis of hierarchically nested designs. The model (17.4), with A = lot, B = box, and C = preparation, is an appropriate model for the polymerization process data; however, since staggered nested designs are unbalanced, observations are not taken for all the responses indicated by the subscripts in the model (17.4). A consequence of the imbalance is that the ANOVA table is constructed using the principle of reduction in error sums of squares.

Three models are fitted: one with lots as the only factor, one with lots and boxes, and one with all three factors. The first fit provides the sum of squares for the lots, the difference in error sums of squares for the first and second models provides the sum of squares for boxes, and the difference in error sums of squares for the last two fits gives the sum of squares for preparations. The error sum of squares for the ANOVA table is obtained from the last fit. The ANOVA table for the polymerization process study is shown in Table 17.8. Note

Table 17.8 ANOVA **Table for Polymerization-Process Data**

Source	df	SS	MS	F	p-Value
Lots	29	855.96	29.52	17.67	0.000
Boxes (lots)	30	50.09	1.67	0.73	0.803
Preparations	30	68.44	2.28	3.52	0.000
(lots, boxes)					
Error	30	19.44	0.65		
Total	119	993.92			

from this table the nearly equal numbers of degrees of freedom for the three experimental factors. An ordinary hierarchical design would have many more degrees of freedom for the preparations than for the lots.

Estimation of variance components with staggered nested designs is more complicated than with hierarchically nested designs. The difficulty arises because of the imbalance in the staggered designs. The interested reader is referred to the references on staggered nested designs at the end of Chapter 10.

17.6 SPLIT-PLOT DESIGNS

Split-plot designs were introduced in Section 10.3 as designs in which crossed factors are nested within levels of other factors. Split-plot designs can be used when experimental units have different sizes, as in agricultural field trials in which some factors are assigned to entire farms and others to fields within the farms. Split-plot designs can also be used when there are different levels of randomization, as in the cylinder-wear example of Section 10.4. Since split-plot designs contain both crossed and nested factors, the analysis of such designs includes features of both factorial experiments and nested experiments.

In Section 10.3 (see also Section 8.1), an investigation of lawnmower automatic-cutoff times was described. The experiment was conducted in a split-plot design. Responses for each of two speeds were taken on each of three lawnmowers from each of the two manufacturers. Figure 10.3 shows the nested and crossed relationships among the factors.

There are two fixed factors in this study (manufacturer and speed), and one random factor (lawnmower). The two fixed factors are crossed, and an interaction between the factors is possible. An ANOVA model for this experiment is

$$y_{ijkl} = \mu + \alpha_i + b_{j(i)} + \gamma_k + (\alpha\gamma)_{ik} + e_{ijkl} \qquad i = 1, 2, \quad j = 1, 2, 3, \quad k = 1, 2, \quad l = 1, 2.$$
$$(17.6)$$

It is important to note that the above model indicates the random effects for lawnmowers are only nested within the levels of the manufacturers. The lawnmowers are not nested within the levels of speed, because all lawnmowers are tested at both speeds. Had some lawnmowers only been tested at the high speed and others only at the low speed, lawnmowers would be nested within the joint levels of manufacturer and speed. In that case the model would be written as

$$y_{ijkl} = \mu + \alpha_i + \beta_j + (\alpha\beta)_{ij} + c_{k(ij)} + e_{ijkl},$$

where factor B (fixed) would represent speed and factor C (random) would represent lawnmowers.

The mean of a response for model (17.5) is $\mu + \alpha_i + \gamma_k + (\alpha\gamma)_{ik}$, where the α_i and γ_k model terms correspond to the manufacturer and speed factors, respectively. The random lawnmower and error terms are assumed to be independently normally distributed with zero means. The $b_{j(i)}$ terms have standard deviation σ_b. The experimental error terms, e_{ijkl}, have standard deviation σ. If there is reason to suspect that the lawnmowers and the speed factors interact, another set of random terms representing the interaction could be added to the above model.

Note that the lawnmowers constitute the whole plots for this split-plot design. The repeat tests on the speeds for each lawnmower constitute the split plots. The whole-plot error, the variability due to replicate test runs on each lawnmower, consists of variation due to different lawnmowers, and to the uncontrollable experimental error. The split-plot error in this example is simply the uncontrollable experimental error. These components of variability are quantified in the expected mean squares.

Table 17.9 displays a symbolic ANOVA table for the experiment on lawnmower automatic-cutoff times. Note that both fixed and random effects and crossed and nested factors occur in this split-plot analysis. The expected mean squares identify the proper F-ratios for testing the factor effects. Observe that the manufacturers are tested against the whole-plot (lawnmower) error mean square, and the speed effects against the split-plot (experimental) error mean square.

Table 17.10 lists the cutoff times for the individual test runs. The ANOVA table for these

Table 17.9   Symbolic ANOVA Table for Split-Plot Analysis of Automatic-Cutoff-Time Experiment

Source	df	SS	MS	F	Expected MS
Manufacturers (M)	$a-1$	SS_A	MS_A	$MS_A/MS_{B(A)}$	$\sigma^2 + cr\sigma_b^2$ $+ bcrQ_\alpha$
Lawnmowers	$a(b-1)$	$SS_{B(A)}$	$MS_{B(A)}$	$MS_{B(A)}/MS_E$	$\sigma^2 + cr\sigma_b^2$
Speed (S)	$c-1$	SS_C	MS_C	MS_C/MS_E	$\sigma^2 + abrQ_\gamma$
$M \times S$	$(a-1)(c-1)$	SS_{AC}	MS_{AC}	MS_{AC}/MS_E	$\sigma^2 + brQ_{\alpha\beta}$
Error	$a(bcr - b - c + 1)$	SS_E	MS_E		σ^2
Total	$abcr - 1$	TSS			

Table 17.10   Lawnmower Automatic Cutoff Times

Manufacturer	Lawnmower	Time (10^{-2} sec)			
		Low	Speed	High	Speed	Average
A	1	211	230	278	278	249.25
	2	184	188	249	272	223.25
	3	216	232	275	271	248.50
B	4	205	217	247	251	230.00
	5	169	168	239	252	207.00
	6	200	187	261	242	222.50
Average		197.50	203.67	258.17	261.00	230.08

Table 17.11 ANOVA **Table for Split-Plot Analysis of Automatic Cutoff Times**

Source	df	SS	MS	F	p-Value
Manufacturer (M)	1	2,521.50	2,521.50	3.73	0.071
Lawnmower	4	2,852.83	713.21	5.39	0.006
Speed (S)	1	20,886.00	20,886.00	157.87	0.000
M × S	1	10.67	10.67	0.08	0.780
Error	16	2,116.83	132.30		
Total	23	23,387.83			

data is shown in Table 17.11. The significance probabilities in the ANOVA table indicate that the interaction between manufacturers and speeds is not statistically significant. The main effect for manufacturers is not statistically significant at the 5% significance level; nevertheless, one may wish to examine the average cutoff times and compute a confidence interval for the difference in the mean cutoff times for the two manufacturers.

In the construction of a confidence interval for the difference of the means for the two manufacturers, the mean square for the lawnmowers is used in the confidence-interval formula, not the error mean square. The mean square used in confidence-interval procedures is always the same mean square that would be used in the denominator of an F-test for the significance of the effect.

Both the main effect for the speeds and the variability due to lawnmowers are statistically significant. Interestingly, the estimates of the two components of variability are approximately equal:

$$s_e = (132.30)^{1/2} = 11.50,$$

$$s_b = \left(\frac{713.21 - 132.30}{4}\right)^{1/2} = 12.05.$$

Thus, the variability incurred by using several lawnmowers in the experiment is of approximately the same magnitude as that from all the sources of uncontrolled variability.

If only one lawnmower had been obtained per manufacturer, no test of the manufacturer main effect would be possible unless the assumption of no whole-plot error could be made. From the above analysis, we see that such an assumption is unreasonable for this experiment, since the lawnmower effect is substantial. In experiments where combinations of levels of two or more factors are assigned to whole plots, tests of factor effects of interest can be obtained without assuming that there is no whole-plot error if the assumption of negligible higher-order interactions among the whole-plot factors is reasonable. Note that in the case of a two-factor factorial experiment in the whole plots, it would be necessary to assume that the two-factor interaction was negligible if no replication of the whole plot was included in the experiment.

The speed factor has only two levels in this experiment. The significant F-ratio for this effect indicates that the two factor levels produce significantly different effects on the cutoff times. For situations in which a factor that is studied at more than two levels is significant, or where a significant interaction is found, it may not be clear which pairs of factor–level averages are significantly different. Procedures such as those discussed in Chapter 16 can be used to determine which pairs of levels are statistically significant. The key to using these procedures is an estimate of the standard error of the difference of two averages.

Obtaining a standard-error estimate is more complicated for a split-plot design than for the other designs discussed, since the proper standard error depends on the averages being compared. There are four situations for distinguishing among averages that need to be considered: factor–level averages for a whole-plot factor, factor–level averages for a split-plot factor, interaction averages from a common level of a whole-plot factor, and interaction averages from different whole-plot factor levels. It was mentioned above that in general, the correct estimate of the error variance is the mean square used to test the significance of a factor effect. In Exhibit 17.5, we make this statement more explicit for the comparison of averages for a split-plot design.

EXHIBIT 17.5 STANDARD ERRORS FOR SPLIT-PLOT AVERAGES

1. For the comparison of two whole-plot averages,

$$\widehat{SE}(\bar{y}_i - \bar{y}_j) = [MS_{WP}(n_i^{-1} + n_j^{-1})]^{1/2},$$

where MS_{WP} is the whole-plot error mean square.

2. For the comparison of two split-plot averages,

$$\widehat{SE}(\bar{y}_i - \bar{y}_j) = [MS_{SP}(n_i^{-1} + n_j^{-1})]^{1/2},$$

where MS_{SP} is the split-plot error mean square, ordinarily MS_E.

3. For the comparison of two interaction averages, two cases must be considered.

  (a) If the two averages have the same level of the whole-plot factor, the standard error is estimated as in instruction 2.

  (b) If the two averages do not have the same level of the whole-plot factor, no exact standard-error formula exists. In place of MS_{SP} in instruction 2, insert a Satterthwaite-type estimate of the sum of the subplot and whole-plot error variances.

The estimated standard errors in Exhibit 17.5 can be used in conjunction with any of the multiple-comparison procedures discussed in Chapter 16. The general approach outlined here can be extended to models that have additional levels of nesting, such as in split-split-plot experiments.

APPENDIX: DETERMINING EXPECTED MEAN SQUARES

Expected mean squares for most balanced experimental designs can be determined using the following rules. These rules apply to crossed factors when there are an equal number of repeat tests for each combination. They also apply to nested factors when there are an equal number of levels at each stage of nesting.

  1. Write out an ANOVA model for the experimental design. Impose constraints that the sum over any subscript of the fixed-effect parameters is zero. Interactions of fixed and random factors are considered random effects.

  2. Label a two-way table with
    (a) A column for each of the subscripts in the ANOVA model and
    (b) A row for each term of the model (except the constant), with the last row being the error term expressed as a nested factor.

3. Proceed down each column labeled by a subscript that represents a fixed effect.

   (a) Enter a 0 if the column subscript appears in a fixed row effect (or a 1 if the column subscript appears in a random row effect) and if no other subscripts are nested within it.

   (b) Enter a 1 if the column subscript appears in the row effect and one or more other subscripts are nested within it.

   (c) If the column subscript does not appear in the row effect, enter the number of levels of the factor corresponding to the subscript.

4. Proceed down each column labeled by a subscript that represents a random effect.

   (a) Enter a 1 if the column subscript appears in the row effect, regardless of whether any other effects are nested within it.

   (b) If the column subscript does not appear in the row effect, enter the number of levels of the factor corresponding to the subscript.

5. For a main effect or interaction consisting of only fixed effects factors, let $\phi = Q$, the mean square of the effects parameters. For a main effect or interaction consisting of one or more random effects factors, let $\phi = \sigma^2$, the variance component for the random effect. List the ϕ parameters in a separate column to the right of the two-way table, with each ϕ parameter on the same line as its corresponding model term.

6. Select a mean square from the ANOVA table. Let MS denote the mean square and C the set of subscripts on the corresponding model term.

   (a) Identify the ϕ parameters whose corresponding model terms contain all of the subscripts in C.

   (b) Determine the multiplier for each ϕ parameter by

      (i) eliminating the columns in the two-way table that correspond to each of the subscripts in C, and

      (ii) taking the product of the remaining elements in the row of the two-way table in which the ϕ parameter occurs.

   (c) The expected mean square for MS is the linear combination of ϕ values identified in (a) with coefficients of the linear combination determined by (b). The expected mean square for MS_E is σ^2.

7. Repeat Step 6 for each mean square in the ANOVA table.

To illustrate the application of the above rules, consider the following model

$$Y_{ijkl} = \mu + \alpha_i + b_{j(i)} + \gamma_k + (\alpha\gamma)_{ik} + e_{ijkl},$$

$$i = 1,\ldots,a \quad j = 1,\ldots,b \quad k = 1,\ldots,c \quad l = 1,\ldots,r.$$

This model has two fixed effects factors (A and C) and one random nested factor (B). The levels of factor B are nested within the levels of factor A. The two-way table of multipliers for the determination of the expected mean squares is as follows:

	i	j	k	l	ϕ
α_i	0	b	c	r	Q_α
$\beta_{j(i)}$	1	1	c	r	σ_b^2
γ_k	a	b	0	r	Q_γ
$(\alpha\gamma)_{ik}$	0	b	0	r	$Q_{\alpha\gamma}$
$e_{l(ijk)}$	1	1	1	1	σ^2

Using these multipliers, the expected mean squares are:

Effect	Expected Mean Square
A	$\sigma^2 + cr\sigma_b^2 + bcrQ_\alpha$
$B(A)$	$\sigma^2 + cr\sigma_b^2$
C	$\sigma^2 + abrQ_\gamma$
AC	$\sigma^2 + brQ_{\alpha\gamma}$
Error	σ^2

REFERENCES

Text References

Most of the references at the end of Chapter 15 provide discussions of random and mixed models. Expected mean squares are discussed in the texts by Ostle and Malone and by Johnson and Leone (cited at end of Chapter 15). The book by Milliken and Johnson provides a comprehensive discussion of the estimation of fixed factor effects and of variance components for unbalanced designs. The rules for expected mean squares are adapted from

Bennett, C. A. and Franklin N. L. (1954). *Statistical Analysis in Chemistry and the Chemical Industry*, New York: John Wiley and Sons, Inc.

Cornfield, J. and Tukey, J. W. (1956). "Average Values of Mean Squares in Factorials," *Annals of Mathematical Statistics*, **27**, 907–949.

Satterthwaite approximations are also discussed in most of the references listed at the end of Chapter 15. Satterthwaite's version of this approximation was first introduced in the following article:

Satterthwaite, F. E. (1946). "An Approximate Distribution of Estimates of Variance Components," *Biometrics Bulletin*, **2**, 110–114.

EXERCISES

1   Use the rules given in the appendix to this chapter to derive expected mean squares for

(a) the torque study decribed in Section 7.1 (see Figure 7.1 and Table 7.3);

(b) the polyethylene density study described in Section 10.1 (see Figure 10.2); and

(c) the connector-pin study described in Section 10.3 (see Table 10.2).

2   Assume that the time effect for the experiment described in Exercise 9 of Chapter 15 is random. Derive the expected mean squares for the factor main effects. Reanalyze the data and compute a confidence interval for the ratio of the standard deviation of the time effects to that of the uncontrolled experimental error.

3   Examine the data on the chemical-analysis variation study in Table 1.1. Which factors are fixed and which are random? Which factors are crossed and which are nested? Write an ANOVA model for the responses. Derive the expected mean squares for all the effects listed in your model.

4   Analyze the data in Table 1.1 using the ANOVA model from Exercise 3. Perform mean comparisons on any fixed effects that are statistically significant. Estimate all the variance components in the model, regardless of whether they are statistically

significant. Do the magnitudes of the estimated variance components conform to what should be expected from the ANOVA results? Explain your conclusion.

5 Conduct separate analyses on the fuel-economy data in Tables 3.1 and 3.2. Include mean comparisons and variance-component estimation where appropriate. Are the expected mean squares for the two analyses identical? Why (not)? Reanalyze the data in Table 3.2 with the outlier on the Volkswagen discarded. How did the deletion of the outlier affect the results of the analysis?

6 An extension of the skin sensitivity study described in Section 17.4 (see Table 17.4) was conducted using two products, one a placebo (control) and the other a potentially allergenic product (treatment). Three animals were exposed to the control product and three to the treatment product. Each animal was administered the products on the ear and on the back. Duplicate tests were run. The data are tabulated below. Analyze these data by constructing an ANOVA table, including the expected mean squares. Perform mean comparisons and estimate variance components where appropriate.

| Group | Animal | Measurements | |
		Ear	Back
Control	1	73, 70	90, 92
	2	66, 66	83, 83
	3	71, 70	77, 74
Treatment	4	77, 74	89, 88
	5	67, 70	70, 78
	6	76, 77	96, 96

7 Plastic viscosity is an important characteristic of a lubricant used to reduce friction between a rotating shaft and the casing that surrounds it. Measurements of plastic viscosity were taken on two lubricants. Four batches of each fluid were used in the tests, with three samples of fluid taken from each batch. The data are given below. Analyze these data with specific reference to how the consistency of the plastic-viscosity measurements is affected by each of the design factors.

| Lubricant | Sample | Viscosity | | | |
		Batch 1	2	3	4
Xtra Smooth	1	4.86	5.87	6.07	4.97
	2	4.34	5.07	5.76	5.42
	3	4.64	5.38	5.22	5.77
EZ Glide	1	1.45	1.78	2.08	2.07
	2	1.26	1.69	2.41	2.19
	3	1.67	1.35	2.27	2.44

8 The following data are a portion of the responses collected during an interlaboratory

study involving several laboratories. Each laboratory was sent a number of materials that were carefully chosen to have different measurement values on the characteristic of interest. The laboratories were required to perform three separate analyses of the test procedure. Devise an appropriate ANOVA model for these data, assuming no laboratory-by-material interaction. Analyze the statistical significance of the design factors by constructing an ANOVA table, including the expected mean squares.

Laboratory	Material	Repeat Measurements
1	A	12.20 12.28 12.16
	B	15.51 15.02 15.29
	C	18.14 18.08 18.21
2	A	12.59 12.30 12.67
	B	14.98 15.46 15.22
	C	18.54 18.31 18.60
3	A	12.72 12.78 12.66
	B	15.33 15.19 15.24
	C	18.00 18.15 17.93

9   Obtain estimates of repeatability and reproducibility for the interlaboratory study described in Exercise 8.

10 Refer to the experiment in Chapter 7, Exercises 5 and 6, where the mold temperature is being measured during the formation of drinking glasses. Five temperatures are sampled from each mold in the experiment. The following data are collected on this hierarchically nested design:

Factory	Disk	Mold	Temperature
1	1	1	459 467 465 472 464
1	1	2	464 468 461 464 469
1	2	1	465 466 469 463 465
1	2	2	466 464 463 466 466
2	1	1	472 471 472 471 470
2	1	2	472 468 468 473 470
2	2	1	471 469 473 471 470
2	2	2	470 468 472 465 472

Analyze these data, assuming that factory is a fixed factor and disk and mold are random factors.

11 Refer to the experiment designed to study the dye quality in a synthetic fiber in Chapter 10, Exercise 7. The following data were collected in a staggered nested design that involved 35 bobbins. Analyze these data and determine which of the factors significantly affect the dye rating.

Dye Quality Rating	Shift	Operator	Package	Bobbin
87	1	1	2	3
65	3	6	17	34
34	2	4	10	19
54	1	2	5	10
86	2	4	12	23
91	3	6	16	31
3	1	1	1	2
65	3	6	18	36
29	2	3	9	18
83	3	6	16	32
75	2	4	11	22
33	1	1	1	1
26	1	1	3	5
74	3	5	13	26
35	2	4	11	21
75	3	6	18	35
81	1	1	3	6
29	2	4	12	24
44	1	1	2	3
28	3	6	17	34
34	2	4	10	19
64	1	2	5	10
79	2	4	12	23
94	3	6	16	31
10	1	1	1	2
53	3	6	18	36
38	2	3	9	18
82	3	6	16	32
70	2	4	11	22
41	1	1	1	1
28	1	1	3	5
72	3	5	13	26
39	2	4	11	21
59	3	6	18	35
87	1	1	3	6
26	2	4	12	24

12 An experiment was performed to assess the effects of three catalysts on the yield of a chemical process. Four replications of the experiment were performed. In each replication, three supposedly identical process runs were begun, samples from each of the processes were collected, and analyses of the chemical yields were performed. After three and five days, samples were again taken and the chemical yields were again determined. Analyze the data below to determine whether the catalysts or the sampling periods (three or five days) significantly affect chemical yields of the process.

Replicate	Catalyst A		Catalyst B		Catalyst C	
	3 Days	5 Days	3 Days	5 Days	3 Days	5 Days
1	83	93	76	82	67	81
2	65	77	65	72	82	78
3	67	88	63	75	65	74
4	81	89	68	74	63	75

CHAPTER 18

Analysis of Blocking Designs and Fractional Factorials

In this chapter the analysis of experiments conducted in block designs and the analysis of fractional factorial experiments are detailed. Analysis-of-variance procedures are detailed, with reference to interval-estimation and multiple-comparison procedures as appropriate. The specific topics included in this chapter are:

- *the analysis of randomized complete block and balanced incomplete block designs,*
- *the analysis of special blocking designs, especially latin-square and crossover designs, and*
- *the analysis of fractional factorial designs, including main effects and screening designs.*

Analysis-of-variance procedures are detailed in Chapters 15 and 17. The interval estimation and hypothesis-testing procedures recommended in those chapters for fixed, random, and mixed models can be applied to data obtained from many different types of experimental designs. In this chapter we utilize ANOVA procedures to analyze data arising from several different types of blocking designs and from fractional factorial experiments.

Although ANOVA procedures are highlighted in this chapter, we stress that a comprehensive analysis of the data from an experimental design requires the estimation of fixed factor effects and of variance components for factors with random levels. For most of the analyses discussed in this chapter, the estimation procedures are identical to those discussed in Chapters 15 to 17. For designs such as balanced incomplete block designs, there are differences from the procedures discussed in Chapters 15 to 17. In such cases, we detail how to perform the necessary analyses.

18.1 RANDOMIZED COMPLETE BLOCK DESIGNS

Randomized complete block (RCB) designs (Chapter 8) are used to control known sources of experimental variability. Variability of test conditions and of experimental units are two primary contributors to the imprecision of test results and therefore to the assessment of factor–level effects. RCB designs are effective in controlling such extraneous sources of variability when a block consists of a sufficient number of test runs or homogeneous experimental units so that all factor–level combinations of interest can be tested in each block.

A RCB design is analyzed like a factorial experiment in which one of the factors, the block factor, does not interact with any of the other factors. Thus the analyses presented in Chapters 15 and 17, depending on whether the factors are all fixed or some are random, can be directly applied to RCB designs.

An ANOVA model for a randomized complete block design in which there are two fixed experimental factors and one random blocking factor can be written as follows:

$$y_{ijkl} = \mu + a_i + \beta_j + \gamma_k + (\beta\gamma)_{jk} + e_{ijkl}, \tag{18.1}$$

where the first factor represents the random blocking factor. The ANOVA table for the model (18.1) would contain three main effects and one interaction. Since the block factor and the fixed experimental factors do not interact, all the effects are tested against the experimental-error mean square. This can be confirmed by determining the expected mean squares using the rules in the appendix to Chapter 17.

Table 18.1 Skin-Thickness Measurements (10^{-2} mm) for Allergic-Reaction Study

	Initial		45 Minutes	
Animal	Back	Shoulder	Back	Shoulder
1	69, 68	75, 80	70, 86	77, 76
2	66, 66	77, 77	66, 82	74, 83
3	77, 77	82, 87	77, 76	86, 87
4	76, 76	78, 76	70, 67	74, 82
5	66, 66	80, 80	86, 62	63, 79
6	82, 82	96, 91	82, 82	95, 94
7	65, 62	74, 77	70, 80	71, 88
8	61, 62	73, 70	63, 63	72, 71
9	78, 77	94, 90	79, 80	92, 88
10	72, 74	78, 76	66, 66	82, 82

Table 18.2 Analysis of Skin-Thickness Measurements

Source	df	SS	MS	F	p-Value
Animals	9	1635.525	181.725	21.28	0.000
Locations (L)	1	0.625	0.625	0.07	0.789
Times (T)	1	893.025	893.025	104.57	0.000
L × T	1	0.025	0.025	0.00	0.957
Error	27	230.575	8.540		
Total	39	2759.775			

Time	Average
Initial	71.10
45 Minutes	80.55

$$s_e = 2.922 \qquad s_a = 4.653$$

An experiment for the which the model (18.1) is appropriate is the allergic-reaction design presented in Table 8.2. A related analysis for a single experimental factor was described and analyzed in Section 17.4 (see Tables 17.4 and 17.5). The animals in this latter analysis also constitute random block effects.

Table 18.1 contains the experimental results for the RCB design in Table 8.2. The ANOVA table for these data is shown in Table 18.2. Only the main effect due to time and the block effect due to animals are statistically significant. From the averages for the two times, it is apparent that the measurements taken 45 minutes after the administration of the solution containing the pollutant do evidence a reaction. The skin-thickness measurements taken 45 minutes after the injection average approximately 0.09 mm more than the average of the initial measurements. Note too that the estimated standard deviation for the block effect due to the laboratory animals is approximately 50% larger than that of the uncontrolled experimental error.

18.2 BALANCED INCOMPLETE BLOCK DESIGNS

Balanced incomplete block (BIB) designs are balanced in the sense that each factor–level combination occurs in exactly r blocks and each pair of factor–level combinations occurs together in exactly p blocks. Viewing the blocking factor as an additional factor in the design, however, results in the design not being balanced. This is because all factor–level combinations do not occur in every block. The principle of reduction in error sum of squares (Section 15.2) is used to determine appropriate sums of squares for assessing the block and factor effects.

Data from an experiment conducted using a BIB design are presented in Table 18.3. This experiment was conducted to develop a method of assessing the deterioration of

Table 18.3 Balanced Incomplete Block Design for Asphalt-Pavement Rating Study*

		District Engineer															
		1	2	3	4	5	6	7	8	9	10	11	12	13	14	15	16
	1				57	70		55	85	70							72
	2		65						60	55			55		74	70	
	3	65						72	60			85		80		78	
	4				52		66	50		62	48					50	
	5		68					60			63			68	75		61
	6	55		58	57					62				73	55		
	7			56			78		55				53	55			72
Road	8			56		55						50			38	74	56
Segment	9	57	59		50		64									70	58
	10	84				95	96	80					80		95		
	11		58			68	60			70		54		49			
	12	50	54	52		45			57		39						
	13		61	57	57			45				56	56				
	14	60								55	68	88	60				76
	15				76		68		60			58	55		75		
	16				35	40						32		35	38	30	

*Values are ratings for each road segment. Each engineer rated only six road segments. The data were provided by Professor Edward R. Mansfield, Department of Management Science and Statistics, The University of Alabama, Tuscaloosa, AL.

asphalt pavement on a state highway. The proposed measurement scale, a quantitative measurement from 0 (no asphalt left) to 100 (excellent condition), needed to yield comparable results when used by different pavement experts on the same segment of the highway. In addition, the scale needed to be broad enough to cover a wide range of asphalt conditions. Sixteen state district engineers and sixteen 200-foot sections of highway were randomly selected. Each expert traveled six preassigned sites and rated the road segment using the proposed measurement scale. Each site was rated by six different experts.

The BIB design model used in this experiment is a main-effect model in which there are $f = 16$ levels of a random experimental factor and $b = 16$ levels of a random blocking factor:

$$y_{ij} = \mu + f_i + b_j + e_{ij}, \qquad i = 1, 2, \ldots, 16, \quad j = 1, 2, \ldots, 16, \qquad (18.2)$$

where f_i denotes the effect of the ith engineer and b_j denotes the effect of the jth road segment. Although the model (18.2) is similar to a two-factor random-effects model for a factorial experiment, only the $N = 96$ combinations shown in Table 18.3 are actually included in the model; in particular, the design is not a complete factorial experiment. Note that we again assume that there is no interaction between the block and experimental factors. We denote this model by M_1 in the calculation of the reduction in error sums of squares.

Denote the reduced models that only have the block (road segments) or experimental (engineers) factors by M_2 and M_3, respectively:

$$
\begin{aligned}
M_2: & \quad y_{ij} = \mu + b_j + e_{ij}, \\
M_3: & \quad y_{ij} = \mu + f_i + e_{ij}.
\end{aligned}
\qquad (18.3)
$$

The statistical significance of the engineer and road-segment effects is gauged by obtaining the reductions in sums of squares due to fitting the full model (18.2) instead of the reduced models in (18.3). Denote the error sums of squares for the three models by SSE_1, SSE_2, and SSE_3, respectively. Then the reductions in error sums of squares of interest are:

Road Segments: $R(M_1 | M_3) = SSE_3 - SSE_1,$

Engineers: $R(M_1 | M_2) = SSE_2 - SSE_1.$

The numbers of degrees of freedom associated with each of these reductions in sums of squares are the differences in those for the error terms:

$$
\begin{aligned}
\mathrm{df}\{R(M_1 | M_3)\} &= \mathrm{df}(SSE_3) - \mathrm{df}(SSE_1) \\
&= (N - f) - (N - b - f + 1) = b - 1, \\
\mathrm{df}\{R(M_1 | M_2)\} &= \mathrm{df}(SSE_2) - \mathrm{df}(SSE_1) \\
&= (N - b) - (N - b - f + 1) = f - 1.
\end{aligned}
$$

The three ANOVA tables for the above fits are shown in Table 18.4. The "model" sum of squares for the complete model M_1 in Table 18.4(a) measures the combined effects of the engineers and the road segments. Table 18.4(b) and (c) show the fits to the reduced models M_2 and M_3. The above reductions and degrees of freedom are readily obtained from Table 18.4. The two reductions in sums of squares are often summarized in one ANOVA

Table 18.4 ANOVA **Tables for Three Fits to Asphalt-Pavement Rating Data**

Source	df	SS	MS	F	p-Value
(a) Full Model: M_1					
Model (engineers and road segments)	30	13,422.13	447.40	7.10	0.000
Error	65	4,098.84	63.06		
Total	95	17,520.97			
(b) Blocks (Road Segments) Only: M_2					
Road segments	15	11,786.96	785.80	10.96	0.000
Error	80	5,734.00	71.86		
Total	95	17,520.97			
(c) Experimental Factor (Engineers) Only: M_3					
Engineers	15	2,416.96	161.13	0.85	0.621
Error	80	15,104.01	188.80		
Total	95	17,520.97			

table such as the one shown in Table 18.5. Note that the reductions in sums of squares are not ordinarily additive; i.e., the quantities in the sum-of-squares column do not usually add up to the total sum of squares. This is a consequence of the imbalance in the design and the use of reductions in sums of squares to measure each of the factor effects.

Testing the significance of the experimental and block factor effects is accomplished through the procedure given in Exhibit 18.1. This procedure is valid regardless of whether the factors are fixed or random. The quantities needed are obtainable from the ANOVA table, Table 18.5.

EXHIBIT 18.1   TESTING THE SIGNIFICANCE OF EXPERIMENTAL AND BLOCK FACTOR EFFECTS

1. Obtain the reductions in error sums of squares for fitting the reduced models M_2 (blocks) and M_3 (experimental factor) by calculating $R(M_1|M_2)$ and $R(M_1|M_3)$.

2. Form the following F ratios:

$$\text{(blocks)} \quad F = \frac{R(M_1|M_2)/(b-1)}{MSE_1},$$

$$\text{(factor)} \quad F = \frac{R(M_1|M_3)/(f-1)}{MSE_1}.$$

3. If either of the F-statistics exceeds an upper $100\alpha\%$ critical point of the F-distribution with numerator degrees of freedom $v_1 = b-1$ or $f-1$, respectively, and denominator degrees of freedom $v_2 = N-f-b+1$, then reject the corresponding hypothesis of no block or factor effects.

Table 18.5 Analysis of Variance Table for BIB Analysis of Asphalt Pavement Rating Data

Source	df	SS	MS	F	p-Value
Road segments	15	11,005.17	733.68	11.63	0.000
Engineers	15	1,635.17	109.01	1.73	0.067
Error	65	4,098.84	63.06		
Total	95	17,520.97			

Table 18.6 Expected Mean Squares for Random Effects in Balanced Incomplete Block Designs

Effect	Mean Square	Expected Mean Square*
Blocks (adj.)	$\dfrac{R(M_1 \mid M_2)}{b-1}$	$\sigma^2 + \dfrac{bk-f}{b-1}\sigma_b^2$
Factor (adj.)	$\dfrac{R(M_1 \mid M_3)}{f-1}$	$\sigma^2 + \dfrac{pf}{k}\sigma_f^2$

*For symmetric BIB designs, $b = f$ and $k = r$. Then $(bk - f)/(b - 1) = pf/k$.

From the F-statistics and the corresponding significance probabilities shown in Table 18.5, it is apparent that the road segments are highly significant, confirming that the asphalt conditions of the selected road surfaces varied widely. On the other hand, the engineer effects are not statistically significant, indicating that there is not significant disagreement among the engineers in their assessments of the condition of the asphalt in the various road segments.

Expected mean squares for BIB designs cannot be determined from the rules in the appendix to Chapter 17. This is because the designs are not balanced in the way described in the appendix; i.e., each factor–level combination does not have an equal number of repeat tests. In this context, a factor–level combination is a block–factor combination. Since some factor levels do not appear in each block, some combinations have one repeat test and others have no repeat tests.

Table 18.6 displays the expected mean squares for random block and factor effects. The multipliers on the variance components are not equal because of the restriction $b \geqslant f$; if $b = f$—the case of *symmetric* BIB designs—the two multipliers are equal. Using these expected mean squares, the estimates of the model standard deviations for the asphalt-pavement rating data are:

$$s_e = 7.94, \qquad s_f = 2.94, \qquad s_b = 11.21.$$

Thus the estimated variability due to the different road surfaces is approximately 40% larger than that due to the uncontrolled measurement errors.

When the experimental factor and the block factor are fixed effects, the usual response averages, $\bar{y}_{i.}$, are biased estimators of the average responses for the factor levels. This is because each factor level (or factor–level combination) does not occur in every block. Thus the responses for the factor levels contain the effects of some, but not all, of the blocks. Even if the block effects are random, it is desirable to adjust the averages for fixed-factor effects in order to eliminate differences due to block effects from comparisons of the factor levels. These adjusted factor–level averages are calculated as shown in Exhibit 18.2.

EXHIBIT 18.2   ADJUSTED FACTOR–LEVEL AVERAGES

$$m_i = \bar{y}.. + c(\bar{y}_{i\cdot} - \bar{y}^{(i)}), \tag{18.4}$$

where

$c = (N - r)/(N - b)$,

$\bar{y}_{i\cdot}$ = average response for factor–level combination i;

$\bar{y}^{(i)}$ = average of all observations from blocks that contain treatment i (this is an average of r block averages);

$\bar{y}..$ = overall average.

The multiplication by c in equation (18.4) can be viewed as correcting for the number of times a factor–level combination appears (i.e., the number of blocks in which a factor–level combination occurs) relative to the total number of blocks in the experiment, b. Aside from this correction term, equation (18.4) is the overall average plus the difference between the usual factor–level response average and the average response for the blocks in which the factor level occurs. If the raw response average for a factor level is greater (less) than the average for the blocks in which it occurs, the overall average is adjusted upward (downward). If each factor–level combination appears in every block (that is, an RCB design) then $c = 1$, $\bar{y}^{(i)} = \bar{y}..$ for all i, and consequently $m_i = \bar{y}_{i\cdot}$, the usual response average.

Table 18.7 shows the values of $\bar{y}._j$, $\bar{y}_{i\cdot}$, and m_i for the data from the asphalt-

Table 18.7   Adjusted Factor Averages for Asphalt Paving Scale Study

Blocking Factor		Experimental Factor		
Road Segment	Average Rating	Engineer	Average Rating	Adjusted Rating
1	68.17	1	61.83	57.30
2	63.17	2	60.83	63.12
3	73.33	3	55.17	62.11
4	54.67	4	55.33	59.69
5	65.83	5	62.17	63.96
6	60.00	6	72.00	67.72
7	61.50	7	60.33	54.93
8	54.33	8	62.83	60.80
9	59.67	9	62.33	61.45
10	88.33	10	51.33	56.93
11	59.83	11	64.67	63.10
12	49.50	12	56.50	56.63
13	55.33	13	60.50	62.51
14	67.83	14	68.67	63.54
15	65.33	15	62.00	66.04
16	35.00	16	65.83	63.94

measurement experiment. Although the experimental and block factors in this experiment are random effects, we present these summary statistics in order to illustrate the changes that can occur in some of the averages due to the imbalance of the BIB design. Note that some of the adjusted factor–level averages m_i differ by as much as 10% from the corresponding raw averages $\bar{y}_{i.}$. Were these factor levels fixed, the adjusted averages would be the appropriate statistics to use to compare the factor–level effects.

The overall F-ratio using the reduction in sum of squares $R(M_1|M_3)$ can be used to ascertain that all the fixed-effect means are not equal. The experimenter will also be keenly interested in identifying which individual factor–level means are different from one another. The estimated standard errors of the adjusted averages can be used with the procedures discussed in Chapter 16 to make this determination. The standard-error estimates can be calculated according to the following formulas:

$$SE(m_i) = s_e\left(N^{-1} + \frac{(k-1)c^2}{rk}\right)^{1/2},$$

$$SE(m_i - m_k) = s_e(2c/r)^{1/2},$$

(18.5)

where $s_e^2 = MSE_1$ is the mean squared error for the full model fit and c is defined in (18.4). The standard errors given in (18.5) can also be used to construct confidence intervals.

18.3 LATIN-SQUARE AND CROSSOVER DESIGNS

Latin-square designs (Section 8.4) are blocking designs that control two extraneous sources of variability. A condition for the use of latin-square designs is that there must be no interactions between the design factor(s) and the two blocking factors. Because of this restriction, the analysis of data from latin-square designs utilizes main-effect ANOVA models.

The layout for a latin-square design used to road-test four different brands of commercial truck tires is given in Table 8.4. The mileage-efficiency data collected for this study are given in Table 18.8. An ANOVA model containing only main effects for these data can be expressed as follows:

$$y_{ijk} = \mu + a_i + b_j + \gamma_k + e_{ijk}, \quad i = 1, 2, 3, 4, \quad j = 1, 2, 3, 4, \quad k = 1, 2, 3, 4. \quad (18.6)$$

Note that although each of the subscripts in the above model ranges from 1 to 4, only the specific test runs shown in Tables 8.4 and 18.8 are included in the data set; in particular, this is not a complete factorial experiment.

In this example, the trucks (a_i) and days (b_j) are random effects. Neither the specific trucks nor the specific days on which the testing was conducted are of primary interest to the experimenters. Both sets of effects can be considered randomly selected from conceptually infinite populations of truck and day effects. Inferences are desired, however, on the specific tire brands (γ_k) used in the tests. For this reason, tire brands are considered fixed effects.

The ANOVA table for these data using the model (18.6) is given in Table 18.9. The sums of squares are calculated in the usual way for main effects; e.g., $SS_A = k^{-1}\sum y_{i..}^2 - (k^2)^{-1}y_{...}^2$, where k is the number of levels of each of the factors (see Table 15.4). Under either fixed or random-effects model assumptions, the expected mean squares for the main effects are

Table 18.8   Mileage Efficiencies for Commercial-Truck-Tire Experiment

Truck	Day	Tire Brand	Efficiency (mi/gal)
1	1	2	6.63
	2	1	6.26
	3	4	7.31
	4	3	6.57
2	1	4	7.06
	2	2	6.59
	3	3	6.64
	4	1	6.71
3	1	1	6.31
	2	3	6.73
	3	2	7.00
	4	4	7.38
4	1	3	6.55
	2	4	7.13
	3	1	6.54
	4	2	7.11

Table 18.9   ANOVA Table for Truck-Tire Experiment

Source	df	SS	MS	F	p-Value
Trucks	3	0.0676	0.0225	1.46	0.317
Days	3	0.2630	0.0877	5.67	0.034
Tires	3	1.3070	0.4357	28.17	0.000
Error	6	0.0928	0.0155		
Total	15	1.7305			

the experimental-error variance plus a nonnegative function of the respective model parameters or variance components, i.e., $\sigma^2 + k\sum\alpha_i^2/(k-1)$ or $\sigma^2 + k\sigma_a^2$. The expected mean square for error is the experimental-error variance. Therefore, the appropriate F-ratios for assessing factor effects for this example, or for data from any latin-square design, are formed by taking the ratio of each of the main-effect mean squares to the error mean square.

From Table 18.9 it is seen that the tire-brand effects are statistically significant ($p < 0.001$), as are the day effects ($p = 0.034$). One would now proceed to estimate the standard deviations of the error and day effects (Section 17.3) and determine which specific tire brands are significantly affecting the mileage efficiency (Chapter 16). We leave the inferences to the exercises.

Crossover designs (Section 8.4) are appropriate when different levels of a factor are given or applied to an experimental unit in a time sequence. A crossover design is in effect a blocking design. Each experimental unit is a block and receives every factor–level combination in a time sequence. In this manner each experimental unit acts as its own control, and variation among experimental units has no bearing on comparisons of factor effects.

Table 18.10 Clinical Responses for Drug-Testing Experiment Using a Two-Period Crossover Design

Subject Number*	Sequence*	Time Period	Drug	Response
2	1	1	1	7.2
		2	2	9.0
4	1	1	1	7.2
		2	2	8.0
5	3	1	2	10.2
		2	1	9.2
13	3	1	2	20.8
		2	1	15.6
19	3	1	2	11.2
		2	1	9.0
20	1	1	1	16.4
		2	2	20.9

*See Table 8.5 for complete crossover design.

Under the simplifying assumptions given in Section 8.4, the analysis of crossover designs involves fitting an ANOVA model with block terms, time-period terms, factor effect terms, and experimental error. If more than one block receives the same sequence of factor–level combinations over the time periods, then a sequence term is added to the model and blocks are nested within a sequence. If the same number of blocks are nested within each sequence, then the design is balanced; otherwise, it is unbalanced. When the design is balanced, the usual sum-of-squares calculations can be performed on the factors. When it is unbalanced, the principle of reduction in error sums of squares (Section 15.2) is used to obtain the sums of squares.

The analysis of a crossover design is illustrated for the clinical study of three experimental drugs shown in Table 8.5. In this study there are six sequences (see Table 8.5) of the order in which three drugs can be administered. The design is unbalanced in that four blocks (test subjects) are assigned to sequences 2 and 6, and three blocks to the remaining sequences. Table 18.10 displays a portion of the data from this experiment. The data from sequences 1 and 3, using only the first two time periods and the first two drugs, are shown. An ANOVA model for this balanced data set is

$$y_{ijkl} = \mu + \alpha_i + b_{j(i)} + \gamma_k + \delta_l + e_{ijkl}, \qquad i = 1, 2, \quad j = 1, 2, 3, \quad k = 1, 2, \quad l = 1, 2.$$

$$(18.7)$$

The mean of a response is $\mu + \alpha_i + \gamma_k + \delta_l$, where the α_i, γ_k, and δ_l model terms correspond to the fixed-sequence, drug, and time-period factors, respectively. The random-effect terms are assumed to be independently normally distributed with zero mean. The $b_{j(i)}$ terms have standard deviation σ_b, and the experimental error terms e_{ijkl} have standard deviation σ. Table 18.11 is a symbolic ANOVA table, with expected mean squares, for this model. Table 18.12 is an ANOVA table for the data in Table 18.10. There are significant effects due to subjects and drugs, but the sequence and time-period effects are

Table 18.11 Symbolic ANOVA Table for the Two-Period Crossover Design of Table 18.9

Source	df	SS	MS	F	Expected MS
Sequences	$a-1$	SS_A	MS_A	$MS_A/MS_{B(A)}$	$\sigma^2 + d\sigma_b^2 + cdQ_\alpha$
Subjects (seq.)	$a(b-1)$	$SS_{B(A)}$	$MS_{B(A)}$	$MS_{B(A)}/MS_E$	$\sigma^2 + d\sigma_b^2$
Drugs	$c-1$	SS_C	MS_C	MS_C/MS_E	$\sigma^2 + bdQ_\gamma$
Time periods	$d-1$	SS_D	MS_D	MS_D/MS_E	$\sigma^2 + bcQ_\delta$
Error	e	SS_E	MS_E		σ^2
Total	$abd-1$	TSS			

a = Number of sequences
b = Number of subjects in each sequence
c = Number of drugs
d = Number of time periods
$e = abd - ab - c - d + 2$

Table 18.12 ANOVA Table for Clinical Drug-Testing Experiment

Source	df	SS	MS	F	p-Value
Sequences	1	4.44	4.44	0.07	0.802
Subjects (seq.)	4	247.78	61.94	29.70	0.003
Drugs	1	20.02	20.02	9.60	0.036
Time periods	1	0.14	0.14	0.07	0.807
Error	4	8.34	2.09		
Total	11	280.72			

not statistically significant. We again leave as an exercise the estimation of the individual factor effects.

18.4 FRACTIONAL FACTORIAL EXPERIMENTS

Fractional factorial experiments (Chapter 9) are analyzed using the appropriate methodology already discussed in this and the previous three chapters. They can occur in completely randomized designs or a variety of block designs. The specific analysis depends on the design used.

Fractional factorial experiments are, by definition, unbalanced. Some factor–level combinations occur in the experiment; others do not. In general, such experiments can be analyzed using the principle of reduction in error sums of squares for any type of design and any model definition. If all factors are fixed and crossed, each effect is tested against the error mean square. If some factors are fixed and others are random, or if some are nested and others are crossed, no general rules can be given for forming F-statistics. The denominators of the test statistics depend on the design and on which factors are fixed and which are random. In many instances, no exact tests exist (see Section 17.4).

For two-level fractional factorial experiments designed as in Chapter 9, the usual sum-of-squares formulas can be used. The expected-mean-square rules in the appendix to Chapter 17 can also be used. Both in the calculation of the sums of squares and in the determination of expected means squares one must be careful to replace the multipliers in front of the various quantities by the number of observations used in the computation of the respective averages. For example, in a half fraction of a 2^3 factorial experiment with no repeat tests, $a = b = c = 2$ and $r = 0$ or 1. In the main-effects formula SS_A for Factor A, the divisor of $\sum y_{i...}^2$ in Table 15.4 would be 2, not $bcr = 2 \times 2 \times 1 = 4$; similarly for the multipliers of Q_α and σ_a^2 in the expected mean squares in Tables 17.1 and 17.2.

We conclude this chapter with the analysis of two fractional factorial experiments. The first is a screening design that uses one of the main-effect fractional factorials popularized by Plackett and Burman (Section 9.4), which have become widely used in ruggedness testing. The second is a $\frac{1}{8}$ fraction of a 2^7 complete factorial experiment.

Certain types of plated materials (for example, a connector in an electronic assembly) are soldered for strength. The manufacturer of the plated material tests it for solderability

Table 18.13   Factors for Solderability Ruggedness Test

	Factor		Levels	
Number	Name	-1	$+1$
1	Solvent dip	No	Yes
2	Surface area	Small	Large
3	Dip device	Manual	Mechanical
4	Magnification	$10 \times$	$30 \times$
5	Solder age	Fresh	Used
6	Flux time	8 sec	30 sec
7	Drain time	10 sec	60 sec
8	Stir	No	Yes
9	Solder time	2 sec	8 sec
10	Residual flux	Not clean	Clean

Table 18.14   Responses and Coded Factor Levels for Solder-Coverage Ruggedness Testing Data

Solder Coverage (%)	Level									
	Factor 1	2	3	4	5	6	7	8	9	10
91	1	1	1	1	-1	1	-1	1	1	-1
97	1	-1	1	-1	1	1	-1	-1	1	-1
89	1	-1	1	1	-1	-1	1	-1	-1	-1
82	1	-1	-1	-1	1	1	1	1	-1	1
82	1	1	1	-1	1	-1	1	1	-1	-1
74	1	-1	-1	1	-1	-1	-1	1	1	1
54	-1	-1	-1	-1	-1	-1	-1	-1	-1	-1
66	-1	1	1	-1	-1	1	-1	-1	-1	1
79	-1	1	-1	-1	-1	1	1	1	1	-1
25	-1	1	-1	1	1	-1	-1	1	-1	-1
77	-1	-1	-1	1	1	1	1	-1	1	-1
44	1	1	-1	-1	1	-1	-1	-1	1	1
86	-1	-1	1	1	1	1	-1	1	-1	1
97	-1	-1	1	-1	-1	-1	1	1	1	1
84	1	1	-1	1	-1	1	1	-1	-1	1
97	-1	1	1	1	1	-1	1	-1	1	1

prior to releasing it for shipment to the customer. In an experiment conducted to assess the effects of various factors on the soldering process, the written procedure for the solderability test method was reviewed in detail to identify variables that might affect the response—percentage solder coverage. Ten factors were identified and are listed, along with the levels chosen for consideration, in Table 18.13. A sixteen-run (randomized) Plackett–Burman screening design was used, with the results shown in Table 18.14.

Each of the factors listed in Table 18.13 consists of fixed effects; i.e., these are the only factor levels of interest in the experiment. A fixed-effects model containing only main effects was fitted, and the resulting ANOVA table is shown in Table 18.15. F-ratios are formed by comparing each of the factor-effect mean squares with the error mean square.

Several of the factors in Table 18.15 are statistically significant. One's philosophy in analyzing screening designs often is to allow a larger significance level in testing hypotheses in order to (a) select several potentially important factors that will be studied further in future experiments, or (b) identify factors in a process that must be more tightly controlled so a more uniform product will result. If one uses a 10% significance level in Table 18.15, more consistent solderability test results can be obtained by better control of the surface areas, dip device, flux time, drain time, and solder time.

A second example of the use of screening designs in a ruggedness test concerns a laboratory viscosity method. Seven potentially sensitive steps in the viscosity method are

Table 18.15 ANOVA Table for Solder-Coverage Ruggedness Tests

Source	df	SS	MS	F	p-Value
Solvent dip	1	240.25	240.25	2.25	0.194
Surface area	1	484.00	484.00	4.53	0.087
Dip device	1	2162.25	2162.25	20.25	0.007
Magnification	1	30.25	30.25	0.28	0.619
Solder age	1	121.00	121.00	1.13	0.336
Flux time	1	625.00	625.00	5.85	0.061
Drain time	1	1406.25	1406.25	13.17	0.015
Stir	1	4.00	4.00	0.04	0.849
Solder time	1	484.00	484.00	4.53	0.087
Residual flux	1	81.00	81.00	0.76	0.423
Error	5	534.00	106.80		
Total	15	6172.00			

Table 18.16 Factors for Viscosity Ruggedness Tests

	Factor		Levels	
No.	Name	−1	+1
1	Sample preparation	Method 1	Method 2
2	Moisture measure	Volume	Weight
3	Speed	800	1600
4	Mixing time	0.5	3.0
5	Equilibrium time	1	2
6	Spindle	Type 1	Type 2
7	Lid	Absent	Present

shown in Table 18.16 with the levels chosen for investigation. Sixteen experiments, corresponding to a one-eighth fraction of a complete factorial experiment, were run. The factor–level combinations, along with the viscosity responses, are shown in Table 18.17. Since this completely randomized design is of resolution IV (see Table 9A.1), the main effects are not confounded with any two-factor interactions.

The ANOVA table for a main-effects model fitted to these data is given in Table 18.18. Again assuming fixed factor effects and negligible interactions, the appropriate F-ratios are obtained by comparing the factor-effect mean squares with the error mean square. Speed, mixing time, and type of spindle are statistically significant effects.

In each of the above analyses the assumption of negligible interaction effects is critical. In both analyses the presence of interaction effects would inflate the estimate of experimental error, MS_E. Likewise, the presence of interaction effects would bias the factor

Table 18.17 Viscosity Data for Ruggedness Tests

Viscosity	Factor 1	2	3	4	5	6	7
2220	−1	−1	−1	1	1	1	−1
2460	1	−1	−1	−1	−1	1	1
2904	−1	1	−1	−1	1	−1	1
2464	1	1	−1	1	−1	−1	−1
3216	−1	−1	1	1	−1	−1	1
3772	1	−1	1	−1	1	−1	−1
2420	−1	1	1	−1	−1	1	−1
2340	1	1	1	1	1	1	1
3376	1	1	1	−1	−1	−1	1
3196	−1	1	1	1	1	−1	−1
2380	1	−1	1	1	−1	1	−1
2800	−1	−1	1	−1	1	1	1
2320	1	1	−1	−1	1	1	−1
2080	−1	1	−1	1	−1	1	1
2548	1	−1	−1	1	1	−1	1
2796	−1	−1	−1	−1	−1	−1	−1

Factor Level

Table 18.18 ANOVA Table for Viscosity Ruggedness Tests

Source	df	SS	MS	F	p-Value
Sample preparation	1	49	49	0.00	0.973
Moisture measure	1	74,529	74,529	1.91	0.204
Speed	1	859,329	859,329	21.99	0.002
Mixing time	1	361,201	361,201	9.24	0.016
Equilibrium time	1	51,529	51,529	1.32	0.284
Spindle	1	1,723,969	1,723,969	44.12	0.000
Lid	1	1,521	1,521	0.04	0.846
Error	8	312,608	39,076		
Total	15	3,384,735			

effects and could either inflate or reduce the size of the individual factor-effect sums of squares, depending on the nature of the biases. With the fractional factorial experiment for the viscosity ruggedness tests, only the three-factor or higher interaction effects would bias the main effects. Using these designs for screening purposes, however, indicates that interest is only in very large effects of the factors, effects that are expected to be detectable with main-effects models.

REFERENCES

Text References

The following references discuss the analysis of a wide variety of restricted designs. Other useful references appear at the end of Chapters 15 and 17.

Cochran, W. G. and Cox, G. M. (1957). *Experimental Designs*, New York: John Wiley and Sons, Inc.

Federer, W. T. (1955). *Experimental Design*, New York: The Macmillan Co. *Many examples of split-plot, crossover, and similar designs.*

Kirk, R. E. (1968). *Experimental Design: Procedures for the Behavioral Sciences*, Belmont, CA: Brooks/Cole Publishing Co.

Milliken, George A. and Johnson, Dallas E. (1984). *Analysis of Messy Data—Volume I: Designed Experiments*, New York: Van Nostrand Reinhold Co. *Comprehensive discussion of the estimation of fixed factor effects and of variance components from unbalanced designs.*

EXERCISES

1 Complete the analysis of the skin-thickness measurements by using the ANOVA results in Table 18.2 to place confidence intervals on the difference in the mean skin-thickness measurements for the two test times and on the ratio of the standard deviations for the animal and the uncontrolled-error effects.

2 Treat the engineer effects for the asphalt-pavement rating study (Table 18.3) as fixed effects. Use Tukey's TSD procedure on the adjusted averages in Table 18.7 to determine which of the sixteen engineers have unusually high or low mean ratings.

3 Complete the analysis of the truck-tire data by determining which of the tire brands have significantly higher mean efficiencies than the others.

4 Place a 99% confidence interval on the difference of the mean response for the two drugs in the two-period crossover design in Table 18.10. Construct a 95% confidence interval on the ratio of the standard deviations for the subject and the uncontrolled error effects.

5 Use the data from the solderability ruggedness test in Tables 18.14 and 18.15 to determine which factor levels in Table 18.13 should be used to provide a high percentage of solder coverage. Just as importantly, which factors do not seem to affect the coverage? Why is this latter finding important?

6 Answer the same questions as in Exercise 5 for the viscosity ruggedness test data presented in Tables 18.16 and 18.18.

7   A traffic engineer is interested in comparing the total unused red-light time for five
different traffic-light signal sequences. The experiment was conducted with a latin-
square design in which the two blocking factors are (a) five randomly selected
intersections and (b) five time periods. Analyze this data set to determine if the signal
sequences (denoted by Roman numerals) are statistically significant. Values in
parentheses represent unused red-light time, the response variable, in minutes.

Inter-section	Period (1)	(2)	(3)	(4)	(5)
1	I (15.2)	II (33.8)	III (13.5)	IV (27.4)	V (29.1)
2	II (16.5)	III (26.5)	IV (19.2)	V (25.8)	I (22.7)
3	III (12.1)	IV (31.4)	V (17.0)	I (31.5)	II (30.2)
4	IV (10.7)	V (34.2)	I (19.5)	II (27.2)	III (21.6)
5	V (14.6)	I (31.7)	II (16.7)	III (26.3)	IV (23.8)

8   A screening experiment for the displacement-efficiency study described in Chapter 9,
Exercise 11, was run, and the data collected were analyzed using an ANOVA model
having only main effects. The following ANOVA table was generated from the
experimental results:

Source	df	Sum of Squares	Mean Square	F	p-Value
Model	7	224.583	32.083	7.00	0.040
Angle	1	140.083	140.083	30.56	0.005
Viscosity	1	4.083	4.083	0.89	0.399
Flow rate	1	70.083	70.083	15.29	0.017
Rot. speed	1	2.083	2.083	0.45	0.537
Wash mat.	1	2.083	2.083	0.45	0.537
Wash vol.	1	2.083	2.083	0.45	0.537
Eccent.	1	4.083	4.083	0.89	0.399
Error	4	18.333	4.583		
Total	11	242.917			

What can one conclude from this analysis about the effects of the seven factors?

9   A new technique was evaluated to determine if a simple low-cost process could
prevent the generation of hazardous airborne asbestos fibers during the removal of
friable insulation material from buildings. The new technique involved spraying an
aqueous saturating solution in the room as the asbestos was being removed. It was
necessary to determine if different lengths of spraying time had any effects on the fiber
count in the room after the insulation had been removed. Ten buildings were chosen
in which to test the spraying technique. Five spraying times were investigated: 90, 60,
50, 40, and 30 minutes. It was determined that only three rooms from each building
could be used in the experiment. The results are given below, with the spraying times
listed for each room and the fiber count (the response of interest) shown in brackets:

| | Spraying Time [Fiber Count] | | |
Building	Room 1	Room 2	Room 3
1	90 [8]	40 [12]	30 [14]
2	60 [6]	50 [10]	40 [12]
3	30 [11]	60 [8]	90 [9]
4	40 [12]	50 [13]	30 [15]
5	50 [11]	90 [9]	60 [10]
6	50 [8]	40 [9]	90 [7]
7	60 [7]	30 [10]	40 [12]
8	50 [10]	90 [7]	30 [12]
9	60 [9]	40 [12]	90 [10]
10	60 [9]	50 [12]	30 [12]

Analyze this BIB design using the principle of reduction in error sum of squares. Are the spraying times statistically significant? Calculate the adjusted factor averages for the spraying times. Which spraying times have the lowest mean fiber counts?

10   Refer to the fluidized-bed combustor experiment described in Chapter 6, Exercise 5. Since the cost of each test in this experiment is very high, the project manager decided to use a fractional factorial screening design to identify which factors were significant in promoting corrosion on the exterior surface of the tubes. The data collected from the fractional factorial design are given below:

Bed Temp. (°C)	Tube Temp. (°C)	Particle Size (μm)	Environment	Particle Material	Tube Material	Corrosion
700	0	1000	Sulfidizing	Limestone	Iron	14
1000	0	4000	Sulfidizing	Limestone	S. steel	12
700	0	4000	Sulfidizing	Dolomite	S. steel	16
700	400	4000	Sulfidizing	Limestone	Iron	14
1000	0	1000	Sulfidizing	Dolomite	Iron	4
700	400	4000	Oxidizing	Limestone	S. steel	13
1000	0	1000	Oxidizing	Dolomite	S. steel	4
1000	400	1000	Oxidizing	Limestone	Iron	14
1000	400	4000	Sulfidizing	Dolomite	Iron	6
1000	400	4000	Oxidizing	Dolomite	S. steel	2
1000	400	1000	Sulfidizing	Limestone	S. steel	15
700	0	1000	Oxidizing	Limestone	S. steel	13
700	0	4000	Oxidizing	Dolomite	Iron	5
700	400	1000	Sulfidizing	Dolomite	S. steel	3
700	400	1000	Oxidizing	Dolomite	Iron	5
1000	0	4000	Oxidizing	Limestone	Iron	11

Define an appropriate ANOVA model for this experiment. Perform an analysis of variance on these data. What factor levels should be chosen to minimize corrosion? What factors do not appear to affect the corrosion rate?

11 Studies are being conducted by a major automobile manufacturer to determine the
relationship between spray angle of an injector in an automobile engine and various
factors concerning the orifice through which fuel is being injected. Six factors are of
interest to the manufacturer's design engineers: number of orifices (7 or 9), total orifice
area (small or large), orifice length/depth ratio (high or low), orifice diameter (small or
large), amount of fuel injected (small or large), and swirl ratio (low or high). The test
lab can only perform 32 test runs for this particular project. A fractional factorial was
run with the following results:

No. of Orifices	Orifice Area	L/D Ratio	Orfice Diam	Amount Fuel	Swirl Ratio	Spray Cone Angle (deg)
9	Large	Low	Large	Small	High	12
7	Large	Low	Small	Large	Low	110
9	Large	Low	Small	Small	Low	35
9	Small	High	Large	Small	High	90
9	Small	Low	Large	Small	Low	87
7	Large	High	Large	Large	Low	93
7	Large	Low	Small	Small	High	45
7	Large	High	Small	Large	High	85
9	Large	High	Small	Small	High	33
9	Small	High	Large	Large	Low	42
9	Large	Low	Large	Large	Low	85
7	Large	Low	Large	Large	High	115
9	Small	High	Small	Large	High	31
9	Large	High	Small	Large	Low	92
7	Small	Low	Small	Large	High	53
7	Small	Low	Small	Small	Low	93
7	Large	High	Small	Small	Low	46
7	Small	High	Small	Large	Low	37
7	Large	High	Large	Small	High	25
9	Small	High	Small	Small	Low	87
9	Large	High	Large	Large	High	78
9	Small	Low	Small	Large	Low	21
7	Small	Low	Large	Small	High	103
7	Small	High	Large	Small	Low	99
9	Small	Low	Small	Small	High	84
9	Large	Low	Small	Large	High	93
7	Small	Low	Large	Large	Low	12
9	Large	High	Large	Small	Low	31
7	Small	High	Small	Small	High	79
9	Small	Low	Large	Large	High	41
7	Small	High	Large	Large	High	25
7	Large	Low	Large	Small	Low	46

Analyze the effects of these factors on the spray cone angles under the assumption that
three-factor and higher-order interactions are zero. Which factor–level combinations
produce the highest (lowest) mean spray cone angle? To which factors is the mean
spray cone angle relatively insensitive?

12 Data were collected in the crossover design experiment described in Chapter 8, Exercise 11. Analyze the data and draw appropriate conclusions.

Jeep	Sequence	Month	Methanol	Iron Wear Rate
1	4	1	2	.73
		2	1	.30
		3	1	.25
2	2	1	1	.27
		2	2	.56
		3	2	.72
3	2	1	1	.35
		2	2	.43
		3	2	.38
4	1	1	1	.22
		2	2	.56
		3	1	.17
5	4	1	2	.67
		2	1	.15
		3	1	.08
6	3	1	2	.82
		2	1	.26
		3	2	.73
7	1	1	1	.30
		2	2	.64
		3	1	.61
8	3	1	2	.42
		2	1	.20
		3	2	.30
9	2	1	1	.13
		2	2	.56
		3	2	.69

CHAPTER 19

Analysis of Covariance

In contrast to experimental factors, covariates are variables that may affect the response but have not been controlled. In this chapter the analysis of experimental-design models that include covariates is discussed. In particular, this chapter highlights

- *models that relate a response variable to both experimental factors and covariates, and*
- *the analysis of data from completely randomized designs and from randomized complete block designs for models that include a single covariate.*

In many experimental situations, responses are affected not only by the controllable experimental factors, but also by uncontrollable variates. Models can be defined that relate the response to both the controllable and the uncontrolled variates, and the influence of all the variates on the response can be investigated. To distinguish controlled experimental factors from uncontrolled experimental variables, the latter are called *covariates.* (see Exhibit 19.1).

EXHIBIT 19.1

Covariate. A covariate is an uncontrolled experimental variable that influences the response but is itself unaffected by the experimental factors.

It is important to note that the definition in Exhibit 19.1 requires that the covariate be unaffected by the experimental factors. If covariates are so affected, they become additional responses and the analytic procedures in this chapter are inappropriate. Covariates are clearly unaffected by the experimental factors if the covariates can be measured on experimental units prior to the assignment of the factor–level combinations. In other cases, the covariate can only be measured at the same time the response is measured. In this latter situation it may be more difficult, but equally important, to determine whether the experimental factors affect the covariates.

If covariates are excluded from the analysis of the response variable, estimates of the factor effects are biased. When covariates are ignored in an analysis, tests for the statistical significance of the factors will suffer from a loss of power. In other words, if covariates are ignored in the analysis, their influences on the response will masquerade as experimental error. This leads to less sensitive factor comparisons, because the estimate of experimental error is inflated by the effects of the covariate.

The analysis of experimental data when both covariates and factor variables are present is termed *analysis of covariance*. This analysis combines features of analysis of variance (Chapters 15 to 18) and regression (Chapters 21 and 22). The first section of this chapter describes an analysis-of-covariance model. The second section provides procedures for the analysis of data from a completely randomized design having one covariate. The final section gives the analysis-of-covariance procedure for a randomized complete block design having one covariate.

19.1 ANALYSIS-OF-COVARIANCE MODEL

In Table 7.4, data were presented on the lifetimes of cutting tools used with lathes. Primary interest focused on the comparison of two different types of cutting tools, labeled type A and type B for simplicity. As is clearly evident from Figure 7.2, however, the tool lifetimes in Table 7.4 cannot be directly compared using either a two-sample t-test or a one-factor analysis of variance. This is because the lifetimes for each type of cutting tool are linear functions of the lathe speed. Although the lathe speeds could have been preset at selected values, they were not in this experiment. Thus lathe speed should be treated as a covariate.

Note that lathe speed and tool type are not affected by one another, so lathe speed can be considered a covariate and not another response. Note too that the trends in Figure 7.2 suggest that the linear relationships between tool life and lathe speed have a common slope for the two tool types. This is not a necessary requirement for analysis-of-covariance models, but a common slope is a reasonable assumption for many applications. Fitted lines with a common slope have been superimposed on Figure 7.2.

A general analysis-of-covariance (ANACOVA) model having one experimental factor of interest and one covariate can be written as

$$y_{ij} = \mu_i + \beta_i x_{ij} + e_{ij}, \qquad i = 1, 2, \ldots, a, \quad j = 1, 2, \ldots, r.$$

In this model, y_{ij} and x_{ij} are the values of the response variable and the covariate, respectively, for the jth observation associated with the ith level of the experimental factor. The mean μ_i denotes the effect of the ith level of the factor on the response, β_i denotes the linear effect (slope) of the covariate for those observations associated with the ith level of the factor, and e_{ij} denotes the random error. As in Chapter 15, μ_i is often expressed as $\mu + \alpha_i$ with the constraint $\sum \alpha_i = 0$ imposed.

Assuming that there is a common slope for the covariate effect with all levels of the experimental factor, the ANACOVA model can be written as

$$y_{ij} = \mu + \alpha_i + \beta x_{ij} + e_{ij}. \tag{19.1}$$

A labeled scatterplot (Section 4.1) is an important guide to whether this assumption of a common slope is reasonable. If a common slope is not a reasonable assumption, general regression techniques (e.g., Chapter 21) can be used to model the effects of the factor and the covariate on the response.

The model (19.1) allows one to estimate the factor effects through estimation of the μ_i. One can also test for significant differences in the effects due to the factor levels by testing

$$H_0 : \mu_1 = \mu_2 = \cdots = \mu_\alpha \quad \text{vs} \quad H_\alpha : \text{at least two } \mu_i \text{ differ}$$

or, equivalently,

$$H_0: \alpha_1 = \alpha_2 = \cdots = \alpha_a = 0 \quad \text{vs} \quad H_a: \text{at least one } \alpha_i \neq 0.$$

Likewise, one can estimate the effect of the covariate on the response through the estimation of β. One can also test whether the covariate has a statistically significant effect on the response by testing

$$H_0: \beta = 0 \quad \text{vs} \quad H_a: \beta \neq 0.$$

ANACOVA models need not have a single covariate, nor need they be used only when there is a single experimental factor of interest. While this chapter explicitly covers only the single-covariate, single-experimental-factor models in either completely randomized or randomized block designs, ANACOVA models can be generalized to several factors and several covariates. Analysis of these more general models would utilize the general regression models and methodology contained in Part IV of this text.

19.2 ANALYSIS OF COVARIANCE FOR COMPLETELY RANDOMIZED DESIGNS

The ANACOVA model for data from a completely randomized design having a single experimental factor of interest and a single covariate is given by (19.1). In order to analyze the data arising from this model, the assumptions listed in Exhibit 19.2 must be valid.

EXHIBIT 19.2   ANALYSIS-OF-COVARIANCE ASSUMPTIONS FOR COMPLETELY RANDOMIZED DESIGNS

1. The experimental data are obtained using a completely randomized statistical design (Section 7.1).
2. The response can be represented as an additive function of the factor effects, a linear covariate effect, and a random error component as in equation (19.1).
3. The covariate is unaffected by the factor levels.
4. The errors can be considered independently normally distributed with mean zero and constant standard deviation σ.

The primary goal of a covariance analysis often is to determine if the factor levels produce different mean responses. In order to make this determination, a related question often must first be answered: is there statistical, theoretical, or previous experimental evidence that the mean of the response depends on the covariate x? If the answer to this latter question is yes, we proceed with the covariance analysis: otherwise, ANOVA procedures (Chapter 15) are appropriate.

The statistical significance of the covariate can be assessed by comparing two models, one with and one without the covariate. The full model (M_1) is (19.1). The reduced model (M_2) is

$$y_{ij} = \mu + \alpha_i + e_{ij}. \tag{19.2}$$

The statistical significance of the covariate is indicated in different ways in the output of different computer programs. Some computer programs report a t-statistic, others an F-

statistic. Either way, if the program is coded correctly, the results will be the same as those using the principle of reduction in error sums of squares described in Section 15.1.

Denote the error sum of squares for the full model (19.1) by SSE_1. Denote the error sum of squares for the reduced model (19.2) by SSE_2. Then

$$R(M_1 | M_2) = SSE_2 - SSE_1.$$

There is one more parameter (β) in the full model than in the reduced one. Therefore, using $MS_E = SSE_1/(n - a - 1)$ from the full model where $n = ar$,

$$F = R(M_1 | M_2)/MS_E$$

has an F-distribution with $v_1 = 1$ and $v_2 = n - a - 1$ degrees of freedom under the hypothesis $H_0: \beta = 0$. This procedure is summarized in Exhibit 19.3.

EXHIBIT 19.3   TESTING THE SIGNIFICANCE OF THE COVARIATE

1. Calculate the reduction in sum of squares for the full and reduced models, $R(M_1 | M_2) = SSE_2 - SSE_1$.

2. Form the F-statistic $F = R(M_1 | M_2)/MS_E$.

3. If $F > F_\alpha(1, n - a - 1)$, then conclude that $\beta \neq 0$, i.e., that the covariate explains a statistically significant amount of the variation of the response variable

---

Table 19.1 exhibits two analysis of variance tables for the tool-life data of Table 7.4. The first ANOVA table is for the model containing the tool-life factor and the lathe-speed covariate. The second ANOVA table eliminates the covariate. Note that the degrees of freedom and the sum of squares for the error term in the reduced model have been increased by the respective quantities for the covariate in the ANOVA table for the full model. This illustrates that failure to include important model terms increases the variability attributable to the error component of the model.

Table 19.2 symbolically illustrates how one ANOVA table can be used to summarize the information from the two fits in Table 19.1. The lines in the table labeled "Model," "Error,"

Table 19.1   Analysis of Variance Tables for Tool Life Data

Source	df	SS	MS	F	p-Value
(a) Model 1: Full Model, Including Covariate					
Tool type & speed	2	1418.00	709.00	76.74	.000
Error	17	157.05	9.24		
Total	19	1575.05			
(b) Model 2: Reduced Model, Excluding Covariate					
Tool type	1	1097.72	1097.72	41.39	.000
Error	18	477.33	26.52		
Total	19	1575.05			

Table 19.2 Symbolic Analysis-of-Variance Table

Source	df	SS*	MS	F	p-Value
Model	a	MSS	MSM	F_M	p_M
Factor A	$a-1$	SS_A	MS_A	F_A	p_A
Covariate	1	SS_C	MS_C	F_C	p_C
Error	$n-a-1$	SSE_1	MS_E		
Total	$n-1$	TSS			

$*SS_C = R(M_1 | M_2)$.

Table 19.3 Analysis of Covariance for Tool Life Data

Source	df	SS	MS	F	p-Value
Model	2	1418.00	709.00	76.74	.000
Tool type	1	1097.72	1097.72	118.80	.000
Speed	1	320.28	320.28	34.67	.000
Error	17	157.05	9.24		
Total	19	1575.05			

and "Total" are the same as those in Table 19.1(a); i.e., they correspond to the full fit to the model, including the covariate. The line labeled "Factor A" includes the degrees of freedom, sum of squares, and mean square for the experimental factor from the second fit, excluding the covariate. The sum of squares in this line measures the effect of only factor A and not the covariate.

The line labeled "Covariate" in Table 19.2 measures the effect of the covariate using the principle of reduction in sum of squares. The number of degrees of freedom for the covariate and the sum of squares, SS_C, can be obtained by subtracting the respective quantities for the error term of the full model from those of the reduced model:

$$df(\text{covariate}) = df(SSE_2) - df(SSE_1) = (n-a) - (n-a-1) = 1,$$
$$SS_C = SSE_2 - SSE_1.$$

This type of ANOVA table is frequently printed out by computer programs in lieu of two separate ANOVA tables. The indenting of the factor and covariate lines in Table 19.2 is used to stress that these sums of squares and degrees of freedom total to those for the "Model" line, the line under which the factor and covariate lines are indented.

Table 19.3 is such an ANOVA table for the tool life example. Note that the covariate is highly significant ($F = 34.67$, $p < 0.001$). This indicates that the linear trends seen in Figure 7.2 are statistically significant; i.e., the lifetimes of the tools are affected by the lathe speed in addition to the tool type.

Had the covariate been judged nonsignificant, one could proceed to test the statistical significance of the main effect for tool type using the F-statistic in Table 19.3. However, since the speed covariate has been judged statistically significant, that cannot be done. Likewise, if an experimenter knows from theoretical considerations or past experimentation that a covariate is important to the satisfactory modeling of a response, the usual F-statistic for testing the significance of factor effects should not be used. A second example will illustrate the reason for this statement.

Table 19.4   Residence Times and Impurity Measurements for Two Chemical Reactors

Reactor	Residence Time (min)	Impurity Measurement*
Production	33.9	−4.6011
	32.9	−4.5791
	32.7	−3.9769
	33.9	−4.7702
	35.4	−6.1052
	32.7	−4.4565
	33.1	−5.0066
	35.1	−6.3283
	37.3	−7.3091
	37.3	−6.2105
Experimental	30.4	−2.8473
	30.0	−2.6173
	31.0	−2.7969
	29.8	−2.5383
	31.7	−3.5066
	31.9	−3.5066
	29.1	−2.3645
	32.5	−4.0174
	30.5	−3.4122
	30.2	−3.3242

*Logarithm of percent impurity concentration.

Table 19.4 lists data collected from an experiment that was conducted to compare the residence times of two chemical reactors. One of the reactors, the *production* model, is currently used to make an intermediate product for a synthetic fiber. The other reactor, the *experimental* model, incorporates new design features and is being compared with the production model. Short residence times are desirable, since they result in increased throughput and consequently increased production.

A straightforward comparison of the average residence times using a two-sample t-statistic ($t = 5.68$, $p < 0.001$) indicates that the average residence times for the two reactors are significantly different. On the basis of this test one would conclude that the experimental reactor ($\bar{y}_E = 30.71$ min) has a significantly lower average residence time than the production reactor ($\bar{y}_P = 34.45$ min).

It is known from past experience with the production reactor that the residence times are affected by the amount of impurities in the reactor. Data on the amount of impurities were collected for each of the test runs and are recorded in Table 19.4 along with the residence times. Figure 19.1 is a labeled scatterplot of the residence times versus the impurity measurements. This scatterplot suggests that the observed differences in the average residence times for the two reactors might be due to differences in the amount of impurities in the reactors and not necessarily due to the reactor designs.

It is important to note that the experimenters did not believe that the reactors would affect impurity formation. If the reactors were thought to affect the amount of impurities in the test runs, both the residence times and the impurity levels would be response variables and an analysis of covariance would not be appropriate. In the following analysis we

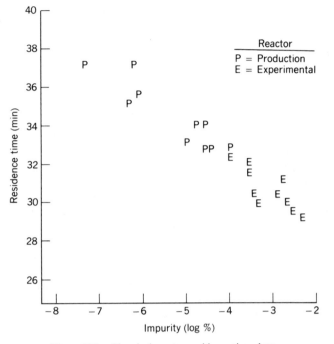

Figure 19.1 Chemical-reactor residence-time data.

Table 19.5 Analysis-of-Variance Tables for Chemical-Reactor Residence Times

Source	df	SS	MS	F	p-Value
(a) *Impurity Effect Adjusted for Reactor Effect*					
Model	2	100.424	50.212	116.96	.000
Reactors	1	69.192	69.192	161.17	.000
Impurity	1	31.232	31.232	72.75	.000
Error	17	7.298	0.429		
Total	19	107.722			
(b) *Reactor Effect Adjusted for Impurity Effect*					
Model	2	100.424	50.212	116.96	.000
Impurity	1	100.325	100.325	233.69	.000
Reactors	1	0.099	0.099	0.23	.638
Error	17	7.298	0.429		
Total	19	107.722			

assume that a covariance analysis is appropriate, i.e., that the level of impurities is not materially affected by the reactor design.

The statistical significance of the effect of the amount of impurities on the residence times is confirmed by the analysis shown in Table 19.5(a). A question that now arises is whether the reactors produce significantly different average residence times when the amount of impurities is the same for both. In order to answer this question the roles of the factor and covariate are reversed in the analysis of variance.

The principle of reduction in sums of squares can provide an appropriate test statistic to examine the effects of the factor on the response after adjusting for the effect of the covariate. Instead of the model (19.2), the reduced model M_3 is now

$$y_{ij} = \mu + \beta x_{ij} + e_{ij}. \tag{19.3}$$

The reduction in sums of squares, $R(M_1 | M_3)$, now measures the incremental effect on the response of the factor above that which is provided by the covariate alone. Thus if it is the covariate that is providing the major effect on the response, $R(M_1 | M_3)$ will be small compared to the error mean square MSE_1 from the full model.

Table 19.5(b) reveals that the reactors do not have a statistically significant effect on the average residence times once the covariate has been accounted for. The difference observed in the average residence times is due to the difference in the amount of impurities in the two reactors. The production reactor has a lower average impurity measurement, -5.33, than does the experimental reactor, -3.09. Once this difference is accounted for by fitting the covariate, the remaining differences in residence times are minor.

Figure 19.2 shows the fitted model (19.1) for the two reactors. The common slope is estimated to be $b = -1.55$. One can obtain the two unadjusted residence-time averages by evaluating the fitted model for each reactor at its average impurity level. This is

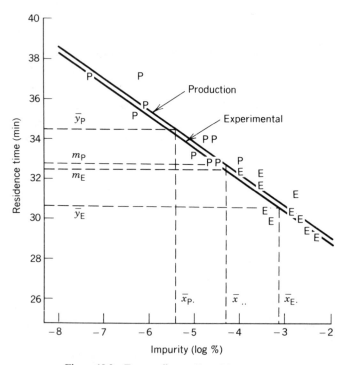

Figure 19.2   Factor effects adjusted for covariate.

graphically shown in Figure 19.2 by proceeding vertically from each average impurity $(\bar{x}_P.$ and $\bar{x}_E.)$ value to the line for the corresponding reactor and then reading the residence time $(\bar{y}_P$ and $\bar{y}_E)$ from the vertical scale.

Fitting the ANACOVA model allows one to obtain adjusted averages, namely, the residence times corresponding to a common level of impurity. Any level of impurity could be chosen to adjust the residence-time averages. Although the adjusted averages will depend on the level of impurity chosen, the *difference* of any two adjusted averages does not depend on the level of impurity. It is common, when no preferred choice such as a trade or industry standard is available, to adjust the averages at the overall average covariate value, $\bar{x}..$, as indicated in Figure 19.2.

Adjusting the response averages to the common covariate value $\bar{x}..$ can be accomplished with the following formula (see the appendix to this chapter):

$$m_i = \bar{y}_i - b(\bar{x}_i. - \bar{x}..). \tag{19.4}$$

Note that if the covariate averages for the different factor levels are all equal (i.e., $\bar{x}_i. = \bar{x}..$), the adjusted factor averages equal the ordinary averages $\bar{y}_i$. The cutting-tool lifetime example (see Figure 7.2) illustrates a situation where this is approximately true. When the covariate averages are not equal, as with the reactor residence times, the adjusted averages can be much different from the ordinary averages. For the reactor residence times,

$$m_E = 30.71 + 1.55(-3.09 + 4.21) = 32.45,$$
$$m_P = 34.43 + 1.55(-5.33 + 4.21) = 32.69.$$

The difference in these adjusted factor averages, 0.24 min, more accurately indicates the differences in the average residence times of the two reactors than does the difference in the ordinary averages, 3.72 min. Figure 19.2 graphically portrays the adjustment of the reactor averages for the effect of the difference in average impurities.

When there are more than two levels of the factor of interest, merely calculating the difference in adjusted factor averages will not suffice to determine which of the pairs of factor averages are significantly different. The overall F-test using the principle of reduction in error sums of squares as in Table 19.5(b) can be used to determine that one or more pairs of means are different from one another, but it cannot identify which pairs are different when there are three or more factor levels.

Multiple-comparison procedures discussed in Chapter 16 can be used for this purpose once the estimated standard errors of the differences have been calculated. The estimated standard error of the difference between two adjusted factor averages, $m_i - m_k$, is given by

$$\widehat{SE}(m_i - m_k) = \left[MS_E \left(n_i^{-1} + n_k^{-1} + \frac{(\bar{x}_i. - \bar{x}_k.)^2}{SS_{xx}} \right) \right]^{1/2}, \tag{19.5}$$

where MS_E is the mean squared error from the fit to the full model (19.1), n_i and n_k are the numbers of observations for the two factor levels, $\bar{x}_i.$ and $\bar{x}_k.$ are the covariate averages for the two factor levels, and SS_{xx} is the error sum of squares from an ANOVA fit to the covariate:

$$SS_{xx} = \sum_i \sum_j (x_{ij} - \bar{x}_i.)^2.$$

The estimated standard errors (19.5) can also be used to construct confidence intervals. The ANACOVA procedure for completely randomized designs is summarized in Exhibit 19.4.

EXHIBIT 19.4   ANALYSIS-OF-COVARIANCE PROCEDURES FOR COMPLETELY RANDOMIZED DESIGNS

1. Using appropriate computer software or calculating formulas, compute the error sums of squares for the models (19.1), (19.2), and (19.3).

2. Calculate the ANOVA table as in Table 19.2, and test the statistical significance of the covariate. If the covariate is not statistically significant, analyze the data using appropriate ANOVA procedures, ignoring the covariate.

3. If the covariate is statistically significant or if the experimenter knows from theoretical considerations that the covariate is required to satisfactorily model the response,

   (a) calculate the ANOVA table adjusting the factor effects for the covariate [e.g., Table 19.5(b)];

   (b) test the statistical significance of the adjusted factor averages, using the estimated standard errors of the differences, as appropriate, to assess which pairs of factor levels are statistically different.

19.3   ANALYSIS OF COVARIANCE FOR RANDOMIZED COMPLETE BLOCK DESIGNS

ANACOVA procedures can be extended to most of the experimental designs discussed in Part II of this book. In this section we illustrate the extension to randomized block designs. For simplicity and clarity of discussion we again consider only a single factor and one covariate.

Suppose that the raw materials used in the reactor example were delivered in batches and that only two test runs could be made from the raw material in each batch. One might decide to split the batches and make one run from each of the reactors from the raw material in a batch. The randomized block designs of Chapter 8 would then be appropriate for this experiment.

Table 19.6   Randomized Block Design for Reactor Residence Times

Block	Production Reactor		Experimental Reactor	
	Residence Times	Impurity Measurement	Residence Times	Impurity Measurement
1	33.9	−4.6011	30.4	−2.8473
2	32.9	−4.5791	30.0	−2.6173
3	32.7	−3.9769	31.0	−2.7969
4	33.9	−4.7702	29.8	−2.5383
5	35.4	−6.1052	31.7	−3.5066
6	32.7	−4.4565	31.9	−3.5066
7	33.1	−5.0066	29.1	−2.3645
8	35.1	−6.3283	32.5	−4.0174
9	37.3	−7.3091	30.5	−3.4122
10	37.3	−6.2105	30.2	−3.3242

Table 19.6 is an illustration of how the experimental results might appear in such a randomized block design. The intent of the covariance analysis in this design would be the same as for the completely randomized design: to determine whether the two reactor types (experimental and production) have a significantly different mean residence time at a specified value of the impurity measurement. The analysis procedure must now take into account the fact that the experiment was blocked.

The ANACOVA model for data from a randomized complete block design in which no repeat tests have been made can be expressed as

$$y_{ij} = \mu + \alpha_i + c_j + \beta x_{ij} + e_{ij}, \qquad i = 1, 2, \ldots, a, \quad j = 1, 2, \ldots, b, \qquad (19.6)$$

where c_j now represents the (random) effect of the jth block (batch of raw material) on the response. The assumptions for ANACOVA models used with randomized complete block designs are given in Exhibit 19.5. The last assumption is only critical when there are no repeat tests from which the experimental-error variation can be estimated. If there are repeat tests available, block-by-factor interaction terms can be added to the model.

EXHIBIT 19.5 ANACOVA **ASSUMPTIONS FOR RANDOMIZED COMPLETE BLOCK DESIGNS**

1. The experimental data are obtained using a randomized complete block design (Section 8.2).
2. Block effects can be either fixed or random. If random, they are assumed to be independent normal variates having mean zero and constant standard deviation σ_c.
3. The response can be represented as an additive function of the factor effects, the block effects, a linear covariate effect, and a random error component as in equation (19.6).
4. The errors can be considered independently normally distributed with mean zero and constant standard deviation σ. The error components are independent of the block components (if the blocks are random).
5. The covariate is unaffected by either the blocks or the factor levels.
6. The block and factor–level components of the model do not interact.

A variety of interesting questions can be asked about the ANACOVA model (19.6). Does the covariate influence the response? If so, are the factor levels significantly different from one another after adjusting for the covariate? Are the block effects statistically significant?

The complete model and three reduced models will be used to answer these questions. We denote the four models as follows:

M_1: The complete model, (19.6);
M_2: $y_{ij} = \mu + \alpha_i + \beta x_{ij} + e_{ij}$;
M_3: $y_{ij} = \mu + \alpha_i + c_j + e_{ij}$;
M_4: $y_{ij} = \mu + c_j + \beta x_{ij} + e_{ij}$.

Note that M_1 is the full ANACOVA model for a randomized complete block design. Models M_2 through M_4 are reduced models: M_2 has no block components, M_3 has no covariate, and M_4 has no factor–level components.

To investigate the questions raised above, three reductions in sums of squares are calculated: $R(M_1|M_2)$, $R(M_1|M_3)$, and $R(M_1|M_4)$. These reductions can be used, respectively, to test the following hypotheses:

$R(M_1|M_2)$: $H_0: \sigma_c = 0$ (blocks random) or
             $H_0: \gamma_1 = \gamma_2 = \cdots = \gamma_b = 0$ (blocks fixed);

$R(M_1|M_3)$: $H_0: \beta = 0$;

$R(M_1|M_4)$: $H_0: \alpha_1 = \alpha_2 = \cdots = \alpha_a = 0$.

To test any one of these hypotheses, one could form ANOVA tables similar to that of Table 19.2. The mean squares for these hypotheses would be formed by dividing the reductions in sums of squares by their numbers of degrees of freedom, which are respectively $b - 1$, 1 and $a - 1$.

Many computer programs summarize the testing of these three hypotheses in a single ANOVA table similar to that shown symbolically in Table 19.7. Note that in this table the sums of squares for the three hypotheses do not add up to the model sum of squares. This is because this type of ANOVA table is summarizing three individual ANOVA tables, each of which (like Table 19.2) does constitute an additive partitioning of the model sum of squares.

Other computer programs do provide an additive partitioning of the model sum of squares according to how the various model terms are listed in the computer control statements. For example, in some programs if the terms are listed in the order (block, factor, covariate), the program fits the following models in sequence and then forms reductions in sums of squares from each succeeding pair of models:

$$Y_{ij} = \mu + e_{ij},$$

$$Y_{ij} = \mu + c_j + e_{ij},$$

$$Y_{ij} = \mu + \alpha_i + c_j + e_{ij},$$

$$Y_{ij} = \mu + \alpha_i + c_j + \beta x_{ij} + e_{ij}.$$

In this sequence of tests, each test is conditional on the outcome of the preceding test being nonsignificant. For example, the test for factor effects uses the reduction in sums of squares between the second and third models listed above. The validity of this test depends

Table 19.7 Symbolic Analysis-of-Variance Table

Source	df	SS	MS	F	p-Value
Model	m	MSS	MSM	F_M	p_M
Blocks	$b - 1$	SS_B	MS_B	F_B	p_B
Factor A	$a - 1$	SS_A	MS_A	F_A	p_A
Covariate	1	SS_C	MS_C	F_C	p_C
Error	e	SS_E	MS_E		
Total	$n - 1$	TSS			

$SS_B = R(M_1 : M_2)$ $MS_B = SS_B/(b-1)$ $F_B = MS_B/MS_E$

$SS_A = R(M_1 : M_4)$ $MS_A = SS_A/(a-1)$ $F_A = MS_A/MS_E$

$SS_C = R(M_1 : M_3)$ $MS_C = SS_C$ $F_C = MS_C/MS_E$

$SS_E = SSE_1$ $MS_E = SS_E/e$ $e = ab - a - b$

                                                     $m = a + b - 1$

Table 19.8 ANACOVA for **Residence-Time Data in a Randomized Block Design**

Source	df	SS	MS	F	p-Value
Model	11	103.865	9.442	19.59	.000
Batches	9	3.442	0.382	0.79	.634
Reactors	1	0.478	0.478	0.99	.348
Impurity	1	14.001	14.001	29.04	.000
Error	8	3.857	0.482		
Total	19	107.722			

Table 19.9 Sequential ANACOVA for **Residence Times in a Randomized Block Design**

Source	df	SS	MS	F	p-Value
Model	11	103.865	9.442	19.59	.000
Batches	9	20.672	2.297	4.76	.019
Reactors	1	69.192	69.192	143.53	.000
Impurity	1	14.001	14.001	29.04	.000
Error	8	3.857	0.482		
Total	19	107.722			

on there being no covariate effect, since these two models ignore the covariate components in the specification of the models.

Table 19.8 is an ANOVA table for the reactor residence times under the assumption that the data in Table 19.6 were obtained from a randomized complete block design. This ANOVA table follows the pattern in Table 19.7 in presenting the analysis of the effects using the reductions in sums of squares. The results indicate that neither batches of raw materials nor the reactor effects are statistically significant once the adjustment for impurities is made. At this point the adjusted factor averages should be calculated as described in the appendix to this chapter. The formula for the adjusted factor averages is the same as equation (19.4), since the blocks are assumed to be random; however, the slope coefficient b is calculated from (19.A.5) in the appendix rather than from (19.A.2). The adjusted averages are

$$m_E = 30.71 + 2.07(-3.09 + 4.21) = 33.03,$$
$$m_P = 34.43 + 2.07(-5.33 + 4.21) = 32.11.$$

These averages are not significantly different from one another, as indicated by the nonsignificant F-value in Table 19.8. Any two adjusted averages can also be compared using t-statistics or confidence intervals. The standard error of the difference between adjusted averages is given by (19.5) with

$$SS_{xx} = \sum\sum(x_{ij} - \bar{x}_{i.} - \bar{x}_{.j} + \bar{x}_{..})^2.$$

In order to illustrate the dangers of using a sequential method of testing the model components, Table 19.9 was constructed from a sequential fitting of models by adding block, factor, and covariate components as described above. Note that the sequential F-tests now indicate that all the model components are statistically significant, contrary to the inferences drawn using the principle of reduction in error sums of squares, which is the correct analysis.

APPENDIX: CALCULATION OF ADJUSTED FACTOR AVERAGES

A. Completely Randomized Design, One Factor

Denote estimates of the model parameters in (19.1) by m, a_i and b. The estimates m and a_i are not unique (see Section 15.3). Using the constraint $\sum \alpha_i = 0$, the estimates are

$$m = \bar{y}_{..}, \qquad a_i = \bar{y}_{i.} - \bar{y}_{..} - b\bar{x}_i, \qquad (19.A.1)$$

where b is the estimate of the slope parameter for the covariate in the full model. The estimate of β is unique, regardless of the solutions used for the other model parameters:

$$b = \frac{\sum_i \sum_j (x_{ij} - \bar{x}_{i.})(y_{ij} - \bar{y}_{i.})}{\sum_i \sum_j (x_{ij} - \bar{x}_{i.})^2} \qquad (19.A.2)$$

Using three estimates of the model parameters, the response averages for the factor levels are adjusted using the following expressions:

$$m_i = m + a_i + b\bar{x}_{..}$$
$$= \bar{y}_{i.} - b(\bar{x}_{i.} - \bar{x}_{..}). \qquad (19.A.3)$$

B. Randomized Block Design, One Factor

Adjusted factor averages for randomized block designs with fixed block effects are computed similarly to those for completely randomized designs. The estimates m, a_i, and c_j again are not unique. One set of solutions is

$$m = \bar{y}_{..}, \qquad a_i = \bar{y}_{i.} - \bar{y}_{..} - b\bar{x}_{i.},$$
$$c_j = \bar{y}_{.j} - \bar{y}_{..} - b\bar{x}_{.j}. \qquad (19.A.4)$$

The estimate b is again unique, regardless of the solutions used for the other model parameters:

$$b = \frac{\sum_i \sum_j (x_{ij} - \bar{x}_{i.} - \bar{x}_{.j})(y_{ij} - \bar{y}_{i.} - \bar{y}_{.j})}{\sum_i \sum_j (x_{ij} - \bar{x}_{i.} - \bar{x}_{.j})^2}. \qquad (19.A.5)$$

Adjusted factor–level averages are given by

$$m_i = m + a_i + b\bar{x}_{..},$$

which reduces to (19.A.3) with the estimated slope (19.A.5) inserted.

REFERENCES

Text References

The following texts provide fairly elementary coverage of ANACOVA techniques, including basic calculation formulas:

Anderson, R. L. and Bancroft, T. A. (1952). *Statistical Theory in Research,* New York: McGraw-Hill, Inc.

Cochran, W. G. and Cox, G. M. (1957). *Experimental Designs,* New York: John Wiley and Sons, Inc.

Ostle, B. and Malone, L. C. (1988). *Statistics in Research, Fourth Edition,* Ames, IA: Iowa State University Press.

Winer, B. J. (1971). *Statistical Principles in Experimental Design,* New York: McGraw-Hill, Inc.

More theoretical coverage is provided in

Searle, S. R. (1971). *Linear Models,* New York: John Wiley and Sons, Inc.

Seber, G. A. F. (1977). *Linear Regression Models,* New York: John Wiley and Sons, Inc.

EXERCISES

1 An experiment was conducted to study the friction properties of lubricants. A key constituent of lubricants that is of interest to the researchers is the additive that is mixed with the base lubricant. In order to ascertain whether two competing additives produce a different effect on the friction properties of lubricants, several mixtures of a base lubricant and each of the additives were made. The mixtures of base lubricant cannot be made sufficiently uniform to ensure that all batches have identical physical properties. Consequently, the plastic viscosity, an important characteristic of the base lubricant that is related to its friction-reducing capability, was measured for each mixture prior to the addition of the additives. Analyze the data below to determine whether the additives differ in the mean friction measurements associated with each.

	Additive 1		Additive 2	
Batch	**Plastic Viscosity**	**Friction Measurement**	**Plastic Viscosity**	**Friction Measurement**
1	12	27.1	15	28.6
2	13	26.6	13	37.1
3	15	28.9	14	37.9
4	14	27.1	14	30.6
5	10	23.6	13	33.6
6	10	26.4	13	34.9
7	13	28.1	13	33.1
8	14	26.1	14	34.4
9	12	24.4	14	32.6
10	14	29.1	17	35.6

2 An experiment was conducted to assess the effects of three new high-nutrient feeds for producing accelerated weight gain in guinea pigs. Pigs from three strains were

administered the feeds, and their weight gains y in grams were recorded. As is common in studies of this type, the initial weights x of the pigs were also recorded for use as a covariate. Analyze these data and determine which, if any, of the feeds produce higher mean weight gains than the others.

	Feed A		Feed B		Feed C	
Strain	x	y	x	y	x	y
1	152	301	160	298	162	336
2	177	337	168	310	181	379
3	114	241	125	234	122	258

3   An automobile manufacturer is interested in determining factors that are important to the stopping of vehicles on wet roads. Three factors were chosen for this study: car type, driver, and road surface material. Two cars, a light-weight and a medium-weight, were chosen. Two drivers were selected at random from a pool of experienced test car drivers. Two road surfaces were selected: asphalt and concrete. The response values measured are the stopping distances (feet) for vehicle traveling at a constant speed of 60 mph. All three factors are crossed in this study. Assume that no three-factor interaction exists. The data collected from this study are given below:

STOPPING DISTANCE (FT) FOR CARS TRAVELING 60 MPH

	Driver 1		Driver 2	
Car	Asphalt	Concrete	Asphalt	Concrete
1	194.1 (49)	197.3 (48)	185.0 (45)	175.7 (45)
	184.1 (47)	188.4 (46)	183.2 (45)	176.2 (51)
	189.0 (45)	187.6 (43)	184.0 (47)	180.9 (49)
2	188.7 (58)	186.3 (62)	191.8 (60)	182.9 (61)
	190.2 (60)	197.0 (63)	194.2 (60)	186.0 (63)
	189.4 (61)	197.5 (61)	190.5 (59)	184.3 (60)

The road surface temperature (°F) was measured in this study, and the values are given in the parentheses in the table above. Analyze this data set by using the road surface temperature as a covariate. Calculate adjusted factor–level means for the factors.

4   A rating procedure is being developed in an engine laboratory that will better quantify the subjective measurement of carbon deposits on engine parts. Two different methods of teaching this complex procedure to newly hired technicians are under investigation. Three senior technicians were chosen to instruct new technicians. Eighteen newly hired technicians were chosen for this study. Each technician rated an engine part before and after completing the training program with instruction from one of the two methods. The results are as follows:

Senior Technician	Method A		Method B	
	Before-Class Rating	**After-Class Rating**	**Before-Class Rating**	**After-Class Rating**
1	42	94	55	87
	59	81	69	94
	80	104	45	77
2	62	97	35	95
	47	99	50	92
	79	98	69	104
3	53	95	68	90
	30	77	37	85
	52	73	48	87

Perform an analysis on this data set to determine if there are significant differences in the teaching methods, the senior technicians, or the interaction between the methods and the senior technicians.

5   A study was conducted to determine the cold-startability of three different makes of diesel engines with a methanol test fuel. Fuel was run through each of three test engines, and the start times (seconds) were recorded. Five tests were run on each engine. It was hypothesized that the amount of time (minutes) the engine was left idle between the tests affected the resulting start times. Suppose that the methanol fuel was received in five 20-gallon drums and that samples were taken from each drum for testing in the three engines. Analyze these data in order to determine whether the starting times differ for the three engines.

Engine	Block (Drum)	Start Time	Idle Time
1	1	24	40
1	2	19	36
1	3	10	41
1	4	17	36
1	5	29	46
2	1	38	31
2	2	31	30
2	3	24	31
2	4	31	34
2	5	15	21
3	1	14	32
3	2	34	36
3	3	25	37
3	4	21	32
3	5	34	36

CHAPTER 20

Analysis of Count Data

Count data consist of frequencies or proportions based on the number of observations that occur in two or more classes. The classes depend on some criterion such as class limits or a set of attributes. The methodology discussed in this chapter includes:

- *the analysis of proportions when a single sample is available,*
- *comparison of proportions taken from two or more samples, and*
- *the analysis of two-way tables of counts (contingency tables).*

Situations often are encountered in experimentation where the values of a response variable are attributes rather than measurements. Examples include agreement ratings (*definitely disagree,..., definitely agree*) on questionnaires, reliability ratings (e.g., *satisfactory* or *unsatisfactory*) of electrial components, the presence or absence of infection after treatment with a postoperative antiinfection drug regimen, and the names of vendors of sampled manufactured products that do not meet design specifications. In each of these instances one would observe the number of responses in each of the response categories and analyze the results.

Data obtained from categorical response variable (see Exhibit 20.1) often are called *count data*, since the data usually are summarized and presented as frequencies or counts of the number of observations for each category of the response. Count data also can be obtained using continuous responses. With continuous responses class limits are used to divide the variable into classes, or categories. Interest is focused on the frequency of observations in each of the classes. When response variables are naturally categorical or have been made categorical by grouping into classes, statistical analyses often must utilize inferential techniques that differ from those for continuous response variables.

EXHIBIT 20.1

Categorical Variable. A variable that is intrinsically nonnumerical is referred to as a categorical, qualitative, or attribute variable.

414

20.1 ANALYSIS OF PROPORTIONS: SINGLE SAMPLE

One of the most useful theoretical distributions for characterizing count data is the binomial probability distribution. The binomial distribution can be used when the sample data have been collected in a *binomial experiment* (see Exhibit 20.2).

EXHIBIT 20.2 BINOMIAL EXPERIMENT

1. There are a fixed number *n* of observations.

2. Only one of two outcomes is possible for each observation; i.e., the response has only two possible values.

3. The probability of each outcome is constant for every observation.

4. The observations are mutually independent.

It is common to denote one of the possible outcomes of the response in a binomial experiment as "success" and the other as "failure." This terminology is used mainly for convenience in discussion; neither of the outcomes need actually be regarded as a success or a failure. The third condition listed above for binomial experiments can now be represented algebraically as follows:

$$\Pr\{\text{success}\} = \theta \quad \text{and} \quad \Pr\{\text{failure}\} = 1 - \theta \tag{20.1}$$

for each observation.

Experiments to test new products often satisfy the requirements for binomial experiments. For example, a cosmetic manufacturer might wish to test a new product designed to reduce the number of severe acne lesions that often accompany the use of facial products by persons with sensitive skin. The design of the experiment would require the selection of *n* persons with the type of sensitive skin the product is intended to treat. The choice of sample size would depend on the known incidence of acne lesions among persons with this type of skin who use facial products of the type being tested. The lower the incidence of acne lesions, the larger the sample size.

In a carefully controlled clinical study of this type, the response for each subject (success = appearance of lesions, failure = no lesions) can be considered independent of the responses for every other subject in the study. The probability of a lesion occurring, θ, can be considered constant for all subjects because of the screening of the participants. There is a fixed sample size, *n*, selected in advance of the initiation of the experiment. Further discussion of the choice of the sample size will be given later in this section.

The response measured in a binomial experiment is the frequency, say *y*, of success in the *n* observations. This random variable has a probability distribution termed the *binomial distribution*. In the above illustration, *y* is the number of subjects who develop acne lesions during the course of the experiment. If 30% ($\theta = 0.30$) of all persons with sensitive skin typically develop acne lesions when using facial products, of interest is whether this new product can reduce the percentage below 30%.

Probabilities for the binomial distribution can be determined directly from computing formulas or from tables based on these formulas. The formula for the binomial probability distribution is as shown in Exhibit 20.3.

EXHIBIT 20.3   BINOMIAL PROBABILITY DISTRIBUTION

Let y denote the number of successes in a binomial experiment having n observations and a constant probability θ of success for each response. The probability of observing exactly y successes is given by

$$Pr(y) = \frac{n!}{y!(n-y)!}\, \theta^y\,(1-\theta)^{n-y}, \qquad y = 0, 1, \ldots, n, \tag{20.2}$$

where $r! = 1 \times 2 \times \cdots \times r$. The mean and standard deviation of a binomial variable are $\mu = n\theta$ and $\sigma = [n\theta(1-\theta)]^{1/2}$.

The rationale for this formula is explained in the appendix to this chapter. Tables of binomial probabilities for select values of n and θ are given in Table A14 in the appendix to this book.

In many applications the sample size is sufficiently large that the exact binomial distribution (20.2) need not be used to calculate probabilities for a binomial random variable. An application of the central limit property (Section 12.1) allows a normal approximation to the binomial distribution to be used. Conditions needed for this normal approximation to be sufficiently accurate are given in Exhibit 20.4.

EXHIBIT 20.4   NORMAL APPROXIMATION TO THE BINOMIAL DISTRIBUTION

Probabilities for a binomial random variable can be approximated by a normal probability distribution if

$$\text{(a)}\ \ n\theta \geqslant 5 \quad \text{and} \quad \text{(b)}\ \ n(1-\theta) \geqslant 5.$$

If these conditions hold, the binomial distribution for y is approximated by the distribution of a normal random variable x that has a mean $n\theta$ and a standard deviation $[n\theta(1-\theta)]^{1/2}$.

When approximating the binomial distribution with the normal distribution, one must compensate for the fact that the binomial distribution is discrete (only has values $0, 1, \ldots, n$) while the normal distribution is continuous. As discussed in the appendix to this chapter, this compensation merely replaces the integer values for the binomial variable y with interval values for the normal variable x. Thus $y = a$ is replaced by $a - \frac{1}{2} < x < a + \frac{1}{2}$. So, for example, to approximate $Pr\{5 \leqslant y \leqslant 8\}$ for a binomial variable having $n = 20$ and $\theta = 0.25$, the normal approximation would be

$$Pr\{4.5 \leqslant x \leqslant 8.5\} = Pr\left\{\frac{4.5 - 5}{(3.75)^{1/2}} \leqslant z \leqslant \frac{8.5 - 5}{(3.75)^{1/2}}\right\}$$

$$= Pr\{-0.26 \leqslant z \leqslant 1.81\} = 0.5675,$$

where z follows a standard normal distribution.

The value of the parameter θ in the calculation of binomial probabilities sometimes is chosen to be a hypothetical value of interest. Other times, as in determining probabilities for games of chance, it is known from theoretical considerations. In practice, however, one often does not have the benefit of knowing the value of θ. One then estimates it from a sample of observations from the population of interest. This is accomplished by determining the proportion of successes among the n observations:

$$f = \frac{y}{n} = \frac{\text{number of successes}}{\text{number of observations}}, \qquad (20.3)$$

where f, the fraction of successes, is an estimator of the parameter θ. The estimator of θ in equation (20.3) is the average of the count data, where each success is coded as a one and each failure is coded as a zero.

When the probability of success, θ, is very close to zero or one, or the sample size is small, the number of successes can equal zero or n. When this occurs, the binomial parameter would be estimated to be 0 or 1, respectively. These estimates are unrealistic because they suggest that either no successes or no failures exist. Rather than estimate the binomial parameter using the formula $f = y/n$ in these instances, adjustments are usually made to prevent the estimates from being so extreme. Many suggestions have been proposed, one of which is to let $f = \frac{3}{8}/(n + \frac{3}{4})$ if $y = 0$ or $f = (n + \frac{3}{8})/(n + \frac{3}{4})$ if $y = n$.

While it is useful to be able to estimate the proportion of successes or failures in selected populations, it is just as important to be able to form confidence intervals on and test hypotheses about the unknown value of θ. For example, in the acne study it is of great interest to be able not only to estimate the values of θ for the test product but also to be able to compare it with the historical value of 0.30.

Inferences on the true value of θ can be made using the statistic f from (20.3) and either the binomial distribution or the normal approximation to the binomial distribution. Observe that f is simply y divided by a constant, n. The random portion of f is just y. Thus, exact binomial probabilities which are computed for y likewise hold for f. The normal approximation for f can be used if the requirements on y given above are satisfied. If so, the normal approximation to the distribution of f has a mean θ and standard deviation $[\theta(1 - \theta)/n]^{1/2}$.

When the normal approximation to the binomial distribution is applicable, sample sizes for binomial experiments can be determined by using the procedures detailed in Section 12.5. The value of θ used in the expression for the standard deviation of f can be based on past estimates of the proportion, or one can insert several hypothetical values to determine sample sizes for several possible values of θ. One might then choose to use the worst-case (i.e., largest) sample size to ensure adequate power for tests on the

---

EXHIBIT 20.5   CONFIDENCE INTERVAL FOR BINOMIAL PARAMETERS

1. If the normal approximation to the binomial distribution is not applicable, obtain values of $\theta_L(\alpha/2)$ and $\theta_U(\alpha/2)$ from Table A23 in the appendix. A two-sided confidence interval for θ having confidence coefficient at least $100(1 - \alpha)\%$ is

$$\theta_L(\alpha/2) \leqslant \theta \leqslant \theta_U(\alpha/2). \qquad (20.4)$$

2. If the normal approximation is applicable, determine the values of $\pm z_{\alpha/2}$ from Table A4 in the Appendix corresponding to the desired confidence coefficient $100(1 - \alpha)\%$. Then an approximate $100(1 - \alpha)\%$ two-sided confidence interval for θ is given by (20.5) with

$$\theta_L(\alpha/2) = f - z_{\alpha/2}[f(1 - f)/n]^{1/2},$$
$$\theta_U(\alpha/2) = f + z_{\alpha/2}[f(1 - f)/n]^{1/2}. \qquad (20.5)$$

parameter or adequate precision for point and interval estimates of θ. Since the standard deviation $[n\theta(1-\theta)]^{1/2}$ is maximized for any n when $\theta = \frac{1}{2}$, this choice of θ, if plausible for the proposed experiment, would yield the worst-case sample size.

Since the exact binomial distribution is often laborious to use, tables and nomograms for obtaining confidence limits on the binomial parameter θ have been constructed (see Exhibit 20.5). One such table is included in Table A23 in the appendix to this book. If the central limit property can be invoked, the normal approximation can be used to construct confidence intervals as described in Section 12.2.

In an acne study similar to the one described above, 43 subjects participated in the experiment. If the test product is similar to other products currently available, $\theta = 0.30$. The normal approximation to the binomial distribution is then applicable, since $n\theta = 12.9$ and $n(1-\theta) = 30.1$. Nine of the subjects developed acne lesions during the course of the study; consequently, the estimate of θ for this new product is $f = 9/43 = 0.21$. Since $nf = 9$ and $n(1-f) = 34$, the normal approximation to the binomial distribution still appears to be appropriate for these data. Thus, an approximate 95% confidence interval for θ is:

$$0.21 - 1.96\left(\frac{(0.21)(0.79)}{43}\right)^{1/2} \leqslant \theta \leqslant 0.21 + 1.96\left(\frac{(0.21)(0.79)}{43}\right)^{1/2},$$

or

$$0.09 \leqslant \theta \leqslant 0.33.$$

EXHIBIT 20.6   SIGNIFICANCE TESTS FOR BINIOMIAL PARAMETERS

In order to test the hypothesis

$$H_0 : \theta = \theta_0 \quad \text{vs} \quad H_a : \theta \neq \theta_0 \qquad (0 < \theta_0 < 1):$$

1. Determine whether the normal approximation to the binomial distribution is applicable by checking the following conditions:

$$n\theta_0 \geqslant 5 \quad \text{and} \quad n(1-\theta_0) \geqslant 5.$$

2. If the normal approximation is not applicable, reject H_0 if

$$y < b_L(\alpha/2) \quad \text{or} \quad y > b_U(\alpha/2),$$

where $b_L(\alpha/2)$ is the largest value of r in the binomial distribution for n and θ_0 (Table A14 in the Appendix) that has a cumulative probability less than or equal to $\alpha/2$, and $b_U(\alpha/2)$ is the smallest value of r that has a cumulative probability greater than or equal to $1 - \alpha/2$.

3. If the normal approximation is applicable, calculate

$$z_1 = \frac{y - \frac{1}{2} - n\theta_0}{[n\theta_0(1-\theta_0)]^{1/2}}$$

and

$$z_2 = \frac{y + \frac{1}{2} - n\theta_0}{[n\theta_0(1-\theta_0)]^{1/2}} \tag{20.6}$$

If $z_1 < -z_{\alpha/2}$ or $z_2 > z_{\alpha/2}$, then reject H_0.

One-Sided Tests

Replace $\alpha/2$ with α in the appropriate upper or lower critical value and make either an upper- or a lower-tail test.

Note that this interval includes the value $\theta = 0.30$, indicating that there is not sufficient evidence to conclude that the new product decreases the incidence of acne lesions over current products. Note too that had the same proportion, $\theta = 0.21$, been obtained with a sample of $n = 100$ subjects, the confidence interval would not have included 0.30 ($0.13 \leqslant \theta \leqslant 0.29$ using $n = 100$).

Tests of hypotheses can also be developed for binomial parameters using either the exact binomial distribution or the normal approximation. These procedures are outlined in Exhibit 20.6.

One can show that the two-sided test of $H_0 : \theta = \theta_0$ versus $H_a : \theta \neq \theta_0$ using the normal approximation and the rejection region defined by (20.6) is equivalent to rejecting H_0 if $X^2 > X_\alpha^2$, where

$$X^2 = \sum (|O_i - E_i| - 0.5)^2 / E_i \tag{20.7}$$

with

$O_1 = y,$	the observed number of successes,
$O_2 = n - y,$	the observed number of failures,
$E_1 = n\theta_0,$	the expected number of successes,
$E_2 = n(1 - \theta_0),$	the expected number of failures,

and X_α^2 is an upper $100\alpha\%$ point of the chi-square distribution with one degree of freedom. The conditions for the use of this statistic are identical to those for the use of the normal approximation (20.6). The statistic is generalized to several possible outcomes, rather than just success or failure, in Section 20.3.

Since the interest in the acne-product study is whether the new product reduces the incidence of acne lesion formation from that with current products, consider testing the hypothesis $H_0 : \theta \geqslant 0.30$ vs $H_a : \theta < 0.30$. Rejection of H_0 would provide statistical evidence that the new product does reduce the incidence of acne lesions. Since the normal approximation was shown above to be applicable and only a lower-tail test is desired, calculate z_1:

$$z_1 = \frac{8.5 - 12.9}{[(12.9)(0.70)]^{1/2}} = -1.46.$$

Since z_1 is not less than $-z_{0.05} = -1.645$, there is not sufficient evidence ($p = 0.0721$) at the 5% significance level to conclude that the new product significantly lessens the formation of acne lesions from that with current products.

In some applications of binomial experiments the sample size is large but the probability of success on any trial is very small. In such situations the calculations for the binomial distribution can be tedious and the conditions needed for the normal approximation to the binomial distribution may not be satisfied. An alternative approximation to the binomial distribution is the Poisson distribution (see Exhibit 20.7). The Poisson distribution is an important probability distribution in its own right as well as a useful approximation to the binomial distribution when n is large and θ is small. Tables of the Poisson distribution, for selected values of λ, are given in Table A24 of the Appendix.

EXHIBIT 20.7   POISSON PROBABILITY DISTRIBUTION

Let y denote the number of occurrences of an event over a fixed interval of time or space. The distribution of y is a Poison probability distribution if:

1. Each occurrence of the event is rare; i.e. $\theta = \Pr\{$event occurrence$\}$ is small.

2. The probability of occurrence of two or more events in a sufficiently small interval of time or space is zero.

3. A large number n of independent observations are made.

4. The average number of occurrences, λ, over a fixed interval of time or space is constant.

Poisson probabilities are given by the following formula:

$$\Pr\{y\} = \lambda^y e^{-\lambda}/y!, \qquad y = 0, 1, 2, \ldots, \tag{20.8}$$

where $y! = 1 \times 2 \times \cdots \times y$. The mean of the Poisson distribution is $\mu = \lambda$; its standard deviation is $\sigma = \lambda^{1/2}$.

If n is large, say $n > 50$, and θ is small, say $\theta < 0.1$, the Poisson distribution often can be used to approximate the binomial distribution by using $\lambda = n\theta$ in (20.8). Poisson approximations have been used to model the incidence of automobile collisions with highway light poles. The probability of a specific pole along a highway being hit in an automobile accident is very small; however, there are thousands of light poles along major highways. If the probability of a specific pole being hit during a fixed time interval is 0.0025 and there are 10,000 poles along a stretch of highway, the average number of poles hit during that time interval is $\lambda = 25$. While the probability of hitting any light pole on a major highway is not the same as that for every other pole, studies have shown that the Poisson distribution is a good model because the average λ remains fairly constant over time. Thus, the Poisson distribution can be used to model the actual number of hits, y.

The parameter λ is the mean of the Poisson distribution. It is estimated by y, the number of events that occur. Confidence intervals and tests for λ can be obtained using the procedure in Exhibit 20.8.

EXHIBIT 20.8   CONFIDENCE INTERVALS FOR POISSON PARAMETERS

Let λ_L be the largest value of λ for which the cumulative Poisson distribution evaluated at the observed Poisson count y has a lower-tail probability less than or equal to $\alpha/2$. Let λ_U be the smallest value of λ for which the cumulative Poisson distribution evaluated at y has an upper-tail probability less than or equal to $\alpha/2$. Then a $100(1-\alpha)\%$ confidence interval for λ is $\lambda_L < \lambda < \lambda_U$.

A Poisson distribution can be approximated by the normal, binomial, or other probability distributions depending on the magnitudes of the Poisson parameter λ and the sample size n. In general, if λ is sufficiently large, a normal variable x having mean $\mu = \lambda$ and standard deviation $\sigma = \lambda^{1/2}$ can be used to approximate the distribution of the Poisson random variable y.

Data from designed experiments involving one or more controlled factors often have binomial or Poisson response variables. The ANOVA procedures discussed in Chapters 15 to 18 cannot be directly applied to responses from such experiments. The procedures reported in those chapters are based on the assumption that the responses are normally distributed and have constant standard deviations. Even if the large-sample normal approximations to the binomial and the Poisson distributions can be invoked, the responses for different factor–level combinations not only have potentially different means, they also have different standard deviations. For both distributions the mean and the variances of the normal approximations depend on the unknown distribution parameters θ and λ, respectively.

In Section 20.3, transformations of binomial and Poisson random variables are introduced. These transformations are termed *variance-stabilizing* transformations because the variances (and standard deviations) of the approximating normal distribution do not depend on the unknown distributional parameters. These transformations allow the ANOVA procedures of Chapters 15 to 18 to be used with count data.

20.2 ANALYSIS OF PROPORTIONS: TWO SAMPLES

As with the comparison of means in Chapter 13, one can compare two binomial proportions when either paired or independent samples of data are available from the populations or processes of interest. We first consider the comparison of binomial proportions when samples are paired.

Suppose independent pairs (y_{1j}, y_{2j}), $j = 1, 2, \ldots, n$, are obtained from two binomial populations. Each of the individual observations in a pair can be labelled a "success" or a "failure." For example, suppose two electronic sensors are proposed for use to screen individuals for the presence of a sufficiently high amount of radiation. One sensor is very expensive but is advertised as having very high sensitivity to radiation at the levels desired. The other sensor is less expensive but also advertised as being of acceptable sensitivity at the levels desired. An experiment is to be conducted to compare the sensitivities of the two sensors. Several objects emitting different amounts of radiation are to be screened by both sensors. A "success" for either sensor is simply the determination that an object has a radiation level that exceeds a specified threshold value.

In this example the observations are paired, because both sensors are used on the same n objects. The pairs of observations are the two determinations of the radiation levels on each of the objects. Each pair of observations can have one of the following four outcomes:

Sensor 1		Sensor 2		Difference	
Label	Code	Label	Code	Code 1–Code 2	Number
Success	1	Success	1	0	a
Success	1	Failure	0	1	b
Failure	0	Success	1	-1	c
Failure	0	Failure	0	0	d
				Total	n

In this description of the possible outcomes, it may be that the most important information is contained in the number of times $(a + d)$ that the two sensors agree, i.e. the number of times that the difference in the above codes is zero. One can use the sample proportion $(a + d)/n$ to form a confidence interval on the probability of agreement of the two sensors. This confidence interval would be calculated using the procedures described in the previous section, i.e., the interval (20.4) or the normal approximation to it. If a one-sided lower confidence interval has a lower limit that is acceptably high, one might be willing to consider the two sensors of equivalent sensitivity.

Note, however, that the only information about any disagreement between the two sensors is contained in the second and third categories, i.e., the number of times $(b + c)$ the difference in codes is either -1 or $+1$. From this information one could construct a test of the hypothesis that the two modes of disagreement are equally likely, i.e. that one is equally likely to have the first sensor indicate a success and the second indicate failure as one is to have the determinations reversed.

If one ignores the categories of agreement, the sample size is reduced to $m = b + c$. Under the hypothesis that differences of $+1$ are as likely to occur as differences of -1, the probability of observing $+1$ is $\theta = \frac{1}{2}$. One can test the hypothesis $H_0: \theta = \frac{1}{2}$ vs $H_a: \theta \neq \frac{1}{2}$ using the binomial testing procedures of the previous section. This application of the test for binomial population proportions would use a sample size of m, and one would now label the $+1$ differences as "successes." This test is frequently referred to as a *sign test*, (see Exhibit 20.9) because the only information used is the number of positive and the number of negative differences.

EXHIBIT 20.9   SIGN TEST

Denote the coded (success = 1, failure = 0) responses from independent pairs of observations from two binomial populations by y_{ij}, $i = 1, 2$, $j = 1, 2, \ldots, n$. Let $d_j = y_{1j} - y_{2j}$.

1. Let b equal the number of negative d_j, i.e., the number of d_j equal to -1. Let c equal the number of positive d_j; i.e., the number of d_j equal to $+1$. Denote the total number of nonzero differences by $m = b + c$.

2. Use the binomial probability distribution (20.2) or the normal approximation to the binomial distribution, as appropriate, to test

$$H_0: \theta = \frac{1}{2} \quad \text{vs} \quad H_a: \theta \neq \frac{1}{2}.$$

When using the binomial distribution to test for equally likely differences in paired outcomes, one is only analyzing a portion of the available data. One should not disregard the information contained in the categories of agreement when comparing two binomial populations with paired data.

The above procedures also can be used when the responses are each continuous (or discrete) but one is only interested in whether one response is greater or less than the other. For such data one would again determine the differences in the pairs of observations and simply count the positive and the negative differences. If one has a choice between using the two-sample paired t-test (Section 13.3) and the sign test, the former is generally preferred. This is because the t-test has greater power, since it uses not only the signs of the differences but their magnitudes as well (i.e., the actual values of the d_j, not just the signs).

If there are two independent samples available from two binomial distributions, a different methodology is required (see Exhibits 20.10 and 20.11). The normal approximation to the binomial distribution is the most frequently used distribution for these inferential techniques on binomial proportions from independent samples.

EXHIBIT 20.10   LARGE-SAMPLE CONFIDENCE INTERVAL FOR $\theta_1 - \theta_2$

Let y_1 be the number of successes from a sample of n_1 observations from a binomial population having θ_1 as the probability of success. Similarly define y_2, n_2, and θ_2 for an independent sample from a second binomial population. Assume that the normal approximation is valid for both samples. An approximate $100(1 - \alpha)\%$ confidence interval on $\theta_1 - \theta_2$ is

$$L \leqslant \theta_1 - \theta_2 \leqslant U, \tag{20.9}$$

where

$$L = (f_1 - f_2) - z_{\alpha/2}\left(\frac{f_1(1 - f_1)}{n_1} + \frac{f_2(1 - f_2)}{n_2}\right)^{1/2}$$

$$U = (f_1 - f_2) + z_{\alpha/2}\left(\frac{f_1(1 - f_1)}{n_1} + \frac{f_2(1 - f_2)}{n_2}\right)^{1/2}$$

and $f_1 = y_1/n_1$, $f_2 = y_2/n_2$.

EXHIBIT 20.11   LARGE-SAMPLE SIGNIFICANCE TEST FOR $\theta_1 - \theta_2$

Assume the same conditions as listed in Exhibit 20.10 for large-sample confidence intervals on $\theta_1 - \theta_2$.

Calculate

$$z = \frac{f_1 - f_2}{[f_p(1 - f_p)(n_1 + n_2)/n_1 n_2]^{1/2}} \tag{20.10}$$

where $f_p = (y_1 + y_2)/(n_1 + n_2)$ is a pooled estimator of the (hypothesized) common value $\theta = \theta_1 = \theta_2$. To test

$$H_0 : \theta_1 = \theta_2 \quad \text{vs} \quad H_a : \theta_1 \neq \theta_2,$$

reject H_0 if $z < -z_{\alpha/2}$ or if $z > z_{\alpha/2}$. To perform one-sided tests, replace $\alpha/2$ with α and compare z with either $-z_\alpha$ or z_α.

20.3   ANALYSIS OF PROPORTIONS: MULTIPLE SAMPLES

When comparing more than two population proportions, the inference procedures presented in the last two sections are no longer applicable. We now turn to an alternative technique (see Exhibit 20.12) using the chi-square probability distribution.

The data in Table 20.1 show the incidence of injuries for 998 accidents involving breakaway light poles. Of interest to the project investigators is whether there are different probabilities of injury in different parts of the country. Assuming that these are independent samples from five binomial populations, the investigators wish to test the hypotheses

$$H_0 : \theta_1 = \theta_2 = \theta_3 = \theta_4 = \theta_5$$

vs

$$H_a : \text{at least two of the } \theta_i \text{ differ.}$$

Table 20.1   Incidence of Injuries by Geographic Location for Breakaway Light Poles

	Geographic Location					
	1	2	3	4	5	Total
Injuries	35	6	16	10	4	71
No injuries	47	280	265	268	67	927
Total	82	286	281	278	71	998

Table 20.2   Symbolic Crosstabulation of Successes and Failures for Binomial Samples

	Sample Number					
	1	2	3	$\cdots$	c	Total
Success	O_{11}	O_{12}	O_{13}	$\cdots$	O_{1c}	R_1
Failure	O_{21}	O_{22}	O_{23}	$\cdots$	O_{2c}	R_2
Total	C_1	C_2	C_3	$\cdots$	C_c	N

Table 20.2 is a symbolic crosstabulation similar to that of Table 20.1. The observed frequencies in the table are denoted by O_{ij}, the row totals by R_i, and the column totals (samples sizes) by C_j. Denote the overall total of successes by R_1, of failures by R_2, and the total sample size by N. Note that the overall proportion of successes is $f = R_1/N$. If the true proportion of successes in all the populations is the same, then $C_j f$ is an estimate of the number of observations that should be expected in the success cell of sample j, and $C_j(1 - f)$ is an estimate of the number of observations that should be expected in its failure cell. Note that $C_j f = R_1 C_j/N$ and $C_j(1 - f) = R_2 C_j/N$. The chi-square statistic compares the observed frequencies in the table with the corresponding expected frequencies, calculated under the assumption that there is a common probability of success θ for all the populations.

EXHIBIT 20.12   COMPARISON OF SEVERAL BINOMIAL POPULATION PROPORTIONS

1. Denote the observed cell frequencies by O_{ij} and the total sample size by N. Denote the row totals by R_i $(i = 1, 2)$, and the column totals by C_j $(j = 1, \ldots, c)$.

2. Calculate expected cell frequencies:

$$E_{ij} = (\text{row total})(\text{column total})/N$$

$$= R_i C_j/N.$$

3. Ensure that all $E_{ij} > 1$ and at least 80% of the $E_{ij} > 5$. If this condition is not met, combine columns of the table until it is satisfied.

4. Calculate the chi-square statistic

$$X^2 = \sum\sum (O_{ij} - E_{ij})^2/E_{ij}. \qquad (20.11)$$

5. If $X^2 > X_\alpha^2$, where X_α^2 is a chi-square critical value corresponding to an upper-tail probability of α and degrees of freedom $v = (c - 1)$, reject $H_0 : \theta_1 = \theta_2 = \cdots = \theta_c$.

Note: If the table has only two columns, use the adjustment discussed in the next section.

Critical values for chi-square random variables are given in Table A6 of the appendix. The third condition listed in Exhibit 20.12 is similar to the condition needed for the normal approximation to the binomial distribution to be valid: it ensures that the sample size is sufficiently large for each of the populations.

The chi-square statistic (20.11) is the c-sample generalization of the chi-square statistic (20.7). The statistic (20.11) has $v = c - 1$ degrees of freedom rather than c because the hypothesized common value of the binomial parameter θ must be estimated from the data. If the common value of θ, say θ_0, is stipulated in the null hypothesis, the expected frequencies are calculated as $E_{ij} = R_i \theta_0$, and the chi-square statistic (20.11) has $v = c$ degrees of freedom.

Table 20.3(a) displays the expected frequencies for each cell of Table 20.1. All the expected frequencies are greater than 5, indicating that the chi-square statistic can be used to test whether the probabilities of injury in the five regions of the country are equal. Table 20.3(b) shows the contribution of each cell to the calculated value of the chi-square statistic in equation (20.11). The chi-square statistic, $X = 174.33$, is highly significant ($p < 0.001$), leading to the conclusion that the five regions of the country significantly differ in the probability of injury.

The expected frequencies in Table 20.3(a) can be compared with the observed frequencies in Table 20.1 to determine which of the regions differ most from what is expected under the hypothesis that all the probabilities are equal. Likewise the sample proportions can be examined to determine which differ the most from the overall proportion. A third technique for isolating differences in the regions is to determine which cells of the table contribute most to the overall chi-square statistic, $X = 174.33$. The individual cell contributions are shown in Table 20.3(b). Any of these analyses will point to the first region as having a much higher injury rate than the other four.

The chi-square statistic (20.11) can be used to investigate the effect of the levels of one controllable factor on a binomial response variable. Experiments are sometimes designed to investigate the effects of two or more controllable factors on binomial or Poisson count data. While one might attempt to model the counts using an ANOVA model, the counts are not normally distributed. Even if the sample sizes for each

Table 20.3   Expected Frequencies and Cell Contributions to the Chi-Square Statistic

(a) *Expected Cell Frequencies*

			Geographic Location			
	1	2	3	4	5	Total
Injuries	5.83	20.35	19.99	19.78	5.05	71
No injuries	76.17	265.65	261.01	258.22	65.95	927
Total	82	286	281	278	71	998

(b) *Cell Contributions to the Chi-Square Statistic*

			Geographic Location			
	1	2	3	4	5	Total
Injuries	145.95	10.12	0.80	4.84	0.22	161.93
No injuries	11.17	0.78	0.06	0.37	0.02	12.40
Total	157.12	10.90	0.86	5.21	0.24	174.33

factor–level combination are sufficiently large for the normal approximations to the distributions of the counts to be applied, the standard deviations of the counts are not all equal. Thus the ANOVA procedures developed in Chapters 15 to 18 are applicable.

Transformations of the binomial or Poisson counts are of great importance in investigations of this type if the transformations produce variates that are approximately normally distributed and have constant variances. Two such transformations for binomial and Poisson counts are, respectively, the arc-sine transformation and the square-root transformation (see Exhibit 20.13).

EXHIBIT 20.13   TRANSFORMATIONS FOR COUNT DATA

Arc-Sine Transformation. Let θ denote the probability of success for a binomial random variable. Then if y denotes the number of successes,

$$x = \sin^{-1}\left\{ \frac{y + \frac{3}{8}}{n + \frac{3}{4}} \right\}^{1/2} \tag{20.12}$$

is approximately normally distributed with mean $\mu = \sin^{-1}(\theta)$ and standard deviation $\sigma \doteq 1/(2n^{1/2})$.

Square-Root Transformation. Let λ denote the mean of a Poisson random variable. If y denotes the number of Poisson events in n observations,

$$z = (y + \tfrac{3}{8})^{1/2} \tag{20.13}$$

is approximately a normal variable with mean $\lambda^{1/2}$ and standard deviation $1/2$.

In order to apply the above transformation for binomial variables, each factor–level combination in an experiment should be based on approximately the same number of observations, n. For Poisson variables, the number of Poisson counts should be based on approximately the same temporal or spatial range of observation. Any of the fixed ANOVA models studied in the previous chapters can be used to model the transformed responses.

Since the standard deviation of the transformed responses is known, each of the sums of squares in the ANOVA table can be divided by the known variance of the responses and is distributed as a chi-square random variable with as many degrees of freedom as has the effect. If some of the interactions can be assumed to be zero, an error sum of squares can be formed from the sums of squares for the interactions that are assumed to be zero. A test for adequacy of the model can be performed using a chi-square statistic that is the ratio of the error sum of squares to the variance of the responses. The degrees of freedom of this chi-square equal the degrees of freedom for error.

20.4   CONTINGENCY-TABLE ANALYSIS

One form of count data that often occurs in experiments is obtained when observations are categorized according to two or more response variables. The example in Table 20.1 categorizes each recorded accident according to the region of the country in which it occurred and whether an injury occurred. Table 20.4(a) categorizes 991 of the accidents in this study according to the location on the vehicle where the impact with the light

Table 20.4 Distribution of Accidents by Vehicle Impact Location and Severity of Injury

(a) *Observed Frequencies*

Injury Severity

		None	Minor	Severe	Total
	Front	244	340	141	724
Location of	Left	59	46	18	124
Impact	Right	66	41	18	126
	Rear	11	5	2	18
	Total	380	432	179	991

(b) *Expected Frequencies*

Injury Severity

		None	Minor	Severe	Total
	Front	277.61	315.61	130.77	724
Location of	Left	47.55	54.05	22.40	124
Impact	Right	48.31	54.93	22.76	126
	Rear	6.90	7.85	3.25	18
	Total	380	432	179	991

(c) *Cell Contributions to Chi-square Statistic*

Injury Severity

		None	Minor	Severe	Total
	Front	4.07	1.88	0.80	6.75
Location of	Left	2.76	1.20	0.86	4.82
Impact	Right	0.28	3.53	1.00	4.81
	Rear	2.44	1.03	0.48	3.95
	Total	9.55	7.64	3.14	20.33

pole occurred and the highest (if more than one injury) severity of injury that occurred in the vehicle. Interest in analyzing these data would focus on whether there is a relationship between the severity of injury and the location of impact.

Contingency-table analyses investigate relationships between two categorizations (responses) of an observation. The statistical hypothesis that usually is of interest is whether the row and column categorizations are independent. Relationships between the categorizations are established by rejecting the hypothesis of independence.

The test for independence in a contingency table (see Exhibit 20.14) is very similar to the test for equal population proportions given in the last section. It is actually a generalization of that procedure in that "success" and "failure" can be replaced by three or more categorizations.

Table 20.4(b) displays the expected frequencies, and Table 20.4(c) the cell contributions to the chi-square statistic. The chi-square statistic is $X = 20.33$ with $v = (3)(2) = 6$ degrees of freedom. Since $p < 0.001$, we conclude that the row and column categorizations are not independent, i.e., that the severity of injury is related to the location of the impact. Table 20.5 displays row percentages for the data in Table 20.4(a). From the cell

---

EXHIBIT 20.14   TEST FOR INDEPENDENCE IN A CONTINGENCY TABLE

1. Denote the observed cell frequencies by O_{ij} and the total sample size by N. Denote the row totals by R_i ($i = 1, \ldots, r$) and the column totals by C_j ($j = 1, \ldots, c$).

2. Calculate expected cell frequencies:

$$E_{ij} = \text{(row total) (column total)}/N$$

$$= R_i C_j / N.$$

3. Ensure that all $E_{ij} > 1$ and at least 80% of the $E_{ij} > 5$. If this condition is not met, combine rows or columns of the table until it is satisfied.

4. Calculate the chi-square statistic

$$X^2 = \sum \sum (O_{ij} - E_{ij})^2 / E_{ij}. \tag{20.14}$$

5. If $X^2 > X_\alpha^2$, where the degrees of freedom of the chi-square statistic are $v = (r-1)(c-1)$, reject the hypothesis that the classifications are independent.

Note: If $v = 1$ (i.e., $r = c = 2$), replace $(O_{ij} - E_{ij})$ by $(|O_{ij} - E_{ij}| - \frac{1}{2})$ in (20.14).

Table 20.5   Row Percentages for Vehicle-Accident Data

		Injury Severity			
		None	Minor	Severe	Total
	Front	34	47	19	100
Location of	Left	48	37	15	100
Impact	Right	53	33	14	100
	Rear	61	28	11	100

contributions to the chi-square statistic in Table 20.4(c) and the row percentages in Table 20.5 a reason for the lack of independence is suggested: there is a higher incidence of minor to severe injuries from frontal impacts than from impacts on any other location on the vehicle.

Categorical responses that are influenced by three or more experimental factors are usually analyzed by taking logarithms of the cell frequencies and applying inferential techniques that are similar in concept to ANOVA modeling. This approach, using "log-linear" models, is beyond the scope of this text. The interested reader is referred to the references for information on this topic.

APPENDIX: BINOMIAL PROBABILITY DISTRIBUTION

The formula for binomial probabilities, equation (20.2), is based on the following argument. Assume that on n independent observations in a binomial experiment the probability of obtaining a success is θ for each observation. Then the probability of obtaining exactly y successes and exactly $n - y$ failures is $\theta^y(1 - \theta)^{n-y}$. This probability is the same for every possible ordering of the successes and the failures. The total number of possible ways of obtaining y successes in n observations is, from combinatorial algebra,

$$\binom{n}{y} = \frac{n!}{y!(n-y)!}.$$

Multiplying $\theta^y(1-\theta)^{n-y}$ by the number of ways y successes can be obtained yields the binomial probability distribution (20.2).

A histogram of the binomial probability distribution would consist of vertical bars, each having height $\Pr\{y\}$ and centered on the value of y. The base of each bar would extend from $y - \frac{1}{2}$ to $y + \frac{1}{2}$, thereby making the area of the bars equal to their respective probabilities. A normal approximation to $\Pr\{y\}$ is obtained by finding the area under the normal curve between the limits $y - \frac{1}{2}$ and $y + \frac{1}{2}$. In general, if y follows a binomial distribution with parameters n and θ, and x follows a normal distribution with mean $\mu = n\theta$ and standard deviation $\sigma = \{n\theta(1-\theta)\}^{1/2}$, the normal approximation to $\Pr\{a \leqslant y \leqslant b\}$ is $\Pr\{a - \frac{1}{2} \leqslant x \leqslant b + \frac{1}{2}\}$.

REFERENCES

Text References

Most textbooks on general statistical methodology cover the techniques discussed in this chapter. The texts referenced in Chapter 12 all cover this material. A few additional references that summarize these procedures are:

Dixon, W. J. and Massey, F. J. (1969). *Introduction to Statistical Analysis*, New York: McGraw-Hill, Inc.

Natrella, M. G. (1966). *Experimental Statistics*, National Bureau of Standards Handbook 91, reprinted by John Wiley and Sons, Inc., New York.

Ostle, B. and Malone, L. C. (1988). *Statistics in Research, Fourth Edition*, Ames, IA: Iowa State University Press.

Snedecor, G. W. and Cochran, W. G. (1967). *Statistical Methods, Sixth Edition*, Ames, IA: Iowa State University Press.

Many types of count data can be analyzed using log-linear models. Log-linear models express the logarithms of the observed frequencies in terms of main effects and interactions as in the definition of ANOVA *models. A treatment of the analysis of count data using log-linear models can be found in the following texts:*

Agresti, A (1984). *Analysis of Ordinal Categorical Data*, New York: John Wiley and Sons, Inc.

Bishop, Y. M. M. Fienberg, S. E., and Holland, P. W. (1975). *Discrete Multivariate Analysis*, Cambridge, MA: The MIT Press.

Feinberg, S. E. (1977). *The Analysis of Cross-Classified Categorical Data*, Cambridge, MA: The MIT Press.

EXERCISES

1 In a study of liver infection, 35 laboratory mice were infected by inserting a specified quantity of parasites in the blood stream. Ten days later, 21 of the mice developed serious levels of infection in their livers. Use these sample data and the normal approximation to the binomial distribution to estimate the unknown proportion of

mice that can be expected to evidence infection of the liver when exposed to parasites as the mice in this experiment were.

2 In experiments similar to the one described in Exercise 1, small portions of the liver are treated and examined under a microscope in order to study the effects of the infection. It is known that in suitably small areas of the liver the chance of detecting a parasite is extremely small. Consequently, larger areas are examined and the parasites observed are carefully counted. Suppose that in one such examination, 15 parasites are counted. Use the square-root transformation (20.13) to place a confidence interval on the mean number of parasites that can be expected to be found in similar examinations of the livers.

3 In order to obtain better precision in the estimation of the mean parasite count, suppose the parasite counts on 12 of the mice are taken using the procedure described in Exercise 2. Use the counts below to construct a confidence interval on the mean parasite count.

Counts: 18, 23, 26, 11, 28, 22, 19, 15, 31, 25, 21, 24.

4 In a study of the presense of pollutants in the fish population of a lake near a metropolitan area, 70 fish were captured and examined. A total of 21 of these fish contained measurable amounts of a specific pollutant. Construct a confidence interval to estimate the proportion of fish in the lake that contain measurable amounts of this pollutant.

5 Using the data in Exercise 4, test the hypothesis that the proportion of fish in the lake that have measurable levels of the pollutant is no greater than 0.10. State all assumptions needed to make this test, and draw conclusions in the context of the exercise.

6 Electronic components for use in copying equipment are required to have a maintenance-free lifetime in excess of 2 years. The service department of a copier manufacturer surveyed its maintenance records for 100 components from each of six vendors and found that the component in question failed to meet the maintenance-free requirement as indicated below. Is there sufficient evidence to conclude that the proportion of electronics components that do not meet the maintenance-free requirement differs for the six vendors?

Vendors	No. of Failures
1	21
2	13
3	4
4	16
5	9
6	7

7 A major manufacturer is concerned with the quality of materials supplied by vendors. On one particular item, the manufacturer monitored the seven shipments from each

of two vendors. Each shipment consisted of 60 items. The number of defective items in each shipment from each of the manufacturers is shown below. Is there sufficient evidence in these data to conclude that proportions of defective products shipped by the two vendors differ?

	No. of Defective Items	
Shipment	Vendor 1	2
1	12	4
2	5	3
3	7	2
4	11	6
5	8	4
6	8	1
7	10	3

8 The following table classifies the numbers of tomato seeds that germinated in harsh soil conditions after being treated with one of five different types of soil nutrients. Is there a statistically significant difference in the effects of the soil treatments on the number of seeds that germinate?

Soil Nutrient	No. of Seeds	
	Germinated	Failed to Germinate
A	12	38
B	26	24
C	8	41
D	35	13
E	17	30

9 In a study of consumer satisfaction, an automobile manufacturer commissioned a survey of 60 owners of one of the manufacturer's automobiles and of a similar number of owners of vehicles of a comparable type manufactured by two of its competitors. Vehicle owners were asked to cite the problem area that caused the most dissatisfaction. The results are shown below. Are there any statistically significant differences in the areas of dissatisfaction for the owners of the three vehicle types?

		Manufacturer		
		A	B	C
	Body	2	11	8
Problem	Engine	16	21	38
Area	Transmission	35	17	6
	Interior	7	11	8
	Total	60	60	60

10 A safety study was conducted to assess the irritation due to a new cosmetic product proposed for use around the eyes. The product was administered in a clinical trial to each of 40 subjects. Two commercially available products currently in use were also tested as controls. The results are tabulated below. Analyze these data and assess whether any of the products significantly differ in the amount of irritation experienced by the participants in the study.

	No. of subjects		
	Product A	B	C
Irritation	8	3	12
No irritation	32	37	28
Total	40	40	40

11 An experiment was conducted to assess the ability of three chemical treatments to produce germination in hybicuramaglobumar seeds. The chemicals were each applied to 200 seeds and the number of seeds that germinated were counted. The respective manufacturers claim that 95% (Chemical A), 85% (Chemical B), and 90% (Chemical C) of the seeds to which their treatments are applied will germinate. Do the data below support the manufacturers' claims? Why (not)?

Chemical:	A	B	C
Number of Seeds Germinating:	176	154	191

PART IV

Fitting Data

CHAPTER 21

Linear Regression with One Variable

Linear regression is used to model one quantitative variable as a function of one or more other variables. In this chapter we introduce regression modeling with the fitting of a response variable as a linear function of one predictor variable. As with the introductory chapters to the previous parts of this book, general principles are introduced, principles which are elaborated in later chapters. The topics covered in this chapter include:

- *uses and misuses of regression modeling,*
- *a strategy for regression modeling,*
- *least-squares parameter estimation, and*
- *procedures for drawing inferences.*

Regression modeling is one of the most widely used statistical modeling techniques for fitting a quantitative response variable y as a function of one or more predictor variables $x_1, x_2, \ldots, x_p$. Regression models can be used to fit data obtained from a statistically designed experiment in which the predictor variables either are quantitative or are indicators of factor levels. All the ANOVA models described in earlier chapters of this book are special types of regression models. In ANOVA models the predictor variables are specially coded *indicator variables*, e.g., the effects representation in Tables 7.5 to 7.7, in which the upper level of a factor is indicated by a $+1$ and the lower level by a -1. ANACOVA models are also special types of regression models in which some of the variables are indicator variables and others are quantitative.

One of the reasons for the widespread popularity of regression models is their use with nonexperimental data, such as observational or historical data. Another reason is that the regression analysis procedures contain diagnostic techniques for (a) identifying incorrect specifications of the model (Chapter 25), (b) assessing the influence of outliers on the fit (Chapter 24), and (c) evaluating whether redundancies (collinearities) among the predictor variables are adversely affecting the estimation of the model parameters (Chapter 27). Perhaps most pragmatically, regression models are widely used because they often provide excellent fits to a response variable when the true functional relationship, if any, between the response and the predictors is unknown.

In this chapter we introduce regression modeling for the special case in which the response is to be modeled as a function of a single predictor variable. In the next section we introduce the particular form of regression model that we shall stress in this part of the text: the linear regression model. Nonlinear regression models will be discussed in Chapter 26.

21.1 USES AND MISUSES OF REGRESSION

Linear regression models are, apart from the random error component, linear in the unknown parameters (see Exhibit 21.1). The coefficients in the model (21.1) appear as either additive constants (β_0) or as multipliers on the predictor variable (β_1). This requirement extends to multiple linear regression models (see Chapter 22). Note too that the predictor variable can be a function, linear or nonlinear, of other predictor variables, e.g., $x = \ln z$ or $x = \sin z$ for some variable z. The predictor variable cannot, however, be a function of unknown parameters, e.g., $x = \ln(z + \phi)$ or $x = \sin(z - \phi)$ for some unknown ϕ. Models which are nonlinear functions of unknown model parameters are referred to as nonlinear models.

EXHIBIT 21.1

Linear Regression Models. Linear regression models that relate a response variable (y) to one predictor variable (x) are defined as

$$y_i = \beta_0 + \beta_1 x_i + e_i, \qquad i = 1, 2, \ldots, n, \tag{21.1}$$

where e_i represents a random error component of the model.

As with all the models discussed in this text, several assumptions usually accompany the model. These assumptions are listed in Table 21.1. It is important to note that violation of one or more of these assumptions can invalidate the inferences drawn on parameter estimates and can result in other, often not apparent, difficulties with fitting linear regression models using least-squares estimators (Section 21.4). Techniques that can be used to assess the reasonableness of these model assumptions are discussed in Chapter 25. Alternative estimators that are appropriate when some of these assumptions are violated are presented in Chapter 28.

The regression texts referenced at the end of this chapter contain much discussion on the appropriate uses and abuses of regression modeling. Table 21.2 lists a few of the more common of each. The uses of regression modeling will be amply illustrated throughout the next several chapters of this text. In the remainder of this section, we wish to comment on some of the more prevalent misuses listed in the table.

Figure 21.1 is a sequence plot for rotational velocities of a bearing as a function of

Table 21.1 Assumptions for Linear Regression Models

1. Over the range of applicability, the true relationship between the response and the predictor variable(s) is well approximated by a linear regression model.
2. The predictor variable(s) is (are) nonstochastic and measured without error.
3. The model errors are statistically independent and satisfactorily modeled by a normal distribution with mean zero and constant, usually unknown, standard deviation σ.

Table 21.2   Common Uses and Misuses of Regression Modeling

Common Uses

- Prediction (forecasting)
- Interpolation
- Data fitting
- Testing for significant factor or predictor variable effects on a response
- Determination of predictor values that maximize or minimize a response
- Control of the response variable by selection of appropriate predictor values

Common Misuses

- Extrapolation
- Misinterpreting spurious relationships
- Overreliance on automated-regression results
- Exaggerated claims about the validity of empirical models

time. Initially, the bearing was spun at a rate of 44.5 revs/sec. The data plotted in Figure 21.1 are the velocities recorded at 10-second intervals following the termination of the spinning of the bearing. The experiment was replicated six times; each plotted point is therefore the average of six velocities. Note that the data appear to follow a straight line; hence, a linear regression model of the form (21.1) should provide an excellent fit to the averages.

An appropriate use of a linear fit to the data in Figure 21.1 would be to describe the rotational velocities (the response) as a function of time (the predictor). The slope of the fit would be a useful estimate of the deceleration rate of the bearing. One could also use the fit to interpolate rotational velocities for any times between 10 and 100 seconds.

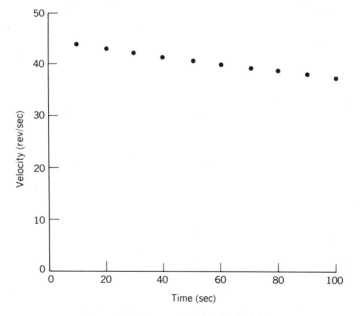

Figure 21.1   Average rotational velocities.

An inappropriate use of the fit would be to extrapolate the rotational velocities at, say, 120 or 150 seconds. Based solely on the data plotted in Figure 21.1, one cannot conclude that the rotational velocities remain linear outside the time frame plotted. One cannot tell from the data in the figure whether the rotational velocities continue to decelerate at a constant rate and then abruptly stop or the deceleration rate itself begins to slow down so that the velocities decrease in a nonlinear fashion. Thus, extrapolation of velocities is highly speculative because of the unknown nature of the deceleration rate beyond 100 seconds. Another extrapolation would be to conclude that the relationship depicted in Figure 21.1 holds for bearing types that are different from the one tested.

Another potential difficulty with the application of regression analysis to the data in Figure 21.1 concerns the possible violation of one of the model assumptions. Since the averages are calculated from individual velocities that are measured 10 seconds apart, it is likely that the measurement errors for the velocities on a single replicate of the experiment are correlated. If so, the averages plotted in Figure 21.1 are also correlated. Before either statistical interval estimates for model parameters or predictions are calculated or tests of hypotheses are conducted, the model assumptions must be investigated. If the correlations are found to be sufficiently large, time-series models or other alternative analyses should be conducted, rather than a regression analysis.

Another potential misuse of regression modeling is illustrated with the scatterplot in Figure 21.2. These data represent measurements taken on 21 operating days of a

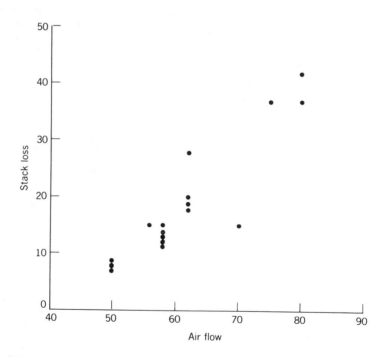

Figure 21.2   Stack loss versus air flow. Data from Brownlee, K. A. (1965). *Statistical Theory and Methodology in Science and Engineering*, New York: John Wiley & Sons, Inc. Copyright 1965 John Wiley & Sons, Inc. Used by permission.

chemical plant. The "stack loss" is ten times the percentage of ammonia that is left un-converted in an oxidation process for converting ammonia to nitric acid. The predictor variable in this plot is the rate of air flow in the process. It is apparent that the stack loss is increasing with the air flow. Having just discussed the dangers of extrapolation, it should be clear that a straight-line fit to these data should not be used to infer the effects of plant operating conditions outside the range of air-flow values plotted in the figure.

Often studies are conducted with nonexperimental data such as these with the intent to determine reasons for the trends. When analyzing experimental or nonexperimental data in which some of the variables are not controlled, extreme care must be taken to ensure that proper inferences are drawn when statistically significant results are obtained. In particular, one must be concerned about confounded predictor variables and spurious relationships.

Confounding was introduced in Section 9.1 in the context of designing fractional factorial experiments. In general, a factor is said to be confounded whenever its effect on the response cannot be uniquely ascribed to the factor. In the present context, a predictor variable is said to be confounded whenever its relationship with the response variable is unclear because of the predictor variable's relationship with one or more other predictors. Failure to recognize the potential for confounding of predictor variables is one of the ways in which exaggerated claims can be made about a fitted model.

Researchers sometimes conclude from trends such as that in Figure 21.2 or from

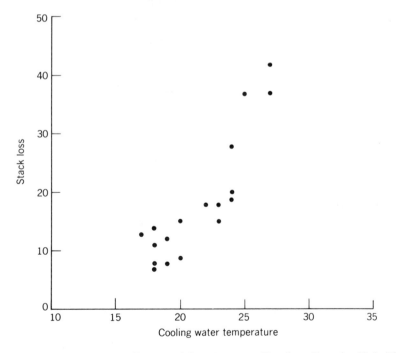

Figure 21.3 Stack loss versus cooling-water inlet temperature. Data from Brownlee, K. A. (1965), *Statistical Theory and Methodology in Science and Engineering*, New York: John Wiley & Sons, Inc. Copyright 1965 John Wiley & Sons, Inc. Used by permission.

regression fits to the data that increases in air flow rates are a major contributor to increases in the stack loss. While such a relationship may be true, the conclusion is not warranted solely on the basis of a regression fit to the data in the figure.

Figure 21.3 is a plot of the stack-loss values versus the inlet water temperature. Note that this plot suggests that increases in inlet water temperature rates are associated with increases in the stack loss. Without a controlled experiment that includes both air flow and inlet water temperature, as well as any other predictor variables that are believed to affect stack loss, no definitive conclusion can be drawn about which of the variables truly affects the stack loss.

With controlled experimentation, many of the abuses listed in Table 21.2 are less likely to occur than with observational or historical data. Since variables are controlled and assigned predetermined values in designed experiments, confounding of variables is usually lessened, sometimes completely absent.

21.2   REGRESSION VERSUS CORRELATION

Regression analysis is appropriate when the relationship between two or more variables is clearly in one direction—i.e., the design factors or the predictor variables influence but are not influenced by the response variable. Situations in which there are two or

Table 21.3   Estimates of Tread Life Using Two Measurement Methods

| | Tread Life (10^2 mi) | |
Tire No.	Weight-Loss Method	Groove-Depth Method
1	459	357
2	419	392
3	375	311
4	334	281
5	310	240
6	305	287
7	309	259
8	319	233
9	304	231
10	273	237
11	204	209
12	245	161
13	209	199
14	189	152
15	137	115
16	114	112

Sample Correlation Coefficient Calculation

x = weight-loss measurement
y = groove-depth measurement

$S_{xx} = 9073.59$ $S_{yy} = 6306.93$
$S_{xy} = 7170.07$ $r = 0.948$

more variables that are related, none of which can be said to cause or determine the values of the others, are analyzed using multivariate statistical methods. Multivariate ANOVA models and multivariate regression models are appropriate when the design factors or predictor variables simultaneously influence the values of two or more response variables.

A correlation analysis is used to assess linear relationships between two variables when both are random and neither variable is believed to cause or determine values of the other. An example of the type of data for which a correlation analysis is appropriate is the tire-wear data plotted in Figure 6.5 and listed in Table 21.3. The linear scatter observable in Figure 6.5 is not due to one of the methods of measuring wear influencing the other method. Rather, both measurements are affected by differences in tires and the road conditions under which each tire is tested.

A measure of the strength of a linear association between two random variables is the sample *correlation coefficient*—more precisely, Pearson's product-moment correlation coefficient, or Pearson's r. It is defined in Exhibit 21.2.

EXHIBIT 21.2   CALCULATION OF PEARSON'S r
(THE SAMPLE CORRELATION COEFFICIENT)

1. Calculate the two sample means, $\bar{x}$ and $\bar{y}$; e.g.,

$$\bar{x} = n^{-1}\sum x_i.$$

2. Calculate the two sample variances, s_{xx} and s_{yy}; e.g.,

$$s_{xx} = (n-1)^{-1}\sum (x_i - \bar{x})^2$$
$$= (n-1)^{-1}\left(\sum x_i^2 - n\bar{x}^2\right).$$

3. Calculate the sample covariance, s_{xy}:

$$s_{xy} = (n-1)^{-1}\sum (x_i - \bar{x})(y_i - \bar{y})$$
$$= (n-1)^{-1}\left(\sum x_i y_i - n\bar{x}\bar{y}\right).$$

4. Calculate the sample correlation coefficient:

$$r = s_{xy}/(s_{xx}s_{yy})^{1/2}. \tag{21.2}$$

The value of Pearson's r for the tire-wear data can be calculated using equation (21.2) and the summary statistics given in Table 21.3. Doing so, one obtains $r = 0.948$. This value of Pearson's r can be interpreted with the aid of the properties listed in Exhibit 21.3.

EXHIBIT 21.3   PROPERTIES OF PEARSON'S r

- Pearson's r measures the strength of a linear association between two quantitative variables.
- The closer r is to zero, the weaker is the linear association.
- The closer r is to -1 or $+1$, the stronger is the linear association, with the sign of r indicating a decreasing ($r < 0$) or an increasing ($r > 0$) linear association.
- The sample correlation coefficient can only equal $+1$ (-1) when the observations lie exactly on a straight line having positive (negative) slope.

Figure 21.4 is a series of scatterplots of pairs of variates that have different positive values of the sample correlation coefficient. Variates having negative correlations would have similar scatterplots with downward rather than upward trends. Thus, the size of the sample correlation coefficient for the tire-wear data indicates that the linear relationship between the two measures of tire wear is extremely strong.

The statistical significance of the sample correlation coefficient can be assessed using a t-statistic. Let ρ denote the correlation coefficient for a population of pairs of quantitative observations. Alternatively, ρ could denote the correlation coefficient for a conceptual population of pairs of observations arising from measurements on some process. The following t statistic can be used to test $H_0 : \rho = 0$ vs $H_a : \rho \neq 0$,

$$t = \frac{r(n-2)^{1/2}}{(1-r^2)^{1/2}}. \tag{21.3}$$

Values of $|t|$ that exceed a two-tailed t-value from Table A5 in the appendix lead to rejection of the null hypothesis. The degrees of freedom of the t-statistic are $v = n - 2$. One-sided tests can also be made using (21.3) and the appropriate one-sided t-value from this table.

The t-statistic for the tire-wear measurements is $t = 11.14$. The t value from Table A5 corresponding to a two-sided 5% significance level and $v = 14$ degrees of freedom is 2.145. Thus the correlation coefficient is statistically significant, leading to the conclusion that there exists a linear relationship between the two tire-wear measurements. The magnitude of the t-statistic and the scatterplot in Figure 6.5 indicate that the linear relationship between the variates is a strong one.

In order to test for nonzero values of a population correlation coefficient, a different test statistic must be used. If one wishes to test $H_0 : \rho = c$ vs $H_a : \rho \neq c$, where $|c| < 1$, the following test statistic should be used:

$$z = (n-3)^{1/2}(\tanh^{-1} r - \tanh^{-1} c), \tag{21.4}$$

where $\tanh^{-1}(r)$ is the inverse hyperbolic-tangent function,

$$\tanh^{-1} r = \tfrac{1}{2} \ln \frac{1+r}{1-r}.$$

The statistic z in (21.4) is approximately distributed as a standard normal random variable under the null hypothesis. Using critical values for two-tailed tests from Table A4 of the appendix, one can test the above hypothesis. One-tailed tests are conducted in a similar fashion.

Confidence intervals for correlation coefficients can be obtained using the statistic z in equation (21.4). One uses the standard normal distribution of z to place confidence limits on $\tanh^{-1} \rho$ and then uses the hyperbolic-tangent function to transform these limits into limits on ρ. The interested reader is referred to the exercises for further details.

A nonparametric alternative to Pearson's r, which can be used when variables are not normally distributed, is called *Spearman's rank correlation coefficient* (see Exhibit 21.4). This measure of association between two variables is calculated using the formula for Pearson's r, but the numerical values of the variables are replaced by their ranks. The calculation of the rank correlation coefficient is not restricted to numerical variables. It can be calculated on any variables for which ranks can be meaningfully assigned to the

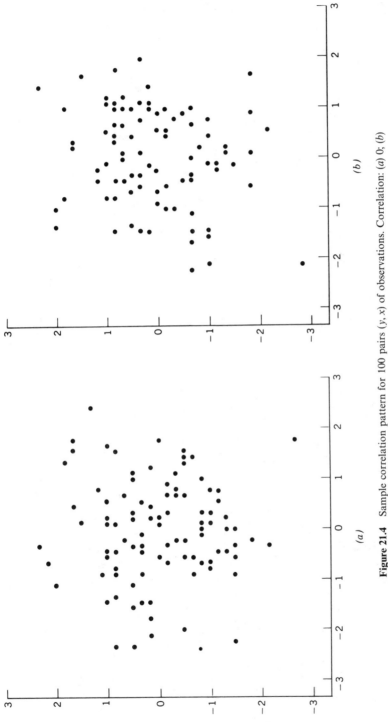

Figure 21.4 Sample correlation pattern for 100 pairs (y, x) of observations. Correlation: (*a*) 0; (*b*) 0.25; (*c*) 0.50; (*d*) 0.75; (*e*) 0.90; (*f*) 0.99.

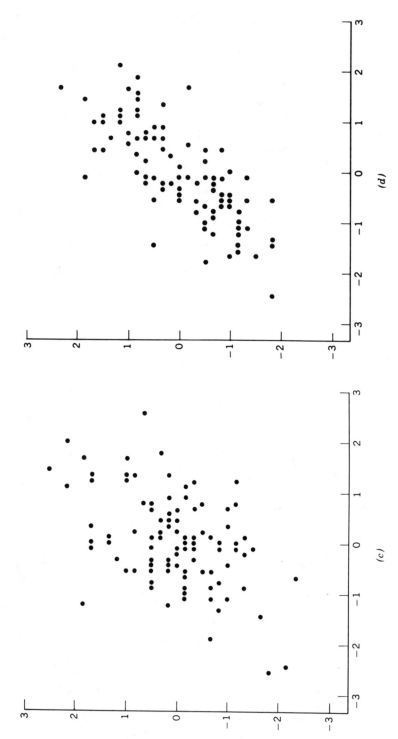

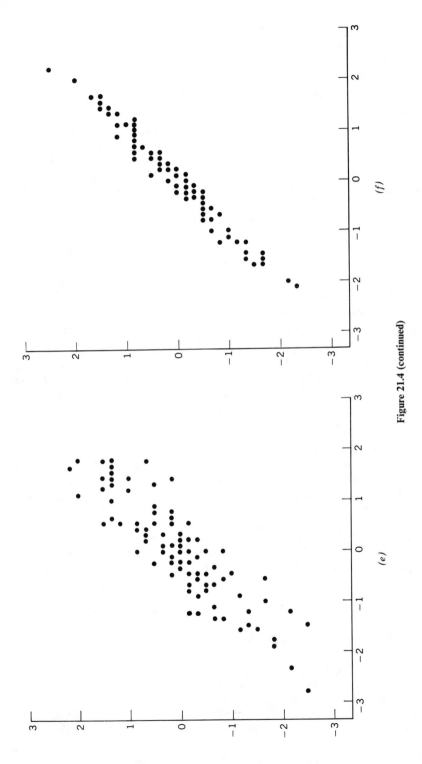

(e)

(f)

Figure 21.4 (continued)

variable values. Data from preference tests or rankings from expert panels are just two examples of nonnumerical ranked data for which the rank correlation coefficient can be calculated.

EXHIBIT 21.4   RANK CORRELATION COEFFICIENT

1. Order the observations on one of the variables (the x's) from smallest to largest. Replace the ordered observations by their ranks from 1 to n. Do the same for the observations on the second variable (the y's).

2. For tied observations on each variable, assign the average of the ranks that would have been assigned had there been no ties.

3. For each pair of observations (x_i, y_i) calculate $d_i = R(x_i) - R(y_i)$, where $R(x_i)$ denotes the rank of x_i among the x's and $R(y_i)$ denotes the rank of y_i among the y's.

4. If there are no ties, Spearman's rank correlation coefficient is

$$r_s = 1 - \frac{6\sum d_i^2}{n(n^2 - 1)}. \tag{21.5}$$

If there are ties, calculate r_s using the formula for Pearson's r, equation (21.2), using the assigned ranks instead of the actual variable values.

Table A25 in the appendix contains upper-tail critical values for the rank correlation coefficient r_s. Lower-tail critical values are the negatives of the tabulated ones. These critical values can be used with r_s to test for the independence of the two sets of ranks. This is a nonparametric test of independence of the two sets of observations. When the sample size exceeds the values in Table 25, an approximate standard normal test statistic is

$$z = r_s(n - 1)^{1/2}.$$

To illustrate the use of the rank correlation coefficient, consider testing the independence of the tread-life measurements in Table 21.3. If each set of measurements is replaced by ranks, $r_s = 0.926$. From Table A25, this value of the rank correlation coefficient exceeds the 0.001 upper-tail critical value for $n = 16$ observations. The null hypothesis of independence of the two methods of measuring tread life is therefore rejected.

As mentioned above, the use of Pearson's r is appropriate only when both of the variates are random variables. Pearson's r can be calculated when one or both of the variables are not random, but none of the inference procedures discussed in this section should be used to draw conclusions about a linear relationship between the variables. The calculated value of r is simply a measure of the degree of linearity in the observed data values.

21.3   THE STRATEGY OF REGRESSION ANALYSIS

A comprehensive regression analysis involves several different phases, each of which is important to the successful fitting of regression models. We describe these phases by the acronym ISEAS: investigate, specify, estimate, assess, select. Table 21.4 summarizes some of the components of each of these phases. In this section we briefly describe them. Subsequent chapters explore each more fully.

The first step in a regression analysis is to investigate (I) the data base. A listing of the data base, especially if it is entered into a computer file, should be scanned in order to detect any obvious errors in the data entry—e.g., negative values for variates that can only be positive, or decimal points entered in the wrong position. Erroneous or incorrectly coded data values can have a serious effect on the estimation of the model parameters; consequently, the detection of outliers at this stage of the analysis can save much time and effort later on and can lessen the chance of erroneous inferences due to outlier effects.

Along with the scanning of the data base, summary statistics for each of the variables should be computed. Incorrectly coded or erroneous data values often can be identified by examining the maximum and minimum data values for each variable. This is especially important when data bases are so large that scanning the individual data values is difficult.

Plotting response and predictor variables is another important step in the first phase of a regression analysis. A sequence plot of the response variable against its run number or against a time index can often identify nonrandom patterns in the experiment. For example, abrupt shifts could indicate unforeseen or unplanned changes in the operating conditions of an experiment. Cyclic trends could indicate a time dependence among the responses.

The response and predictor variables should also be plotted against one another. Such scatterplots may indicate that one or more of the variates need to be reexpressed prior to the fitting of a regression model. For example, plotting the response variable against each of the predictor variables may indicate the need to reexpress the response variable if several of the plots show the same nonlinear trend. If only one or two of the plots show nonlinear trends, the predictors involved should be reexpressed.

Box plots and stem-and-leaf plots are valuable aids for identifying possible outliers in a single variable. Scatterplots are useful for identifying combinations of data values that might not be apparent from an examination of each variable separately. In a scatterplot an observation might appear in a corner without having an extremely large or small value on either of the variates.

The second stage of a regression analysis is to initially specify (S) the form of the regression model. Not only must functional forms for each of the variables be specified, but one must consider whether interaction terms, polynomial terms, or nonlinear functions of the predictor variables should be included in the model.

The specification of a regression model relies heavily on one's knowledge of the experiment or process being studied. Not only does such knowledge aid one in selecting the variables to be included in the experiment, and thus in the model, but it is also important when questions concerning the functional form of the model are discussed. Plots of the variables, as suggested above, can aid in the initial specification of the model, especially when an experiment is being conducted or a process is being studied for which little is known about the effects of variables on the response.

Ordinarily, if no indications to the contrary are shown by plots or by one's knowledge of the problem being studied, all variables are entered into the model in a linear fashion with no transformations. The model (21.1) is an example of the specification for a single predictor variable.

The next stage in a regression model is to estimate (E) the model parameters, ordinarily using appropriate computer software. In addition to just the estimates of the model parameters, regression programs usually provide statistics that assist one in assessing the fit. Thus the estimation stage leads directly to the assessment stage of the analysis.

When assessing (A) the adequacy of the model fit one must consider several questions.

Table 21.4   Components of a Comprehensive Regression Analysis

---

Investigate.   Investigate the data base; calculate summary statistics; plot the variables.

Specify.   Specify a functional form for each variable; formulate an initial model, reexpressing variables as needed.

Estimate.   Estimate the parameters of the model; calculate statistics which summarize the adequacy of the fit.

Assess.   Assess the model assumptions; examine diagnostics for influential observations and collinearities.

Select.   Select statistically significant predictor variables.

Are the model assumptions reasonable (Chapter 25), including the functional form of the response and predictor variables (Sections 22.1, 25.2, 26.1)? If so, does the model adequately fit the data (Sections 21.4, 22.2)? Can the errors be considered to be normally distributed with zero means and constant standard deviations (Sections 25.1, 25.3)? Are any observations unduly influencing the fit (Chapter 24)?

Following the assessment of the specified model, perhaps following a respecification of the model or some other remedial action, one proceeds to a selection (s) of the individual predictors (Chapter 27). This is done because often one is unsure, prior to the experiment, which of the predictor variables affect the response.

As indicated throughout this discussion, each of these topics is explored more fully in later sections of this chapter and in subsequent chapters. Table 21.4 is intended to provide a general scenario for the analysis of data using regression models. One's decision at one stage of the analysis can cause the reevaluation of conclusions reached at a previous stage. For example, if an observation is deleted because it has an undue influence on the coefficient estimate of a predictor variable and then that variable is deleted from the model at a later stage of the analysis, one should reevaluate the need to delete the observation. Thus the process is often an iterative one.

21.4 LEAST-SQUARES ESTIMATION

The linear regression model used to relate a response variable to a single predictor variable was defined in equation (21.1). The assumptions that ordinarily accompany this model are given in Table 21.1. We now wish to consider the estimation of the model parameters. Since we focus on several different aspects of fitting regression models in this and the next section, we break each into several subsections.

21.4.1 Intercept and Slope Estimates

The model (21.1) can be written in a form similar to the ANOVA models that were discussed in Chapters 15–19:

$$y_i = \mu_i + e_i, \qquad i = 1, 2, \ldots, n. \tag{21.6}$$

This expression of the model coupled with the assumptions in Table 21.1 imply that the responses are independent normal random variables with means μ_i and common standard deviation σ. On comparing the model (21.6) with the model (21.1) it is apparent that the means of the responses are

$$\mu_i = \beta_0 + \beta_1 x_i. \tag{21.7}$$

Figure 21.5 depicts the linear relationship between the response μ and the predictor variable x, as well as the variability in the observed response y associated with the normally distributed errors for each value of x.

Let b_0 and b_1 denote some estimators of the intercept and slope parameters. Insertion of these estimators into (21.7) provides estimators of the means for each of the n responses. We term the equation

$$\hat{y} = b_0 + b_1 x \tag{21.8}$$

the *fitted* regression model. Note that the fitted model is an estimator of the mean of the regression model, i.e., the deterministic portion of the model (21.1). When the n predictor-variable values x_i are inserted into the prediction equation (21.8), the resulting quantities are not only estimates of the respective means μ_i, but also of the actual responses y_i. The latter use of the fitted model results in the estimates being referred to as *predicted* responses. The differences between actual and fitted responses are termed *residuals*, r_i:

$$r_i = y_i - \hat{y}_i = y_i - b_0 - b_1 x_i. \tag{21.9}$$

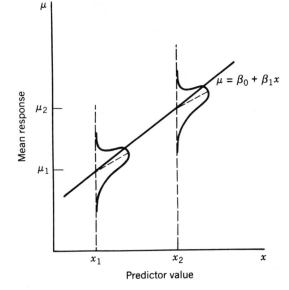

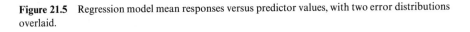

Figure 21.5 Regression model mean responses versus predictor values, with two error distributions overlaid.

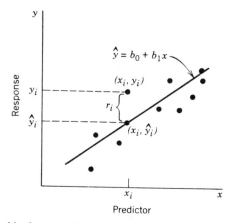

Figure 21.6 Relationships between observed responses (y_i), predicted responses ($\hat{y}_i$), and residuals (r_i).

Figure 21.6 shows the relationship among the actual responses, fitted responses, and residuals.

The residuals (21.9) provide a measure of the closeness of agreement of the actual and the fitted responses; hence, they provide a measure of the adequacy of the fitted model. The sum of the squared residuals, SS_E, is an overall measure of the adequacy of the fitted model:

$$SS_E = \sum r_i^2 = \sum (y_i - b_0 - b_1 x_i)^2. \tag{21.10}$$

Note that SS_E is a function of the intercept and slope parameter estimators. A reasonable criteria for selecting intercept and slope estimators is to choose those that minimize the residual sum of squares. This is the principle of *least squares*, and the resulting estimators are called *least-squares estimators* (see Exhibit 21.5).

EXHIBIT 21.5   LEAST-SQUARES ESTIMATORS

1. Assume a linear regression model of the form (21.1).

2. Let $\hat{y}_i = b_0 + b_1 x_i$ denote predicted responses using some estimators b_0 and b_1 of the intercept and slope. Denote the residuals by $r_i = y_i - \hat{y}_i$.

3. Least squares estimators b_0 and b_1 minimize the sum of the squared residuals, $SS_E = \sum r_i^2$. The least-squares intercept and slope estimators are

$$b_0 = \bar{y} - b_1 \bar{x} \quad \text{and} \quad b_1 = s_{xy}/s_{xx}, \tag{21.11}$$

where $\bar{y}$ and $\bar{x}$ are the sample means of the response and the predictor variables, respectively, and

$$s_{xy} = (n-1)^{-1} \sum (y_i - \bar{y})(x_i - \bar{x}),$$
$$s_{xx} = (n-1)^{-1} \sum (x_i - \bar{x})^2.$$

Table 21.5 Acid-Content Data and Least-Squares Fit*

Acid Number x	Acid Content y	Predicted Acid Content	Residual
123	76	75.01	0.98
109	70	70.51	−0.51
62	55	55.40	−0.40
104	71	68.91	2.09
57	55	53.79	1.21
37	48	47.36	0.64
44	50	49.61	0.39
100	66	67.62	−1.62
16	41	40.60	0.40
28	43	44.46	−1.46
138	82	79.84	2.16
105	68	69.23	−1.23
159	88	86.59	1.41
75	58	59.58	−1.58
88	64	63.76	0.24
164	88	88.20	−0.20
169	89	89.81	−0.81
167	88	89.17	−1.17
149	84	83.38	0.62
167	88	89.17	−1.17

*$b_0 = 35.458$, $b_1 = 0.322$.
Data from Daniel, C. and Wood, F. S. (1971), *Fitting Equations to Data*, New York: John Wiley & Sons, Inc., Copyright 1971 John Wiley & Sons, Inc. Used by permission.

Table 21.5 lists two measurements of the acid content of a chemical. The measurements of the response variable are obtained by a fairly expensive extraction and weighing procedure. The measurements of the predictor variable are obtained from an inexpensive titration method. The measurement error in the titration method is believed to be negligible. A scatterplot of the two variables indicates that a straight line should provide an excellent fit to the data.

Least-squares intercept and slope parameter estimates are also shown in Table 21.5. Because the linear fit approximates the observed data points so well, the residuals are all very small. Small residuals are one important indicator of the adequacy of a regression fit.

21.4.2 Interpreting Least-Squares Estimates

The least-squares intercept estimate is usually interpreted as the predicted response associated with the zero level of the predictor variable. Geometrically, it is the point where the fitted line crosses the vertical axis, when $x = 0$. Interpreting the intercept in this manner is an extrapolation when the predictor-variable values in the data set do not include the origin.

The slope is usually interpreted as the change in the fitted response associated with a unit (1) change in the predictor variable:

$$[b_0 + b_1(x + 1)] - [b_0 + b_1 x] = b_1.$$

This interpretation is usually adequate, but sometimes requires modification. We point out one occasional misinterpretation in this section and another one in Section 27.4.

The sample correlation coefficient between the response variables y_i and the least-squares predicted responses $\hat{y}_i = b_0 + b_1 x_i$ is related to the least-squares slope estimator (21.11). If r is calculated as in (21.2) using $\hat{y}_i$ instead of x_i, the relationship is as follows:

$$b_1 = r(s_{yy}/s_{xx})^{1/2}. \tag{21.12}$$

Furthermore, since $\hat{y}_i$ and x_i are linearly related, the calculation of r using $\hat{y}_i$ gives the same result as the calculation using x_i in (21.2). Note that the slope estimate and the correlation coefficient are equal when the standard deviations of the two variables are the same. Whenever the standard deviations are the same the least-squares slope estimate will be less than 1 (in absolute value), since $b_1 = r$.

Occasionally researchers draw incorrect conclusions because they do not understand or know about the relationship (21.12). For example, in pre- and posttesting of the same subjects prior to and following the administration of medical or psychological treatment, researchers often fit regression lines in order to assess the effectiveness of the treatment. An observed slope that is less than one is sometimes interpreted to mean that the treatment has lowered the posttest measurements from the pretest measurements.

Such an effect is to be expected, regardless of whether the treatment is effective, since the variability of the test scores is likely to be similar for the two sets of scores. Thus the slope is less than one only because it is approximately equal to the sample correlation coefficient. The supposed effect occurs because of the mathematical relationship (21.12) and not because of a real treatment effect. This phenomenon is referred to as the "regression effect" or "regression to the mean effect," and its (erroneous) interpretation as the "regression fallacy."

21.4.3 No-Intercept Models

Least-squares estimators can be calculated for models that do not have intercepts (see Exhibit 21.6). Only minor modification of equation (21.11) is required. These models are appropriate when (a) the response is believed to be zero when the predictor variable is zero and (b) the true model is believed to be linear from the origin to the range covered by the values of the predictor variable contained in the data set. The second condition

---

EXHIBIT 21.6   LEAST SQUARES ESTIMATORS: NO-INTERCEPT MODELS

1. Assume a linear regression model of the form (21.1), but without the constant term b_0.
2. Let $\hat{y}_i = b_1 x_i$ denote predicted responses using some estimator b_1 of the slope. Denote the residuals by $r_i = y_i - \hat{y}_i$.
3. The least-squares estimator b_1 minimizes the sum of the squared residuals, $SS_E = \sum r_i^2$. The least-squares slope estimator is

$$b_1 = s_{xy}/s_{xx},$$

where s_{xy} and s_{xx} are not adjusted for the averages $\bar{y}$ and $\bar{x}$:

$$s_{xy} = n^{-1}\sum y_i x_i \quad \text{and} \quad s_{xx} = n^{-1}\sum x_i^2.$$

---

is sometimes difficult to ensure when theoretical models are not known in advance of the data collection. In such cases it is wise to include an intercept term in the model and then test for its statistical significance.

21.4.4   Model Assumptions

The calculation of least-squares estimates does not require the use of any of the assumptions listed in Table 21.1. Least-squares estimates can be obtained for any data set, regardless of whether the model is correctly specified. Statistical properties of the intercept and slope estimators, however, do depend on the assumptions that accompany the model. In particular, if one wishes to draw inferences about the model parameters or to make predictions using the fitted model, the model assumptions may be of critical importance.

The last abuse of regression modeling listed in Table 21.2 is of great concern because of the ability to calculate least-squares estimates for any regression data. In the next section, a number of inferential techniques, from examination of simple descriptive statistics to testing hypotheses about model parameters, will be discussed. All too frequently only the descriptive statistics are examined when the adequacy of the fit is assessed. The references at the end of this chapter contain numerous examples of how such a superficial examination can lead to misleading conclusions. We again stress the need to perform a comprehensive regression analysis using all the steps (ISEAS) listed in Table 21.4 in order to ensure that proper conclusions about the relationship between the response and the predictor variables are drawn.

21.5   INFERENCE

Inference on regression models can take the form of simply interpreting the numerical values of the intercept and slope estimates or the construction of tests of hypotheses or confidence intervals. In this section we present several inferential techniques that can be used with regression models.

21.5.1   Analysis-of-Variance Table

The same procedures used to derive the ANOVA table in Section 15.1. can be used to derive similar tables for regression models. We again use as a measure of the total variation in the response variable the total sum of squares:

$$\text{TSS} = \sum (y_i - \bar{y})^2. \tag{21.13}$$

The error sum of squares is again equal to

$$\text{SS}_E = \sum (y_i - \hat{y}_i)^2. \tag{21.14}$$

The model sum of squares is also the same, but it is now referred to as the regression sum of squares:

$$\text{SS}_R = \sum (\hat{y}_i - \bar{y})^2 = b_1^2 \sum (x_i - \bar{x})^2. \tag{21.15}$$

Table 21.6   Symbolic ANOVA Table for Regression Models Having One Predictor Variable

Source of Variation	df	Sum of Squares	Mean Squares	F- Value	p- Value
Regression	1	SS_R	$MS_R = SS_R$	$F = MS_R/MS_E$	p
Error	$n-2$	SS_E	$MS_E = SS_E/(n-2)$		
Total	$n-1$	TSS			

A symbolic ANOVA table for regression models having a single predictor variable is shown in Table 21.6. The source of variation labeled "Regression" reflects the variability in the response variable that can be attributed to the predictor variable. There is only one degree of freedom for regression, because only one coefficient, namely β_1, must be estimated to obtain the regression sum of squares (21.15).

The estimate of the standard deviation, $s_e = (MS_E)^{1/2}$, is a measure of the adequacy of the fitted regression line. It is important to realize that the standard deviation of the error terms in the model (21.1) measures the variability of the observations about the regression line. Thus an estimated standard deviation that is small relative to the average response indicates that the observed responses are tightly clustered around the fitted line, just as a small model standard deviation σ indicates that the response values are clustered tightly around the mean regression line (21.7).

The estimate s_e is model-dependent. The value of s_e measures not only the uncontrolled variation of the responses but also contributions due to any model misspecification that may occur. This *lack of fit* of the data to the assumed model can be a substantial portion of the magnitude of SS_E and hence of s_e. In Section 22.3 we describe a technique for assessing possible lack of fit of a regression model when repeat observations are available for each of several values of the predictor variable. Residual-based techniques for assessing model misspecification when repeat observations are not available are detailed in Section 25.2.

The sample correlation r between the actual and fitted responses is another measure of the adequacy of the fit. The square of this sample correlation is readily computed from the statistics in the analysis of variance table and is termed the *coefficient of determination* (R^2):

$$R^2 = [\mathrm{corr}(y, \hat{y})]^2 = 1 - \frac{SS_E}{TSS}. \tag{21.16}$$

The coefficient of determination could be calculated directly by calculating Pearson's r between the observed and predicted responses, but the above formula is simpler and less subject to roundoff error. Coefficients of determination are usually expressed as a percentage by multiplying (21.16) by 100. Then R^2 values lie between 0 and 100%. The closer they are to the upper bound, the better is the fit.

Although coefficients of determination are very popular measures of the adequacy of the fitted model, there is a tendency to overrely on them because of their straightforward interpretation. Coefficients of determination can be made arbitrarily large even when the fit to the data is poor. For example, a single observation, if sufficiently removed from the bulk of the data, can produce very large R^2 values even though there is no linear association for the majority of the observations in the data base. For this reason, we

Table 21.7 ANOVA Table for the Acid-Content Data

Source of Variation	df	Sum of Squares	Mean Squares	F-Value	p-Value
Regression	1	5071.57	5071.57	3352.33	0.000
Error	18	27.23	1.51		
Total	19	5098.80			

again stress that a comprehensive regression analysis (ISEAS) is required before satisfactory conclusions can be drawn about the adequacy of the fit.

The F-statistic in the ANOVA table allows one to test the statistical significance of the slope parameter. Unlike tests for the significance of factor effects in the analysis of designed experiments, the test for significance of the slope parameter is ordinarily conducted using a large significance level, say $\alpha = 0.25$. The change in the significance level is due to the desire to protect against Type II errors (see Section 12.4), i.e., erroneously concluding that the predictor variable is not useful in modeling the response. Even though the Type II error probability is unknown (because the true value of β_1 is unknown), we seek to reduce it by using a large significance level.

The ANOVA table for the acid content data in Table 21.5 is shown in Table 21.7. Note that the F-statistic is highly significant. The estimated model standard deviation is $s_e = 1.23$. The coefficient of determination, $R^2 = 99.47\%$, also indicates that the prediction equation fits the observed responses well.

21.5.2 Tests and Confidence Intervals

A t-statistic can be constructed for testing $H_0:\beta_1 = c$ vs $H_a:\beta_1 \neq c$, where c is a specified constant, using the general procedures outlined in Section 12.1. From the assumptions listed in Table 21.1, it follows that the least-squares slope estimator b_1 has a normal distribution with mean β_1 and standard error $\sigma/[\sum(x_i - \bar{x})^2]^{1/2}$. The following statistic has a Student t-distribution with $n - 2$ degrees of freedom:

$$t = \frac{b_1 - \beta_1}{s_e/s_{xx}^{1/2}}, \tag{21.17}$$

$$s_{xx} = \sum(x_i - x)^2.$$

Inserting c for β_1 in (21.17) provides the t-statistic that is suitable for testing the above hypotheses.

Inserting $c = 0$ in (21.17) and squaring the resulting t-statistic yields the F-statistic from the ANOVA table, Table 21.6. This t-variate can also be used to form confidence intervals on the slope parameter. Using the same type of procedures discussed in Section 12.2, the following limits for a $100(1 - \alpha)\%$ confidence interval for β_1 can be derived:

$$b_1 - t_{\alpha/2}\frac{s_e}{s_{xx}^{1/2}} \leq \beta_1 \leq b_1 + t_{\alpha/2}\frac{s_e}{s_{xx}^{1/2}}, \tag{21.18}$$

where $t_{\alpha/2}$ is a $100(\alpha/2)\%$ upper-tail t critical value having $n - 2$ degrees of freedom.

For the acid-content data, a 95% confidence interval for β_1 using (21.18) is

$$-0.322 \pm 2.101 \frac{1.230}{(54,168.2)^{1/2}},$$

or

$$-0.311 \leqslant \beta_1 \leqslant -0.333.$$

The tightness of this interval is in part due to the small estimated model standard deviation, $s_e = 1.230$.

Tests of hypotheses and confidence intervals for the intercept parameter can be made using the following t-variate:

$$t = \frac{(b_0 - \beta_0)}{s_e(n^{-1} + \bar{x}^2/s_{xx})^{1/2}}. \qquad (21.19)$$

21.5.3   No-Intercept Models

When the regression model does not contain an intercept term, modifications of several of the above statistics are required. The total sum of squares is now unadjusted, and the regression sum of squares is not adjusted for the average of the predictor variables:

$$\text{TSS} = \sum y_i^2 \quad \text{and} \quad \text{SS}_R = b_1^2 \sum x_i^2. \qquad (21.20)$$

Likewise the number of degrees of freedom for error and the total number of degrees of freedom are now $n-1$ and n, respectively. The t-variate that is used to test for the significance of the slope parameter and to form confidence intervals is adjusted by replacing s_{xx} in (21.17) with

$$s_{xx} = \sum x_i^2 \qquad (21.21)$$

and using $n-1$ degrees of freedom rather than $n-2$.

21.5.4   Intervals for Responses

It is often of interest to form confidence intervals around $\mu = \beta_0 + \beta_1 x$ for a fixed value of x. Under the assumptions listed in Table 21.1, the predicted response $\hat{y}$ has a normal probability distribution with mean $\mu = \beta_0 + \beta_1 x$ and standard deviation $\sigma[a_1 + (x - a_2)^2/s_{xx}]^{1/2}$, where

$$a_1 = n^{-1}, \quad a_2 = \bar{x}, \text{ and } s_{xx} = \sum(x_i - \bar{x})^2 \qquad \text{for intercept models}$$

and

$$a_1 = 0, \quad a_2 = 0, \text{ and } s_{xx} = \sum x_i^2 \qquad \text{for no-intercept models.}$$

Thus the following t-variate can be used to form confidence intervals for an expected response μ:

$$t = \frac{\hat{y} - \mu}{s_e[a_1 + (x - a_2)^2/s_{xx}]^{1/2}}. \tag{21.22}$$

If one wishes to form prediction intervals for an actual future response, not the expected value of a response, equation (21.22) can again be used if one replaces μ by y_f and a_1 by $1 + a_1$. In this formulation y_f represents the future response and $\hat{y}$ its predicted value. The change from a_1 to $1 + a_1$ in the formula for the standard error occurs because the future response has a standard deviation σ, an added component of variability to that of the predicted response $\hat{y}$. Thus the standard deviation of $\hat{y} - y_f$ is $\sigma\{1 + a_1 + (x - a_2)^2/s_{xx}\}^{1/2}$.

REFERENCES

Text References

The are many excellent texts on regression analysis. Among the more current, comprehensive texts are the following:

Daniel, C. and Wood, F. S. (1971). *Fitting Equations to Data*, New York: John Wiley and Sons, Inc.

Draper, N. R. and Smith, H. (1981). *Applied Regression Analysis, Second Edition*, New York: John Wiley and Sons, Inc.

Gunst, R. F. and Mason, R. L. (1980). *Regression Analysis and Its Application: A Data-Oriented Approach*, New York: Marcel Dekker, Inc.

Montgomery, D. C. and Peck, E. (1982). *Introduction to Linear Regression Analysis*, New York: John Wiley and Sons, Inc.

Myers, R. H. (1986). *Classical and Modern Regression with Applications*, Boston, MA: PWS and Kent Publishing Co.

Neter, J., Wasserman, W., and Kutner, M. H. (1983). *Applied Regression Analysis*, Homewood, IL: Irwin Publishing Co.

Younger, M.S. (1985). *A First Course in Linear Regression, Second Edition*, Boston, MA: Duxbury Press.

Further information on inference procedures for correlation coefficients can be found in the general statistics texts referenced at the end of our Chapter 13, e.g., Ostle and Malone (1988).

Data References

The tire-wear data are taken from Natrella (1963); see the references at the end of Chapter 6. The stack-loss data are taken from:

Brownlee, K. A. (1965). *Statistical Theory and Methodology in Science and Engineering, Second Edition*, New York: John Wiley and Sons, Inc.

The acid content data are taken from Daniel and Wood's text, referenced above.

EXERCISES

1 An electronics firm is interested in documenting a relationship between sales price (y) of one of its videocassette recorders (VCR) and its factory-authorized discount (x) used in promotions of the VCR. The firm authorizes different discounts over a 15-week period and determines the average sales price of the recorders sold during each week. The data are shown below:

Week	Price ($)	Discount ($)	Week	Price ($)	Discount ($)
1	320	40	9	340	15
2	320	46	10	335	25
3	315	48	11	305	50
4	310	49	12	350	13
5	340	14	13	325	40
6	315	45	14	350	15
7	335	20	15	330	25
8	335	30			

Investigate the data set. Pay special attention to the presence of extreme data values and the possible need for transformations of the variables.

2 Using the data in Exercise 1, estimate the least-squares intercept and slope coefficients. Plot the fit on a scatterplot of the data values. Is the fit visually satisfactory? Why (not)?

3 An engineering research firm is conducting an experiment to compare the strength of various metals under dynamic load conditions. A cylindrical weight is allowed to drop freely down a rod and hit a spring at the base of the rod. Thin rectangular strips of different metals are attached to the weight so that they extend radially outward from the weight perpendicular to the direction of fall. These cantilever beams each experience permanent deformation, depending on the drop height. The deformation values tabulated below are from beams of 5052-0 aluminum having a free length of 10 inches, a width of 0.5 inches, and a thickness of 0.04 inches:

Drop Height (in.)	Deformation
20	4.00
20	4.85
25	5.45
25	5.60
30	6.25
30	6.10
35	7.00
35	7.40
40	8.00
40	7.30

Investigate the suitability of performing a linear regression analysis by making a scatterplot of the observations. Should a no-intercept model be fit to the data? Why (not)? Place a confidence interval on the intercept parameter. Does this interval confirm your conclusion about the desirability of using a no-intercept model?

4 The following measurements are of beta-particle emissions of a several-hundred-year-old geological artifact. The counts are taken from two channels of the same decay counter, the counts differing because one counter counts emissions over a wider energy range than the other. Calculate Pearson's r for these data. Place a confidence interval

on the true correlation between measurements from these two counters. State two reasons why a correlation analysis of these data is more appropriate than a regression analysis.

Channel A	Channel B	Channel A	Channel B
3287	2418	3184	2325
3130	2328	3129	2318
3183	2366	3147	2359
3168	2405	3181	2350
3294	2445	3246	2439
3300	2454	3216	2389
3213	2352	3229	2394
3263	2416	3294	2420
3195	2401	3212	2416
3252	2412	3252	2418
3288	2404	3254	2389

5 Construct an ANOVA table for the beam deformation data in Exercise 3. Does the R^2 value suggest that the fit is adequate? Construct a confidence interval for the mean response for a drop height of 30 in. Construct a second confidence interval for the mean response, this time for a drop height of 50 in. Based on the widths of these two confidence intervals, what is the effect on the confidence interval of moving from the center to the extremes of the predictor variable values?

6 In a tire-treadwear study, a factor believed to influence the amount of wear is the ambient temperature. The data below represent wear rates of one brand of tires that were measured in an experiment. Each of the test runs was made at a different temperature, and all were made using the same vehicle. Perform a comprehensive regression analysis on these data. Are the intercept and slope coefficients statistically significant? Do you consider the fit to the data satisfactory? Why (not)?

Temperature ($^\circ$F)	Tire Wear (mm)
66.0	1.17
70.3	1.00
53.2	1.15
53.3	1.07
88.1	0.97
89.6	0.94
78.4	1.33
76.9	1.36

7 A study was conducted to investigate factors that might contribute to tooth discoloration in humans. Of particular interest was the possible relationship between flouride levels and tooth discoloration. Dental examinations were given to 20 participants at the beginning and the end of the year-long study. The participants were selected from communities that differed in the fluoride levels in the public water supply. The response of interest was the percentage increase in discoloration over the course of

study. Perform a regression analysis of the data given below and assess the adequacy of the fit. In particular, evaluate whether fluoride level alone is satisfactory in predicting the response variable.

Fluoride level (ppm)	Discolor- ation (%)	Fluoride level (ppm)	Discolor- ation (%)
0.7	12	0.5	11
1.3	10	1.8	15
1.5	14	2.1	22
2.4	20	0.4	21
2.6	18	3.4	27
2.9	18	2.9	25
3.0	25	2.1	18
3.6	21	3.4	21
3.8	36	3.0	18
4.0	44	1.7	20

8 An investigation was performed to determine whether two measures of the strength of a specific type of metal could be considered to be equivalent. One measure is a tensile strength measurement and the other is a Brinell hardness number. Two questions are of importance in this investigation:

(a) Are the measures linearly related?

(b) If so, how strong is the relationship?

Use the data below to answer these questions.

Sample:	1	2	3	4	5	6	7
Brinell Hardness:	105	107	106	107	101	106	100
Tensile Strength:	38.9	40.4	39.9	40.8	33.7	39.5	33.0

CHAPTER 22

Linear Regression with Several Variables

The methods presented in Chapter 21 for regression on a single predictor variable can be generalized to the regression of a response variable on several predictors. Due to interrelationships among the predictor variables, both the fitting of multiple regression models and the interpretation of the fitted models require additional specialized techniques that are not ordinarily necessary with single-variable models. In this chapter attention is concentrated on the estimation of model parameters, including the following topics:

- *model specification,*
- *least-squares parameter estimation, and*
- *inferential techniques for model parameters.*

A multiple linear regression analysis involves the fitting of a response to more than one predictor variable. For example, an experimenter may be interested in modeling automobile emissions (y) as a function of several fuel properties, including viscosity (x_1), cetane number (x_2), and a distillation temperature (x_3). Additional complications are introduced into such an analysis beyond the single-variable analyses discussed in the last chapter, because of the interrelationships that are possible among the predictors. As with ANOVA models, two or more factors can have synergistic effects on the response, leading to the need to include joint functions of the several predictors in order to model the response adequately.

A benefit of the modeling of a response as a function of several predictor variables is flexibility for the analyst. A wider variety of response variables can be satisfactorily modeled with multiple regression models than can be with single-variable models. In a single-variable analysis one is confined to using functions of only one predictor. In multiple regression analyses, different individual and joint functional forms for each of the predictors is permitted. Direct comparisons of alternative choices of predictors also can be made.

In this chapter we begin a comprehensive development of the use of multiple-regression modeling methodology. The three sections of this chapter examine, respectively, model-specification issues, least-squares parameter estimation, and statistical inference on model parameters.

22.1   MODEL SPECIFICATION

Multiple linear regression models can be defined as follows:

$$y_i = \beta_0 + \beta_1 x_{i1} + \beta_2 x_{i2} + \cdots + \beta_p x_{ip} + e_i, \qquad i = 1, 2, \ldots, n. \tag{22.1}$$

This model is a "linear" regression model because the unknown coefficients appear in linear forms, i.e. as additive constants or multipliers of the values of the predictor variables. The predictor variables x_j may be functions of other variables so long as there are no unknown constants needed to determine their values. The predictor variables may be intrinsically numerical or they may be categorical variables. The predictor variables can be powers or products of other predictors, e.g., $x_3 = x_1 x_2$. Because of the close relationships between the fitting of regression models and ANOVA models, two of the connections between these models are now examined: categorical variables and interactions.

22.1.1   Categorical Variables: Variables with Two Values

Categorical variables were used extensively in Chapters 6–19, where they were defined as factors in designed experiments. Categorical response variables were defined in the introduction to Chapter 20 to be variables that are intrinsically nonnumerical. In a regression model, categorical variables are any qualitative variables that could be important in the prediction of response-variable values. Categorical variables, in this context, can be obtained as controllable factors in designed experiments or as uncontrollable covariates (Chapter 19) in either designed or observational studies.

ANOVA models and the regression model (22.1) can be shown to be equivalent if categorical variables (factors) are assigned appropriate numerical codes. There are many numerical codes that can be used and are equivalent to one another. Effects coding was used in Chapters 7 (e.g., Tables 7.5 to 7.7) and 9 to represent main effects and interactions. This type of coding can be used in a regression analysis. The following example introduces an alternative coding that clarifies the connection between ANOVA and regression models.

Consider an experiment that was conducted to study the effects of ambient temperature on tire treadwear for four brands of tires, one of which was used as a control. Two convoys, each consisting of four cars of the same type, were driven day and night using replicate tire sets during two seasons of the year, winter and summer. Rotation of the vehicles in the convoy and of tire brands on the vehicles ensured that no systematic bias would be incurred on any tire brand due to the experimental procedure. Various forms of randomization were also included in the design to eliminate systematic vehicle and driver biases on the comparison of brand effects. Thirty-two tires of each of the four tire brands were exposed to 8000 miles of wear. One response of interest in this study was the relative wear rate, obtained by dividing the average wear rate (treadwear loss in miles per 1000 miles of exposure) of a test tire by the average wear rate of the control tire.

Suppose that interest lies in a comparison of only two of the test-tire brands. If one wishes to account for the effects of temperature (x_1 = Fahrenheit temperature/100) and tire brand (x_2) on the relative wear rate (y), the following *indicator variable* for tire brand can be used:

$$x_2 = \begin{cases} 1 & \text{if brand A} \\ 0 & \text{if brand B.} \end{cases}$$

A regression model fitted to the 64 observations for these two tire brands using the method of least squares (Section 22.2) resulted in the following fit:

$$\hat{y} = 1.68 - 0.30x_1 - 0.31x_2. \tag{22.2}$$

In a regression analysis the effect a predictor variable has on the response equals the difference in predicted responses for two values of the predictor variable(s). If one fixes the temperature variable at any value, the effect of the tire brands is the difference in $\hat{y}$ at $x_2 = 1$ and $x_2 = 0$. This difference is the estimated regression coefficient $b_2 = -0.31$. Thus the effect of the tire brands is obtainable from the estimated coefficient for the tire-brand variable. With the definition of the tire-brand categorical variable given above, the estimated coefficient b_2 indicates that brand A results in a lower estimated relative wear rate than brand B. Note in the above discussion that the difference in predicted responses estimates the difference in two model means.

Another coding scheme which could be used to designate the tire brands is the following:

$$x_2^* = \begin{cases} 1 & \text{if brand A,} \\ -1 & \text{if brand B.} \end{cases}$$

The resulting fitted prediction equation is

$$\hat{y} = 1.525 - 0.30x_1 - 0.155x_2^*.$$

By noting that $x_2 = (x_2^* + 1)/2$, the equivalence between the above fitted model and (22.2) is readily established:

$$\begin{aligned} \hat{y} &= 1.525 - 0.30x_1 - 0.155x_2^* \\ &= 1.525 - 0.30x_1 - 0.155(2x_2 - 1) \\ &= 1.68 - 0.30x_1 - 0.31x_2. \end{aligned}$$

Other choices of values for the coding on x_2 result in changes in the estimates of the intercept and slope coefficients, as did these two choices for the tire-brand coding. The predicted responses, however, remain the same regardless of the coding used. In any coding, the predicted response at a specified temperature for tire brand A is 0.31 units lower than that for tire brand B.

When reporting a fitted regression equation involving categorical variables, it is generally preferable to list separate equations for each value of the categorical variable. This is done to preclude confusion over the use of the categorical variable and because ready comparisons may be made of the different equations. For the tire treadwear data, one simply inserts the two values for the categorical variable in equation (22.2):

$$\hat{y}_A = 1.37 - 0.30x_1, \qquad \text{brand A,}$$
$$\hat{y}_B = 1.68 - 0.30x_1, \qquad \text{brand B.}$$

As mentioned above, the effect of the tire brands is the difference between the average relative wear rates for the two tires. This difference is the same as would be obtained in an ANOVA model for a factorial experiment in which two controllable factors, temperature and tire brand, were included at preselected levels. In this example, the reason the estimated slope coefficient for tire brands exactly equals the difference in the tire-brand averages, $\bar{y}_A - \bar{y}_B = -0.31$, is that x_2 was coded with the 0–1 convention and the two sets of tire brand measurements, y_{Aj} and y_{Bj}, were obtained from a test run on one vehicle. Because both measurements were taken on the same vehicle, the ambient temperature for each tire-brand measurement in a pair is the same. This experimental layout is identical to that of a factorial experiment in two factors. The major difference is that for this experiment the temperature values are not preselected; they are measured ambient values. Table 22.1 displays the portion of the data set relevant to this discussion.

Had the ambient temperature values for each set of tire-brand measurements been

Table 22.1 Tire Treadwear Rates for Tire Brands A and B

| Obs. | Temp. | Relative Wear | |
No.	(°F)	Brand A	Brand B
1	66.0	1.17272	1.58489
2	66.0	1.43768	1.76644
3	66.0	1.16290	1.65358
4	66.0	1.51047	1.56035
5	70.3	1.00000	1.44000
6	70.3	1.33778	1.37333
7	70.3	0.98667	1.40000
8	70.3	1.28889	1.41778
9	53.2	1.15254	1.63983
10	53.2	1.11864	1.50000
11	53.2	1.01695	1.42797
12	53.2	1.32627	1.34322
13	53.3	1.07066	1.50321
14	53.3	1.38758	1.44754
15	53.3	1.05353	1.43897
16	53.3	1.19058	1.37045
17	88.1	0.97074	1.48138
18	88.1	1.25000	1.42819
19	88.1	0.98670	1.47606
20	88.1	0.80319	1.41223
21	89.6	0.94475	1.33149
22	89.6	0.81215	1.40055
23	89.6	0.94751	1.41436
24	89.6	1.32044	1.37569
25	78.4	1.33430	1.47093
26	78.4	0.87500	1.44477
27	78.4	1.20640	1.47674
28	78.4	1.22384	1.50872
29	76.9	1.36145	1.40060
30	76.9	1.34036	1.48494
31	76.9	1.22289	1.48193
32	76.9	1.21687	1.43675

different, the estimated coefficient for x_2 would not exactly equal the difference in the averages for the two tire brands. The amount of disagreement between the estimated slope coefficient (b_2) and the difference in the tire brand averages ($\bar{y}_A - \bar{y}_B$) would depend on how similar were the temperature values for the two sets of measurements, i.e., on how close the combinations of values of tire brands and temperatures were to the layout of a complete factorial experiment.

This example illustrates the equivalence of a two-factor ANOVA model and a regression model when one of the factors is categorical and has two levels. One can algebraically derive the equivalence between the estimated slope coefficient and the main effect for the categorical factor, but we leave that exercise to the interested reader. Note, however, that this discussion was restricted to two-level factors and models without interaction terms. These results do not necessarily generalize, as we demonstrate in the next subsection.

22.1.2 Categorical Variables: Variables with More than Two Values

When a categorical variable has more than two values, it is more difficult to isolate the influence of the predictor variable on the response. Care must be taken in model specification so that an ordering or a constant difference between categories is not arbitrarily imposed on the model when no such effect exists on the response variable. This can happen if an arbitrary assignment of values is made to the various categories.

As an example, consider extending the analysis of the above tire treadwear data to include all three test tire brands A, B, and C. One possible coding for brand type is

$$x_2 = \begin{cases} 1 & \text{if brand A,} \\ 2 & \text{if brand B,} \\ 3 & \text{if brand C.} \end{cases}$$

The resulting prediction equation is

$$\hat{y} = 0.76 - 0.30x_1 + 0.54x_2.$$

This fitted model is very similar to the equation obtained using only brands A and B, and the results appear plausible. A major interpretive problem with this fit is that the above definition of x_2 imposes an arbitrary ordering of the effects of the three tire brands. The estimated coefficient for x_2 implies that the estimated relative wear rate for brand B is 0.54 units above that for brand A, and the estimated relative wear rate for brand C is 0.54 units above that for brand B.

This estimated constant difference and fixed ordering is strictly the result of the coding chosen for x_2 and not necessarily of the relationship of brand type to mean treadwear. Using the above coding with any rearrangement of the tire brands will result in a similar interpretation: the effects of the tire brands will always be ordered according to the order specified in the definition of x_2, and the effects of the three tire brands will always be increasing or decreasing (depending on the sign of b_2) by a constant amount.

This problem can be overcome by using two indicator variables to designate the tire type. For example, let

$$x_2 = \begin{cases} 1 & \text{if brand A,} \\ 0 & \text{otherwise} \end{cases} \quad \text{and} \quad x_3 = \begin{cases} 1 & \text{if brand B,} \\ 0 & \text{otherwise.} \end{cases}$$

Using these predictor variables, tire brand C is indicated by $x_2 = x_3 = 0$. This specification imposes no ordering or fixed differences in tire-brand effects on treadwear. The resulting prediction equation for the treadwear data is

$$\hat{y} = 2.45 - 0.30x_1 - 1.08x_2 - 0.77x_3.$$

The arbitrariness of the fit when using a single categorical variable to specify three or more levels of a categorical factor is well illustrated with this example. While the signs on b_2 and b_3 in this fit are both negative, the latter estimated coefficient is almost half the size of the former one. The effect of tire brand on treadwear is that brand A decreases the estimated relative wear rate by 1.08 units over brand C, while brand B decreases the estimated relative wear rate by 0.77 units over brand C. The use of two predictor variables leads to correct conclusions about the influence of the tire brands on treadwear.

Many different specifications could be used to code a categorical variable with three or more values. One of the most common specifications of categorical variables in regression analyses is given in Exhibit 22.1.

EXHIBIT 22.1   CODING CATEGORICAL VARIABLES

If a categorical variable can take on any one of k different values, use $k - 1$ indicator variables of the form

$$x_j = \begin{cases} 1 & \text{if } j\text{th value,} \\ 0 & \text{otherwise} \end{cases} \tag{22.3}$$

for $k - 1$ values of the categorical variable. It is arbitrary which $k - 1$ values are chosen.

While it is arbitrary which of the values of the categorical variable are assigned to the indicator variables, there is sometimes a category that is preferable. Note that the kth category is identified by $x_1 = x_2 = \cdots = x_{k-1} = 0$. The estimated regression coefficient for x_j then measures the effect of changing from the kth category to the jth one. This was the interpretation used in the above example, and it is the interpretation used with all two-level predictor variables. Hence if one of the values of the categorical variable represents a standard, it should represent the kth level.

The above tire-treadwear example illustrates a secondary use of categorical variables, namely in specifying an ANACOVA model (see Chapter 19). An ANACOVA model is a regression model containing both quantitative variables and categorical variables. Often, the categorical variable is of most interest, but uncontrollable quantitative factors also affect the response. In such circumstances ANACOVA models are defined and analyzed in a manner similar to that in the above example.

22.1.3   Interactions

In Chapter 9, interactions were defined as joint factor effects in an ANOVA model. Interactions have a similar interpretation in terms of the joint effects of predictor variables in regression models. Most often interaction terms are formed as products of two or more predictor variables, although they can be specified in any manner felt to be reasonable for the model under investigation.

Algebraically one can show that if a factor in an ANOVA model is represented in a regression model in terms of indicator variables as defined by equation (22.3), the ANOVA sum of squares for its main effect is obtainable as the total of the regression sums of squares for the effects represented by the individual indicator variables. There will be $k - 1$ such effects, one for each indicator variable, since there will be $k - 1$ of these terms in the regression model (22.1). Note too that the total number of degrees of freedom for these indicator variables, $k - 1$, is the same as for a main effect in an ANOVA model.

Continuing with this representation, an interaction between two categorical factors can be represented in regression models by products of the individual indicator variables for each factor. There are $(a - 1)(b - 1)$ such products for two factors having, respectively, a and b levels. The sum of squares for the interaction between the two factors would be

Table 22.2   Tire Treadwear Rates for Tire Brand C

Relative Wear	Temp. (°F)	Wet Miles (mi)
2.25847	53.2	388
2.19915	53.2	388
2.19068	53.2	388
1.99153	53.2	388
2.27837	53.3	438
2.22698	53.3	438
2.03854	53.3	438
2.05139	53.3	438
2.68891	66.0	58
2.63003	66.0	58
2.59078	66.0	58
2.58587	66.0	58
2.35556	70.3	7
2.41333	70.3	7
2.34222	70.3	7
2.42667	70.3	7
2.15361	76.9	28
2.20181	76.9	28
2.22892	76.9	28
2.16867	76.9	28
2.09884	78.4	25
2.25000	78.4	25
2.04360	78.4	25
2.08721	78.4	25
2.07979	88.1	275
2.23404	88.1	275
2.07713	88.1	275
2.21011	88.1	275
2.01934	89.6	324
2.26796	89.6	324
2.10221	89.6	324
2.01105	89.6	324

obtained as the total of the individual regression sums of squares for the product terms. The total interaction degrees of freedom are $(a-1)(b-1)$, the number of degrees of freedom stated in Chapter 15 for two-factor interactions.

This development can be extended to any number of factors having an arbitrary number of levels and to interactions involving more than two factors. Thus, the assignment of sums of squares to main effects and interactions in ANOVA models results from a consideration of their expression as indicator variables in regression models. We now wish to consider interactions in regression models in a slightly more general framework for two quantitative variables.

Consider again the tire-treadwear example. For a single tire brand, we wish to examine relative wear rate y using temperature (x_1 = Fahrenheit temperature/100) and miles traveled on wet pavement (x_2 = distance/100), referred to as "wet miles," as the predictor variables. Table 22.2 contains the data for this example in their original units of measurement, using tire brand C. Least-squares estimates of the model parameters (see Section 22.2) result in an estimated relationship of the form

$$\hat{y} = 2.75 - 0.55x_1 - 0.06x_2. \tag{22.4}$$

The graph of this fitted model is a plane in a three-dimensional data space, Figure 22.1.

There is no interaction term between temperature and wet miles in equation (22.4). Graphically one can see from Figure 22.1 that a change in temperature (x_1) for a fixed value of wet miles (x_2) produces the same change in the relative wear rate regardless of the value of wet miles used (and vice versa). Thus the fitted model is a plane in three

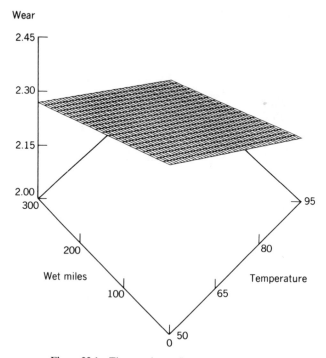

Figure 22.1   Tire-treadwear fit without interaction.

dimensions. The contour lines (curves representing constant values of the fitted response variable) of the surface of the plane are parallel for fixed values of x_1 as well as for fixed values of x_2. Models of this type are called *first-order* models, since the exponent of each x-term is one.

As will be shown in subsequent sections of this chapter, the relative wear rates are fitted better if an interaction term between temperature and wet miles is added to the model. This implies that the effect of the two factors on the response is not satisfactorily modeled by the individual terms in the above prediction equation.

Suppose that an interaction term, the product of temperature and wet miles, is included in the specification of the model. Symbolically, an interaction of this form is denoted by defining a variable $x_3 = x_1x_2$, where the notation indicates that for each observation $(i = 1, 2, \ldots, n)$ the interaction x_{i3} is the product of the observed values for the two variables, $x_{i1}x_{i2}$. Least-squares estimates of the model parameters produce the following fit:

$$\hat{y} = 5.29 - 4.04x_1 - 0.80x_2 + 1.06x_1x_2. \tag{22.5}$$

The response surface corresponding to equation (22.5) is plotted in Figure 22.2. Temperature and wet miles interact in equation (22.5) because of the presence of the product term. The change in the relative wear rate corresponding to a specified change in temperature depends on how many wet miles the vehicle is driven. Consequently, the contour lines on the surface of Figure 22.2 are not parallel. Models of this type are a special type of *second-order* regression models, since the sum of the exponents of the highest-order term (x_1x_2) is two. A complete second-order model would contain linear, cross-product, and pure quadratic $(x_1^2$ and $x_2^2)$ terms.

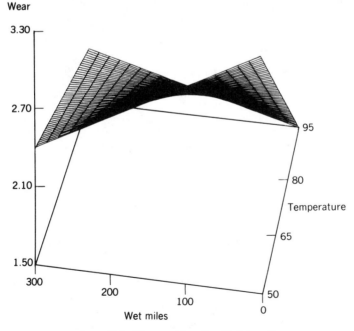

Figure 22.2 Tire-treadwear fit with interaction.

Scatterplots are often useful in determining whether an interaction term might be beneficial in a regression model. If one of the variables is categorical, a labeled scatterplot of y versus the quantitative variable might show different slopes for some values of the categorical variable. If both predictors are quantitative, a plot of the residuals from a fitted model without the interaction term [e.g., equation (22.4)] versus the product $x_1 x_2$ might show a strong linear trend. If so, an interaction term in equation (22.5) would be appropriate. If the plot is nonlinear, an interaction component to the model is still suggested, but it might be a nonlinear function of the product.

Ordinarily, one should not routinely insert products of all the predictors in a regression model. To do so might create unnecessary complications in the analysis and interpretation of the fitted models due to collinear predictor variables (Section 23.1). The purpose of including interaction terms in regression models is to improve the fit either because theoretical considerations require a modeling of the joint effects or because an analysis of regression data indicates that joint effects are needed in addition to the linear terms of the individual variables.

22.2   LEAST-SQUARES ESTIMATION

The multiple linear regression model used to relate a response variable to one or more predictor variables was defined in equation (22.1). As with the simple linear regression model described in Chapter 21, several assumptions usually accompany this model. These assumptions are listed in Table 21.1 and discussed in Chapter 25. None of these model assumptions need be valid in order to calculate least-squares estimates of the model parameters. The model assumptions are of critical importance if one wishes to make statistical inferences on the true model parameters or on the distribution of the model errors.

22.2.1   Coefficient Estimates

As observed in Chapter 21, the model in equation (22.1) can be written similarly to an ANOVA model:

$$y_i = \mu_i + e_i, \qquad i = 1, 2, \ldots, n, \tag{22.6}$$

where the means, or *expected values*, of the response variables y_i are given by

$$\mu_i = \beta_0 + \beta_1 x_1 + \beta_2 x_2 + \cdots + \beta_p x_p. \tag{22.7}$$

To estimate these n means, we insert estimates of the coefficients into the right side of equation (22.7):

$$m_i = b_0 + b_1 x_1 + b_2 x_2 + \cdots + b_p x_p. \tag{22.8}$$

Equation (22.8) can be used to estimate not only the means μ_i of the response variables but also the actual responses y_i. When used for this latter purpose the resulting estimates are termed *predicted* responses, notationally indicated by replacing m_i with $\hat{y}_i$, and equation (22.8) is called a *prediction equation*. Again these comments parallel those given in Chapter 21 for linear regression with a single predictor variable.

One method of obtaining the coefficient estimates in equation (22.8) is to use the principle of least squares. The residuals between the observed and predicted responses are

$$r_i = y_i - \hat{y} = y_i - b_0 - b_1 x_{i1} - b_2 x_{i2} - \cdots - b_p x_{ip}. \tag{22.9}$$

The sum of the squared residuals, SSE, is given by

$$SS_E = \sum r_i^2 = \sum (y_i - b_0 - b_1 x_{i1} - b_2 x_{i2} - \cdots - b_p x_{ip})^2. \tag{22.10}$$

The coefficient estimates, $b_0, b_1, \ldots, b_p$, that minimize SS_E are termed the least-squares coefficient estimates (see Exhibit 22.2).

EXHIBIT 22.2   LEAST-SQUARES ESTIMATORS FOR MULTIPLE LINEAR REGRESSION MODELS

1. Assume a multiple linear regression model of the form (22.1).

2. Least-squares estimators $b_0, b_1, \ldots, b_p$ minimize the sum of the squared residuals,

$$SS_E = \sum (y_i - b_0 - b_1 x_{i1} - \cdots - b_p x_{ip})^2.$$

The process of minimizing SS_E and obtaining coefficient estimates involves the differentiation of equation (22.10) with respect to each of the $p + 1$ unknown regression coefficients. The $p + 1$ derivatives are then set equal to zero, and the least-squares estimates are obtained by simultaneously solving the equations. We leave the algebraic details as an exercise for the interested reader.

The $p + 1$ equations that must be simultaneously solved to find the least-squares estimators are termed *normal equations*. The normal equations for a model having $p = 2$ predictor variables are

$$\bar{y} = b_0 + b_1 \bar{x}_1 + b_2 \bar{x}_2,$$
$$\sum x_{i1} y_i = b_0 \sum x_{i1} + b_1 \sum x_{i1}^2 + b_2 \sum x_{i1} x_{i2}, \tag{22.11}$$
$$\sum x_{i2} y_i = b_0 \sum x_{i2} + b_1 \sum x_{i1} x_{i2} + b_2 \sum x_{i2}^2.$$

There are three equations and three unknowns in (22.11). One can simultaneously solve them and obtain the unique least-squares coefficient estimates for β_0, β_1, and β_2. Ordinarily, the existence of a unique solution for the least-squares estimators requires that the sample size n be greater than the number p of predictor variables.

As the number of predictor variables increases, the solutions to the normal equations are ordinarily obtained using computer software. The least-squares fits presented in the last section were all obtained using computer algorithms that solve the normal equations. The algebraic solutions to the normal equations are most easily expressed in matrix notation. The interested reader is referred to the appendix to this chapter for the algebraic solutions.

Least-squares estimators also can be calculated for models that do not have an intercept term β_0. Whether to include an intercept term in the model is an option with most computer regression algorithms. When one is uncertain about including an intercept term, it should be kept in the model and then tested for statistical significance (see Section 22.3).

22.2.2 Interpreting Least-Squares Estimates

The least-squares coefficient estimates for the regression model in (22.1) have a slightly modified interpretation from that given for the slope estimate in Chapter 21. The estimates generally are said to measure the change in the predicted response variable due to a unit change in one predictor variable while all remaining predictor variables are held constant. For this reason the estimates often are termed *partial* regression coefficient estimates.

Because of interrelationships among predictor variables, it is not always possible to change one predictor variable while holding the others fixed. The magnitudes of the coefficient estimates themselves can depend heavily on which other predictor variables are included in the regression model. Adding or deleting predictor variables can cause the estimated coefficients to drastically change in size as well as sign. These problems become acute when severe collinearity is present among the predictor variables (see Sections 23.1 and 27.4).

EXHIBIT 22.3   INTERPRETATION OF LEAST-SQUARES ESTIMATES

The least-squares estimates b_j measure the change in the predicted response $\hat{y}_i$ associated with a unit change in x_j after adjusting for its common variation with the other predictor variables.

An explicit interpretation of the least-squares coefficient estimates, one that takes into account the above difficulties, is given in Exhibit 22.3. According to this interpretation, if one desires to determine the effect on $\hat{y}$ due to changes in x_j, the predictor variable values for all predictor variables must be specified. The net changes in $\hat{y}$ due to the change in x_j are then a function of the changes not only of x_j but also of the specific changes that occur in the other predictor variables. Only if x_j does not vary systematically with the other predictor variables can the usual interpretation that b_j measures the effect of a unit change in x_j be considered appropriate.

The specific adjustment referred to in the above interpretation for least-squares estimators can be obtained as follows. Just as linear regression can be used to determine a linear relationship among the response and the predictor variables, it also can be used to relate the predictors to one another. Regress x_j on the other $p - 1$ predictor variables and a constant term. Denote the residuals from this fit by r^*:

$$r_i^* = x_{ij} - \hat{x}_{ij}. \tag{22.12}$$

The least-squares estimate b_j measures the change in the response due to a unit change in r^*, not x_j. Note that if x_j cannot be well predicted by the other predictor variables, r^* is approximately equal to $x_j - \bar{x}_j$ and hence b_j can be interpreted in the usual manner as measuring the change in $\hat{y}$ associated with a unit change in x_j.

Figure 22.3 shows this relationship for the tire treadwear data with the relative wear plotted against the temperature residuals r^*. The temperature residuals were obtained by regressing temperature x_1 on a constant term and on the other two predictors, x_2 and $x_3 = x_1 x_2$, in equation (22.5). The slope of the fitted line in Figure 22.3 is the least-squares estimate b_1 from the fit to all three predictor variables, equation (22.5).

The magnitudes of least-squares estimates often are used as indicators of the importance of the predictor variables. Because predictor variables generally are measured on different scales, it is inappropriate to compare coefficient magnitudes directly. The concern about systematic variation among the predictors is another reason why regression coefficients should not be directly compared. In its extreme form, collinearities

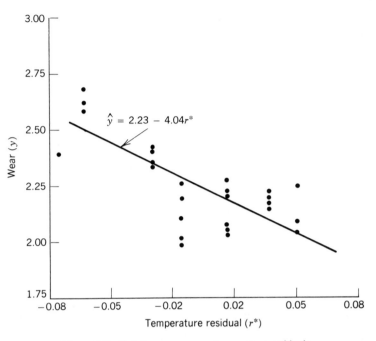

Figure 22.3 Relative wear versus temperature residuals.

can render such comparisons completely meaningless—worse yet, erroneous.

If one is convinced that such comparisons don't suffer from possible misinterpretation due to systematic variation among the predictor variables, *standardized* coefficient estimates, also termed *beta-weight coefficients*, should be used. These standardized estimates, $\hat{\beta}_j$, remove the scaling problem and produce dimensionless estimates that are directly comparable. They are defined as

$$\hat{\beta}_j = b_j(s_{jj}/s_{yy})^{1/2}, \tag{22.13}$$

where $s_{yy} = (n-1)^{-1}\sum(y_i - \bar{y})^2$ and $s_{jj} = (n-1)^{-1}\sum(x_{ij} - \bar{x}_j)^2$.

22.3 INFERENCE

Inferences on multiple linear regression models are needed to assess not only the overall fit of the prediction equation but also the contributions of individual predictor variables to the fit. In most of the inferential procedures discussed in this section we impose the model assumptions listed in Table 21.1, including the assumption that the error terms are normally distributed with a constant standard deviation. Investigating the validity of these assumptions for specific data sets is discussed in Chapter 25.

22.3.1 Analysis of Variance

Many different statistical measures are available for assessing the adequacy of the fitted model. While we stress the need to perform a comprehensive regression analysis (ISEAS)

Table 22.3 Symbolic ANOVA Table for Multiple Regression Models

Source of Variation	df	Sum of Squares	Mean Squares	F-Statistic
Regression	p	SS_R	$MS_R = SS_R/p$	$F = MS_R/MS_E$
Error	$n - p - 1$	SS_E	$MS_E = SS_E/(n - p - 1)$	
Total	$n - 1$	TSS		

before claiming a fit is adequate, we examine below a few of the measures commonly reported with regression analyses. These generally are based on the use of the ANOVA table and the estimated regression coefficients.

The derivation of an ANOVA table for a multiple regression analysis is similar to the derivation for a single predictor variable in Chapter 21. The total and error sum of squares can again be expressed as

$$TSS = \sum (y_i - \bar{y})^2, \qquad SS_E = \sum (y_i - \hat{y}_i)^2.$$

The regression sum of squares is

$$SS_R = \sum (\hat{y}_i - \bar{y})^2.$$

In multiple regression models there are p degrees of freedom for the sum of squares due to regression, because p coefficients, namely $\beta_1, \beta_2, \ldots, \beta_p$, must be estimated to obtain the regression sum of squares. One can show this algebraically (see the exercises). A symbolic ANOVA table for a regression model having p predictor variables is shown in Table 22.3.

One useful measure of the adequacy of the fitted model is the estimated error standard deviation, $s_e = (MS_E)^{1/2}$, where $MS_E = SS_E/(n - p - 1)$. A small value for this statistic indicates that the predicted responses closely approximate the observed responses, while a large value may indicate either a large random error component or improper specification of the model.

The coefficient of determination, R^2, is another extensively used measure of the goodness of fit of a regression model. It can be defined in many similar ways, and because of this it is sometimes used inappropriately. Care should be exercised when using R^2 values to compare fits between (a) models with and without an intercept term, (b) models in which the response variable is not in exactly the same functional form, and (c) linear and nonlinear regression models. The references at the end of this chapter should be consulted regarding the appropriate calculation of R^2 in some of the alternatives to multiple linear regression models.

A preferred choice for calculating R^2 is as follows:

$$R^2 = 1 - \frac{SS_E}{TSS}. \tag{22.14}$$

For least-squares estimates of the model parameters the value of R^2 lies between 0 and 1; the closer it is to 1, the closer the predicted responses are to the observed responses. The coefficient of determination for models fitted by least squares can also be interpreted as the square of the ordinary correlation coefficient between the observed and the predicted responses [see equation (21.16)].

Table 22.4 ANOVA **Table for Tire-Treadwear Data**

Source of Variation	df	Sum of Squares	Mean Squares	F-Statistic	p-Value
		(a) *No Interaction*			
Regression	2	0.378	0.189	7.60	0.002
Error	29	0.720	0.025		
Total	31	1.098			
		(b) *Interaction Added*			
Regression	3	0.835	0.278	29.58	0.000
Error	28	0.263	0.009		
Total	31	1.098			

The coefficient of determination is often adjusted to take account of the number of observations and the number of predictor variables. This adjustment is made because R^2 can be arbitrarily close to 1 if the number of predictor variables is too close to the number of observations. In fact, $R^2 = 1$ if $n = p + 1$ and no two responses have exactly the same set of predictor-variable values, regardless of whether there is any true relationship among the response and the predictor variables, i.e., regardless of whether equation (22.1) is the correct model. The adjusted R^2 is calculated from the following formula:

$$R_a^2 = 1 - a\frac{SS_E}{TSS}, \qquad (22.15)$$

where $a = (n-1)/(n-p-1)$. This adjusted coefficient of determination is always less than (22.14), but the difference between the two is usually minor if n is sufficiently large relative to p. When n is not much larger than p, the adjusted coefficient of determination should be used.

It is possible to find data sets where the R^2 value is near 1 but the estimated standard deviation is large. This usually is the result of model misspecification, although large error variability also might be a cause. Hence, caution should be used in relying on a single measure of the fit, such as R^2 values.

The ANOVA table for the tire-treadwear data in Table 22.2 using the fitted model without an interaction term, equation (22.4), is shown in Table 22.4(a). The estimated model standard deviation is $s_e = 0.158$, and the coefficient of determination of the fitted model is R^2 ($\times 100\%$) $= 34.4\%$, $R_a^2 = 29.9\%$. Addition of the interaction term as in equation (22.5) improves the fit dramatically [see Table 22.4(b)]: $s_e = 0.097$, $R^2(\times 100\%) = 76.0\%$, $R_a^2 = 73.4\%$. Note that the adjusted R_a^2 values do not differ appreciably from the unadjusted values, because the number of observations, $n = 32$, is sufficiently large relative to the number of predictors in the model, $p = 2$ or 3.

22.3.2 Lack of Fit

The ability to determine when a regression fit adequately models a response variable is greatly enhanced when repeat observations for several combinations of the predictor variables are available. The error sum of squares can then be partitioned into a component due to *pure* error and a component due to *lack-of-fit* error:

$$SS_E = SSE_P + SSE_{LOF}. \tag{22.16}$$

The pure error sum of squares, SSE_P, is computed using the responses at the repeat observations:

$$SSE_P = \sum_{i=1}^{m} \left(\sum_{j=1}^{n_i} (y_{ij} - \bar{y}_{i.})^2 \right). \tag{22.17}$$

In equation (22.17), we have temporarily changed to a double subscript for clarity. It is only in the calculation of SSE_P that this is needed. We assume that there are m combinations of values of variables with repeat observations. For the ith combination, n_i repeats are available, the average of which is denoted by $\bar{y}_i$. The pure error sum of squares is identical to the numerator of the pooled variance discussed in Sections 14.3 and 15.3 for the estimation of a common variance. The total number of degrees of freedom for this estimate of error is

$$f_P = \sum_{i=1}^{m} (n_i - 1) = q - m, \tag{22.18}$$

where $q = n_1 + n_2 + \cdots + n_m$. The lack-of-fit error sum of squares, SSE_{LOF}, is obtained as the difference between SS_E and SSE_P, with $f_{LOF} = (n - p - 1) - (q - m)$ degrees of freedom.

The test procedure for checking model specification is based on comparing the lack-of-fit error mean square with the pure error mean square. The symbolic ANOVA table in Table 22.3 is expanded in Table 22.5 to include this partitioning. This table includes the lack-of-fit F-statistic, $F = MSE_{LOF}/MSE_P$, to test the hypothesis that the model is correctly specified versus the alternative that the model is misspecified. The degrees of freedom for this F statistic are $v_1 = f_{LOF}$ and $v_2 = f_P$.

A statistically significant lack-of-fit F-statistic implies that the terms in the model do not capture all of the assignable-cause variation of the response variable. Since the pure error mean square measures the uncontrolled error variation only through calculations on repeat observations, a large F-ratio indicates that the lack-of-fit error mean square is measuring variation that exceeds that due to uncontrollable repeat-test error. When a

Table 22.5 Symbolic ANOVA Table for Lack-of-Fit Test

Source of Variation	df*	Sum of Squares	Mean Squares	F- Statistic
Regression	p	SS_R	MS_R	$F = MS_R/MS_E$
Error	$n - p - 1$	SS_E	MS_E	
Lack of fit	f_{LOF}	SSE_{LOF}	MSE_{LOF}	$F = MS_{LOF}/MSE_P$
Pure	f_P	SSE_P	MSE_P	
Total	$n - 1$	TSS		

*$f_P = q - m$, $f_{LOF} = n - p - 1 - f_P$.

Table 22.6 ANOVA **Table for Lack of Fit of Treadwear Data**

Source of Variation	df	Sum of Squares	Mean Squares	F-Value	p-Value
Regression	2	0.378	0.189	7.60	0.002
Error	29	0.720	0.025		
Lack of fit	5	0.532	0.106	13.56	0.000
Pure	24	0.188	0.008		
Total	31	1.098			

large F-ratio occurs, one should consider alternative model specifications (e.g. Section 25.2), including transformations of the response and predictors, additional polynomial terms for the predictors, etc.

The treadwear data in Table 22.2 have $m = 8$ values of temperature and wet miles for which there are 4 responses each. Hence, it is possible to test for lack of fit of the proposed model. The partitioned ANOVA table is given in Table 22.6 for a regression fit containing only the linear terms involving the two predictor variables [e.g. Table 22.4(a)]. The lack-of-fit F-statistic is statistically significant ($p < 0.001$).

When the test for lack of fit is rerun after adding the interaction term, the F-statistic is much smaller, $F = 2.39 \ (0.05 < p < 0.10)$. The fitted interaction model (22.5) is illustrated in Figure 22.2, and its no-interaction counterpart (22.4) in Figure 22.1. The effect of the interaction term is clearly evident in the twisted shape of Figure 22.2.

An investigator, whenever possible, should plan to include repeat predictor-variable values in a regression data base. This will facilitate the use of the above lack-of-fit test. When it is not possible to collect repeat points, nearby points (i.e., *nearest neighbors*) can be used to perform an approximate test for misspecification. It should be noted that model misspecification resulting from the exclusion of useful predictor variables may not be detected by the lack-of-fit test. Its sensitivity is directed mainly to reexpressions of current variables in the model.

22.3.3 Tests on Parameters

The F-statistic in the ANOVA table in Table 22.3 allows one to simultaneously test that all $\beta_j = 0$ versus the alternative that at least one β_j is not zero. Specifically, to test

$$H_0 : \beta_1 = \beta_2 = \cdots = \beta_p = 0 \quad \text{vs} \quad H_a : \text{at least one } \beta_j \neq 0,$$

use $F = \text{MS}_R / \text{MS}_E$ as the test statistic, with $v_1 = p$ and $v_2 = n - p - 1$ degrees of freedom. As in Chapter 21, we recommend that a large significance level, say $\alpha = 0.25$, be used in this test in order to reduce the chance of incorrectly deleting a useful predictor variable.

The F-statistic in the ANOVA table for the tire-treadwear data [Table 22.4(b)] is highly significant ($p < 0.001$). This indicates that at least one of the three predictor variables (temperature, wet miles, and their interaction) is useful in predicting the relative wear rate of this brand of tire.

Testing hypotheses on individual regression coefficients often is of primary interest to an experimenter performing a regression analysis. As with the single-variable model in

Chapter 21, a t-statistic can be constructed for testing

$$H_0: \beta_j = c \quad \text{vs} \quad H_a: \beta_j \neq c$$

for some specified constant c using the general procedures given in Section 12.1. The test statistic used for this purpose is

$$t = \frac{b_j - c}{s_e c_{jj}^{1/2}}, \tag{22.19}$$

where $s_e = (\text{MS}_E)^{1/2}$ is the estimated error standard deviation and

$$c_{jj} = [(n-1)s_j^2(1 - R_j^2)]^{-1}. \tag{22.20}$$

In equation (22.20), s_j^2 is the sample variance of the n values of the jth predictor variable and R_j^2 is the coefficient of determination for the regression of x_j on the constant term and the $p - 1$ other predictor variables.

The t-statistics (22.19) with $c = 0$ are used to test the statistical significance of the individual model parameters; i.e., the usefulness of x_j as a predictor of the response variable. Note that since β_j is a partial regression coefficient and both b_j and c_{jj} are functions of the values of the other predictor variables, this test determines the importance of the jth predictor variable only *conditionally*, i.e., conditioned on the other predictor variables being in the model. Thus it can be considered a conditional test and should not be interpreted as a determinant of the significance of the jth predictor variable without regard to the presence or absence of the other predictor variables.

The above approach also can be used to test the significance of the intercept term. To test $H_0: \beta_0 = c$ vs $H_a: \beta_0 \neq c$ we use a t-statistic similar in form to (22.19). The appendix to this chapter outlines the algebraic details for the formation of this t-statistic. Most regression computer programs provide the t-statistic for testing whether the intercept or the individual regression coefficients are zero.

Table 22.7 contains additional information on the regression of treadwear on temperature, wet miles, and their interaction. The intercept term is significantly different from zero, and each individual predictor variable contributes significantly to the fits, given that the other two predictor variables are also included in the model.

Table 22.7 Summary Statistics for Tire-Treadwear Regression

Variable	Coefficient Estimate	t Statistic*	95% Confidence Interval
Intercept	5.288	13.94	(4.512, 6.064)
Temperature[†]	−4.044	−7.79	(−5.102, −2.982)
Wet miles[‡]	−0.801	−7.52	(−1.018, −0.583)
Temperature × Wet miles	1.061	6.97	(0.539, 1.372)

*All are statistically significant ($p < 0.001$).
[†]Fahrenheit temperature/100.
[‡]Distance in 10^{-2} mi.

The contention that t-tests on individual parameters are conditional tests can be appreciated by relating the t-statistic to an equivalent F-statistic derived using the principle of reduction in sums of squares, described in Section 15.1.

Consider the full regression model in (22.1) and a reduced model containing any subset of $k < p$ predictor variables. Denote the full-model (M_1) error sum of squares by SSE_1 and the reduced-model (M_2) error sum of squares by SSE_2. Denote the reduction in error sum of squares resulting from the fit of the additional terms in the full model by

$$R(M_1|M_2) = SSE_2 - SSE_1. \tag{22.21}$$

There are $p - k$ more predictor variables in the full model than the reduced one. Therefore the F-statistic for determining the statistical significance of this subset of predictor variables is

$$F = \frac{MSR(M_1|M_2)}{MSE_1}, \tag{22.22}$$

where $MSR(M_1|M_2) = R(M_1|M_2)/(p - k)$. Under the null hypothesis that the $p - k$ additional predictor variables in the full model have regression coefficients equal to zero, this statistic has an F-distribution with $v_1 = p - k$ and $v_2 = n - p - 1$ degrees of freedom. This test is called a "partial F-test" in many textbooks, since it measures the contribution of a subset of predictor variables conditioned on other predictor variables being in the model. If $k = p - 1$, the F-statistic (22.22) is the square of the t-statistic from the full model corresponding to the term left out of the reduced model.

To apply this principle reconsider the two models for the tire-treadwear data, i.e., the fits with and without the interaction term. Using the two error sums of squares from Table 22.4(a) and (b), the reduction in error sum of squares is

$$R(M_1|M_2) = 0.720 - 0.263 = 0.457,$$

and the corresponding partial F-statistic is

$$F = \frac{0.457/1}{0.263/28} = 48.65.$$

Since this statistic is highly significant ($p < 0.001$), the temperature-by-wet-miles interaction is a statistically significant predictor of wear rate in addition to the contributions of the individual linear terms for the two variables. The equivalence of this procedure and the use of the t-statistic to test the significance of the interaction term is readily established, since $t = F^{1/2} = (48.65)^{1/2} = 6.97$ is the value of the interaction t-statistic in Table 22.7.

22.3.4   Confidence Intervals

Confidence intervals for the regression coefficients in equation (22.1) can be constructed using the same type of procedures discussed in Sections 21.5 and 12.2. A $100(1 - \alpha)\%$ confidence interval for β_j is given by

$$b_j - t_{\alpha/2} s_e c_{jj}^{1/2} \leqslant \beta_j \leqslant b_j + t_{\alpha/2} s_e c_{jj}^{1/2}, \tag{22.23}$$

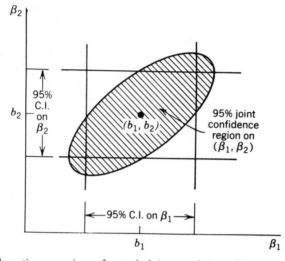

Figure 22.4 Schematic comparison of a typical (rectangular) confidence region formed from individual 95% confidence intervals with the true joint (elliptical) 95% confidence region.

where $t_{\alpha/2}$ is a two-tailed $100\alpha\%$ t critical value having $n - p - 1$ degrees of freedom. A $100(1 - \alpha)\%$ confidence interval for β_0 can be defined in a similar fashion (see the appendix to this chapter). 95% confidence intervals for the coefficients in the tire treadwear data are shown in the last column of Table 22.6.

Simultaneous confidence intervals for all the coefficients in equation (22.1) cannot be computed using the individual coefficient intervals for β_0 and the β_j. The individual intervals are useful for estimating ranges for the individual coefficients, but they ignore the systematic variation of the predictor variables and consequent correlation among the coefficient estimators. Even if the estimated coefficients were statistically independent, the construction of several individual confidence intervals would not result in an overall confidence region with the stated confidence. The reasons for this are similar to those discussed in Chapter 16 and are related to the distinction between experimentwise and comparisonwise error rates. Thus, the chance that one particular interval covers the corresponding regression coefficient is as stated, $1 - \alpha$, but the probability that all the intervals simultaneously cover all the regression coefficients can be much less than $1 - \alpha$.

Figure 22.4 illustrates the difference between joint and individual confidence statements. The individual $100(1 - \alpha)\%$ interval for β_1 and β_2 create a rectangular region, while the simultaneous interval is elliptical. The stronger the systematic variation between the two involved predictor variables, the narrower will be the joint confidence ellipsoid. The algebraic details of the construction of simultaneous confidence regions are outlied in the appendix to this chapter.

APPENDIX: MATRIX FORM OF LEAST-SQUARES ESTIMATORS

The multiple linear regression model (22.1) can be represented in matrix form as follows:

$$\mathbf{y} = X_c \boldsymbol{\beta}_c + \mathbf{e},\qquad(22.A.1)$$

where

$$\mathbf{y} = \begin{pmatrix} y_1 \\ y_2 \\ \vdots \\ y_i \\ \vdots \\ y_n \end{pmatrix},$$

$$X_c = \begin{bmatrix} 1 & x_{11} & x_{12} & \cdots & x_{1p} \\ 1 & x_{21} & x_{22} & \cdots & x_{2p} \\ \vdots & \vdots & \vdots & & \vdots \\ 1 & x_{i1} & x_{i2} & \cdots & x_{ip} \\ \vdots & \vdots & \vdots & & \vdots \\ 1 & x_{n1} & x_{n2} & \cdots & x_{np} \end{bmatrix},$$

$$\mathbf{e} = \begin{pmatrix} e_1 \\ e_2 \\ \vdots \\ e_i \\ \vdots \\ e_n \end{pmatrix},$$

and $\boldsymbol{\beta}_c = (\beta_0, \beta_1, \beta_2, \ldots, \beta_p)$. We use the notation $\mathbf{z}'$ to denote the transpose of a vector (similarly for a matrix).

The subscript c used with the X-matrix and the vector of regression coefficients is intended to stress that the constant term is included in the model; consequently, β_0 appears in the coefficient vector $\boldsymbol{\beta}_c$, and the first column of X_c is a column of ones. In subsequent chapters properties of the least-squares estimators of the coefficients of the nonconstant predictor variables $x_1, x_2, \ldots, x_p$ will be discussed. When such discussions occur, of interest will be the X-matrix without a column of ones and the $\boldsymbol{\beta}$-vector without the constant β_0. These quantities will be denoted by dropping the subscript c.

Using the model (22.A.1), the general expression for the normal equations in matrix form is

$$X_c' X_c \mathbf{b}_c = X_c' \mathbf{y}.\qquad(22.A.2)$$

The solutions of the normal equations, the least-squares estimators for multiple linear regression models, are obtained by multiplying both sides of equation (22.A.2) by the matrix inverse of $X_c'X_c$, denoted $(X_c'X_c)^{-1}$:

$$\mathbf{b}_c = (X_c'X_c)^{-1}X_c'\mathbf{y}. \tag{22.A.3}$$

The elements of the vector $X_c'\mathbf{y}$ are the sums of the cross-products of the elements of the columns of X and the observations in $\mathbf{y}$. The diagonal elements of $X_c'X_c$ are the sums of the squares of the elements in the columns of X_c, and the off-diagonal elements are the sums of the cross-products of the elements in the columns of X_c.

The t-statistic and confidence intervals for the intercept parameter and the regression coefficients are based on the least-squares estimators of $\boldsymbol{\beta}_c$, equation (22.A.3), and the matrix $C = (X_c'X_c)^{-1}$. Denote the diagonal elements of this matrix by $c_{00}, c_{11}, \ldots, c_{pp}$. The c_{jj}-values used in equations (22.19) and (22.23) are the last p diagonal elements of C. To test hypotheses and form confidence intervals on β_0, these same equations can be used when b_0 is substituted for b_j and when c_{00}, the first diagonal element of C, is inserted for c_{jj}.

The formula for the $100(1 - \alpha)\%$ confidence region for $\boldsymbol{\beta}_c$ is

$$\frac{(\mathbf{b}_c - \boldsymbol{\beta}_c)'X_c'X_c(\mathbf{b}_c - \boldsymbol{\beta}_c)}{(p + 1)\,\mathrm{MS}_E} < F_\alpha, \tag{22.A.4}$$

where F_α is the upper-tail $100\alpha\%$ point of an F-distribution with $v_1 = p + 1$ and $v_2 = n - p - 1$ degrees of freedom. The references at the end of this chapter include detailed discussions on these types of intervals and how to construct them. They also discuss the construction of simultaneous confidence regions for the regression coefficients in $\boldsymbol{\beta}$, i.e., excluding the constant term.

Confidence intervals for mean responses [i.e., equation (22.7)] and prediction intervals for future responses can also be constructed for fixed values of the predictor variables. A $100(1 - \alpha)\%$ confidence interval for the mean response μ, at a fixed point $\mathbf{u}' = (1, u_1, \ldots, u_p)$, where u_j is the value of the jth predictor variable for which the mean response is desired, is

$$\hat{y} \pm t_{\alpha/2}\{\mathrm{MS}_E\,[\mathbf{u}'(X_c'X_c)^{-1}\mathbf{u}]\}^{1/2}, \tag{22.A.5}$$

where $\hat{y} = \mathbf{b}_c'\mathbf{u} = b_0 + \sum b_j u_j$. To construct a $100(1 - \alpha)\%$ prediction interval for a future response y_f, we modify equation (22.A.5) as follows:

$$\hat{y} \pm t_{\alpha/2}\{\mathrm{MS}_E\,[1 + \mathbf{u}'(X_c'X_c)^{-1}\mathbf{u}]\}^{1/2}, \tag{22.A.6}$$

where, again, $\hat{y} = \mathbf{b}_c'\mathbf{u}$.

Simultaneous confidence and prediction intervals corresponding to the single intervals (22.A.5) and (22.A.6) also can be constructed. These have forms similar to the joint region defined in equation (22A.4) for the coefficient parameters. The references at the end of this chapter include more details on these procedures.

Note that the uncertainty limits in the confidence intervals (22.A.5) and (22.A.6) depend on the value of $\mathbf{u}$ through the term $\mathbf{u}'(X_c'X_c)^{-1}\mathbf{u}$. While it is not obvious, the limits are smallest when $\mathbf{u}' = (1, \bar{x}_{.1}, \ldots, \bar{x}_{.p})$, where $\bar{x}_{.j}$ is the average of the predictor values for x_j. The magnitude of $\mathbf{u}'(X_c'X_c)^{-1}\mathbf{u}$ increases as $\mathbf{u}'$ deviates from $(1, \bar{x}_{.1}, \ldots, \bar{x}_{.p})$.

REFERENCES

Text References

The texts referenced at the end of Chapter 21 also provide coverage of the topics discussed in this chapter. These texts rely on matrix algebra in the discussions of multiple linear regression.

Useful texts for simultaneous confidence and prediction limits include those by Draper and Smith and by Montgomery and Peck (also referenced in Chapter 21). An advanced text that includes a chapter on this topic is

Seber, G. A. F. (1977). *Linear Regression Analysis*, New York: John Wiley and Sons, Inc.

A discussion of various definitions of the coefficient of determination, including recommendations for which ones to use, is found in the following article:

Kvalseth, T. O. (1985). "Cautionary Note About R^2," *The American Statistician*, **39**, 279–285.

Data References

The treadwear data were taken from the following report:

R. N. Pierce, R. L. Mason, K. E. Hudson, and H. E. Staph (1985). "An Investigation of a Low-Variability Tire Treadwear Test Procedure and of Treadwear Adjustment for Ambient Temperature," Report No. SwRI EFL-7928-1, Southwest Research Institute, San Antonio, TX.

The data in Exercise 6 are extracted from

Hare, C. T. (1977). "Light Duty Diesel Emission Correction Factors for Ambient Conditions," Final Report to the Environmental Protection Agency under Contract No. 68-02-1777, Southwest Research Institute, San Antonio, TX.

The data in Exercise 7 were provided by Professor Ladislav P. Novak, Department of Anthropology, Southern Methodist University.

EXERCISES

1 Fit a complete second-order regression model to the brand-A tire-treadwear rates in Table 22.1, using both the temperature and the wet-miles predictor variables from Table 22.2. Test for lack of fit. If no significant lack of fit is present, investigate the statistical significance of the individual model terms.

2 Compare an ANACOVA model in which the design factor has two levels with a two-factor regression model in which one of the predictors is an indicator variable. Algebraically relate the regression coefficient for the indicator variable to the effects parameters for the ANACOVA model. Algebraically relate the least-squares estimators of the regression parameters to the estimators of the ANACOVA model parameters (see the appendix to Chapter 19).

3 Model the weight-gain data in Exercise 2, Chapter 19, as a regression model using indicator variables. Compute the total of the sums of squares for the categorical variables using the principle of reduction in sums of squares. Compare this total with the corresponding total for the main effects sum of squares from the analysis of covariance.

4   Model the engine start-time data in Exercise 5, Chapter 19, as a regression model using
    (a) values 1, 2, and 3 in a single predictor to designate the engines, and (b) two indicator
    variables to identify the engines. Calculate the least-squares estimates for the two
    models. How do the regression-coefficient estimates affect predictions for the two fits?

5   Algebraically show that the least-squares coefficient estimators minimize the error sum
    of squares (22.10).

6   The following data are taken from an experiment designed to investigate the effects of
    three environmental variables on exhaust emissions of light-duty diesel trucks.
    Investigate regression fits to the nitrogen oxide (NO_x) data listed below. Do any of
    your fits to the data appear to be satisfactory? The researchers expected that humidity
    would have a substantial negative effect on NO_x emissions and that temperature might
    not be an important predictor. Does your analysis tend to confirm or contradict these
    expectations?

NO_x (ppm)	Humidity (%)	Temperature (°F)	Barometric Pressure (in. Hg)
0.70	96.50	78.10	29.08
0.79	108.72	87.93	28.98
0.95	61.37	68.27	29.34
0.85	91.26	70.63	29.03
0.79	96.83	71.02	29.05
0.77	95.94	76.11	29.04
0.76	83.61	78.29	28.87
0.79	75.97	69.35	29.07
0.77	108.66	75.44	29.00
0.82	78.59	85.67	29.02
1.01	33.85	77.28	29.43
0.94	49.20	77.33	29.43
0.86	75.75	86.39	29.06
0.79	128.81	86.83	28.96
0.81	82.36	87.12	29.12
0.87	122.60	86.20	29.15
0.86	124.69	87.17	29.09
0.82	120.04	87.54	29.09
0.91	139.47	87.67	28.99
0.89	105.44	86.12	29.21

7   The following data are physical measurements (all in cm) taken on female applicants to
    the police department of a metropolitan police force. They are a portion of a much
    larger data base that was compiled to investigate physical requirements for police
    officers. Fit a linear regression model to these data, using the overall height as the
    response variable. Assess the adequacy of the fit.

Measurements of Female Police Department Applicants (Exercise 7)

Appli-cant	Overall Height	Sitting Height	Upper Arm Length	Forearm Length	Hand Length	Upper Leg Length	Lower Leg Length	Foot Length
1	165.8	88.7	31.8	28.1	18.7	40.3	38.9	6.7
2	169.8	90.0	32.4	29.1	18.3	43.3	42.7	6.4
3	170.7	87.7	33.6	29.5	20.7	43.7	41.1	7.2
4	170.9	87.1	31.0	28.2	18.6	43.7	40.6	6.7
5	157.5	81.3	32.1	27.3	17.5	38.1	39.6	6.6
6	165.9	88.2	31.8	29.0	18.6	42.0	40.6	6.5
7	158.7	86.1	30.6	27.8	18.4	40.0	37.0	5.9
8	166.0	88.7	30.2	26.9	17.5	41.6	39.0	5.9
9	158.7	83.7	31.1	27.1	18.3	38.9	37.5	6.1
10	161.5	81.2	32.3	27.8	19.1	42.8	40.1	6.2
11	167.3	88.6	34.8	27.3	18.3	43.1	41.8	7.3
12	167.4	83.2	34.3	30.1	19.2	43.4	42.2	6.8
13	159.2	81.5	31.0	27.3	17.5	39.8	39.6	4.9
14	170.0	87.9	34.2	30.9	19.4	43.1	43.7	6.3
15	166.3	88.3	30.6	28.8	18.3	41.8	41.0	5.9
16	169.0	85.6	32.6	28.8	19.1	42.7	42.0	6.0
17	156.2	81.6	31.0	25.6	17.0	44.2	39.0	5.1
18	159.6	86.6	32.7	25.4	17.7	42.0	37.5	5.0
19	155.0	82.0	30.3	26.6	17.3	37.9	36.1	5.2
20	161.1	84.1	29.5	26.6	17.8	38.6	38.2	5.9
21	170.3	88.1	34.0	29.3	18.2	43.2	41.4	5.9
22	167.8	83.9	32.5	28.6	20.2	43.3	42.9	7.2
23	163.1	88.1	31.7	26.9	18.1	40.1	39.0	5.9
24	165.8	87.0	33.2	26.3	19.5	43.2	40.7	5.9
25	175.4	89.6	35.2	30.1	19.1	45.1	44.5	6.3
26	159.8	85.6	31.5	27.1	19.2	42.3	39.0	5.7
27	166.0	84.9	30.5	28.1	17.8	41.2	43.0	6.1
28	161.2	84.1	32.8	29.2	18.4	42.6	41.1	5.9
29	160.4	84.3	30.5	27.8	16.8	41.0	39.8	6.0
30	164.3	85.0	35.0	27.8	19.0	47.2	42.4	5.0
31	165.5	82.6	36.2	28.6	20.2	45.0	42.3	5.6
32	167.2	85.0	33.6	27.1	19.8	46.0	41.6	5.6
33	167.2	83.4	33.5	29.7	19.4	45.2	44.0	5.2

CHAPTER 23

Polynomial Models

Polynomial models provide sufficient flexibility to adequately approximate many complicated but unknown relationships between a response and one or more predictor variables. In this chapter the uses and the fitting of polynomial models are discussed. Specific topics covered include:

- *appropriate uses and potential misuses of polynomial models,*
- *use of polynomial models to fit data from response-surface designs, including the optimization of responses, and*
- *using polynomial models to analyze mixture experiments.*

The basic concepts and procedures for fitting linear regression models were introduced in Chapters 21 and 22. A necessary requirement for regression modeling and a key component of the ISEAS comprehensive regression analysis approach (Chapter 21) is the specification of the functional relationship between the response and the predictor variables. If this relationship is known to the experimenter on the basis of theoretical arguments or previous empirical results, it should be used; however, frequently the model is not known prior to the data analysis. In many important problems in engineering and science the underlying mechanism that generates the data is not well understood, due to the complexity of the problem and to lack of sufficient theory. In these cases polynomial models often can provide adequate approximations to the unknown functional relationship.

This chapter focuses on uses of polynomial models. Particular attention is given to specifying and fitting polynomial response-surface models to data collected according to the designs discussed in Chapter 11.

23.1 USES AND PERILS OF POLYNOMIALS

The polynomial models discussed in this chapter are defined in Exhibit 23.1. Many theoretical models are polynomial. For example, the physical law describing the volume expansion of a rectangular solid is a third-order polynomial in temperature t given by

$$V = L_0 W_0 H_0 (1 + 3\beta t + 3\beta^2 t^2 + \beta^3 t^3),$$

486

where L_0, W_0, and H_0 are the dimensions of the solid at $0°C$ and β is the coefficient of linear expansion. Another example is the second-order relationship between temperature T, pressure P, and volume V in ideal gas laws:

$$RT = PV,$$

where R is a constant that depends on the type of gas.

EXHIBIT 23.1 POLYNOMIAL MODELS

- A polynomial model for p predictor variables has the form

$$y = \beta_0 + \beta_1 t_1 + \beta_2 t_2 + \cdots + \beta_m t_m + e, \tag{23.1}$$

  where the ith variable, t_i, is either a single predictor variable or a product of at least two of the predictors; each variable in t_i can be raised to a (positive) power.
- The order of a polynomial model is determined by the maximum (over the m terms) of the sum of the powers of the predictor variables in each term of the model.

(a) A first-order model is of the form

$$y = \beta_0 + \beta_1 x_1 + \beta_2 x_2 + \cdots + \beta_p x_p + e. \tag{23.2}$$

(b) A complete second-order model (also called a quadratic model) is of the form

$$y = \beta_0 + \sum_i \beta_i x_i + \sum\sum_{i<j} \beta_{ij} x_i x_j + \sum_i \beta_{ii} x_i^2 + e. \tag{23.3}$$

Note that for a first-order model $m = p$ and for a second-order model $m = p(p+3)/2$.

Our primary concern in this section is the use of polynomial models to approximate unknown relationships between responses and predictor variables. A motivation for this is contained in calculus-based formulations of Maclaurin and Taylor series expansions of functions: any suitably well-behaved function of a mathematical variable x can be written as an infinite sum of terms involving increasing powers of x. A Taylor series expansion often is used to approximate a complicated function by expressing it as a polynomial, perhaps containing an infinite number of terms, and retaining only a few low-order terms of the series. The number of terms necessary to give an adequate approximation depends on the complexity of the function, the range of interest of x, and the use of the approximation.

The polynomial models considered in this chapter exhibit the same functional form in the predictor variables as the corresponding Taylor series approximation. The unknown coefficients of the powers of the predictor variables $[\beta_j, j = 1, 2, \ldots, m$ in equation (23.1)] are estimated as described in Chapter 22.

In many respects a polynomial model in several predictor variables can be viewed as a multidimensional french curve. A french curve is a drawing instrument that is used to draw a smooth curve through points that do not fall on such regular curves as an ellipse or a circle. In the same way, polynomial models are used to smooth response data over a region of the predictor variables. As higher-order terms are included in the model, the "french curve" acquires more twists and bends. An important operating principle for determining the order of the model to fit to experimental data (that is, how elaborate a french curve should be used) is that of parsimony.

Table 23.1   Synthetic-Rubber Process Data: Production Rates and Solvent Weights

Rate (lb/hr)	Solvent Weight (%)	Rate (lb/hr)	Solvent Weight (%)	Rate (lb/hr)	Solvent Weight (%)
400	0.02	450	0.03	460	0.06
480	0.01	480	0.03	640	0.03
760	0.06	890	0.09	1020	0.16
1090	0.16	1100	0.12	1100	0.10
1120	0.14	1140	0.12	1140	0.14
1150	0.11	1140	0.13	1150	0.13
1190	0.19	1220	0.17	1280	0.12
1290	0.12	1290	0.15	1290	0.19
1320	0.13	1320	0.16	1320	0.21
1330	0.25	1340	0.13	1350	0.20
1380	0.16	1380	0.23	1410	0.30
1420	0.43	1610	0.41	1660	0.40
1750	0.26				

One should start with the simplest model warranted by what is known about the physical mechanism under study and by suggestions obtained from plots of the response variable versus the predictor variables. If a lack-of-fit test (Section 22.3) or a residual analysis (Section 25.2) indicates that a proposed model is an inadequate approximation to the observed responses, one can either add the next higher-order terms into the model or investigate nonlinear models. In many experimental situations, a first- or second-order polynomial is adequate to describe a response.

The data shown in Table 23.1 are 37 observations from a synthetic-rubber process. The data were collected in order to investigate the relationship between the weight (in percent) of a solvent and the corresponding production rate of the rubber process. Figure 23.1 shows a plot of the solvent content versus the production rate. The plot indicates a strong linear component in the relationship, with an indication of possible curvature. A linear regression model was fit to the data. The ANOVA table is presented in Table 23.2. Note that there is evidence of lack of fit with the straight-line model. A quadratic model provides an improved fit, as will be demonstrated below.

Although polynomial models are useful tools for regression analysis, their use has some potential drawbacks. One such drawback is that in order to model some response variables satisfactorily, a complicated polynomial model may be needed, when simpler nonlinear models (Chapter 26) or a linear model with one or more predictors reexpressed may be equally satisfactory. This is especially likely when polynomial models are used routinely without proper thought about the physical nature of the system being studied.

The relationship between volume V and pressure P in the expansion of gases at a constant temperature is known to follow Boyle's law, derivable from the kinetic theory of gases:

$$PV = k, \quad \text{or} \quad V = kP^{-1},$$

where k depends primarily on the mass of the gas and on the constant temperature. If one were to attempt to fit the volume with a polynomial (positive powers) in pressure, at least a quadratic or cubic polynomial would be needed to fit the data adequately over a reasonable range of pressure values. A simpler model is the inverse relationship shown above.

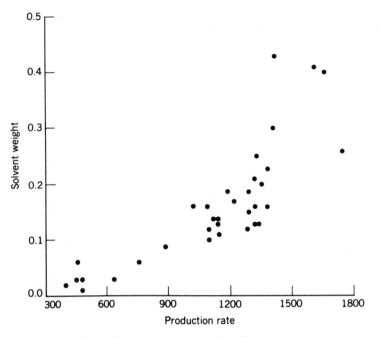

Figure 23.1 Data on a synthetic-rubber process.

Table 23.2 ANOVA Table for the Synthetic-Rubber Process Data

Source	df	SS	MS	F	p-Value
Regression	1	.2433	.2433	64.0	.000
Error	35	.1329	.0038		
Lack of fit	25	.1239	.0050	5.52	.004
Pure	10	.0090	.0009		
Total	36	.3762			

A second risk in fitting polynomial models is the temptation to use them to extrapolate outside the experimental region. Remembering the french-curve analogy, the fitted polynomial does no more than describe the response in a smooth manner over the region where data were obtained. Like a french curve, polynomial models can be used to extrapolate past the range of the data, but the behavior of the actual system may be entirely different.

Another peril of polynomial models is collinear predictor variables. In mathematics courses, notably linear algebra, two lines are said to be collinear if they have the same intercept and the same slope. Three lines are said to be coplanar if they all lie in a two-dimensional plane. Four or more lines are said to be coplanar if they reside in a subspace of dimension at most one less than the number of lines ("coplanar" is used even though the subspace may be of dimension greater than two). In each of these instances the lines are said to be linearly dependent. Rather than distinguish collinear from coplanar situations, we will refer to all of these situations as being collinear.

In regression analyses, collinear predictor variables supply redundant information (see Exhibit 23.2). Frequently variables are not exactly redundant but are extremely close to being redundant. The closer the linear dependence among two or more of the predictor variables is to being exact, the stronger the redundancy among the variables. Methods for detecting collinearities and their effects on regression analyses are described in Section 28.3.

EXHIBIT 23.2

Collinear Predictor Variables. Two or more predictor variables are collinear if there exists an approximate linear dependence among them.

The collinearity problem is not confined to polynomial models, but they can display it in special ways. A common way in which collinearities arise with polynomial models is through the failure to standardize the predictor variables before forming products and powers (see Exhibit 23.3). The form of standardization given in Exhibit 23.3 is sometimes called the *normal-deviate* standardization because of its similarity to the formation of standard normal variables. An equivalent form of standardization is

$$z_{ij} = \frac{x_{ij} - \bar{x}_j}{d_j},\qquad(23.5)$$

where $d_j = (n-1)^{1/2}s_j$. This form of standardization is usually called *correlation-form* standardization, because

$$\sum_{i=1}^{n} z_{ij}z_{ik} = r_{jk},$$

the correlation coefficient for the observations on x_j and x_k.

EXHIBIT 23.3

Standardized Predictor Variable. A standardized predictor variable, denoted z_j, is obtained from the corresponding raw predictor variable x_j by subtracting the average $\bar{x}_j$ from each observation and then dividing the result by the standard deviation s_j:

$$z_{ij} = \frac{x_{ij} - \bar{x}_j}{s_j}.\qquad(23.4)$$

These forms of standardization result in the standardized variables having an average of zero and standard deviations of 1 or $(n-1)^{1/2}$, respectively. The normal-deviate standardization scales predictor variables so they have a range of approximately -2.5 to $+2.5$, regardless of the original units. One major benefit of standardization is that collinearity problems may be lessened, perhaps even eliminated, simply by standardizing prior to forming products and powers of the predictor variables. A second benefit is that inaccuracies in computations due to roundoff error are lessened when computing the regression coefficients.

Table 23.3 Correlation Coefficients for Powers of x Equally Spaced in Increments of 0.1 from 1 to 5

Raw Predictors		Standardized Predictors	
Variables	Correlation	Variables	Correlation
x, x^2	0.985	z, z^2	0.000
x, x^3	0.951	z, z^3	0.917
x^2, x^3	0.990	z^2, z^3	0.000

Two other benefits are derived from standardizing the predictors. First, the magnitude of the regression coefficients are comparable and can be used to assess the relative effects of the predictor variables on the response. If the response and the predictor variables are all standardized, the resulting coefficient estimates are the beta weights of equation (22.13). Second, the constant term has a meaningful interpretation when fitting response surfaces. It is the estimated value of the response at the centroid (center) of the experimental region.

A straightforward example provides a quantitative illustration of the collinearity problems that can arise with polynomial models and the potential benefits of standardization. Suppose one wishes to model a response variable as a polynomial function of a single predictor variable. Suppose further that the response variable is observed at values of the predictor variable from 1 to 5 in increments of 0.1:

$$x_i = 1 + 0.1(i-1), \qquad i = 1, 2, \ldots, 41.$$

If this predictor is not standardized, the correlations between pairs of the linear, quadratic, and cubic powers of x are extremely highly correlated as shown in Table 23.3.

The average of the 41 values of x is 3.0, and the standard deviation is 1.198. Standardized values of the predictor variable are then obtainable from the following equations:

$$z_i^k = \left(\frac{x_i - 3.0}{1.198} \right)^k, \qquad k = 1, 2, 3.$$

Correlations [Pearson's r-values, equation (21.2)] between pairs of these standardized powers are also shown in Table 23.3. Observe that the correlations between the linear and quadratic and between the quadratic and cubic powers are zero. Very high correlation remains between the linear and cubic terms.

Since many response-surface models are at most quadratic functions of several variables, this illustration demonstrates the potential for reduction, even elimination, of collinearity problems with such polynomial fits. The illustration also indicates that collinearity problems are not always eliminated by standardization. The benefits of standardization are dependent on the polynomial terms used in the model and the range of values of the predictors. Note, however, that one can always calculate correlation coefficients as one indicator of whether collinearities may remain a serious problem after standardization.

Table 23.4 contains two fits to the solvent data listed in Table 23.1. Table 23.4(a) summarizes a quadratic fit when the production rate is not standardized, and Table 23.4(b) provides a similar summary for the standardized production rate. The regression

Table 23.4 Summary of a Quadratic Fit to the Synthetic-Rubber Data

Model Term	Coefficient Estimate	t-Value
(a) Raw Rate Values		
Intercept	0.0558	0.82
Linear (rate)	-0.0147×10^{-2}	-1.05
Quadratic (rate^2)	0.0193×10^{-5}	2.80
(b) Standardized Rate Values		
Intercept	0.1358	11.13
Linear (rate)	0.0995	8.85
Quadratic (rate^2)	0.0229	2.80

Table 23.5 ANOVA Table for Quadratic Fit to Synthetic-Rubber Data

Source	df	SS	MS	F	p-Value
Regression	2	.2682	.1341	42.2	.000
Error	34	.1080	.0032		
Lack of Fit	24	.0990	.0041	4.59	.008
Pure	10	.0090	.0009		
Total	36	.3762			

equations produce the same ANOVA table (see Table 23.5) and identical predicted values. For these reasons the fitted models are considered to be equivalent.

Interpreting the coefficients in Table 23.4(a) is troublesome. The estimated regression coefficient for the linear term is negative and not statistically significant; yet the trend in Figure 23.1 is clearly increasing in the production rate. On standardizing the production rate prior to forming the linear and quadratic terms, the interpretation is more reasonable. Both the linear and the quadratic coefficient estimates are positive and statistically significant. Although the coefficients for the nonstandardized fit are obtainable from the standardized coefficient estimates (see the exercises), the reasonable interpretation of the standardized estimates render this fit more appealing.

The differences in the coefficient estimates for the nonstandardized and standardized fits can be explained by referring to the interpretation of least-squares coefficient estimates in Section 22.2. The linear-coefficient estimate can be obtained by regressing the solvent weights y on the residuals r^* from a least-squares fit of the production-rate values x on the squares of these values x^2; i.e., we regress y on $r^* = x - \hat{x}$, where $\hat{x} = a_1 + a_2 x^2$ and a_1 and a_2 are least-squares estimates from the regression of x on x^2. The high correlation between x and x^2 $(r = 0.98)$ results in all the residuals r^* being close to zero. The regression of y on r^* thus yields a nonsignificant estimated coefficient, which happens to be negative.

Standardizing the predictor variable prior to forming the quadratic term produces standardized linear and quadratic variables that are much less correlated $(r = -0.55)$. Consequently the regression of y on r^* is much better able to indicate the effect of the linear term of the model. This is why the estimated coefficient is positive and statistically significant.

Significant lack of fit (see Table 23.5) is detected in the ANOVA table for both of the quadratic fits shown in Table 23.4. A reasonable question to ask is whether a cubic term will help better describe the relationship between solvent content and production rate. The answer is no. This lack of fit is caused by poorly fitting the observations corresponding to high solvent contents, in particular observations 34 and 37. The poor fitting of these two points is symptomatic of the real problem with this fit: the increasing variability of the solvent weights with the magnitude of the weights. A transformation of the solvent weights that satisfactorily accounts for both its apparent nonlinear relationship with the production rate and its nonconstant variability is presented in Section 26.2.

23.2 RESPONSE-SURFACE MODELS

A response-surface model represents the functional form of a response surface. Response surface models can be based on either theoretical or empirical considerations. When a theoretical model cannot be specified in an experimental investigation (the usual case), polynomial models often are used to approximate the response surface. A quadratic polynomial [equation (23.3)] can provide a useful approximation for a broad range of applications.

A quadratic response-surface model in two predictor variables consists of the following terms:

$$y = \beta_0 + \beta_1 x_1 + \beta_2 x_2 + \beta_{12} x_1 x_2 + \beta_{11} x_1^2 + \beta_{22} x_2^2 + e. \tag{23.6}$$

As in equation (23.3), the regression coefficients in this model for the second-order terms have double subscripts. The number of times a 1 occurs in the subscripts indicates the power of x_1, and similarly for x_2. In addition to a sufficient number of observations, at least three distinct values of a predictor variable are needed to fit linear and quadratic powers of the predictor.

Response-surface models, and regression models in general, can be fitted to basically two types of data: observational data and data collected in a prescribed manner according to a designed experiment. Observational data are known by several names, such as "historical data," "old process data," and "happenstance data." Whatever the name, fitting observational data with response-surface models has several potential pitfalls, as shown in Table 23.6.

The list of potential problems with observational data is presented in part to point out the strengths of statistical designs that meet the criteria presented in Table 6.9. Data from factorial and fractional factorial experiments (Chapters 7 and 9) and from central

Table 23.6 Frequent Defects in Observational Data

- Collinearities among predictor variables are common. This partial confounding among the predictors is especially a problem among polynomial functions of a predictor variable.
- Important predictor effects may go undiscovered because the predictors vary over too narrow a range.
- When significant effects are identified, causation cannot be confirmed. The variable(s) that are significant may be surrogates for other unobserved or uncontrolled variables.
- Excessive effort may be spent dealing with gross data errors, missing values, and inconsistent data-collection periods.

Table 23.7   Responses from a Central Composite Design for a Polymer Density Study

No.	Location	Annealing Time (min)	Annealing Temperature (°C)	Polymer Density g/ml
1	Star point	60	190	101
2	Corner point	50	155	72
3	Corner point	50	225	101
4	Star point	30	140	70
5	"Extra" point	30	170	91
6	Center point	30	190	98
7	Star point	30	240	Missing
8	Corner point	10	155	70
9	Corner point	10	225	83
10	Star point	2	190	70

Data from Snee, R. D. (1985). "Computer-Aided Design of Experiments—Some Practical Examples," *Journal of Quality Technology,* **17,** 222–236. Copyright American Society for Quality Control, Inc., Milwaukee, WI. Used with permission.

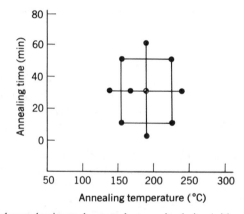

Figure 23.2   Polymer-density study: central composite design (with extra point added).

composite and Box–Behnken designs (Chapter 11) can be efficiently used to fit a response-surface model. Three-level factorial experiments, central composite designs, and Box–Behnken designs will provide data that can be used to fit a full quadratic response-surface model. Two-level factorial experiments and their fractions will yield data to fit a more limited model, linear in all the factors with some product terms.

Table 23.7 displays data from an experiment that used a central composite design to study the affects of annealing time and annealing temperature on the density of a polymer. Run 5 was an additional time and temperature condition included in the experiment by the investigator. Figure 23.2 shows the central–composite design for this experiment.

A quadratic response-surface model will be fit to these data using the standardized form of the predictor variables shown in equation (23.4). Table 23.8(a) lists the two standardized predictor variables, the products of these two variables, and the two squared variables for the complete quadratic model, equation (23.6). The correlations in

Table 23.8 Standardized Predictor Values for a Quadratic Fit to the Polymer Density Data

(a) *Standardized Predictors*

No.	Time t	Temp. T	tT	t^2	T^2
1	1.5601	0.0592	0.0924	2.4340	0.0035
2	1.0366	−0.9774	−1.0132	1.0745	0.9553
3	1.0366	1.0958	1.1360	1.0745	1.2009
4	−0.0105	−1.4216	0.0149	0.0001	2.0210
5	−0.0105	−0.5331	0.0056	0.0001	0.2842
6	−0.0105	0.0592	−0.0006	0.0001	0.0035
7	−0.0105	1.5401	−0.0161	0.0001	2.3719
8	−1.0575	−0.9774	1.0336	1.1184	0.9553
9	−1.0575	1.0959	−1.1589	1.1184	1.2009
10	−1.4764	0.0592	−0.0875	2.1797	0.0035

(b) *Correlations*

	t	T	tT	t^2	T^2
T	0.001	1			
tT	0.082	−0.014	1		
t^2	0.054	0.065	0.003	1	
T^2	−0.012	0.184	−0.004	−0.488	1

Table 23.9 Summary of the Quadratic Fit (Standardized Predictors) to the Polymer Density Data

(a) *Coefficient Estimates*

Model Term	Estimate	t-Value
Intercept	97.59	21.32
Time (z_1)	7.55	3.79
Temperature (z_2)	10.05	4.00
$z_1 z_2$	3.69	1.34
z_1^2	−6.09	−2.33
z_2^2	−7.88	−2.44

(b) ANOVA

Source	df	SS	MS	F	p-Value
Regression	5	1449.78	289.96	8.19	0.057
Error	3	106.22	35.41		
Total	8	1556.00			

Table 23.8(b) among the linear, product, and squared terms are modest and in many cases near zero. This indicates that each model term is essentially contributing independent information about the response. The small correlations among these model terms is a property of most of the statistical experimental designs presented in Chapters 7 to 11.

A summary of the quadratic fit is given in Table 23.9. Even though one of the

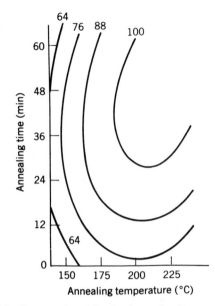

Figure 23.3   Contour plot of fitted polymer-density response surface.

experimental conditions resulted in no density measurement (the polymer melted at the highest temperature), the remaining data allowed the fit without any appreciable confounding of the effects of the terms in the model. This is another advantage of well-designed experiments: loss of one or two observations usually does not impair fitting of response surfaces or meaningful interpretation of the fitted models.

The estimated response function for the quadratic fit to the polymer density data is shown graphically in the contour plot displayed in Figure 23.3. The quadratic nature of the response surface is evident from this fit, a fit that required only nine well-chosen observations.

23.3   OPTIMIZATION

When a polynomial representation of a response surface is obtained, the experimenter often is interested in determining the factor levels that provide an optimum response. The preferred approach for accomplishing this task is the contour-plotting method described in Section 11.1. This use of contour plots is preferred for a variety of reasons. The foremost of these is that in most practical situations there is more than one response of interest in an experimental program. Thus, identification of an optimum in any one response is not as important as is knowing the tradeoffs available among several responses. Whether there is a set of predictor-variable conditions that will result in the optimum of all responses simultaneously becomes apparent when contour plots are overlaid. This approach is discussed in terms of the gentamicin study example in Section 11.1.

A second reason for employing this graphical method is that in addition to the optimum condition being apparent in a contour plot (it is often on the edge of an experimental region), there may also be portions of the experimental region where the response is insensitive to changes in the predictor variables. Where a response is robust to changes in predictor-variable values is important, especially in designing for product quality.

A drawback of the contour-plot approach is that as the number of responses and particularly the number of predictor variables increase, the number of plots to review may appear overwhelming. This can be minimized by eliminating nonsignificant variables from the model and putting predictor variables involved in interactions and curvature on the two axes available while holding other predictor variables at constant values. A contour plot is generated for each response at each combination of the levels of the predictor variables that are being held fixed. A rule of thumb that keeps the number of plots within bounds while making it large enough to allow the salient features of the system under study to be seen is to have each predictor variable being held constant take on three equally spaced values.

For example, if there are four responses of interest and five predictor variables, three of the predictor variables will have to be held constant. If they are each held constant at three different levels, the number of contour plots that result for study is $3 \times 3 \times 3 = 27$ per response, for a grand total of 108. After studying the initial set of plots, the investigator may decide to interchange the fixed predictor variables with one or both of those on the axes; in addition, one or more of the fixed variables may be held at different levels. A new set of contour plots can then be generated.

The question of an optimum response can be mathematically investigated. If an maximum or minimum response value exists, it corresponds to the predictor variable values that are a solution of the set of equations obtained by setting each of the partial derivatives of the estimated response function with respect to the predictor variables equal to zero. The predictor-variable values that satisfy this set of equations are called a *stationary point* (in a p-dimensional space). Canonical analysis is a mathematical approach that can be used to determine the stationary point and whether it represents a maximum, minimum, or saddle point. Due to the mathematical background required to understand and use canonical analysis (a facility with matrix algebra and calculus), a comprehensive treatment of this topic is beyond the scope of this book. The appendix to this chapter outlines algebraic methods for locating predictor-variable values that optimize a response function.

23.4 MIXTURE MODELS

The mixture or formulation problem was introduced in Section 11.4. Just as the mixture constraint $(x_1 + x_2 + \cdots + x_p = 1)$ places restrictions on the factor space, it also affects the form of polynomial models fitted to data from mixture experiments. Using the constraint $\sum x_i = 1$, the linear polynomial model given in equation (23.2) can be reexpressed as

$$y = \beta_0 + \beta_1 x_1 + \beta_2 x_2 + \cdots + \beta_p x_p + e$$
$$= \beta_1^* x_1 + \beta_2^* x_2 + \cdots + \beta_p^* x_p + e,$$

where $\beta_i^* = \beta_0 + \beta_i$, $i = 1, 2, \ldots, p$. Using this same constraint and noting that it implies $x_i = 1 - \sum_{j \neq i} x_j$, the quadratic polynomial model shown in equation (23.3) can be reexpressed as

$$y = \beta_0 + \sum_i \beta_i x_i + \sum \sum_{i<j} \beta_{ij} x_i x_j + \sum_i \beta_{ii} x_i^2 + e$$

$$= \sum_i \beta_i^* x_i + \sum \sum_{i<j} \beta_{ij}^* x_i x_j + e,$$

where $\beta_i^* = \beta_0 + \beta_i + \beta_{ii}$ and $\beta_{ij}^* = \beta_{ij} - \beta_{ii} - \beta_{jj}$.

For the remainder of this section, the reexpressed polynomial models will be discussed. In doing so, we drop the asterisks for convenience. Thus, the above two models will be written as

$$y = \beta_1 x_1 + \beta_2 x_2 + \cdots + \beta_p x_p + e \qquad (23.7)$$

and

$$y = \sum_i \beta_i x_i + \sum \sum_{i<j} \beta_{ij} x_i x_j + e \qquad (23.8)$$

respectively. The designs in the appendix to Chapter 11 enable the fitting of linear mixture models. The designs in Table 11.A.3 can be used to fit linear models of the form (23.7). The designs in Table 11.A.2 allow the fitting of a quadratic model of the form (23.8).

The coefficients in the models specified in equations (23.7) and (23.8) have a different interpretation than do the response-surface model coefficients given in equations (23.2) and (23.3). For response-surface models in which predictor variable values are not constrained by the values of the other predictors, the β_i are slopes, the β_{ij} describe an interaction, and the β_{ii} convey the direction and magnitude of curvature. This is not true for the coefficients of the mixture models, due to the constraint on the predictor variables. The β_i correspond to the expected response for formulations that are 100% of a single component (or at the maximum value of that component in the case of pseudocomponents—see Section 11.4). Alternatively, β_i is the height of the response surface at one of the vertices of a simplex design or, if a simplex design is not used, when $x_i = 1$ and $x_j = 0, j \neq i$. The β_{ij} describe the nonlinear behavior of the response; i.e., these coefficients describe the departure of the response surface from a linear representation.

An illustration of the interpretation of mixture-model coefficients for a two-component mixture system is shown in Figure 23.4. Two response surfaces are depicted in this figure, one linear and the other quadratic. The two surfaces are related in that they have identical coefficients for the linear terms in the model. That is, the theoretical mean response value is the same for the pure mixtures ($x_1 = 1.0$ and $x_2 = 0.0$ or $x_1 = 0.0$ and $x_2 = 1.0$) for both surfaces. The quadratic surface depicts synergistic nonlinear mixing of the two components. The maximum difference between the responses for the two surfaces equals $0.25\beta_{12}$ at the 50:50 mixture ($x_1 = x_2 = 0.5$).

The hypotheses of interest to the experimenter in mixture experiments differ from those described for regression models in Chapters 21 and 22. One null hypothesis of interest is that the response surface is linear; i.e., no curvature. This hypothesis, $H_0: \beta_{ij} = 0$ for all i and j, can be tested using the principle of reduction in error sums of squares (Sections 15.1 and 21.3) and fits to the reduced (M_2) linear model in equation (23.7) and the full (M_1) quadratic model in equation (23.8).

A second hypothesis of interest is that the mixture components do not affect the response. In terms of the mixture-model coefficients for the quadratic model, this

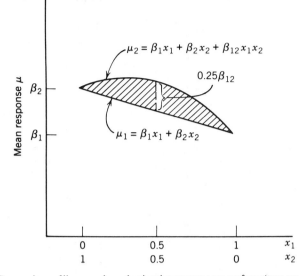

Figure 23.4   Comparison of linear and quadratic mixture response surfaces (two-component model).

hypothesis is $H_0: \beta_1 = \beta_2 = \cdots = \beta_p$ and $\beta_{ij} = 0$. For a linear mixture model the hypothesis is $H_0: \beta_1 = \beta_2 = \cdots = \beta_p$. These hypotheses can again be tested using the principle of reduction in error sums of squares. The reduced model in either case is the mean model (13.1):

$$M_2 : y = \mu + e,$$

where μ is the common value of the β_i; and the full model (M_1) is either equation (23.7) or (23.8).

Caution should be exercised when using computer software to test hypotheses with mixture models. While equations (23.7) and (23.8) appear to be ordinary regression models without intercept terms, the statistics appropriate for testing the above hypotheses are not identical to those from ordinary least-squares calculations. Using the principle of reduction in error sums of squares, correct statistics are derived. One can also use that principle to calculate sums of squares that are appropriate for testing the significance of individual model coefficients.

ANOVA tables can be formed to summarize the fit to mixture models. Using the principle of reduction in error sums of squares, the regression sum of squares has $m - 1$ degrees of freedom and the error sum of squares has $n - m$ degrees of freedom, where $m = p$ for linear mixture models and $m = p(p + 1)/2$ for quadratic mixture models. The degrees of freedom for regression point out another difference between the fitting of mixture models and the fitting of unconstrained regression models. There are m terms in the model, and one can calculate a t-statistic for each one; however, the overall regression fit has only $m - 1$ degrees of freedom.

This result implies that the individual tests on components are not all statistically independent. The problem is similar to that of forming individual tests for each level of a main effect in an ANOVA model. Note that if one were to use individual t-statistics to test for

Table 23.10   Simplex Mixture Experiment Data for Animal-Feed Additive Example

Monomer	Oligomer	Water	ln(viscosity)
0.30	0.45	0.25	4.11
0.60	0.30	0.10	6.18
0.30	0.60	0.10	6.80
0.30	0.45	0.25	4.35
0.30	0.30	0.40	1.24
0.45	0.30	0.25	3.76
0.40	0.40	0.20	4.86
0.35	0.50	0.15	5.87
0.30	0.60	0.10	6.91
0.45	0.45	0.10	6.35
0.50	0.35	0.15	5.04
0.45	0.30	0.25	3.47
0.40	0.40	0.20	5.02
0.60	0.30	0.10	5.83
0.35	0.50	0.15	6.10
0.50	0.35	0.15	5.51
0.35	0.35	0.30	2.99
0.45	0.45	0.10	6.26
0.30	0.30	0.40	1.33
0.35	0.35	0.30	3.45

the statistical significance of a main effect using only the observations on one level of a factor, the null hypothesis would be $H_0: \mu + \alpha_i = 0$ (ignoring interaction parameters) and not $H_0: \alpha_i = 0$. So too with mixture models. Hypotheses corresponding to individual t-statistics involve several of the original model parameters because of the definition of the parameters in the reexpressed models.

As with the testing of hypotheses in mixture models, one must exercise caution with the formulas used to calculate summary statistics describing the adequacy of the fit. In particular, the coefficient of determination should be calculated as in equations (22.14) and (22.15), with $a = (n-1)/(n-m)$ in the latter, rather than with alternative formulas that only may be appropriate for least-squares estimates from unrestricted models.

Standardization of mixture experiment data is unnecessary when the designs advocated in the appendix to Chapter 11 are used. These designs are specified in terms of coded component values that are suitably standardized. Likewise, transforming from raw components to pseudocomponents [equation (11.4)] automatically standardizes the original component values.

The data in Table 23.10 are the responses for the simplex mixture experiment given in Table 11.6 for the animal-feed additive example. The response is the natural logarithm of the viscosity. In addition to the response, both the pseudocomponents and the raw component values for each of the twenty experiments are shown. The lower bounds for the raw component are $l_1 = 0.30$, $l_2 = 0.30$, and $l_3 = 0.10$. Figure 23.5 is a plot of the average responses for the two repeat tests at each design point superimposed on the simplex of the original components.

Table 23.11 summarizes the fit of a quadratic mixture model to ln(viscosity). The coefficients of the individual model terms with their associated t-values indicate that

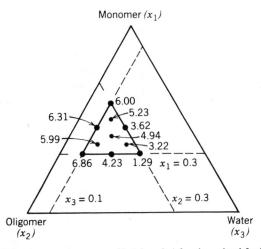

Figure 23.5 Average responses of ln(viscosity) for the animal-feed study.

Table 23.11 Summary of the Quadratic Fit (Pseudocomponents) to the Animal-Feed Additive Data

(a) *Coefficient Estimates*

Model Term*	Estimate	t-Value
s_1	5.95	45.19
s_2	6.86	52.10
s_3	1.29	9.80
$s_1 s_2$	−0.17	−0.27
$s_1 s_3$	0.21	0.34
$s_2 s_3$	1.09	1.79

(b) ANOVA

Source	df	SS	MS	F	p-Value
Regression	5	52.204	10.441	282	0.000
Error	14	0.522	0.037		
Total	19	52.726			

*$s_1 = (\text{monomer} - 0.3)/0.3$, $s_2 = (\text{oligomer} - 0.3)/0.3$, $s_3 = (\text{water} - 0.1)/0.3$.

ln(viscosity) mixes essentially as a linear function of the three components. Using the fitted model, the effects of the mixture components can be calculated (see Exhibit 23.4). These effects are not simply differences in averages.

EXHIBIT 23.4 MIXTURE COMPONENT EFFECTS

The effect of the ith component on the response variable in a mixture model is the difference between the maximum and the minimum predicted response over the range of the ith component values in the experimental region.

In general, contour plots on response traces (see below) must be used to identify the maximum and minimum predicted responses for each component. This is especially true when mixture models are nonlinear or when the experimental region is constrained so that at least one of the components cannot attain the lower bound of zero or the upper bound of one (i.e., for at least one component $a_i \leqslant x_i \leqslant b_i$ and $a_i > 0$ or $b_i < 1$). When only the original mixture constraint $\sum x_i = 1$ and $0 \leqslant x_i \leqslant 1$ is imposed and the mixture model is linear, mixture component effects can be calculated from the estimated model coefficients (see Exhibit 23.5).

EXHIBIT 23.5   CALCULATING LINEAR MIXTURE COMPONENT EFFECTS

1. The linear effect for the ith component is calculated as the difference between the predicted response value for $x_i = 1$ and $x_j = 0$ ($j \neq i$), and the predicted response value for $x_j = 0$ and $x_j = (p-1)^{-1}(j \neq i)$ (i.e., the mixture equally split among the remaining components), where the predicted value is based on the fitted linear model:

$$\hat{y} = b_1 x_1 + b_2 x_2 + \cdots + b_p x_p.$$

2. The linear effect can be written in terms of the linear coefficients from the fit as

$$\text{Effect}_i = b_i - (p-1)^{-1} \sum_{j \neq 1} b_j.$$

Response traces are plots of the predicted responses versus the component or pseudocomponent values, as appropriate, for each individual component as it varies from 0 to 1. As the individual component is varied from 0 to 1, the other $p-1$

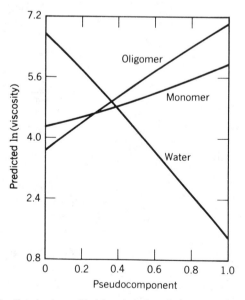

Figure 23.6   Predicted values of ln(viscosity) plotted along pseudocomponent axis.

(pseudo)components are assigned values $(1 - x_i)/(p - 1)$ or $(1 - s_i)/(p - 1)$, as appropriate. The p different plots can be overlaid to show the relative effects of the components.

Figure 23.6 is a plot of the predicted responses for each of the pseudocomponents in the animal feed study along a common axis. From the figure it is apparent that water has the greatest effect on ln(viscosity), since the response change from approximately 6.4 to 1.5 is greater than the maximum changes due to either of the other two components. It is also apparent that along the pseudocomponent axes water has a negative effect on ln (viscosity), whereas the other two components have positive effects. The exact nature of the joint effects of these three components as well as optimum formulations for achieving desired values of ln(viscosity) can be obtained from an evaluation of contour plots.

Based on the preceding discussion, a procedural outline for fitting polynomial mixture models is given in Exhibit 23.6.

EXHIBIT 23.6 FITTING POLYNOMIAL MIXTURE MODELS

1. Specify either a linear [equation (23.7)] or a quadratic [equation (23.8)] model. Ensure that the experimental design used will allow the fitting of the specified model.

2. Standardize mixture components to pseudocomponents, if appropriate.

3. Using a regression computer program that allows suppression of the intercept term, fit the specified model to the data. Assess the adequacy of the fit using the techniques described in Chapter 25, with the modifications suggested in this section. If the initial model is not adequate, consider higher-order models or transformations of the response.

4. Calculate component effects or plot the predicted responses for the components along their (pseudo)component axis.

5. Construct contour plots of the estimated response surface over the experimental region.

There will be occasions when the polynomial mixture models introduced in this chapter do not provide the necessary flexibility to describe adequately the response surface under study. Other classes of models are available that may ameliorate this situation. Discussions of these alternative models can be found in the references at the end of this chapter.

APPENDIX: LOCATING OPTIMUM RESPONSES

1 Path of Steepest Ascent

Graphical or algebraic techniques for identifying predictor-variable values that correspond to a maximum or a minimum of a response surface are most effective when the predictor variables are equivalently centered or scaled. This can be accomplished by using coded predictor or factor levels as in Section 11.4 or by standardization as in equations (23.4) or (23.5).

The path of steepest ascent through the experimental region follows the direction of maximal increase or decrease in the predicted response variable. It is most frequently used with prediction equations that are linear in the predictors. Canonical analysis is usually used with second-order prediction equations.

The path of steepest ascent is most satisfactorily implemented when the predictor variables or factor levels are coded so that the center of the experimental region corresponds to the value 0 for each predictor and the limits on each factor are ± 1 (see

Section 11.4). Denoting the scaled predictors by x_j, $j = 1, 2, \ldots, k$, the fitted prediction equation is

$$\hat{y} = b_0 + b_1 x_1 + b_2 x_2 + \cdots + b_k x_k. \tag{23.A.1}$$

The path of steepest ascent is determined by finding the values of x_j that either maximize or minimize (23.A.1) subject to the constraint

$$\sum x_j^2 = c, \tag{23.A.2}$$

where c is any constant. The constraint (23.A.2) simply requires that the solution to the maximization or minimization of (23.A.1) lie on a sphere of radius $c^{1/2}$. The path of steepest ascent is then a line from the origin through the point on the sphere (23.A.2) for which the response (23.A.1) is the largest or smallest, as appropriate.

One can show that the solution to this problem is

$$x_j = q b_j, \tag{23.A.3}$$

where q is a constant that depends on the radius of the sphere in (23.A.2), and b_j is the coefficient of x_j in (23.A.1). Since choices of the radius of the sphere are arbitrary, so are values of q. Because of this, the procedure in Exhibit 23.A.1 can be used to locate an optimum response using the path of steepest ascent.

EXHIBIT 23.A.1   PATH OF STEEPEST ASCENT

1. Code the factor or response variable so that the center of the experimental region is at the origin $x_1 = \cdots = x_k = 0$ and the extremes in each variable are -1 and $+1$.

2. Fit a linear response surface of the form (23.A.1) to the response.

3. Choose a value for one of the predictors, say x_1, at which a new test run or observation is to be made. This new value should be some fractional change in the largest or smallest coded value, depending on the sign of b_1 and on whether the optimum response is a maximum or a minimum of the response surface. For example, if b_1 is positive and a maximum response is sought, choose x_1 to be 1.1, 1.2, or some other convenient value.

4. With a selected value of x_1, solve (23.A.3) for q: $q = b_1 / x_1$. Insert this value of q into (23.A.3), and solve for the other x_j. The values of $x_1, x_2, \ldots, x_k$ determine a set of predictor-variable values or factor–level combinations at which the next observation is to be taken.

5. Determine the value of the response for the x_j-values from step 4. If a maximum (minimum) is sought and the new response is an increase (a decrease) over the previous one, return to step 3. Otherwise, terminate experimentation along the path of the steepest ascent.

Once the path of steepest ascent no longer leads to increases or decreases, as appropriate, or the changes become small, this procedure should be discontinued in favor of a more elaborate experiment. The response-surface designs discussed in Chapter 11 should then be used to collect data from which a satisfactory response-surface model can be fitted.

2   Canonical Analysis

When second-order regression models such as equation (23.3) are used to characterize a response surface, the path of steepest ascent cannot be used to identify the location of the optimum response. A *canonical* analysis of the fitted prediction equation is used.

For a canonical analysis, it is convenient to use standardized predictor variables, since polynomial models are being fitted. A complete second-order model in matrix notation (see the appendix to Chapter 22) can be written as

$$\hat{y} = b_0 + \mathbf{x}'\mathbf{b} + \mathbf{x}'B\mathbf{x}. \tag{23.A.4}$$

Here $\mathbf{b}' = (b_1, b_2, \ldots, b_k)$ is a vector that contains the least-squares estimates of the linear terms of the model (23.3), $\mathbf{x}$ is a vector of the values of the k predictor variables, and B is a symmetric matrix containing the estimates of the second-order terms of the model (23.3) in the following form:

$$B = \begin{bmatrix} b_{11} & b_{12}/2 & \cdots & b_{1k}/2 \\ b_{12}/2 & b_{22} & \cdots & b_{2k}/2 \\ \vdots & \vdots & & \vdots \\ b_{1k}/2 & b_{2k}/2 & \cdots & b_{kk} \end{bmatrix}.$$

A stationary point of the fixed response surface (23.A.4) is obtained by setting the partial derivatives of the fitted surface with respect to each of the predictors equal to zero. The result is

$$\mathbf{x}_s = -B^{-1}\mathbf{b}/2. \tag{23.A.5}$$

The nature of the stationary point—whether it locates a maximum, a minimum, or a saddle point—is determined by expressing the prediction equation (23.A.4), evaluated at the stationary point (23.A.5), in canonical form.

Denote the eigenvalues of B by $l_1, l_2, \ldots, l_k$ and the corresponding eigenvectors by $\mathbf{v}_1$, $\mathbf{v}_2, \ldots, \mathbf{v}_k$. The eigenvalues and eigenvectors satisfy the matrix equation $B\mathbf{v}_j = l_j\mathbf{v}_j$, $j = 1$, $2, \ldots, k$. The canonical form of the prediction equation can then be expressed as follows:

$$\hat{y} = \hat{y}_s + l_1 u_1^2 + l_2 u_2^2 + \cdots + l_k u_k^2, \tag{23.A.6}$$

where $\hat{y}_s$ is the value of the prediction equation (23.A.4) at the stationary point $\mathbf{x}_s$, and $u_j = \mathbf{v}_j'(\mathbf{x} - \mathbf{x}_s)$ is a transformation of the original k predictor variables to k new canonical variables u_j.

If all the l_j are positive, (23.A.6) shows that any departure from the stationary point $\mathbf{x}_s$ results in an increase in the predicted response. Thus, the stationary point locates a minimum of the response surface. If all the l_j are negative, the stationary point locates a maximum of the fitted response surface, since any departure from the stationary value decreases the predicted response. If some of the l_j are positive and some are negative, the stationary point is a saddle point: the fitted response increases in directions corresponding to positive l_j, and it decreases in directions corresponding to negative l_j. These directions emanate from the stationary point and are defined by the transformations $\mathbf{v}_j'\mathbf{x}$ of the original coordinate axes.

The magnitudes of the l_j indicate the relative sensitivity of the predicted response variable to changes in the direction defined by u_j. Large values of l_j indicate [see equation (23.A.6)] that unit changes in the corresponding u_j-direction result in large increases or decreases in the predicted response relative to those directions corresponding to small l_j.

REFERENCES

Text References

The following texts discuss the role of empirical models, including polynomial models, in describing response surfaces. They also illustrate and discuss the method of canonical analysis for describing the fitted surface mathematically.

Box, G. E. P. and Draper, N. R. (1987). *Empirical Model-Building and Responses Surfaces*, New York: John Wiley and Sons, Inc.

Box, G. E. P., Hunter, W. G., and Hunter, J. S. (1978). *Statistics for Experimenters*, New York: John Wiley and Sons, Inc.

Myers, R. H. (1976). *Response Surface Methodology*, Boston: Allyn and Bacon, Inc.

The texts referenced in Chapter 21 cover regression analysis in general, including polynomial models. While each text is comprehensive, the depth of coverage of polynomial models varies.

The following text covers a broad range of mixture data-analysis topics, including the design of mixture experiments, the fitting of polynomial mixture models, and alternative models to the polynomial mixture models:

Cornell, J. A. (1981). *Experiments with Mixtures: Designs, Models, and the Analysis of Mixture Data.* New York: John Wiley and Sons, Inc.

The first of the following articles discusses mixture-model hypotheses and appropriate test statistics. The next two describe plotting the predicted response versus the (pseudo)components. The fourth provides a computer code to construct contour plots that take into account the mixture constraints. The third article and the references therein are especially helpful for determining component effects.

Marquardt, D. W. and Snee, R. D. (1974). "Test Statistics for Mixture Models," *Technometrics*, **16**, pp. 533–537.

Hare, L. B. (1985). "Graphical Display of the Results of Mixture Experiments," in *Experiments in Industry: Design, Analysis, and Interpretation of Results* (R. D. Snee, L. B. Hare, and J. R. Trout, eds.), Milwaukee, WI: American Society for Quality Control.

Snee, R. D. (1975). "Discussion: The Use of Gradients to Aid in the Interpretation of Mixture Response Surfaces," (by J. A. Cornell and L. Ott), *Technometrics*, **17**, pp. 425–430.

Koons, G. F. and Heasley, R. H. (1981). "Response Surface Contour Plots for Mixture Problems," *Journal of Quality Technology*, **13**, 207–214.

EXERCISES

1 Refer to Exercise 6, Chapter 22. Define second-order cross-product and quadratic terms for each of the three predictor variables in the nitrogen oxide data set. Do so without centering or scaling the raw humidity, temperature, or barometric-pressure variables. Calculate correlations between all pairs of the nine second-order terms. Are there any obvious collinearities among the second-order terms? Comment on the advisability of including all the second-order terms in a model for the nitrogen oxide emissions.

2 Calculate least-squares coefficient estimates for the complete nine-term second-order fit to the nitrogen oxide emissions using the untransformed second-order terms for Exercise 1. Compare the least-squares coefficient estimates of the linear parameters from this fit with those from a fit using only the linear terms. How have the coefficient estimates changed? Are the effects of the collinearities apparent? If so, how? Which fit would you prefer to use? Why?

3 Redo Exercises 1 and 2 using standardized predictor variables. How do the conclusions from the two previous exercises change?

4 Refer to the data in Exercise 7, Chapter 22. Two indices that are often important in studies of this type are the brachial index and the tibiofemural index. The brachial index is the ratio of the upper-arm length to the forearm length. The tibiofemural index is the ratio of the upper-leg length to the lower-leg length. Fit a regression model to the overall height using the seven original predictors and the two indices. Compare the coefficient estimates of the nine-predictor fit with those of the fit to the seven original variables. Examine the t-statistics for the two fits. Are there any important differences in the fits, especially among the four arm and leg predictors? (We return to this data set for a comprehensive examination of its collinearities in the exercises in Chapter 28.)

5 Consider the algebraic form of a first-order multiple linear regression model, equation (23.2). Use suitable transformations of the predictor variables and the regression coefficients to rewrite the original model in terms of the standardized predictors (23.5). How do the original model parameters relate to the parameters in the transformed model? Assume that least-squares estimates are available for the original or for the transformed model. The least-squares estimates for the two models satisfy the same relationships as the two sets of model parameters. Express the least-squares estimates for the parameters of the transformed model as a function of the least-squares estimates of the original model parameters.

6 Refer to Exercise 4, Chapter 11. Using the methodology described in the appendix to this chapter, design a sequence of test runs along the path of steepest ascent to locate the maximum temperature. List at least five test runs.

7 The data below are coded wear measurements for the second design in Exercise 5 of Chapter 11. Fit a second-order model to these data. Do the t-statistics from the model indicate that a second-order model is necessary for an adequate fit to the wear measurements?

	Coded Factor Levels	
	-1	$+1$
Temperature (°F)	500	1000
Contact Pressure (psi)	15	21

Run No	Temperature	Pressure	Wear
1	-1	-1	0.1014
2	-1	1	0.5009
3	1	-1	0.8152
4	1	1	0.4026
5	0	-1.5	0.7001
6	0	1.5	0.1995
7	-1.5	0	0.5753
8	1.5	0	0.8747
9	0	0	0.4893
10	0	0	0.5031
11	0	0	0.5118

8 Use the coefficient estimates in Table 23.9 to locate the stationary point for the polymer-density fitted response surface. Transform the coded annealing-time and annealing-temperature values at the stationary point to values in the original scales of the two predictors. Using a canonical analysis, determine whether the stationary point locates a maximum, a minimum, or a saddle point for the response.

9 Use a canonical analysis to investigate the nature of a second-order response surface fitted to the data in Exercise 7. Is there a maximum or a minimum within the experimental region? If so, identify the factor levels that correspond to the optimum response.

10 The data below are measures of peak separation (larger is better) on a liquid chromatograph using various solvents that are mixtures of three different base solvents. The mixture components are expressed as volume proportions. Fit a quadratic mixture model and find the solvent combination that gives the best resolution.

Base Solvent			
A	*B*	*C*	**Peak Separation**
1	0	0	0.6
0	1	0	3.1
0	0	1	0.2
0.5	0.5	0	1.3
0.5	0	0.5	1.6
0	0.5	0.5	6.9
0.33	0.33	0.34	3.5
0.66	0.17	0.17	1.9
0.17	0.66	0.17	4.0
0.17	0.17	0.66	0.8

11 The equilibrium temperature deviations from baseline for the formation of gas hydrates in drilling fluids are given below for various combinations of six different drilling-fluid constituents. The mixtures for this screening test are given in Table 11A.3.

Kind of Point	No.	Temperature Deviation
Vertex	1	14.5
	2	3.7
	3	9.8
	4	7.5
	5	6.0
	6	0.6
Interior	1	7.9
	2	6.7
	3	2.8
	4	7.9
	5	0.5
	6	3.6
Centroid	1	9.2, 7.5, 9.5

Due to constraints on the minimum amount needed for some of the constituents, pseudocomponents were used. The minimum constraints were:

Constituent	Minimum Needed (Volume %)
x_1	50
x_2	5
x_3	1
x_4	5
x_5	0
x_6	2

Determine the effects that each of the six drilling fluid components has on the temperature deviation.

CHAPTER 24

Outlier Detection

Outliers are observations that have extreme values relative to other observations observed under the same conditions. Observations may be outliers because of a single large or small value of one variable or because of an unusual combination of values of two or more variables. In this chapter we discuss outlier detection techniques for:

- *response variables collected under one experimental condition,*
- *response variables observed for several factor–level combinations in a designed experiment, and*
- *response and predictor variables used to fit regression models.*

Outliers are important for at least two reasons. First, their presence in a data set may obscure characteristics about the phenomena being studied that are present in the bulk of the other data values. Second, outliers may provide unique information about the phenomenon of interest that is not contained in the other observations. Outliers were introduced in Chapter 3 in the context of the deleterious effects that they have on traditional summary statistics. Outliers can also influence the results obtained in the analysis of designed experiments, and in fitting regression models.

Statistical techniques for dealing with the possible presence of outliers in a data set fall into two general categories: identification and accommodation. Identification refers to techniques used to determine whether any outliers exist in a data set and, if so, which observations are outliers. Accommodation of outliers refers to techniques used to mitigate their effects. These techniques include the deletion of outliers, trimming extreme observations to less extreme values, using outlier-resistant estimators, respecifying the assumed model, or collecting additional data.

While the focus of this chapter is on the identification of outliers, a few comments are in order about accommodation. Accommodation need not be preceded by the finding of outliers. Often outlier-resistant estimators are used without regard to a prior determination that outliers exist. Identification and accommodation are not competing concepts; rather, they reinforce one another. In fact, many outlier-resistant estimators provide statistics that are valuable for identifying outliers. The M-estimator of location and the other robust estimators introduced in Chapter 3 are outlier-resistant forms of accommodation. Robust regression alternatives presented in Section 28.2 are also outlier-resistant. In both of these settings, the weights used with the respective estimators can be used to identify outliers.

Identifying outliers is a key to both remedial action and discovering potential new information. Regardless of whether any form of accommodation is used, outliers should be identified and studied, since they may arise from experimental conditions that differ from those that were intended. Information obtained in this manner may lead to a more fundamental understanding of the physical mechanism under investigation.

24.1 UNIVARIATE TECHNIQUES

The graphical techniques introduced in Chapters 3 and 4 are useful for identifying potential outliers in one variable. These techniques include stem-and-leaf plots, box plots, and normal quantile–quantile plots. We recommend the use of normal quantile–quantile plots, in part because many of the statistical procedures recommended in this book are based on an assumption of normality.

Table 3.2 in Chapter 3 contains fuel-economy measurements for four vehicles, each using eight different fuels. The scatterplot in Figure 3.2 and the box plots in Figure 4.7 highlight the one very low measurement on the Volkswagen as an outlier. By comparison, the plots for the Mercedes and the Oldsmobile do not show any discrepant points. The plots for the Peugeot exhibit one moderately large value, which might be labeled an outlier, but the plots are not as conclusive as those for the Volkswagen.

Figure 24.1 displays normal quantile–quantile plots of the fuel-economy data for the Volkswagen and the Peugeot. Outliers in a quantile–quantile plot usually appear as one or more values at the extremes that depart from a line that passes through the bulk of the remaining observations. The argument that the extreme values that depart from a line through the remaining points are outliers is offered cautiously, because there are often a few points that depart modestly from a straight line at the extremes of quantile–quantile plots even when the data do conform to the reference distribution.

The plot of the Peugeot values in Figure 24.1a is not sufficient to label the largest observation an outlier, because its departure is not large enough to justify an unequivocal conclusion. Contrast the uncertainty about the largest observation in Figure 24.1a with the gross departure of the smallest fuel-economy value for the Volkswagen from the remaining seven values in Figure 24.1b. This large a deviation from a straight line through the other seven points is beyond the modest departures which might be expected of extreme values.

Statistical tests for outlying observations are available to help confirm patterns seen in plots of the data. These statistical tests are especially valuable when extreme observations exhibit modest departures from the bulk of the other observations in the various plots mentioned above.

Commonly used statistics for detecting outliers among observations on a single variable are of two general types. *Dixon* tests for outliers use ratios of ranges and subranges of the data. *Grubbs* tests use ratios of two sums of squares. The numerator sum of squares does not contain the suspect observations. The denominator sum of squares contains all the observations, including the suspect values. Both Dixon and Grubbs tests are based on an assumption of normally distributed model errors (other than the outliers) in equation (13.1), $y = \mu + e$.

Dixon tests are of primary benefit when small samples are involved and only one or two observations are suspected as outliers. An appealing feature of Dixon tests is that they only

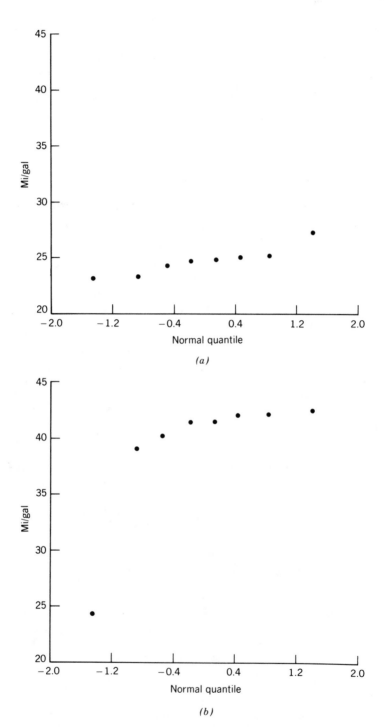

Figure 24.1   Fuel-economy quantile–quantile plots: (*a*) Peugeot, (*b*) Volkswagen.

involve the calculation of ranges, which is quick and easy to perform. Grubbs tests (see Exhibit 24.1) require more calculational effort but are generally more powerful and can be used to test any number of outliers. In the remainder of this section we discuss the use of Grubbs tests more fully. The interested reader is referred to the references for further information on Dixon tests.

EXHIBIT 24.1   TEST FOR OUTLIERS

1. Let $y_{(i)}$ denote the ith smallest observation; i.e., let the ordered responses be denoted by

$$y_{(1)} \leqslant y_{(2)} \leqslant \cdots \leqslant y_{(n)}.$$

2. If the k largest values in a data set are suspected as outliers, calculate

$$L_k = \frac{1}{S_{yy}} \sum_{i=1}^{n-k} (y_{(i)} - \bar{y}_L)^2, \qquad (24.1)$$

where $\bar{y}_L = \sum_{i=1}^{n-k} y_{(i)}/(n-k)$ and $s_{yy} = \sum_{i=1}^{n}(y_i - \bar{y})^2$. If the k smallest values in a data set are suspected as outliers, calculate

$$S_k = \frac{1}{S_{yy}} \sum_{i=k+1}^{n} (y_{(i)} - \bar{y}_S)^2, \qquad (24.2)$$

where s_{yy} is as defined above and $\bar{y}_S = (\sum_{i=k+1}^{n} y_{(i)}/(n-k)$

3. Conclude that the group of k observations are outliers if the calculated value of L_k or S_k is less than the critical value given in Table A26 of the appendix.

4. If the k most extreme observations are suspected as outliers but some are the largest values in the data set while others are the smallest ones, calculate

$$E_k = \frac{1}{S_{yy}} \sum_{i=1}^{n-k} (z_{(i)} - \bar{z}_E)^2, \qquad (24.3)$$

where $z_{(i)}$ is the y_i corresponding to the ith smallest value of the ordered absolute values of the deviations, i.e., the y_i corresponding to

ith smallest $|y_i - \bar{y}|$.

The value of $\bar{z}_E$ is the average of the y_i corresponding to the $n-k$ smallest deviations. Conclude that the group of k observations are outliers if the calculated value of E_k is less than the critical value given in Table A27 in the appendix.

The calculated value of S_k ($k=1$) for the Volkswagen fuel-economy data that corresponds to the lowest test result is

$$S_1 = 8.714/265.429 = 0.033.$$

The p-value associated with this statistic (see Table A26) is less than 0.01. Thus the lowest observed fuel-economy value is not consistent with the remainder of the data. As noted in Chapter 3, this low test result is not an error in the experimentation; it was confirmed in subsequent test runs. Robust measures of center and scale were used in Chapter 3 to summarize the data (accommodation).

The value of L_1 associated with the highest fuel-economy test result for the Peugeot is

$$L_1 = 4.000/11.875 = 0.337.$$

This value of the test statistic is not significant at the 0.05 level of significance ($0.05 < p < 0.10$).

The procedures just discussed for outlier testing should not be used without an accompanying plot of the data. If the statistics L_k, S_k, and E_k are routinely used as outlier diagnostics without any visual inspection of data sets, the researcher may be misled due to problems such as *masking*.

Masking occurs when two or more outliers have similar values. If a data set contains, say, two large values that are similar in magnitude, an outlier test for one of the observations ordinarily will not be statistically significant. This is because both the numerator and the denominator of the statistics (24.1) to (24.3) will be large, due to the presence of one outlier in the numerator and both outliers in the denominator. Only a test for both the two largest observations will be statistically significant. Plotting the data provides a measure of protection against this difficulty by identifying sets of outlying observations.

One other caution is important to understand when considering the use of either graphical or computational outlier-detection techniques. If all model assumptions (e.g., fixed-effects models with independent normal errors) are correct, test statistics for single outliers will identify observations as being outliers $100\alpha\%$ of the time; i.e., the proportion of observations erroneously identified as outliers will equal the significance level of the test. Tests for groups of outliers also will erroneously lead to the conclusion that the group of observations significantly deviates from the bulk of the remaining data $100\alpha\%$ of the time. For these reasons, a small significance level should be used with these tests, e.g., 0.01 or smaller.

With the above two cautions in mind, a general procedure for identifying outliers is outlined in Exhibit 24.2.

EXHIBIT 24.2   OUTLIER DETECTION

1. Plot responses using point plots, sequence plots, box plots, or normal quantile–quantile plots. Note any unusual trends or clustering of observations, as well as all extreme observations.

2. Based on the plots constructed in step 1, a scan of the data set, or any other analysis deemed appropriate, select the observation or group of observations that are suspected to be outliers.

3. Use a statistical test for outliers with a small significance level to determine whether the group of observations deviate significantly from the bulk of the observations.

24.2   RESPONSE OUTLIERS

Outliers in response variables can occur in the analysis of data from designed experiments and in regression analyses from either designed experiments or observational studies. In designed experiments outliers usually do not occur in the factor variables. This is because the levels of the factor variables are generally selected by the experimenter according to a balanced arrangement in the factor space. In regression analyses using either observational data or predictor variables that cannot satisfactorily be controlled, outliers can occur in the predictor variables. This section presents outlier detection techniques for response variables from designed experiments or regression analyses. Outlier diagnostic techniques for predictor variables in regression models are discussed in the next section.

Outlier-detection techniques for response variables are based on residuals: differences between observed (y) and predicted ($\hat{y}$) responses, $r = y - \hat{y}$. Residuals were introduced in Section 21.4 in the context of fitting regression models. Residuals can be defined similarly for data from designed experiments that are analyzed using ANOVA techniques. In designed experiments the predicted values are obtained by inserting estimates of all fixed effects for the corresponding model parameters [e.g., estimates of the parameters in equation (15.8)]. Any solution to the normal equations can be inserted for the model parameters to obtain the predicted response. See Section 25.1 for further discussion of residuals for univariate models and ANOVA models.

Residual plots are very effective for detecting outliers. Recommended techniques include plotting residuals in scatterplots versus the corresponding predicted responses, each factor or predictor variable, experimental run order, or any other meaningful variable. Outliers generally occur as points far above or below the bulk of the plotted residuals. Other graphical techniques include plotting the residuals in stem-and-leaf plots, box plots, and normal quantile–quantile plots as discussed in the previous section.

Table 24.1 contains 58 observations on yield and four operating conditions (catalyst

Table 24.1 Yield and Operating Conditions for a Chemical Process

Obs.	Yield	Catalyst	Conversion	Flow	Ratio
1	55.45	0	11.79	118.9	0.155
2	54.83	0	11.95	105.0	0.089
3	52.21	0	12.14	97.0	0.094
4	50.40	0	12.06	101.0	0.108
5	49.32	0	12.04	44.0	0.100
6	41.36	0	12.28	30.0	0.036
7	49.28	0	12.36	38.0	0.113
8	47.89	0	12.22	32.0	0.123
9	52.40	0	11.90	220.0	0.135
10	52.62	0	11.34	350.0	0.183
11	59.34	0	11.20	160.0	0.166
12	45.32	0	12.03	328.0	0.221
13	48.97	0	12.04	325.0	0.192
14	49.87	0	12.02	330.0	0.188
15	47.06	0	12.02	330.0	0.201
16	40.05	0	12.05	322.0	0.153
17	51.29	0	12.05	322.0	0.194
18	43.73	0	12.14	326.0	0.097
19	51.44	0	12.05	366.0	0.136
20	49.33	0	12.13	350.0	0.143
21	51.85	0	11.94	510.0	0.116
22	52.33	0	12.02	513.0	0.195
23	47.72	0	12.02	523.0	0.160
24	46.65	0	11.80	430.0	0.164
25	49.83	0	11.40	325.0	0.197

Table 24.1 (continued)

Obs.	Yield	Catalyst	Conversion	Flow	Ratio
26	51.36	0	11.19	380.0	0.233
27	51.19	0	11.19	380.0	0.211
28	48.10	0	11.18	383.0	0.222
29	48.28	0	11.21	375.0	0.223
30	51.28	0	11.21	375.0	0.229
31	47.81	0	11.88	410.0	0.170
32	46.78	0	12.09	485.0	0.163
33	44.24	0	11.97	445.0	0.153
34	53.07	0	11.30	232.0	0.180
35	54.08	0	10.90	220.0	0.126
36	54.19	0	10.90	223.0	0.152
37	54.26	0	10.90	222.0	0.184
38	55.94	0	10.80	240.0	0.225
39	55.85	0	10.90	238.0	0.169
40	56.57	0	10.80	258.0	0.161
41	53.85	0	10.75	272.0	0.197
42	54.81	0	10.72	280.0	0.201
43	54.14	0	11.10	426.0	0.221
44	53.74	0	11.12	404.0	0.215
45	58.40	1	10.90	466.0	0.343
46	51.02	1	11.00	490.0	0.278
47	49.53	1	10.90	483.0	0.280
48	48.50	1	10.90	427.0	0.296
49	51.97	1	10.96	465.0	0.339
50	52.35	1	11.42	527.0	0.306
51	51.83	1	11.60	480.0	0.274
52	44.51	1	11.65	465.0	0.243
53	49.46	1	11.60	150.0	0.159
54	49.58	1	11.50	160.0	0.165
55	51.42	1	11.50	166.0	0.185
56	48.85	1	11.30	166.0	0.165
57	47.42	1	11.50	167.0	0.160
58	48.33	1	11.50	167.0	0.161

type, percent conversion, flow of raw material, and ratio of reactants) for a chemical process. The experimenters believed that yield would be linearly related to conversion, flow, and the reciprocal of the reactant ratio ("invratio"). In addition, a conversion-by-flow interaction term has physical meaning, and catalyst type is suspected to cause a shift in the mean response. Since an interaction term is to be included in the model, we follow the recommendation of Section 23.1 and standardize the continuous predictor variables (conversion, flow, invratio) using the normal-deviate form, equation (23.4), prior to forming the interaction term. The fit of the five predictor variables to the yield is shown in Table 24.2.

Figure 24.2 displays a plot of the residuals versus the fitted values, one of the residual plots suggested above. One large negative residual (observation 16) appears to be separated from the bulk of the remaining residuals. This residual corresponds to the observation having the smallest response in Table 24.1.

Table 24.2 Regression Analysis for Chemical-Process Yield

| | | | *ANOVA* | | |
Source of Variation	df	Sum of Squares	Mean Square	F	p-Value
Regression	5	379.218	75.843	8.34	0.000
Error	52	472.728	9.091		
Total	57	851.946			

Variable*	Coefficient Estimate	t-Statistic	p-Value
Intercept	51.097	110.77	0.000
Catalyst	−2.202	−2.15	0.036
Conversion	−2.170	−4.73	0.000
Flow	−1.107	−2.24	0.030
Conversion × flow	−0.046	−0.10	0.919
Invratio	−1.088	−1.96	0.055

*Normal-deviate standardization for conversion, flow, and invratio.

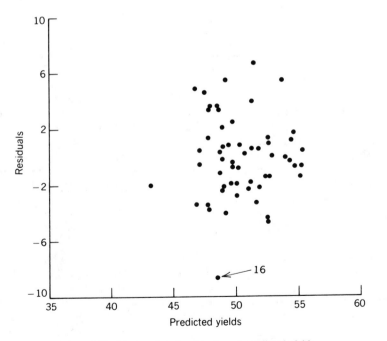

Figure 24.2 Scatterplot of residuals and predicted yields.

As with the univariate outlier-plotting techniques discussed in the last section, plots of raw residuals lead to subjective decisions about outliers. More objective decisions are available with the use of diagnostic statistics that are functions of the residuals. The general procedures that were recommended in the last section (plotting and calculating appropriate statistics) are also recommended for outlier detection with regression and ANOVA models. Instead of using Dixon or Grubbs tests, however, other test statistics are available for these models.

Studentized deleted residuals are popular outlier-detection statistics. The computation of studentized deleted residuals is based on deleting an observation from the data set, fitting the regression model with the remaining $n - 1$ observations, and using the fitted model to predict the observation that was deleted from the data set. If the deleted residual is large, an outlier is indicated. Studentized deleted residuals are deleted residuals that have been scaled by independent estimates of their standard errors so they individually follow Student t-distributions (see Exhibit 24.3).

EXHIBIT 24.3   STUDENTIZED DELETED RESIDUALS

Let $\hat{y}_{(-i)}$ denote the predicted response for the ith observation from a fit to a regression or ANOVA model when the ith observation has been deleted from the data set. The ith studentized deleted residual is

$$t_{(-i)} = r_{(-i)}/\widehat{SE}(r_{(-i)}), \tag{24.4}$$

where

$$r_{(-i)} = y_i - \hat{y}_{(-i)}, \tag{24.5}$$

and $\widehat{SE}(r_{(-i)})$ is a statistically independent estimate of the standard error of the ith deleted residual $r_{(-i)}$.

The studentized deleted residual $t_{(-i)}$ follows Student's t-distribution with $v - 1$ degrees of freedom, where v equals the number of degrees of freedom for error from the fitted model using all the data.

Calculation of studentized deleted residuals is outlined in the appendix to this chapter. Studentized deleted residuals can always be found for regression models once the least-squares estimates are calculated. For some designed experiments, Studentized deleted residuals cannot be calculated. This occurs when there is a single response for a factor–level combination. Deletion of that factor–level combination for some designs does not allow the estimation of all model effects. If all the effects corresponding to the (deleted) combination of interest cannot be estimated, its response cannot be predicted and the studentized deleted residual cannot be calculated.

Table 24.3 lists the raw and the studentized deleted residuals for a subset of the 58 observations for the chemical-process yield example of Table 24.1. The three other statistics (h_i, DFFITS$_i$, DFBETAS) in this table are defined later in this section or in Section 24.3. The large studentized deleted residual ($t_{(-16)} = -3.14$, $p < 0.001$) for observation 16 confirms the visual impression left by the residual plot, Figure 24.2. Note too the large value for observation 45 ($t_{(-45)} = 2.56$, $p = 0.012$). This observation has the second largest response value in Table 24.1 and is the first observation taken with the second catalyst type. There are now two outliers which the experimenter may wish to study further.

Residual plots are often made with studentized deleted residuals rather than raw residuals r_i. One reason for this is that studentized deleted residuals are conveniently scaled; since they behave like Student t-statistics, most of them should have values between -3 and 3. Another reason for plotting studentized deleted residuals is that they have equal standard errors, unlike raw residuals, which are not equally variable. Both types of plots generally display the same trends and highlight the same outliers.

Another conveniently scaled residual is termed a *studentized residual*:

$$t_i = r_i / \widehat{\text{SE}}(r_i).$$

The standard error used in t_i, unlike that for a studentized deleted residual, is not independent of r_i; therefore, the statistic t_i does not follow a Student t-distribution. It is, however, sufficiently similar to a Student's t-statistic that it can be considered to have an approximate t-distribution with v degrees of freedom, where v is the number of degrees of freedom for MS_E from the fit to the complete data set. The derivation of this statistic is outlined in the appendix to this chapter.

One of the key concerns of experimenters when deciding how to accommodate the presence of outliers is whether the outliers would seriously affect statistical procedures. Outliers that do affect statistical estimation or inference procedures are termed *influential observations* (see Exhibit 24.4).

EXHIBIT 24.4

Influential Observations. An outlier (extreme observation) is termed an influential observation if its presence in a data set strongly affects parameter estimates or statistical inference procedures.

The definition of influential observations given in Exhibit 24.4 does not provide explicit rules for the determination of whether outliers are influential. While the determination of when outliers constitute influential observations is often subjective, there are many outlier diagnostics that can aid in this decision.

Studentized deleted residuals are useful diagnostics for influential observations because they are measures of when a data set cannot satisfactorily predict a particular response. A related statistic, referred to by the mnemonic DFFITS, measures the change in the predicted value of the ith response using two fits to the data, one with and one without the ith response:

$$\text{DFFITS}_i = \frac{\hat{y}_i - \hat{y}_{(-i)}}{\widehat{\text{SE}}(\hat{y}_i)} \tag{24.6}$$

$$= \frac{h_i}{1 - h_i} t_{(-i)},$$

where $\hat{y}_i$ is the predicted response from the full (all n observations) fit to the model, $\hat{y}_{(-i)}$ is the predicted response from the fit in which the ith observation is deleted, and $\widehat{\text{SE}}(\hat{y}_i)$ is an independent estimate of the standard error of $\hat{y}_i$. As shown in the second expression in equation (24.6), DFFITS$_i$ is a multiple of the ith studentized deleted residual. The multiplier of $t_{(-i)}$ in (24.6) is a simple function of the ith "leverage value" h_i (see Section 24.3).

While similar in form to studentized deleted residuals, DFFITS values measure change in predicted responses. As such they are another measure of influence and are important in

their own right. Criteria for assessing whether an observation is influential according to the
DFFITS statistic depend on both the sample size and the number of predictor variables in
the model. One useful cutoff for regression models with intercept terms is $|\text{DFFITS}_i| >
3[(p + 1)/n]^{1/2}$, roughly corresponding to a Student t-value of 3. This cutoff would use p
rather than $p + 1$ for no-intercept models.

Another outlier diagnostic that is especially useful for assessing the influence of the ith
observation on the estimation of the jth regression coefficient in a regression analysis is
referred to by the mnemonic DFBETAS:

$$\text{DFBETAS}_{ij} = \frac{b_j - b_{j(i)}}{\widehat{\text{SE}}(b_j)}, \tag{24.7}$$

where the two estimates of β_j are from fits with and without the ith observation and $\widehat{\text{SE}}(b_j)$ is
an independent estimate of the standard error of the least-squares estimator b_j. A cutoff for
DFBETAS_{ij} is $3/n^{1/2}$.

Table 24.3 lists the outlier diagnostics DFFITS_i and DFBETAS_{ij} for selected observations.
The cutoff values for DFFITS and DFBETAS are, respectively, $3[(p + 1)/n]^{1/2} = 3(6/58)^{1/2} =
0.96$ and $3/n^{1/2} = 3/(58)^{1/2} = 0.39$. Observation 16 does not have a large value on DFFITS
but it does on DFBETAS(intercept, conversion). Observation 45 has a relatively large value
for both DFFITS and DFBETAS(flow, conversion × flow). Note too that observation 6 has
large values for DFFITS and DFBETAS(catalyst, flow, invratio) and the following observations
have large values for DFBETAS on at least one predictor variable: observation 11 (flow),
observation 21 (flow, invratio).

Since the diagnostics presented in Table 24.3 are from a regression analysis, the
influential observations that were identified could be the result of outliers in the response
or in the predictor variables. In the next section we discuss techniques for identifying
outlying observations in the predictors. We shall return to this example at that time.

**Table 24.3 Diagnostics for Influential Observations: Selected Observations, Chemical-Process
Yield Data**

Obs.	h_i	r_i	$t_{(-i)}$	DFFITS_i
6	0.834	−1.933	−1.594	−3.566
7	0.250	0.407	0.154	0.089
8	0.207	−1.856	−0.688	−0.351
11	0.084	5.565	1.982	0.559
16	0.043	−8.562	−3.141	−0.668
21	0.134	4.925	1.792	0.705
45	0.144	6.799	2.564	1.050

			DFBETAS			
Obs.	Int.	Cat.	Conv.	Flow	$C \times F$	Invratio
6	−0.164	−0.425	0.842	−0.999	0.261	−3.158
7	0.020	−0.021	0.037	−0.026	−0.062	−0.039
8	−0.091	0.084	−0.140	0.142	0.220	0.173
11	0.344	−0.099	−0.199	−0.397	0.311	−0.054
16	−0.447	0.132	−0.446	−0.020	−0.078	0.150
21	0.266	0.010	0.083	0.442	0.291	0.398
45	0.033	0.311	−0.293	0.448	−0.606	−0.046

Each of the outlier diagnostics, (24.4) to (24.6), appears to require $n + 1$ fits to a data set, one for the full data set and one for each reduced data set in which one observation is deleted. In fact, one can show that all these statistics can be calculated from quantities available after only one fit to the data, the fit to the complete data set. The interested reader is referred to the references and the appendix to this chapter for details.

As in the univariate case, the outlier diagnostics presented in this section can fail to detect extreme observations due to the clustering of two or more observations in the $(p + 1)$-dimensional space of response and predictor (or factor) variables. The techniques described in this section are most beneficial when outliers occur singly. The most effective outlier diagnostics for multiple outliers make use of group deletion of observations. To date there are few economically feasible procedures for identifying an arbitrary number of clustered outliers. The references cite several procedures that have been suggested for identifying multiple outliers.

24.3 PREDICTOR-VARIABLE OUTLIERS

The outlier diagnostics discussed in the last section can exceed their cutoff values not only because a response value is extreme but also because of outliers in the predictor variables. These diagnostics are thus indicators of influential observations, but they do not identify the cause of the influence. In this section we present diagnostic techniques that explicitly identify outliers in the predictor variables. These techniques do identify extreme (combinations of) predictor-variable values, but they are not measures of influence. They should be used in conjunction with the diagnostics for influential observations in order to assess whether the predictor-variable outliers are affecting estimates or inference procedures.

Response–predictor and predictor–predictor scatterplots are useful tools for discovering extreme values of the predictor variables. Figure 24.3 shows a scatterplot of yield versus one of the predictor variables, the inverse of the reactant ratio, for the chemical-process example. Observation 6 appears to be an outlier in this plot. This observation has an extremely large value for invratio, a value that removes it from the bulk of the other observations plotted in Figure 24.3. On occasion, a very large value for a predictor variable occurs, but the plotted point is consistent with a trend in the data. The extreme value for invratio does not appear to be consistent with any trend among the other points plotted.

Figure 24.4 is a plot of two of the predictor variables, conversion versus invratio. Again, observation 6 is an outlier on this plot. This observation appears to be an outlier on any plot involving the invratio. It is now clear why the extremely large DFBETAS value (the largest in the entire data set) for the invratio on observation 6 exists. This large value is distorting the estimated regression coefficient for invratio and is also the cause of the large DFFITS value for this observation.

At this juncture in a statistical analysis, the question of accommodation arises—in particular, the deletion of observations. While we urge caution in choosing to delete observations, if a decision is to be based solely on outlier diagnostics and if the invratio is to be included in the fitted model, observation 6 should be deleted. There is dramatic evidence that observation 6 is affecting the fit. If the invratio is ultimately removed as a predictor variable, one can reinsert observation 6 into the data set and reevaluate its influence on the fit. An underlying principle in this decision is that a single observation should not dictate the fit to a data set.

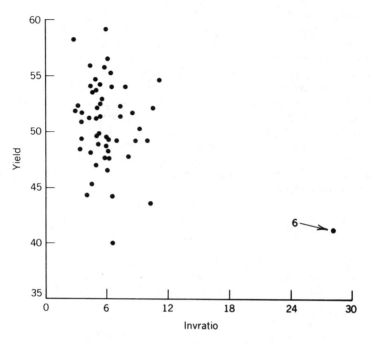

Figure 24.3   Scatterplot of yield and invratio.

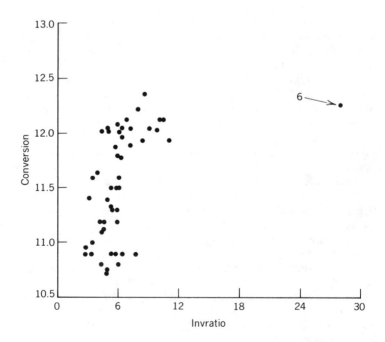

Figure 24.4   Scatterplot of conversion and invratio.

A decision to delete observation 6 is based in part on the premise that the data for this experiment were collected under controlled conditions and that observation 6 is not representative of other data that would be collected for this same value of invratio. If a researcher believes this is a valid observation in spite of the outlier diagnostics, it would be extremely worthwhile to collect a few more observations with the same predictor-variable values as observation 6. Then the influence of this observation on the fit could be unequivocally assessed.

When three or more predictor variables are used in a regression model, extreme combinations of the predictors may not be apparent from inspecting response–predictor and predictor–predictor scatterplots. Unusual combinations of values on response and predictor variables might involve three or more variables, thereby not appearing extreme on any single variable or any combination of two variables. The computation of *leverage values* aids in the identification of unusual combinations of predictor–variable values (see Exhibit 24.5). The calculation of leverage values is described using matrix algebra in the appendix to this chapter.

EXHIBIT 24.5   LEVERAGE VALUES

Leverage values h_{ii} (or h_i) are constants calculated from the values of the predictor variables. Leverage values have the following properties:

(a) $y_i = h_{ii}y_i + \sum_{j \ne i} h_{ij}y_j$, $i = 1, 2, \ldots, n$, where the h_{ij} are also constants calculated from the predictor variables;

(b) $0 \le h_{ii} \le 1$, $-1 \le h_{ij} \le 1$;

(c) $h_{ii} = \sum_{j=1}^{n} h_{ij}^2$; and

(d) h_{ii} measures the distance of the ith set of predictor-variable values from the point of averages of all $p + 1$ predictors (including the constant).

The algebraic properties described above can be summarized as follows. Property (a) states that a predicted response is a multiple of its corresponding observed response, added to a linear combination of the other responses. The importance of this property is that the multiplier on the observed response is the leverage value. Together, properties (b) and (c) state that leverage values are fractions between 0 and 1, and the closer the leverage value is to 1, the closer the h_{ij} are to 0. In the limit, if $h_{ii} = 1$, then $h_{ij} = 0$ ($j \ne i$). Thus properties (a) to (c) state that if a leverage value is close to 1, a predicted response is almost completely determined by its observed response, regardless of the adequacy of the fit of the model or the value of the observed response.

An implication of these properties is that leverage values close to 1 force the fit to pass through the observed response for the ith observation. Geometrically, the fitted line or plane must pass through the plotted point for the ith observation, regardless of any trend in the bulk of the remaining observations. Property (d) states that this will occur whenever the predictor-variable values for the ith observation are sufficiently extreme from the remaining observations.

Leverage values are important not only because of the above interpretations, but also because they can be calculated for any number of predictor variables. Leverage values can identify extreme observations when they occur because of an unusually large value on one variable or because of an unusual combination of values on two or more variables.

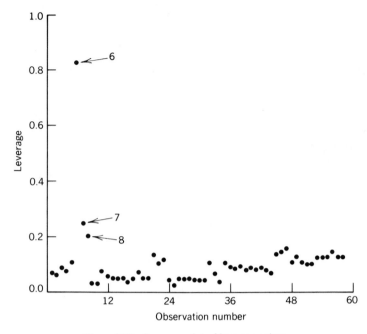

Figure 24.5   Sequence plot of leverage values.

Leverage values thereby aid in the detection of predictor-variable outliers even when such observations are not apparent from listings or plots of the variables in a data set.

Figure 24.5 shows a sequence plot of the leverage values h_i for the chemical process example. The cutoff value typically used to identify outliers in the predictors is $h_i > 2(p + 1)/n$ for models with intercepts and $h_i > 2p/n$ for no-intercept models. For this example,

$$2\frac{p + 1}{n} = 2\frac{6}{58} = 0.21.$$

Table 24.4   Comparison of Regression Coefficients with Influential Observations Removed: Chemical-Yield Data

Variable*	Complete Data Set	Observation 6 Deleted	Observations 6, 7, 8 Deleted
Intercept	51.097	51.171	51.217
Catalyst	−2.202	−1.173	−1.877
Conversion	−2.170	−2.565	−2.486
Flow	−1.107	−0.620	−0.703
Conversion × flow	−0.046	−0.165	−0.281
Invratio	−1.088	0.636	0.508

*Normal-deviate standardization for conversion, flow, and invratio.

A scan of either Table 24.3 or Figure 24.5 reveals that observation 6 clearly stands out as an outlier with a leverage value of 0.83. This information reinforces the conclusions made from the scatterplots and the outlier diagnostics discussed earlier in this section and in the last one. Figure 24.5 also shows that observations 7 and 8 are potentially influential ($h_7 = 0.25$ and $h_8 = 0.21$).

Table 24.4 shows a comparison of the regression coefficients from a fit of (1) all the data, (2) the data with observation 6 removed, and (3) the data with observations 6, 7, and 8 removed. Dramatic changes occur with the least-squares estimate of the invratio when observation 6 is removed. Only small changes occur when the other two observations are removed. This finding is consistent with the suggestions of the scatterplots, leverage values, and DFBETAS.

APPENDIX: CALCULATION OF LEVERAGE VALUES AND OUTLIER DIAGNOSTICS

Using the matrix form of the multiple linear regression model given in the appendix to Chapter 22, the least-squares estimator of the regression coefficients can be written as

$$\mathbf{b}_c = (X'_c X_c)^{-1} X'_c \mathbf{y}, \tag{24.A.1}$$

where the subscript c is again used to denote the presence of a constant term in the model; i.e., the first column of X_c is a column of ones. The predicted values of the elements in the response vector $\mathbf{y}$ are given by

$$\hat{\mathbf{y}} = X_c \mathbf{b}_c = H \mathbf{y}, \tag{24.A.2}$$

where $H = X_c (X'_c X_c)^{-1} X'_c$. The matrix H is commonly called the *hat matrix* (notice that this is the matrix that when postmultiplied by $\mathbf{y}$ yields $\hat{\mathbf{y}}$). The leverage values, h_{ii} (or h_i), introduced in Section 24.3, are the diagonal elements of H.

The vector of raw residual values can be obtained by subtracting the predicted responses from the observed responses:

$$\mathbf{r} = \mathbf{y} - \hat{\mathbf{y}} = (I - H)\mathbf{y}. \tag{24.A.3}$$

From equation (24.A.3) one can show that the variances and covariances of the elements of $\hat{\mathbf{y}}$ and $\mathbf{r}$ are contained on the diagonals and off-diagonals, respectively, of the following matrices:

$$\text{var}(\hat{\mathbf{y}}) = H \sigma^2 \tag{24.A.4}$$

and

$$\text{var}(\mathbf{r}) = (I - H)\sigma^2. \tag{24.A.5}$$

Studentized residuals t_i are the ratios of the raw residuals r_i and an estimate of the standard error of r_i. From (24.A.5), an estimated standard error of r_i is

$$\widehat{\text{SE}}(r_i) = (1 - h_i)^{1/2} s_e,$$

where $s_e = (\text{MS}_E)^{1/2}$ and MS_E is the mean squared error from the least-squares fit to the complete data set. Thus

$$t_i = \frac{r_i}{(1 - h_i)^{1/2} s_e}. \tag{24.A.6}$$

Using matrix equations similar to (24.A.1)–(24.A.5), one can derive expressions for least-squares estimators and prediction equations for a fit in which the ith observation is deleted. Straightforward matrix-algebra derivations (e.g., Gunst and Mason 1980) leads to the expression for a deleted residual:

$$r_{(-i)} = \frac{r_i}{1 - h_i}. \tag{24.A.7}$$

From equation (24.A.7) and the ith diagonal element of (24.A.5), it follows that the standard error of a deleted residual is

$$\widehat{SE}(r_{(-i)}) = \frac{\sigma}{(1 - h_i)^{1/2}}. \tag{24.A.8}$$

Again using matrix algebra and the matrix expressions for residuals from the complete data set and the data set with the ith observation deleted, one can show that an estimator of the error standard deviation that is independent of the predicted responses and the residuals from both the complete and the reduced data sets is

$$s_{(-i)} = \left(\frac{SS_E - r_i^2/(1 - h_i)}{\nu - 1} \right)^{1/2}, \tag{24.A.9}$$

where SS_E is the error sum of squares from the complete data set and is based on ν degrees of freedom.

REFERENCES

Text References

The following references provide additional information on univariate outlier-detection techniques. The second reference also provides a history of the study of outliers.

"Standard Practice for Dealing with Outlying Observations," in *Annual Book of ASTM Standards*, Designation E 178-80, Philadelphia: American Society for Testing Materials.

Beckman, R. J. and Cook, R. D. (1983). "Outlier.........s," *Technometrics*, **25**, 119–149.

Tietjen, G. L. and Moore, R. H. (1972). "An Extension of Some Grubbs-Type Statistics for the Detection of Several Outliers," *Technometrics*, **14**, 583–598.

The following texts provide specific coverage of the diagnostics for influential observations that were discussed in Sections 24.2 and 24.3. All these texts are at an advanced mathematical level.

Belsley, D. A., Kuh, E., and Welsch, R. E. (1980). *Regression Diagnostics*, New York: John Wiley and Sons, Inc. This text provides extensive coverage of outlier diagnostics such as DFFITS and DFBETAS.

Cook, R. D. and Weisberg, S. (1982). *Residuals and Influence in Regression*, New York: Chapman and Hall. Excellent coverage of statistical basis for the identification of outliers. Extensive treatment of alternative outlier diagnostic procedures.

Gnust, R. F. and Mason, R. L. (1980). *Regression Analysis and Its Application: A Data-Oriented Approach*. New York: Marcel Dekker, Inc. Comprehensive treatment of deleted residuals and leverage values.

Meyers, R. H. (1986). *Classical and Modern Regression with Applications*. Boston, MA: PWS and Kent Publishing Co.

Weatherill, G. B. (1986). *Regression Analysis with Applications*. New York: Chapman and Hall.

EXERCISES

1 A study of engine-induced structural-borne noise in a single-engine light aircraft was carried out to determine the level to which structural-borne noise influences interior levels of noise. Cabin noise was recorded for 20 ground tests for engine-attached aircraft configurations. Construct a quantile–quantile plot for this data set. Check suspect observations using Grubbs test. Do the data appear to contain any outliers?

CABIN NOISE [DB(A)]

186	263	201	211	199	198	189	257	204	203
185	209	225	193	166	175	210	205	182	223

2 A study was conducted to identify factors that may contribute to tooth discoloration in humans. Twenty subjects were randomly chosen from different communities to participate in the year-long study. Dental examinations were given to each at the beginning of the study and again one year later. Percentage increases in tooth discoloration were measured. Data gathered from each individual included:

(a) Age (years).

(b) Fluoride level (ppm) of public water supply in the subject's community.

(c) Average number of tooth brushings per week.

DATA SET FOR TOOTH DISCOLORATION (EXERCISE 2)

Tooth Discoloration (%)	Fluoride Level (ppm)	Age (years)	Avg. No. of Brushings per Week
12	0.7	46	5
10	1.3	44	8
14	1.5	34	7
20	2.4	72	6
18	2.6	22	9
18	2.9	40	11
25	3.0	52	5
21	3.6	37	10
36	3.8	11	7
44	4.0	10	6
11	0.5	24	8
15	1.8	39	7
22	2.1	42	6
21	0.4	6	5
27	3.4	68	5
25	2.9	59	7
18	2.1	18	10
21	3.4	32	12
18	3.0	15	10
20	1.7	59	9

The standardized regression equation for this data set is

tooth discoloration $= 20.8 + 6.944 \times$ fluoride $- 2.674 \times$ age $- 4.297 \times$ brush.

Calculate the residuals and produce a scatterplot of residuals versus predicted values. Also plot the residuals versus each of the predictor variables. Comment on any interesting features of the plots.

3 Calculate the studentized deleted residuals from the fit to the data in Exercise 2. Calculate leverage, DFFITS, and DFBETAS values for each observation in the data set. From the plots and the statistics calculated in this exercise and the previous one, what conclusions can be drawn about the presence of outliers?

4 Refer to the videocassette sales data in Exercise 1, Chapter 21. Perform a comprehensive analysis of the possible presence of outliers in the raw data. Construct appropriate plots, and calculate the outlier diagnostics discussed in this chapter. What conclusions can be drawn about the presence of outliers?

5 Calculate outlier diagnostics for the synthetic-rubber process data in Table 23.1 for two models, one having only the linear term in the production rates and the other having both the linear and the quadratic terms. Which observations appear to be outliers? Redo this analysis using the natural logarithm of solvent weight as the response variable. How do the outlier diagnostics compare? Construct plots of the residuals to interpret any differences noted. Which model fit do you prefer? Why?

6 Calculate outlier diagnostics for the height data in Exercise 7, Chapter 22. Use appropriate plots to interpret conclusions drawn from the investigation of these diagnostics.

7 An experiment was conducted to study the effects of four factors on an engine-knock measurement.

(a) Investigate the observations on knock number using the techniques described in Section 24.1. Are any outliers identified in this analysis?

(b) Fit a regression model, linear in the four predictors, to the knock-number measurements. Are any observations identified as influential? If these results differ from those of (a), determine the reasons for the differences.

DATA FOR ENGINE KNOCKS EXPERIMENT (EXERCISE 7)

Spark Timing	Air–Fuel Ratio	Intake Temp.	Exhaust Temp.	Knock Number
13.3	13.9	31	697	84.4
13.3	14.1	30	697	84.1
13.4	15.2	32	700	88.4
12.7	13.8	31	669	84.2
14.4	13.6	31	631	89.8
14.4	13.8	30	638	84.0
14.5	13.9	32	643	83.7
14.2	13.7	31	629	84.1
12.2	14.8	36	724	90.5
12.2	15.3	35	739	90.1
12.2	14.9	36	722	89.4
12.0	15.2	37	743	90.2
12.9	15.4	36	723	93.8
12.7	16.1	35	649	93.0
12.9	15.1	36	721	93.3
12.7	15.9	37	696	93.1

CHAPTER 25

Assessment of Model Assumptions

Assessing the validity of model assumptions is often critical to the successful application of statistical inference procedures. In this chapter techniques for investigating the validity of model assumptions are presented. The topics discussed include:

- *graphics and test statistics for the assessment of normally distributed errors,*
- *procedures for evaluating the adequacy of model specifications,*
- *tests for the independence of errors, and*
- *a discussion of robustness to violations of model assumptions.*

The methodology presented in this chapter concerns techniques for verifying the common model assumptions used in a variety of statistical inference procedures. It will be assumed throughout these discussions that the data have already been checked for outliers using the procedures recommended in Chapter 24. Further, it is assumed that initial model specifications have been made and, if appropriate, a cursory inspection of the reasonableness of those assumptions has been made through an examination of simple plots such as scatterplots. Thus, we consider in this chapter the evaluation of a specific statistical model for which assumptions of normality, independence of errors, or correct specification of predictor variables is of concern.

25.1 NORMALLY DISTRIBUTED ERRORS

One of the central assumptions used in many statistical procedures is that the model errors are normally distributed. Normally distributed errors were assumed for single-sample models, ANOVA models, and regression models in earlier chapters of this book. While many statistical procedures (e.g., nonparametric tests) do not require an assumption of normality and others are fairly insensitive to such assumptions (see Section 25.4), the assumption of normality is required for the exact sampling distributions of many statistics to be valid. Some statistics (e.g., F-tests for the equality of two population variances) are extremely sensitive to departures from normality.

The error terms in statistical models are unobservable. Most statistical procedures for assessing the assumption of normally distributed errors thus rely on using residuals in place of the model errors. Residuals were introduced in Chapter 21 for regression models as the differences between the actual response values and those predicted from the fitted model, $r = y - \hat{y}$. Residuals can be calculated for any statistical model, including univariate models and ANOVA models.

For single-sample models of the form (13.1), $y = \mu + e$, the deterministic portion of the model only contains the unknown population mean. Residuals for this model are $r_i = y_i - \bar{y}$. Since these residuals only differ from the original responses by an additive amount $\bar{y}$, the raw responses are generally used to assess model assumptions instead of the residuals; however, residuals could also be used.

ANOVA models involve one or more factors, each having two or more levels. When all factor levels are fixed, residuals can be calculated by inserting solutions to the normal equations for the fixed-effects (assignable-cause) components of the model in order to obtain predicted responses. Any solution to the normal equations will produce the same residuals.

For a one-factor model, one set of solutions, m and a_i, to the normal equations for μ and α_i is

$$m = \bar{y}_{..} \quad \text{and} \quad a_i = \bar{y}_{i.} - \bar{y}_{...} \tag{25.1}$$

The resulting residuals are

$$r_{ij} = y_{ij} - \bar{y}_{i.}. \tag{25.2}$$

For balanced, complete factorial experiments involving two or more factors conducted in completely randomized designs, predicted responses depend on whether an interaction term is included in the model. For example, a two-factor model without interaction would result in predicted responses equal to

$$\hat{y}_{ijk} = \bar{y}_{i..} + \bar{y}_{.j.} - \bar{y}_{...}, \tag{25.3}$$

whereas if an interaction term is included in the model,

$$\hat{y}_{ijk} = \bar{y}_{ij}. \tag{25.4}$$

In general, as indicated by equation (25.4) for two factors, predicted responses for complete factorial experiments in which all interactions are included in the model are simply the average responses for all test runs having each specified factor–level combination. Residuals are then obtained by subtracting the predicted responses from each of the responses for that combination.

For unbalanced designs or models that do not have a complete factorial structure, the fitted model is obtained by finding a solution to the normal equations. Any solution can be used. The residuals are again the differences between the actual responses and the predicted responses using the fitted model.

One of the easiest and most direct evaluations of the assumption of normality is the normal quantile–quantile plot. This plotting technique was introduced in Section 5.3 and discussed further in Section 24.1 in the context of outlier identification. A normal quantile–quantile plot consists of plotting the ordered residuals from a fitted model against the corresponding ordered quantiles from a standard normal reference distribution. Approximate linearity indicates that the sample data are consistent with an assumption of normally distributed errors. These plots can be supplemented with other data displays such as stem-and-leaf plots (Section 4.2) or relative-frequency histograms (Section 4.2) of the residuals.

In Section 24.2 normal quantile–quantile plots for regression models were introduced. In order to demonstrate their use with ANOVA models, we reconsider the torque study that was designed in Section 7.1 (Table 7.3) and analyzed in Section 16.3. The complete data set, including the duplicate responses for each factor–level combination, is shown in Table 25.1.

Since this is a balanced, complete factorial experiment and a complete three-factor

Table 25.1 Predicted Responses for Torque Study

Obs.	Sleeve	Lubricant	Torque	Predicted Torque
Alloy: Steel				
1	Porous	1	82	79.0
2	Porous	1	76	79.0
3	Porous	2	75	77.5
4	Porous	2	80	77.5
5	Porous	3	77	77.0
6	Porous	3	77	77.0
7	Porous	4	76	74.5
8	Porous	4	73	74.5
9	Nonporous	1	78	78.5
10	Nonporous	1	79	78.5
11	Nonporous	2	65	67.0
12	Nonporous	2	69	67.0
13	Nonporous	3	79	77.5
14	Nonporous	3	76	77.5
15	Nonporous	4	81	79.0
16	Nonporous	4	77	79.0
Alloy: Aluminum				
17	Porous	1	79	78.0
18	Porous	1	77	78.0
19	Porous	2	72	72.0
20	Porous	2	72	72.0
21	Porous	3	77	78.5
22	Porous	3	80	78.5
23	Porous	4	72	73.5
24	Porous	4	75	73.5
25	Nonporous	1	71	69.5
26	Nonporous	1	68	69.5
27	Nonporous	2	73	70.5
28	Nonporous	2	68	70.5
29	Nonporous	3	69	69.0
30	Nonporous	3	69	69.0
31	Nonporous	4	70	70.0
32	Nonporous	4	70	70.0

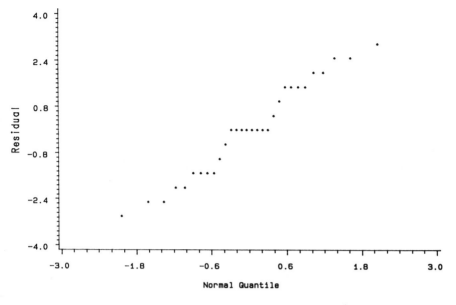

Figure 25.1 Normal quantile–quantile plot for torque residuals.

interaction model was used to fit the data, the predicted responses are simply the averages of the duplicate measurements for each alloy–sleeve–lubricant combination. These averages are shown in Table 16.3 and are listed next to the observed-torque values in Table 25.1. The residuals are listed in order in Table 25.2 and are plotted in Figure 25.1. The residual plot exhibits reasonable fidelity to a straight line. Thus, the assumption of normally distributed errors for these data appears to be reasonable.

Since graphical evaluations are subjective, it is useful to supplement normal quantile

---

EXHIBIT 25.1 SHAPIRO–WILK TEST FOR NORMALITY

1. Let $y_{(i)}$ denote the ordered observations, either response values or residuals, as appropriate:

$$y_{(1)} \leqslant y_{(2)} \leqslant \cdots \leqslant y_{(n)}.$$

2. Compute the numerator of the sample variance of the observations:

$$s^2 = \sum (y_{(i)} - \bar{y})^2.$$

3. Using the values of a_i found in Table A28 of the appendix, calculate

$$b = \sum a_i y_{(i)}.$$

4. Compute the Shapiro–Wilk statistic

$$W = b^2 / s^2. \tag{25.5}$$

5. Reject the assumption of normality if W is less than the value in Table A29 of the appendix.

Table 25.2 Ordered Torque Residuals and Constants for the Shapiro–Wilk Test for Normally Distributed Errors

Ordered Residual	a_i	Ordered Residual	a_i
−3.0	−0.4188	0.0	0.0068
−2.5	−0.2898	0.0	0.0206
−2.5	−0.2463	0.0	0.0344
−2.0	−0.2141	0.0	0.0485
−2.0	−0.1878	0.5	0.0629
−1.5	−0.1651	1.0	0.0777
−1.5	−0.1449	1.5	0.0931
−1.5	−0.1265	1.5	0.1093
−1.5	−0.1093	1.5	0.1265
−1.5	−0.0931	1.5	0.1449
−1.0	−0.0777	1.5	0.1651
−0.5	−0.0629	2.0	0.1878
0.0	−0.0485	2.0	0.2141
0.0	−0.0344	2.5	0.2463
0.0	−0.0206	2.5	0.2898
0.0	−0.0068	3.0	0.4188

plots with appropriate tests of statistical hypotheses. Generally regarded as the most powerful statistical test of normality is the Shapiro–Wilk test (see Exhibit 25.1).

The assumptions underlying the use of the Shapiro–Wilk statistic (25.5) require independent observations. The statistic is appropriate for use when a single sample of independent observations is to be tested for normality. When observations are not independent, as with the use of residuals from analysis of variance and regression models, the statistic provides only an approximate test for normality. Its use is still recommended as a valuable statistical measure to augment the use of residual plots.

Table 25.2 lists the constants from Table A28 needed to compute the Shapiro–Wilk statistic (25.5) for the torque data. For this data set, $b = 8.9359$, $s^2 = SS_E = 84.00$, and $W = 0.9506$. From Table A29, this value of W is not statistically significant ($0.10 < p < 0.50$). Hence, the hypothesis of normality is not rejected, and statistical inference procedures that require the assumption of normally distributed observations may be used.

When the sample size exceeds 50, several alternatives to the Shapiro–Wilk test are available. Two popular tests are the Kolmogorov–Smirnov test and the Anderson–Darling test. Information on these tests can be found in the references listed at the end of the chapter.

As with most statistical procedures, sample-size considerations have an important effect on the performance of statistical tests for normality. Very small sample sizes may lead to nonrejection of tests for normality because of little power in the test statistics. So too, very large sample sizes may lead to rejection when only minor departures from normality are exhibited by a data set. This occurs because of the extremely high power levels for the tests. These test procedures should always be accompanied by residual plots, especially quantile–quantile plots and stem-and-leaf plots (or histograms). An assessment using all of these procedures will lead to an informed evaluation of the departure, if any, from normal assumptions and the need for accommodation due to nonnormality.

We mention in closing this section that when ANOVA models are saturated (i.e., contain all main effects and interactions), as in the torque example, and only two repeat tests are available, the normal quantile–quantile plot and tests for normality will not be sensitive to many types of skewed error distributions. This is because the residuals for each factor–level combination will be equal in magnitude and opposite in sign. Thus, the distribution of the calculated residuals will be symmetric, regardless of the true error distribution. We used the torque data to illustrate the above procedures for ANOVA models, not to draw a definitive conclusion about its error distribution.

There are several alternative courses of action that may be taken when statistical tests or graphical procedures indicate that data are not normally distributed. Some statistical procedures are not highly sensitive to the effects of nonnormality (Section 25.4). For such procedures only extreme nonnormality, such as excessive skewness or bimodal distributions, call for corrective action. Reexpression of response variables (Section 26.2) is often an effective strategy when departures from normality are severe or when test statistics are extremely sensitive to even small departures from normality. Use of nonlinear estimation procedures (Section 26.1) may be appropriate if transformations do not statisfactorily rectify problems associated with nonnormality. Finally, nonparametric inference procedures may be suitable for use with nonnormal data.

25.2 MODEL SPECIFICATION

Correct model specification implies that all relevent design factors, covariates, or predictor variables are included in a statistical model and that all model variables, including the response, are expressed in an appropriate functional form. Model specification is most acutely a problem in regression models, especially when the functional form of an appropriate theoretical model is unknown.

In this section we discuss techniques for detecting when the functional form of a response or a predictor variable in a regression model is incorrectly defined. Some of these techniques can also be used with ANOVA models when interest is in the correct functional form for the response variable. Although the selection of variables is an important component of correct model specification, we delay consideration of it until Chapter 27.

Correcting a misspecified variable requires an ability to recognize a variety of algebraic curves. A discussion of various functional forms for use in building linear and nonlinear models is given in Section 26.1. Recognizing these algebraic forms is aided through the use of residual plots. The numerical values of least-squares residuals can be shown to be uncorrelated with the predictor-variable values in a regression model. For this reason, a plot of residuals against any predictor variable should result in a random scatter of points about the line $r = 0$. Any systematic trend in the plot indicates the need for some reexpression of either the response variable or the predictor variable.

The most frequently occurring patterns indicating a need for reexpression of variables in residual plots are either wedge-shaped or curvilinear trends. A wedge shape in a plot of the residuals versus the predictor variables usually indicates that the error standard deviation is not constant for all values of the predictor variable. The question of whether to transform the response or the predictor variable cannot be unequivocally answered. A useful guideline is that if the wedge shape occurs in only one of the plots, the predictor variable should be transformed. If it occurs in two or more plots, the response should probably be transformed.

A curvilinear trend in a plot of residuals versus a predictor variable often indicates the need for additional variables in the model or a reexpression of one or more of the current model variables. For example, a quadratic trend in a plot may indicate the need for a quadratic term in x_j. Such a trend might also be indicative of the need to add a variable to the model because the error terms are reflecting a systematic, not random, pattern. The latter would especially be true if patterns were present in several of the residual plots.

Residual plots of this type also are helpful in visually assessing whether an interaction term $x_j x_k$ should be added to the model. A plot of the residuals from a fit excluding the interaction term $x_j x_k$ versus the interaction term should be a random scatter about zero. Any systematic departure from a random scatter suggests the need for an interaction variable in the model.

Another useful residual plot for detecting model misspecification is termed the *partial-residual plot*. A partial residual is defined in Exhibit 25.2. Partial residuals measure the linear effect of a predictor variable relative to the random error component of the model. As suggested by equation (25.6), if a predictor variable has a strong linear effect on a response variable, the partial residuals for that predictor should be dominated by the linear term $b_j x_{ij}$. If the linear effect is weak or negligible, the randomness present in the

---

EXHIBIT 25.2   PARTIAL RESIDUALS

Partial residuals corresponding to the predictor variable x_j adjust the least-squares residuals for the portion of the regression fit attributable to the linear effect of x_j:

$$r_i^* = r_i + b_j x_{ij}. \tag{25.6}$$

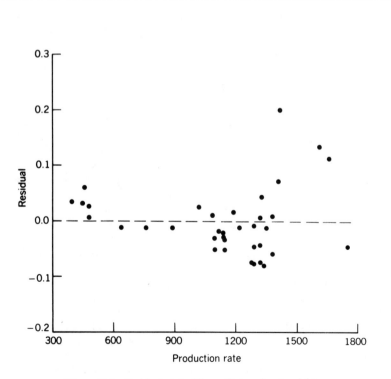

Figure 25.2   Residual plot of linear fit to solvent weights.

least-squares residuals should dominate the partial residual. If some nonlinear function of the predictor variable is needed, the partial residual will often reflect the nonlinear function because the least-squares residuals will contain that portion of the misspecification.

By plotting the partial residuals versus the predictor-variable values x_{ij}, the direction and strength of the linear effect of x_j on the response can be visually gauged. The regression of r_i^* versus x_{ij} is a line with an intercept of zero and a slope equal to b_j, the least-squares estimate of the coefficient of x_j in the full model. This is in contrast to an ordinary residual plot of r_i versus x_{ij}, in which the slope is always zero. This property of partial residuals enhances their ability to portray the extent of nonlinearity in a predictor variable.

An example of an ordinary residual plot is given in Figure 25.2. The residuals plotted in this figure are from a straight-line fit of the solvent weights to the production rates using the data in Table 23.1. Observe that the first few plotted points are all positive in Figure 25.2, and most of the last several are also positive. In the middle, the predominance of plotted points have negative residuals. This type of curvilinear trend suggests that a quadratic term in the rate variable be added to the model.

Figure 25.3 is a partial-residual plot for the fit to the solvent weights. Observe the strong linear component to the plot and the variation of the points around the linear trend. The clarity of the linear trend relative to the variation around it attests to the strength of the linear effect of production rate on solvent weights.

It is also clear from Figure 25.3 that the plotted points are curving upward. The upward trend is not as strong as the linear trend; nevertheless, a linear trend does not adequately capture the entire relationship between solvent weights and production rates. The ability to visually assess the strengths of nonlinear and linear effects relative to one another and to the random error variation is the primary reason for the use of partial-residual plots.

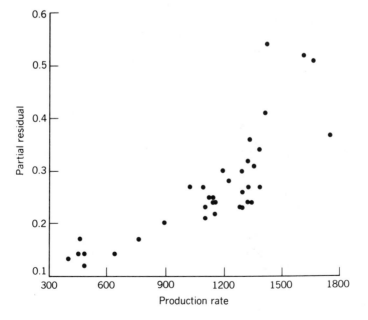

Figure 25.3 Partial-residual plot of solvent-weight data.

A careful examination of Figures 25.2 and 25.3 indicates that not only a nonlinear trend but also increasing variability of the residuals as a function of the production rates is evident in the plots. This characteristic, like a wedge-shaped plot of residuals, suggests the need for a transformation of the response variable. We confirm this need in Section 26.2.

A type of plot that is closely related to a partial residual plot is termed a *partial-regression* or *partial-regression leverage* plot. In this type of plot the response variable and one of the predictor variables are regressed on the other $p - 1$ predictor variables in two separate regressions. The response residuals are then plotted against the residuals of the predictor variable. A least-squares fit to the plotted points, as with the partial-residual plot, has a zero intercept and a slope equal to the least-squares estimate b_j from a fit to the complete data set.

25.3 ADDITIONAL ERROR ASSUMPTIONS

Randomness of the error terms in statistical models often validates the use of sampling distributions even when the stated assumptions are not strictly correct. For example, a random sample of observations often allows one to use Student's t-distribution even though the responses are not normally distributed. On the other hand, correlated model errors are a particularly severe violation of the usual statistical model assumptions. Multivariate statistical analyses are specifically designed to accommodate this problem. The procedures recommended in this book, however, generally are not valid when errors are correlated.

In the experimental settings stressed in this book, randomization of experimental units or of the test sequence does much to alleviate the effects of correlated errors on statistical inference procedures. However, many experiments are conducted in time sequence, a common source of correlated errors. In addition, observational studies are frequently conducted in time sequence. In such circumstances an examination of time-ordered responses or residuals can provide important insight into correlated errors.

One simple diagnostic for detecting correlations between time-consecutive observations is to plot the responses or residuals versus their time sequence. A nonrandom scatter of points indicates that the data are time-ordered and not independent. An interesting example of a time-sequence plot of residuals is shown in Figure 25.4. In this figure the residuals from a regression fit to a data set used in the calibration of a cryogenic flowmeter (see the appendix to Chapter 27) are plotted. It is apparent from the plot that the residuals are not randomly scattered.

The residuals in Figure 25.4 are drifting upward with time. Upon viewing this plot the researchers noted that the data were taken for various runs on several different days, and the reading taken later in the day were higher than those taken earlier in the day. The problem causing the drift was noted during the experiment and corrective action was taken, accounting for the removal of the upward drift in the last few plotted points.

In addition to time-ordered sequence plots, there are several statistical tests that are used to test for randomness. One procedure, termed the *runs* test, involves examining the sign arrangement ($+$ or $-$) of time-ordered residuals. An unusually large or small number of runs (see Exhibit 25.3) is indicative of correlated observations.

EXHIBIT 25.3

Run. A run is a sequence (of any length) of observations all of which have the same sign.

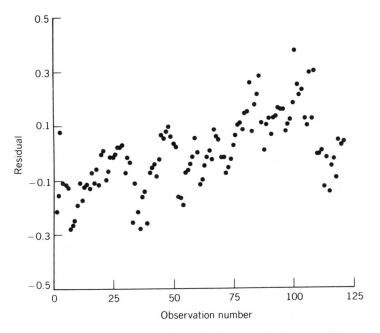

Figure 25.4   Time-sequence residual plot for cryogenic-flowmeter data.

The runs test is a nonparametric statistical procedure defined in Exhibit 25.4.

EXHIBIT 25.4   RUNS TEST

1. Calculate the residuals from a fit to an ANOVA or regression model. For a single sample of observations from a population, use the deviations $y_i - \bar{y}$.

2. Order the residuals according to their time sequence.

3. Count the number of runs (r), the number of positive residuals (n_1), and the number of negative residuals (n_2).

4. If $n_1 \leqslant 20$ and $n_2 \leqslant 20$, reject the hypothesis of randomness if $r < r_L$ or if $r > r_U$, where r_L and r_U are the upper and lower critical values given in Table A30 in the appendix.

5. For larger samples sizes, reject the hypothesis of randomness if $z > z_{\alpha/2}$, where

$$z = \frac{|r - \mu| - 0.5}{\sigma},$$

$$\mu = 1 + \frac{2n_1 n_2}{n_1 + n_2}, \qquad (25.7)$$

$$\sigma = \frac{2n_1 n_2 (2n_1 n_2 - n_1 - n_2)}{(n_1 + n_2)^2 (n_1 + n_2 - 1)},$$

and $z_{\alpha/2}$ is an upper $100(\alpha/2)\%$ critical point of the standard normal distribution corresponding to a two-tailed significance level of α.

Table 22.7 displays the least-squares coefficient estimates for a fit to the tire treadwear data of Table 22.2. If one computes the residuals from this fit, the signs on the time-ordered residuals are

$$(+)(---)(++)(--)(++++)(-----)(++)$$

$$(--)(+)(---)(+)(-)(+)(-)(+)(--).$$

Each set of parentheses shown above encompasses one run. There are $r = 16$ runs among these time-ordered residuals, with $n_1 = 13$ positive residuals and $n_2 = 19$ negative ones. From Table A30 of the appendix, a two-tailed 10%-level test of the hypothesis of randomness would not reject it, since $r_L = 10$ and $r_U = 23$. In contrast, the residuals for the cryogenic-flowmeter data plotted in Figure 25.4 have $r = 20$ runs with $n_1 = 56$ and $n_2 = 64$. Using the normal approximation (25.7), an unusually low number of runs is indicated, since $z = (|20 - 60.73| - 0.5)/5.43 = 7.41$ is greater than $z_{0.025} = 1.96$.

If errors are correlated, it is sometimes of interest to test for a specific form of correlational pattern among the errors. A pattern often of interest with time-ordered data is that of first-order serial correlation, in which consecutive errors are related as follows:

$$e_i = \rho e_{i-1} + u_i, \qquad i = 2, 3, \ldots, n.$$

This relationship states that each error is the sum of a multiple of the previous error and an independent error (u_i). The coefficient, ρ, is the *autocorrelation* between consecutive errors ($|\rho| < 1$). A test for the hypothesis $H_0: \rho = 0$ is the *Durbin–Watson test* (see Exhibit 25.5).

EXHIBIT 25.5   DURBIN–WATSON TEST

1. Calculate the test statistic

$$d = \sum_{i=2}^{n} (r_i - r_{i-1})^2 \left/ \sum_{i-1}^{n} r_i^2 \right. \tag{25.8}$$

2. Determine upper (d_U) and lower (d_L) bounds for d from Table A31 in the appendix.
3. To test $H_0: \rho = 0$ vs $H_a: \rho > 0$:

    (a) Reject H_0 if $d < d_L$.
    (b) Do not reject H_0 if $d > d_U$.
    (c) Draw no conclusion if $d_L < d < d_U$.

4. To test $H_0: \rho = 0$ vs $H_a: \rho < 0$, replace d by $d^* = 4 - d$ and use criteria (a)–(c) of step 3.

One difficulty with using the Durbin–Watson test is the inconclusive region where no decision can be made about the null hypothesis. Several approximate procedures have been suggested when this occurs. The most common approach is to include the inconclusive region in the nonrejection decision; i.e., only reject H_0 if $d < d_L$.

Although sequence plots, the runs test, and the Durbin–Watson test have been presented for time-ordered data, other logical data sequences can be used. Data could be ordered according to location, position, or any other meaningul variable that is not affected by the experimentation.

Finally, many statistical procedures assume that the model errors all have the same standard deviation σ. In Section 14.3, Bartlett's test and the F-max test were proposed for use to test this assumption. Each of these tests requires a number of observations for which repeat tests are available. Each is also very sensitive to departures from the normality assumption for the model errors. When no repeat tests are available, plotting the residuals versus the predicted responses should result in a horizontal band of points around the line $r_i = 0$. Wedge-shaped and other systematic departures from a horizontal band indicate nonconstant standard deviations.

25.4 ROBUSTNESS

The consequences of violating statistical model assumptions should be understood in order to be able to evaluate whether certain types of violations seriously affect conclusions that are drawn from inferential procedures. In many instances, the procedures employed are *robust* to some model assumptions. Robustness implies that violations must be severe before they will appreciably effect inferences. When statistical procedures are not robust with respect to violations of model assumptions and such violations are detected, alternative techniques such as nonparametric procedures or some of the alternatives discussed in Chapter 28 may be needed.

25.4.1 Normality

Moderate departures from normality will have little effect on the procedures recommended in previous chapters for one- or two-sample inferences on population means, for inferences on fixed-effect ANOVA models, or for regression coefficients. In particular, the t-statistics and F-statistics used in these procedures are known to be robust to departures from normality. This is a consequence of the central limit property (Section 12.1). Very skewed error distributions are an example of a departure from normality that can seriously affect these statistics.

Inference procedures for variances are greatly affected by nonnormality. F-statistics used to compare two population variances, inferences on variances in random-effect models, and the tests for equality of variances discussed in the Section 14.3 are examples. Alternative courses of action when data are not normally distributed were discussed at the end of Section 25.1.

25.4.2 Constant Standard Deviations

The consequences of moderate departures from the assumption of constant error standard deviations are generally not too serious for tests on means, fixed-effect ANOVA parameters, and regression coefficients. In ANOVA or regression tests the overall F-test of significance is least affected, but individual t-tests are only moderately more susceptible. Often departures from the assumption of constant error standard deviations can be corrected by a reexpression of the data. A weighted-regression approach, which is briefly described in Section 26.2, is another alternative.

25.4.3 Independence

Correlated errors pose a much more serious problem for tests on means, fixed effects in ANOVA, and regression parameters than do nonnormality or nonconstant error standard deviations. Substantial biases in standard-error estimation can result from violations of this assumption. Further, the accuracy of stated significance probabilities can be severely impaired. Lack of independence rarely can be corrected through data reexpression. Multivariate statistical procedures are the best solution to this problem.

25.4.4 Nonstochastic Predictor Variables

Random values for predictor variables arise in many experiments in which regression models are used. Random predictor variables occur in observational studies in which the predictor variables are not controlled and in designed experiments in which the predictor variables are subject to measurement errors. Measurement errors can occur, for example, when the values of the predictor variables are obtained from equipment settings or chemical analyses.

As an illustration, consider the design layout given in Figure 25.5. The design was constructed for a study of the effects of fuel properties on vehicle emissions. Illustrated are the target design properties (the circles of the cube) and the measured fuel properties achieved by the blending process (the points connected by the dashed lines). Although this was to be a designed experiment with controlled observations, there was variation in the factor levels due to the blending process and the measurement of the fuel properties.

Accommodation of random factor or predictor variables is highly problem-dependent. We offer a few suggestions, realizing that they may not be appropriate in all circumstances.

In the analysis of fixed-effect ANOVA models, key considerations are how much measured factor levels differ from target values and the size of the likely effect on the response variable. If the difference between measured and target values is small relative to the difference between two adjacent target values, the effect of using the target factor levels instead of the measured ones in a statistical analysis is likely to be very small. Likewise, if changes anticipated in the response variable between the target and actual factor levels are likely to be small relative to changes expected between two target levels, use of the target levels should not substantially affect inferences. Two-level factorial and fractional factorial experiments are known to be robust with respect to measurement errors in the predictor variables.

In both of the settings described in the last paragraph, one could use the measured

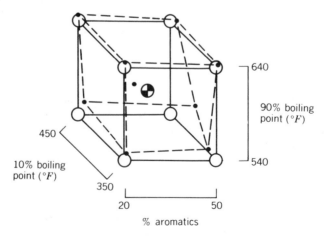

Figure 25.5   Target and measured factor–level values for fuel blends.

factor levels in an ANOVA model or in a regression model, as appropriate. The recommendation to use the target values is based on a presumption that the experiment was designed and, therefore, that the target factor levels were chosen in part to provide a suitable analysis with unconfounded factors. Use of measured factor levels may introduce confounding (Section 9.1) or the need to use the principle of reduction in sums of squares (Section 15.2) to analyze the data, whereas use of target factor levels would not add this complexity.

In regression analyses, measured predictor-variable values are most often used instead of target values. This is especially true when random predictor variables are a result of data being collected in an observational study. If the observations on the predictor variables represent typical values for the predictors, then the random nature of the predictor variables is not a problem. Inferences are simply conditioned on the observed predictor values. In order to use this line of argument, the burden is on the researcher to ensure that these predictor-variable values are indeed representative of the phenomenon under study.

When the levels of the predictor variables are subject to measurement error, standard least-squares analysis procedures often can still be used provided the response is a linear function of the predictor variables. For nonlinear models the effect depends heavily on the observed values of the predictors. An alternative to least-squares estimation for linear measurement-error regression models is given in Section 28.1.

REFERENCES

Text References

The references listed at the end of Chapters 14, 15, 21, and 22 provide informative coverage on various aspects of statistical model assumptions and their assessment. The following books are among many that comprehensively cover multivariate statistical procedures. These books can be consulted for inferential procedures when model errors are correlated.

Anderson, T. W. (1958). *An Introduction to Multivariate Statistical Analysis*, New York: John Wiley and Sons, Inc.

Johnson, R. A. and Wichern, D. W. (1982). *Applied Multivariate Statistical Analysis*, Englewood Cliffs, NJ: Prentice-Hall, Inc.

Morrison, D. F. (1976). *Multivariate Statistical Methods*, New York: McGraw-Hill Book Company.

Some useful articles on procedures for detecting model-assumption violations are given below. The first concerns measurement error in experimental design. The second discusses assumption violations in ANOVA models. The last two present various test procedures for normality.

Box, G. E. P. (1963). "The Effects of Errors in the Factor Levels and Experimental Design," *Technometrics*, **5**, 247–262.

Cochran, W. G. (1947). "Some Consequences When the Assumptions for the Analysis of Variance are not Satisfied," *Biometrics*, **3**, 22–38.

Shapiro, S. S. (1980). *How to Test Normality and Other Distributional Assumptions, Volume 3*, Milwaukee, WI: ASQC Basic References in Quality Control.

Shapiro, S. S. and Wilk, M. B. (1965). "An Analysis of Variance Test for Normality (Complete Samples)," *Biometrika*, **52**, 591–611.

Data References

The cryogenic-flowmeter data are taken from the following article:

Joiner, B. (1977). "Evaluation of Cryogenic Flow Meters: An Example in Non-standard Experimental Design and Analysis," *Technometrics*, **19**, 353–380. Copyright American Statistical Association, Alexandria, VA. Reprinted by permission.

EXERCISES

1 Refer to the experiment described in Exercise 5, Chapter 7, regarding the study of glass-mold temperatures. Calculate the residuals from an analysis of variance using molten-glass temperature and glass type as factors. Include an interaction term in the analysis. Assess the assumption of normality using (a) graphical techniques and (b) the Shapiro–Wilk test. On the basis of these analyses, is the assumption of normally distributed errors reasonable?

2 Refer to Exercise 2 in Chapter 7, which describes an experiment in which temperatures and engine speeds are varied in order to study the effect they may have on engine stability. The data collected for this experiment are given in the accompanying table.

DATA FOR ENGINE STABILITY EXPERIMENT

Engine Speed (rpm)	Temperature (°F)	Time (sec)
1000	70	34
1200	40	36
1200	70	20
1200	20	43
1200	10	44
1400	20	24
1600	10	19
1400	20	23
1600	40	15
1600	70	11
1000	40	37
1400	70	14
1200	70	18
1400	10	27
1600	20	18
1400	40	18
1000	20	37
1000	40	42
1000	20	39
1400	70	12
1600	70	10
1400	40	17
1200	20	40
1000	10	39
1200	40	35
1000	70	30
1000	40	32
1600	20	12
1600	40	10
1600	10	14
1200	10	40
1400	10	23

Calculate the residuals from a regression analysis in which engine speed and temperature are the only predictors. Plot the residuals against the predicted values. Is curvature evident in the plot? If so, reexpress the model by either adding additional model terms or transforming variables to obtain a fit in which the residual plot does not display curvature.

3   An investigation was carried out to study the hazards posed by chemical vapors released during cargo-tanker operations. Specifically, the vapor concentration at breathing height downwind from vapor vents on the ship's deck was measured. The data in the accompanying table represent vapor concentrations collected while the ship was carrying benzene in the cargo tank. Fit a regression model to their data using distance as the predictor variable. Construct a partial-residual plot. Interpret the results of this analysis. Reexpress the model, if appropriate, to rectify any difficulties noted in the plot.

VAPOR CONCENTRATIONS

In[Vapor Conc. (kg/m^3)]	Downwind Distance (m)
-2.0	2.0
-2.3	4.0
-1.9	2.5
-2.2	7.5
-2.3	6.0
-1.0	0.1
-3.0	5.5
-2.0	3.5
-2.3	3.0
-1.1	0.5
-2.3	9.0
-2.4	6.5
-2.1	4.5
-2.1	1.0
-2.4	5.0
-3.0	8.0
-2.0	1.5
-2.2	10.5
-2.9	7.0

4   The data in Exercise 2 were collected sequentially in time in the order they are listed. Plot the residuals versus the time sequence (observation number) after correcting for any curvature that was observed. From the plot and a Durbin–Watson test for correlated errors, what conclusion can be drawn about the independence assumption of the model errors?

5   Perform a runs test on the residuals from the fit to the data in Exercise 2. Compare this result with the one obtained in Exercise 4.

6   Assess the normality assumption for the VCR sales data in Exercise 1 of Chapter 21. If the normality assumption does not appear to be reasonable, identify a suitable

reexpression of the model that does permit an assumption of normally distributed errors.

7 Assess the normality assumption for the nitrogen oxide data in Exercise 6 of Chapter 22.

8 Assess the normality assumption for the wear data in Exercise 7 of Chapter 23.

9 Investigate the effects of outliers on graphical and numerical evaluations of normality by assessing the assumption of normality for the cabin-noise data of Exercise 1, Chapter 24, (a) using the complete set, and (b) excluding the two largest observations. Is your conclusion about the reasonableness of an assumption of normally distributed errors affected by the removal of the two observations? Explain your conclusion.

CHAPTER 26

Model Respecification

The regression procedures discussed in Chapters 21 and 22 rely on several key assumptions. Two of these assumptions are that the deterministic component of the model is a linear function of the unknown regression coefficients, and the model errors are additive and follow a normal probability distribution with mean zero and a constant standard deviation. This chapter describes model respecification procedures which are useful when there is evidence that one or both of these assumptions are not appropriate. Topics discussed in this chapter include:

- *reexpression of a nonlinear model to satisfy linearity assumptions,*
- *parameter estimation for nonlinear models when the specified model is not reexpressible as a linear model, and*
- *reexpression to better satisfy the assumptions that the model errors are additive, have constant standard deviations, and are normally distributed.*

A necessary phase of a comprehensive regression analysis (ISEAS, Section 21.3) is the formulation and specification of the regression model. Two basic assumptions (see Table 21.1) about the specified model are that the functional relationship between the response and the predictor variables is linear in the unknown regression coefficients and that the model errors are additive and normally distributed with mean zero and constant standard deviation. Chapters 24 and 25 present techniques for assessing the validity of these model assumptions. Information from such an assessment or known behavior of the phenomenon being studied may indicate that an initial model specification violates one or both of these assumptions.

Models that are nonlinear in the unknown model parameters sometimes can be respecified as a linear model by suitable reexpressions of the response or (a subset of) the predictor variables or both. One value of such reexpressions is that they allow the procedures developed for linear models to be applied to nonlinear relationships between a response and a set of predictors. Reexpression of the model may also be necessary when the model is assumed to be linear but the errors do not have constant standard deviations (for example, when the magnitude of the error standard deviation is proportional to the mean of the response), or the error distribution is believed to be nonnormal.

Respecification of a nonlinear model is not, however, always a viable option. Reexpression of a nonlinear model might result in a complicated linear form for the deterministic component of the model and loss of the original measurement metric. Thus, at times a nonlinear model is the candidate model of choice because it provides a more parsimonious description of the relationship between response and predictor variables. Likewise, nonlinear models are sometimes the models chosen for fitting purposes because of the physical nature of the problem. For example, in engineering experiments the response of interest may be represented by the solution of a differential equation that is nonlinear in the model parameters.

In this chapter we discuss reexpressions of response and predictor variables. We consider the effects these reexpressions have on inferential procedures and interpreting the regression results. We also introduce the methodology necessary for fitting nonlinear regression models.

26.1 NONLINEAR MODELS

Regression models of the form (21.1) and (22.1) are termed linear, because the unknown regression coefficients are required to appear as either additive constants or multipliers of the predictor variables. Nonlinear regression models are another important and useful family of regression models.

In the last several chapters, several examples of linear regression models were used to model response variables. It is important to emphasize that the term "linear" applied to these models refers to the unknown regression coefficients, not to the response and predictor variables. Thus, the following models are all linear regression models:

$$y = \beta_0 + \beta_1 x_1 + \cdots + \beta_p x_p + e,$$

$$y = \beta_0 + \beta_1 x + \beta_2 x^2 + \cdots + \beta_p x^p + e,$$

$$\ln y = \beta_0 + \beta_1 x_1 + \beta_2 \sin x_2 + \beta_3 \cos x_3 + e.$$

When one or more of the unknown regression coefficients do not appear as additive constants or multipliers, the models are nonlinear (see Exhibit 26.1). Some examples are

$$y = \beta_0 + \beta_1 x_1^{\beta_2} + e,$$

$$y = \beta_0 + \beta_1 (x - \beta_3)^{-1} + \beta_2 (x - \beta_3)^{-2} + e,$$

$$\ln y = \beta_0 + \beta_1 x_1 + \beta_2 \sin \frac{x_2}{\beta_4} + \beta_3 \cos \frac{x_2}{\beta_4} + e.$$

EXHIBIT 26.1 NONLINEAR REGRESSION MODELS

Nonlinear regression models relate a response variable y to a set of predictor variables $x_1, x_2, \ldots, x_p$ using a functional form that is not linear in the unknown regression coefficients:

$$y_i = f(x_{ij}, \beta_k, e_i), \qquad i = 1, 2, \ldots n, \quad j = 1, 2, \ldots p, \quad k = 0, 1, \ldots q, \qquad (26.1)$$

where β_k is the kth unknown regression coefficient, x_{ij} is the ith value of the jth predictor variable, $f(x_{ij}, \beta_k, e_i)$ is a nonlinear response function [i.e., f cannot be expressed as in equation (22.1)], and e_i is the ith random error component of the model.

For notational convenience, the model (26.1) is symbolically represented as a function of only a typical predictor-variable value x_{ij} and a typical regression coefficient β_k. It should be understood, however, that the response y_i is a function of the ith values of all p predictor variables, all $q + 1$ regression coefficients, and the ith error.

The strategies for fitting nonlinear regression models are similar to those outlined in Section 21.3; however, the procedures used to obtain the model coefficient estimates and the techniques used to draw inferences on the model parameters are different for nonlinear models. Before introducing these topics, we discuss model respecification to achieve linearity or to stabilize the error standard deviation.

26.1.1 Reexpressing Nonlinear Models

Knowledge of theoretical relationships between the response and the predictor variables may indicate what reexpressions would be helpful in linearizing a response function $f(x_{ij}, \beta_k, e_i)$. In addition, scatterplots of the response versus the predictor variables may suggest forms of reexpression.

Figure 26.1 shows plots of several common functional forms of a response y versus a single predictor variable x. The different shapes displayed in Figure 26.1 are rich in variety.

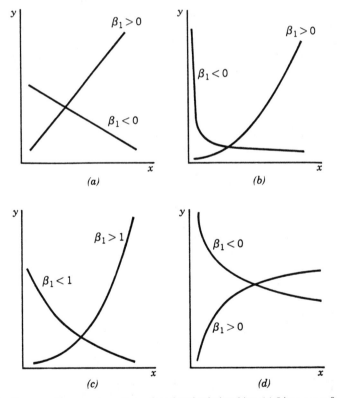

Figure 26.1 Common linear and nonlinear functional relationships. (a) Linear: $y = \beta_0 + \beta_1 x$. (b) Power: $y = \beta_0 x^{\beta_1}$. (c) Exponential: $y = \beta_0 \beta_1^x$. (d) Logarithmic: $y = \beta_0 + \beta_1 \ln x$.

Minor modifications to the expressions, such as location changes, allow even more flexibility. Figures 26.1a and d are linear in the regression coefficients, while b and c are nonlinear. Both of the nonlinear relationships can be linearized by taking logarithms of both sides of the functional equation.

Logarithmic reexpression of response or predictor variables is perhaps the most frequently used transformation of nonlinear models. While the natural logarithm (base e) is often used, the common logarithm (base 10) or a logarithm to any convenient base can be used, since all such transformations are multiples of one another:

$$\log_a y = \frac{\ln y}{\ln a},$$

where $\log_a y$ denotes the logarithm to the base a and $\ln y$ denotes the natural logarithm. Thus, any logarithmic reexpression has the same ability to linearize as does the natural logarithm; the investigator can use any base that is convenient.

There is a connection between reexpression and plotting on logarithmic graph paper. Engineering and scientific data are often plotted on either semilog graph paper (one of the axes has a logarithmic scale and the other has an arithmetic scale) or log–log paper (both axes have a logarithmic scale). The intent of such plots is to determine a combination of axis scales that result in a linear relationship. This is equivalent to reexpressing the variables using logarithmic transformations of one or both of the variables and plotting them on axes that have arithmetic scales. Working with the reexpressed values makes graph paper with special scales unnecessary and facilitates using computer-constructed scatterplots.

The data displayed in Table 26.1 were collected as part of a testing program to measure

Table 26.1   Hydrocarbon-Emissions Data

Obs.	Emissions	Wind Speed
1	4.8	5.0
2	8.2	7.2
3	10.0	7.3
4	14.6	12.0
5	22.1	14.7
6	21.1	15.3
7	20.5	18.3
8	16.5	18.3
9	21.4	22.2
10	34.5	23.6
11	32.3	28.2
12	31.9	30.5
13	48.0	31.3
14	34.0	31.9
15	37.6	31.9
16	67.0	33.7
17	32.7	34.3
18	47.4	38.2
19	53.2	40.5
20	49.4	42.5
21	81.3	45.3

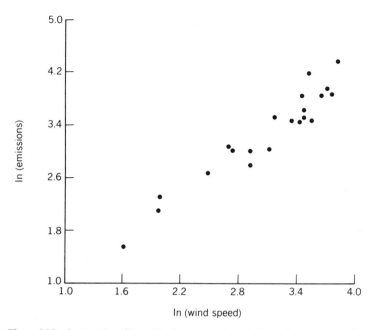

Figure 26.2   Scatterplot of logarithmic reexpressions: hydrocarbon-emission data.

hydrocarbon emissions from internal floating roof tanks used to store petroleum products. The emission and wind speed data were originally plotted using log–log graph paper. A satisfactory linear plot using log–log graph paper implies that a power relationship (Figure 26.1*b*) may be an adequate deterministic model component for relating emissions to wind speed. Figure 26.2 shows a scatterplot of ln (emission) versus ln (wind speed). The relationship appears linear; hence, the least-squares procedures of Chapter 21 can be used to fit a model to the reexpressed data.

Logarithmic reexpressions are not only used to linearize a nonlinear model. Another frequent use of them is to stabilize (make constant) the error standard deviation. ANOVA and regression models sometimes have error components whose variability increases with the size of the response. Plots of standard deviations versus average responses for ANOVA models having repeat observations can reveal such trends. Residual plots showing increased variability in the residuals as a function of the size of the predicted response is an indication of nonconstant error standard deviations for regression models (see Section 25.3).

Figure 26.3 is a plot of the standard deviations versus averages from an interlaboratory study on measuring the smoothness of paper. The standard deviations and averages were calculated from eight repeat observations on each of fourteen different samples of paper, all observations taken at a single testing laboratory. The increase in the standard deviations as a function of the average paper smoothness is apparent from the figure. This increase violates the constant-standard-deviation assumption of ANOVA and regression models. A logarithmic transformation of the original measurements removes this trend from a plot of the standard deviations and averages for the transformed responses.

Reexpressing one or more of the predictor variables is the preferred method (when possible) of linearizing a response function in a regression model. This approach leaves the

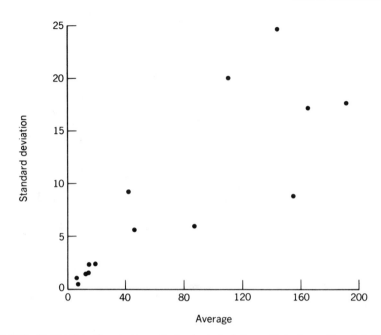

Figure 26.3 Scatterplot of averages and standard deviations. Data from Chemical Division Technical Supplement (1978), "Interlaboratory Testing Techniques," American Society for Quality Control. Copyright American Society for Quality Control, Inc., Milwaukee, WI. Reprinted by permission.

original scale of the response variable and the error structure of the model intact. When linearizing $f(x_{ij}, \beta_k, e_i)$ involves reexpressing the response y, a word of caution is in order.

If the true form of the relationship between y and $x_1, x_2, \ldots, x_p$ has an additive error component, then any reexpression of y that linearizes the deterministic portion of the model will result in a transformed model that may not have an additive error structure. For example, if the relationship between a response and a predictor variable is

$$y = \beta_0 x^{\beta_1} + e,$$

a logarithmic transformation will result in a nonlinear relationship between y and both x and the error component e. This violates the linearity assumption of a linear regression model.

In practice, the form of the model error structure is usually not known. Thus, when the response is reexpressed and a linear regression model is used to fit the data, it is incumbent on the investigator to check carefully that the assumptions made on the model errors are reasonable. The residual-based techniques discussed in Chapters 24 and 25 can be used for this purpose.

26.1.2 Fitting Nonlinear Models

When reexpression is not possible due to the functional form of the regression model, nonlinear model-fitting procedures are used to estimate the model parameters. Nonlinear model fitting can give a parsimonious specification of the relationship between the response and the predictor variables, a relationship in which the response scale has meaning to the experimenter and the model parameters are easily interpreted. Nonlinear models should be used when ease in interpretation offsets the more difficult nonlinear model-fitting procedures.

Nonlinear least-squares estimation procedures can be used to estimate the unknown regression coefficients when model errors are additive. Additivity of the error terms requires that the nonlinear model (26.1) be expressible as:

$$y_i = f(x_{ij}, \beta_k) + e_i. \tag{26.2}$$

As with linear regression models, several assumptions accompany this model. These assumptions are listed in Table 26.2.

The model parameters β_k in equation (26.2) are estimated using the principle of least squares. The nonlinear least-squares estimates are those values $b_0, b_1, \ldots, b_q$ that minimize

$$SS_E = \sum_i [y_i - f(x_{ij}, b_k)]^2, \tag{26.3}$$

the sum of the squared residuals. As with a linear regression model, SS_E is minimized and the parameter estimates are obtained by constructing and solving a set of $q + 1$ normal equations.

The normal equations are obtained by differentiating equation (26.3) with respect to each of the regression coefficients and setting each of these partial derivatives equal to zero. The second assumption in Table 26.2 ensures that the derivatives exist and that solutions to the normal equations can be found. The least-squares estimates of β_k are the values b_k that solve the normal equations. The solution to the normal equations cannot in general be expressed in a closed form; an iterative estimation procedure must be used (see below).

Consider again the data on hydrocarbon emissions from storage tanks given in Table 26.1. In order for least-squares estimation of the logarithmically reexpressed model used above to be appropriate, the original model errors must be multiplicative. A power model with multiplicative errors is expressible as

$$y_i = \beta_0 x_i^{\beta_1} e_i. \tag{26.4}$$

A logarithmic transformation of both sides of equation (26.4) then results in the linear regression model

$$z_i = \theta_0 + \theta_1 w_i + u_i, \tag{26.5}$$

Table 26.2 Nonlinear-Regression-Model Assumptions

---

1. Over the range of applicability, the true relationship between the response and the predictor variable(s) is well approximated by a nonlinear model of the form (26.2).
2. The functional form $f(x_j, \beta_k)$ is known. The function f is continuous in both x_j and β_k. The first and second partial derivatives of f with respect to β_k are continuous in both x_j and β_k.
3. The predictor variable(s) is (are) nonstochastic and measured without error.
4. The model errors are normally distributed with mean zero and constant standard deviation σ.

---

where $z = \ln y$, $\theta_0 = \ln \beta_0$, $\theta_1 = \beta_1$, $w = \ln x$, and $u_i = \ln e$. The model (26.5) is in the form of a linear regression model, and least squares can be used to estimate its parameters. This is the model that would be fitted to the data in Figure 26.2.

If the errors in the model (26.4) are additive rather than multiplicative, logarithmic transformations do not produce a linear model. Logarithmic transformations result in a more complicated nonlinear model. A nonlinear least-squares estimation procedure should be used to fit this model.

A power model with additive errors is expressible as

$$y_i = \beta_0 x_i^{\beta_1} + e_i. \tag{26.6}$$

The normal equations for this model are

$$\sum(y_i - b_0 x_i^{b_1})x_i^{b_1} = 0,$$
$$\sum(y_i - b_0 x_i^{b_1})b_0 x_i^{b_1} \ln x_i = 0. \tag{26.7}$$

There are two equations and two unknowns in (26.7). The values b_0 and b_1 that solve these equations are the least-squares estimates of β_0 and β_1.

Computer algorithms for nonlinear regression are used to obtain the least-squares estimates of the model parameters. These programs employ efficient iterative techniques to find a solution to the normal equations. Although the computer algorithm may depend on the software used, the same input is typically required (see Exhibit 26.2).

EXHIBIT 26.2   INPUT REQUIRED FOR NONLINEAR-REGRESSION COMPUTER PROGRAMS

1. The form of the nonlinear model, including explicit identification of parameters, the response, and the predictor variables.

2. The partial derivatives of the deterministic portion of the nonlinear model with respect to each of the parameters.

3. Initial estimates of the parameters to be used as starting values for the iterative routine.

Item 1 is obtained once the nonlinear model is specified. Providing the derivatives of the response function (item 2) is optional with some nonlinear-regression software that utilizes a "derivative-free" method. Depending on the particular problem, the initial estimates of the model parameters (item 3) can be obtained in a variety of ways. These include approximating the nonlinear model with a linear one to obtain starting values; choosing $q + 1$ of the observations, substituting these values into the specified model (ignoring the error component), and solving the resulting system of equations; and using limiting properties or boundary conditions of the deterministic portion of the model to determine starting values.

Table 26.3 gives details of two fits for the hydrocarbon-emission data. The first fit is to a linear regression model using the logarithmic respecification (26.5). The intercept and slope estimates c_0 and c_1 of the parameters θ_0 and θ_1 provide estimates of the original model parameters β_0 and β_1:

$$b_0 = e^{c_0} = e^{-0.028} = 0.97 \quad \text{and} \quad b_1 = c_1 = 1.08.$$

The second fit in Table 26.3 is to the nonlinear model (26.6) using these additive model

Table 26.3   Regression Fits to Hydrocarbon-Emissions Data

(a) *Linear Fit to* ln*(Emissions)*

Predictor Variable	Estimated Coefficient
Constant	$-0.0282 = c_0$
ln(speed)	$1.0770 = c_1$

$$b_0 = e^{c_0} = 0.9722 \qquad b_1 = c_1 = 1.0770$$

(b) *Nonlinear Fit to Emissions*

Model Parameter	Estimated Parameter
β_0	0.5630
β_1	1.2435

estimates as initial values. The solutions to the normal equations (26.7) yield the estimates

$$b_0 = 0.56 \quad \text{and} \quad b_1 = 1.24.$$

There are several evaluations that should be conducted in order to determine which, if either, of the above fits is suitable for these data. Residual plots should be made to assess the error assumptions. For the linearized model, partial residual plots aid in an evaluation of the linearity of the transformed response with respect to the transformed predictor. Contour plots of SS_E can be made as a function of the two model parameters, β_0 and β_1. These plots will aid in the evaluation of how sensitive SS_E is to the solution of the normal equations.

Estimates of parameters in nonlinear models often are highly correlated. When high correlations occur there will be a narrow valley in the SS_E surface, running through the minimum SS_E value at (b_0, b_1), within which the parameter estimates can jointly change without a substantial change in SS_E. For example, the estimated correlation between the two estimated parameters for the nonlinear hydrocarbon emissions fit is -0.997. Because of this high negative correlation the least-squares estimates are not stable: values of b_0 and b_1 for which $b_1 = 1.5 - 0.4b_0$ produce SS_E values close to the minimum.

Inference techniques for model parameters differ between nonlinear regression models and linear models. These methods are briefly discussed in Section 26.3.

26.2   POWER REEXPRESSIONS

Logarithmic transformations are special cases of more general power families of transformations. One of the simplest power families for positive response values can be expressed as

$$z = \begin{cases} y^\lambda, & \lambda \neq 0, \\ \ln y, & \lambda = 0. \end{cases} \tag{26.8}$$

This power family includes the reciprocal ($\lambda = -1$), square-root ($\lambda = \frac{1}{2}$), and logarithmic ($\lambda = 0$) reexpressions as special cases. It is a continuous function of the parameter λ. As noted in the previous section, logarithms to any base are constant multiples of the natural logarithms used in (26.8), so any convenient logarithm can be used in place of the natural logarithm.

For estimation purposes, it is preferable to use a power family that is expressed on a common scale for all values of the parameter λ. The following power family, often referred to as the *Box–Cox* family of transformations, is generally used when estimation of the parameter λ is to be included in the modeling procedure:

$$z = \begin{cases} (y^\lambda - 1)/(\lambda h^{\lambda - 1}), & \lambda \neq 0, \\ (\ln y)h, & \lambda = 0, \end{cases} \tag{26.9}$$

where h is the geometric mean of the responses,

$$h = (y_1 y_2 \cdots y_n)^{1/n}.$$

Once λ is estimated using (26.9), the responses are reexpressed using (26.8) and the estimated value for λ. The regression coefficients are then estimated using ordinary least squares with reexpressed responses (see Exhibit 26.3).

A modification of the transformation (26.9) changes the origin of the reexpressed variable. If y and y_i are replaced by $y - \theta$ and $y_i - \theta$, the origin of the transformation is shifted to θ. Through this two-parameter reexpression, responses that have negative values can be respecified. A change of origin is also useful when the original responses are all several orders of magnitude greater than zero.

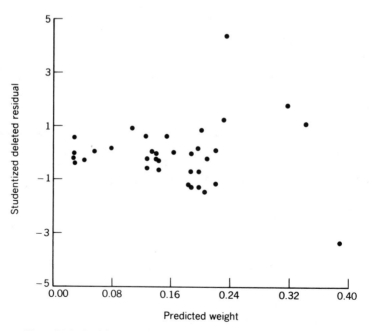

Figure 26.4 Residual plot for a quadratic fit to the solvent-weight data.

The estimation of λ is accomplished by fitting linear regression models to transformed responses (26.9) for several values of λ. The residual sums of squares are then plotted versus the corresponding values of λ. The minimum value of SS_E determines the value of the transformation parameter λ. Usually the value of λ that minimizes SS_E is rounded to a nearby convenient decimal, fraction, or integer in order to facilitate a reasonable interpretation of the transformation.

When plotting SS_E versus λ, scaling may be needed because SS_E can change by several orders of magnitude as λ is varied. When such large changes occur, either a replotting of only those SS_E values near the minimum or a scaling of SS_E can be used. A convenient scaling is to plot $L = -(n/2)\ln(SS_E/n)$. The values of L are *likelihood* values corresponding to λ; the value of λ that minimizes SS_E also maximizes L.

EXHIBIT 26.3   REGRESSION ESTIMATION WITH POWER-FAMILY REEXPRESSIONS

1. Initially choose several values of λ in the interval $[-2, 2]$.

2. For each value of λ reexpress the response values using equation (26.9).

3. Fit the reexpressed responses z_i to a linear regression model, and plot the residual sum of squares SS_E from each of the fits versus λ. Denote the value of λ that yields the minimum value of SS_E by λ^*. (If the interval $[-2, 2]$ or the increment of λ used to determine the SS_E values is too imprecise, refine the interval or increment values as needed.)

4. Determine a $100(1-\alpha)\%$ approximate confidence interval for λ by drawing a horizontal line on the plot of SS_E versus λ from the vertical axis at

$$SS_E(\lambda^*)\left(1 + \frac{t_{\alpha/2}^2}{v}\right),$$

where $t_{\alpha/2}$ is a value from the Student-t table corresponding to v degrees of freedom and v equals the number of the degrees of freedom for SS_E. The values of λ where this line intersects the graph are the approximate $100(1-\alpha)\%$ lower and upper confidence limits for λ.

5. Using λ^* or a nearby rounded value, fit the regression model using the reexpressed responses (26.8) to obtain final estimates of the regression coefficients.

Data collected on a synthetic-rubber process were introduced in Table 23.1. A fit of a quadratic polynomial model to the data, summarized in Tables 23.4 and 23.5, appears to adequately describe the curvilinear trend between the solvent weights and the production rates shown in Figure 23.1. However, a plot of the studentized deleted residuals in Figure 26.4 from the quadratic fit displays a wedge shape. This pattern is suggestive of a nonconstant standard deviation. The power transformation (26.8) is now investigated in an effort to find a transformation that will both linearize the relationship between the solvent weights and the production rates and also stabilize the standard deviations.

The quadratic response function is only one of many that could possibly characterize the curvilinear relationship between the response and the predictor variable. A range of transformations using equation (26.9) and values of λ from -2 to 2 in increments of 0.25 were constructed. After plotting the SS_E values, a refined interval of values from -0.5 to 1.0 in increments of 0.1 was used. Figure 26.5 shows a plot of the SS_E values versus λ over this latter range with the 95% confidence interval noted.

Since $\lambda = 1$ is well outside the confidence interval shown in Figure 26.5, the need for reexpression of the response variable is confirmed. Even though the logarithmic transformation $(\lambda = 0)$ is slightly outside the approximate confidence interval, the

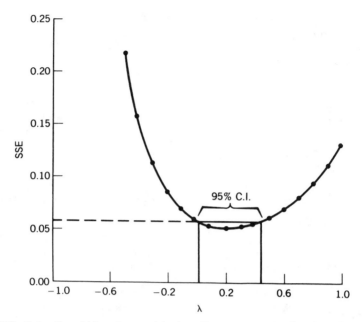

Figure 26.5 Estimation of λ for solvent-weight data. $\lambda^* = 0.2$; $\text{SSE}(\lambda)[1 + t_{0.025}^2(35)/35] = 0.06$; 95% confidence interval: $0.05 \leqslant \lambda \leqslant 0.44$.

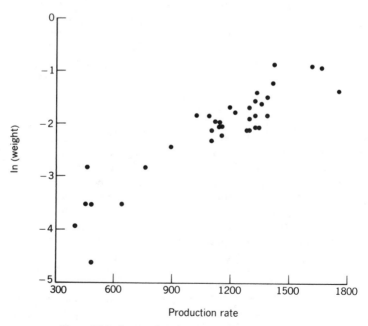

Figure 26.6 Scatterplot of reexpressed solvent weights.

experimenter opted to use the logarithmic reexpression for ease of interpretation and justification.

Figure 26.6 displays a scatterplot of the reexpressed solvent weights using the logarithmic transformation $z = \ln y$. No substantive departure from a straight line is apparent. The regression fit is summarized in Table 26.4 along with a lack-of-fit test (Section 22.3). The lack-of-fit test does not indicate significant misspecification. Apart from the presence of an outlier, the plot of the studentized deleted residuals for the reexpressed model in Figure 26.7 shows no anomalies; in particular, the wedge shape in

Table 26.4   Summary of Linear Fit of ln(Weight) for Solvent Weight Data

Predictor Variable			Estimated Coefficient		
Constant			−4.5841		
Rate			0.0022		

ANOVA

Source	df	Sum of Squares	Mean Squares	F	p-Value
Regression	1	20.61	20.61	173.19	0.000
Error	35	4.17	0.12		
Lack of Fit	25	2.24	0.13	1.40	0.297
Pure	10	0.93	0.09		
Total	36				

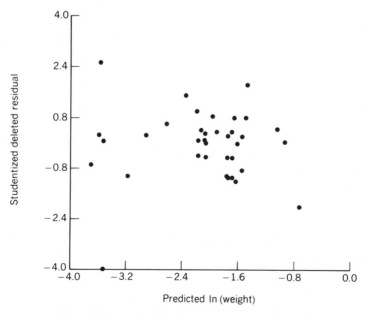

Figure 26.7   Residual plot for reexpressed solvent weights.

Figure 26.4 has been eliminated. Thus, in this example the reexpression of the solvent weights using the power transformation (26.8) results in a parsimonious model fit that satisfies the regression-model assumptions better than the quadratic model fit of the original responses.

The methodology for estimating λ in nonlinear regression models is a straightforward extension of the methodology for linear models. One simply fits z_i to the nonlinear model instead of y_i. The confidence-interval procedure described above can also be used, but the exact confidence level is more uncertain.

Power reexpressions can be extended to include applying the same reexpression to both the response variable y and the response function $f(x_j, \beta_k)$. The following model transformations result:

$$z = \begin{cases} y^\lambda = [f(x_j, \beta_k)]^\lambda + e, & \lambda \neq 0, \\ \ln y = \ln[f(x_j, \beta_k)] + e, & \lambda = 0. \end{cases} \tag{26.10}$$

This extension is referred to as *reexpressing both sides* and can be used to preserve linearity or a meaningful physical model.

Suppose that the response function $f(x_j, \beta_k)$ has been specified, based on the nature of the process or phenomenon being studied, and that the model parameters have meaningful interpretations to the investigator. When the original model is respecified using equation (26.10) in order to remedy a violation of the error-distribution assumptions, the form of the resulting prediction equation is

$$\hat{z} = [f(x_j, b_k)]^l,$$

where l is the estimated value of λ and $l \neq 0$. Since this transformation is a unique function, the inverse reexpression can be applied to both sides of the prediction equation. In doing so, the original response metric is retained, and the original form and parametrization of the response function is preserved. Note that if $f(x_j, \beta_k)$ is linear in the unknown coefficients, then linearity is also preserved. Similar results are true for the logarithmic transformation using $l = 0$. Inverting the reexpression will be discussed further in Section 26.3.

Estimation of λ when reexpressing both sides of the model requires modification similar to the use of (26.9) instead of (26.8). When reexpressing both sides the following transformed response is fitted:

$$z = \begin{cases} \dfrac{[f(x_j, \beta_k)]^\lambda - 1}{\lambda h^{\lambda-1}} + e, & \lambda \neq 0, \\ [\ln f(x_j, \beta_k)] h + e, & \lambda = 0. \end{cases} \tag{26.11}$$

The nonlinear estimation procedure described in the last section is used to estimate the model parameters for each value of λ and to calculate SS_E. Once a suitable estimate of λ is obtained by minimizing SS_E, equation (26.10) is used to reexpress both sides of the equation in order to obtain the final estimates of the regression coefficients.

Reexpression of predictor variables has been discussed several times in previous chapters, e.g., Sections 22.1, 23.1, and 25.2. Most of those discussions involved the use of various types of scatterplots to suggest transformations of the predictors. The procedures mentioned above for reexpressing both sides of a model assume that the functional form of the model is known. There are a few analytical procedures for determining the form of the predictor-variable specifications when the underlying model is unknown. One procedure, using power transformations, is the Box–Tidwell transformation.

Box–Tidwell transformations assume that the relationship between the response and the predictors is a power function of the form

$$y = \beta_0 + \beta_1 x_1^{\lambda_1} + \beta_2 x_2^{\lambda_2} + \cdots + \beta_p x_p^{\lambda_p} + e. \tag{26.12}$$

The selection of the powers $\lambda_1, \lambda_2, \ldots, \lambda_p$ is accomplished by expanding the terms in (26.12) in a first-order Taylor series and approximating each term by two terms, x_j and $x_j \ln x_j$. One can show that the coefficients of the x_j terms in this approximation do not involve the power parameters, and those of the $x_j \ln x_j$ terms equal $\beta_j(\lambda_j - 1)$. The derivations lead to the procedure (see Exhibit 26.4) for approximating the powers in the model (26.12). This estimation procedure should be used when the power model (26.12) is the correct model for the responses, i.e., when each predictor appears in the model as a single term raised to a power. This power transformation procedure is not intended to select the correct polynomial when several powers of one or more of the predictors are needed to model the response satisfactorily. Collinearities between linear and polynomial powers of predictor variables, such as between x_j and $x_j \ln x_j$, can seriously affect the estimates of the powers λ_j.

To illustrate the use of this procedure, reconsider the hydrocarbon-emission data in Table 26.1. A transformation that satisfactorily models the relationship between the hydrocarbon emissions and the wind speed is a linear model relating the natural logarithms of the two variables (see Figure 26.2). Suppose now that the power family (26.12) is used to relate ln(emissions) to the untransformed wind-speed measurements.

EXHIBIT 26.4 POWER REEXPRESSION OF PREDICTOR VARIABLES

1. Fit a linear regression of the form (22.1) in the original predictor variables. Denote the estimated regression coefficients by $a_1, a_2, \ldots, a_p$.
2. Add to the linear terms of the model one or more terms of the form $x_j \ln x_j$, and fit the expanded linear regression model. Denote the estimated regression coefficients for the linear terms by $b_1, b_2, \ldots, b_p$, and the estimated coefficients for the added terms by c_j.
3. Estimate the power parameters λ_j by $1 + c_j/a_j$.
4. Assess the adequacy of the transformation(s) using the procedures recommended in Chapter 25.

A linear fit of $y = \ln(\text{emissions})$ to $x = \text{wind speed}$ results in coefficient estimates $a_0 = 1.9122$ and $a_1 = 0.0542$. A linear fit of ln(emissions) to x and $x \ln x$ results in coefficient estimates $b_0 = 1.0000$, $b_1 = 0.2499$, and $c_1 = -0.0476$. The estimate of the power parameter λ_1 is $1 + (-0.0476/0.0542) = 0.12$. Rounding this value suggests $\lambda_1 = 0$, or the logarithmic transformation. If one performs this procedure on a power-family model of $y = \ln(\text{emissions})$ versus $x = \ln(\text{speed})$, the estimated power parameter λ_1 is 1.09, suggesting that no transformation is needed.

The approaches discussed in the last two sections to remedying model violations are not the only ones. One may use reexpressions other than power transformations, and weighted least-squares regression. Information on these possibilities is available in the references at the end of this chapter.

26.3 INFERENCE

The inference techniques discussed in this section are distinguished by whether (a) the original response is used with one or more reexpressed predictor variables in a linear model, (b) the original response is used with a nonlinear model, and (c) the response is reexpressed and the model is either linear or nonlinear, including the possibility of transformations that reexpress both sides of the model. All the inference procedures outlined below are approximate. No general statement can be made about the adequacy of the approximations, since such statements are heavily model- and hypothesis-dependent. Where possible, inferences drawn on fitted models should be validated by using the same transformations and inference procedures on additional data sets collected under identical experimental conditions.

When the original response variable is fit with a linear model in which one or more of the predictor variables have been reexpressed, the inference procedures described in Chapter 22 are still the most appropriate. These procedures are not theoretically exact if the transformation(s) used were determined by examining plots or calculating statistics from the data set used to estimate the model parameters. The same caveat applies to any analysis in which more than one inferential procedure is used, e.g., variable selection. In practice, the inferential procedures of Chapter 22 are used as though they were exact.

Inferences for models in which the original responses are nonlinear functions of the predictor variables are analogous to those presented in Section 22.3. They are, however, explicitly based on the functional form of equation (26.2). Specifically, the c_{jj}-values in equations (22.19) and (22.23) are not calculated from the original predictor-variable values x_{ij}. The predictor-variable values are replaced by the partial derivatives of $f(x_{ij}, \beta_k)$ with respect to the regression coefficients β_k. Referring to the appendix to Chapter 22, the elements x_{ij} ($i = 1, 2, \ldots, n$, $j = 1, 2, \ldots, p$) of the matrix X_c are replaced by

$$w_{ir} = \frac{\partial f(x_{ij}, \beta_k)}{\partial \beta_r}, \qquad r = 0, 1, \ldots, q,$$

where $f(x_{ij}, \beta_k)$ denotes the nonlinear response function evaluated at the predictor-variable values for the ith observation. The partial derivatives of the response function are evaluated at the nonlinear least-squares estimates $b_0, b_1, \ldots, b_q$.

If the w_{ir} are substituted for the x_{ij} in the calculation of the statistics that make up c_{jj} in equation (22.20), and the degrees of freedom associated with the estimated error standard deviation are changed from $n - p - 1$ to $n - q - 1$, then the t-statistics (22.19) and the confidence intervals (22.23) can be used. Similar substitutions are made in the reductions in error sums of squares (22.21) and F-statistics (22.22).

As noted in Section 26.1, the estimates of parameters in nonlinear models often are highly correlated. When this occurs, care must be taken when interpreting confidence intervals for individual model parameters. Simultaneous confidence regions, introduced at the end of Section 22.3 for linear models, are helpful in understanding the interrelationship among the parameter estimates.

Reexpression of the response often leaves the results of the model fit in an undesirable metric. This is especially true for predicted values from the fit. To aid in interpretation, the reexpression can be inverted. This inversion results in a prediction equation having the following general form:

$$\hat{y}_i = g(\hat{z}_i), \qquad i = 1, 2, \ldots, n, \tag{26.13}$$

where $\hat{z}_i$ denotes the predicted value from the fit with the reexpressed response z_i, and g is a function that transforms the reexpressed responses back to the original metric. The units of $\hat{y}$ are the same as those of the original response y.

A subtle change has occurred in what is being estimated by the inverted prediction equation (26.13). The distribution of the model error component associated with z is assumed to be satisfactorily approximated by the normal distribution. Consequently, the distribution for $g(z)$ is not generally a normal distribution, because of the transformation.

For responses that follow a normal distribution, the mean and the median coincide. The mean and the median of the distribution of a function g of a normal distribution normally do not coincide. In that case, $\hat{y}$ ordinarily estimates the median of the distribution $g(z)$.

When the technique of reexpressing both sides is used to better satisfy the assumptions on the regression error distribution, the form of the inverted prediction equation is

$$\hat{y} = f(x_j, b_k),$$

where b_k is the estimated coefficient from the fit to the model (26.10). The original response scale is retained, and the original form and parametrization of the response function is preserved. For the same reasons as noted above—the distribution of y is likely to be nonnormal—predicted responses $\hat{y}$ obtained from inverting both sides of the reexpression estimate medians of the response distribution corresponding to fixed values of the predictor variables.

As a rule of thumb, inferential procedures are carried out in the reexpressed scale. The inverse reexpression is then used as necessary for prediction purposes and the formation of confidence intervals on median responses. Since approximate $100(1 - \alpha)\%$ prediction limits for z are estimated percentiles for future values of z, if the inverse reexpression is a unique transformation, these limits and the resulting limits in the original measurement scale have the proper interpretation for new values of y. This property does not hold for a confidence interval on the mean response, since in general the mean of y is not equal to g(mean of z).

REFERENCES

Text References

The texts referenced at the end of Chapter 21 provide coverage of many of the topics discussed in Chapter 26. Particularly useful for a more detailed discussion of nonlinear models are the books by Daniel and Wood and by Draper and Smith. Cook and Weisberg's text, referenced in Chapter 24, and the text by Box, Hunter, and Hunter, referenced in Chapters 6 and 7, detail the use of power-family reexpressions. Most references on nonlinear model fitting and inference are at an advanced mathematical level. This includes the text by Draper and Smith and the following references:

Milliken, G. A. (1982). *Nonlinear Statistical Models*, Arlington, Virginia: The Institute for Professional Education.

Ratkowsky, D. A. (1983). *Nonlinear Regression Modeling*, New York: Marcel Dekker, Inc.

Gallant, A. R. (1987). *Nonlinear Statistical Models*, New York: John Wiley and Sons, Inc.

Additional information on reexpression is contained in the following articles:

Box, G. E. P. and Tidwell, P. W. (1962). "Transformation of the Independent Variables," *Technometrics*, **4**, 531–550.

Patterson, R. L. (1966). "Difficulties Involved in the Estimation of a Population Mean Using Transformed Sample Data," *Technometrics*, **8**, 535–537.

Snee, R. D. (1986). "An Alternative Approach to Fitting Models When Re-expression of the Response is Useful," *Journal of Quality Technology*, **18**, 211–225.

EXERCISES

1 An experimental study was made to determine the frequency responses to longitudinal excitations of thin-walled hemispherical–cylindrical tanks containing liquid. The experimenter would like to express natural frequency in terms of the liquid height ratio in the tanks. The data in the accompanying table were collected from twelve experiments. Construct a scatterplot of these data using the liquid height ratio as the predictor variable. Reexpress the data, if necessary, to a linear form. Examine both Box–Cox and Box–Tidwell power transformations.

DATA FROM FREQUENCY RESPONSE EXPERIMENTS

Liquid Height Ratio $h/1$	Frequency (Hz)
0.2	2010
0.3	1876
0.4	1720
0.5	1650
0.6	1584
0.7	1532
0.8	1413
0.9	1482
1.0	1406
1.1	1385
1.2	1371
1.3	1350

2 The feasibility of using a portable instrument to determine the physical condition of fabric-reinforced polyurethane fuel tanks was studied. This instrument measured the surface resistivity of seven polyurethane samples in varying degrees of degradation due to humid aging. It is known that the data plot as a straight line on semilog paper. Confirm this fact by plotting the data. Replot the data on arithmetic paper by making the appropriate variable reexpression. Examine Box–Cox and/or Box–Tidwell power transformations as a means of estimating the power transformation.

Exposure Time (days)	Surface Resistivity $(10^{10}\ \Omega/cm^2)$
0	4500
7	2050
15	3100
25	1000
42	225
57	32
65	21

3   Different fuels being used in certain military aircraft have caused problems in the
    engine's fuel pump by the presence of peroxides in the turbine fuel. To avoid these
    problems in the future, it would be advantageous to predict the potential peroxide
    content of a fuel before the fuel has any measurable peroxide content. A study was
    made of a particular fuel in which peroxides were measured as a function of the storage
    time of the fuel at 43°C. (See accompanying table.) Produce a scatterplot of the data
    and linearize them (if necessary) using appropriate reexpressions.

PEROXIDES AND STORAGE TIME

Storage Time (weeks)	Peroxides (ppm)	Storage Time (weeks)	Peroxides (ppm)
1	5	12	372
2	26	16	612
5	92	20	1105
8	151	24	1500
10	197	4	75
7	172	18	801
14	625	22	1401

4   Determine which of the nonlinear models below can be reexpressed as linear models.
    Show the reexpressions, where applicable.

(a) $y = ae^{bx}$

(b) $y = \dfrac{1}{a + bx}$

(c) $y = \dfrac{1}{1 + be^x}$

(d) $y = ax_1^b x_2^c$

(e) $y = ab^x$

(f) $y = a - be^x$

(g) $y = \dfrac{1}{1 + e^{a + bx}}$

(h) $y = a + bx_1 + c^{x_2}$

(i) $y = (a + b \sin x)^{-1}$

5   A heavily insulated and cooled diesel engine was used to measure the starting
    temperature of 15 test fuels. Multiple regression analysis will be applied to relate the
    minimum starting temperature of the diesel engine to several fuel properties:
    autoignition temperature, cloud point, flash point, and viscosity. Investigate the need
    for any reexpression among the data in the accompanying table.

DATA FOR STARTING TEMPERATURE EXPERIMENT (EXERCISE 5)

Minimum Starting Temp. (°C)	Auto-ignition Temp. (°C)	Cloud Point (°C)	Flash Point (°C)	Viscosity
−4.0	180	−6	77	1.40
−1.0	185	−46	64	1.50
−8.6	180	−21	83	1.95
2.8	190	−65	36	1.07
−9.4	200	−38	55	1.56
1.0	180	−68	−24	0.78
−4.2	191	−54	−21	1.12
−7.6	185	−67	43	1.39
−6.0	185	−48	64	2.07
−9.8	179	−16	84	2.57
9.0	202	−64	−21	0.76
12.0	210	3	32	0.23
−4.0	190	−60	−22	0.82
4.0	190	−59	−2	0.78
3.5	204	0	37	0.20

6 Examine the chemical-vapor data in Exercise 3 of Chapter 25. Use the techniques described in this chapter, as well as alternatives such as adding additional powers of the predictor variable, to obtain a satisfactory fit to the data. Seek a fitted model that is parsimonious (i.e., has a simple form) and for which the usual error assumptions appear reasonable.

7 Investigate whether second-order terms in the predictor variables for the engine-knock data in Exercise 7, Chapter 24, improve the fit to the knock measurements. Assess the reasonableness of the usual error assumptions, perhaps after deleting influential observations.

CHAPTER 27

Variable Selection Techniques

Regression analyses often are conducted with a large set of candidate predictor variables, only a subset of which are useful for predicting the response. In this chapter techniques for identifying important predictor variables are discussed. Specific topics to be covered include:

- *criteria for the selection of important predictor variables,*
- *subset selection methods, including the assessment of all possible regressions and stepwise selection procedures, and*
- *collinearity effects on selection methods.*

One of the last components of a comprehensive regression analysis (ISEAS, Table 21.4) is the selection of the most suitable predictor variables. This step in a comprehensive analysis usually is performed last because variable selection methods can be affected by model specification, influential observations, and collinearities. Variable selection is a concern in any regression analyses for which no complete theoretical model is available that stipulates the relationship between the response and the predictor variables.

Two competing goals face an investigator in the fitting of regression models that do not have theoretical bases for the specification of the model. On one hand is the desire for an adequate fit to explain the relationship between the response and the predictors. This usually is the primary concern, and it often mandates that many candidate predictors be included in an initial fit to the data.

The second concern is simplicity. One seeks the simplest expression of the relationship between the response and the predictors. Elimination of unnecessary predictor variables and a parsimonious expression of the relationship are sought. This goal must, however, be secondary to the first one cited above. For this reason, we have delayed consideration of variable selection until discussions of model specification and the accommodation of influential observations have been completed.

In this chapter we consider variable selection techniques. In the next section several criteria for identifying important subsets of variables are introduced. In Sections 27.2 and 27.3, methods for selecting these subsets using the criteria of Section 27.1 are discussed. The final section of this chapter illustrates the effects that collinearities can have on least-squares estimates and on variable selection procedures. Methods for accommodating collinearities are discussed in Chapter 28.

27.1 COMPARING FITTED MODELS

In the general discussions of variable selection criteria presented in this section, two models are compared: the complete model and a model with a reduced number of predictor variables. For convenience we write these two models as follows:

$$M_1: \quad y_i = \beta_0 + \beta_1 x_{i1} + \beta_2 x_{i2} + \cdots + \beta_p x_{ip} + e_i \tag{27.1}$$

and

$$M_2: \quad y_i = \beta_0 + \beta_1 x_{i1} + \beta_2 x_{i2} + \cdots + \beta_k x_{ik} + e_i^*. \tag{27.2}$$

The first model, M_1, contains the complete set of p predictor variables and is assumed to adequately relate the response variable to the predictor variables, although there may be some predictors that are unnecessary. Specifically, the model is assumed to be correctly specified, but there might be some predictor variables included in the model for which $\beta_j = 0$. The intercept term can be excluded from the models (27.1) and (27.2) with obvious changes in the discussions which follow; however, if the reduced model (27.2) contains an intercept term, the complete model (27.1) must also contain one.

The second model, M_2, consists of only the first $k < p$ predictor variables. The first k of the p variables are chosen, rather than an arbitrary subset of k, only for ease of presentation. The variable selection methods presented in subsequent sections do not require that the first k variables be the subset of interest. Note that on reducing the model the error term may change to reflect the exclusion of important predictor variables. If important predictor variables, those for which $\beta_j \neq 0$, are erroneously deleted from the model, their effects on the response are included in the model error terms. Thus, we change the notation on the error term from e_i to e_i^* in the reduced model to reflect the bias which may occur. Likewise, coefficient estimates may change dramatically when important predictors are eliminated, reflecting biases incurred by eliminating these variables.

Many criteria have been proposed for assessing whether a reduced model is an adequate representation of the relationship between the response and predictor variables. Most of these criteria involve a comparison of the complete and the reduced models, since it is assumed that the complete model does adequately relate the response and predictors. The F-statistic in (22.22) provides a statistical test of the significance of the deleted parameters:

$$F = \frac{\text{MSR}(M_1 | M_2)}{\text{MSE}_1}. \tag{27.3}$$

This statistic is based on the principle of reduction in error sums of squares (Section 15.2) and can be calculated from the error sums of squares for the two models:

$$\text{MSR}(M_1 | M_2) = \frac{\text{SSE}_2 - \text{SSE}_1}{p - k} \quad \text{and} \quad \text{MSE}_1 = \frac{\text{SSE}_1}{n - p - 1}.$$

This statistic is used to test the hypotheses

$$H_0: \beta_{k+1} = \cdots = \beta_p = 0$$

vs

$$H_a: \text{at least one } \beta_j \neq 0, \quad j = k + 1, \ldots, p.$$

Under the null hypothesis the F-statistic (27.3) follows an F probability distribution with $v_1 = p - k$ and $v_2 = n - p - 1$.

An often used criterion for assessing the predictive ability of a reduced model relative to the complete model is a comparison of the coefficients of determination, R^2. Since the reduced models do not contain the same number of predictor variables as the complete model, adjusted coefficients of determination, equation (22.15), should be compared:

$$R_a^2 = 1 - \frac{n-1}{n-k-1} \frac{\text{SSE}_2}{\text{TSS}}. \tag{27.4}$$

For the complete model, k is replaced by p, and SSE_2 by SSE_1. If the adjusted coefficients of determination are approximately equal for the full and a reduced model, the two fits are regarded as having equal predictive ability.

A third criterion for assessing reduced models relative to the complete model is the following statistic:

$$C_m = \frac{\text{SSE}_2}{\text{MSE}_1} - (n - 2m), \tag{27.5}$$

where $m = k + 1$. An adequate reduced model is one for which C_m is approximately equal to m. When predictors for which $\beta_j = 0$ are correctly eliminated from the model, SSE_2 estimates $(n - k - 1)\sigma^2$ and MSE_1 estimates σ^2. Thus C_m is approximately equal to m.

C_m can be much greater than m when important predictor variables are deleted from the model. In such circumstances, SSE_2 contains the bias due to the incorrect deletion of the predictors and it overestimates $(n - k - 1)\sigma^2$, while MSE_1 still estimates σ^2. When unimportant predictors remain in a reduced model, C_m can be much less than m. In such cases there are usually subsets with fewer variables for which C_m is approximately equal to m.

Use of the statistic (27.5) is generally accompanied by a plot of the C_m-values versus m for the better subsets, i.e., those with small values of m and for which $C_m \leq m$. Such plots allow a visual comparison of alternative subsets.

There are many other alternative criteria for the assessment of reduced models. The above three criteria are presented because they are among the most widely used and they illustrate the important features of subset selection criteria. In applications they are combined with a methodology for selecting candidate subsets. In the next two sections we present some of the more popular subset selection methods.

It is important to note that the material in this section and the next two are based solely on statistical criteria. As with all other statistical procedures, they are not intended to replace expert judgment or subject-matter theoretical considerations. In many applications variable selection procedures are not even appropriate, since theoretical considerations define both the important predictors and the functional relationship between the response and the predictors.

27.2 ALL-POSSIBLE-SUBSET COMPARISONS

The most comprehensive assessment of candidate reduced models is a comparison of the selection criteria (27.3)–(27.5) for all possible subsets of predictor variables. By comparing all possible subsets, an investigator can not only determine the "best" reduced model(s) according to the above criteria, but also identify alternatives to the "best" ones.

The term "best" is placed in quotes because of the frequent misconception that there is a single reduced model that outperforms all others. When using the above selection criteria, one should realize that they are all random variables and therefore subject to variability. The reduced model that has the highest R_a^2, for example, is not necessarily the best one if there are other reduced models that have R_a^2 close to the maximum. The selection criteria should be used as guides to select all of the better subsets, and whenever possible, engineering or scientific judgment should be used to select from among the better candidate reduced models.

To illustrate these points, we return to the chemical-yield data introduced in Table 24.1. In Chapter 24 this data set was used to illustrate the detection of influential observations. Prior to the calculation of subset selection criteria, we examined diagnostics

Table 27.1   Subset Selection Criteria for All Possible Reduced Models: Chemical-Yield Data

Subset	R_a^2	m	C_m
INT	0*	2	39.4
INV	0.026	2	39.1
FLOW	0.021	2	35.9
CAT	0.031	2	35.1
CON	0.237	2	17.2
INV, INT	0*	3	40.9
FLOW, INV	0.009	3	37.4
CAT, INT	0.012	3	37.1
CAT, INV	0.012	3	37.1
FLOW, INT	0.018	3	36.6
CAT, FLOW	0.045	3	34.2
CON, INT	0.225	3	19.0
CON, FLOW	0.302	3	12.4
CAT, CON	0.347	3	8.6
CON, INV	0.386	3	5.2
CAT, INV, INT	0*	4	39.1
FLOW, INV, INT	0.005	4	37.9
CAT, FLOW, INT	0.034	4	35.6
CAT, FLOW, INV	0.060	4	33.4
CON, FLOW, INT	0.294	4	13.8
CAT, CON, INT	0.345	4	9.6
CON, FLOW, INV	0.374	4	7.2
CON, INV, INT	0.375	4	7.1
CAT, CON, FLOW	0.402	4	4.8
CAT, CON, INV	0.433	4	2.3
CAT, FLOW, INV, INT	0.047	5	34.8
CON, FLOW, INV, INT	0.363	5	9.0
CAT, CON, FLOW, INT	0.390	5	6.8
CAT, CON, INV, INT	0.421	5	4.3
CAT, CON, FLOW, INV	0.425	5	4.0
CAT, CON, FLOW, INV, INT	0.412	6	6.0

*Negative values set to zero.

for influential observations and, after several fits to the model with various observations deleted, decided to eliminate five observations from the data set: observations 6, 11, 16, 18, and 45. We examine the issue of outliers in this data set further in Section 28.2.

Table 27.1 exhibits subset selection criteria for all possible reduced models using the data set in Table 24.1 with the five influential observations mentioned above eliminated. The five candidate predictor variables are catalyst (CAT), conversion (CON), flow of raw materials (FLOW), the inverse of a ratio of reactants (INV), and the interaction of conversion and flow (INT).

We focus attention in Table 27.1 on the use of the adjusted coefficient of determination (27.4) and the C-statistic (27.5), since computer programs that perform all subset regressions generally print out these statistics but do not always print out the F-statistics, equation (27.3). Once screening for the better subsets is completed, the F-statistics should be evaluated for the selected subsets.

Screening for the better subsets on the basis of large R_a^2-values leads to several candidate reduced models. If one screens by using a cutoff of, say, $R_a^2 > 0.35$, then one subset with two predictors, four subsets with three predictors, and four subsets with four predictors will be selected. The choice of $R_a^2 > 0.35$ is arbitrary but is reasonably close to the maximum value of $R_a^2 = 0.433$. The important issue here is that several candidates have been identified, not simply the one with the highest value for the coefficient of determination. We also would not hesitate to include one or more of the reduced models that have slightly smaller values of R_a^2 if these models have physical or experimental meaning to the investigator.

A further delineation of the better reduced models occurs when the criterion $C_m \approx m$ is examined. A plot of C_m versus k, the number of predictor variables (excluding the constant term, which is included in all the models), is shown in Figure 27.1 for all reduced

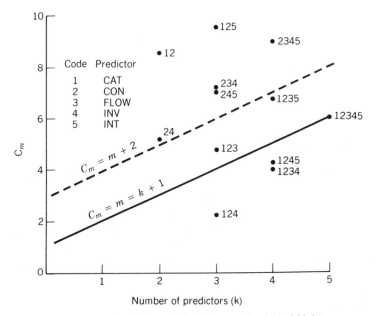

Figure 27.1   Plot of C_m statistics ($C_m \leqslant 10$), chemical-yield data.

models having C_m less than 10. Code numbers are used in the figure to identify the subsets of predictors. Again, somewhat arbitrarily, we choose a cutoff of $C_m = m + 2$ for identification of the better subsets. With this criterion, two subsets with three predictors and three subsets with four predictors remain as candidates for reduced models.

If one had to select a single subset solely on the basis of a statistical analysis of this data set, one of the two subsets with $C_m < m + 2 = 6$ having three predictors would be selected. Each of these subsets has an acceptable R_a^2-value, their C_m-values are approximately equal to m, and each subset contains one fewer variable than the three better reduced models having four predictors. The choice between the two three-predictor reduced models is not unequivocal on the sole basis of statistical criteria. However, in the four-variable reduced model that has all four variables that are included in these two three-variable models (CAT, CON, FLOW, INV), FLOW is not statistically significantly ($p = 0.570$) at the recommended significance level of $\alpha = 0.25$ (see Sections 21.5, 22.3). Based on this information, the subset CAT, CON, and INV would be selected.

The examination of all possible reduced models allows the greatest scrutiny of alternative candidates. Statistical criteria can be used to screen the better subset models and, with engineering and scientific judgment, to select one or more for use as final prediction equations. This approach to the identification of reduced models is recommended whenever the number of candidate predictor variables is not prohibitively large. When it is, the procedures discussed in the next section should be used.

27.3   STEPWISE SELECTION METHODS

Stepwise selection methods sequentially add or delete predictor variables to the prediction equation, generally one at a time. These methods involve many fewer model fits than all possible subsets, since each step in the procedure leads directly to the next one. Three popular subset selection methods will be discussed in this section: forward selection, backward elimination, and stepwise iteration.

The selection methods presented in this section are especially useful when a large number of candidate predictors are available for possible inclusion in the final reduced model. The major disadvantage of these methods is that they generally isolate only one reduced model and do not identify alternative candidate subsets of the same size. These methods also do not necessarily identify the reduced model that is optimal according to the selection criteria discussed in Section 27.1.

27.3.1   Forward Selection

The forward selection (FS) procedure for variable selection begins with no predictor variables in the model. Variables are added one at a time until a satisfactory fit is achieved or until all predictors have been added. The decision when to terminate the procedure can be made using any of the selection criteria discussed in Section 27.1; however, F-statistics generally are used. Since these F-statistics are based on the principle of reduction in error sums of squares, they measure the incremental contribution of a predictor variable above that provided by the variables already in the model. When the addition of a predictor does not result in a statistically significant F-statistic, the procedure is terminated (see Exhibit 27.1).

EXHIBIT 27.1   FORWARD SELECTION PROCEDURE

1. Fit p single-variable regression models, each having one of the predictor variables in it. Calculate the overall model F-statistic for each of the p fits. Select the fit with the largest F-statistic.

2. Using a significance level of $\alpha = 0.25$, test the significance of the predictor variable. If it is not significant, terminate the procedure and conclude that none of the predictors are useful in predicting the response. If it is significant, retain the predictor variable, set $r = 1$, and proceed to step 3.

3. Fit $p - r$ reduced models, each having the r predictor variables from the previous stage of the selection process and one of the remaining candidate predictors. Calculate the overall model F-statistic for each of the fits. Select the fit with the largest F-statistic.

4. Using a significance level of $\alpha = 0.25$, test the significance of the additional predictor variable using the F-statistic (27.3) with the model M_1 containing the $r + 1$ predictors and the reduced model M_2 containing the r predictors from the previous stage of the selection process. The degrees of freedom for this F-statistic are $v_1 = 1$ and $v_2 = n - r - 2$.

5. If the F-statistic is not significant, terminate the procedure and retain the r predictors from the previous stage. If it is significant, add the additional predictor to the r previously selected, increment r by 1, and return to step 3.

The F-statistic used in step 4 of the FS procedure tests the hypothesis

$$H_0 : \beta_{r+1} = 0 \quad \text{vs} \quad H_a : \beta_{r+1} \neq 0$$

in the model

$$M_1: \quad y_i = \beta_0 + \beta_1 x_{i1} + \beta_2 x_{i2} + \cdots + \beta_{r+1} x_{i,r+1} + e_i^*. \tag{27.6}$$

The procedure selects as the $(r + 1)$st predictor variable the one in the remaining $p - r$ that has the largest F-value. If this predictor is not statistically significant, the procedure terminates and no further predictors are added to the model. If it is statistically significant, the procedure continues with the remaining $p - r - 1$ candidate predictors.

Table 27.2 shows the application of the forward selection procedure to the chemical-yield data of Table 24.1 with the five outliers mentioned in the last section deleted. At stage 1, conversion is added to the model. At stage 2, the inverse of the reactant ratio variable is added. At stage 3, catalyst is added. The procedure terminates at stage 4. In this instance, the method produced the optimal subset of size $r = 3$, the one ultimately selected in the examination of all possible reduced models in the last section. We again note that selection of the optimal subset cannot be guaranteed with this procedure; however, the optimal subset is often selected when predictor variables are not highly collinear.

When there are many candidate predictors, application of FS is frequently allowed to proceed two or three steps beyond the one that suggests termination. This is because the estimate of the uncontrolled error variance, MS_E, is heavily biased in the early steps of this procedure. The effects of any predictors that should be in the model but have not yet been included are contained in the error term e_i^*. When possible, the final reduced model should be compared with the complete model using the F-statistic (27.3) to ensure that the procedure did not terminate prematurely due to an inflated error component.

Interpretation of the results of FS should be done with some care. The order of entry of

Table 27.2   Forward Selection Procedure for the Chemical-Yield Data

Stage	Reduced Model	F-Statistic for Added Predictor	p-Value
1	CAT	2.67	0.109
	CON*	17.19	0.000
	FLOW	2.13	0.150
	INV	0.23	0.633
	INT	0.03	0.874
2	CON, CAT	9.54	0.003
	CON, FLOW	5.72	0.021
	CON, INV*	13.33	0.001
	CON, INT	0.15	0.701
3	CON, INV, CAT*	5.10	0.028
	CON, INV, FLOW	0.01	0.915
	CON, INV, INT	0.13	0.717
4	CON, INV, CAT, FLOW	0.33	0.570
	CON, INV, CAT, INT	0.03	0.858

*Variable selected

the predictor variables should not be interpreted as indicating an order of importance of the variables. Interrelationships among the predictors, synergistic effects of the variables, and the biases due to poor model specification in the early steps of the procedure can all affect the order of entry of the predictors.

27.3.2   Backward Elimination

The backward elimination (BE) procedure for variable selection begins with all the predictor variables in the model. Variables are deleted one at a time until an unsatisfactory fit is encountered (see Exhibit 27.2). As with forward selection, the decision of when to terminate the procedure is most frequently based on the F-statistic (27.3).

EXHIBIT 27.2   BACKWARD ELIMINATION PROCEDURE

1. Fit the complete model having all p predictor variables in it. Select the variable having the smallest t or F statistic. The F-statistic (27.3) is the square of the t-statistic (22.19) used to test the significance of β_j in the complete model.

2. Using a significance level of $\alpha = 0.25$, test the significance of the predictor variable. If it is significant terminate the procedure and retain all the predictor variables. If it is not significant delete the predictor variable, set $r = 1$, and proceed to step 3.

3. Fit the reduced model having $p - r$ predictor variables remaining from the previous step. Calculate F-statistics for testing the significance of each of the remaining variables. Select the predictor variable having the smallest F-statistic. The degrees of freedom of this F-statistic are $v_1 = 1$ and $v_2 = n - p + r - 1$.

4. Using a significance level of $\alpha = 0.25$, test the significance of the selected predictor variable. If the F-statistic is significant, terminate the procedure and retain all of the remaining $p - r$ predictor variables from the previous stage of the selection process. If it is not significant, delete the selected variable, increment r by 1, and return to step 3.

Table 27.3 Backward Elimination Procedure for the Chemical-Yield Data

Stage	Candidate Variable for Deletion	F-Statistic for Candidate Variable	p-Value
1	CAT	5.03	0.030
	CON	30.83	0.000
	FLOW	0.29	0.593
	INV	2.84	0.099
	INT*	0.00	0.978
2	CAT	5.35	0.025
	CON	32.04	0.000
	FLOW*	0.33	0.570
	INV	2.91	0.095
3	CAT	5.10	0.028
	CON	38.05	0.000
	INV	8.55	0.005

*Variable deleted

At each step of BE the statistical significance of the predictor variable having the smallest t or F statistic is examined. Only if the predictor is not statistically significant is it deleted from the model. Unlike forward selection, the error term in the denominator of the F-statistic (27.3) should not be heavily biased in backward elimination, because the analysis begins with what is assumed to be a correctly specified model, apart from possibly some extraneous predictor variables. At each step of the selection procedure only one model is fitted: the reduced model containing all predictors not yet deleted.

Table 27.3 illustrates the application of BE to the chemical-yield data. At stage 1 the interaction between conversion and flow is deleted. At stage 2, flow is deleted. No further predictors are deleted at the next stage, so the procedure terminates. The optimal subset that was obtained from an examination of all possible reduced models has also been selected by BE.

As with forward selection, backward elimination cannot guarantee the selection of optimal subsets. Likewise, only one reduced model is selected; no alternative subsets are identified. Backward elimination begins with the complete model and a good estimate of the error standard deviation. For these reasons, BE is preferred to FS unless the number of predictors is so large that BE is inefficient or, if $n < p + 1$, BE cannot calculate estimates of the regression coefficients. We again caution that no order of importance should be attached to the predictor variables based on the size of the t or F statistics.

27.3.3 Stepwise Iteration

The stepwise iteration (SI) procedure adds predictor variables one at a time like a forward selection procedure, but at each stage of the procedure the deletion of variables is permitted (see Exhibit 27.3). It thus combines features of both forward selection and backward elimination.

EXHIBIT 27.3   STEPWISE ITERATION PROCEDURE

1. Initiate the selection procedure as in forward selection.
2. After a predictor variable is added, apply steps 3 and 4 of the backward elimination procedure to the predictor variables in the model.
3. Continue the selection procedure as in forward selection. At each stage of the process, if a predictor variable is added, return to step 2. Terminate the procedure when no predictor variables are added at a stage of forward selection.

Table 27.4 illustrates the application of the SI procedure. Stages 1 and 2 are identical to the same stages of forward selection shown in Table 27.2. Once two predictors are in the model, backward elimination is performed as each variable is entered. Thus, stage 3 is a backward-elimination stage for the reduced model containing the two variables entered at stage 2. Since each of the significance probabilities at stage 3 is less than 0.25, both predictors are retained. At stage 4 forward selection is again performed with the remaining predictors. Stage 5 is a backward elimination procedure for the reduced model containing the three predictors resulting from stage 4. Again, none of the predictors is deleted, so stage 6 is the next stage of forward selection. Since no further variables are entered at stage 6, the process terminates with the optimal subset of three predictors.

More computational effort is involved with stepwise iteration than with either forward selection or backward elimination. The tradeoff for the additional computational effort is the ability to delete nonsignificant predictors as variables are added and the ability to add

Table 27.4   Stepwise Iteration Procedure for the Chemical-Yield Data

	Predictor Variables				
Stage	In	Added	Deleted	F-Statistic	p-Value
1	None	CAT		2.67	0.109
	None	CON*		17.19	0.000
	None	FLOW		2.13	0.150
	None	INV		0.23	0.633
	None	INT		0.03	0.874
2	CON	CAT		9.54	0.003
	CON	FLOW		5.72	0.021
	CON	INV*		13.33	0.001
	CON	INT		0.15	0.701
3	CON, INV		CON	34.30	0.000
	CON, INV		INV	13.33	0.001
4	CON, INV	CAT*		5.10	0.029
	CON, INV	FLOW		0.01	0.915
	CON, INV	INT		0.13	0.717
5	CON, INV, CAT		CAT	5.10	0.029
	CON, INV, CAT		CON	38.05	0.000
	CON, INV, CAT		INV	8.55	0.005
6	CON, INV, CAT	FLOW		0.33	0.570
	CON, INV, CAT	INT		0.03	0.858

*Variable added.

new predictors following deletion. In this sense it is a compromise between the strict addition or deletion of FS and BE and the major expenditure of effort involved with the evaluation of all possible reduced models. We again stress that the advantages of being able to assess all possible subsets of predictors outweighs the additional computational effort involved unless the number of predictor variables renders this procedure impractical.

27.4 COLLINEAR EFFECTS

In the last section it was stressed that different subset selection methods need not result in the selection of the same subset of predictors, nor need the subsets selected be optimal according to the selection criteria used. The chemical-yield data were used to illustrate the stepwise procedures in the last section in part because they do yield the optimal subset regardless of which of the methods is employed. Suitably well-behaved data sets can be expected to yield consistent results on the various subset selection methods. Many designed experiments, notably complete factorials, yield identical subsets regardless of the selection method used.

One property of data sets that is known often to produce results depending on the subset selection method is collinearity among predictor variables. Collinearities (redundancies) were introduced in Section 23.1 in the context of fitting polynomial regression models. They are discussed further in Section 28.3, at which point the detection and accommodation of collinearities will be detailed. In this section we wish to illustrate some of the effects of collinearities, especially their effects on subset selection methods.

The cryogenic-flowmeter data that were introduced in Section 25.3 consist of 120 observations on a response variable (PD) and seven predictor variables: TEMP, PRESS, DENS, COOL, TIME, PULSE, and ULLAGE. A description of these variables and a listing of this data set is appended to this chapter. After reviewing residual plots, there appears to be a need to add a quadratic term for temperature. We add such a term, TEMP2, after first standardizing the temperature as in equation (23.5).

An initial assessment of the resulting fit, ignoring the questions of model specification that were raised in Section 25.3, suggests that the fit is satisfactory. The coefficient of determination in Table 27.5 is 78%; the estimated error standard deviation is 0.112. Several of the t-statistics are nonsignificant, leading to a consideration of variable selection to reduce the model.

Forward selection selects six predictor variables, all those listed in Table 27.5 except TEMP and TIME. Backward elimination also selects six predictors, deleting DENS and TIME. Stepwise iteration agrees with forward selection. An examination of all possible reduced models indicates that both of these six-variable subsets are acceptable, the FS subset having $R^2 = 0.775$ and $C_7 = 6.08$, and the BE subset having the same R^2 with $C_7 = 6.17$. What is disturbing about these results is that the t-statistics in Table 27.5 suggest that both DENS and TEMP can be deleted but neither of the above subsets deletes both.

The reason for these apparently contradictory results is that DENS and TEMP are highly collinear. The correlation coefficient (Pearson's r) between these two predictors is -0.999. As explained in the reference from which these data are taken, the process under study is such that the density measurements of the liquid nitrogen used in the experiment are solely a function of temperature and pressure, and the dependence on pressure was thought to be minimal throughout the experimental region.

Collinearities such as this one often lead to inconsistent variable-selection results when

Table 27.5   Regression Fit to Cryogenic-Flowmeter Data

			ANOVA		
Source	df	Sums of Squares	Mean Squares	F	p-Value
Regression	8	4.908	0.614	48.48	0.000
Error	111	1.405	0.013		
Total	119	6.313			
Predictor Variable*	Estimate	Standard Error	t	p-Value	VIF
Constant	−2.993	35.514	−0.08	0.933	
TEMP	−0.066	0.156	−0.42	0.674	1,541.1
TEMP2	4.258	1.157	3.68	0.000	2.3
PRESS	0.023	0.009	2.62	0.010	101.3
DENS	1.128	3.558	0.32	0.752	1,255.7
COOL	−0.150	0.061	−2.44	0.016	145.3
TIME	−0.000	0.000	−0.99	0.325	1.2
PULSE	0.120	0.003	5.89	0.000	1.6
ULLAGE	−0.002	0.001	−2.94	0.004	1.2

*TEMP2 formed using the correlation-form standardization of TEMP.

different methods are applied. Other problems are also associated with collinearities. The standard errors of least-squares coefficient estimators, Section 22.2, are proportional to their *variance inflation factors* (VIF):

$$VIF_j = (1 - R_j^2)^{-1}, \qquad (27.7)$$

where R_j^2 is the coefficient of determination of the regression of x_j on the constant term and the other predictor variables in the model. A predictor variable that is highly collinear with the constant term or one or more other predictors will have a R_j^2-value close to 1. This in turn will produce a large variance inflation factor and, unless the estimated error standard deviation s_e is compensatingly small, a large estimated standard error for the least-squares estimate b_j.

The dramatic effect of the collinearity between DENS and TEMP on their variance inflation factors can be seen in the last column of Table 27.5. Each of their variance inflation factors is over 1000. This means that the estimators' standard errors are over 30 times larger than they would be if the two predictor variables were completely uncorrelated with all other predictors in the model ($R_j^2 = 0$, VIF = 1). It is interesting to note that there appear to be other collinear variables in this data set, since both PRESS and COOL have variance inflation factors that exceed 100. We return for a more complete assessment of collinearities in this data set in Section 28.3.

Among the effects known to result from collinearities are the following. Coefficient estimates often tend to be too large in magnitude and occasionally of the wrong sign. Variance inflation factors can be orders of magnitude larger than their minimum value of 1; consequently, standard errors of coefficient estimators tend to be much larger than for estimators of noncollinear predictors. Finally, variable selection procedures, including *t*-statistics from the full model fit, can erroneously indicate addition or deletion of collinear variables.

Although strong collinearities cause difficulties such as these, their effects can sometimes be minimized. Proper consideration of variables, including the elimination of variables known to be always redundant, and standardization prior to the formation of polynomial terms can reduce the occurrence of collinearities. Variable selection techniques can be used at times to screen for collinear predictors. Biased estimators of the regression coefficients are another alternative. These topics are explored more fully in Section 28.3.

REFERENCES

Text References

Most of the regression texts referenced at the end of Chapters 21, 22, and 24 contain discussions devoted to variable selection techniques and the effects of collinearities. Three additional articles that specifically address this topic are:

Gunst, R. F. (1983). "Regression Analysis with Multicollinear Predictor Variables: Definition, Detection, and Effects," *Communications in Statistics*, **12**, 2217–2260.

Henderson, H. V. and Velleman, P. F. (1981). "Building Multiple Regression Models Interactively," *Biometrics*, **37**, 391–441.

Hocking, R. R. (1976). "The Analysis and Selection of Variables in Linear Regression," *Biometrics*, **32**, 1–49.

APPENDIX: CRYOGENIC-FLOWMETER DATA

This data set contains 120 observations collected during an experiment to evaluate the accuracy and precision of a new facility for calibrating cryogenic flowmeters. The reference for the original data set is provided at the end of Chapter 25. The variable definitions are:

PD   Percent difference between the weight recorded by the flowmeter under test and the weight provided by an independent weighing process.

TEMP   Temperature of the liquid nitrogen at the meter (K).

PRESS   Pressure at the flow meter (lb/in.^{2}).

DENS   Density of the liquid nitrogen (lb/gal).

COOL   Subcooling temperature (amount below the boiling point of the liquid) (K).

TIME   Elapsed time of the accumulation at the meter (sec).

PULSE   Number of pulses of the photoelectric cell counting revolutions of the flowmeter shaft.

ULLAGE Ullage temperature of the gaseous helium above the liquid nitrogen in the weighing tank (K).

Cryogenic-Flowmeter Data

TEMP	PRESS	DENS	COOL	TIME	PULSE	ULLAGE	PD
78.88	82.3	6.69	17.27	78.93	59	120.55	−0.4861
79.02	83.0	6.68	17.24	85.79	65	106.15	−0.4252
79.00	83.2	6.68	17.29	84.92	65	103.07	−0.1933
79.35	86.7	6.67	17.47	84.73	65	84.84	−0.3660
78.51	85.2	6.70	18.08	84.79	65	88.64	−0.3951
77.89	83.7	6.73	18.48	83.22	64	89.25	−0.4165
78.00	80.0	6.72	17.79	78.38	58	85.03	−0.5655
78.07	82.0	6.72	18.03	74.64	58	87.98	−0.5542
78.10	82.4	6.72	18.07	75.26	58	87.87	−0.5382
78.95	89.7	6.69	18.29	72.50	56	82.82	−0.4643
79.07	90.5	6.68	18.28	73.51	57	86.80	−0.3763
79.21	90.8	6.68	18.18	72.35	56	87.54	−0.4361
87.31	97.0	6.35	10.89	92.59	74	95.46	0.3343
87.77	98.6	6.33	10.63	93.71	74	98.38	0.4132
87.89	98.0	6.32	10.44	95.24	75	97.39	0.4158
87.75	106.3	6.33	11.55	159.48	74	99.41	0.4500
87.36	109.1	6.35	12.25	157.83	74	96.54	0.3557
87.13	108.5	6.36	12.42	158.85	74	99.55	0.3706
79.76	104.0	6.66	19.27	100.14	70	99.87	−0.3618
79.42	105.7	6.67	19.82	101.44	70	101.82	−0.2609
79.39	106.5	6.67	19.93	101.60	70	103.06	−0.2493
79.51	107.0	6.67	19.87	79.08	55	102.46	−0.3516
79.53	107.6	6.67	19.91	80.41	56	100.39	−0.3213
79.49	107.5	6.67	19.95	80.25	56	96.75	−0.2731
79.51	107.3	6.67	19.90	99.88	69	90.74	−0.2688
79.46	108.2	6.67	20.05	98.50	68	96.52	−0.2587
79.46	107.9	6.67	20.01	99.21	69	92.53	−0.2354
79.51	108.0	6.67	19.98	99.91	69	91.38	−0.2296
79.76	108.3	6.66	19.76	95.99	67	87.91	−0.2166
81.13	109.5	6.60	18.52	88.92	68	83.64	−0.2512
79.56	104.8	6.67	19.57	88.60	67	85.15	−0.2688
78.61	102.2	6.70	20.22	88.24	67	85.85	−0.3154
79.69	94.1	6.66	18.13	91.09	67	115.55	−0.4985
79.81	94.3	6.65	18.04	90.76	67	109.69	−0.3540
79.90	94.5	6.65	17.98	89.79	66	102.59	−0.4536
81.22	87.8	6.60	15.75	87.29	67	143.33	−0.4524
81.50	88.2	6.59	15.53	88.58	68	131.99	−0.3122
81.73	88.6	6.58	15.36	87.49	67	122.64	−0.2805
82.03	90.4	6.57	15.31	212.45	68	110.55	−0.3796
81.82	90.5	6.57	15.53	199.38	67	107.80	−0.2091
81.87	91.0	6.57	15.55	210.06	68	104.87	−0.1859
81.80	88.4	6.57	15.26	108.84	68	101.56	−0.1738
81.64	88.4	6.58	15.42	109.81	68	100.05	−0.2340
81.59	88.7	6.58	15.51	109.40	68	98.90	−0.1695
81.59	92.4	6.58	16.02	151.85	68	86.51	−0.0828
81.71	94.2	6.58	16.14	148.84	67	90.39	−0.0902
81.78	94.9	6.58	16.16	151.62	68	93.55	−0.0586
81.78	98.6	6.58	16.62	98.48	67	85.08	−0.0431
82.05	101.0	6.57	16.63	94.06	68	89.86	−0.0608

Cryogenic-Flowmeter Data
(continued)

TEMP	PRESS	DENS	COOL	TIME	PULSE	ULLAGE	PD
82.19	101.3	6.56	16.53	93.71	68	89.41	−0.0730
82.26	101.3	6.56	16.46	92.57	68	89.68	−0.0811
81.24	73.8	6.59	13.43	88.62	69	99.26	−0.3291
81.41	73.9	6.59	13.29	88.92	69	98.79	−0.3254
81.52	73.8	6.58	13.15	87.48	68	97.69	−0.3425
78.65	75.6	6.70	16.36	91.27	69	89.23	−0.3474
78.51	75.9	6.70	16.55	91.02	69	91.82	−0.3491
78.37	76.2	6.71	16.75	91.41	69	92.15	−0.3235
78.98	81.9	6.68	17.11	92.21	69	83.65	−0.2837
79.09	82.9	6.68	17.15	92.32	69	87.15	−0.2155
79.18	83.5	6.68	17.15	91.02	68	87.51	−0.2624
82.33	90.5	6.55	15.02	90.72	70	131.25	−0.2146
82.56	90.5	6.54	14.79	90.98	70	118.27	−0.1794
82.63	90.9	6.54	14.78	90.75	70	111.47	−0.1204
80.97	83.5	6.61	15.37	88.53	69	86.94	−0.2040
81.08	84.3	6.60	15.38	88.59	69	88.95	−0.1725
81.13	85.0	6.60	15.44	89.83	70	89.51	−0.1974
81.66	87.1	6.58	15.21	89.15	70	84.50	−0.0589
81.82	88.4	6.57	15.24	88.55	70	87.86	−0.0766
82.01	89.5	6.57	15.21	88.59	70	88.10	−0.0748
82.88	94.6	6.53	15.01	90.77	70	86.50	−0.0712
83.23	92.3	6.52	14.36	90.99	70	99.83	−0.0397
81.54	93.1	6.59	16.15	91.91	69	106.76	−0.2226
81.54	93.1	6.59	16.15	93.22	70	106.76	−0.2038
81.57	93.0	6.58	16.11	93.46	70	102.15	−0.1725
81.57	94.5	6.58	16.31	94.82	70	85.86	−0.1255
81.59	95.3	6.58	16.39	95.21	70	88.64	−0.0863
81.68	96.0	6.58	16.38	95.25	70	88.68	−0.0432
82.49	99.4	6.55	16.00	94.35	70	85.31	0.0217
82.61	100.7	6.54	16.04	91.14	70	87.62	0.0099
82.63	101.1	6.54	16.06	91.14	70	88.10	0.0692
84.80	115.8	6.46	15.57	112.44	71	87.90	0.2890
85.12	120.0	6.45	15.75	97.59	72	91.02	0.4289
85.21	105.4	6.44	13.98	95.60	72	98.78	0.2603
85.84	109.8	6.41	13.84	95.42	72	88.88	0.4350
86.00	111.3	6.41	13.84	95.09	72	92.06	0.4909
79.97	92.7	6.65	17.67	90.89	69	111.00	0.0493
80.37	92.0	6.63	17.18	92.92	70	106.57	−0.1090
80.37	91.7	6.63	17.14	91.92	69	102.66	−0.2092
80.62	88.6	6.62	16.46	91.01	69	85.34	−0.1009
80.67	89.3	6.62	16.51	92.19	70	88.50	−0.0781
80.78	89.6	6.62	16.44	92.42	70	88.91	−0.1347
81.61	92.7	6.58	16.02	91.16	69	84.70	−0.0199
81.78	94.4	6.58	16.09	92.08	70	88.11	−0.0020
81.91	95.5	6.57	16.09	93.57	71	89.06	0.0349
83.09	99.2	6.52	15.37	92.91	71	87.05	0.1232
83.23	99.8	6.52	15.30	92.64	71	91.03	0.1351

Cryogenic-Flowmeter Data
(continued)

TEMP	PRESS	DENS	COOL	TIME	PULSE	ULLAGE	PD
83.32	99.8	6.52	15.21	92.99	71	90.60	0.0646
77.47	68.9	6.74	16.25	88.86	68	147.56	−0.1911
77.61	68.9	6.74	16.11	89.03	68	143.52	−0.1736
77.65	69.3	6.73	16.15	89.00	68	142.36	−0.1105
77.98	73.6	6.72	16.66	90.82	69	88.38	0.0859
78.17	74.7	6.72	16.69	88.38	68	90.19	−0.0376
78.19	75.3	6.71	16.78	89.54	69	90.22	−0.0702
78.98	77.3	6.68	16.35	89.73	69	82.64	−0.0333
77.93	75.7	6.72	17.10	88.44	68	84.31	−0.1601
77.17	74.6	6.75	17.67	88.62	68	85.35	−0.1908
77.00	79.6	6.76	18.72	88.76	68	82.01	0.0020
77.07	80.7	6.76	18.83	88.94	68	85.01	−0.1671
77.10	81.4	6.76	18.91	90.09	69	85.59	0.0097
77.91	65.0	6.72	14.99	89.46	68	111.92	−0.2960
77.98	65.7	6.72	15.07	90.10	68	107.46	0.2922
78.05	66.2	6.72	15.11	89.44	68	103.85	−0.2805
76.91	69.4	6.76	16.90	87.41	67	85.88	−0.4197
76.80	69.7	6.77	17.08	89.07	68	86.64	−0.3098
76.84	77.9	6.77	18.59	88.90	68	87.54	−0.4386
77.21	78.3	6.75	18.29	88.72	68	82.52	−0.3418
77.21	79.4	6.75	18.47	88.99	68	86.01	−0.3202
77.31	80.0	6.75	18.47	87.54	67	86.30	−0.3887
78.19	84.2	6.72	18.25	88.60	68	85.42	−0.2431
78.30	84.5	6.71	18.18	89.41	68	86.35	−0.2571

Data from Joiner, B. (1977), "Evaluation of Cryogenic Flow Meters: An Example in Non-standard Experimental Design and Analysis," *Technometrics*, **19**, 353–380. Copyright American Statistical Association, Alexandria, VA. Reprinted by permission.

EXERCISES

1   The effects of fuel properties on startability were evaluated in a diesel engine using 18 test fuels. The minimum starting temperature was recorded as each test fuel flowed through the engine. Analysis of the fuel yielded five fuel properties: cetane number, viscosity, distillation temperature at 90% boiling point, autoignition temperature, and flash point (see accompanying table). Perform all possible subset comparisons using the five fuel properties as predictor variables. Produce a C_m-plot of the C_m-statistics. Which subset(s) do you feel may be candidate models for the engine starting temperature?

FUEL PROPERTIES AND STARTABILITY (EXERCISE 1)

Minimum Starting Temperature	Cetane Number	Viscosity	90% Distillation	Auto-ignition Temp.	Flash Point
−9.6	57	1.95	304.8	252	82.6
−2.0	45	1.50	257.8	257	63.6
0.0	35	0.78	246.7	252	−24.0
8.0	28	0.76	236.7	209	−21.0
11.0	35	3.74	361.0	217	32.0
1.8	41	1.07	221.3	262	36.0
−10.4	47	1.56	258.4	272	55.0
−5.0	50	3.00	343.0	252	77.0
−5.2	43	1.12	481.2	198	−21.0
−8.6	49	1.39	474.0	192	43.0
−7.0	48	2.07	469.6	192	64.0
−10.8	60	2.57	471.8	186	84.0
−5.0	39	0.82	250.5	197	−22.0
3.0	40	0.78	234.0	197	−2.0
2.5	37	3.73	357.0	211	37.0
−10.4	42	1.46	251.0	192	55.0
−5.5	39	1.80	329.3	197	50.0
−5.5	44	1.49	281.4	190	60.5

2 Analyze the data in Exercise 1 using the forward selection, backward elimination, and stepwise iteration procedures with a level of significance $\alpha = 0.25$. Compare the results of these procedures with the conclusions drawn in Exercise 1.

3 An experiment was conducted to determine which factors affect the knock intensity of an engine during the combustion process. Twelve different fuels were chosen in this study, and varying levels of speed, load, fuel flow, manifold pressure, and brake horsepower were applied as each fuel flowed through the engine. A listing of the resultant knock-intensity values is given in the accompanying table. Identify one or more better subsets of the predictor variables by summarizing all possible reduced models in a C_m-plot. If all subsets can't be computed, see Exercise 4.

Response variable
 KNOCK (knock intensity, psi/angular degree)
Predictor variables
 SPEED (rpm)
 LOAD (lb/ft)
 FUELFLOW (fuel flow, lb/hr)
 EXTEMP (exhaust temperature, °F)
 MANPRESS (manifold pressure, inches of mercury)
 BHP (brake horsepower, hp)
 GRAVITY (specific gravity, PAPI)
 VISCOS (viscosity, cS, at 40°C)
 D50 (distillation, 50% recovery, °C)
 FBP (final boiling point, °C)

Knock-Intensity Values (Exercise 3)

OBS	SPEED	LOAD	FUELFLOW	EXTEMP	MANPRESS	BHP	GRAVITY	VISCOS	D50	FBP	KNOCK
1	1404	885.9	90.2	1278	4.0	236.8	30.9	2.20	256.3	418.3	148.90
2	1800	1010.9	123.6	1217	9.0	346.5	30.9	2.20	256.3	418.3	130.50
3	2200	1016.8	154.8	1175	14.0	444.8	30.9	2.20	256.3	418.3	121.60
4	2600	981.8	165.0	1239	15.8	486.0	30.9	2.20	256.3	418.3	123.70
5	1400	869.4	84.5	1237	4.0	231.8	37.5	2.51	278.9	486.4	123.80
6	1800	990.0	110.2	1148	8.0	339.3	37.5	2.51	278.9	486.4	111.20
7	2200	1032.8	137.8	1095	13.0	432.6	37.5	2.51	278.9	486.4	95.81
8	2600	954.3	157.5	1133	15.0	472.4	37.5	2.51	278.9	486.4	95.48
9	1400	871.8	87.3	1223	4.0	232.4	33.7	1.76	249.4	421.3	228.30
10	1800	969.8	109.9	1158	8.0	332.4	33.7	1.76	249.4	421.3	175.70
11	2200	1000.4	132.7	1097	12.5	419.1	33.7	1.76	249.4	421.3	150.50
12	2600	930.4	150.8	1137	15.0	460.6	33.7	1.76	249.4	421.3	156.60
13	1401	886.4	90.8	1295	4.0	236.5	35.8	3.20	283.3	481.2	90.03
14	1800	1013.8	124.4	1212	9.0	347.5	35.8	3.20	283.3	481.2	86.25
15	2200	1070.3	153.7	1152	13.9	448.3	35.8	3.20	283.3	481.2	87.50
16	2600	1018.9	178.9	1234	15.5	504.4	35.8	3.20	283.3	481.2	84.32
17	1403	860.8	88.8	1213	4.0	230.0	35.9	1.56	239.1	421.3	232.70
18	1803	955.2	105.5	1144	8.0	327.9	35.9	1.56	239.1	421.3	183.90
19	2201	994.2	139.5	1105	13.0	416.6	35.9	1.56	239.1	421.3	161.60
20	2601	908.9	147.1	1126	14.5	450.1	35.9	1.56	239.1	421.3	162.20
21	1401	840.1	85.8	1204	4.0	224.1	38.8	1.29	222.8	421.3	266.70
22	1801	940.2	110.8	1136	8.0	322.4	38.8	1.29	222.8	421.3	191.40
23	2206	980.2	130.2	1097	12.3	411.7	38.8	1.29	222.8	421.3	168.80
24	2604	870.7	159.4	1101	14.0	431.7	38.8	1.29	222.8	421.3	174.30
25	1402	829.4	86.4	1168	3.5	221.4	39.5	1.33	224.8	414.7	356.00

26	1805	930.5	108.5	1129	9.0	319.8	39.5	1.33	224.8	414.7	246.20
27	2201	965.4	130.4	1100	12.0	404.6	39.5	1.33	224.8	414.7	212.50
28	2600	873.2	146.2	1112	13.8	432.3	39.5	1.33	224.8	414.7	233.60
29	1404	830.4	84.1	1182	3.5	222.0	35.6	1.65	241.3	421.3	304.60
30	1801	949.1	111.7	1148	8.0	325.5	35.6	1.65	241.3	421.3	216.20
31	2202	960.3	123.2	1095	12.0	402.6	35.6	1.65	241.3	421.3	176.20
32	2602	874.8	145.3	1123	14.5	433.4	35.6	1.65	241.3	421.3	194.60
33	1402	813.8	84.1	1132	3.5	217.2	41.4	1.19	213.9	422.6	386.80
34	1801	927.8	106.9	1119	8.0	318.2	41.4	1.19	213.9	422.6	255.00
35	2202	943.2	122.4	1079	12.2	395.5	41.4	1.19	213.9	422.6	241.90
36	2602	850.7	144.9	1097	13.8	421.5	41.4	1.19	213.9	422.6	251.60
37	1400	847.2	85.0	1230	2.5	225.8	43.2	1.08	197.3	412.9	238.90
38	1800	940.4	104.8	1151	8.9	322.3	43.2	1.08	197.3	412.9	376.60
39	2200	990.2	140.7	1127	13.8	414.8	43.2	1.08	197.3	412.9	331.70
40	2600	851.2	142.5	1125	14.4	421.4	43.2	1.08	197.3	412.9	401.10
41	1400	804.8	90.7	1360	3.3	214.5	40.1	2.00	271.1	475.2	113.40
42	1800	893.7	115.2	1342	6.0	306.3	40.1	2.00	271.1	475.2	74.00
43	2200	965.1	153.8	1315	9.0	404.3	40.1	2.00	271.1	475.2	78.11
44	2600	879.6	154.3	1302	10.0	435.5	40.1	2.00	271.1	475.2	93.30
45	1400	824.9	86.0	1312	4.0	219.9	42.7	1.60	263.0	466.0	80.71
46	1800	952.6	111.9	1245	7.6	326.5	42.7	1.60	263.0	466.0	67.05
47	2200	1004.1	145.7	1208	11.5	420.6	42.7	1.60	263.0	466.0	79.17
48	2600	890.6	156.5	1202	12.4	440.9	42.7	1.60	263.0	466.0	81.20

4 Perform the forward selection, backward elimination, and stepwise iteration procedures on the knock-intensity data in Exercise 3. If all three procedures do not result in the same preferred subset, examine the data further to identify the reason for the differences. Which subset(s) do you recommend? Why?

5 Identify the better subsets for the complete quadratic model for the oxides-of-nitrogen data in Exercise 6 of Chapter 22. Do not center or scale the raw predictors. Compare the results of the stepwise selection methods (FS, BE, SI). Do the strong collinearities among the polynomial terms have a noticeable effect on the selection procedures?

6 Repeat Exercise 5 with standardized predictors. How do the results of the stepwise methods change?

7 Identify the better subsets for the height data in Exercise 7, Chapter 22. Next, add two predictors, the brachial index and the tibiofemural index defined in Exercise 4 of Chapter 23. Redo the variable selection process. Are any of the previous results altered? Are either of the two additional predictors included in the better subsets? (See the exercises in Chapter 28 for an assessment of collinearities in this data set.)

8 Standardize the predictors for the engine-knock data in Exercise 7 of Chapter 24. Form a complete second-order model from the standardized predictors. Investigate whether the second-order terms aid in the ability to predict the knock measurements.

CHAPTER 28

Alternative Regression Estimators

Characteristics of a data base sometimes require that estimators other than least squares be used to fit regression models. In this chapter three alternatives to least squares are introduced. These estimation procedures are not covered comprehensively, but references are provided for the reader desiring more information. The estimation procedures introduced in this chapter are:

- *regression estimators for measurement error models, appropriate when both the response and the predictor variables are subject to measurement errors,*
- *robust regression estimators, appropriate when outliers are known or suspected, and*
- *biased regression estimators, appropriate when predictor variables are highly collinear.*

Least-squares estimators are overwhelmingly the most popular and most often used estimators of the parameters of regression models. However, certain characteristics of data bases can have serious effects on least-squares estimators, leading to distortion of the resulting parameter estimates. It is well known, for example, that one strategically placed data point can force the least-squares fit to pass through that data point regardless of any existing trend in the other $n - 1$ observations. If this point is not representative of the true linear relationship between the response and the predictor variable(s), the intercept and slope estimates can be greatly distorted from the true values of the parameters, values that might be well estimated by the least-squares fit to the other $n - 1$ observations.

In this chapter we introduce three classes of alternatives to least-squares estimators. Each of the alternatives presented either is not susceptible to specific undesirable effects of data sets or compensates for these effects. Each class of alternatives also provides diagnostic information about the presence of these undesirable data characteristics or their effects on least-squares estimators.

The three classes of alternatives discussed in this chapter are: (1) estimators that can be used when measurement errors occur in the predictor variables in addition to those occurring in the response variable, (2) robust regression estimators, which can be used when outliers occur in a data set, and (3) biased regression estimators, which can be used when predictor variables are collinear. Since the literature on each of these classes of alternatives is substantial, we introduce only one specific alternative from each class. The references may be used as a guide to further information or additional alternatives.

28.1 MEASUREMENT-ERROR MODEL ESTIMATION

Measurement-error models differ from the regression models discussed previously in this book primarily through the violation of assumption 2 in Table 21.1; i.e., the predictor variables are subject to measurement errors and are therefore random variables. Measurement errors in predictor variables can arise in a variety of ways, most often when the predictors are not controlled in an experiment. For example, rather than being preselected values, predictor variables may be measured during the course of an experiment through chemical analyses or by various types of instrumentation.

In order to focus on the key issues involved in measurement-error model estimation, the model used in this section is the single-variable linear regression model discussed under the usual regression assumptions in Chapter 21. Specifically, we assume there is a true linear relationship between unobservable variables Y and X:

$$Y_i = \beta_0 + \beta_1 X_i, \qquad i = 1, 2, \ldots, n. \tag{28.1}$$

We refer to Y and X as unobservable because both variables are subject to measurement errors; hence, their exact values are not observable. The observable variables are denoted by y and x, so that

$$y_i = Y_i + v_i \quad \text{and} \quad x_i = X_i + u_i, \tag{28.2}$$

where v_i and u_i denote random measurement errors.

Equations (28.1) and (28.2) can be combined to form a single equation that is similar in form to the usual linear regression model (21.1):

$$y_i = \beta_0 + \beta_1 x_i + e_i, \qquad i = 1, 2, \ldots, n. \tag{28.3}$$

While equation (28.3) is similar in form to equation (21.1), there are important differences in the specifications of the two models. The predictor-variable values in the model (21.1) are assumed to be known constants; those in (28.3) are random, due to the random errors in (28.2). Also, the error terms e_i in (28.3), unlike those in (21.1), are correlated with the predictor-variable values x_i, since both contain the random errors u_i:

$$x_i = X_i + u_i \quad \text{and} \quad e_i = v_i - \beta_1 u_i. \tag{28.4}$$

Either of two assumptions could be made about the true values of the predictor variable X in (28.1): the X_i are unknown constants, or the X_i are random variables. In the former case, the model is referred to as a *functional* measurement-error model; in the later, a *structural* measurement-error model. In addition, a variety of assumptions could be made about the distributional properties of the error terms, v_i and u_i. Often, it is assumed that the errors are normally distributed. Table 28.1 summarizes the model assumptions that will be used in this section.

Of particular importance about the assumptions in Table 28.1 is that the model (28.1) is assumed to be the correct theoretical relationship between Y and X; that the X_i, if random, are normally distributed; and that the errors are normally distributed. We stress that there are other model assumptions that may be more appropriate for some applications.

Although measurement-error models often are far more realistic than ordinary regression models, the estimation procedures for measurement-error models are theoreti-

Table 28.1 Measurement-Error Model Assumptions

1. Over the range of applicability, the true relationship between the response variable Y and the predictor variable X is well approximated by a linear regression with negligible model error; i.e., equation (28.1) is the correct theoretical relationship between Y and X.

2. Both the response and the predictor variables are subject to additive measurement errors as in the equations (28.2).

3. The values of the predictor variable X are either

   (a) constants (functional model), or

   (b) independently normally distributed with mean μ_X and standard deviation σ_X (structural model).

4. The model errors, v_i and u_i, are jointly normally distributed with mean zero, standard deviations σ_v and σ_u, respectively, and correlation ρ. Errors for different observations (different i) are independent.

cally more difficult and computationally more complex than are those for least-squares estimation. One can show that under the assumptions listed in Table 28.1, the parameters of the model (28.3) cannot all be estimated unless additional information is known about the model. This additional information can be either knowledge of the values of two of the model parameters, data on additional variables that are correlated with the predictor-variable values X_i but not with the errors u_i, or replicated observations x_{ij} corresponding to fixed values of X_i.

The need for additional information about the model is a direct consequence of the normality assumptions in Table 28.1. If the measurement errors are not normally distributed or, for structural models, if the X_i are not normally distributed, estimators of all the model parameters can be derived without any additional knowledge about the model. The reason for the difficulty with the normal-error assumptions is most easily appreciated by considering the structural model with the assumptions listed in Table 28.1. Under these assumptions the observable predictor variables are sums of two independent normally distributed components, X_i and u_i. Thus the x_i are normally distributed with mean $\mu_x = \mu_X$ and standard deviation $\sigma_x = (\sigma_X^2 + \sigma_u^2)^{1/2}$. The sample mean and the sample standard deviation of the x_i estimate the corresponding population values. The sample standard deviation s_x can only estimate σ_x; it cannot estimate the individual standard deviations σ_X and σ_u.

Through theoretical considerations, knowledge of measurement processes, or perhaps calibration data taken on instruments, a researcher may be willing to assume knowledge of certain model parameters. One might be willing, for example, to assume that the measurement errors v_i and u_i are independent of one another ($\rho = 0$) and that the variabilities of the two sets of errors are approximately equal ($\lambda = \sigma_v^2/\sigma_u^2 = 1$). Alternatively, one might be willing to assume that the errors are independent and that one of the error standard deviations (e.g., σ_u) is known. In either of these situations, estimates of all the remaining model parameters can be obtained.

For the remainder of this section we assume that the error correlation ρ is known and that the square of the ratio of the error standard deviations, $(\sigma_v/\sigma_u)^2$, is known. With these assumptions, the estimates of the model parameters for both functional and structural models can be obtained using an estimation technique known as *maximum likelihood*. The theoretical basis for maximum-likelihood estimators is beyond the scope of this book. Maximum-likelihood estimators are known to possess many optimum statistical properties, and they reduce to least-squares estimators when the usual model assumptions of Table 21.1 are valid.

EXHIBIT 28.1   MEASUREMENT-ERROR MODEL: INTERCEPT AND SLOPE ESTIMATORS

1. Assume a linear measurement-error model of the form (28.1) with errors as in (28.2) and satisfying either the functional or the structural assumptions of Table 28.1.

2. Maximum-likelihood estimators of the intercept and slope are

$$b_0 = \bar{y} - b_1\bar{x},$$
$$b_1 = s(\lambda, \theta) + [s(\lambda, \theta)^2 + t(\lambda, \theta)]^{1/2}, \tag{28.5}$$

where $\theta = \rho\lambda^{1/2}$,

$$s(\lambda, \theta) = \frac{s_{yy} - \lambda s_{xx}}{2(s_{xy} - \theta s_{xx})},$$

$$t(\lambda, \theta) = \frac{\lambda s_{xy} - \theta s_{yy}}{s_{xy} - \theta s_{xx}},$$

and $s_{yy} = n^{-1}\sum(y_i - \bar{y})^2$, $s_{xx} = n^{-1}\sum(x_i - \bar{x})^2$, $s_{xy} = n^{-1}\sum(y_i - \bar{y})(x_i - \bar{x})$. The sample statistics s_{yy}, s_{xx}, and s_{xy} can be replaced by unbiased estimators of the corresponding population variances if n^{-1} is replaced by $(n-1)^{-1}$. This replacement does not affect the intercept and slope estimates (28.5).

Maximum-likelihood estimators for the intercept and slope parameters (see Exhibit 28.1) of equation (28.1) are the same for both functional and structural measurement-error models. The intercept estimator b_0 has the same form as that for the

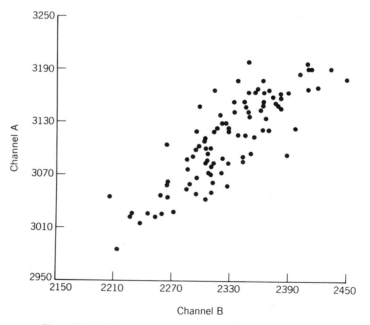

Figure 28.1   Carbon-14 decay counts for a charcoal specimen.

least-squares estimator. The slope estimator b_1 is a function of the two sample variances s_{xx} and s_{yy} and of the sample covariance s_{xy}. When the correlation ρ is zero and the measurement-error standard deviation σ_u for the errors in x is also zero, the intercept and slope estimators (28.5) reduce to the least-squares estimators (21.11).

Estimators of the remaining model parameters depend on which of the two assumptions about the X_i is made (see Exhibit 28.2). For functional models the individual X_i are estimated. For structural models the mean μ_X and the standard deviation σ_X are estimated; the X_i need not be estimated, since they differ from the mean μ_X only by random measurement error.

EXHIBIT 28.2   MEASUREMENT-ERROR MODEL: ADDITIONAL PARAMETER ESTIMATORS

Functional Models

$$\hat{X}_i = \bar{x} + \frac{(y_i - \bar{y})(b_1 - \theta) + (x_i - \bar{x})(\lambda - b_1\theta)}{g},$$

$$\hat{\sigma}_u^2 = \frac{s_{yy} - 2b_1 s_{xy} + b_1^2 s_{xx}}{g}, \tag{28.6}$$

$$\hat{\sigma}_v^2 = \lambda \hat{\sigma}_u^2, \qquad g = (b_1 - \theta)^2 + (\lambda - \theta^2).$$

Structural Models

$$\hat{\mu}_X = \bar{x}, \qquad \hat{\sigma}_X^2 = s_{xx} - \hat{\sigma}_u^2, \qquad \hat{\sigma}_v^2 = \lambda \hat{\sigma}_u^2,$$

$$\hat{\sigma}_u^2 = \frac{s_{yy} - 2b_1 s_{xy} + b_1^2 s_{xx}}{g}, \tag{28.7}$$

$$g = (b_1 - \theta)^2 + (\lambda - \theta^2).$$

Figure 28.1 is a scattergram of $n = 96$ 100-minute carbon-14 radiation counts from a 650-year-old specimen of charcoal (see Section 1.2). The plotted points represent counts taken from two channels of the same decay counter. The energy window over which counts on Channel A are taken is larger than that of Channel B; consequently, the counts for Channel A are larger than those for Channel B. If it is reasonable to assume that the true radiocarbon counts for the two windows, apart from measurement error, are linearly related, then the model (28.1) will express the theoretical relationship between the two sets of counts. Since the counts from each of the channels are subject to measurement errors, a linear measurement-error model is more appropriate than the usual linear regression model.

The true decay counts X_i are random; hence, a structural measurement-error model with the assumptions listed in Table 28.1 will be assumed. It is reasonable to assume that the errors associated with the two decay counters are independent ($\rho = 0$) and that the variation in the errors is similar for the two counters ($\lambda = 1$). Figure 28.2 displays box plots for the two sets of counts. Note that the variation in the observed counts is similar. This supports the assumption that the errors have similar variation, even though Figure 28.2 plots the observed counts and not the errors in the counts.

Table 28.2 exhibits summary statistics and the least-squares and maximum-likelihood estimates of the intercept and slope parameters. The least-squares slope estimate is smaller than the maximum-likelihood one. One can show that the least-squares estimate is always

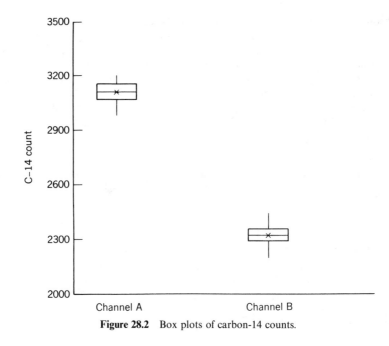

Figure 28.2   Box plots of carbon-14 counts.

Table 28.2   Parameter Estimates for the Radiocarbon Dating Example

Summary Statistics

$n = 96$	$\bar{y} = 3111.95$	$\bar{x} = 2326.47$
$s_{xy} = 2103.02$	$s_{yy} = 2525.65$	$s_{xx} = 2467.94$

Least-Squares Estimates

$$b_0 = 1129.47 \qquad b_1 = 0.85$$

Maximum-Likelihood Estimates

$$b_0 = 753.34 \qquad b_1 = 1.01$$

$$\hat{\sigma}_u^2 = \hat{\sigma}_v^2 = 393.57 \qquad \hat{\mu}_X = 2326.47 \qquad \hat{\sigma}_X^2 = 2074.36$$

smaller (in magnitude) than the maximum-likelihood one when the errors are independent. One can see the relationship between the two fits in Figure 1.3.

Apart from the more realistic modeling of the relationship between the two sets of decay counts, there are other arguments that lead to a preference for the maximum-likelihood fit over the least-squares. Least-squares estimators are known to be biased when both response and predictor variables are subject to measurement errors. (Least-squares estimators of the slope parameter are biased downward under the conditions assumed in this example.) Not only are they biased; they are not consistent—that is, they do not converge (in probability) to the true parameter value as the sample size is increased. It is not known whether maximum-likelihood estimators are unbiased; however, it is known that they are consistent.

Least-squares estimators are appropriate when the measurement error in x is negligible. Note that the estimates in Table 28.2 suggest that the error variation in x, as measured by $\hat{\sigma}_u$, is not negligible relative to the error variation in y, as measured by $\hat{\sigma}_v$, since it is assumed that $\lambda = 1$. The error variation in x is also not negligible relative to the variation in the true predictor variable X, as measured by $\hat{\sigma}_X$, since $\hat{\sigma}_u/\hat{\sigma}_X = 0.44$. This is another argument for preferring the maximum-likelihood estimates.

The references cited at the end of this chapter list books and journal articles that present both theoretical and empirical justification for the use of measurement-error models. Several of these references provide information on inference procedures for maximum-likelihood estimators and on alternative estimation schemes. These references should be consulted for more detailed information on estimation with measurement-error models.

28.2 ROBUST REGRESSION

Influential observations (Section 24.2) can seriously affect the estimation of regression coefficients and resulting inferences. As was stressed in Chapter 24, investigation of outliers can lead to the most important conclusions in a research project. There are many instances, however, when the presence of outliers impairs the satisfactory fitting of a regression model. In addition, it is not always clear which specific observations are influential, even when diagnostics for influential observations are examined. This is because frequently there are observations that are only slightly below or slightly above the suggested cutoff values for the diagnostics.

One form of accommodation of influential observations is to use a regression-fitting procedure that is less susceptible than least squares to the effects of outliers. Many such procedures have been proposed. The estimation technique we introduce in this section is similar to the M-estimator of the center of a distribution (Section 3.2).

For linear regression models, the normal equations [see (22.11) for $p = 2$] can be concisely expressed as follows, where $x_{i0} = 1$ for the constant term:

$$\sum_{i=1}^{n} x_{ij} \left(y_i - \sum_{k=0}^{p} b_k x_{ik} \right) = 0, \qquad j = 0, 1, \ldots, p. \tag{28.8}$$

Noting that the expression in parentheses is the ith residual r_i, these equations can be rewritten as

$$\sum_{i=1}^{n} x_{ij} r_i = 0, \qquad j = 0, 1, \ldots, p,$$

where $r_i = y_i - b_0 x_{i0} - b_1 x_{i1} - b_2 x_{i2} - \cdots - b_p x_{ip}$.

Robust M-estimators for regression coefficients can be viewed as weighted least-squares estimators (see Exhibit 28.3). The weights used in the estimation formulas are determined iteratively from the sizes of the residuals r_i obtained from the previous fit in the iteration scheme. Usually, a robust estimator of the variability of the model errors is also estimated (see Exhibit 28.4).

EXHIBIT 28.3   REGRESSION *M*-ESTIMATORS

M-estimators of the regression coefficients of a linear regression model are found by simultaneously solving the following weighted normal equations:

$$\sum_{i=1}^{n} w_i x_{ij} r_i = 0, \qquad j = 0, 1, \ldots, p, \tag{28.9}$$

where $x_{i0} = 1$, $r_i = y_i - b_0 x_{i0} - b_1 x_{i1} - \cdots - b_p x_{ip}$,

$$w_i = \begin{cases} -tv/r_i & \text{if} & r_i < -tv, \\ 1 & \text{if} & -tv \leqslant r_i \leqslant tv, \\ tv/r_i & \text{if} & tv < r_i, \end{cases} \tag{28.10}$$

the tuning constant $t = 1.345$ or 1.5, and v is a robust measure of the variability of the model errors.

EXHIBIT 28.4   *M*-ESTIMATOR OF SPREAD

An *M*-estimator of spread is

$$v = \left(\frac{\sum_{i=1}^{n} w_i^2 r_i^2}{2a} \right)^{1/2}, \tag{28.11}$$

where $a = 0.355v$ if $t = 1.345$, $a = 0.389v$ if $t = 1.50$, and v equals the number of degrees of freedom for the least-squares mean squared error.

Regression *M*-estimators are solved for iteratively. At each stage of the iteration large residuals r_i are weighted, with the weights inversely proportional to how much larger the residuals are than tv. The iteration scheme has many variations, depending on the starting values and the choice for the tuning constant t. One useful iteration scheme is given in Exhibit 28.5.

EXHIBIT 28.5   *M*-ESTIMATION ITERATION PROCEDURE

1. Using the outlier diagnostics discussed in Chapter 24, delete clearly influential observations from the data set.
2. Select a tuning constant (e.g., $t = 1.345$), the maximum number of iterations (e.g., $N = 50$), and the relative accuracy desired (e.g., $\max\{|b_j^{(k+1)} - b_j^{(k)}|/b_j^{(k)}\} < 10^{-6}$).
3. Fit the remaining observations by least squares. Denote these estimates by $b_j^{(1)}$. Let $v^{(1)}$ be the median absolute deviation (Section 3.2) of the residuals:

$$v^{(1)} = \frac{\text{median}[|r_i - \text{median}(r_i)|]}{0.6745}.$$

4. Using the current estimates $b_j^{(k)}$ and $v^{(k)}$ and the current residuals $r_i^{(k)}$, update the weights w_i using (28.10). From these updated weights, calculate new estimates $b_j^{(k+1)}$ by solving the equations (28.9). Calculate updated residuals $r_i^{(k+1)}$ using the updated coefficient estimates and then update the estimate of scale by calculating $v^{(k+1)}$ from equation (28.11).
5. If the relative accuracy is achieved or the maximum number of iterations is exceeded, stop the iteration and report $k + 1$, $b_j^{(k+1)}$, $v^{(k+1)}$, and $\max\{|b_j^{(k+1)} - b_j^{(k)}|/b_j^{(k)}\}$; otherwise, return to step 4.

Initial estimates (starting values) are extremely important for the satisfactory application of robust regression estimation procedures. If initial estimates are severely affected by the presence of outliers, robust procedures may not be able to compensate for their effects. For example, least-squares residuals for high leverage points can be small due to the distortion caused by influential observations. Throughout the iteration procedure the residuals can be so small that they are left unweighted, thereby leaving the distortion unaffected by the robust estimator.

For these reasons, the first step in the iteration scheme for robust estimation is to eliminate from the data base observations that are clearly influential. Observations having extremely large leverage values and DFBETAS should be removed from the data base. Note that one can examine the effects of these observations by examining robust fits with and without the removal of the influential observations. If the two robust fits are substantially different, the fit in which the influential observations are not used in determining the initial estimates should be preferred.

Once the clearly influential observations are removed, the least-squares coefficient estimates and the MAD estimator of variability are the initial estimates in the iteration. From the least-squares residuals, the M-estimator weights are determined using equation (28.10). From these weights, updated estimates of the regression coefficients and the estimate of error variation are calculated by solving equations (28.9) and (28.11). This process continues until convergence or the maximum number of iterations is reached.

The chemical-process yield data in Table 24.1 were used in Chapter 24 to illustrate the application of diagnostics for influential observations. In Table 24.3 several diagnostics are displayed for the seven observations in this data set that have large values on one or more of the diagnostics. From these diagnostics it appears the observations 6, 16, and 45 are having a major influence on the least-squares fit. It is less certain whether the other observations shown are highly influential. In Section 27.2, several analyses in which subsets of the observations were deleted from the data set resulted in the conclusion that observations 11 and 18 should be deleted in addition to 6, 16, and 45.

While it may be clear that three of the observations should be eliminated from this data set, it is much less certain that the other two should be. One of the benefits of a robust regression estimation scheme is that only the clearly influential observations need be deleted.

Table 28.3 displays the least-squares estimates from the complete data set and those

Table 28.3   Comparison of Fits to the Chemical-Yield Data

	Least-Squares Fits		M-Estimator Fit[a]
Standardized Predictor Variable[b]	Complete Data Set	Obsns. 6, 11, 16, 18, 45 Removed	Obsns. 6, 16, 45 Removed
Constant	51.10	51.22	51.30
Catalyst	− 2.20	− 1.85	− 2.25
Conversion	− 2.18	− 2.22	− 2.26
Flow	− 1.11	− 0.26	− 0.61
Interaction[c]	− 0.05	0.01	0.08
Invratio[d]	− 1.09	0.93	0.59

[a] $t = 1.345$.
[b] Normal-deviate standardization, except for catalyst.
[c] Conversion × flow.
[d] (Ratio)$^{-1}$.

Table 28.4   *M*-Estimator Weights for Chemical-Yield Data

Obs. No.	Weight
1	0.77
11	0.60
18	0.58
22	0.79
28	0.86
29	0.93
33	0.74
50	0.74
51	0.75

from the reduced data set in which the above five observations were deleted. This table also contains the robust *M*-estimator estimates in which the initial least-squares estimates are obtained from the fit with observations 6, 16, and 45 eliminated. The least-squares estimates for the full data set differ substantially on the last three variables from those for the reduced data set and the robust fit. The estimates for the latter two fits are more similar for these three predictors and are quite similar overall.

The similarity between the reduced least-squares estimates and those for the robust fit is partly explainable by examining the *M*-estimator weights in Table 28.4. Both observations 11 and 18, the two extra observations eliminated in the reduced least squares fit, are heavily weighted in the robust fit.

28.3   COLLINEARITY AND BIASED ESTIMATION

Collinearities occur in a data set when two or more predictor variables repeat information. Two-variable collinearities are indicated by large correlations between the variables. Three- or higher-variable collinearities are indicated by approximate linear dependencies among the variables (see Section 23.1). The damaging effects of collinearities on least-squares estimation and inference procedures is illustrated in Section 27.4. In this section we discuss the detection of collinearities and introduce a biased alternative to least squares.

Prior to an investigation of the presence of collinearities in a regression data set, all predictor variables should be standardized, especially if products or powers of individual variables are to be included in the model. We assume in this section that all predictor variables are standardized using the correlation form of standardization defined in equation (23.5); i.e., the standardized form of the predictors is

$$z_{ij} = \frac{x_{ij} - \bar{x}_j}{d_j}, \qquad i = 1, 2, \ldots, n, \quad j = 1, 2, \ldots, p, \tag{28.12a}$$

where

$$d_j^2 = \sum_{i=1}^{n} (x_{ij} - \bar{x}_j)^2.$$

One consequence of standardization is that all collinearities with the constant term are eliminated. In some settings it is important to assess whether collinearities do involve the constant term. If so, the standardization is modified as follows:

$$z_{ij} = \frac{x_{ij}}{d_j}, \qquad d_j^2 = \sum_{i=1}^{n} x_{ij}^2. \tag{28.12b}$$

There are many techniques for detecting collinearities. We discuss three of them in this section (see Exhibit 28.6) and refer the reader to the references listed at the end of this chapter for further details and additional collinearity diagnostics.

EXHIBIT 28.6   DIAGNOSING COLLINEARITIES

1. Standardize all predictor variables using either equation (28.12a) or (28.12b). Products or powers formed from standardized variables must themselves be standardized.

2. For each pair of predictor variables, including products or powers formed from the standardized predictors, calculate correlation coefficients. If the correlation between two predictors exceeds 0.95 (in absolute value), the predictor variables are collinear.

3. Determine the coefficients of determination R_j^2 of each predictor variable regressed on all the other predictor variables. Equivalently, calculate the variance inflation facors

$$\text{VIF}_j = (1 - R_j^2)^{-1}. \tag{28.13}$$

The jth predictor variable is collinear with one or more other predictors if $R_j^2 > 0.90$; equivalently, if $\text{VIF}_j > 10$.

4. Examine the eigenvalues, l_s, and the elements of the corresponding eigenvectors, v_{rs}, of the correlation matrix of the predictor variables. Any eigenvalue l_s that is less than 0.05 identifies a collinearity. Large elements v_{rs} of the corresponding eigenvectors indicate which predictors x_r are involved in the collinearity.

Pairwise correlations were used in Table 23.3 to show the effects of standardization on collinearities involving polynomial terms in regression models. Pairwise correlations can be calculated for any pair of variables in a regression data set. The variables do not have to be standardized prior to the calculation of the correlations unless the regression model includes polynomial terms of second or higher order.

Pairwise correlations are useful for detecting two-variable collinearities, but they may be unable to detect collinearities involving three or more variables. For example, it is easy to construct exact collinearities $x_1 + x_2 + \cdots + x_p = 0$ involving $p = 3$ or more variables for which none of the pairwise correlations exceed 0.95. For this reason, additional collinearity diagnostics that can detect collinearities involving any number of predictor variables are needed.

Any collinearity involving a predictor variable can be detected by regressing the variable of interest on the other predictor variables. The resulting coefficient of determination can only be close to one if there exists an approximate linear dependence (a collinearity) involving x_j and the other predictors—i.e., only if x_j can be well predicted by some linear combination of the constant term and the other predictor variables. Variance inflation factors were introduced in Section 27.4 as measures of one of the ill effects of collinearities. Variance inflation factors can be related to the coefficients of determination as indicated in equation (28.13).

As shown in the appendix to this chapter, variance inflation factors VIF_j and coefficients of determination R_j^2 can be obtained from the least-squares fit to the complete data set. One need not regress each predictor variable on all the other variables—p

regression fits—in order to obtain these quantities. The cutoffs $VIF_j > 10$ and $R_j^2 > 0.90$, like the cutoff of 0.95 for correlations between pairs of predictors, indicate strong redundancies in a data set, redundancies that may result in severe distortions of least-squares estimates.

Although large coefficients of determination and variance inflation factors identify which predictor variables are involved in collinearities, they do not specify how many collinearities exist in a data set, nor which predictors are involved in each collinearity. A widely used diagnostic procedure that satisfies these last requirements is an examination of the eigenvalues and eigenvectors of the correlation matrix of the predictor variables.

A thorough discussion of eigenvalues and eigenvectors requires the use of matrix algebra and is beyond the scope of this book. A brief discussion is contained in the appendix to this chapter. We simply note that they are quantities that can be calculated from the correlations among pairs of the predictor variables. Small values of the eigenvalues identify collinearities; large elements in the corresponding eigenvectors identify the predictor variables involved in each collinearity.

Table 28.5 lists collinearity diagnostics for the cryogenic-flowmeter data set that was introduced in Section 25.3 and discussed further in Section 27.4. The correlations shown in Table 28.5 reveal an extremely strong negative association between temperature and density. It was noted during the experimentation that this strong association is unavoidable, since the density of the liquid nitrogen used in the experiment is solely a function of temperature and pressure, and the pressure dependence was thought to be minimal throughout the experimental region.

Table 28.5   Collinearity Diagnostics for Cryogenic-Flowmeter Data

Predictor-Variable Correlations

	TEMP2	PRESS	DENS	COOL	TIME	PULSE	ULLAGE
TEMP	0.536	0.625	−0.999	−0.746	0.374	0.528	0.354
TEMP2		0.175	−0.553	−0.565	0.190	0.396	−0.026
PRESS			−0.605	0.048	0.222	0.151	−0.154
DENS				0.762	−0.375	−0.536	−0.037
COOL					−0.289	−0.559	−0.190
TIME						0.289	0.097
PULSE							0.010

	VIF	R_j^2	$l_1 = 0.0004$ v_1	$l_2 = 0.0031$ v_2	$l_3 = 0.4538$ v_3
TEMP	1541.11	0.999	−0.745	−0.312	−0.230
TEMP2	2.31	0.567	−0.015	−0.019	0.536
PRESS	101.28	0.990	0.073	0.517	0.269
DENS	1255.65	0.999	−0.659	0.475	0.228
COOL	145.27	0.993	−0.067	−0.640	0.469
TIME	1.20	0.167	0.000	0.003	−0.039
PULSE	1.60	0.375	−0.002	−0.010	0.477
ULLAGE	1.16	0.138	−0.000	−0.014	0.294

The variance inflation factors and coefficients of determination also reveal the involvement of temperature and density in collinearities. In addition, pressure and subcooling temperature are seen to be involved in collinearities. Note that an examination of only the pairwise correlations would not identify this latter pair of collinear variables. Note too that at this point it is not known how many collinearities there are nor which variables are involved in each collinearity.

The eigenvalues reveal that there are two strong collinearities, since $l_1 = 0.0004$ and $l_2 = 0.0031$, both less than 0.05. Two relatively large elements occur in the first eigenvector v_1, those corresponding to temperature (-0.745) and density (-0.659), the strong pairwise relationship identified by the correlations. In the second eigenvector v_2, large elements occur for both of the temperature variables (TEMP and COOL), pressure, and density. This collinearity is probably also due to the relationship between density, pressure, and temperature cited above.

Two of the most effective means of accommodating collinearities are the respecification of regression models and the collection of additional data. Respecification of models is effective when collinearities are induced by the specification of the predictor variables. In the discussion of the fitting of polynomial models in Section 23.1, it was shown that quadratic and higher powers of a predictor variable are often collinear with one another and with the linear term. Since these collinearities are created through the inclusion of several functions of a single variable, superfluous terms should be eliminated.

Standardization of predictors prior to the formation of the higher-order terms was also shown to reduce, or even eliminate, collinearities between certain powers (e.g., linear and quadratic). In the cryogenic-flowmeter data set, temperature was standardized prior to the formation of the quadratic term. In Table 28.5, the linear and quadratic functions of temperature are not collinear.

A special case of model respecification is variable selection. When collinear predictor variables are clearly redundant and would always be expected to be so, one or more of the collinear variables can be deleted from the model to eliminate the collinearity problem. One must be sure that the variables are not arbitrarily deleted just because of the occurrence of collinearities, because one might eliminate important predictors. If temperature and density can be shown to always be strongly collinear in experiments with cryogenic flowmeters, one of the two should be deleted from the model.

Respecification of predictor variables may not be a satisfactory response to collinearities if the variates involved are not functionally or experimentally related. Collinearities can arise because of the data values themselves and not because of the definitions of the variables. Such collinearities are sometimes an artifact of the sample and would not be expected to occur in every experiment or data set collected. In this instance, supplementing the data base with additional observations can eliminate the problem.

When collinearities cannot be eliminated through respecification of the model or the collection of additional observations, estimation procedures other than least squares can be effective alternatives. One estimation procedure that has been used effectively as an alternative to least squares with collinear data is termed *ridge regression*. (see Exhibit 28.7). Ridge-regression estimators are motivated by several different considerations. One is that least-squares estimators are so unstable when predictor variables are collinear that very minor perturbations of the data set result in drastically different coefficient estimates. The modified normal quations (28.14) are one way to perturb the data set. Ridge estimates using a suitably small value of k ($k > 0$) are stable solutions to these equations.

EXHIBIT 28.7   RIDGE-REGRESSION ESTIMATORS

1. Standardize each of the predictor variables as in (28.12a).

2. For a fixed, positive value of the *ridge parameter k*, solve the following modified normal equations for the standardized ridge regression coefficients b_j^s:

$$b_0^s = \bar{y},$$

$$\sum z_{i1} y_i = b_1^s \left(\sum z_{i1}^2 + k \right) + b_2^s \sum z_{i1} z_{i2} + \cdots + b_p^s \sum z_{i1} z_{ip},$$

$$\sum z_{i2} y_1 = b_1^s \sum z_{i1} z_{i2} + b_2^s \left(\sum z_{i2}^2 + k \right) + \cdots + b_p^s \sum z_{i2} z_{ip},$$

$$\vdots \hspace{8cm} (28.14)$$

$$\sum z_{ip} y_i = b_1^s \sum z_{i1} z_{ip} + b_2^s \sum z_{i2} z_{ip} + \cdots + b_p^s \left(\sum z_{ip}^2 + k \right).$$

3. The solutions to these modified normal equations are the standardized ridge-regression coefficient estimates corresponding to the ridge parameter k. Ridge estimates of the original regression coefficients are

$$b_j = b_j^s / d_j, \qquad j = 1, 2, \ldots, p, \hspace{3cm} (28.15)$$

$$b_0 = \bar{y} - b_1 \bar{x}_1 - b_2 \bar{x}_2 - \cdots - b_p \bar{x}_p.$$

Another motivation for ridge-regression estimation is the often observed tendency of least-squares estimates to have unreasonably large magnitudes and incorrect signs. This tendency is a direct result of the unstable nature of the least-squares solutions to the normal equations. Ridge estimates, on the other hand, tend to have generally more reasonable magnitudes, and often the signs are different from those of least squares and more consistent with known theory or past empirical evidence.

A final motivation for ridge estimators is the extremely large variance inflation factors caused by collinearities. One can calculate equivalent ridge-regression variance inflation factors and show that any nonzero value of the ridge parameter will decrease the variance inflation factors from those of least squares.

Standardized ridge estimates b_j^s are generally calculated for a range of values of the ridge parameter k and then plotted versus k. Such a plot is termed a *ridge trace*. Because of the subjective nature of plots, one frequently makes two or three ridge traces using equally spaced values of k. Traces using k-values from 0 to 1 followed by a narrower range, say 0 to 0.1 or 0.2, are commonly made.

Coefficient estimates for collinear variables usually change dramatically when k is initially incremented a small amount from 0. The trace soon stabilizes, and changes in the coefficient estimates become small. A value of k is chosen for which all the coefficient estimates have stabilized. This value of k determines the final ridge estimates.

Because of the subjectivity of choosing k from a ridge trace, many studies have been conducted to evaluate specific methods for chosing the ridge parameter. An estimator of k that has been repeatedly cited as one of the better techniques for selecting the ridge parameter is

$$k = \frac{p s_e^2}{\sum_{j=1}^{p} (b_j^s)^2}, \hspace{3cm} (28.16)$$

where the regression estimates b_j^s are least-squares estimates for standardized predictor variables, and $s_e^2 = MS_E$, the least squares mean squared error.

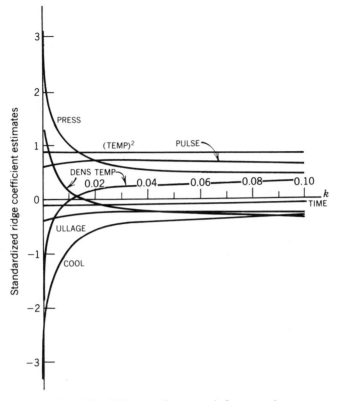

Figure 28.3   Ridge trace for cryogenic-flowmeter data.

Ridge regression estimators are biased. The tradeoff for variance reduction and the more stable properties of the ridge estimator is some bias in the coefficient estimators. Many studies have shown, however, that under a wide range of problems likely to be encountered in practice that the bias incurred is small relative to the increase in precision of the estimator. In other words, the net effect of the reduction in the standard errors of the estimators and the inclusion of bias is that the ridge estimates generally tend to be closer to the true parameter values than with least squares.

Figure 28.3 shows a ridge trace for the cryogenic-flowmeter data with all the predictor variables included. Initially the trace was made using the interval $0 \leqslant k \leqslant 1$, but the changes in the coefficient estimates occur in a range much closer to the origin. The trace was replotted using the range $0 \leqslant k \leqslant 0.1$.

Sharp changes in the standardized coefficient estimates occur for temperature, pressure, density, and subcooling temperature. The ridge trace appears to stabilize by approximately $k = 0.02$. The estimate of k from equation (28.16) is $k = 0.004$. This appears to be slightly too low for this example. We recommend a value in the interval $0.02 \leqslant k \leqslant 0.04$.

Table 28.6 lists the standardized coefficient estimates for least squares, the ridge estimates using the estimator from equation (28.16), and the ridge estimates using $k = 0.02$.

Table 28.6 Comparison of Fits to the Cryogenic-Flowmeter Data

Standardized Predictor Variable*	Least Squares	Ridge Regression	
		$k = 0.004$	0.02
TEMP	−1.86	−0.36	0.16
TEMP2	0.63	0.69	0.71
PRESS	2.97	1.51	0.74
DENS	1.26	0.58	−0.09
COOL	−3.30	−1.53	−0.58
TIME	−0.12	−0.13	−0.13
PULSE	0.84	0.86	0.86
ULLAGE	−0.36	−0.32	−0.29

*Correlation-form standardization.

Note that the ridge coefficient estimates for all the collinear variables have smaller magnitudes than the corresponding estimates for least squares. The signs on all estimates are the same for the first two estimators (least squares and ridge with $k = 0.004$), but two of the signs change with the ridge estimator using $k = 0.02$.

The references at the end of this chapter provide information on many alternatives to least squares when predictor variables are collinear. Included in these references are discussions of the merits of ridge and other biased estimators relative to least squares. Inference procedures for individual regression coefficients as well as overall assessments of model fits are also presented.

APPENDIX: COLLINEARITY DIAGNOSTICS

Write the standardized form of the linear regression model in matrix form as

$$y = b_0^s 1 + Zb^s + e, \qquad (28.A.1)$$

where 1 is an $n \times 1$ vector of ones and Z contains the p columns of standardized predictor variables. Least-squares estimators of the standardized regression coefficients are [cf. equation (22.A.3)]

$$b_0^s = \bar{y} \quad \text{and} \quad b^s = (Z'Z)^{-1}Z'y. \qquad (28.A.2)$$

Pairwise correlations for all pairs of the predictor variables are contained on the off-diagonal elements of $Z'Z$. Using properties of the inverse of partitioned matrices, one can show that the variance inflation factors, VIF_j, are the diagonal elements of $(Z'Z)^{-1}$. From the variance inflation factors the respective coefficients of determination R_j^2 can be determined from equation (28.13). All these quantities are obtained from a single fit to the full model.

The eigenvalues l_k and the corresponding p-dimensional orthonormal eigenvectors v_k are solutions to the system of equations

$$Z'Zv_k = l_k v_k, \qquad k = 1, 2, \ldots, p. \qquad (28.A.3)$$

Since the eigenvalues are normalized, it follows that

$$\mathbf{v}'_k Z'Z\mathbf{v}_k = l_k, \qquad k = 1, 2, \ldots, p. \tag{28.A.4}$$

From this it follows that if $l_k \approx 0$, then $Z\mathbf{v}_k \approx \mathbf{0}$; i.e., the columns of Z are approximately linearly dependent—the definition of a collinearity. Thus, eigenvalues that are near zero, say less than 0.05, identify collinearities. The elements of the corresponding eigenvectors identify the nature of the linear dependence. In particular, large elements of the eigenvectors identify the predictor variables involved in the collinearities.

REFERENCES

Text References

Most of the literature on measurement-error models, also known as errors-in-variables models, is contained in journal articles. A major work on this subject, written at a high mathematical level, is:

Fuller, W. A. (1987). *Measurement Error Models.* New York: John Wiley and Sons, Inc.
  The Fuller text is comprehensive, has extensive references, includes multiple regression and multivariate multiple regression, and discusses the relationship between measurement-error models and other models widely used in the economic and social-science literature.

Important surveys on topics dealing with measurement-error models are:

Kendall, M. G. and Stuart, A. (1979). *The Advanced Theory of Statistics,* Chapter 27, New York: Hafner Publishing Co.

Malinvaud, E. (1970). *Statistical Methods of Econometrics,* Chapter 10, Amsterdam: North Holland Publishing Co.

Robust regression procedures are also widely available, primarily in journal articles. Important theoretical references on topics dealing with robust regression are:

Huber, P. J. (1981). *Robust Statistics,* New York: John Wiley and Sons, Inc.

Hampel, F. R., Ronchetti, E. M., Rousseeuw, P. J., and Stahel, W. A. (1986). *Robust Statistics,* New York: John Wiley and Sons, Inc.
  Each of these works lists numerous references to various topics dealing with robust estimation in general and robust regression in particular.

Biased regression estimation is discussed in many of the regression texts listed at the end of Chapters 21 and 22. Among the more comprehensive treatments are those in the following texts:

Draper, N. R. and Smith, H. (1981). *Applied Regression Analysis, Second Edition,* New York: John Wiley and Sons, Inc.

Gunst, R. F. and Mason, R. L. (1980). *Regression Analysis and Its Application: A Data-Oriented Approach,* New York: Marcel Dekker, Inc.

Montgomery, D. C. and Peck, E. (1982). *Introduction to Linear Regression Analysis,* New York: John Wiley and Sons, Inc.

Meyers, R. H. (1986). *Classical and Modern Regression with Applications,* Boston, MA: PWS and Kent Publishing Co.

Data References

The radiocarbon-dating example was provided by Drs. Herbert Haas and James Devine of the Radiocarbon Dating Laboratory of Southern Methodist University.

EXERCISES

1   Refer to Exercise 2, Chapter 24, in which a study was conducted to identify factors that may contribute to tooth discoloration in humans. Determine the measurement-error model slope and intercept estimates for a regression model in which the percentage of tooth discoloration is the response variable and fluoride level is the predictor variable. Assume it is known from past experimental work that $\rho = 0$ and $\lambda = 25$.

2   Compute the least-squares fit to the data given in Exercise 1. Compare the results of the two regression models. How do the assumptions differ for the two models being fit? Which set of assumptions appears more reasonable? Which fit appears to represent the data best? Why?

3   After reexpressing, if needed, the data on the chemical vapors in Exercise 3 of Chapter 25, use an M-estimator to estimate the intercept and slope parameters. For this exercise do not add polynomial terms to the model; simply regress (an appropriate function of) the log(concentration) on (an appropriate function of) the distance. What do the M-estimator weights suggest about the presence of outliers? Compare the information obtained from these weights with diagnostics for influential observations from the least-squares fit to these same two variables.

4   Examine collinearity diagnostics for the height data in Exercise 7, Chapter 22, using the seven predictors shown and the two indices defined in Exercise 4, Chapter 23. Use these collinearity diagnostics to explain the behavior of the estimates noted in the latter exercise and in Exercise 7, Chapter 27. Find suitable ridge-regression estimates for the regression coefficients in the nine-predictor model. Compare the estimates from this fit with least-squares estimates from fits to the original seven-predictor model and the enlarged nine-predictor model.

5   The petroleum-tank emission data in Table 26.1 consists of two variables, emissions and wind speed, that are both subject to measurement errors. Model ln(emissions) as a linear function of ln(wind speed) using a measurement-error model with $\rho = 0$. Suppose that from previous measurements on these variates it is believed that an error variance ratio of $\lambda = 1$ is reasonable. Do the measurement-error model estimates differ appreciably from the least-squares estimates? Plot the data, and superimpose the two fits on the scatterplot. Where would predictions from the two fits differ the most?

6   Compute M-estimates for the regression coefficients of the fit to the engine-knock data in Exercise 7, Chapter 24. Use the weights w_i to identify influential observations. How does the identification of outliers using these weights compare with the least-squares diagnostics for influential observations? Delete any observations that have weights that are less than 0.8. How do the coefficients compare for (a) the least-squares fit to the complete data set, (b) the robust fit to the complete data set, and (c) the least-squares fit to the reduced data set?

7   After discarding any influential observations identified in Exercise 6, standardize the predictors and fit a complete second-order model to the knock measurements. Evaluate whether any collinearities exist in the data set. If strong collinearities exist,

compute ridge regression coefficient estimates and compare the results with least squares. Perform variable selection on the least-squares fit using the techniques discussed in Chapter 27. Use the ridge trace to perform variable selection on the ridge parameter estimates (i.e., delete those predictors that have coefficient estimates close to zero). How do the results of the two procedures compare? Which do you prefer? Why?

Appendices

Table A1 Table of Random Numbers

46	96	85	77	27	92	86	26	45	21	89	91	71	42	64	64	58	22	75	81	74	91	48	46	18
44	19	15	32	63	55	87	77	33	29	45	00	31	34	84	05	72	90	44	27	78	22	07	62	17
34	39	80	62	24	33	81	67	28	11	34	79	26	35	34	23	09	94	00	80	55	31	63	27	91
74	97	80	30	65	07	71	30	01	84	47	45	89	70	74	13	04	90	51	27	61	34	63	87	44
22	14	61	60	86	38	33	71	13	33	72	08	16	13	50	56	48	51	29	48	30	93	45	66	29
40	03	96	40	03	47	24	60	09	21	21	18	00	05	86	52	85	40	73	73	57	68	36	33	91
52	33	76	44	56	15	47	75	78	73	78	19	87	06	98	47	48	02	62	03	42	05	32	55	02
37	59	20	40	93	17	82	24	19	90	80	87	32	74	59	84	24	49	79	17	23	75	83	42	00
11	02	55	57	48	84	74	36	22	67	19	20	15	92	53	37	13	75	54	89	56	73	23	39	07
10	33	79	26	34	54	71	33	89	74	68	48	23	17	49	18	81	05	52	85	70	05	73	11	17
67	59	28	25	47	89	11	65	65	20	42	23	96	41	64	20	30	89	87	64	37	93	36	96	35
93	50	75	20	09	18	54	34	68	02	54	87	23	05	43	36	98	29	97	93	87	08	30	92	98
24	43	23	72	80	64	34	27	23	46	15	36	10	63	21	59	69	76	02	62	31	62	47	60	34
39	91	63	18	38	27	10	78	88	84	42	32	00	97	92	00	04	94	50	05	75	82	70	80	35
74	62	19	67	54	18	28	92	33	69	98	96	74	35	72	11	68	25	08	95	31	79	11	79	54
91	03	35	60	81	16	61	97	25	14	78	21	22	05	25	47	26	37	80	39	19	06	41	02	00
42	57	66	76	72	91	03	63	48	46	44	01	33	53	62	28	80	59	55	05	02	16	13	17	54
06	36	63	06	15	03	72	38	01	58	25	37	66	48	56	19	56	41	29	28	76	49	74	39	50
92	70	96	70	89	80	87	14	25	49	25	94	62	78	26	15	41	39	48	75	64	69	61	06	38
91	08	88	53	52	13	04	82	23	00	26	36	47	44	04	08	84	80	07	44	76	51	52	41	59
68	85	97	74	47	53	90	05	90	84	87	48	25	01	11	05	45	11	43	15	60	40	31	84	59
59	54	13	09	13	80	42	29	63	03	24	64	12	43	28	10	01	65	62	07	79	83	05	59	61
39	18	32	69	33	46	58	19	34	03	59	28	97	31	02	65	47	47	70	39	74	17	30	22	65
67	43	31	09	12	60	19	57	63	78	11	80	10	97	15	70	04	89	81	78	54	84	87	83	42
61	75	37	19	56	90	75	39	03	56	49	92	72	95	27	52	87	47	12	52	54	62	43	23	13
78	10	91	11	00	63	19	63	74	58	69	03	51	38	60	36	53	56	77	06	69	03	89	91	24
93	23	71	58	09	78	08	03	07	71	79	32	25	19	61	04	40	33	12	06	78	91	97	88	95
37	55	48	82	63	89	92	59	14	72	19	17	22	51	90	20	03	64	96	60	48	01	95	44	84
62	13	11	71	17	23	29	25	13	85	33	35	07	69	25	68	57	92	57	11	84	44	01	33	66
29	89	97	47	03	13	20	86	22	45	59	98	64	53	89	64	94	81	55	87	73	81	58	46	42
16	94	85	82	89	07	17	30	29	89	89	80	98	36	25	36	53	02	49	14	34	03	52	09	20
04	93	10	59	75	12	98	84	60	93	68	16	87	60	11	50	46	56	58	45	88	72	50	46	11
95	71	43	68	97	18	85	17	13	08	00	50	77	50	46	92	45	26	97	21	48	22	23	08	32
86	05	39	14	35	48	68	18	36	57	09	62	40	28	87	08	74	79	91	08	27	12	43	32	03
59	30	60	10	41	31	00	69	63	77	01	89	94	60	19	02	70	88	72	33	38	88	20	60	86
05	45	35	40	54	03	98	96	76	27	77	84	80	08	64	60	44	34	54	24	85	20	85	77	32
71	85	17	74	66	27	85	19	55	56	51	36	48	92	32	44	40	47	10	38	22	52	42	29	96
80	20	32	80	98	00	40	92	57	51	52	83	14	55	31	99	73	23	40	07	64	54	44	99	21
13	50	78	02	73	39	66	82	01	28	67	51	75	66	33	97	47	58	42	44	88	09	28	58	06
67	92	65	41	45	36	77	96	46	21	14	39	56	36	70	15	74	43	62	69	82	30	77	28	77
72	56	73	44	26	04	62	81	15	35	79	26	99	57	28	22	25	94	80	62	95	48	98	23	86
28	86	85	64	94	11	58	78	45	36	34	45	91	38	51	10	68	36	87	81	16	77	30	19	36
69	57	40	80	44	94	60	82	94	93	98	01	48	50	57	69	60	77	69	60	74	22	05	77	17
71	20	03	30	79	25	74	17	78	34	54	45	04	77	42	59	75	78	64	99	37	03	18	03	36
89	98	55	98	22	45	12	49	82	71	57	33	28	69	50	59	15	09	25	79	39	42	84	18	70
58	74	82	81	14	02	01	05	77	94	65	57	70	39	42	48	56	84	31	59	18	70	41	74	60
50	54	73	81	91	07	81	26	25	45	49	61	22	88	41	20	00	15	59	93	51	60	65	65	63
49	33	72	90	10	20	65	28	44	63	95	86	75	78	69	24	41	65	86	10	34	10	32	00	93
11	85	01	43	65	02	85	69	56	88	34	29	64	35	48	15	70	11	77	83	01	34	82	91	04
34	22	46	41	84	74	27	02	57	77	47	93	72	02	95	63	75	74	69	69	61	34	31	92	13

05	57	23	06	26	23	08	66	16	11	75	28	81	56	14	62	82	45	65	80	36	02	76	55	63
37	78	16	06	57	12	46	22	90	97	78	67	39	06	63	60	51	02	07	16	75	12	90	41	16
23	71	15	08	82	64	87	29	01	20	46	72	05	80	19	27	47	15	76	51	58	67	06	80	54
42	67	98	41	67	44	28	71	45	08	19	47	76	30	26	72	33	69	92	51	95	23	26	85	76
05	83	03	84	32	62	83	27	48	83	09	19	84	90	20	20	50	87	74	93	51	62	10	23	30
60	46	18	41	23	74	73	51	72	90	40	52	95	41	20	89	48	98	27	38	81	33	83	82	94
32	80	64	75	91	98	09	40	64	89	29	99	46	35	69	91	50	73	75	92	90	56	82	93	24
79	86	53	77	78	06	62	37	48	82	71	00	78	21	65	65	88	45	82	44	78	93	22	78	09
45	13	23	32	01	09	46	36	43	66	37	15	35	30	46	15	90	53	19	13	91	59	81	81	87
20	60	97	48	21	41	84	22	72	77	99	81	83	30	46	15	90	26	51	73	66	34	99	40	60
67	91	44	83	43	25	56	33	28	80	99	53	27	56	19	80	76	32	53	95	07	53	09	61	98
86	50	76	93	86	35	68	45	37	83	47	44	92	57	66	59	64	16	48	39	26	94	54	66	40
66	73	38	38	23	36	10	95	16	01	10	01	59	71	55	99	24	88	31	41	00	73	13	80	62
55	11	50	29	17	73	97	04	20	39	20	22	71	11	43	00	15	10	12	35	09	11	00	89	05
23	54	33	87	92	92	04	49	73	96	57	53	57	08	93	09	69	87	83	07	46	39	50	37	85
41	48	67	79	44	57	40	29	10	34	58	63	51	18	07	41	02	39	79	14	40	68	10	01	61
03	97	71	72	43	27	36	24	59	88	82	87	26	31	11	44	28	58	99	47	83	21	35	22	88
90	24	83	48	07	41	56	68	11	14	77	75	48	68	08	90	89	63	87	00	06	18	63	21	91
98	98	97	42	27	11	80	51	13	13	03	42	91	14	51	22	15	48	67	52	09	40	34	60	85
74	20	94	21	49	96	51	69	99	85	43	76	55	81	36	11	88	68	32	43	08	14	78	05	34
94	67	48	87	11	84	00	85	93	56	43	99	21	74	84	13	56	41	90	96	30	04	19	68	73
58	18	84	82	71	23	66	33	19	25	65	17	90	84	24	91	75	36	14	83	86	22	70	86	89
31	47	28	24	88	49	28	69	78	62	23	45	53	38	78	65	87	44	91	93	91	62	76	09	20
45	62	31	06	70	92	73	27	83	57	15	64	40	57	56	54	42	35	40	93	55	82	08	78	87
31	49	87	12	27	41	07	91	72	64	63	42	06	66	82	71	28	36	45	31	99	01	03	35	76
69	37	22	23	46	10	75	83	62	94	44	65	46	23	65	71	69	20	89	12	16	56	61	70	41
93	67	21	56	98	42	52	53	14	86	24	70	25	18	23	23	56	24	03	86	11	06	46	10	23
77	56	18	37	01	32	20	18	70	79	20	85	77	89	28	17	77	15	52	47	15	30	35	12	75
37	07	47	79	60	75	24	15	31	63	25	93	27	66	19	53	52	49	98	45	12	12	06	00	32
72	08	71	01	73	46	39	60	37	58	22	25	20	84	30	02	03	62	68	58	38	04	06	89	94
55	22	48	46	72	50	14	24	47	67	84	37	32	84	82	64	97	13	69	86	20	09	80	46	75
69	24	98	90	70	29	34	25	33	23	12	69	90	50	38	93	84	32	28	96	03	65	70	90	12
01	86	77	18	21	91	66	11	84	65	48	75	26	94	51	40	51	53	36	39	77	69	06	25	07
51	40	94	06	80	61	34	28	46	28	11	48	48	94	60	65	06	63	71	06	19	35	05	32	56
58	78	02	85	80	29	67	27	44	07	67	23	20	28	22	62	97	59	62	13	41	72	70	71	07
33	75	88	51	00	33	56	15	84	34	28	50	16	65	12	81	56	43	54	14	63	37	74	97	59
58	60	37	45	62	09	95	93	16	59	35	22	91	78	04	97	98	80	20	04	38	93	13	92	30
72	13	12	95	32	87	99	32	83	65	40	17	92	57	22	68	98	79	16	23	53	56	56	07	47
22	21	13	16	10	52	57	71	40	49	95	25	55	36	95	57	25	25	77	05	38	05	62	57	77
97	94	83	67	90	68	74	88	17	22	38	01	04	33	49	38	47	57	61	87	15	39	43	87	00
09	03	68	53	63	29	27	31	66	53	39	34	88	87	04	35	80	69	52	74	99	16	52	01	65
29	95	61	42	65	05	72	27	28	18	09	85	24	59	46	03	91	55	38	62	51	71	47	37	38
81	96	78	90	47	41	38	36	33	95	05	90	26	72	85	23	23	30	70	51	56	93	23	84	80
44	62	20	81	21	57	57	85	00	47	26	10	87	22	45	72	03	51	75	23	38	38	56	77	97
68	91	12	15	08	02	18	74	56	79	21	53	63	41	77	15	07	39	87	11	19	25	62	19	30
29	33	77	60	29	09	25	09	42	28	07	15	40	67	56	29	58	75	84	06	19	54	31	16	53
54	13	39	19	29	64	97	73	71	61	78	03	24	02	93	86	69	76	74	28	08	98	84	08	23
75	16	85	64	64	93	85	68	08	84	15	41	57	84	45	11	70	13	17	60	47	80	10	13	00
36	47	17	08	79	03	92	85	18	42	95	48	27	37	99	98	81	94	44	72	06	95	42	31	17
29	61	08	21	91	23	76	72	84	98	26	23	66	54	86	88	95	14	82	57	17	99	16	28	99

Source: The RAND Corporation, *A Million Random Degits with 100,100 Normal Deviates*, New York: The Free Press, 1965. Copyright 1955 and 1983, the RAND Corporation. Used by permission.

Table A2 Factors for Determining Control Limits for Means and Ranges

Number of Observations in Subgroup n	Factor for $\bar{X}$-Chart, A_2	Factors for R-Chart	
		Lower Control Limit D_3	Upper Control Limit D_4
2	1.88	0	3.27
3	1.02	0	2.57
4	0.73	0	2.28
5	0.58	0	2.11
6	0.48	0	2.00
7	0.42	0.08	1.92
8	0.37	0.14	1.86
9	0.34	0.18	1.82
10	0.31	0.22	1.78
11	0.29	0.26	1.74
12	0.27	0.28	1.72
13	0.25	0.31	1.69
14	0.24	0.33	1.67
15	0.22	0.35	1.65
16	0.21	0.36	1.64
17	0.20	0.38	1.62
18	0.19	0.39	1.61
19	0.19	0.40	1.60
20	0.18	0.41	1.59

Source: Grant, E. L. and Leavenworth, R. S. (1980). *Statistical Quality Control*, New York: McGraw-Hill Book Company. Copyright McGraw-Hill, New York. Reprinted with permission.

Table A3 Factors for Determining Control Limits for Standard Deviations (S.D.)

| | Factors for S.D. Chart | |
| | Lower Control Limit | Upper Control Limit |
Number of Observations in Subgroup, n	B_3	B_4
2	0	3.27
3	0	2.57
4	0	2.27
5	0	2.09
6	0.03	1.97
7	0.12	1.88
8	0.19	1.81
9	0.24	1.76
10	0.28	1.72
11	0.32	1.68
12	0.35	1.65
13	0.38	1.62
14	0.41	1.59
15	0.43	1.57
16	0.45	1.55
17	0.47	1.53
18	0.48	1.52
19	0.50	1.50
20	0.51	1.49
21	0.52	1.48
22	0.53	1.47
23	0.54	1.46
24	0.55	1.45
25	0.56	1.44
30	0.60	1.40
35	0.63	1.37
40	0.66	1.34
45	0.68	1.32
50	0.70	1.30
55	0.71	1.29
60	0.72	1.28
65	0.73	1.27
70	0.74	1.26
75	0.75	1.25
80	0.76	1.24
85	0.77	1.23
90	0.77	1.23
95	0.78	1.22
100	0.79	1.21

Source: Adapted from Grant, E. L and Leavenworth, R. S. (1980). *Statistical Quality Control*, New York: McGraw-Hill Book Company. Copyright McGraw-Hill, New York. Reprinted with permission.

Table A4 Standard Normal Cumulative Probabilities*

z_c	.00	.01	.02	.03	.04	.05	.06	.07	.08	.09
0.0	.5000	.5040	.5080	.5120	.5160	.5199	.5239	.5279	.5319	.5359
0.1	.5398	.5438	.5478	.5517	.5557	.5596	.5636	.5675	.5714	.5753
0.2	.5793	.5832	.5871	.5910	.5948	.5987	.6026	.6064	.6103	.6141
0.3	.6179	.6217	.6255	.6293	.6331	.6368	.6406	.6443	.6480	.6517
0.4	.6554	.6591	.6628	.6664	.6700	.6736	.6772	.6808	.6844	.6879
0.5	.6915	.6950	.6985	.7019	.7054	.7088	.7123	.7157	.7190	.7224
0.6	.7257	.7291	.7324	.7357	.7389	.7422	.7454	.7486	.7517	.7549
0.7	.7580	.7611	.7642	.7673	.7704	.7734	.7764	.7794	.7823	.7852
0.8	.7881	.7910	.7939	.7967	.7995	.8023	.8051	.8078	.8106	.8133
0.9	.8159	.8186	.8212	.8238	.8264	.8289	.8315	.8340	.8365	.8389
1.0	.8413	.8438	.8461	.8485	.8508	.8531	.8554	.8577	.8599	.8621
1.1	.8643	.8665	.8686	.8708	.8729	.8749	.8770	.8790	.8810	.8830
1.2	.8849	.8869	.8888	.8907	.8925	.8944	.8962	.8980	.8997	.9015
1.3	.9032	.9049	.9066	.9082	.9099	.9115	.9131	.9147	.9162	.9177
1.4	.9192	.9207	.9222	.9236	.9251	.9265	.9279	.9292	.9306	.9319
1.5	.9332	.9345	.9357	.9370	.9382	.9394	.9406	.9418	.9429	.9441
1.6	.9452	.9463	.9474	.9484	.9495	.9505	.9515	.9525	.9535	.9545
1.7	.9554	.9564	.9573	.9582	.9591	.9599	.9608	.9616	.9625	.9633
1.8	.9641	.9649	.9656	.9664	.9671	.9678	.9686	.9693	.9699	.9706
1.9	.9713	.9719	.9726	.9732	.9738	.9744	.9750	.9756	.9761	.9767
2.0	.9772	.9778	.9783	.9788	.9793	.9798	.9803	.9808	.9812	.9817
2.1	.9821	.9826	.9830	.9834	.9838	.9842	.9846	.9850	.9854	.9857
2.2	.9861	.9864	.9868	.9871	.9875	.9878	.9881	.9884	.9887	.9890
2.3	.9893	.9896	.9898	.9901	.9904	.9906	.9909	.9911	.9913	.9916
2.4	.9918	.9920	.9922	.9925	.9927	.9929	.9931	.9932	.9934	.9936
2.5	.9938	.9940	.9941	.9943	.9945	.9946	.9948	.9949	.9951	.9952
2.6	.9953	.9955	.9956	.9957	.9959	.9960	.9961	.9962	.9963	.9964
2.7	.9965	.9966	.9967	.9968	.9969	.9970	.9971	.9972	.9973	.9974
2.8	.9974	.9975	.9976	.9977	.9977	.9978	.9979	.9979	.9980	.9981
2.9	.9981	.9982	.9982	.9983	.9984	.9984	.9985	.9985	.9986	.9986
3.0	.9987	.9987	.9987	.9988	.9988	.9989	.9989	.9989	.9990	.9990
3.1	.9990	.9991	.9991	.9991	.9992	.9992	.9992	.9992	.9993	.9993
3.2	.9993	.9993	.9994	.9994	.9994	.9994	.9994	.9995	.9995	.9995
3.3	.9995	.9995	.9995	.9996	.9996	.9996	.9996	.9996	.9996	.9997
3.4	.9997	.9997	.9997	.9997	.9997	.9997	.9997	.9997	.9997	.9998
3.5	.9998	.9998	.9998	.9998	.9998	.9998	.9998	.9998	.9998	.9998
3.6	.9998	.9998	.9999	.9999	.9999	.9999	.9999	.9999	.9999	.9999
3.7	.9999	.9999	.9999	.9999	.9999	.9999	.9999	.9999	.9999	.9999
3.8	.9999	.9999	.9999	.9999	.9999	.9999	.9999	.9999	.9999	.9999
3.9	1.000	1.0000	1.0000	1.0000	1.0000	1.0000	1.0000	1.0000	1.0000	1.0000

z_c		0.675	1.282	1.645	1.960	2.326	2.576
Probability		.75	.90	.95	.975	.99	.995
Upper Tail		.25	.10	.05	.025	.01	.005

*The entries in this table show the probability that a standard normal variate is less than or equal to z_c.

Table A5 Student _t_ Cumulative Probabilities*

ν	Cumulative Probability					
	.75	.90	.95	.975	.99	.995
1	1.000	3.078	6.314	12.706	31.821	63.657
2	0.816	1.886	2.920	4.303	6.965	9.925
3	.765	1.638	2.353	3.182	4.541	5.841
4	.741	1.533	2.132	2.776	3.747	4.604
5	0.727	1.476	2.015	2.571	3.365	4.032
6	.718	1.440	1.943	2.447	3.143	3.707
7	.711	1.415	1.895	2.365	2.998	3.499
8	.706	1.397	1.860	2.306	2.896	3.355
9	.703	1.383	1.833	2.262	2.821	3.250
10	0.700	1.372	1.812	2.228	2.764	3.169
11	.697	1.363	1.796	2.201	2.718	3.106
12	.695	1.356	1.782	2.179	2.681	3.055
13	.694	1.350	1.771	2.160	2.650	3.012
14	.692	1.345	1.761	2.145	2.624	2.977
15	0.691	1.341	1.753	2.131	2.602	2.947
16	.690	1.337	1.746	2.120	2.583	2.921
17	.689	1.333	1.740	2.110	2.567	2.898
18	.688	1.330	1.734	2.101	2.552	2.878
19	.688	1.328	1.729	2.093	2.539	2.861
20	0.687	1.325	1.725	2.086	2.528	2.845
21	.686	1.323	1.721	2.080	2.518	2.831
22	.686	1.321	1.717	2.074	2.508	2.819
23	.685	1.319	1.714	2.069	2.500	2.807
24	.685	1.318	1.711	2.064	2.492	2.797
25	0.684	1.316	1.708	2.060	2.485	2.787
26	.684	1.315	1.706	2.056	2.479	2.779
27	.684	1.314	1.703	2.052	2.473	2.771
28	.683	1.313	1.701	2.048	2.467	2.763
29	.683	1.311	1.699	2.045	2.462	2.756
30	0.683	1.310	1.697	2.042	2.457	2.750
60	.679	1.296	1.671	2.000	2.390	2.660
90	.678	1.291	1.662	1.987	2.368	2.632
120	.677	1.289	1.658	1.980	2.358	2.617
∞	.674	1.282	1.645	1.960	2.326	2.576
Upper Tail	.25	.10	.05	.025	.01	.005

*The entries in this table show the value of the student _t_ variate with _v_ degrees of freedom corresponding to the designated cumulative probability.

Table A6 Chi-Square Cumulative Probabilities*

ν	.005	.010	.025	.050	.100	.250	.500	.750	.900	.950	.975	.990	.995	.999
								Cumulative Probability						
1					0.02	0.10	0.45	1.32	2.71	3.84	5.02	6.63	7.88	10.83
2	0.01	0.02	0.05	0.10	0.21	0.58	1.39	2.77	4.61	5.99	7.38	9.21	10.60	13.82
3	0.07	0.11	0.22	0.35	0.58	1.21	2.37	4.11	6.25	7.81	9.35	11.34	12.84	16.27
4	0.21	0.30	0.48	0.71	1.06	1.92	3.36	5.39	7.78	9.49	11.14	13.28	14.86	18.47
5	0.41	0.55	0.83	1.15	1.61	2.67	4.35	6.63	9.24	11.07	12.83	15.09	16.75	20.52
6	0.68	0.87	1.24	1.64	2.20	3.45	5.35	7.84	10.64	12.59	14.45	16.81	18.55	22.46
7	0.99	1.24	1.69	2.17	2.83	4.25	6.35	9.04	12.02	14.07	16.01	18.48	20.28	24.32
8	1.34	1.65	2.18	2.73	3.49	5.07	7.34	10.22	13.36	15.51	17.53	20.09	21.96	26.12
9	1.73	2.09	2.70	3.33	4.17	5.90	8.34	11.39	14.68	16.92	19.02	21.67	23.59	27.88
10	2.16	2.56	3.25	3.94	4.87	6.74	9.34	12.55	15.99	18.31	20.48	23.21	25.19	29.59
11	2.60	3.05	3.82	4.57	5.58	7.58	10.34	13.70	17.28	19.68	21.92	24.73	26.76	31.26
12	3.07	3.57	4.40	5.23	6.30	8.44	11.34	14.85	18.55	21.03	23.34	26.22	28.30	32.91
13	3.57	4.11	5.01	5.89	7.04	9.30	12.34	15.98	19.81	22.36	24.74	27.69	29.82	34.53
14	4.07	4.66	5.63	6.57	7.79	10.17	13.34	17.12	21.06	23.68	26.12	29.14	31.32	36.12
15	4.60	5.23	6.26	7.26	8.55	11.04	14.34	18.25	22.31	25.00	27.49	30.58	32.80	37.70
16	5.14	5.81	6.91	7.96	9.31	11.91	15.34	19.37	23.54	26.30	28.85	32.00	34.27	39.25
17	5.70	6.41	7.56	8.67	10.09	12.79	16.34	20.49	24.77	27.59	30.19	33.41	35.72	40.79
18	6.26	7.01	8.23	9.39	10.86	13.68	17.34	21.60	25.99	28.87	31.53	34.81	37.16	42.31
19	6.84	7.63	8.91	10.12	11.65	14.56	18.34	22.72	27.20	30.14	32.85	36.19	38.58	43.82
20	7.43	8.26	9.59	10.85	12.44	15.45	19.34	23.83	28.41	31.41	34.17	37.57	40.00	45.32
21	8.03	8.90	10.28	11.59	13.24	16.34	20.34	24.93	29.62	32.67	35.48	38.93	41.40	46.80
22	8.64	9.54	10.98	12.34	14.04	17.24	21.34	26.04	30.81	33.92	36.78	40.29	42.80	48.27
23	9.26	10.20	11.69	13.09	14.85	18.14	22.34	27.14	32.01	35.17	38.08	41.64	44.18	49.73
24	9.89	10.86	12.40	13.85	15.66	19.04	23.34	28.24	33.20	36.42	39.36	42.98	45.56	51.18
25	10.52	11.52	13.12	14.61	16.47	19.94	24.34	29.34	34.38	37.65	40.65	44.31	46.93	52.62
26	11.16	12.20	13.84	15.38	17.29	20.84	25.34	30.43	35.56	38.89	41.92	45.64	48.29	54.05
27	11.81	12.88	14.57	16.15	18.11	21.75	26.34	31.53	36.74	40.11	43.19	46.96	49.64	55.48
28	12.46	13.56	15.31	16.93	18.94	22.66	27.34	32.62	37.92	41.34	44.46	48.28	50.99	56.89
29	13.12	14.26	16.05	17.71	19.77	23.57	28.34	33.71	39.09	42.56	45.72	49.59	52.34	58.30
30	13.79	14.95	16.79	18.49	20.60	24.48	29.34	34.80	40.26	43.77	46.98	50.89	53.67	59.70
40	20.71	22.16	24.43	26.51	29.05	33.66	39.34	45.62	51.81	55.76	59.34	63.69	66.77	73.40
50	27.99	29.71	32.36	34.76	37.69	42.94	49.33	56.33	63.17	67.50	71.42	76.15	79.49	86.66
60	35.53	37.48	40.48	43.19	46.46	52.29	59.33	66.98	74.40	79.08	83.30	88.38	91.95	99.61
70	43.28	45.44	48.76	51.74	55.33	61.70	69.33	77.58	85.53	90.53	95.02	100.42	104.22	112.32
80	51.17	53.54	57.15	60.39	64.28	71.14	79.33	88.13	96.58	101.88	106.63	112.33	116.32	124.84
90	59.20	61.75	65.65	69.13	73.29	80.62	89.33	98.65	107.56	113.14	118.14	124.12	128.30	137.21
100	67.33	70.06	74.22	77.93	82.36	90.13	99.33	109.14	118.50	124.34	129.56	135.81	140.17	149.45
Upper Tail	.995	.990	.975	.950	.900	.750	.500	.250	.100	.050	.025	.010	.005	.001

*The entries in this table are values of a chi-square variate with ν degrees of freedom corresponding to the designated cumulative probability.

Table A7(a)   F Cumulative Probabilities: 0.75 (Upper Tail: 0.25)*

v_2 \ v_1	1	2	3	4	5	6	7	8	9	10	12	15	20	24	30	40	60	120	∞
1	5·83	7·50	8·20	8·58	8·82	8·98	9·10	9·19	9·26	9·32	9·41	9·49	9·58	9·63	9·67	9·71	9·76	9·80	9·85
2	2·57	3·00	3·15	3·23	3·28	3·31	3·34	3·35	3·37	3·38	3·39	3·41	3·43	3·43	3·44	3·45	3·46	3·47	3·48
3	2·02	2·28	2·36	2·39	2·41	2·42	2·43	2·44	2·44	2·44	2·45	2·46	2·46	2·46	2·46	2·47	2·47	2·47	2·47
4	1·81	2·00	2·05	2·06	2·07	2·08	2·08	2·08	2·08	2·08	2·08	2·08	2·08	2·08	2·08	2·08	2·08	2·08	2·08
5	1·69	1·85	1·88	1·89	1·89	1·89	1·89	1·89	1·89	1·89	1·89	1·89	1·88	1·88	1·88	1·88	1·87	1·87	1·87
6	1·62	1·76	1·78	1·79	1·79	1·78	1·78	1·78	1·77	1·77	1·77	1·76	1·76	1·75	1·75	1·75	1·74	1·74	1·74
7	1·57	1·70	1·72	1·72	1·71	1·71	1·70	1·70	1·69	1·69	1·68	1·68	1·67	1·67	1·66	1·66	1·65	1·65	1·65
8	1·54	1·66	1·67	1·66	1·66	1·65	1·64	1·64	1·63	1·63	1·62	1·62	1·61	1·60	1·60	1·59	1·59	1·58	1·58
9	1·51	1·62	1·63	1·63	1·62	1·61	1·60	1·60	1·59	1·59	1·58	1·57	1·56	1·56	1·55	1·54	1·54	1·53	1·53
10	1·49	1·60	1·60	1·59	1·59	1·58	1·57	1·56	1·56	1·55	1·54	1·53	1·52	1·52	1·51	1·51	1·50	1·49	1·48
11	1·47	1·58	1·58	1·57	1·56	1·55	1·54	1·53	1·53	1·52	1·51	1·50	1·49	1·49	1·48	1·47	1·47	1·46	1·45
12	1·46	1·56	1·56	1·55	1·54	1·53	1·52	1·51	1·51	1·50	1·49	1·48	1·47	1·46	1·45	1·45	1·44	1·43	1·42
13	1·45	1·55	1·55	1·53	1·52	1·51	1·50	1·49	1·49	1·48	1·47	1·46	1·45	1·44	1·43	1·42	1·42	1·41	1·40
14	1·44	1·53	1·53	1·52	1·51	1·50	1·49	1·48	1·47	1·46	1·45	1·44	1·43	1·42	1·41	1·41	1·40	1·39	1·38
15	1·43	1·52	1·52	1·51	1·49	1·48	1·47	1·46	1·46	1·45	1·44	1·43	1·41	1·41	1·40	1·39	1·38	1·37	1·36
16	1·42	1·51	1·51	1·50	1·48	1·47	1·46	1·45	1·44	1·44	1·43	1·41	1·40	1·39	1·38	1·37	1·36	1·35	1·34
17	1·42	1·51	1·50	1·49	1·47	1·46	1·45	1·44	1·43	1·43	1·41	1·40	1·39	1·38	1·37	1·36	1·35	1·34	1·33
18	1·41	1·50	1·49	1·48	1·46	1·45	1·44	1·43	1·42	1·42	1·40	1·39	1·38	1·37	1·36	1·35	1·34	1·33	1·32
19	1·41	1·49	1·49	1·47	1·46	1·44	1·43	1·42	1·41	1·41	1·40	1·38	1·37	1·36	1·35	1·34	1·33	1·32	1·30
20	1·40	1·49	1·48	1·47	1·45	1·44	1·43	1·42	1·41	1·40	1·39	1·37	1·36	1·35	1·34	1·33	1·32	1·31	1·29
21	1·40	1·48	1·48	1·46	1·44	1·43	1·42	1·41	1·40	1·39	1·38	1·37	1·35	1·34	1·33	1·32	1·31	1·30	1·28
22	1·40	1·48	1·47	1·45	1·44	1·42	1·41	1·40	1·39	1·39	1·37	1·36	1·34	1·33	1·32	1·31	1·30	1·29	1·28
23	1·39	1·47	1·47	1·45	1·43	1·42	1·41	1·40	1·39	1·38	1·37	1·35	1·34	1·33	1·32	1·31	1·30	1·28	1·27
24	1·39	1·47	1·46	1·44	1·43	1·41	1·40	1·39	1·38	1·38	1·36	1·35	1·33	1·32	1·31	1·30	1·29	1·28	1·26
25	1·39	1·47	1·46	1·44	1·42	1·41	1·40	1·39	1·38	1·37	1·36	1·34	1·33	1·32	1·31	1·29	1·28	1·27	1·25
26	1·38	1·46	1·45	1·44	1·42	1·41	1·39	1·38	1·37	1·37	1·35	1·34	1·32	1·31	1·30	1·29	1·28	1·26	1·25
27	1·38	1·46	1·45	1·43	1·42	1·40	1·39	1·38	1·37	1·36	1·35	1·33	1·32	1·31	1·30	1·28	1·27	1·26	1·24
28	1·38	1·46	1·45	1·43	1·41	1·40	1·39	1·38	1·37	1·36	1·34	1·33	1·31	1·30	1·29	1·28	1·27	1·25	1·24
29	1·38	1·45	1·45	1·43	1·41	1·40	1·38	1·37	1·36	1·35	1·34	1·32	1·31	1·30	1·29	1·27	1·26	1·25	1·23
30	1·38	1·45	1·44	1·42	1·41	1·39	1·38	1·37	1·36	1·35	1·34	1·32	1·30	1·29	1·28	1·27	1·26	1·24	1·23
40	1·36	1·44	1·42	1·40	1·39	1·37	1·36	1·35	1·34	1·33	1·31	1·30	1·28	1·26	1·25	1·24	1·22	1·21	1·19
60	1·35	1·42	1·41	1·38	1·37	1·35	1·33	1·32	1·31	1·30	1·29	1·27	1·25	1·24	1·22	1·21	1·19	1·17	1·15
120	1·34	1·40	1·39	1·37	1·35	1·33	1·31	1·30	1·29	1·28	1·26	1·24	1·22	1·21	1·19	1·18	1·16	1·13	1·10
∞	1·32	1·39	1·37	1·35	1·33	1·31	1·29	1·28	1·27	1·25	1·24	1·22	1·19	1·18	1·16	1·14	1·12	1·08	1·00

(a)

*The entries in this table are values of the F variate having v_1 numerator and v_2 denominator degrees of freedom corresponding to the designated cumulative probability.

Table A7(b)   F Cumulative Probabilities: 0.90 (Upper Tail: 0.10)

ν_2\\ν_1	1	2	3	4	5	6	7	8	9	10	12	15	20	24	30	40	60	120	∞
1	39.86	49.50	53.59	55.83	57.24	58.20	58.91	59.44	59.86	60.19	60.71	61.22	61.74	62.00	62.26	62.53	62.79	63.06	63.33
2	8.53	9.00	9.16	9.24	9.29	9.33	9.35	9.37	9.38	9.39	9.41	9.42	9.44	9.45	9.46	9.47	9.47	9.48	9.49
3	5.54	5.46	5.39	5.34	5.31	5.28	5.27	5.25	5.24	5.23	5.22	5.20	5.18	5.18	5.17	5.16	5.15	5.14	5.13
4	4.54	4.32	4.19	4.11	4.05	4.01	3.98	3.95	3.94	3.92	3.90	3.87	3.84	3.83	3.82	3.80	3.79	3.78	3.76
5	4.06	3.78	3.62	3.52	3.45	3.40	3.37	3.34	3.32	3.30	3.27	3.24	3.21	3.19	3.17	3.16	3.14	3.12	3.10
6	3.78	3.46	3.29	3.18	3.11	3.05	3.01	2.98	2.96	2.94	2.90	2.87	2.84	2.82	2.80	2.78	2.76	2.74	2.72
7	3.59	3.26	3.07	2.96	2.88	2.83	2.78	2.75	2.72	2.70	2.67	2.63	2.59	2.58	2.56	2.54	2.51	2.49	2.47
8	3.46	3.11	2.92	2.81	2.73	2.67	2.62	2.59	2.56	2.54	2.50	2.46	2.42	2.40	2.38	2.36	2.34	2.32	2.29
9	3.36	3.01	2.81	2.69	2.61	2.55	2.51	2.47	2.44	2.42	2.38	2.34	2.30	2.28	2.25	2.23	2.21	2.18	2.16
10	3.29	2.92	2.73	2.61	2.52	2.46	2.41	2.38	2.35	2.32	2.28	2.24	2.20	2.18	2.16	2.13	2.11	2.08	2.06
11	3.23	2.86	2.66	2.54	2.45	2.39	2.34	2.30	2.27	2.25	2.21	2.17	2.12	2.10	2.08	2.05	2.03	2.00	1.97
12	3.18	2.81	2.61	2.48	2.39	2.33	2.28	2.24	2.21	2.19	2.15	2.10	2.06	2.04	2.01	1.99	1.96	1.93	1.90
13	3.14	2.76	2.56	2.43	2.35	2.28	2.23	2.20	2.16	2.14	2.10	2.05	2.01	1.98	1.96	1.93	1.90	1.88	1.85
14	3.10	2.73	2.52	2.39	2.31	2.24	2.19	2.15	2.12	2.10	2.05	2.01	1.96	1.94	1.91	1.89	1.86	1.83	1.80
15	3.07	2.70	2.49	2.36	2.27	2.21	2.16	2.12	2.09	2.06	2.02	1.97	1.92	1.90	1.87	1.85	1.82	1.79	1.76
16	3.05	2.67	2.46	2.33	2.24	2.18	2.13	2.09	2.06	2.03	1.99	1.94	1.89	1.87	1.84	1.81	1.78	1.75	1.72
17	3.03	2.64	2.44	2.31	2.22	2.15	2.10	2.06	2.03	2.00	1.96	1.91	1.86	1.84	1.81	1.78	1.75	1.72	1.69
18	3.01	2.62	2.42	2.29	2.20	2.13	2.08	2.04	2.00	1.98	1.93	1.89	1.84	1.81	1.78	1.75	1.72	1.69	1.66
19	2.99	2.61	2.40	2.27	2.18	2.11	2.06	2.02	1.98	1.96	1.91	1.86	1.81	1.79	1.76	1.73	1.70	1.67	1.63
20	2.97	2.59	2.38	2.25	2.16	2.09	2.04	2.00	1.96	1.94	1.89	1.84	1.79	1.77	1.74	1.71	1.68	1.64	1.61
21	2.96	2.57	2.36	2.23	2.14	2.08	2.02	1.98	1.95	1.92	1.87	1.83	1.78	1.75	1.72	1.69	1.66	1.62	1.59
22	2.95	2.56	2.35	2.22	2.13	2.06	2.01	1.97	1.93	1.90	1.86	1.81	1.76	1.73	1.70	1.67	1.64	1.60	1.57
23	2.94	2.55	2.34	2.21	2.11	2.05	1.99	1.95	1.92	1.89	1.84	1.80	1.74	1.72	1.69	1.66	1.62	1.59	1.55
24	2.93	2.54	2.33	2.19	2.10	2.04	1.98	1.94	1.91	1.88	1.83	1.78	1.73	1.70	1.67	1.64	1.61	1.57	1.53
25	2.92	2.53	2.32	2.18	2.09	2.02	1.97	1.93	1.89	1.87	1.82	1.77	1.72	1.69	1.66	1.63	1.59	1.56	1.52
26	2.91	2.52	2.31	2.17	2.08	2.01	1.96	1.92	1.88	1.86	1.81	1.76	1.71	1.68	1.65	1.61	1.58	1.54	1.50
27	2.90	2.51	2.30	2.17	2.07	2.00	1.95	1.91	1.87	1.85	1.80	1.75	1.70	1.67	1.64	1.60	1.57	1.53	1.49
28	2.89	2.50	2.29	2.16	2.06	2.00	1.94	1.90	1.87	1.84	1.79	1.74	1.69	1.66	1.63	1.59	1.56	1.52	1.48
29	2.89	2.50	2.28	2.15	2.06	1.99	1.93	1.89	1.86	1.83	1.78	1.73	1.68	1.65	1.62	1.58	1.55	1.51	1.47
30	2.88	2.49	2.28	2.14	2.05	1.98	1.93	1.88	1.85	1.82	1.77	1.72	1.67	1.64	1.61	1.57	1.54	1.50	1.46
40	2.84	2.44	2.23	2.09	2.00	1.93	1.87	1.83	1.79	1.76	1.71	1.66	1.61	1.57	1.54	1.51	1.47	1.42	1.38
60	2.79	2.39	2.18	2.04	1.95	1.87	1.82	1.77	1.74	1.71	1.66	1.60	1.54	1.51	1.48	1.44	1.40	1.35	1.29
120	2.75	2.35	2.13	1.99	1.90	1.82	1.77	1.72	1.68	1.65	1.60	1.55	1.48	1.45	1.41	1.37	1.32	1.26	1.19
∞	2.71	2.30	2.08	1.94	1.85	1.77	1.72	1.67	1.63	1.60	1.55	1.49	1.42	1.38	1.34	1.30	1.24	1.17	1.00

(b)

*The entries in this table are values of the F variate having ν_1 numerator and ν_2 denominator degrees of freedom corresponding to the designated cumulative probability.

Table A7(c)   *F* Cumulative Probabilities: 0.95 (Upper Tail: 0.05)

v_2 \ v_1	1	2	3	4	5	6	7	8	9	10	12	15	20	24	30	40	60	120	∞
1	161·4	199·5	215·7	224·6	230·2	234·0	236·8	238·9	240·5	241·9	243·9	245·9	248·0	249·1	250·1	251·1	252·2	253·3	254·3
2	18·51	19·00	19·16	19·25	19·30	19·33	19·35	19·37	19·38	19·40	19·41	19·43	19·45	19·45	19·46	19·47	19·48	19·49	19·50
3	10·13	9·55	9·28	9·12	9·01	8·94	8·89	8·85	8·81	8·79	8·74	8·70	8·66	8·64	8·62	8·59	8·57	8·55	8·53
4	7·71	6·94	6·59	6·39	6·26	6·16	6·09	6·04	6·00	5·96	5·91	5·86	5·80	5·77	5·75	5·72	5·69	5·66	5·63
5	6·61	5·79	5·41	5·19	5·05	4·95	4·88	4·82	4·77	4·74	4·68	4·62	4·56	4·53	4·50	4·46	4·43	4·40	4·36
6	5·99	5·14	4·76	4·53	4·39	4·28	4·21	4·15	4·10	4·06	4·00	3·94	3·87	3·84	3·81	3·77	3·74	3·70	3·67
7	5·59	4·74	4·35	4·12	3·97	3·87	3·79	3·73	3·68	3·64	3·57	3·51	3·44	3·41	3·38	3·34	3·30	3·27	3·23
8	5·32	4·46	4·07	3·84	3·69	3·58	3·50	3·44	3·39	3·35	3·28	3·22	3·15	3·12	3·08	3·04	3·01	2·97	2·93
9	5·12	4·26	3·86	3·63	3·48	3·37	3·29	3·23	3·18	3·14	3·07	3·01	2·94	2·90	2·86	2·83	2·79	2·75	2·71
10	4·96	4·10	3·71	3·48	3·33	3·22	3·14	3·07	3·02	2·98	2·91	2·85	2·77	2·74	2·70	2·66	2·62	2·58	2·54
11	4·84	3·98	3·59	3·36	3·20	3·09	3·01	2·95	2·90	2·85	2·79	2·72	2·65	2·61	2·57	2·53	2·49	2·45	2·40
12	4·75	3·89	3·49	3·26	3·11	3·00	2·91	2·85	2·80	2·75	2·69	2·62	2·54	2·51	2·47	2·43	2·38	2·34	2·30
13	4·67	3·81	3·41	3·18	3·03	2·92	2·83	2·77	2·71	2·67	2·60	2·53	2·46	2·42	2·38	2·34	2·30	2·25	2·21
14	4·60	3·74	3·34	3·11	2·96	2·85	2·76	2·70	2·65	2·60	2·53	2·46	2·39	2·35	2·31	2·27	2·22	2·18	2·13
15	4·54	3·68	3·29	3·06	2·90	2·79	2·71	2·64	2·59	2·54	2·48	2·40	2·33	2·29	2·25	2·20	2·16	2·11	2·07
16	4·49	3·63	3·24	3·01	2·85	2·74	2·66	2·59	2·54	2·49	2·42	2·35	2·28	2·24	2·19	2·15	2·11	2·06	2·01
17	4·45	3·59	3·20	2·96	2·81	2·70	2·61	2·55	2·49	2·45	2·38	2·31	2·23	2·19	2·15	2·10	2·06	2·01	1·96
18	4·41	3·55	3·16	2·93	2·77	2·66	2·58	2·51	2·46	2·41	2·34	2·27	2·19	2·15	2·11	2·06	2·02	1·97	1·92
19	4·38	3·52	3·13	2·90	2·74	2·63	2·54	2·48	2·42	2·38	2·31	2·23	2·16	2·11	2·07	2·03	1·98	1·93	1·88
20	4·35	3·49	3·10	2·87	2·71	2·60	2·51	2·45	2·39	2·35	2·28	2·20	2·12	2·08	2·04	1·99	1·95	1·90	1·84
21	4·32	3·47	3·07	2·84	2·68	2·57	2·49	2·42	2·37	2·32	2·25	2·18	2·10	2·05	2·01	1·96	1·92	1·87	1·81
22	4·30	3·44	3·05	2·82	2·66	2·55	2·46	2·40	2·34	2·30	2·23	2·15	2·07	2·03	1·98	1·94	1·89	1·84	1·78
23	4·28	3·42	3·03	2·80	2·64	2·53	2·44	2·37	2·32	2·27	2·20	2·13	2·05	2·01	1·96	1·91	1·86	1·81	1·76
24	4·26	3·40	3·01	2·78	2·62	2·51	2·42	2·36	2·30	2·25	2·18	2·11	2·03	1·98	1·94	1·89	1·84	1·79	1·73
25	4·24	3·39	2·99	2·76	2·60	2·49	2·40	2·34	2·28	2·24	2·16	2·09	2·01	1·96	1·92	1·87	1·82	1·77	1·71
26	4·23	3·37	2·98	2·74	2·59	2·47	2·39	2·32	2·27	2·22	2·15	2·07	1·99	1·95	1·90	1·85	1·80	1·75	1·69
27	4·21	3·35	2·96	2·73	2·57	2·46	2·37	2·31	2·25	2·20	2·13	2·06	1·97	1·93	1·88	1·84	1·79	1·73	1·67
28	4·20	3·34	2·95	2·71	2·56	2·45	2·36	2·29	2·24	2·19	2·12	2·04	1·96	1·91	1·87	1·82	1·77	1·71	1·65
29	4·18	3·33	2·93	2·70	2·55	2·43	2·35	2·28	2·22	2·18	2·10	2·03	1·94	1·90	1·85	1·81	1·75	1·70	1·64
30	4·17	3·32	2·92	2·69	2·53	2·42	2·33	2·27	2·21	2·16	2·09	2·01	1·93	1·89	1·84	1·79	1·74	1·68	1·62
40	4·08	3·23	2·84	2·61	2·45	2·34	2·25	2·18	2·12	2·08	2·00	1·92	1·84	1·79	1·74	1·69	1·64	1·58	1·51
60	4·00	3·15	2·76	2·53	2·37	2·25	2·17	2·10	2·04	1·99	1·92	1·84	1·75	1·70	1·65	1·59	1·53	1·47	1·39
120	3·92	3·07	2·68	2·45	2·29	2·17	2·09	2·02	1·96	1·91	1·83	1·75	1·66	1·61	1·55	1·50	1·43	1·35	1·25
∞	3·84	3·00	2·60	2·37	2·21	2·10	2·01	1·94	1·88	1·83	1·75	1·67	1·57	1·52	1·46	1·39	1·32	1·22	1·00

(c)

*The entries in this table are values of the *F* variate having v_1 numerator and v_2 denominator degrees of freedom corresponding to the designated cumulative probability.

Table A7(d) F Cumulative Probabilities: 0.975 (Upper Tail: 0.025)

v_2 \ v_1	1	2	3	4	5	6	7	8	9	10	12	15	20	24	30	40	60	120	∞
1	647.8	799.5	864.2	899.6	921.8	937.1	948.2	956.7	963.3	968.6	976.7	984.9	993.1	997.2	1001	1006	1010	1014	1018
2	38.51	39.00	39.17	39.25	39.30	39.33	39.36	39.37	39.39	39.40	39.41	39.43	39.45	39.46	39.46	39.47	39.48	39.49	39.50
3	17.44	16.04	15.44	15.10	14.88	14.73	14.62	14.54	14.47	14.42	14.34	14.25	14.17	14.12	14.08	14.04	13.99	13.95	13.90
4	12.22	10.65	9.98	9.60	9.36	9.20	9.07	8.98	8.90	8.84	8.75	8.66	8.56	8.51	8.46	8.41	8.36	8.31	8.26
5	10.01	8.43	7.76	7.39	7.15	6.98	6.85	6.76	6.68	6.62	6.52	6.43	6.33	6.28	6.23	6.18	6.12	6.07	6.02
6	8.81	7.26	6.60	6.23	5.99	5.82	5.70	5.60	5.52	5.46	5.37	5.27	5.17	5.12	5.07	5.01	4.96	4.90	4.85
7	8.07	6.54	5.89	5.52	5.29	5.12	4.99	4.90	4.82	4.76	4.67	4.57	4.47	4.42	4.36	4.31	4.25	4.20	4.14
8	7.57	6.06	5.42	5.05	4.82	4.65	4.53	4.43	4.36	4.30	4.20	4.10	4.00	3.95	3.89	3.84	3.78	3.73	3.67
9	7.21	5.71	5.08	4.72	4.48	4.32	4.20	4.10	4.03	3.96	3.87	3.77	3.67	3.61	3.56	3.51	3.45	3.39	3.33
10	6.94	5.46	4.83	4.47	4.24	4.07	3.95	3.85	3.78	3.72	3.62	3.52	3.42	3.37	3.31	3.26	3.20	3.14	3.08
11	6.72	5.26	4.63	4.28	4.04	3.88	3.76	3.66	3.59	3.53	3.43	3.33	3.23	3.17	3.12	3.06	3.00	2.94	2.88
12	6.55	5.10	4.47	4.12	3.89	3.73	3.61	3.51	3.44	3.37	3.28	3.18	3.07	3.02	2.96	2.91	2.85	2.79	2.72
13	6.41	4.97	4.35	4.00	3.77	3.60	3.48	3.39	3.31	3.25	3.15	3.05	2.95	2.89	2.84	2.78	2.72	2.66	2.60
14	6.30	4.86	4.24	3.89	3.66	3.50	3.38	3.29	3.21	3.15	3.05	2.95	2.84	2.79	2.73	2.67	2.61	2.55	2.49
15	6.20	4.77	4.15	3.80	3.58	3.41	3.29	3.20	3.12	3.06	2.96	2.86	2.76	2.70	2.64	2.59	2.52	2.46	2.40
16	6.12	4.69	4.08	3.73	3.50	3.34	3.22	3.12	3.05	2.99	2.89	2.79	2.68	2.63	2.57	2.51	2.45	2.38	2.32
17	6.04	4.62	4.01	3.66	3.44	3.28	3.16	3.06	2.98	2.92	2.82	2.72	2.62	2.56	2.50	2.44	2.38	2.32	2.25
18	5.98	4.56	3.95	3.61	3.38	3.22	3.10	3.01	2.93	2.87	2.77	2.67	2.56	2.50	2.44	2.38	2.32	2.26	2.19
19	5.92	4.51	3.90	3.56	3.33	3.17	3.05	2.96	2.88	2.82	2.72	2.62	2.51	2.45	2.39	2.33	2.27	2.20	2.13
20	5.87	4.46	3.86	3.51	3.29	3.13	3.01	2.91	2.84	2.77	2.68	2.57	2.46	2.41	2.35	2.29	2.22	2.16	2.09
21	5.83	4.42	3.82	3.48	3.25	3.09	2.97	2.87	2.80	2.73	2.64	2.53	2.42	2.37	2.31	2.25	2.18	2.11	2.04
22	5.79	4.38	3.78	3.44	3.22	3.05	2.93	2.84	2.76	2.70	2.60	2.50	2.39	2.33	2.27	2.21	2.14	2.08	2.00
23	5.75	4.35	3.75	3.41	3.18	3.02	2.90	2.81	2.73	2.67	2.57	2.47	2.36	2.30	2.24	2.18	2.11	2.04	1.97
24	5.72	4.32	3.72	3.38	3.15	2.99	2.87	2.78	2.70	2.64	2.54	2.44	2.33	2.27	2.21	2.15	2.08	2.01	1.94
25	5.69	4.29	3.69	3.35	3.13	2.97	2.85	2.75	2.68	2.61	2.51	2.41	2.30	2.24	2.18	2.12	2.05	1.98	1.91
26	5.66	4.27	3.67	3.33	3.10	2.94	2.82	2.73	2.65	2.59	2.49	2.39	2.28	2.22	2.16	2.09	2.03	1.95	1.88
27	5.63	4.24	3.65	3.31	3.08	2.92	2.80	2.71	2.63	2.57	2.47	2.36	2.25	2.19	2.13	2.07	2.00	1.93	1.85
28	5.61	4.22	3.63	3.29	3.06	2.90	2.78	2.69	2.61	2.55	2.45	2.34	2.23	2.17	2.11	2.05	1.98	1.91	1.83
29	5.59	4.20	3.61	3.27	3.04	2.88	2.76	2.67	2.59	2.53	2.43	2.32	2.21	2.15	2.09	2.03	1.96	1.89	1.81
30	5.57	4.18	3.59	3.25	3.03	2.87	2.75	2.65	2.57	2.51	2.41	2.31	2.20	2.14	2.07	2.01	1.94	1.87	1.79
40	5.42	4.05	3.46	3.13	2.90	2.74	2.62	2.53	2.45	2.39	2.29	2.18	2.07	2.01	1.94	1.88	1.80	1.72	1.64
60	5.29	3.93	3.34	3.01	2.79	2.63	2.51	2.41	2.33	2.27	2.17	2.06	1.94	1.88	1.82	1.74	1.67	1.58	1.48
120	5.15	3.80	3.23	2.89	2.67	2.52	2.39	2.30	2.22	2.16	2.05	1.94	1.82	1.76	1.69	1.61	1.53	1.43	1.31
∞	5.02	3.69	3.12	2.79	2.57	2.41	2.29	2.19	2.11	2.05	1.94	1.83	1.71	1.64	1.57	1.48	1.39	1.27	1.00

(*d*)

*The entries in this table are values of the *F* variate having v_1 numerator and v_2 denominator degrees of freedom corresponding to the designated cumulative probability.

Table A7(e)   F Cumulative Probabilities: 0.99 (Upper Tail: 0.01)

$v_2 \backslash v_1$	1	2	3	4	5	6	7	8	9	10	12	15	20	24	30	40	60	120	∞
1	4052	4999·5	5403	5625	5764	5859	5928	5981	6022	6056	6106	6157	6209	6235	6261	6287	6313	6339	6366
2	98·50	99·00	99·17	99·25	99·30	99·33	99·36	99·37	99·39	99·40	99·42	99·43	99·45	99·46	99·47	99·47	99·48	99·49	99·50
3	34·12	30·82	29·46	28·71	28·24	27·91	27·67	27·49	27·35	27·23	27·05	26·87	26·69	26·60	26·50	26·41	26·32	26·22	26·13
4	21·20	18·00	16·69	15·98	15·52	15·21	14·98	14·80	14·66	14·55	14·37	14·20	14·02	13·93	13·84	13·75	13·65	13·56	13·46
5	16·26	13·27	12·06	11·39	10·97	10·67	10·46	10·29	10·16	10·05	9·89	9·72	9·55	9·47	9·38	9·29	9·20	9·11	9·02
6	13·75	10·92	9·78	9·15	8·75	8·47	8·26	8·10	7·98	7·87	7·72	7·56	7·40	7·31	7·23	7·14	7·06	6·97	6·88
7	12·25	9·55	8·45	7·85	7·46	7·19	6·99	6·84	6·72	6·62	6·47	6·31	6·16	6·07	5·99	5·91	5·82	5·74	5·65
8	11·26	8·65	7·59	7·01	6·63	6·37	6·18	6·03	5·91	5·81	5·67	5·52	5·36	5·28	5·20	5·12	5·03	4·95	4·86
9	10·56	8·02	6·99	6·42	6·06	5·80	5·61	5·47	5·35	5·26	5·11	4·96	4·81	4·73	4·65	4·57	4·48	4·40	4·31
10	10·04	7·56	6·55	5·99	5·64	5·39	5·20	5·06	4·94	4·85	4·71	4·56	4·41	4·33	4·25	4·17	4·08	4·00	3·91
11	9·65	7·21	6·22	5·67	5·32	5·07	4·89	4·74	4·63	4·54	4·40	4·25	4·10	4·02	3·94	3·86	3·78	3·69	3·60
12	9·33	6·93	5·95	5·41	5·06	4·82	4·64	4·50	4·39	4·30	4·16	4·01	3·86	3·78	3·70	3·62	3·54	3·45	3·36
13	9·07	6·70	5·74	5·21	4·86	4·62	4·44	4·30	4·19	4·10	3·96	3·82	3·66	3·59	3·51	3·43	3·34	3·25	3·17
14	8·86	6·51	5·56	5·04	4·69	4·46	4·28	4·14	4·03	3·94	3·80	3·66	3·51	3·43	3·35	3·27	3·18	3·09	3·00
15	8·68	6·36	5·42	4·89	4·56	4·32	4·14	4·00	3·89	3·80	3·67	3·52	3·37	3·29	3·21	3·13	3·05	2·96	2·87
16	8·53	6·23	5·29	4·77	4·44	4·20	4·03	3·89	3·78	3·69	3·55	3·41	3·26	3·18	3·10	3·02	2·93	2·84	2·75
17	8·40	6·11	5·18	4·67	4·34	4·10	3·93	3·79	3·68	3·59	3·46	3·31	3·16	3·08	3·00	2·92	2·83	2·75	2·65
18	8·29	6·01	5·09	4·58	4·25	4·01	3·84	3·71	3·60	3·51	3·37	3·23	3·08	3·00	2·92	2·84	2·75	2·66	2·57
19	8·18	5·93	5·01	4·50	4·17	3·94	3·77	3·63	3·52	3·43	3·30	3·15	3·00	2·92	2·84	2·76	2·67	2·58	2·49
20	8·10	5·85	4·94	4·43	4·10	3·87	3·70	3·56	3·46	3·37	3·23	3·09	2·94	2·86	2·78	2·69	2·61	2·52	2·42
21	8·02	5·78	4·87	4·37	4·04	3·81	3·64	3·51	3·40	3·31	3·17	3·03	2·88	2·80	2·72	2·64	2·55	2·46	2·36
22	7·95	5·72	4·82	4·31	3·99	3·76	3·59	3·45	3·35	3·26	3·12	2·98	2·83	2·75	2·67	2·58	2·50	2·40	2·31
23	7·88	5·66	4·76	4·26	3·94	3·71	3·54	3·41	3·30	3·21	3·07	2·93	2·78	2·70	2·62	2·54	2·45	2·35	2·26
24	7·82	5·61	4·72	4·22	3·90	3·67	3·50	3·36	3·26	3·17	3·03	2·89	2·74	2·66	2·58	2·49	2·40	2·31	2·21
25	7·77	5·57	4·68	4·18	3·85	3·63	3·46	3·32	3·22	3·13	2·99	2·85	2·70	2·62	2·54	2·45	2·36	2·27	2·17
26	7·72	5·53	4·64	4·14	3·82	3·59	3·42	3·29	3·18	3·09	2·96	2·81	2·66	2·58	2·50	2·42	2·33	2·23	2·13
27	7·68	5·49	4·60	4·11	3·78	3·56	3·39	3·26	3·15	3·06	2·93	2·78	2·63	2·55	2·47	2·38	2·29	2·20	2·10
28	7·64	5·45	4·57	4·07	3·75	3·53	3·36	3·23	3·12	3·03	2·90	2·75	2·60	2·52	2·44	2·35	2·26	2·17	2·06
29	7·60	5·42	4·54	4·04	3·73	3·50	3·33	3·20	3·09	3·00	2·87	2·73	2·57	2·49	2·41	2·33	2·23	2·14	2·03
30	7·56	5·39	4·51	4·02	3·70	3·47	3·30	3·17	3·07	2·98	2·84	2·70	2·55	2·47	2·39	2·30	2·21	2·11	2·01
40	7·31	5·18	4·31	3·83	3·51	3·29	3·12	2·99	2·89	2·80	2·66	2·52	2·37	2·29	2·20	2·11	2·02	1·92	1·80
60	7·08	4·98	4·13	3·65	3·34	3·12	2·95	2·82	2·72	2·63	2·50	2·35	2·20	2·12	2·03	1·94	1·84	1·73	1·60
120	6·85	4·79	3·95	3·48	3·17	2·96	2·79	2·66	2·56	2·47	2·34	2·19	2·03	1·95	1·86	1·76	1·66	1·53	1·38
∞	6·63	4·61	3·78	3·32	3·02	2·80	2·64	2·51	2·41	2·32	2·18	2·04	1·88	1·79	1·70	1·59	1·47	1·32	1·00

(e)

*The entries in this table are values of the F variate having v_1 numerator and v_2 denominator degrees of freedom corresponding to the designated cumulative probability.

620

Table A7(f) F Cumulative Probabilities: 0.995 (Upper Tail: 0.005)

v_2 \ v_1	1	2	3	4	5	6	7	8	9	10	12	15	20	24	30	40	60	120	∞
1	16211	20000	21615	22500	23056	23437	23715	23925	24091	24224	24426	24630	24836	24940	25044	25148	25253	25359	25465
2	198·5	199·0	199·2	199·2	199·3	199·3	199·4	199·4	199·4	199·4	199·4	199·4	199·4	199·5	199·5	199·5	199·5	199·5	199·5
3	55·55	49·80	47·47	46·19	45·39	44·84	44·43	44·13	43·88	43·69	43·39	43·08	42·78	42·62	42·47	42·31	42·15	41·99	41·83
4	31·33	26·28	24·26	23·15	22·46	21·97	21·62	21·35	21·14	20·97	20·70	20·44	20·17	20·03	19·89	19·75	19·61	19·47	19·32
5	22·78	18·31	16·53	15·56	14·94	14·51	14·20	13·96	13·77	13·62	13·38	13·15	12·90	12·78	12·66	12·53	12·40	12·27	12·14
6	18·63	14·54	12·92	12·03	11·46	11·07	10·79	10·57	10·39	10·25	10·03	9·81	9·59	9·47	9·36	9·24	9·12	9·00	8·88
7	16·24	12·40	10·88	10·05	9·52	9·16	8·89	8·68	8·51	8·38	8·18	7·97	7·75	7·65	7·53	7·42	7·31	7·19	7·08
8	14·69	11·04	9·60	8·81	8·30	7·95	7·69	7·50	7·34	7·21	7·01	6·81	6·61	6·50	6·40	6·29	6·18	6·06	5·95
9	13·61	10·11	8·72	7·96	7·47	7·13	6·88	6·69	6·54	6·42	6·23	6·03	5·83	5·73	5·62	5·52	5·41	5·30	5·19
10	12·83	9·43	8·08	7·34	6·87	6·54	6·30	6·12	5·97	5·85	5·66	5·47	5·27	5·17	5·07	4·97	4·86	4·75	4·64
11	12·23	8·91	7·60	6·88	6·42	6·10	5·86	5·68	5·54	5·42	5·24	5·05	4·86	4·76	4·65	4·55	4·44	4·34	4·23
12	11·75	8·51	7·23	6·52	6·07	5·76	5·52	5·35	5·20	5·09	4·91	4·72	4·53	4·43	4·33	4·23	4·12	4·01	3·90
13	11·37	8·19	6·93	6·23	5·79	5·48	5·25	5·08	4·94	4·82	4·64	4·46	4·27	4·17	4·07	3·97	3·87	3·76	3·65
14	11·06	7·92	6·68	6·00	5·56	5·26	5·03	4·86	4·72	4·60	4·43	4·25	4·06	3·96	3·86	3·76	3·66	3·55	3·44
15	10·80	7·70	6·48	5·80	5·37	5·07	4·85	4·67	4·54	4·42	4·25	4·07	3·88	3·79	3·69	3·58	3·48	3·37	3·26
16	10·58	7·51	6·30	5·64	5·21	4·91	4·69	4·52	4·38	4·27	4·10	3·92	3·73	3·64	3·54	3·44	3·33	3·22	3·11
17	10·38	7·35	6·16	5·50	5·07	4·78	4·56	4·39	4·25	4·14	3·97	3·79	3·61	3·51	3·41	3·31	3·21	3·10	2·98
18	10·22	7·21	6·03	5·37	4·96	4·66	4·44	4·28	4·14	4·03	3·86	3·68	3·50	3·40	3·30	3·20	3·10	2·99	2·87
19	10·07	7·09	5·92	5·27	4·85	4·56	4·34	4·18	4·04	3·93	3·76	3·59	3·40	3·31	3·21	3·11	3·00	2·89	2·78
20	9·94	6·99	5·82	5·17	4·76	4·47	4·26	4·09	3·96	3·85	3·68	3·50	3·32	3·22	3·12	3·02	2·92	2·81	2·69
21	9·83	6·89	5·73	5·09	4·68	4·39	4·18	4·01	3·88	3·77	3·60	3·43	3·24	3·15	3·05	2·95	2·84	2·73	2·61
22	9·73	6·81	5·65	5·02	4·61	4·32	4·11	3·94	3·81	3·70	3·54	3·36	3·18	3·08	2·98	2·88	2·77	2·66	2·55
23	9·63	6·73	5·58	4·95	4·54	4·26	4·05	3·88	3·75	3·64	3·47	3·30	3·12	3·02	2·92	2·82	2·71	2·60	2·48
24	9·55	6·66	5·52	4·89	4·49	4·20	3·99	3·83	3·69	3·59	3·42	3·25	3·06	2·97	2·87	2·77	2·66	2·55	2·43
25	9·48	6·60	5·46	4·84	4·43	4·15	3·94	3·78	3·64	3·54	3·37	3·20	3·01	2·92	2·82	2·72	2·61	2·50	2·38
26	9·41	6·54	5·41	4·79	4·38	4·10	3·89	3·73	3·60	3·49	3·33	3·15	2·97	2·87	2·77	2·67	2·56	2·45	2·33
27	9·34	6·49	5·36	4·74	4·34	4·06	3·85	3·69	3·56	3·45	3·28	3·11	2·93	2·83	2·73	2·63	2·52	2·41	2·29
28	9·28	6·44	5·32	4·70	4·30	4·02	3·81	3·65	3·52	3·41	3·25	3·07	2·89	2·79	2·69	2·59	2·48	2·37	2·25
29	9·23	6·40	5·28	4·66	4·26	3·98	3·77	3·61	3·48	3·38	3·21	3·04	2·86	2·76	2·66	2·56	2·45	2·33	2·21
30	9·18	6·35	5·24	4·62	4·23	3·95	3·74	3·58	3·45	3·34	3·18	3·01	2·82	2·73	2·63	2·52	2·42	2·30	2·18
40	8·83	6·07	4·98	4·37	3·99	3·71	3·51	3·35	3·22	3·12	2·95	2·78	2·60	2·50	2·40	2·30	2·18	2·06	1·93
60	8·49	5·79	4·73	4·14	3·76	3·49	3·29	3·13	3·01	2·90	2·74	2·57	2·39	2·29	2·19	2·08	1·96	1·83	1·69
120	8·18	5·54	4·50	3·92	3·55	3·28	3·09	2·93	2·81	2·71	2·54	2·37	2·19	2·09	1·98	1·87	1·75	1·61	1·43
∞	7·88	5·30	4·28	3·72	3·35	3·09	2·90	2·74	2·62	2·52	2·36	2·19	2·00	1·90	1·79	1·67	1·53	1·36	1·00

(f)

*The entries in this table are values of the F variate having v_1 numerator and v_2 denominator degrees of freedom corresponding to the designated cumulative probability.

Source: E. S. Pearson and H. O. Hartley, eds., *Biometrika Table for Statisticians*, Volume I, (1966). Copyright Biometrika Trustees. Reprinted with permission.

621

Table A8 Factors for Determining One-sided Tolerance Limits

| | γ = 0.90 | | | γ = 0.95 | | | γ = 0.99 | | |
| | p | | | p | | | p | | |
n	0.90	0.95	0.99	0.90	0.95	0.99	0.90	0.95	0.99
2	10.253	13.090	18.500	20.581	26.260	37.094	103.029	131.426	185.617
3	4.258	5.311	7.340	6.155	7.656	10.553	13.995	17.370	23.896
4	3.188	3.957	5.438	4.162	5.144	7.042	7.380	9.083	12.387
5	2.742	3.400	4.666	3.407	4.203	5.741	5.362	6.578	8.939
10	2.066	2.568	3.532	2.355	2.911	3.981	3.048	3.738	5.074
15	1.867	2.329	3.212	2.068	2.566	3.520	2.521	3.102	4.222
20	1.765	2.208	3.052	1.926	2.396	3.295	2.276	2.808	3.832
25	1.702	2.132	2.952	1.838	2.292	3.158	2.129	2.633	3.601
30	1.657	2.080	2.884	1.777	2.220	3.064	2.030	2.515	3.447
35	1.624	2.041	2.833	1.732	2.167	2.995	1.957	2.430	3.334
40	1.598	2.010	2.793	1.697	2.125	2.941	1.902	2.364	3.249
45	1.577	1.986	2.761	1.669	2.092	2.898	1.857	2.312	3.180
50	1.559	1.965	2.735	1.646	2.065	2.862	1.821	2.269	3.125
60	1.532	1.933	2.694	1.609	2.022	2.807	1.764	2.202	3.038
70	1.511	1.909	2.662	1.581	1.990	2.765	1.722	2.153	2.974
80	1.495	1.890	2.638	1.559	1.964	2.733	1.688	2.114	2.924
90	1.481	1.874	2.618	1.542	1.944	2.706	1.661	2.082	2.883
100	1.470	1.861	2.601	1.527	1.927	2.684	1.639	2.056	2.850
150	1.433	1.818	2.546	1.478	1.870	2.611	1.566	1.971	2.740
200	1.411	1.793	2.514	1.450	1.837	2.570	1.524	1.923	2.679
250	1.397	1.777	2.493	1.431	1.815	2.542	1.496	1.891	2.638
300	1.386	1.765	2.477	1.417	1.800	2.522	1.475	1.868	2.608
350	1.378	1.755	2.466	1.406	1.787	2.506	1.461	1.850	2.585
400	1.372	1.748	2.456	1.398	1.778	2.494	1.448	1.836	2.567
450	1.366	1.742	2.448	1.391	1.770	2.484	1.438	1.824	2.553
500	1.362	1.736	2.442	1.385	1.763	2.475	1.430	1.814	2.540
550	1.358	1.732	2.436	1.380	1.757	2.468	1.422	1.806	2.530
600	1.355	1.728	2.431	1.376	1.752	2.462	1.416	1.799	2.520
650	1.352	1.725	2.427	1.372	1.748	2.456	1.411	1.792	2.512
700	1.349	1.722	2.423	1.368	1.744	2.451	1.406	1.787	2.505
750	1.347	1.719	2.420	1.365	1.741	2.447	1.401	1.782	2.499
800	1.344	1.717	2.417	1.363	1.737	2.443	1.397	1.777	2.493
850	1.343	1.714	2.414	1.360	1.734	2.439	1.394	1.773	2.488
900	1.341	1.712	2.411	1.358	1.732	2.436	1.390	1.769	2.483
950	1.339	1.711	2.409	1.356	1.729	2.433	1.387	1.766	2.479
1000	1.338	1.709	2.407	1.354	1.727	2.430	1.385	1.762	2.475
∞	1.282	1.645	2.326	1.282	1.645	2.326	1.282	1.645	2.326

Source: Adapted and reprinted from Odeh, R. E. and Owen, D. B. (1980). *Tables for Normal Tolerance Limits, Sampling Plans, and Screening*. New York: Marcel Dekker, Inc., pp. 22–25 and 30–37, by courtesy of Marcel Dekker, Inc.

Table A9 Factors for Determining Two-sided Tolerance Limits

| | $\gamma = 0.90$ | | | $\gamma = 0.95$ | | | $\gamma = 0.99$ | | |
| | p | | | p | | | p | | |
n	0.900	0.950	0.990	0.900	0.950	0.990	0.900	0.950	0.990
2	15.512	18.221	23.423	31.092	36.519	46.944	155.569	182.720	234.877
3	5.788	6.823	8.819	8.306	9.789	12.647	18.782	22.131	28.586
4	4.157	4.913	6.372	5.368	6.341	8.221	9.416	11.118	14.405
5	3.499	4.142	5.387	4.291	5.077	6.598	6.655	7.870	10.220
10	2.546	3.026	3.958	2.856	3.393	4.437	3.617	4.294	5.610
15	2.285	2.720	3.565	2.492	2.965	3.885	2.967	3.529	4.621
20	2.158	2.570	3.372	2.319	2.760	3.621	2.675	3.184	4.175
25	2.081	2.479	3.254	2.215	2.638	3.462	2.506	2.984	3.915
30	2.029	2.417	3.173	2.145	2.555	3.355	2.394	2.851	3.742
35	1.991	2.371	3.114	2.094	2.495	3.276	2.314	2.756	3.618
40	1.961	2.336	3.069	2.055	2.448	3.216	2.253	2.684	3.524
45	1.938	2.308	3.032	2.024	2.412	3.168	2.205	2.627	3.450
50	1.918	2.285	3.003	1.999	2.382	3.129	2.166	2.580	3.390
60	1.888	2.250	2.956	1.960	2.335	3.068	2.106	2.509	3.297
70	1.866	2.224	2.922	1.931	2.300	3.023	2.062	2.457	3.228
80	1.849	2.203	2.895	1.908	2.274	2.988	2.028	2.416	3.175
90	1.835	2.186	2.873	1.890	2.252	2.959	2.001	2.384	3.133
100	1.823	2.172	2.855	1.875	2.234	2.936	1.978	2.357	3.098
150	1.786	2.128	2.796	1.826	2.176	2.859	1.905	2.271	2.985
200	1.764	2.102	2.763	1.798	2.143	2.816	1.866	2.223	2.921
250	1.750	2.085	2.741	1.780	2.121	2.788	1.839	2.191	2.880
300	1.740	2.073	2.725	1.767	2.106	2.767	1.820	2.169	2.850
350	1.732	2.064	2.713	1.757	2.094	2.752	1.806	2.152	2.828
400	1.726	2.057	2.703	1.749	2.084	2.739	1.794	2.138	2.810
450	1.721	2.051	2.695	1.743	2.077	2.729	1.785	2.127	2.795
500	1.717	2.046	2.689	1.737	2.070	2.721	1.777	2.117	2.783
550	1.713	2.041	2.683	1.733	2.065	2.713	1.770	2.109	2.772
600	1.710	2.038	2.678	1.729	2.060	2.707	1.765	2.103	2.763
650	1.707	2.034	2.674	1.725	2.056	2.702	1.759	2.097	2.755
700	1.705	2.032	2.670	1.722	2.052	2.697	1.755	2.091	2.748
750	1.703	2.029	2.667	1.719	2.049	2.692	1.751	2.086	2.742
800	1.701	2.027	2.664	1.717	2.046	2.688	1.747	2.082	2.736
850	1.699	2.025	2.661	1.715	2.043	2.685	1.744	2.078	2.731
900	1.697	2.023	2.658	1.712	2.040	2.682	1.741	2.075	2.727
950	1.696	2.021	2.656	1.711	2.038	2.679	1.738	2.071	2.722
1000	1.695	2.019	2.654	1.709	2.036	2.676	1.736	2.068	2.718
∞	1.645	1.960	2.576	1.645	1.960	2.576	1.645	1.960	2.576

Source: Adapted and reprinted from Odeh, R. E. and Owen, D. B. (1980). *Tables for Normal Tolerance Limits, Sampling Plans, and Screening*. New York: Marcel Dekker, Inc., pp. 90–93 and 98–105, by courtesy of Marcel Dekker, Inc.

Table A10(a) Operating characteristic curve for test on mean of a normal population, known standard deviation. Two-sided normal test for a level of significance $\alpha = 0.10$

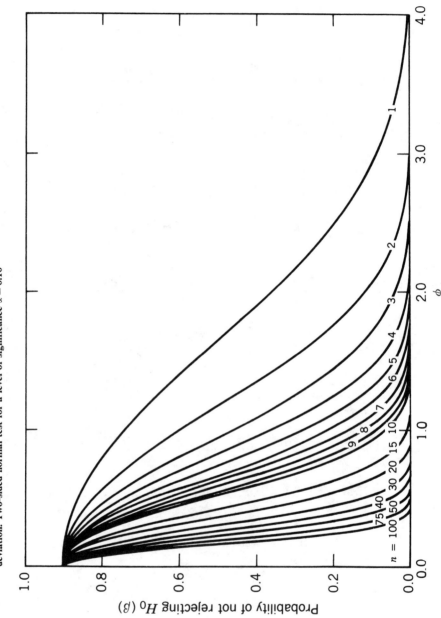

Table A10(b) Operating characteristic curve for test on mean of a normal population, known standard deviation. Two-sided normal test for a level of significance $\alpha = 0.05$

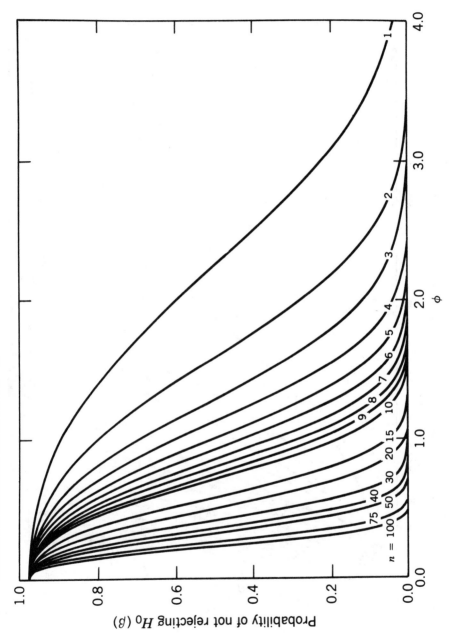

Table A10(c) Operating characteristic curve for test on mean of a normal population, known standard deviation. Two-sided normal test for a level of significance $\alpha = 0.025$

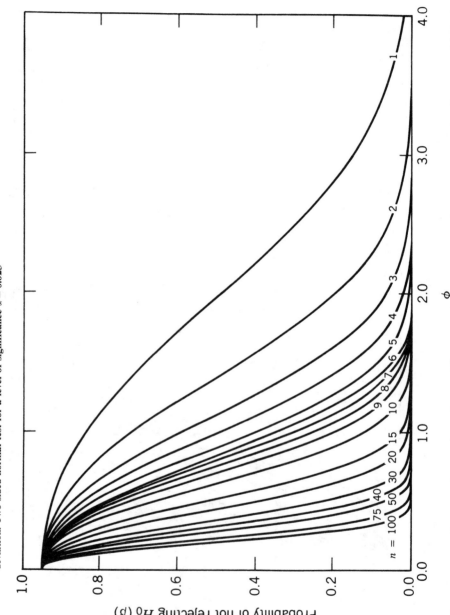

Table A10(d) Operating characteristic curve for test on mean of a normal population, known standard deviation. Two-sided normal test for a level of significance $\alpha = 0.01$

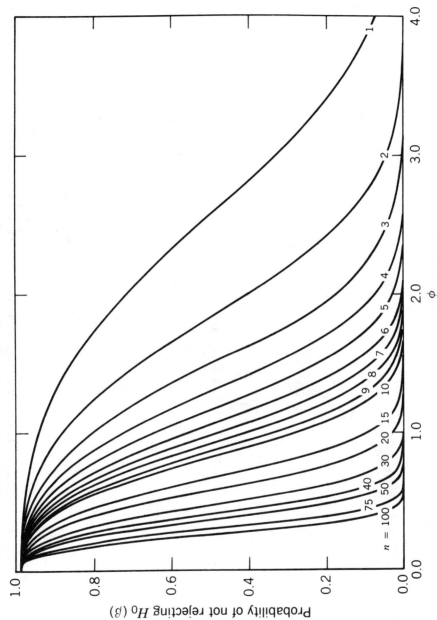

Probability of not rejecting H_0 (β)

ϕ

Table A10(e) Operating characteristic curve for test on mean of a normal population, known standard deviation. One-sided normal test for a level of significance $\alpha = 0.10$.

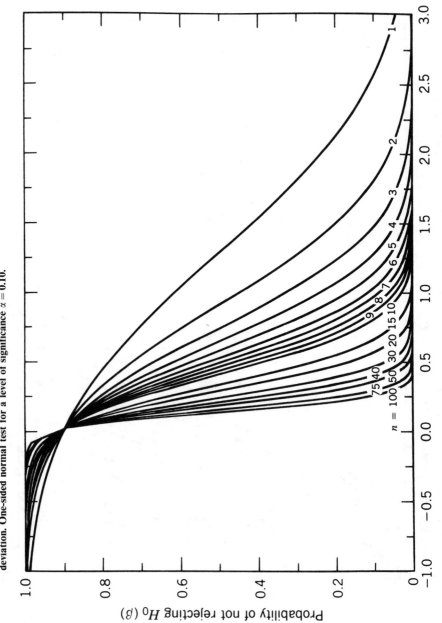

Table A10(f) Operating characteristic curve for test on mean of a normal population, known standard deviation. One-sided normal test for a level of significance $\alpha = 0.05$.

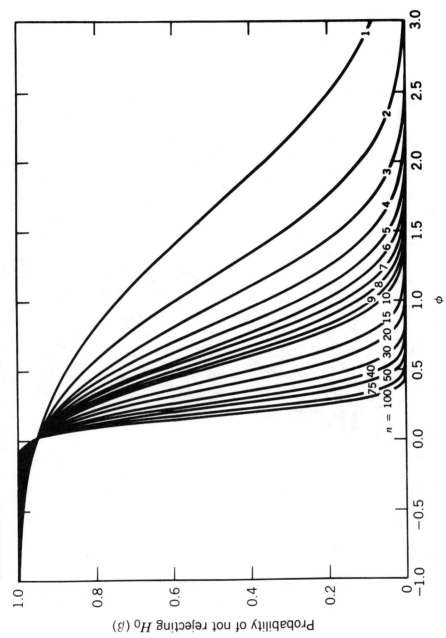

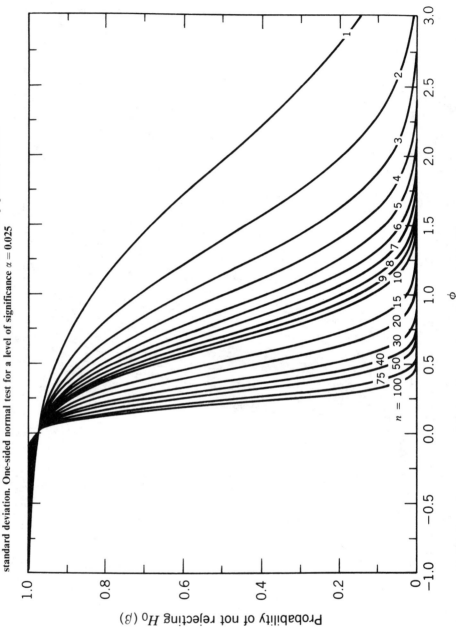

Table A10(g) Operating characteristic curve for test on mean of a normal population, known standard deviation. One-sided normal test for a level of significance $\alpha = 0.025$

Table A10(h) Operating characteristic curve for test on mean of a normal population, known standard deviation. One-sided normal test for a level of significance $\alpha = 0.01$

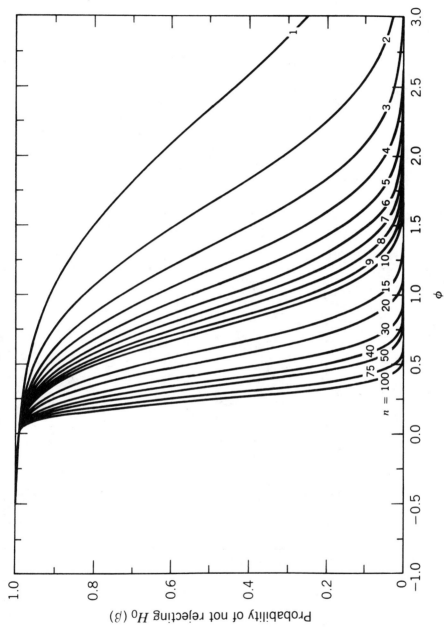

Table A11 Sample-Size Requirements for Tests on the Mean of a Normal Distribution, Unknown Standard Deviation

	Level of t Test																
Single-Sided Test	α = 0.005				α = 0.01				α = 0.025				α = 0.05				
Double-Sided Test	α = 0.01				α = 0.02				α = 0.05				α = 0.1				
β =	0.01	0.05	0.1	0.2	0.01	0.05	0.1	0.2	0.01	0.05	0.1	0.2	0.01	0.05	0.1	0.2	
0.05																	0.05
0.10																	0.10
0.15																	0.15
0.20																	0.20
0.25												128			139	101	0.25
0.30				134				115			119	90		122	97	71	0.30
0.35			125	99			109	85		109	88	67		90	72	52	0.35
0.40		115	97	77		101	85	66	117	84	68	51	101	70	55	40	0.40
0.45		92	77	62	110	81	68	53	93	67	54	41	80	55	44	33	0.45
0.50	100	75	63	51	90	66	55	43	76	54	44	34	65	45	36	27	0.50
0.55	83	63	53	42	75	55	46	36	63	45	37	28	54	38	30	22	0.55
0.60	71	53	45	36	63	47	39	31	53	38	32	24	46	32	26	19	0.60
0.65	61	46	39	31	55	41	34	27	46	33	27	21	39	28	22	17	0.65
0.70	53	40	34	28	47	35	30	24	40	29	24	19	34	24	19	15	0.70
0.75	47	36	30	25	42	31	27	21	35	26	21	16	30	21	17	13	0.75
0.80	41	32	27	22	37	28	24	19	31	22	19	15	27	19	15	12	0.80
0.85	37	29	24	20	33	25	21	17	28	21	17	13	24	17	14	11	0.85
0.90	34	26	22	18	29	23	19	16	25	19	16	12	21	15	13	10	0.90
0.95	31	24	20	17	27	21	18	14	23	17	14	11	19	14	11	9	0.95
1.00	28	22	19	16	25	19	16	13	21	16	13	10	18	13	11	8	1.00
1.1	24	19	16	14	21	16	14	12	18	13	11	9	15	11	9	7	1.1
1.2	21	16	14	12	18	14	12	10	15	12	10	8	13	10	8	6	1.2
1.3	18	15	13	11	16	13	11	9	14	10	9	7	11	8	7	6	1.3
1.4	16	13	12	10	14	11	10	9	12	9	8	7	10	8	7	5	1.4
1.5	15	12	11	9	13	10	9	8	11	8	7	6	9	7	6		1.5
1.6	13	11	10	8	12	10	9	7	10	8	7	6	8	6	6		1.6
1.7	12	10	9	8	11	9	8	7	9	7	6	5	8	6	5		1.7
1.8	12	10	9	8	10	8	7	7	8	7	6		7	6			1.8
1.9	11	9	8	7	10	8	7	6	8	6	6		7	5			1.9
2.0	10	8	8	7	9	7	7	6	7	6	5		6				2.0
2.1	10	8	7	7	8	7	6	6	7	6			6				2.1
2.2	9	8	7	6	8	7	6	5	7	6			6				2.2
2.3	9	7	7	6	8	6	6		6	5			5				2.3
2.4	8	7	7	6	7	6	6		6								2.4
2.5	8	7	6	6	7	6	6		6								2.5
3.0	7	6	6	5	6	5	5		5								3.0
3.5	6	5	5		5												3.5
4.0	6																4.0

Value of $\phi = \dfrac{|\delta|}{\sigma}$

Source: Reproduced from Table E of Owen L. Davies, *The Design and Analysis of Industrial Experiments*, 2nd ed., Longman House, Essex, 1956. By permission of the author and publishers.

Note: The entries in this table show the number of observations needed in a t test of the significance of a mean in order to control the probabilities of errors of the first and second kinds at α and β respectively.

Table A12 Sample-Size Requirements for Tests on the Difference of Two Means from Independent Normal Populations, Equal Population Standard Deviations

	Level of t Test																
Single-Sided Test Double-Sided Test	α = 0.005 α = 0.01				α = 0.01 α = 0.02				α = 0.025 α = 0.05				α = 0.05 α = 0.01				
β =	0.01	0.05	0.1	0.2	0.01	0.05	0.1	0.2	0.01	0.05	0.1	0.2	0.01	0.05	0.1	0.2	
0.05																	0.05
0.10																	0.10
0.15																	0.15
0.20																	0.20
0.25																	0.25
0.30																	0.30
0.35																102	0.35
0.40												100			108	78	0.40
0.45				118				101			105	79		108	86	62	0.45
0.50				96			106	82		106	86	64		88	70	51	0.50
0.55			101	79		106	88	68		87	71	53	112	73	58	42	0.55
0.60		101	85	67		90	74	58	104	74	60	45	89	61	49	36	0.60
0.65		87	73	57	104	77	64	49	88	63	51	39	76	52	42	30	0.65
0.70	100	75	63	50	90	66	55	43	76	55	44	34	66	45	36	26	0.70
0.75	88	66	55	44	79	58	48	38	67	48	39	29	57	40	32	23	0.75
0.80	77	58	49	39	70	51	43	33	59	42	34	26	50	35	28	21	0.80
0.85	69	51	43	35	62	46	38	30	52	37	31	23	45	31	25	18	0.85
0.90	62	46	39	31	55	41	34	27	47	34	27	21	40	28	22	16	0.90
0.95	55	42	35	28	50	37	31	24	42	30	25	19	36	25	20	15	0.95
1.00	50	38	32	26	45	33	28	22	38	27	23	17	33	23	18	14	1.00
1.1	42	32	27	22	38	28	23	19	32	23	19	14	27	19	15	12	1.1
1.2	36	27	23	18	32	24	20	16	27	20	16	12	23	16	13	10	1.2
1.3	31	23	20	16	28	21	17	14	23	17	14	11	20	14	11	9	1.3
1.4	27	20	17	14	24	18	15	12	20	15	12	10	17	12	10	8	1.4
1.5	24	18	15	13	21	16	14	11	18	13	11	9	15	11	9	7	1.5
1.6	21	16	14	11	19	14	12	10	16	12	10	8	14	10	8	6	1.6
1.7	19	15	13	10	17	13	11	9	14	11	9	7	12	9	7	6	1.7
1.8	17	13	11	10	15	12	10	8	13	10	8	6	11	8	7	5	1.8
1.9	16	12	11	9	14	11	9	8	12	9	7	6	10	7	6	5	1.9
2.0	14	11	10	8	13	10	9	7	11	8	7	6	9	7	6	4	2.0
2.1	13	10	9	8	12	9	8	7	10	8	6	5	8	6	5	4	2.1
2.2	12	10	8	7	11	9	7	6	9	7	6	5	8	6	5	4	2.2
2.3	11	9	8	7	10	8	7	6	9	7	6	5	7	5	5	4	2.3
2.4	11	9	8	6	10	8	7	6	8	6	5	4	7	5	4	4	2.4
2.5	10	8	7	6	9	7	6	5	8	6	5	4	6	5	4	3	2.5
3.0	8	6	6	5	7	6	5	4	6	5	4	4	5	4	3		3.0
3.5	6	5	5	4	6	5	4	4	5	4	4	3	4	3			3.5
4.0	6	5	4	4	5	4	4	3	4	4	3		4				4.0

Value of $\phi = \dfrac{|\delta|}{\sigma}$

Source: Reproduced from Table E.1 of Owen L. Davies, *The Design and Analysis of Industrial Experiments*, 2nd ed., Longman House, Essex, 1956. By permission of the author and publishers.

Note: The entries in this table show the number of observations needed (for each of two samples of equal size) in a test of the significance of the difference between two means in order to control the probabilities of the errors of the first and second kinds at α and β respectively.

Table A13(a)   Critical Values for the Analysis-of-Means Procedure: $\alpha = 0.10$

Number of Means, k

ν	3	4	5	6	7	8	9	10	11	12	13	14	15	16	17	18	19	20	ν
3	3.16																		3
4	2.81	3.10																	4
5	2.63	2.88	3.05																5
6	2.52	2.74	2.91	3.03															6
7	2.44	2.65	2.81	2.92	3.02														7
8	2.39	2.59	2.73	2.85	2.94	3.02													8
9	2.34	2.54	2.68	2.79	2.88	2.95	3.01												9
10	2.31	2.50	2.64	2.74	2.83	2.90	2.96	3.02											10
11	2.29	2.47	2.60	2.70	2.79	2.86	2.92	2.97	3.02										11
12	2.27	2.45	2.57	2.67	2.75	2.82	2.88	2.93	2.98	3.02									12
13	2.25	2.43	2.55	2.65	2.73	2.79	2.85	2.90	2.95	2.99	3.03								13
14	2.23	2.41	2.53	2.63	2.70	2.77	2.83	2.88	2.92	2.96	3.00	3.03							14
15	2.22	2.39	2.51	2.61	2.68	2.75	2.80	2.85	2.90	2.94	2.97	3.01	3.04						15
16	2.21	2.38	2.50	2.59	2.67	2.73	2.79	2.83	2.88	2.92	2.95	2.99	3.02	3.05					16
17	2.20	2.37	2.49	2.58	2.65	2.72	2.77	2.82	2.86	2.90	2.93	2.97	3.00	3.03	3.05				17
18	2.19	2.36	2.47	2.56	2.64	2.70	2.75	2.80	2.84	2.88	2.92	2.95	2.98	3.01	3.03	3.06			18
19	2.18	2.35	2.46	2.55	2.63	2.69	2.74	2.79	2.83	2.87	2.90	2.94	2.96	2.99	3.02	3.04	3.06		19
20	2.18	2.34	2.45	2.54	2.62	2.68	2.73	2.78	2.82	2.86	2.89	2.92	2.95	2.98	3.00	3.03	3.05	3.07	20
24	2.15	2.32	2.43	2.51	2.58	2.64	2.69	2.74	2.78	2.82	2.85	2.88	2.91	2.93	2.96	2.98	3.00	3.02	24
30	2.13	2.29	2.40	2.48	2.55	2.61	2.66	2.70	2.74	2.77	2.81	2.84	2.86	2.89	2.91	2.93	2.96	2.98	30
40	2.11	2.27	2.37	2.45	2.52	2.57	2.62	2.66	2.70	2.73	2.77	2.79	2.82	2.85	2.87	2.89	2.91	2.93	40
60	2.09	2.24	2.34	2.42	2.49	2.54	2.59	2.63	2.66	2.70	2.73	2.75	2.78	2.80	2.82	2.84	2.86	2.88	60
120	2.07	2.22	2.32	2.39	2.45	2.51	2.55	2.59	2.62	2.66	2.69	2.71	2.74	2.76	2.78	2.80	2.82	2.84	120
∞	2.05	2.19	2.29	2.36	2.42	2.47	2.52	2.55	2.59	2.62	2.65	2.67	2.69	2.72	2.74	2.76	2.77	2.79	∞

Table A13(b)   Critical Values for the Analysis-of-Means Procedure: $\alpha = 0.05$

Number of Means, k

ν	3	4	5	6	7	8	9	10	11	12	13	14	15	16	17	18	19	20
3	4.18																	
4	3.56	3.89																
5	3.25	3.53	3.72															
6	3.07	3.31	3.49	3.62														
7	2.94	3.17	3.33	3.45	3.56													
8	2.86	3.07	3.21	3.33	3.43	3.51												
9	2.79	2.99	3.13	3.24	3.33	3.41	3.48											
10	2.74	2.93	3.07	3.17	3.26	3.33	3.40	3.45										
11	2.70	2.88	3.01	3.12	3.20	3.27	3.33	3.39	3.44									
12	2.67	2.85	2.97	3.07	3.15	3.22	3.28	3.33	3.38	3.42								
13	2.64	2.81	2.94	3.03	3.11	3.18	3.24	3.29	3.34	3.38	3.42							
14	2.62	2.79	2.91	3.00	3.08	3.14	3.20	3.25	3.30	3.34	3.37	3.41						
15	2.60	2.76	2.88	2.97	3.05	3.11	3.17	3.22	3.26	3.30	3.34	3.37	3.40					
16	2.58	2.74	2.86	2.95	3.02	3.09	3.14	3.19	3.23	3.27	3.31	3.34	3.37	3.40				
17	2.57	2.73	2.84	2.93	3.00	3.06	3.12	3.16	3.21	3.25	3.28	3.31	3.34	3.37	3.40			
18	2.55	2.71	2.82	2.91	2.98	3.04	3.10	3.14	3.18	3.22	3.26	3.29	3.32	3.35	3.37	3.40		
19	2.54	2.70	2.81	2.89	2.96	3.02	3.08	3.12	3.16	3.20	3.24	3.27	3.30	3.32	3.35	3.37	3.40	
20	2.53	2.68	2.79	2.88	2.95	3.01	3.06	3.11	3.15	3.18	3.22	3.25	3.28	3.30	3.33	3.35	3.37	3.40
24	2.50	2.65	2.75	2.83	2.90	2.96	3.01	3.05	3.09	3.13	3.16	3.19	3.22	3.24	3.27	3.29	3.31	3.33
30	2.47	2.61	2.71	2.79	2.85	2.91	2.96	3.00	3.04	3.07	3.10	3.13	3.16	3.18	3.20	3.22	3.25	3.27
40	2.43	2.57	2.67	2.75	2.81	2.86	2.91	2.95	2.98	3.01	3.04	3.07	3.10	3.12	3.14	3.16	3.18	3.20
60	2.40	2.54	2.63	2.70	2.76	2.81	2.86	2.90	2.93	2.96	2.99	3.02	3.04	3.06	3.08	3.10	3.12	3.14
120	2.37	2.50	2.59	2.66	2.72	2.77	2.81	2.84	2.88	2.91	2.93	2.96	2.98	3.00	3.02	3.04	3.06	3.08
∞	2.34	2.47	2.56	2.62	2.68	2.72	2.76	2.80	2.83	2.86	2.88	2.90	2.93	2.95	2.97	2.98	3.00	3.02

Table A13(c) Critical Values for the Analysis-of-Means Procedure: $\alpha = 0.01$

Number of Means, k

ν	3	4	5	6	7	8	9	10	11	12	13	14	15	16	17	18	19	20	ν
3	7.51																		3
4	5.74	6.21																	4
5	4.93	5.29	5.55																5
6	4.48	4.77	4.98	5.16															6
7	4.18	4.44	4.63	4.78	4.90														7
8	3.98	4.21	4.38	4.52	4.63	4.72													8
9	3.84	4.05	4.20	4.33	4.43	4.51	4.59												9
10	3.73	3.92	4.07	4.18	4.28	4.36	4.43	4.49											10
11	3.64	3.82	3.96	4.07	4.16	4.23	4.30	4.36	4.41										11
12	3.57	3.74	3.87	3.98	4.06	4.13	4.20	4.25	4.31	4.35									12
13	3.51	3.68	3.80	3.90	3.98	4.05	4.11	4.17	4.22	4.26	4.30								13
14	3.46	3.63	3.74	3.84	3.92	3.98	4.04	4.09	4.14	4.18	4.22	4.26							14
15	3.42	3.58	3.69	3.79	3.86	3.92	3.98	4.03	4.08	4.12	4.16	4.19	4.22						15
16	3.38	3.54	3.65	3.74	3.81	3.87	3.93	3.98	4.02	4.06	4.10	4.14	4.17	4.20					16
17	3.35	3.50	3.61	3.70	3.77	3.83	3.89	3.93	3.98	4.02	4.05	4.09	4.12	4.14	4.17				17
18	3.33	3.47	3.58	3.66	3.73	3.79	3.85	3.89	3.94	3.97	4.01	4.04	4.07	4.10	4.12	4.15			18
19	3.30	3.45	3.55	3.63	3.70	3.76	3.81	3.86	3.90	3.94	3.97	4.00	4.03	4.06	4.08	4.11	4.13		19
20	3.28	3.42	3.53	3.61	3.67	3.73	3.78	3.83	3.87	3.90	3.94	3.97	4.00	4.02	4.05	4.07	4.09	4.12	20
24	3.21	3.35	3.45	3.52	3.58	3.64	3.69	3.73	3.77	3.80	3.83	3.86	3.89	3.91	3.94	3.96	3.98	4.00	24
30	3.15	3.28	3.37	3.44	3.50	3.55	3.59	3.63	3.67	3.70	3.73	3.76	3.78	3.81	3.83	3.85	3.87	3.89	30
40	3.09	3.21	3.29	3.36	3.42	3.46	3.50	3.54	3.58	3.60	3.63	3.66	3.68	3.70	3.72	3.74	3.76	3.78	40
60	3.03	3.14	3.22	3.29	3.34	3.38	3.42	3.46	3.49	3.51	3.54	3.56	3.59	3.61	3.63	3.64	3.66	3.68	60
120	2.97	3.07	3.15	3.21	3.26	3.30	3.34	3.37	3.40	3.42	3.45	3.47	3.49	3.51	3.53	3.55	3.56	3.58	120
∞	2.91	3.01	3.08	3.14	3.18	3.22	3.26	3.29	3.32	3.34	3.36	3.38	3.40	3.42	3.44	3.45	3.47	3.48	∞

Table A13(d) Critical Values for the Analysis-of-Means Procedure; $\alpha = 0.001$

Number of Means, k

ν	3	4	5	6	7	8	9	10	11	12	13	14	15	16	17	18	19	20	ν
3	16.4																		3
4	10.6	11.4																	4
5	8.25	8.79	9.19																5
6	7.04	7.45	7.76	8.00															6
7	6.31	6.65	6.89	7.09	7.25														7
8	5.83	6.12	6.32	6.49	6.63	6.75													8
9	5.49	5.74	5.92	6.07	6.20	6.30	6.40												9
10	5.24	5.46	5.63	5.76	5.87	5.97	6.05	6.13											10
11	5.05	5.25	5.40	5.52	5.63	5.71	5.79	5.86	5.92										11
12	4.89	5.08	5.22	5.33	5.43	5.51	5.58	5.65	5.71	5.76									12
13	4.77	4.95	5.08	5.18	5.27	5.35	5.42	5.48	5.53	5.58	5.63								13
14	4.66	4.83	4.96	5.06	5.14	5.21	5.28	5.33	5.38	5.43	5.48	5.51							14
15	4.57	4.74	4.86	4.95	5.03	5.10	5.16	5.21	5.26	5.31	5.35	5.39	5.42						15
16	4.50	4.66	4.77	4.86	4.94	5.00	5.06	5.11	5.16	5.20	5.24	5.28	5.31	5.34					16
17	4.44	4.59	4.70	4.78	4.86	4.92	4.98	5.03	5.07	5.11	5.15	5.18	5.22	5.25	5.28				17
18	4.38	4.53	4.63	4.72	4.79	4.85	4.90	4.95	4.99	5.03	5.07	5.10	5.14	5.16	5.19	5.22			18
19	4.33	4.47	4.58	4.66	4.73	4.79	4.84	4.88	4.93	4.96	5.00	5.03	5.06	5.09	5.12	5.14	5.17		19
20	4.29	4.42	4.53	4.61	4.67	4.73	4.78	4.83	4.87	4.90	4.94	4.97	5.00	5.03	5.05	5.08	5.10	5.12	20
24	4.16	4.28	4.37	4.45	4.51	4.56	4.61	4.65	4.69	4.72	4.75	4.78	4.81	4.83	4.86	4.88	4.90	4.92	24
30	4.03	4.14	4.23	4.30	4.35	4.40	4.44	4.48	4.51	4.54	4.57	4.60	4.62	4.64	4.67	4.69	4.71	4.72	30
40	3.91	4.01	4.09	4.15	4.20	4.25	4.29	4.32	4.35	4.38	4.40	4.43	4.45	4.47	4.49	4.50	4.52	4.54	40
60	3.80	3.89	3.96	4.02	4.06	4.10	4.14	4.17	4.19	4.22	4.24	4.27	4.29	4.30	4.32	4.33	4.35	4.37	60
120	3.69	3.77	3.84	3.89	3.93	3.96	4.00	4.03	4.05	4.07	4.09	4.11	4.13	4.15	4.16	4.17	4.19	4.21	120
∞	3.58	3.66	3.72	3.76	3.80	3.84	3.87	3.89	3.91	3.93	3.95	3.97	3.99	4.00	4.02	4.03	4.04	4.06	∞

Source: Nelson, L. (1983). "Exact critical values for use with the analysis of means", *Journal of Quality Technology*, **15**, 40–44. Copyright American Society for Quality Control, Inc., Milwaukee, WI. Reprinted by permission.

Table A14 Individual Terms of the Binomial Distribution*

						θ					
n	x	.05	.10	.15	.20	.25	.30	.35	.40	.45	.50
1	0	.950	.900	.850	.800	.750	.700	.650	.600	.550	.500
	1	.050	.100	.150	.200	.250	.300	.350	.400	.450	.5u0
2	0	.902	.810	.722	.640	.562	.490	.422	.360	.302	.250
	1	.095	.180	.255	.320	.375	.420	.455	.480	.495	.500
	2	.002	.010	.022	.040	.062	.090	.122	.160	.202	.250
3	0	.857	.729	.614	.512	.422	.343	.275	.216	.166	.125
	1	.135	.243	.325	.384	.422	.441	.444	.432	.408	.375
	2	.007	.027	.057	.096	.141	.189	.239	.288	.334	.375
	3	.000	.001	.003	.008	.016	.027	.043	.064	.091	.125
4	0	.814	.656	.522	.410	.316	.240	.178	.130	.091	.062
	1	.171	.292	.368	.410	.422	.412	.384	.346	.299	.250
	2	.013	.049	.097	.154	.211	.265	.310	.346	.367	.375
	3	.000	.004	.011	.026	.047	.076	.111	.154	.200	.250
	4	.000	.000	.000	.002	.004	.008	.015	.026	.041	.062
5	0	.774	.590	.444	.328	.237	.168	.116	.078	.050	.031
	1	.204	.328	.391	.410	.395	.360	.312	.259	.206	.156
	2	.021	.073	.138	.205	.264	.309	.336	.346	.337	.312
	3	.001	.008	.024	.051	.088	.132	.181	.230	.276	.312
	4	.000	.000	.002	.006	.015	.028	.049	.077	.113	.156
	5	.000	.000	.000	.000	.001	.002	.005	.010	.019	.031
6	0	.735	.531	.377	.262	.178	.118	.075	.047	.028	.016
	1	.232	.354	.399	.393	.356	.302	.244	.187	.136	.094
	2	.030	.098	.176	.246	.297	.324	.328	.311	.278	.234
	3	.002	.015	.041	.082	.132	.185	.235	.276	.303	.312
	4	.000	.001	.005	.015	.033	.059	.095	.138	.186	.234
	5	.000	.000	.000	.001	.004	.010	.020	.037	.061	.094
	6	.000	.000	.000	.000	.000	.001	.002	.004	.008	.016
7	0	.698	.478	.321	.210	.133	.082	.049	.028	.015	.008
	1	.257	.372	.396	.367	.311	.247	.185	.131	.087	.055
	2	.041	.124	.210	.275	.311	.318	.298	.261	.214	.164
	3	.004	.023	.062	.115	.173	.227	.268	.290	.292	.273
	4	.000	.003	.011	.029	.058	.097	.144	.193	.239	.273
	5	.000	.000	.001	.004	.011	.025	.047	.077	.117	.164
	6	.000	.000	.000	.000	.001	.004	.008	.017	.032	.055
	7	.000	.000	.000	.000	.000	.000	.001	.002	.004	.008
8	0	.663	.430	.272	.168	.100	.058	.032	.017	.008	.004
	1	.279	.383	.385	.335	.267	.198	.137	.090	.055	.031
	2	.051	.149	.238	.294	.311	.296	.259	.209	.157	.109
	3	.005	.033	.084	.147	.208	.254	.279	.279	.257	.219
	4	.000	.005	.018	.046	.086	.136	.187	.232	.263	.273
	5	.000	.000	.003	.009	.023	.047	.081	.124	.172	.219
	6	.000	.000	.000	.001	.004	.010	.022	.041	.070	.109
	7	.000	.000	.000	.000	.000	.001	.003	.008	.016	.031
	8	.000	.000	.000	.000	.000	.000	.000	.001	.002	.004

*The entries in this table show the probability that a binomial variate with parameter values n and θ equals the value x.

n	x	.05	.10	.15	.20	.25	.30	.35	.40	.45	.50
						θ					
9	0	.630	.387	.232	.134	.075	.040	.021	.010	.005	.002
	1	.298	.387	.368	.302	.225	.156	.100	.060	.034	.018
	2	.063	.172	.260	.302	.300	.267	.216	.161	.111	.070
	3	.008	.045	.107	.176	.234	.267	.272	.251	.212	.164
	4	.001	.007	.028	.066	.117	.171	.219	.251	.260	.246
	5	.000	.001	.005	.016	.039	.073	.118	.167	.213	.246
	6	.000	.000	.001	.003	.009	.021	.042	.074	.116	.164
	7	.000	.000	.000	.000	.001	.004	.010	.021	.041	.070
	8	.000	.000	.000	.000	.000	.000	.001	.003	.008	.018
	9	.000	.000	.000	.000	.000	.000	.000	.000	.001	.002
10	0	.599	.349	.197	.107	.056	.028	.013	.006	.002	.001
	1	.315	.387	.347	.268	.188	.121	.072	.040	.021	.010
	2	.075	.194	.276	.302	.282	.233	.176	.121	.076	.044
	3	.010	.057	.130	.201	.250	.267	.252	.215	.166	.117
	4	.001	.011	.040	.088	.146	.200	.238	.251	.238	.205
	5	.000	.001	.008	.026	.058	.103	.154	.201	.234	.246
	6	.000	.000	.001	.005	.016	.037	.069	.111	.160	.205
	7	.000	.000	.000	.001	.003	.009	.021	.042	.075	.117
	8	.000	.000	.000	.000	.000	.001	.004	.011	.023	.044
	9	.000	.000	.000	.000	.000	.000	.000	.002	.004	.010
11	0	.569	.314	.167	.086	.042	.020	.009	.004	.001	.000
	1	.329	.383	.325	.236	.155	.093	.052	.027	.012	.005
	2	.087	.213	.287	.295	.258	.200	.139	.089	.051	.027
	3	.014	.071	.152	.221	.258	.257	.225	.177	.126	.081
	4	.001	.016	.054	.118	.172	.220	.243	.236	.206	.161
	5	.000	.002	.013	.039	.080	.132	.183	.221	.236	.226
	6	.000	.000	.002	.010	.027	.057	.098	.147	.193	.226
	7	.000	.000	.000	.002	.006	.017	.038	.070	.113	.161
	8	.000	.000	.000	.000	.001	.004	.010	.023	.046	.081
	9	.000	.000	.000	.000	.000	.000	.002	.005	.013	.027
	10	.000	.000	.000	.000	.000	.000	.000	.001	.002	.005
12	0	.540	.282	.142	.069	.032	.014	.006	.002	.001	.000
	1	.341	.377	.301	.206	.127	.071	.037	.017	.007	.003
	2	.099	.230	.292	.283	.232	.168	.109	.064	.034	.016
	3	.017	.085	.172	.236	.258	.240	.195	.142	.092	.054
	4	.002	.021	.068	.133	.194	.231	.237	.213	.170	.121
	5	.000	.004	.019	.053	.103	.158	.204	.227	.222	.193
	6	.000	.000	.004	.015	.040	.079	.128	.177	.212	.226
	7	.000	.000	.001	.003	.011	.029	.059	.101	.149	.193
	8	.000	.000	.000	.000	.002	.008	.020	.042	.076	.121
	9	.000	.000	.000	.000	.000	.001	.005	.012	.028	.054
	10	.000	.000	.000	.000	.000	.000	.001	.002	.007	.016
	11	.000	.000	.000	.000	.000	.000	.000	.000	.001	.003

*The entries in this table show the probability that a binomial variate with parameter values n and θ equals the value x.

Table A14 (continued)   Individual Terms of the Binomial Distribution

n	x	.05	.10	.15	.20	.25	.30	.35	.40	.45	.50
13	0	.513	.254	.121	.055	.024	.010	.004	.001	.000	.000
	1	.351	.367	.277	.179	.103	.054	.026	.011	.004	.002
	2	.111	.245	.294	.268	.206	.139	.084	.045	.022	.009
	3	.021	.100	.190	.246	.252	.218	.165	.111	.066	.035
	4	.003	.028	.084	.153	.210	.234	.222	.184	.135	.087
	5	.000	.005	.027	.069	.126	.180	.215	.221	.199	.157
	6	.000	.001	.006	.023	.056	.103	.155	.197	.217	.209
	7	.000	.000	.001	.006	.019	.044	.083	.131	.177	.209
	8	.000	.000	.000	.001	.005	.014	.034	.066	.109	.157
	9	.000	.000	.000	.000	.001	.003	.010	.024	.049	.087
	10	.000	.000	.000	.000	.000	.001	.002	.006	.016	.035
	11	.000	.000	.000	.000	.000	.000	.000	.001	.004	.009
	12	.000	.000	.000	.000	.000	.000	.000	.000	.000	.002
14	0	.488	.229	.103	.044	.018	.007	.002	.001	.000	.000
	1	.359	.356	.254	.154	.083	.041	.018	.007	.003	.001
	2	.123	.257	.291	.250	.180	.113	.063	.032	.014	.006
	3	.026	.114	.206	.250	.240	.194	.137	.084	.046	.022
	4	.004	.035	.100	.172	.220	.229	.202	.155	.104	.061
	5	.000	.008	.035	.086	.147	.196	.218	.207	.170	.122
	6	.000	.001	.009	.032	.073	.126	.176	.207	.209	.183
	7	.000	.000	.002	.009	.028	.062	.108	.157	.195	.209
	8	.000	.000	.000	.002	.008	.023	.051	.092	.140	.183
	9	.000	.000	.000	.000	.002	.007	.018	.041	.076	.122
	10	.000	.000	.000	.000	.000	.001	.005	.014	.031	.061
	11	.000	.000	.000	.000	.000	.000	.001	.003	.009	.022
	12	.000	.000	.000	.000	.000	.000	.000	.000	.002	.006
	13	.000	.000	.000	.000	.000	.000	.000	.000	.000	.001
15	0	.463	.206	.087	.035	.013	.005	.002	.000	.000	.000
	1	.366	.343	.231	.132	.067	.030	.013	.005	.002	.000
	2	.135	.267	.286	.231	.156	.092	.048	.022	.009	.003
	3	.031	.128	.218	.250	.225	.170	.111	.063	.032	.014
	4	.005	.043	.116	.188	.225	.219	.179	.127	.078	.042
	5	.001	.010	.045	.103	.165	.206	.212	.186	.140	.092
	6	.000	.002	.013	.043	.092	.147	.191	.207	.191	.153
	7	.000	.000	.003	.014	.039	.081	.132	.177	.201	.196
	8	.000	.000	.000	.003	.013	.035	.071	.118	.165	.196
	9	.000	.000	.000	.001	.003	.012	.030	.061	.105	.153
	10	.000	.000	.000	.000	.001	.003	.010	.024	.051	.092
	11	.000	.000	.000	.000	.000	.001	.002	.007	.019	.042
	12	.000	.000	.000	.000	.000	.000	.000	.002	.005	.014
	13	.000	.000	.000	.000	.000	.000	.000	.000	.001	.003
	14	.000	.000	.000	.000	.000	.000	.000	.000	.000	.000

*The entries in this table show the probability that a binomial variate with parameter values n and θ equals the value x.

						θ					
n	x	.05	.10	.15	.20	.25	.30	.35	.40	.45	.50
16	0	.440	.185	.074	.028	.010	.003	.001	.000	.000	.000
	1	.371	.329	.210	.113	.053	.023	.009	.003	.001	.000
	2	.146	.274	.277	.211	.134	.073	.035	.015	.006	.002
	3	.036	.142	.228	.246	.208	.146	.089	.047	.021	.008
	4	.006	.051	.131	.200	.225	.204	.155	.101	.057	.028
	5	.001	.014	.055	.120	.180	.210	.201	.162	.112	.067
	6	.000	.003	.018	.055	.110	.165	.198	.198	.168	.122
	7	.000	.000	.004	.020	.052	.101	.152	.189	.197	.175
	8	.000	.000	.001	.005	.020	.049	.092	.142	.181	.196
	9	.000	.000	.000	.001	.006	.018	.044	.084	.132	.175
	10	.000	.000	.000	.000	.001	.006	.017	.039	.075	.122
	11	.000	.000	.000	.000	.000	.001	.005	.014	.034	.067
	12	.000	.000	.000	.000	.000	.000	.001	.004	.011	.028
	13	.000	.000	.000	.000	.000	.000	.000	.001	.003	.008
	14	.000	.000	.000	.000	.000	.000	.000	.000	.000	.002
17	0	.418	.167	.063	.022	.007	.002	.001	.000	.000	.000
	1	.374	.315	.189	.096	.043	.017	.006	.002	.000	.000
	2	.157	.280	.267	.191	.114	.058	.026	.010	.003	.001
	3	.041	.156	.236	.239	.189	.124	.070	.034	.014	.005
	4	.008	.060	.146	.209	.221	.187	.132	.080	.041	.018
	5	.001	.017	.067	.136	.191	.208	.185	.138	.087	.047
	6	.000	.004	.024	.068	.128	.178	.199	.184	.143	.094
	7	.000	.001	.006	.027	.067	.120	.168	.193	.184	.148
	8	.000	.000	.001	.008	.028	.064	.113	.161	.188	.185
	9	.000	.000	.000	.002	.009	.028	.061	.107	.154	.185
	10	.000	.000	.000	.000	.002	.009	.026	.057	.101	.148
	11	.000	.000	.000	.000	.000	.003	.009	.024	.052	.094
	12	.000	.000	.000	.000	.000	.001	.002	.008	.021	.047
	13	.000	.000	.000	.000	.000	.000	.000	.002	.007	.018
	14	.000	.000	.000	.000	.000	.000	.000	.000	.002	.005
	15	.000	.000	.000	.000	.000	.000	.000	.000	.000	.001
18	0	.397	.150	.053	.018	.006	.002	.000	.000	.000	.000
	1	.376	.300	.170	.081	.034	.013	.004	.001	.000	.000
	2	.168	.283	.256	.172	.096	.046	.019	.007	.002	.001
	3	.047	.168	.241	.230	.170	.105	.055	.025	.009	.003
	4	.009	.070	.159	.215	.213	.168	.110	.061	.029	.012
	5	.001	.022	.079	.151	.199	.202	.166	.115	.067	.033
	6	.000	.005	.030	.082	.144	.187	.194	.165	.118	.071
	7	.000	.001	.009	.035	.082	.138	.179	.189	.166	.121
	8	.000	.000	.002	.012	.038	.081	.133	.173	.186	.167
	9	.000	.000	.000	.003	.014	.039	.079	.128	.169	.185
	10	.000	.000	.000	.001	.004	.015	.038	.077	.125	.167
	11	.000	.000	.000	.000	.001	.005	.015	.037	.074	.121

*The entries in this table show the probability that a binomial variate with parameter values n and θ equals the value x.

Table A14 (continued)   Individual Terms of the Binomial Distribution

n	x	.05	.10	.15	.20	.25	.30	.35	.40	.45	.50
18	12	.000	.000	.000	.000	.000	.001	.005	.014	.035	.071
	13	.000	.000	.000	.000	.000	.000	.001	.004	.013	.033
	14	.000	.000	.000	.000	.000	.000	.000	.001	.004	.012
	15	.000	.000	.000	.000	.000	.000	.000	.000	.001	.003
	16	.000	.000	.000	.000	.000	.000	.000	.000	.000	.001
19	0	.377	.135	.046	.014	.004	.001	.000	.000	.000	.000
	1	.377	.285	.153	.068	.027	.009	.003	.001	.000	.000
	2	.179	.285	.243	.154	.080	.036	.014	.005	.001	.000
	3	.053	.180	.243	.218	.152	.087	.042	.017	.006	.002
	4	.011	.080	.171	.218	.202	.149	.091	.047	.020	.007
	5	.002	.027	.091	.164	.202	.192	.147	.093	.050	.022
	6	.000	.007	.037	.095	.157	.192	.184	.145	.095	.052
	7	.000	.001	.012	.044	.097	.152	.184	.180	.144	.096
	8	.000	.000	.003	.017	.049	.098	.149	.180	.177	.144
	9	.000	.000	.001	.005	.020	.051	.098	.146	.177	.176
	10	.000	.000	.000	.001	.007	.022	.053	.098	.145	.176
	11	.000	.000	.000	.000	.002	.008	.023	.053	.097	.144
	12	.000	.000	.000	.000	.000	.002	.008	.024	.053	.096
	13	.000	.000	.000	.000	.000	.000	.002	.008	.023	.052
	14	.000	.000	.000	.000	.000	.000	.001	.002	.008	.022
	15	.000	.000	.000	.000	.000	.000	.000	.000	.002	.007
	16	.000	.000	.000	.000	.000	.000	.000	.000	.000	.002
20	0	.358	.122	.039	.011	.003	.001	.000	.000	.000	.000
	1	.377	.270	.137	.058	.021	.007	.002	.000	.000	.000
	2	.189	.285	.229	.137	.067	.028	.010	.003	.001	.000
	3	.060	.190	.243	.205	.134	.072	.032	.012	.004	.001
	4	.013	.090	.182	.218	.190	.130	.074	.035	.014	.005
	5	.002	.032	.103	.175	.202	.179	.127	.075	.036	.015
	6	.000	.009	.045	.109	.169	.192	.171	.124	.075	.037
	7	.000	.002	.016	.054	.112	.164	.184	.166	.122	.074
	8	.000	.000	.005	.022	.061	.114	.161	.180	.162	.120
	9	.000	.000	.001	.007	.027	.065	.116	.160	.177	.160
	10	.000	.000	.000	.002	.010	.031	.069	.117	.159	.176
	11	.000	.000	.000	.000	.003	.012	.034	.071	.118	.160
	12	.000	.000	.000	.000	.000	.003	.014	.035	.073	.120
	13	.000	.000	.000	.000	.000	.001	.004	.015	.037	.074
	14	.000	.000	.000	.000	.000	.000	.001	.005	.015	.037
	15	.000	.000	.000	.000	.000	.000	.000	.001	.005	.015
	16	.000	.000	.000	.000	.000	.000	.000	.000	.001	.005

Source: Adapted and reprinted with permission from *CRC Standard Mathematical Tables*, 15th Ed., Edited by S. M. Selby, The Chemical Rubber Company. Copyright CRC Press, Inc., Boca Raton, FL.

*The entries in this table show the probability that a binomial variate with parameter values n and θ equals the value x.

Table A15   Lower-Tail Critical Values for the Wilcoxon Signed-Rank Test

Lower-Tail Probability

One-sided	Two-sided	$n = 5$	6	7	8	9	10	11	12	13	14	15	16
.05	.10	1	2	4	6	8	11	14	17	21	26	30	36
.025	.05		1	2	4	6	8	11	14	17	21	25	30
.01	.02			0	2	3	5	7	10	13	16	20	24
.005	.01				0	2	3	5	7	10	13	16	19

		$n = 17$	18	19	20	21	22	23	24	25	26	27	28
.05	.10	41	47	54	60	68	75	83	92	101	110	120	130
.025	.05	35	40	46	52	59	66	73	81	90	98	107	117
.01	.02	28	33	38	43	49	56	62	69	77	85	93	102
.005	.01	23	28	32	37	43	49	55	61	68	76	84	92

		$n = 29$	30	31	32	33	34	35	36	37	38	39	
.05	.10	141	152	163	175	188	201	214	228	242	256	271	
.025	.05	127	137	148	159	171	183	195	208	222	235	250	
.01	.02	111	120	130	141	151	162	174	186	198	211	224	
.005	.01	100	109	118	128	138	149	160	171	183	195	208	

		$n = 40$	41	42	43	44	45	46	47	48	49	50	
.05	.10	287	303	319	336	353	371	389	408	427	446	466	
.025	.05	264	279	295	311	327	344	361	379	397	415	434	
.01	.02	238	252	267	281	297	313	329	345	362	380	398	
.005	.01	221	234	248	262	277	292	307	323	339	356	373	

Source: Material from *Some Rapid Approximate Statistical Procedures*, Copyright 1949, 1964, Lederle Laboratories Division of American Cyanamid Company, all rights reserved and reprinted with permission.

Table A16(a)   Lower-Tail Critical Values* for the Wilcoxon Rank-Sum Test: $p = 0.05$

n_2	$n_1 = 3$	4	5	6	7	8	9	10	11	12	13	14	15	16	17	18	19	20	21	22	23	24	25
3	6																						
4	7	12																					
5	7	13	19																				
6	8	14	20	28																			
7	9	15	22	30	39																		
8	9	16	24	32	41	52																	
9	10	17	25	33	43	54	66																
10	11	18	26	35	46	57	69	83															
11	11	19	27	37	48	60	72	86	101														
12	12	20	29	39	50	62	75	89	105	121													
13	13	21	30	41	52	65	78	93	109	125	143												
14	13	22	32	42	54	67	81	96	112	129	148	167											
15	14	23	33	44	57	70	84	100	116	134	152	172	192										
16	15	24	34	46	59	73	87	103	120	138	157	177	198	220									
17	15	25	36	48	61	75	90	107	124	142	162	182	203	226	249								
18	16	26	37	50	63	78	93	110	128	147	166	187	209	232	256	280							
19	17	27	39	52	65	80	96	114	132	151	171	192	215	238	262	287	314						
20	17	28	40	53	68	83	100	117	136	155	176	197	220	244	268	294	321	349					
21	18	29	42	55	70	86	103	120	139	159	181	203	226	250	275	301	328	356	386				
22	19	30	43	57	72	88	106	124	143	164	185	208	231	256	281	308	336	364	394	424			
23	19	31	44	59	74	91	109	127	147	168	190	213	237	262	288	315	343	372	402	433	465		
24	20	32	46	61	77	94	112	131	151	172	195	218	242	268	294	322	350	380	410	442	474	508	
25	21	33	47	62	79	96	115	134	155	177	199	223	248	274	301	329	358	387	418	450	483	517	552
$n_1 = 3$	4	5	6	7	8	9	10	11	12	13	14	15	16	17	18	19	20	21	22	23	24	25	

*Upper-tail critical values can be found from the equation $w_{(1-p)} = n_1(n_1 + n_2 + 1) - w_p$.

Source: Adapted and reprinted with permission from CRC Handbook of Tables for Probability and Statistics, Second Edition, Edited by W. H. Beyer, The Chemical Rubber Company. Copyright CRC Press, Inc., Boca Raton, FL.

Table A16(b) Lower-Tail Critical Values* for the Wilcoxon Rank-Sum Test: $p = 0.025$

n_2	$n_1 = 3$	4	5	6	7	8	9	10	11	12	13	14	15	16	17	18	19	20	21	22	23	24	25
3	5																						
4	6	11																					
5	6	12	18																				
6	7	12	19	26																			
7	7	13	20	28	37																		
8	8	14	21	29	39	49																	
9	8	15	22	31	41	51	63																
10	9	16	24	32	43	54	66	79															
11	9	17	25	34	45	56	68	82	96														
12	10	17	26	36	46	58	71	85	100	116													
13	10	18	27	37	48	61	74	88	103	120	137												
14	11	19	29	39	50	63	77	91	107	124	141	160											
15	11	20	30	41	52	65	79	94	110	128	146	165	185										
16	12	21	31	42	54	68	82	97	114	131	150	170	190	212									
17	12	22	32	44	56	70	85	101	118	135	154	174	195	217	240								
18	13	23	33	45	58	72	88	104	121	139	159	179	201	223	246	271							
19	13	24	35	47	60	75	90	107	125	143	163	184	206	229	252	277	303						
20	14	24	36	49	62	77	93	110	128	147	168	189	211	234	258	284	310	337					
21	14	25	37	50	64	80	96	113	132	151	172	194	216	240	264	290	317	345	373				
22	15	26	38	52	66	82	99	117	135	155	176	198	221	245	271	297	324	352	381	411			
23	15	27	40	53	68	84	101	120	139	159	181	203	227	251	277	303	331	359	389	419	451		
24	16	28	41	55	70	87	104	123	143	163	185	208	232	257	283	310	338	367	397	428	460	493	
25	17	29	42	57	72	89	107	126	146	167	190	213	237	262	289	316	345	374	404	436	468	502	536
$n_1 = 3$	4	5	6	7	8	9	10	11	12	13	14	15	16	17	18	19	20	21	22	23	24	25	

*Upper-tail critical values can be found from the equation $w_{(1-p)} = n_1(n_1 + n_2 + 1) - w_p$.

Source: Adapted and reprinted with permission from *CRC Handbook of Tables for Probability and Statistics*, Second Edition, Edited by W. H. Beyer, The Chemical Rubber Company. Copyright CRC Press, Inc, Boca Raton, FL.

Table A16(c) Lower-Tail Critical Values* for the Wilcoxon Rank-Sum Test: $p = 0.01$

n_2	$n_1=3$	4	5	6	7	8	9	10	11	12	13	14	15	16	17	18	19	20	21	22	23	24	25
3	5																						
4	5	10																					
5	6	10	16																				
6	6	11	17	24																			
7	6	12	18	26	34																		
8	7	12	19	27	36	46																	
9	7	13	20	28	38	48	59																
10	7	14	21	30	39	50	62	74															
11	8	14	22	31	41	52	64	77	91														
12	8	15	23	32	43	54	66	80	94	110													
13	9	16	24	34	44	56	69	83	97	113	130												
14	9	16	25	35	46	58	71	85	101	117	134	153											
15	9	17	26	36	48	60	74	88	104	120	138	157	177										
16	10	18	27	38	49	62	76	91	107	124	142	161	181	202									
17	10	18	28	39	51	64	79	94	110	128	146	166	186	208	230								
18	10	19	29	40	53	66	81	97	113	131	150	170	191	213	236	260							
19	11	20	30	42	55	68	83	100	117	135	154	174	196	218	241	266	291						
20	11	20	31	43	56	71	86	102	120	138	158	179	200	223	247	272	297	324					
21	12	21	32	45	58	73	88	105	123	142	162	183	205	228	253	278	304	331	359				
22	12	22	33	46	60	75	91	108	126	146	166	188	210	234	258	284	310	338	367	396			
23	12	23	34	47	61	77	93	111	130	149	170	192	215	239	264	290	317	345	374	404	435		
24	13	23	35	49	63	79	96	114	133	153	174	196	220	244	269	296	323	352	381	412	443	476	
25	13	24	36	50	65	81	98	117	136	157	178	201	225	249	275	302	330	359	388	419	451	484	518
$n_1=3$	4	5	6	7	8	9	10	11	12	13	14	15	16	17	18	19	20	21	22	23	24	25

*Upper-tail critical values can be found from the equation $w_{(1-p)} = n_1(n_1 + n_2 + 1) - w_p$.

Source: Adapted and reprinted with permission from *CRC Handbook of Tables for Probability and Statistics*, Second Edition, Edited by W. H. Beyer, The Chemical Rubber Company. Copyright CRC Press, Inc., Boca Raton, FL.

Table A16(d) Lower-Tail Critical Values* for the Wilcoxon Rank-Sum Test: $p = 0.005$

n_2	$n_1=3$	4	5	6	7	8	9	10	11	12	13	14	15	16	17	18	19	20	21	22	23	24	25
3	5																						
4	5	9																					
5	5	10	15																				
6	5	10	16	23																			
7	6	11	17	24	33																		
8	6	11	18	25	34	44																	
9	6	12	19	27	36	46	57																
10	6	12	19	28	37	47	59	71															
11	7	13	20	29	39	49	61	74	88														
12	7	13	21	30	40	51	63	76	91	106													
13	7	14	22	31	42	53	65	79	94	109	126												
14	8	15	23	32	43	55	68	82	97	113	130	148											
15	8	15	24	34	45	57	70	84	100	116	133	152	171										
16	8	16	25	35	46	59	72	87	102	119	137	156	176	196									
17	9	16	26	36	48	61	74	89	105	123	141	160	180	201	223								
18	9	17	26	37	49	62	77	92	108	126	144	164	184	206	229	252							
19	9	17	27	38	51	64	79	95	111	129	148	168	189	211	234	258	283						
20	9	18	28	40	52	66	81	97	114	133	152	172	194	216	239	264	289	316					
21	10	19	29	41	54	68	83	100	117	136	156	176	198	221	245	269	295	322	350				
22	10	19	30	42	55	70	86	103	120	139	159	180	203	226	250	275	301	329	357	386			
23	10	20	31	43	57	72	88	105	123	143	163	185	207	231	255	281	307	335	364	393	424		
24	11	20	32	45	59	74	90	108	126	146	167	189	212	236	261	287	314	342	371	401	432	464	
25	11	21	33	46	60	76	93	110	129	150	171	193	216	241	266	292	320	348	378	408	440	472	506
$n_1=3$	4	5	6	7	8	9	10	11	12	13	14	15	16	17	18	19	20	21	22	23	24	25	

*Upper-tail critical values can be found from the equation $w_{(1-p)} = n_1(n_1 + n_2 + 1) - w_p$.

Source: Adapted and reprinted with permission from *CRC Handbook of Tables for Probability and Statistics*, Second Edition, Edited by W. H. Beyer, The Chemical Rubber Company. Copyright CRC Press, Inc., Boca Raton, FL.

Table A17 **Number of Observations Required for the Comparison of a Population Variance with a Standard Value Using the Chi-Square Test**

ν	$\alpha = 0.01$			$\alpha = 0.05$		
	$\beta = 0.01$	$\beta = 0.05$	$\beta = 0.1$	$\beta = 0.01$	$\beta = 0.05$	$\beta = 0.1$
1	42,240	1,687	420.2	24,450	977.0	243.3
2	458.2	89.78	43.71	298.1	58.40	28.43
3	98.79	32.24	19.41	68.05	22.21	13.37
4	44.69	18.68	12.48	31.93	13.35	8.920
5	27.22	13.17	9.369	19.97	9.665	6.875
6	19.28	10.28	7.628	14.44	7.699	5.713
7	14.91	8.524	6.521	11.35	6.491	4.965
8	12.20	7.352	5.757	9.418	5.675	4.444
9	10.38	6.516	5.198	8.103	5.088	4.059
10	9.072	5.890	4.770	7.156	4.646	3.763
12	7.343	5.017	4.159	5.889	4.023	3.335
15	5.847	4.211	3.578	4.780	3.442	2.925
20	4.548	3.462	3.019	3.802	2.895	2.524
24	3.959	3.104	2.745	3.354	2.630	2.326
30	3.403	2.752	2.471	2.927	2.367	2.125
40	2.874	2.403	2.192	2.516	2.103	1.919
60	2.358	2.046	1.902	2.110	1.831	1.702
120	1.829	1.661	1.580	1.686	1.532	1.457
∞	1.000	1.000	1.000	1.000	1.000	1.000

Source: Reproduced from Table G of Owen L. Davies. *The Design and Analysis of Industrial Experiments*, 2nd ed., Longman House, Essex, 1956. By permission of the author and publishers.

Note: The entries in this table show the value of the ratio λ of the population variance σ^2 to a standard variance σ_0^2 that will be undetected with probability β in a χ^2 test at the 100α percent significance level. The estimate s^2 is based on ν degrees of freedom.

Table A18 Number of Observations Required for the Comparison of Two Population Variances Using the F-Test

ν	$\alpha = 0.01$			$\alpha = 0.05$		
	$\beta = 0.01$	$\beta = 0.05$	$\beta = 0.1$	$\beta = 0.01$	$\beta = 0.05$	$\beta = 0.1$
1	16,420,000	654,200	161,500	654,200	26,070	6,436
2	9,801	1,881	891.0	1,881	361.0	171.0
3	867.7	273.3	158.8	273.3	86.06	50.01
4	255.3	102.1	65.62	102.1	40.81	26.24
5	120.3	55.39	37.87	55.39	25.51	17.44
6	71.67	36.27	25.86	36.27	18.35	13.09
7	48.90	26.48	19.47	26.48	14.34	10.55
8	36.35	20.73	15.61	20.73	11.82	8.902
9	28.63	17.01	13.06	17.01	10.11	7.757
10	23.51	14.44	11.26	14.44	8.870	6.917
12	17.27	11.16	8.923	11.16	7.218	5.769
15	12.41	8.466	6.946	8.466	5.777	4.740
20	8.630	6.240	5.270	6.240	4.512	3.810
24	7.071	5.275	4.526	5.275	3.935	3.376
30	5.693	4.392	3.833	4.392	3.389	2.957
40	4.470	3.579	3.183	3.579	2.866	2.549
60	3.372	2.817	2.562	2.817	2.354	2.141
120	2.350	2.072	1.939	2.072	1.828	1.710
∞	1.000	1.000	1.000	1.000	1.000	1.000

Source: Reproduced from Table H of Owen L. Davies, *The Design and Analysis of Industrial Experiments*, 2nd ed., Longman House, Essex, 1956. By permission of the author and publishers.

Note: The entries in this table show the value of the ratio λ of two population variances σ_2^2/σ_1^2 that will be undetected with probability β in a variance ratio test at the 100α percent significance level. Both s_1^2 and s_2^2 are based on ν degree of freedom.

Table A19   Upper-Tail Critical Values for the *F*-Max Test

		Critical Value									
v	α	*k* = 3	4	5	6	7	8	9	10	11	12
2	.10	42.48	69.13	98.18	129.1	161.7	195.6	230.7	266.8	303.9	341.9
	.05	87.49	142.5	202.4	266.2	333.2	403.1	475.4	549.8	626.2	704.4
	.01	447.5	729.2	1036	1362	1705	2063	2432	2813	3204	3604
3	.10	16.77	23.95	30.92	37.73	44.40	50.94	57.38	63.72	69.97	76.14
	.05	27.76	39.51	50.88	61.98	72.83	83.48	93.94	104.2	114.4	124.4
	.01	84.56	119.8	153.8	187.0	219.3	251.1	282.3	313.0	343.2	373.1
4	.10	10.38	13.88	17.08	20.06	22.88	25.57	28.14	30.62	33.01	35.33
	.05	15.46	20.56	25.21	29.54	33.63	37.52	41.24	44.81	48.27	51.61
	.01	36.70	48.43	59.09	69.00	78.33	87.20	95.68	103.8	111.7	119.3
5	.10	7.68	9.86	11.79	13.54	15.15	16.66	18.08	19.43	20.71	21.95
	.05	10.75	13.72	16.34	18.70	20.88	22.91	24.83	26.65	28.38	30.03
	.01	22.06	27.90	33.00	37.61	41.85	45.81	49.53	53.06	56.42	59.63
6	.10	6.23	7.78	9.11	10.30	11.38	12.38	13.31	14.18	15.01	15.79
	.05	8.36	10.38	12.11	13.64	15.04	16.32	17.51	18.64	19.70	20.70
	.01	15.60	19.16	22.19	24.89	27.32	29.57	31.65	33.61	35.46	37.22
7	.10	5.32	6.52	7.52	8.41	9.20	9.93	10.60	11.23	11.82	12.37
	.05	6.94	8.44	9.70	10.80	11.80	12.70	13.54	14.31	15.05	15.74
	.01	12.09	14.55	16.60	18.39	20.00	21.47	22.82	24.08	25.26	26.37
8	.10	4.71	5.68	6.48	7.18	7.80	8.36	8.88	9.36	9.81	10.23
	.05	6.00	7.19	8.17	9.02	9.77	10.46	11.08	11.67	12.21	12.72
	.01	9.94	11.77	13.27	14.58	15.73	16.78	17.74	18.63	19.46	20.24
9	.10	4.26	5.07	5.74	6.31	6.82	7.28	7.70	8.09	8.45	8.78
	.05	5.34	6.31	7.11	7.79	8.40	8.94	9.44	9.90	10.33	10.73
	.01	8.49	9.93	11.10	12.11	12.99	13.79	14.52	15.19	15.81	16.39
10	.10	3.93	4.63	5.19	5.68	6.11	6.49	6.84	7.16	7.46	7.74
	.05	4.85	5.67	6.34	6.91	7.41	7.86	8.27	8.64	8.99	9.32
	.01	7.46	8.64	9.59	10.39	11.10	11.74	12.31	12.84	13.33	13.79
12	.10	3.45	4.00	4.44	4.81	5.13	5.42	5.68	5.92	6.14	6.35
	.05	4.16	4.79	5.30	5.72	6.09	6.42	6.72	6.99	7.24	7.48
	.01	6.10	6.95	7.63	8.20	8.69	9.13	9.53	9.89	10.23	10.54
15	.10	3.00	3.41	3.74	4.02	4.25	4.46	4.65	4.82	4.98	5.13
	.05	3.53	4.00	4.37	4.67	4.94	5.17	5.38	5.57	5.75	5.91
	.01	4.93	5.52	5.99	6.37	6.71	7.00	7.27	7.51	7.73	7.93
20	.10	2.57	2.87	3.10	3.29	3.46	3.60	3.73	3.85	3.96	4.06
	.05	2.95	3.28	3.53	3.74	3.92	4.08	4.22	4.35	4.46	4.57
	.01	3.90	4.29	4.60	4.85	5.06	5.25	5.42	5.57	5.70	5.83
30	.10	2.14	2.34	2.50	2.62	2.73	2.82	2.90	2.97	3.04	3.10
	.05	2.40	2.61	2.77	2.90	3.01	3.11	3.19	3.27	3.34	3.40
	.01	2.99	3.23	3.41	3.56	3.68	3.79	3.88	3.97	4.04	4.12
60	.10	1.71	1.82	1.90	1.96	2.02	2.07	2.11	2.14	2.18	2.21
	.05	1.84	1.96	2.04	2.11	2.16	2.21	2.25	2.29	2.32	2.35
	.01	2.15	2.26	2.35	2.42	2.47	2.52	2.57	2.61	2.64	2.67

Source: Nelson, L. (1987). "Upper 10%, 5%, and 1% Points of the Maximum *F*-Ratio," *Journal of Quality Technology*, **19**, 165–67. Copyright American Society for Quality Control, Inc., Milwaukee, WI. Reprinted by permission.

Table A20 Orthogonal Polynomial Coefficients

k = 3

a_1	a_2
-1	+1
0	-2
+1	+1
D: 2	6
λ: 1	3

k = 4

a_1	a_2	a_3
-3	+1	-1
-1	-1	+3
+1	-1	-3
+3	+1	+1
D: 20	4	20
λ: 2	1	10/3

k = 5

a_1	a_2	a_3	a_4
-2	+2	-1	+1
-1	-1	+2	-4
0	-2	0	+6
+1	-1	-2	-4
+2	+2	+1	+1
D: 10	14	10	70
λ: 1	1	5/6	35/12

k = 6

a_1	a_2	a_3	a_4	a_5
-5	+5	-5	+1	-1
-3	-1	+7	-3	+5
-1	-4	+4	+2	-10
+1	-4	-4	+2	+10
+3	-1	-7	-3	-5
+5	+5	+5	+1	+1
D: 70	84	180	28	252
λ: 2	3/2	5/3	7/12	21/10

k = 7

a_1	a_2	a_3	a_4	a_5
-3	+5	-1	+3	-1
-2	0	+1	-7	+4
-1	-3	+1	+1	-5
0	-4	0	+6	0
+1	-3	-1	+1	+5
+2	0	+1	+1	-4
+3	+5	+1	+3	+1
D: 28	84	6	154	84
λ: 1	1	1/6	7/12	7/20

k = 8

a_1	a_2	a_3	a_4	a_5
-7	+7	-7	+7	-7
-5	+1	+5	-13	+23
-3	-3	+7	-3	-17
-1	-5	+3	+9	-15
+1	-5	-3	+9	+15
+3	-3	-7	-3	+17
+5	+1	-5	-13	-23
+7	+7	+7	+7	+7
D: 168	168	264	616	2184
λ: 2	1	2/3	7/12	7/10

k = 9

a_1	a_2	a_3	a_4	a_5
0	-20	0	+18	0
+1	-17	-9	+9	+9
+2	-8	-13	-11	+4
+3	+7	-7	-21	-11
+4	+28	+14	+14	+4
D: 60	2,772	990	2,002	468
λ: 1	3	5/6	7/12	3/20

k = 10

a_1	a_2	a_3	a_4	a_5
+1	-4	-12	+18	+6
+3	-3	-31	+3	+11
+5	-1	-35	-17	+1
+7	+2	-14	-22	-14
+9	+6	+42	+18	+6
D: 330	132	8,580	2,860	780
λ: 2	1/2	5/3	5/12	1/10

k = 11

a_1	a_2	a_3	a_4	a_5
0	-10	0	+6	0
+1	-9	-14	+4	+4
+2	-6	-23	-1	+4
+3	-1	-22	-6	-1
+4	+6	-6	-6	-6
+5	+15	+30	+6	+3
D: 110	858	4,290	286	156
λ: 1	1	5/6	1/12	1/40

Note: For $k \geq 9$, the remainder of the coefficients for column a_i are given by $a_{i,k-j+1} = -a_{i,j}$ for i odd, and $a_{i,k-j+1} = a_{i,j}$ for i even.

Table A20 (continued)   Orthogonal Polynomial Coefficients

k = 12

a_1	a_2	a_3	a_4	a_5
+1	-35	-7	+28	+20
+3	-29	-19	+12	+44
+5	-17	-25	-13	+29
+7	+1	-21	-33	-21
+9	+25	-3	-27	-57
+11	+55	+33	+33	+33
D	572	12,012	5,148	8,008	15,912
λ	2	3	2/3	7/24	3/20

(D row: a_1 572, a_2 12,012, a_3 5,148, a_4 8,008, a_5 15,912; λ row: 2, 3, 2/3, 7/24, 3/20)

k = 13

a_1	a_2	a_3	a_4	a_5
0	-14	0	+84	0
+1	-13	-4	+64	+20
+2	-10	-7	+11	+26
+3	-5	-8	-54	+11
+4	+2	-6	-96	-18
+5	+11	0	-66	-33
+6	+22	+11	+99	+22

D: 182, 2,002, 572, 68,068, 6,188
λ: 1, 1, 1/6, 7/12, 7/120

k = 14

a_1	a_2	a_3	a_4	a_5
+1	-8	-24	-108	+60
+3	-7	-67	+63	+145
+5	-5	-95	-13	+139
+7	-2	-98	-92	+28
+9	+2	-66	-132	-132
+11	+7	+11	-77	-187
+13	+13	+143	+143	+143

D: 910, 728, 97,240, 136,136, 235,144
λ: 2, 1/2, 5/3, 7/12, 7/30

k = 15

a_1	a_2	a_3	a_4	a_5
0	-56	0	+756	0
+1	-53	-27	+621	+675
+2	-44	-49	+251	+1000
+3	-29	-61	-249	+751
+4	-8	-58	-704	-44
+5	+19	-35	-869	-979
+6	+52	+13	-429	-1144
+7	+91	+91	+1001	+1001

D: 280, 37,128, 39,780, 6,466,460, 10,581,480
λ: 1, 3, 5/6, 35/12, 21/20

k = 15

a_1	a_2	a_3	a_4	a_5
+1	-21	-63	+189	+45
+3	-19	-179	+129	+115
+5	-15	-265	+23	+131
+7	-9	-301	-101	+77
+9	-1	-267	-201	-33
+11	+9	-143	-221	-143
+13	+21	+91	-91	-143
+15	+35	+455	+273	+143

D: 1,360, 5,712, 1,007,760, 470,288, 201,552
λ: 2, 1, 10/3, 7/12, 1/10

Table A20 (continued)   Orthogonal Polynomial Coefficients

17

a_1	a_2	a_3	a_4	a_5
0	-24	0	+36	0
+1	-23	-7	+31	+55
+2	-20	-13	+17	+88
+3	-15	-17	-3	+83
+4	-8	-18	-24	+36
+5	+1	-15	-39	-39
+6	+12	-7	-39	-104
+7	+25	+7	-13	-91
+8	+40	+28	+52	+104
D 408	7,752	3,876	16,796	100,776
λ 1	1	1/6	1/12	1/20

18

a_1	a_2	a_3	a_4	a_5
+1	-40	-8	+44	+220
+3	-37	-23	+33	+583
+5	-31	-35	+13	+733
+7	-22	-42	-12	+588
+9	-10	-42	-36	+156
+11	+5	-33	-51	-429
+13	+23	-13	-47	-871
+15	+44	+20	-12	-676
+17	+68	+68	+68	+884
D 1,938	23,256	23,256	28,424	6,953,544
λ 2	3/2	1/3	1/12	3/10

19

a_1	a_2	a_3	a_4	a_5
0	-30	0	+396	0
+1	-29	-44	+352	+44
+2	-26	-83	+227	+74
+3	-21	-112	+42	+79
+4	-14	-126	-168	+54
+5	-5	-120	-354	+3
+6	+6	-89	-453	-58
+7	+19	-28	-388	-98
+8	+34	+68	-68	-68
+9	+51	+204	+612	+102
D 570	13,566	213,180	2,288,132	89,148
λ 1	1	5/6	7/12	1/40

20

a_1	a_2	a_3	a_4	a_5
+1	-33	-99	+1188	+396
+3	-31	-287	+948	+1076
+5	-27	-445	+503	+1441
+7	-21	-553	-77	+1351
+9	-13	-591	-687	+771
+11	-3	-539	-1187	-187
+13	+9	-377	-1402	-1222
+15	+23	-85	-1122	-1802
+17	+39	+357	-102	-1122
+19	+57	+969	+1938	+1938
D 2,660	17,556	4,903,140	22,881,320	31,201,800
λ 2	1	10/3	35/24	7/20

Table A20 (continued) Orthogonal Polynomial Coefficients

	21					22				
k	a_1	a_2	a_3	a_4	a_5	a_1	a_2	a_3	a_4	a_5
	0	-110	0	+594	0	+1	-20	-12	+702	+390
	+1	-107	-54	+540	+1404	+3	-19	-35	+585	+1079
	+2	-98	-103	+385	+2444	+5	-17	-55	+365	+1509
	+3	-83	-142	+150	+2819	+7	-14	-70	+70	+1554
	+4	-62	-166	-130	+2354	+9	-10	-78	-258	+1158
	+5	-35	-170	-406	+1063	+11	-5	-77	-563	+363
	+6	-2	-149	-615	-788	+13	+1	-65	-775	-663
	+7	+37	-98	-680	-2618	+15	+8	-40	-810	-1598
	+8	+82	-12	-510	-3468	+17	+16	0	-570	-1938
	+9	+133	+114	0	-1938	+19	+25	+57	+57	-969
	+10	+190	+285	+969	+3876	+21	+35	+133	+1197	+2261
D	770	201,894	432,630	5,720,330	121,687,020	3,542	7,084	96,140	8,748,740	40,562,340
λ	1	3	5/6	7/12	21/40	2	1/2	1/3	7/12	7/30

Source: Taken from Table XXIII of Fisher, R. A. and Yates, F. (1974) *Statistical Tables for Biological, Agricultural, and Medical Research*, 6th ed. Published by Longman Group UK Ltd. London (previously published by Oliver and Boyd Ltd. Edinburgh). By permission of the authors and publishers.

Note: The two values at the bottom of each column are the values of D, the sum of squares of the coefficients used in the orthogonal polynomial, and λ, the coefficient of the highest power of X_i in the orthogonal polynomial.

Table A21 Percentage Points of the Studentized Range

$\alpha = .05$

ν \ k	2	3	4	5	6	7	8	9	10	11	12	13	14	15	16	17	18	19
1	17.97	26.98	32.82	37.08	40.41	43.12	45.40	47.36	49.07	50.59	51.96	53.20	54.33	55.36	56.32	57.22	58.04	58.83
2	6.085	8.331	9.798	10.88	11.74	12.44	13.03	13.54	13.99	14.39	14.75	15.08	15.38	15.65	15.91	16.14	16.37	16.57
3	4.501	5.910	6.825	7.502	8.037	8.478	8.853	9.177	9.462	9.717	9.946	10.15	10.35	10.53	10.69	10.84	10.98	11.11
4	3.927	5.040	5.757	6.287	6.707	7.053	7.347	7.602	7.826	8.027	8.208	8.373	8.525	8.664	8.794	8.914	9.028	9.134
5	3.635	4.602	5.218	5.673	6.033	6.330	6.582	6.802	6.995	7.168	7.324	7.466	7.596	7.717	7.828	7.932	8.030	8.122
6	3.461	4.339	4.896	5.305	5.628	5.895	6.122	6.319	6.493	6.649	6.789	6.917	7.034	7.143	7.244	7.338	7.426	7.508
7	3.344	4.165	4.681	5.060	5.359	5.606	5.815	5.998	6.158	6.302	6.431	6.550	6.658	6.759	6.852	6.939	7.020	7.097
8	3.261	4.041	4.529	4.886	5.167	5.399	5.597	5.767	5.918	6.064	6.175	6.287	6.389	6.483	6.571	6.653	6.729	6.802
9	3.199	3.949	4.415	4.756	5.024	5.244	5.432	5.595	5.739	5.867	5.983	6.089	6.186	6.276	6.359	6.437	6.510	6.579
10	3.151	3.877	4.327	4.654	4.912	5.124	5.305	5.461	5.599	5.722	5.833	5.935	6.028	6.114	6.194	6.269	6.339	6.405
11	3.113	3.820	4.256	4.574	4.823	5.028	5.202	5.353	5.487	5.606	5.713	5.811	5.901	5.984	6.062	6.134	6.202	6.265
12	3.082	3.773	4.199	4.508	4.751	4.950	5.119	5.265	5.396	5.511	5.615	5.710	5.798	5.878	5.953	6.023	6.089	6.151
13	3.055	3.735	4.151	4.453	4.690	4.885	5.049	5.192	5.318	5.431	5.533	5.625	5.711	5.789	5.862	5.931	5.995	6.065
14	3.033	3.702	4.111	4.407	4.639	4.829	4.990	5.131	5.254	5.364	5.463	5.554	5.637	5.714	5.786	5.852	5.915	5.974
15	3.014	3.674	4.076	4.367	4.596	4.782	4.940	5.077	5.198	5.306	5.404	5.493	5.574	5.649	5.720	5.785	5.846	5.904
16	2.998	3.649	4.046	4.333	4.557	4.741	4.897	5.031	5.150	5.256	5.352	5.439	5.520	5.594	5.662	5.727	5.788	5.843
17	2.984	3.628	4.020	4.303	4.524	4.705	4.858	4.991	5.108	5.212	5.307	5.392	5.471	5.544	5.612	5.675	5.734	5.790
18	2.971	3.609	3.997	4.277	4.495	4.673	4.824	4.956	5.071	5.174	5.267	5.352	5.429	5.501	5.568	5.630	5.688	5.743
19	2.960	3.598	3.977	4.253	4.469	4.645	4.794	4.924	5.038	5.140	5.231	5.315	5.391	5.462	5.528	5.589	5.647	5.701
20	2.950	3.578	3.958	4.232	4.445	4.620	4.768	4.896	5.008	5.108	5.199	5.282	5.357	5.427	5.498	5.553	5.610	5.663
24	2.919	3.532	3.901	4.166	4.373	4.541	4.684	4.807	4.915	5.012	5.099	5.179	5.251	5.319	5.381	5.439	5.494	5.545
30	2.888	3.486	3.845	4.102	4.302	4.464	4.602	4.720	4.824	4.917	5.001	5.077	5.147	5.211	5.271	5.327	5.379	5.429
40	2.858	3.442	3.791	4.039	4.232	4.389	4.521	4.635	4.735	4.824	4.904	4.977	5.044	5.106	5.163	5.216	5.266	5.318
60	2.829	3.399	3.737	3.977	4.163	4.314	4.441	4.550	4.646	4.732	4.808	4.878	4.942	5.001	5.066	5.107	5.154	5.199
120	2.800	3.356	3.685	3.917	4.096	4.241	4.363	4.468	4.560	4.641	4.714	4.781	4.842	4.898	4.960	4.998	5.044	5.096
∞	2.772	3.314	3.633	3.858	4.030	4.170	4.286	4.387	4.474	4.552	4.622	4.686	4.743	4.796	4.846	4.891	4.924	4.974

Table A21 (continued)   Percentage Points of the Studentized Range

$\alpha = .05$

ν \ k	20	22	24	26	28	30	32	34	36	38	40	50	60	70	80	90	100
1	59.56	60.91	62.12	63.22	64.23	65.15	66.01	66.81	67.56	68.26	68.92	71.73	73.97	75.82	77.40	78.77	79.98
2	16.77	17.13	17.45	17.75	18.02	18.27	18.50	18.72	18.92	19.11	19.28	20.06	20.66	21.16	21.59	21.96	22.29
3	11.24	11.47	11.68	11.87	12.06	12.21	12.36	12.50	12.68	12.75	12.87	13.36	13.76	14.08	14.36	14.61	14.82
4	9.233	9.418	9.584	9.736	9.875	10.00	10.12	10.23	10.34	10.44	10.53	10.93	11.24	11.51	11.73	11.92	12.09
5	8.208	8.368	8.512	8.643	8.764	8.875	8.979	9.075	9.166	9.250	9.330	9.674	9.949	10.18	10.38	10.54	10.69
6	7.587	7.730	7.861	7.979	8.088	8.189	8.288	8.370	8.462	8.529	8.601	8.913	9.163	9.370	9.548	9.702	9.839
7	7.170	7.303	7.423	7.538	7.634	7.728	7.814	7.886	7.972	8.043	8.110	8.400	8.632	8.824	8.989	9.133	9.261
8	6.870	6.995	7.109	7.212	7.307	7.395	7.477	7.554	7.625	7.693	7.756	8.029	8.248	8.430	8.586	8.722	8.843
9	6.644	6.763	6.871	6.970	7.061	7.145	7.222	7.296	7.363	7.428	7.488	7.749	7.968	8.132	8.281	8.410	8.526
10	6.467	6.582	6.686	6.781	6.868	6.948	7.023	7.098	7.159	7.220	7.279	7.529	7.730	7.897	8.041	8.166	8.276
11	6.326	6.436	6.536	6.628	6.712	6.790	6.868	6.930	6.994	7.063	7.110	7.352	7.546	7.708	7.847	7.968	8.075
12	6.209	6.317	6.414	6.508	6.585	6.660	6.731	6.796	6.858	6.916	6.970	7.206	7.394	7.552	7.687	7.804	7.909
13	6.112	6.217	6.312	6.398	6.478	6.551	6.620	6.684	6.744	6.800	6.854	7.083	7.267	7.421	7.552	7.667	7.769
14	6.029	6.132	6.224	6.309	6.387	6.459	6.526	6.588	6.647	6.702	6.754	6.979	7.159	7.309	7.438	7.550	7.650
15	5.958	6.059	6.149	6.233	6.309	6.379	6.445	6.506	6.564	6.618	6.669	6.888	7.065	7.212	7.339	7.449	7.546
16	5.897	5.995	6.084	6.166	6.241	6.310	6.374	6.434	6.491	6.544	6.594	6.810	6.984	7.128	7.252	7.360	7.457
17	5.842	5.940	6.027	6.107	6.181	6.249	6.313	6.372	6.427	6.479	6.529	6.741	6.912	7.064	7.176	7.283	7.377
18	5.794	5.890	5.977	6.055	6.128	6.195	6.258	6.316	6.371	6.422	6.471	6.680	6.848	6.989	7.109	7.213	7.307
19	5.752	5.846	5.932	6.009	6.081	6.147	6.209	6.267	6.321	6.371	6.419	6.626	6.792	6.980	7.048	7.152	7.244
20	5.714	5.807	5.891	5.968	6.039	6.104	6.165	6.222	6.275	6.325	6.373	6.576	6.740	6.877	6.994	7.097	7.187
24	5.594	5.683	5.764	5.888	5.906	5.968	6.027	6.081	6.132	6.181	6.226	6.421	6.579	6.710	6.822	6.920	7.008
30	5.475	5.561	5.638	5.709	5.774	5.833	5.889	5.941	5.990	6.037	6.090	6.267	6.417	6.543	6.650	6.744	6.827
40	5.358	5.439	5.513	5.581	5.642	5.700	5.753	5.808	5.849	5.893	5.934	6.112	6.255	6.375	6.477	6.566	6.645
60	5.241	5.319	5.389	5.453	5.512	5.566	5.617	5.664	5.708	5.750	5.789	5.968	6.093	6.206	6.308	6.387	6.462
120	5.126	5.200	5.266	5.327	5.382	5.434	5.481	5.526	5.568	5.607	5.644	5.802	5.929	6.085	6.126	6.205	6.275
∞	5.012	5.081	5.144	5.201	5.253	5.301	5.346	5.388	5.427	5.463	5.498	5.646	5.764	5.963	5.947	6.020	6.085

Table A21 (continued) Percentage Points of the Studentized Range

$\alpha = .01$

k ν	2	3	4	5	6	7	8	9	10	11	12	13	14	15	16	17	18	19
1	90.08	135.0	164.3	185.6	202.2	215.8	227.2	237.0	245.6	253.2	260.0	266.2	271.8	277.0	281.8	286.3	290.4	294.3
2	14.04	19.02	22.29	24.72	26.63	28.20	29.53	30.68	31.69	32.59	33.40	34.13	34.81	35.43	36.00	36.53	37.03	37.50
3	8.261	10.62	12.17	13.33	14.24	15.00	15.64	16.20	16.69	17.13	17.53	17.89	18.22	18.52	18.81	19.07	19.32	19.55
4	6.512	8.120	9.173	9.958	10.58	11.10	11.55	11.98	12.27	12.57	12.84	13.09	13.32	13.53	13.73	13.91	14.08	14.24
5	5.702	6.976	7.804	8.421	8.913	9.321	9.669	9.972	10.24	10.48	10.70	10.89	11.08	11.24	11.40	11.55	11.68	11.81
6	5.243	6.381	7.033	7.556	7.973	8.318	8.613	8.869	9.097	9.301	9.485	9.653	9.808	9.951	10.08	10.21	10.32	10.43
7	4.949	5.919	6.543	7.006	7.373	7.679	7.939	8.166	8.368	8.548	8.711	8.860	8.997	9.124	9.242	9.353	9.456	9.554
8	4.746	5.635	6.204	6.625	6.960	7.237	7.474	7.681	7.863	8.027	8.176	8.312	8.436	8.552	8.659	8.760	8.854	8.943
9	4.596	5.428	5.957	6.348	6.658	6.915	7.134	7.325	7.496	7.647	7.784	7.910	8.025	8.132	8.232	8.325	8.412	8.496
10	4.482	5.270	5.769	6.136	6.428	6.669	6.875	7.055	7.213	7.356	7.485	7.603	7.712	7.812	7.906	7.993	8.076	8.153
11	4.392	5.146	5.621	5.970	6.247	6.476	6.672	6.842	6.992	7.128	7.250	7.362	7.465	7.560	7.649	7.732	7.809	7.883
12	4.320	5.046	5.502	5.836	6.101	6.321	6.507	6.670	6.814	6.943	7.060	7.167	7.265	7.356	7.441	7.520	7.594	7.665
13	4.260	4.964	5.404	5.727	5.981	6.192	6.372	6.528	6.667	6.791	6.903	7.006	7.101	7.188	7.269	7.345	7.417	7.485
14	4.210	4.895	5.322	5.634	5.881	6.085	6.258	6.409	6.543	6.664	6.772	6.871	6.962	7.047	7.126	7.199	7.268	7.333
15	4.168	4.836	5.252	5.556	5.796	5.994	6.162	6.309	6.439	6.555	6.660	6.757	6.845	6.927	7.003	7.074	7.142	7.204
16	4.131	4.786	5.192	5.489	5.722	5.915	6.079	6.222	6.349	6.462	6.564	6.658	6.744	6.823	6.898	6.967	7.032	7.093
17	4.099	4.742	5.140	5.430	5.659	5.847	6.007	6.147	6.270	6.381	6.480	6.572	6.656	6.734	6.806	6.873	6.937	6.997
18	4.071	4.708	5.094	5.379	5.608	5.788	5.944	6.081	6.201	6.310	6.407	6.497	6.579	6.655	6.725	6.792	6.854	6.912
19	4.046	4.670	5.054	5.334	5.554	5.735	5.889	6.022	6.141	6.247	6.342	6.430	6.510	6.585	6.654	6.719	6.780	6.837
20	4.024	4.639	5.018	5.294	5.510	5.688	5.839	5.970	6.087	6.191	6.285	6.371	6.450	6.523	6.591	6.654	6.714	6.771
24	3.956	4.546	4.907	5.168	5.374	5.542	5.685	5.809	5.919	6.017	6.106	6.186	6.261	6.330	6.394	6.453	6.510	6.563
30	3.889	4.455	4.799	5.048	5.242	5.401	5.536	5.653	5.756	5.849	5.932	6.008	6.078	6.143	6.203	6.259	6.311	6.361
40	3.825	4.367	4.696	4.931	5.114	5.265	5.392	5.502	5.599	5.686	5.764	5.835	5.900	5.961	6.017	6.069	6.119	6.166
60	3.762	4.282	4.595	4.818	4.991	5.133	5.253	5.356	5.447	5.528	5.601	5.667	5.728	5.785	5.837	5.886	5.981	5.974
120	3.702	4.200	4.497	4.709	4.872	5.006	5.118	5.214	5.299	5.375	5.443	5.506	5.562	5.614	5.662	5.708	5.750	5.790
∞	3.643	4.120	4.408	4.603	4.757	4.882	4.987	5.078	5.157	5.227	5.290	5.348	5.400	5.448	5.493	5.535	5.574	5.611

Table A21 (continued)   Percentage Points of the Studentized Range

$\alpha = .01$

ν \ k	20	22	24	26	28	30	32	34	36	38	40	50	60	70	80	90	100
1	296.0	304.7	310.8	316.3	321.3	326.0	330.3	334.3	338.0	341.5	344.8	358.9	370.1	379.4	387.3	394.1	400.1
2	37.95	38.76	39.49	40.15	40.76	41.32	41.84	42.33	42.78	43.21	43.61	45.33	46.70	47.83	48.80	49.64	50.38
3	19.77	20.17	20.53	20.86	21.16	21.44	21.70	21.96	22.17	22.39	22.59	23.45	24.13	24.71	25.19	25.62	25.99
4	14.40	14.68	14.93	15.16	15.37	15.57	15.75	15.92	16.08	16.23	16.37	16.98	17.46	17.86	18.20	18.50	18.77
5	11.93	12.16	12.36	12.54	12.71	12.87	13.02	13.15	13.28	13.40	13.52	14.00	14.39	14.72	14.99	15.23	15.45
6	10.54	10.73	10.91	11.06	11.21	11.34	11.47	11.58	11.69	11.80	11.90	12.31	12.65	12.92	13.16	13.37	13.55
7	9.646	9.815	9.970	10.11	10.24	10.36	10.47	10.58	10.67	10.77	10.85	11.23	11.52	11.77	11.99	12.17	12.34
8	9.027	9.182	9.322	9.450	9.569	9.678	9.779	9.874	9.964	10.05	10.13	10.47	10.75	10.97	11.17	11.34	11.49
9	8.573	8.717	8.847	8.966	9.075	9.177	9.271	9.360	9.443	9.521	9.594	9.912	10.17	10.38	10.57	10.73	10.87
10	8.226	8.361	8.483	8.595	8.698	8.794	8.883	8.966	9.044	9.117	9.187	9.486	9.726	9.927	10.10	10.25	10.39
11	7.952	8.080	8.196	8.303	8.400	8.491	8.575	8.654	8.728	8.798	8.864	9.148	9.377	9.568	9.732	9.875	10.00
12	7.731	7.853	7.964	8.066	8.159	8.246	8.327	8.402	8.473	8.539	8.603	8.875	9.094	9.277	9.434	9.571	9.693
13	7.548	7.665	7.772	7.870	7.960	8.043	8.121	8.193	8.262	8.326	8.387	8.648	8.859	9.035	9.187	9.318	9.436
14	7.395	7.508	7.611	7.705	7.792	7.873	7.948	8.018	8.084	8.146	8.204	8.457	8.661	8.832	8.978	9.106	9.219
15	7.264	7.374	7.474	7.566	7.650	7.728	7.800	7.869	7.932	7.992	8.049	8.296	8.492	8.658	8.800	8.924	9.035
16	7.152	7.258	7.356	7.445	7.527	7.602	7.673	7.739	7.802	7.860	7.916	8.154	8.347	8.507	8.646	8.767	8.874
17	7.053	7.158	7.253	7.340	7.420	7.498	7.563	7.627	7.687	7.745	7.799	8.081	8.219	8.377	8.511	8.630	8.735
18	6.968	7.070	7.163	7.247	7.325	7.398	7.465	7.528	7.587	7.643	7.696	7.924	8.107	8.261	8.388	8.508	8.611
19	6.891	6.992	7.082	7.166	7.242	7.313	7.379	7.440	7.498	7.553	7.606	7.828	8.008	8.159	8.298	8.401	8.502
20	6.823	6.922	7.011	7.092	7.168	7.237	7.302	7.362	7.419	7.473	7.523	7.742	7.919	8.067	8.194	8.306	8.404
24	6.612	6.706	6.789	6.865	6.936	7.001	7.062	7.119	7.173	7.223	7.270	7.476	7.642	7.780	7.900	8.004	8.097
30	6.407	6.494	6.572	6.644	6.710	6.772	6.828	6.881	6.932	6.978	7.023	7.215	7.370	7.500	7.611	7.709	7.796
40	6.209	6.289	6.362	6.429	6.490	6.547	6.600	6.660	6.697	6.740	6.782	6.960	7.104	7.225	7.328	7.419	7.500
60	6.015	6.090	6.158	6.220	6.277	6.330	6.378	6.424	6.467	6.507	6.546	6.710	6.843	6.964	7.050	7.133	7.207
120	5.827	5.897	5.969	6.016	6.069	6.117	6.162	6.204	6.244	6.281	6.316	6.467	6.588	6.689	6.776	6.852	6.919
∞	5.645	5.709	5.766	5.818	5.866	5.911	5.952	5.990	6.026	6.060	6.092	6.228	6.338	6.429	6.507	6.575	6.636

Source: Adapted from Harter, H. L. (1960). "Tables of Range and Studentized Range," *Annals of Mathematical Statistics*, **31**, 1122–1147. Used by permission of the Institute of Mathematical Statistics.

Table A22 Critical Values for Duncan's Multiple Range Test

α = .05

ν \ p	2	3	4	5	6	7	8	9	10	11	12	13	14	15	16	17	18	19
1	17.97	17.97	17.97	17.97	17.97	17.97	17.97	17.97	17.97	17.97	17.97	17.97	17.97	17.97	17.97	17.97	17.97	17.97
2	6.085	6.085	6.085	6.085	6.085	6.085	6.085	6.085	6.085	6.085	6.085	6.085	6.085	6.085	6.085	6.085	6.085	6.085
3	4.501	4.516	4.516	4.516	4.516	4.516	4.516	4.516	4.516	4.516	4.516	4.516	4.516	4.516	4.516	4.516	4.516	4.516
4	3.927	4.013	4.033	4.033	4.033	4.033	4.033	4.033	4.033	4.033	4.033	4.033	4.033	4.033	4.033	4.033	4.033	4.033
5	3.635	3.749	3.797	3.814	3.814	3.814	3.814	3.814	3.814	3.814	3.814	3.814	3.814	3.814	3.814	3.814	3.814	3.814
6	3.461	3.587	3.649	3.680	3.694	3.697	3.697	3.697	3.697	3.697	3.697	3.697	3.697	3.697	3.697	3.697	3.697	3.697
7	3.344	3.477	3.548	3.588	3.611	3.622	3.626	3.626	3.626	3.626	3.626	3.626	3.626	3.626	3.626	3.626	3.626	3.626
8	3.261	3.399	3.475	3.521	3.549	3.566	3.575	3.579	3.579	3.579	3.579	3.579	3.579	3.579	3.579	3.579	3.579	3.579
9	3.199	3.339	3.420	3.470	3.502	3.523	3.536	3.544	3.547	3.547	3.547	3.547	3.547	3.547	3.547	3.547	3.547	3.547
10	3.151	3.293	3.376	3.430	3.465	3.489	3.505	3.516	3.522	3.525	3.526	3.526	3.526	3.526	3.526	3.526	3.526	3.526
11	3.113	3.256	3.342	3.397	3.435	3.462	3.480	3.493	3.501	3.506	3.509	3.510	3.510	3.510	3.510	3.510	3.510	3.510
12	3.082	3.225	3.313	3.370	3.410	3.439	3.459	3.474	3.484	3.491	3.496	3.498	3.499	3.499	3.499	3.499	3.499	3.499
13	3.055	3.200	3.289	3.348	3.389	3.419	3.442	3.458	3.470	3.478	3.484	3.488	3.490	3.490	3.490	3.490	3.490	3.490
14	3.033	3.178	3.268	3.329	3.372	3.403	3.426	3.444	3.457	3.467	3.474	3.479	3.482	3.484	3.484	3.484	3.485	3.485
15	3.014	3.160	3.250	3.312	3.356	3.389	3.413	3.432	3.446	3.457	3.465	3.471	3.476	3.478	3.480	3.481	3.481	3.481
16	2.998	3.144	3.235	3.298	3.343	3.376	3.402	3.422	3.437	3.449	3.458	3.465	3.470	3.473	3.477	3.478	3.478	3.478
17	2.984	3.130	3.222	3.285	3.331	3.366	3.392	3.412	3.429	3.441	3.451	3.459	3.465	3.469	3.473	3.475	3.476	3.476
18	2.971	3.118	3.210	3.274	3.321	3.356	3.383	3.405	3.421	3.435	3.445	3.454	3.460	3.465	3.470	3.472	3.474	3.474
19	2.960	3.107	3.199	3.264	3.311	3.347	3.375	3.397	3.415	3.429	3.440	3.449	3.456	3.462	3.467	3.470	3.472	3.473
20	2.950	3.097	3.190	3.255	3.303	3.339	3.368	3.391	3.409	3.424	3.436	3.445	3.453	3.459	3.464	3.467	3.470	3.472
24	2.919	3.066	3.160	3.226	3.276	3.315	3.345	3.370	3.390	3.406	3.420	3.432	3.441	3.449	3.456	3.461	3.465	3.469
30	2.888	3.035	3.131	3.199	3.250	3.290	3.322	3.349	3.371	3.389	3.405	3.418	3.430	3.439	3.447	3.454	3.460	3.466
40	2.858	3.006	3.102	3.171	3.224	3.266	3.300	3.328	3.352	3.373	3.390	3.405	3.418	3.429	3.439	3.448	3.456	3.463
60	2.829	2.976	3.073	3.143	3.198	3.241	3.277	3.307	3.333	3.355	3.374	3.391	3.406	3.419	3.431	3.442	3.451	3.460
120	2.800	2.947	3.045	3.116	3.172	3.217	3.254	3.287	3.314	3.337	3.359	3.377	3.394	3.409	3.423	3.435	3.446	3.457
∞	2.772	2.918	3.017	3.089	3.146	3.193	3.232	3.265	3.294	3.320	3.343	3.363	3.382	3.399	3.414	3.428	3.442	3.454

Table A22 (continued)   Critical Values for Duncan's Multiple Range Test

$\alpha = .05$

$\nu \backslash p$	20	22	24	26	28	30	32	34	36	38	40	50	60	70	80	90	100
1	17.97	17.97	17.97	17.97	17.97	17.97	17.97	17.97	17.97	17.97	17.97	17.97	17.97	17.97	17.97	17.97	17.97
2	6.085	6.085	6.085	6.085	6.085	6.085	6.085	6.085	6.085	6.085	6.085	6.085	6.085	6.085	6.085	6.085	6.085
3	4.516	4.516	4.516	4.516	4.516	4.516	4.516	4.516	4.516	4.516	4.516	4.516	4.516	4.516	4.516	4.516	4.516
4	4.033	4.033	4.033	4.033	4.033	4.033	4.033	4.033	4.033	4.033	4.033	4.033	4.033	4.033	4.033	4.033	4.033
5	3.814	3.814	3.814	3.814	3.814	3.814	3.814	3.814	3.814	3.814	3.814	3.814	3.814	3.814	3.814	3.814	3.814
6	3.697	3.697	3.697	3.697	3.697	3.697	3.697	3.697	3.697	3.697	3.697	3.697	3.697	3.697	3.697	3.697	3.697
7	3.626	3.626	3.626	3.626	3.626	3.626	3.626	3.626	3.626	3.626	3.626	3.626	3.626	3.626	3.626	3.626	3.626
8	3.579	3.579	3.579	3.579	3.579	3.579	3.579	3.579	3.579	3.579	3.579	3.579	3.579	3.579	3.579	3.579	3.579
9	3.547	3.547	3.547	3.547	3.547	3.547	3.547	3.547	3.547	3.547	3.547	3.547	3.547	3.547	3.547	3.547	3.547
10	3.526	3.526	3.526	3.526	3.526	3.526	3.526	3.526	3.526	3.526	3.526	3.526	3.526	3.526	3.526	3.526	3.526
11	3.510	3.510	3.510	3.510	3.510	3.510	3.510	3.510	3.510	3.510	3.510	3.510	3.510	3.510	3.510	3.510	3.510
12	3.499	3.499	3.499	3.499	3.499	3.499	3.499	3.499	3.499	3.499	3.499	3.499	3.499	3.499	3.499	3.499	3.499
13	3.490	3.490	3.490	3.490	3.490	3.490	3.490	3.490	3.490	3.490	3.490	3.490	3.490	3.490	3.490	3.490	3.490
14	3.485	3.485	3.485	3.485	3.485	3.485	3.485	3.485	3.485	3.485	3.485	3.485	3.485	3.485	3.485	3.485	3.485
15	3.481	3.481	3.481	3.481	3.481	3.481	3.481	3.481	3.481	3.481	3.481	3.481	3.481	3.481	3.481	3.481	3.481
16	3.478	3.478	3.478	3.478	3.478	3.478	3.478	3.478	3.478	3.478	3.478	3.478	3.478	3.478	3.478	3.478	3.478
17	3.476	3.476	3.476	3.474	3.476	3.476	3.476	3.476	3.476	3.476	3.476	3.476	3.476	3.476	3.476	3.476	3.476
18	3.474	3.474	3.474	3.474	3.474	3.474	3.474	3.474	3.474	3.474	3.474	3.474	3.474	3.474	3.474	3.474	3.474
19	3.474	3.474	3.474	3.474	3.474	3.474	3.474	3.474	3.474	3.474	3.474	3.474	3.474	3.474	3.474	3.474	3.474
20	3.473	3.474	3.474	3.474	3.474	3.474	3.474	3.474	3.474	3.474	3.474	3.474	3.474	3.474	3.474	3.474	3.474
24	3.471	3.475	3.477	3.477	3.477	3.477	3.477	3.477	3.477	3.477	3.477	3.477	3.477	3.477	3.477	3.477	3.477
30	3.470	3.477	3.481	3.484	3.486	3.486	3.486	3.486	3.486	3.486	3.486	3.486	3.486	3.486	3.486	3.486	3.486
40	3.469	3.479	3.486	3.492	3.497	3.500	3.503	3.504	3.504	3.504	3.504	3.504	3.504	3.504	3.504	3.504	3.504
60	3.467	3.481	3.492	3.501	3.509	3.515	3.521	3.525	3.529	3.531	3.534	3.537	3.537	3.537	3.537	3.537	3.537
120	3.466	3.483	3.498	3.511	3.522	3.532	3.541	3.548	3.555	3.561	3.566	3.585	3.596	3.600	3.601	3.601	3.601
∞	3.466	3.486	3.505	3.522	3.536	3.550	3.562	3.574	3.584	3.594	3.603	3.640	3.668	3.690	3.708	3.722	3.735

Table A22 (continued)   Critical Values for Duncan's Multiple Range Test

$$\alpha = .01$$

v\\p	2	3	4	5	6	7	8	9	10	11	12	13	14	15	16	17	18	19
1	90.03	90.03	90.03	90.03	90.03	90.03	90.03	90.03	90.03	90.03	90.03	90.03	90.03	90.03	90.03	90.03	90.03	90.03
2	14.04	14.04	14.04	14.04	14.04	14.04	14.04	14.04	14.04	14.04	14.04	14.04	14.04	14.04	14.04	14.04	14.04	14.04
3	8.261	8.321	8.321	8.321	8.321	8.321	8.321	8.321	8.321	8.321	8.321	8.321	8.321	8.321	8.321	8.321	8.321	8.321
4	6.512	6.677	6.740	6.756	6.756	6.756	6.756	6.756	6.756	6.756	6.756	6.756	6.756	6.756	6.756	6.756	6.756	6.756
5	5.702	5.893	5.989	6.040	6.065	6.074	6.074	6.074	6.074	6.074	6.074	6.074	6.074	6.074	6.074	6.074	6.074	6.074
6	5.243	5.439	5.549	5.614	5.655	5.680	5.694	5.701	5.703	5.703	5.703	5.703	5.703	5.703	5.703	5.703	5.703	5.703
7	4.949	5.145	5.260	5.334	5.383	5.416	5.439	5.454	5.464	5.470	5.472	5.472	5.472	5.472	5.472	5.472	5.472	5.472
8	4.746	4.939	5.057	5.135	5.189	5.227	5.256	5.276	5.291	5.302	5.309	5.314	5.316	5.317	5.317	5.317	5.317	5.317
9	4.596	4.787	4.906	4.986	5.043	5.086	5.118	5.142	5.160	5.174	5.185	5.193	5.199	5.203	5.205	5.206	5.206	5.206
10	4.482	4.671	4.790	4.871	4.931	4.975	5.010	5.037	5.058	5.074	5.088	5.098	5.106	5.112	5.117	5.120	5.122	5.124
11	4.392	4.579	4.697	4.780	4.841	4.887	4.924	4.952	4.975	4.994	5.009	5.021	5.031	5.039	5.045	5.050	5.054	5.057
12	4.320	4.504	4.622	4.706	4.767	4.815	4.852	4.883	4.907	4.927	4.944	4.958	4.969	4.978	4.986	4.993	4.998	5.002
13	4.260	4.442	4.560	4.644	4.706	4.755	4.793	4.824	4.850	4.872	4.889	4.904	4.917	4.928	4.937	4.944	4.950	4.956
14	4.210	4.391	4.508	4.591	4.654	4.704	4.743	4.775	4.802	4.824	4.843	4.859	4.872	4.884	4.894	4.902	4.910	4.916
15	4.168	4.347	4.463	4.547	4.610	4.660	4.700	4.733	4.760	4.783	4.803	4.820	4.834	4.846	4.857	4.866	4.874	4.881
16	4.131	4.309	4.425	4.509	4.572	4.622	4.663	4.696	4.724	4.748	4.768	4.786	4.800	4.813	4.825	4.835	4.844	4.851
17	4.099	4.275	4.391	4.475	4.539	4.589	4.630	4.664	4.693	4.717	4.738	4.756	4.771	4.785	4.797	4.807	4.816	4.824
18	4.071	4.246	4.362	4.445	4.509	4.560	4.601	4.635	4.664	4.689	4.711	4.729	4.745	4.759	4.772	4.783	4.792	4.801
19	4.046	4.220	4.335	4.419	4.483	4.534	4.575	4.610	4.639	4.665	4.686	4.705	4.722	4.736	4.749	4.761	4.771	4.780
20	4.024	4.197	4.312	4.395	4.459	4.510	4.552	4.587	4.617	4.642	4.664	4.684	4.701	4.716	4.729	4.741	4.751	4.761
24	3.956	4.126	4.239	4.322	4.386	4.437	4.480	4.516	4.546	4.573	4.596	4.616	4.634	4.651	4.665	4.678	4.690	4.700
30	3.889	4.056	4.168	4.250	4.314	4.366	4.409	4.445	4.477	4.504	4.528	4.550	4.569	4.586	4.601	4.615	4.628	4.640
40	3.825	3.988	4.098	4.180	4.244	4.296	4.339	4.376	4.408	4.436	4.461	4.483	4.503	4.521	4.537	4.553	4.566	4.579
60	3.762	3.922	4.031	4.111	4.174	4.226	4.270	4.307	4.340	4.368	4.394	4.417	4.438	4.456	4.474	4.490	4.504	4.518
120	3.702	3.858	3.965	4.044	4.107	4.158	4.202	4.239	4.272	4.301	4.327	4.351	4.372	4.392	4.410	4.426	4.442	4.456
∞	3.643	3.796	3.900	3.978	4.040	4.091	4.135	4.172	4.205	4.235	4.261	4.285	4.307	4.327	4.345	4.363	4.379	4.394

Table A22 (continued) Critical Values for Duncan's Multiple Range Test

α = .01

ν \ p	20	22	24	26	28	30	32	34	36	38	40	50	60	70	80	90	100
1	90.03	90.03	90.03	90.03	90.03	90.03	90.03	90.03	90.03	90.03	90.03	90.03	90.03	90.03	90.03	90.03	90.03
2	14.04	14.04	14.04	14.04	14.04	14.04	14.04	14.04	14.04	14.04	14.04	14.04	14.04	14.04	14.04	14.04	14.04
3	8.321	8.321	8.321	8.321	8.321	8.321	8.321	8.321	8.321	8.321	8.321	8.321	8.321	8.321	8.321	8.321	8.321
4	6.756	6.756	6.756	6.756	6.756	6.756	6.756	6.756	6.756	6.756	6.756	6.756	6.756	6.756	6.756	6.756	6.756
5	6.074	6.074	6.074	6.074	6.074	6.074	6.074	6.074	6.074	6.074	6.074	6.074	6.074	6.074	6.074	6.074	6.074
6	5.703	5.703	5.703	5.703	5.703	5.703	5.703	5.703	5.703	5.703	5.703	5.703	5.703	5.703	5.703	5.703	5.703
7	5.472	5.472	5.472	5.472	5.472	5.472	5.472	5.472	5.472	5.472	5.472	5.472	5.472	5.472	5.472	5.472	5.472
8	5.317	5.317	5.317	5.317	5.317	5.317	5.317	5.317	5.317	5.317	5.317	5.317	5.317	5.317	5.317	5.317	5.317
9	5.206	5.206	5.206	5.206	5.206	5.206	5.206	5.206	5.206	5.206	5.206	5.206	5.206	5.206	5.026	5.206	5.206
10	5.124	5.124	5.124	5.124	5.124	5.124	5.124	5.124	5.124	5.124	5.124	5.124	5.124	5.124	5.124	5.124	5.124
11	5.059	5.061	5.061	5.061	5.061	5.061	5.061	5.061	5.061	5.061	5.061	5.061	5.061	5.061	5.061	5.061	5.061
12	5.006	5.010	5.011	5.011	5.011	5.011	5.011	5.011	5.011	5.011	5.011	5.011	5.011	5.011	5.011	5.011	5.011
13	4.960	4.966	4.970	4.972	4.972	4.972	4.972	4.972	4.972	4.972	4.972	4.972	4.972	4.972	4.972	4.972	4.972
14	4.921	4.929	4.935	4.938	4.940	4.940	4.940	4.940	4.940	4.940	4.940	4.940	4.940	4.940	4.940	4.940	4.940
15	4.887	4.897	4.904	4.909	4.912	4.914	4.914	4.914	4.914	4.914	4.914	4.914	4.914	4.914	4.914	4.914	4.914
16	4.858	4.869	4.877	4.883	4.887	4.890	4.892	4.892	4.892	4.892	4.892	4.892	4.892	4.892	4.892	4.892	4.892
17	4.832	4.844	4.853	4.860	4.865	4.869	4.872	4.873	4.874	4.874	4.874	4.874	4.874	4.874	4.874	4.874	4.874
18	4.808	4.821	4.832	4.839	4.846	4.850	4.854	4.856	4.857	4.858	4.858	4.858	4.858	4.858	4.858	4.858	4.858
19	4.788	4.802	4.812	4.821	4.828	4.833	4.838	4.841	4.843	4.844	4.845	4.845	4.845	4.845	4.845	4.845	4.845
20	4.769	4.786	4.795	4.805	4.813	4.818	4.823	4.827	4.830	4.832	4.833	4.833	4.833	4.833	4.833	4.833	4.833
24	4.710	4.727	4.741	4.752	4.762	4.770	4.777	4.783	4.788	4.791	4.794	4.802	4.802	4.802	4.802	4.802	4.802
30	4.650	4.669	4.685	4.699	4.711	4.721	4.730	4.738	4.744	4.750	4.755	4.772	4.777	4.777	4.777	4.777	4.777
40	4.591	4.611	4.630	4.645	4.659	4.671	4.682	4.692	4.700	4.708	4.715	4.740	4.754	4.761	4.764	4.764	4.764
60	4.530	4.553	4.573	4.591	4.607	4.620	4.633	4.645	4.655	4.665	4.673	4.707	4.730	4.745	4.755	4.761	4.665
120	4.469	4.494	4.516	4.535	4.552	4.568	4.583	4.596	4.609	4.619	4.630	4.673	4.703	4.727	4.745	4.759	4.770
∞	4.408	4.434	4.457	4.478	4.497	4.514	4.530	4.545	4.559	4.572	4.584	4.635	4.675	4.707	4.734	4.756	4.776

Source: Reproduced from H. L. Harter, "Critical Values for Duncan's Multiple Range Test." *Biometrics*, **16**, 671–685 (1960). With permission from the Biometric Society.

Table A23(a)   Confidence Limits ($\times 10^3$) for Proportions* (Confidence Coefficient 0.95)

y Denotes the Numerator and n the Denominator of the Relative Frequency

	n − y	1	2	3	4	5	6	7	8	9
	y				Denominator Minus Numerator					
	0	975	842	708	602	522	459	410	369	336
		000	000	000	000	000	000	000	000	000
	1	987	906	806	716	641	579	527	483	445
		013	008	006	005	004	004	003	003	003
	2	992	932	853	777	710	651	600	556	518
		094	068	053	043	037	032	028	025	023
	3	994	947	882	816	755	701	652	610	572
		194	147	118	099	085	075	067	060	055
	4	995	957	901	843	788	738	692	651	614
		284	223	184	157	137	122	109	099	091
	5	996	963	915	863	813	766	723	684	649
		359	290	245	212	187	167	151	139	128
	6	996	968	925	878	833	789	749	711	677
		421	349	299	262	234	211	192	177	163
	7	997	972	933	891	849	808	770	734	701
		473	400	348	308	277	251	230	213	198
	8	997	975	940	901	861	823	787	753	722
		517	444	390	349	316	289	266	247	230
	9	997	977	945	909	872	837	802	770	740
		555	482	428	386	351	323	299	278	260
	10	998	979	950	916	882	848	816	785	756
		587	516	462	419	384	354	329	308	289
	11	998	981	953	922	890	858	827	797	769
		615	546	492	449	413	383	357	335	315
	12	998	982	957	927	897	867	837	809	782
		640	572	519	476	440	410	384	361	340
	13	998	983	960	932	903	874	846	819	793
		661	595	544	501	465	435	408	384	364
	14	998	984	962	936	909	881	854	828	803
		681	617	566	524	488	457	430	407	385
	15	998	985	964	939	913	887	861	836	812
		698	636	586	544	509	478	451	427	406
	20	999	989	972	953	932	910	889	868	847
		762	708	664	626	593	564	537	513	492
	50	1000	995	988	979	970	960	949	939	928
		896	868	843	821	800	781	763	746	730

Numerator of the Relative Frequency (left vertical axis label)

*The pair of entries for each combination of y and $n - y$ are upper and lower confidence limits on the binomial parameter θ. For example, for $y = 5$ and $n - y = 6$, a 95% confidence interval is $0.167 < \theta < 0.766$.

Table A23(a) (continued)   Confidence Limits ($\times 10^3$) for Proportions (Confidence Coefficient 0.95)

y Denotes the Numerator and n the Denominator of the Relative Frequency

$n-y$ / y	Denominator Minus Numerator							
	10	11	12	13	14	15	20	50
0	308	285	265	247	232	218	168	071
	000	000	000	000	000	000	000	000
1	413	385	360	339	319	302	238	104
	002	002	002	002	002	002	001	001
2	484	454	428	405	383	364	292	132
	021	019	018	017	016	015	011	005
3	538	508	481	456	434	414	336	157
	050	047	043	040	038	036	028	012
4	581	551	524	499	476	456	374	179
	084	078	073	068	064	061	047	021
5	616	587	560	535	512	491	407	200
	118	110	103	097	091	087	068	030
6	646	617	590	565	543	522	436	219
	152	142	133	126	119	113	090	040
7	671	643	616	592	570	549	463	237
	184	173	163	154	146	139	111	051
8	692	665	639	616	593	573	487	254
	215	203	191	181	172	164	132	061
9	711	685	660	636	615	594	508	270
	244	231	218	207	197	188	153	072
10	728	702	678	655	634	614	528	286
	272	257	244	232	221	211	173	083
11	743	718	694	672	651	631	546	300
	298	282	268	256	244	234	192	094
12	756	732	709	687	666	647	563	314
	322	306	291	278	266	255	211	104
13	768	744	722	701	680	661	579	327
	345	328	313	299	287	275	229	115
14	779	756	734	713	694	675	593	340
	366	349	334	320	306	295	247	125
15	789	766	745	725	705	687	607	352
	386	369	353	339	325	313	263	135
20	827	808	789	771	753	737	662	406
	472	454	437	421	407	393	338	184
50	917	906	896	885	875	865	816	602
	714	700	686	673	660	648	594	398

Numerator of the Relative Frequency (row axis label)

*The pair of entries for each combination of y and $n-y$ are upper and lower confidence limits on the binomial parameter θ. For example, for $y = 5$ and $n - y = 6$, a 95% confidence interval is $0.167 < \theta < 0.766$.

y Denotes the Numerator and n the Denominator of the Relative Frequency

$n - y$ / y	Denominator Minus Numerator								
	1	2	3	4	5	6	7	8	9
0	995	929	829	734	653	586	531	484	445
	000	000	000	000	000	000	000	000	000
1	997	959	889	815	746	685	632	585	544
	003	002	001	001	001	001	001	001	001
2	998	971	917	856	797	742	693	648	608
	041	029	023	019	016	014	012	011	010
3	999	977	934	882	830	781	735	693	655
	111	083	066	055	047	042	037	033	030
4	999	981	945	900	854	809	767	728	691
	185	144	118	100	087	077	069	062	057
5	999	984	953	913	872	831	791	755	720
	254	203	170	146	128	114	103	094	087
6	999	986	958	923	886	848	811	777	744
	315	258	219	191	169	152	138	127	117
7	999	988	963	931	897	862	828	795	764
	368	307	265	233	209	189	172	159	147
8	999	989	967	938	906	873	841	811	781
	415	352	307	272	245	223	205	189	176
9	999	990	970	943	913	883	853	824	795
	456	392	345	309	280	256	236	219	205
10	1000	991	972	947	920	891	863	835	808
	491	427	379	342	312	286	265	247	232
11	1000	992	974	951	925	899	872	845	819
	523	459	411	373	342	315	293	274	257
12	1000	992	976	955	930	905	879	854	829
	551	488	439	401	369	342	319	299	282
13	1000	993	978	957	935	910	886	862	838
	576	514	466	427	395	367	343	323	305
14	1000	993	979	960	938	915	892	869	846
	598	537	490	451	418	390	366	345	326
15	1000	994	980	962	942	920	898	875	854
	619	559	512	473	440	412	388	366	347
20	1000	995	985	971	954	936	918	900	881
	696	642	599	562	530	502	478	455	435
50	1000	998	994	987	980	972	963	954	945
	863	834	808	785	763	743	725	708	691

*The pair of entries for each combination of y and $n - y$ are upper and lower confidence limits on the binomial parameter θ. For example, for $y = 5$ and $n - y = 6$, a 95% confidence interval is $0.167 < \theta < 0.766$.

Table A23(b) (continued)   Confidence Limits ($\times 10^3$) for Proportions (Confidence Coefficient 0.99)

y Denotes the Numerator and n the Denominator of the Relative Frequency

y \ n − y	Denominator Minus Numerator							
	10	11	12	13	14	15	20	50
0	411	382	357	335	315	298	233	101
	000	000	000	000	000	000	000	000
1	509	477	449	424	402	381	304	137
	000	000	000	000	000	000	000	000
2	573	541	512	486	463	441	358	166
	009	008	008	007	007	006	005	002
3	621	589	561	534	510	488	401	192
	028	026	024	022	021	020	015	006
4	658	627	599	573	549	527	438	215
	053	049	045	043	040	038	029	013
5	688	658	631	605	582	560	470	237
	080	075	070	065	062	058	046	020
6	714	685	658	633	610	588	498	257
	109	101	095	090	085	080	064	028
7	735	707	681	657	634	612	522	275
	137	128	121	114	108	102	082	037
8	753	726	701	677	655	634	545	292
	165	155	146	138	131	125	100	046
9	768	743	718	695	674	653	565	309
	192	181	171	162	154	146	119	055
10	782	758	734	712	690	670	583	324
	218	205	195	185	176	168	137	065
11	795	771	748	726	705	686	600	339
	242	229	218	207	197	189	155	074
12	805	782	760	739	719	700	616	352
	266	252	240	228	218	209	172	084
13	815	793	772	751	731	713	630	366
	288	274	261	249	238	228	189	094
14	824	803	782	762	743	724	643	379
	310	295	281	269	257	247	206	103
15	832	811	791	772	753	735	655	390
	330	314	300	287	276	265	222	112
20	863	845	828	811	794	778	705	443
	417	400	384	370	357	345	295	158
50	935	926	916	906	897	888	842	631
	676	661	648	634	621	610	557	369

Numerator of the Relative Frequency (vertical axis label)

*The pair of entries for each combination of y and $n − y$ are upper and lower confidence limits on the binomial parameter θ. For example, for $y = 5$ and $n − y = 6$, a 95% confidence interval is $0.167 < \theta < 0.766$.

Source: Adapted and reprinted with permission from *CRC Handbook of Tables for Probability and Statistics*, Second Edition, Edited by W. H. Beyer, The Chemical Rubber Company. Copyright CRC Press, Inc., Boca Raton, FL.

Table A24 Individual Terms of the Poisson Distribution*

	λ									
y	0.1	0.2	0.3	0.4	0.5	0.6	0.7	0.8	0.9	1.0
0	.905	.819	.741	.670	.607	.549	.497	.449	.407	.368
1	.090	.164	.222	.268	.303	.329	.348	.359	.366	.368
2	.005	.016	.033	.054	.076	.099	.122	.144	.165	.184
3		.001	.003	.007	.013	.020	.028	.038	.049	.061
4				.001	.002	.003	.005	.008	.011	.015
5							.001	.001	.002	.003
6										.001

	λ									
y	1.1	1.2	1.3	1.4	1.5	1.6	1.7	1.8	1.9	2.0
0	.333	.301	.273	.247	.223	.202	.183	.165	.150	.135
1	.366	.361	.354	.345	.335	.323	.311	.298	.284	.271
2	.201	.217	.230	.242	.251	.258	.264	.268	.270	.271
3	.074	.087	.100	.113	.126	.138	.150	.161	.171	.180
4	.020	.026	.032	.039	.048	.055	.064	.072	.081	.090
5	.004	.006	.009	.011	.014	.018	.022	.026	.031	.036
6	.001	.001	.002	.003	.004	.005	.006	.008	.010	.012
7				.001	.001	.001	.001	.002	.003	.003
8									.001	.001

	λ									
y	2.5	3.0	3.5	4.0	4.5	5.0	5.5	6.0	6.5	7.0
0	.082	.050	.030	.018	.011	.007	.004	.002	.002	.001
1	.205	.149	.106	.073	.050	.034	.022	.015	.010	.006
2	.257	.224	.185	.147	.112	.084	.062	.045	.032	.022
3	.214	.224	.216	.195	.169	.140	.113	.089	.069	.052
4	.134	.168	.189	.195	.190	.175	.156	.134	.112	.091
5	.067	.101	.132	.156	.171	.175	.171	.161	.145	.128
6	.028	.050	.077	.104	.128	.146	.157	.161	.157	.149
7	.010	.022	.039	.060	.082	.104	.123	.138	.146	.149
8	.003	.008	.017	.030	.046	.065	.085	.103	.119	.130
9	.001	.003	.007	.013	.023	.036	.052	.069	.086	.101
10		.001	.002	.005	.010	.018	.029	.041	.056	.071
11			.001	.002	.004	.008	.014	.023	.033	.045
12				.001	.002	.003	.007	.011	.018	.026
13					.001	.001	.003	.005	.009	.014
14							.001	.002	.004	.007
15								.001	.002	.003
16									.001	.001
17										.001

*An entry in this table is the probability that a Poisson variate with parameter λ equals the value y.

Table A24 (continued)   Individual Terms of the Poisson Distribution

y	7.5	8.0	8.5	9.0	9.5	10	11	12	13	14	15	20
						λ						
0	.001	.000	.000	.000	.000	.000	.000	.000	.000	.000	.000	.000
1	.004	.003	.002	.001	.001	.000	.000	.000	.000	.000	.000	.000
2	.016	.011	.007	.005	.003	.002	.001	.000	.000	.000	.000	.000
3	.039	.029	.021	.015	.011	.008	.004	.002	.001	.000	.000	.000
4	.073	.057	.044	.034	.025	.019	.010	.005	.003	.001	.001	.000
5	.109	.092	.075	.061	.048	.038	.022	.013	.007	.004	.002	.000
6	.137	.122	.107	.091	.076	.063	.041	.025	.015	.009	.005	.000
7	.146	.140	.129	.117	.104	.090	.065	.044	.028	.017	.010	.000
8	.137	.140	.138	.132	.123	.113	.089	.066	.046	.030	.019	.001
9	.114	.124	.130	.132	.130	.125	.109	.087	.066	.047	.032	.003
10	.086	.099	.110	.119	.124	.125	.119	.105	.086	.066	.049	.006
11	.059	.072	.085	.097	.107	.114	.119	.114	.101	.084	.066	.011
12	.037	.048	.060	.073	.084	.095	.109	.114	.110	.098	.083	.018
13	.021	.030	.040	.050	.062	.073	.093	.106	.110	.106	.096	.027
14	.011	.017	.024	.032	.042	.052	.073	.090	.102	.106	.102	.039
15	.006	.009	.014	.019	.027	.035	.053	.072	.088	.099	.102	.052
16	.003	.005	.007	.011	.016	.022	.037	.054	.072	.087	.096	.065
17	.001	.002	.004	.006	.009	.013	.024	.038	.055	.071	.085	.076
18		.001	.002	.003	.005	.007	.015	.026	.040	.055	.071	.084
19			.001	.001	.002	.004	.008	.016	.027	.041	.056	.089
20				.001	.001	.002	.005	.010	.018	.029	.042	.089
21						.001	.002	.006	.011	.019	.030	.085
22							.001	.003	.006	.012	.020	.077
23							.001	.002	.004	.007	.013	.067
24								.001	.002	.004	.008	.056
25									.001	.002	.005	.045
26									.001	.001	.003	.034
27										.001	.002	.025
28											.001	.018
29												.012
30												.008
31												.005
32												.003
33												.002
34												.001
35												.001

Source: Adapted and reprinted with permission from *CRC Standard Mathematical Tables*, 15th Ed., Edited by S. M. Selby, The Chemical Rubber Company. Copyright CRC Press, Inc., Boca Raton, FL.

Table A25 Upper-Tail Critical Values for the Spearman Rank-Correlation Test*

n	p = .900	.950	.975	.990	.995	.999
4	.8000	.8000				
5	.7000	.8000	.9000	.9000		
6	.6000	.7714	.8286	.8857	.9429	
7	.5357	.6786	.7450	.8571	.8929	.9643
8	.5000	.6190	.7143	.8095	.8571	.9286
9	.4667	.5833	.6833	.7667	.8167	.9000
10	.4424	.5515	.6364	.7333	.7818	.8667
11	.4182	.5273	.6091	.7000	.7455	.8364
12	.3986	.4965	.5804	.6713	.7273	.8182
13	.3791	.4780	.5549	.6429	.6978	.7912
14	.3626	.4593	.5341	.6220	.6747	.7670
15	.3500	.4429	.5179	.6000	.6536	.7464
16	.3382	.4265	.5000	.5824	.6324	.7265
17	.3260	.4118	.4853	.5637	.6152	.7083
18	.3148	.3994	.4716	.5480	.5975	.6904
19	.3070	.3895	.4579	.5333	.5825	.6737
20	.2977	.3789	.4451	.5203	.5684	.6586
21	.2909	.3688	.4351	.5078	.5545	.6455
22	.2829	.3597	.4241	.4963	.5426	.6318
23	.2767	.3518	.4150	.4852	.5306	.6186
24	.2704	.3435	.4061	.4748	.5200	.6070
25	.2646	.3362	.3977	.4654	.5100	.5962
26	.2588	.3299	.3894	.4564	.5002	.5856
27	.2540	.3236	.3822	.4481	.4915	.5757
28	.2490	.3175	.3749	.4401	.4828	.5660
29	.2443	.3113	.3685	.4320	.4744	.5567
30	.2400	.3059	.3620	.4251	.4665	.5479

*For n greater than 30 the approximate quantiles of r_s may be obtained from

$$w_p \approx \frac{x_p}{\sqrt{n-1}}$$

where x_p is the pth quantile of a standard normal random variable obtained from Table A4. The entries in this table are selected quantiles of the Spearman rank-correlation coefficient r_s when used as a test statistic. The lower quantiles may be obtained from the equation

$$w_p = -w_{,1-p}$$

The critical region corresponds to values of r_s smaller than (or greater than) but not including the appropriate quantile.

Source: Adapted from Glasser, G. J. and Winter, R. F. (1961). "Critical values of the coefficient of rank correlation for testing the hypothesis of independence," *Biometrika*, **48**, 444–448. Copyright Biometrika Trustees. Reprinted with permission.

Table A26 Critical Values for Outlier Test Using L_k and S_k

Critical Value × 10³

k	α	n = 50	45	40	35	30	25	20	19	18	17	16	15	14	13	12	11	10	9	8	7	6	5	4	3
1	0.01	760	745	722	690	650	607	539	522	504	485	463	440	414	386	355	321	283	241	195	145	93	44	10	
	0.025	796	776	756	732	699	654	594	579	562	544	525	503	479	453	423	390	353	310	262	207	145	81	25	1
	0.05	820	802	784	762	730	692	638	624	610	593	576	556	534	510	482	451	415	374	326	270	203	127	49	3
	0.10	840	826	812	792	766	732	685	673	660	646	631	613	594	573	548	520	488	450	405	350	283	199	98	11
2	0.01	667	641	610	573	527	468	391	373	353	332	310	286	261	233	204	174	141	108	75	44	19	4		
	0.025	697	667	644	610	567	512	439	421	403	382	360	337	311	284	254	221	186	149	110	71	35	9	1	
	0.05	720	698	673	641	601	550	480	464	446	426	405	382	357	330	300	267	230	191	148	102	56	18	3	
	0.10	746	726	702	674	637	591	527	511	494	476	456	435	411	384	355	323	286	245	199	148	92	38		
3	0.01	592	558	522	484	434	377	300	272	260	237	219	194	172	147	120	98	70	48	28	10	2			
	0.025	622	592	561	527	479	417	341	321	299	282	261	239	214	184	162	129	100	73	45	21	5			
	0.05	646	618	588	554	506	450	377	354	337	322	300	276	250	224	196	162	129	99	64	32	10			
	0.10	673	648	622	586	523	489	420	398	384	364	342	322	298	270	240	208	170	134	95	56	20			
4	0.01	531	498	460	418	369	308	231	211	192	171	151	132	113	94	70	52	32	18	8					
	0.025	559	529	491	455	408	342	265	243	226	208	185	167	145	122	96	74	52	30	13					
	0.05	588	556	523	482	434	374	299	277	259	240	219	197	174	150	125	98	70	45	22					
	0.10	614	586	554	516	472	412	339	316	302	282	260	236	212	186	159	128	98	68	38					
5	0.01	483	444	408	364	312	246	175	154	140	126	108	90	72	56	38	26	12							
	0.025	510	473	433	398	352	282	209	189	171	151	135	113	95	77	57	40	23							
	0.05	535	502	468	424	376	312	238	217	200	181	159	140	122	98	76	54	34							
	0.10	562	533	499	458	411	350	273	251	236	216	194	172	150	126	103	74	51							
6	0.01	438	399	364	321	268	204	136	118	104	91	72	57	46	33	19									
	0.025	466	430	387	348	302	233	165	145	129	117	96	78	63	47	31									
	0.05	490	456	421	376	327	262	188	168	154	136	115	97	79	60	42									
	0.10	518	488	451	410	359	296	220	199	184	165	144	124	104	82	62									

k	α	n=50	45	40	35	30	25	20	19	18	17	16	15	14	13	12	11	10	9	8	7	6	5	4	3
7	0.01	400	361	324	282	229	168	104	88	76	64	49	37	27											
	0.025	428	391	348	308	261	192	128	108	95	82	65	51	38											
	0.05	450	417	378	334	283	222	150	130	116	100	82	66	50											
	0.10	477	447	408	365	316	251	176	158	142	125	104	86	68											
8	0.01	368	328	292	250	196	144	78	64	53	44	30													
	0.025	392	356	314	274	226	159	98	80	68	58	45													
	0.05	414	382	342	297	245	184	115	99	86	72	55													
	0.10	442	410	372	328	276	213	140	124	108	92	73													
9	0.01	336	296	262	220	166	112	58	46	36															
	0.025	363	325	283	242	193	132	73	59	48															
	0.05	383	350	310	264	212	154	88	74	62															
	0.10	410	378	338	294	240	180	110	94	80															
10	0.01	308	270	234	194	142	92	42																	
	0.025	334	295	257	213	165	108	54																	
	0.05	356	320	280	235	183	126	66																	
	0.10	380	348	307	262	210	152	85																	

Source: Adapted from Tietjen, G. L. and Moore, R. H. (1972). "Some Grubbs-Type Statistics for the Detection of Several Outliers," *Technometrics*, **14**, 583–598. Copyright American Statistical Association, Alexandria, VA. Reprinted by permission.

Adapted from Grubbs, F. E. (1950). "Sample Criteria for Testing Outlying Observations," *Annals of Mathematical Statistics*, **21**, 27–58. Used by permission of the Institute of Mathematical Statistics.

Adapted from Grubbs, F. E. and Beck, G. (1972). "Extension of Sample Sizes and Percentage Points for Significance Tests of Outlying Observations," *Technometrics*, **14**, 847–854. Copyright American Statistical Association, Alexandria, VA. Reprinted by permission.

Table A27 Critical Values for Outlier Test Using E_k

															Critical Value × 10³											
k	α	n = 50	45	40	35	30	25	20	19	18	17	16	15	14	13	12	11	10	9	8	7	6	5	4	3
1	0.01	748	728	704	669	624	571	499	484	459	440	422	404	374	337	311	274	235	197	156	110	68	29	4	
	0.05	796	776	756	732	698	654	594	579	562	544	525	503	479	453	423	390	353	310	262	207	145	81	25	1
	0.10	820	802	784	762	730	692	638	624	610	593	576	556	534	510	482	451	415	374	326	270	203	127	49	3
2	0.01	636	607	574	533	482	418	339	323	306	290	263	238	207	181	159	134	101	78	50	28	12	2		
	0.05	684	658	629	596	549	493	416	398	382	362	340	317	293	262	234	204	172	137	99	65	34	10	1	
	0.10	708	684	657	624	582	528	460	442	424	406	384	360	337	309	278	250	214	175	137	94	56	22	2	
3	0.01	550	518	480	435	386	320	236	219	206	188	166	146	123	103	83	64	44	26	14	6	1			
	0.05	599	567	534	495	443	381	302	287	267	248	227	206	179	156	133	107	83	57	34	16	4			
	0.10	622	593	562	523	475	417	338	322	304	284	263	240	216	189	162	138	108	80	53	27	9			
4	0.01	482	446	408	364	308	245	170	156	141	122	107	90	72	56	42	30	18	9	4					
	0.05	529	492	458	417	364	298	221	203	187	170	153	134	112	92	73	55	37	21	10					
	0.10	552	522	486	443	391	331	252	234	217	198	182	160	138	116	94	73	52	32	16					
5	0.01	424	386	347	299	250	188	121	108	94	79	68	54	42	31	20	12	6							
	0.05	468	433	395	351	298	236	163	146	132	116	102	84	68	53	39	26	14							
	0.10	492	459	422	379	325	264	188	172	156	140	122	105	86	68	52	36	22							
6	0.01	376	336	298	252	204	146	86	74	62	52	40	32	22	14	8									
	0.05	417	381	343	298	246	186	119	105	91	78	67	52	39	28	18									
	0.10	440	406	367	324	270	210	138	124	110	95	82	67	52	38	26									

k	α	n=50	45	40	35	30	25	20	19	18	17	16	15	14	13	12	11	10	9	8	7	6	5	4	3
7	0.01	334	294	258	211	166	110	58	50	41	32	24	18	12											
	0.05	373	337	297	254	203	146	85	74	62	50	41	30	21											
	0.10	396	360	320	276	224	168	102	89	76	64	53	40	29											
8	0.01	297	258	220	177	132	87	40	32	26	18	14													
	0.05	334	299	259	214	166	114	59	50	41	32	24													
	0.10	355	320	278	236	186	132	72	62	51	42	32													
9	0.01	264	228	190	149	108	66	26	20	14															
	0.05	299	263	223	181	137	89	41	33	26															
	0.10	319	284	243	202	154	103	51	42	34															
10	0.01	235	200	164	124	87	50	17																	
	0.05	268	233	195	154	112	68	28																	
	0.10	287	252	212	172	126	80	35																	

Source: Adapted from Tietjen, G. L. and Moore, R. H. (1972). "Some Grubbs-Type Statistics for the Detection of Several Outliers," *Technometrics*, **14**, 583–598. Copyright American Statistical Association, Alexandria, VA. Reprinted by permission.

Adapted from Grubbs, F. E. (1950). "Sample Criteria for Testing Outlying Observations," *Annals of Mathematical Statistics*, **21**, 27–58. Used by permission of the Institute of Mathematical Statistics.

Table A28 Coefficients Used in the Shapiro–Wilk Test for Normality*

a_{n-i+1}

i	$n=3$	4	5	6	7	8	9	10	11	12	13	14
1	0.7071	0.6872	0.6646	0.6431	0.6233	0.6052	0.5888	0.5739	0.5601	0.5475	0.5359	0.5251
2		0.1677	0.2413	0.2806	0.3031	0.3164	0.3244	0.3291	0.3315	0.3325	0.3325	0.3318
3				0.0875	0.1401	0.1743	0.1976	0.2141	0.2260	0.2347	0.2412	0.2460
4						0.0561	0.0947	0.1224	0.1429	0.1586	0.1707	0.1802
5								0.0399	0.0695	0.0922	0.1099	0.1240
6										0.0303	0.0539	0.0727
7												0.0240

i	$n=15$	16	17	18	19	20	21	22	23	24	25	26
1	0.5150	0.5056	0.4968	0.4886	0.4808	0.4734	0.4643	0.4590	0.4542	0.4493	0.4450	0.4407
2	0.3306	0.3290	0.3273	0.3253	0.3232	0.3211	0.3185	0.3156	0.3126	0.3098	0.3069	0.3043
3	0.2495	0.2521	0.2540	0.2553	0.2561	0.2565	0.2578	0.2571	0.2563	0.2554	0.2543	0.2533
4	0.1878	0.1939	0.1988	0.2027	0.2059	0.2085	0.2119	0.2131	0.2139	0.2145	0.2148	0.2151
5	0.1353	0.1447	0.1524	0.1587	0.1641	0.1686	0.1736	0.1764	0.1787	0.1807	0.1822	0.1836
6	0.0880	0.1005	0.1109	0.1197	0.1271	0.1334	0.1399	0.1443	0.1480	0.1512	0.1539	0.1563
7	0.0433	0.0593	0.0725	0.0837	0.0932	0.1013	0.1092	0.1150	0.1201	0.1245	0.1283	0.1316
8		0.0196	0.0359	0.0496	0.0612	0.0711	0.0804	0.0878	0.0941	0.0997	0.1046	0.1089
9				0.0163	0.0303	0.0422	0.0530	0.0618	0.0696	0.0764	0.0823	0.0876
10						0.0140	0.0263	0.0368	0.0459	0.0539	0.0610	0.0672
11								0.0122	0.0228	0.0321	0.0403	0.0476
12										0.0107	0.0200	0.0284
13												0.0094

*$a_i = -a_{n-i+1}$ for $i = 1, 2, \ldots, k$ where $k = n/2$ if n is even and $k = (n-1)/2$ if n is odd.

	$n=27$	28	29	30	31	32	33	34	35	36	37	38
1	0.4366	0.4328	0.4291	0.4254	0.4220	0.4188	0.4156	0.4127	0.4096	0.4068	0.4040	0.4015
2	0.3018	0.2992	0.2968	0.2944	0.2921	0.2898	0.2876	0.2854	0.2834	0.2813	0.2794	0.2774
3	0.2522	0.2510	0.2499	0.2487	0.2475	0.2463	0.2451	0.2439	0.2427	0.2415	0.2403	0.2391
4	0.2152	0.2151	0.2150	0.2148	0.2145	0.2141	0.2137	0.2132	0.2127	0.2121	0.2116	0.2110
5	0.1848	0.1857	0.1864	0.1870	0.1874	0.1878	0.1880	0.1882	0.1883	0.1883	0.1883	0.1881
6	0.1584	0.1601	0.1616	0.1630	0.1641	0.1651	0.1660	0.1667	0.1673	0.1678	0.1683	0.1686
7	0.1346	0.1372	0.1395	0.1415	0.1433	0.1449	0.1463	0.1475	0.1487	0.1496	0.1505	0.1513
8	0.1128	0.1162	0.1192	0.1219	0.1243	0.1265	0.1284	0.1301	0.1317	0.1331	0.1344	0.1356
9	0.0923	0.0965	0.1002	0.1036	0.1066	0.1093	0.1118	0.1140	0.1160	0.1179	0.1196	0.1211
10	0.0728	0.0778	0.0822	0.0862	0.0899	0.0931	0.0961	0.0988	0.1013	0.1036	0.1056	0.1075
11	0.0540	0.0598	0.0650	0.0697	0.0739	0.0777	0.0812	0.0844	0.0873	0.0900	0.0924	0.0947
12	0.0358	0.0424	0.0483	0.0537	0.0585	0.0629	0.0669	0.0706	0.0739	0.0770	0.0798	0.0824
13	0.0178	0.0253	0.0320	0.0381	0.0435	0.0485	0.0530	0.0572	0.0610	0.0645	0.0677	0.0706
14		0.0084	0.0159	0.0227	0.0289	0.0344	0.0395	0.0441	0.0484	0.0523	0.0559	0.0592
15				0.0076	0.0144	0.0206	0.0262	0.0314	0.0361	0.0404	0.0444	0.0481
16						0.0068	0.0131	0.0187	0.0239	0.0287	0.0331	0.0372
17								0.0062	0.0119	0.0172	0.0220	0.0264
18										0.0057	0.0110	0.0158
19												0.0053

Table A28 (continued)

	n = 39	40	41	42	43	44	45	46	47	48	49	50
1	0.3989	0.3964	0.3940	0.3917	0.3894	0.3872	0.3850	0.3830	0.3808	0.3789	0.3770	0.3751
2	0.2755	0.2737	0.2719	0.2701	0.2684	0.2667	0.2651	0.2635	0.2620	0.2604	0.2589	0.2574
3	0.2380	0.2368	0.2357	0.2345	0.2334	0.2323	0.2313	0.2302	0.2291	0.2281	0.2271	0.2260
4	0.2104	0.2098	0.2091	0.2085	0.2078	0.2072	0.2065	0.2058	0.2052	0.2045	0.2038	0.2032
5	0.1880	0.1878	0.1876	0.1874	0.1871	0.1868	0.1865	0.1862	0.1859	0.1855	0.1851	0.1847
6	0.1689	0.1691	0.1693	0.1694	0.1695	0.1695	0.1695	0.1695	0.1695	0.1693	0.1692	0.1691
7	0.1520	0.1526	0.1531	0.1535	0.1539	0.1542	0.1545	0.1548	0.1550	0.1551	0.1553	0.1554
8	0.1366	0.1376	0.1384	0.1392	0.1398	0.1405	0.1410	0.1415	0.1420	0.1423	0.1427	0.1430
9	0.1225	0.1237	0.1249	0.1259	0.1269	0.1278	0.1286	0.1293	0.1300	0.1306	0.1312	0.1317
10	0.1092	0.1108	0.1123	0.1136	0.1149	0.1160	0.1170	0.1180	0.1189	0.1197	0.1205	0.1212
11	0.0967	0.0986	0.1004	0.1020	0.1035	0.1049	0.1062	0.1073	0.1085	0.1095	0.1105	0.1113
12	0.0848	0.0870	0.0891	0.0909	0.0927	0.0943	0.0959	0.0972	0.0986	0.0998	0.1010	0.1020
13	0.0733	0.0759	0.0782	0.0804	0.0824	0.0842	0.0860	0.0876	0.0892	0.0906	0.0919	0.0932
14	0.0622	0.0651	0.0677	0.0701	0.0724	0.0745	0.0765	0.0783	0.0801	0.0817	0.0832	0.0846
15	0.0515	0.0546	0.0575	0.0602	0.0628	0.0651	0.0673	0.0694	0.0713	0.0731	0.0748	0.0764
16	0.0409	0.0444	0.0476	0.0506	0.0534	0.0560	0.0584	0.0607	0.0628	0.0648	0.0667	0.0685
17	0.0305	0.0343	0.0379	0.0411	0.0442	0.0471	0.0497	0.0522	0.0546	0.0568	0.0588	0.0608
18	0.0203	0.0244	0.0283	0.0318	0.0352	0.0383	0.0412	0.0439	0.0465	0.0489	0.0511	0.0532
19	0.0101	0.0146	0.0188	0.0227	0.0263	0.0296	0.0328	0.0357	0.0385	0.0411	0.0436	0.0459
20		0.0049	0.0094	0.0136	0.0175	0.0211	0.0245	0.0277	0.0307	0.0335	0.0361	0.0386
21				0.0045	0.0087	0.0126	0.0163	0.0197	0.0229	0.0259	0.0288	0.0314
22						0.0042	0.0081	0.0118	0.0153	0.0185	0.0215	0.0244
23								0.0039	0.0076	0.0111	0.0143	0.0174
24										0.0037	0.0071	0.0104
25												0.0035

*$a_i = -a_{n-i+1}$ for $i = 1, 2, \ldots, k$ where $k = n/2$ if n is even and $k = (n-1)/2$ if n is odd.

Source: Shapiro, S. S. and Wilk, M. B. (1965). "An Analysis of Variance Test for Normality (Complete Samples)," *Biometrika*, **52**, 591–611. Copyright Biometrika Trustees. Reprinted with permission.

Table A29 Critical Values for the Shapiro–Wilk Test for Normality

n	$\alpha = 1\%$	2%	Critical Value 5%	10%	50%
3	0.753	0.756	0.767	0.789	0.959
4	0.687	0.707	0.748	0.792	0.935
5	0.686	0.715	0.762	0.806	0.927
6	0.713	0.743	0.788	0.826	0.927
7	0.730	0.760	0.803	0.838	0.928
8	0.749	0.778	0.818	0.851	0.932
9	0.764	0.791	0.829	0.859	0.935
10	0.781	0.806	0.842	0.869	0.938
11	0.792	0.817	0.850	0.876	0.940
12	0.805	0.828	0.859	0.883	0.943
13	0.814	0.837	0.866	0.889	0.945
14	0.825	0.846	0.874	0.895	0.947
15	0.835	0.855	0.881	0.901	0.950
16	0.844	0.863	0.887	0.906	0.952
17	0.851	0.869	0.892	0.910	0.954
18	0.858	0.874	0.897	0.914	0.956
19	0.863	0.879	0.901	0.917	0.957
20	0.868	0.884	0.905	0.920	0.959
21	0.873	0.888	0.908	0.923	0.960
22	0.878	0.892	0.911	0.926	0.961
23	0.881	0.895	0.914	0.928	0.962
24	0.884	0.898	0.916	0.930	0.963
25	0.888	0.901	0.918	0.931	0.964
26	0.891	0.904	0.920	0.933	0.965
27	0.894	0.906	0.923	0.935	0.965
28	0.896	0.908	0.924	0.936	0.966
29	0.898	0.910	0.926	0.937	0.966
30	0.900	0.912	0.927	0.939	0.967
31	0.902	0.914	0.929	0.940	0.967
32	0.904	0.915	0.930	0.941	0.968
33	0.906	0.917	0.931	0.942	0.968
34	0.908	0.919	0.933	0.943	0.969
35	0.910	0.920	0.934	0.944	0.969
36	0.912	0.922	0.935	0.945	0.970
37	0.914	0.924	0.936	0.946	0.970
38	0.916	0.925	0.938	0.947	0.971
39	0.917	0.927	0.939	0.948	0.971
40	0.919	0.928	0.940	0.949	0.972
41	0.920	0.929	0.941	0.950	0.972
42	0.922	0.930	0.942	0.951	0.972
43	0.923	0.932	0.943	0.951	0.973
44	0.924	0.933	0.944	0.952	0.973
45	0.926	0.934	0.945	0.953	0.973
46	0.927	0.935	0.945	0.953	0.974
47	0.928	0.928	0.946	0.954	0.974
48	0.929	0.937	0.947	0.954	0.974
49	0.929	0.937	0.947	0.955	0.974
50	0.930	0.938	0.947	0.955	0.974

Source: Adapted from Shapiro, S. S. and Wilk, M. B. (1965), "An Analysis of Variance Test for Normality (Complete Samples)," *Biometrika*, **52**, 591–611. Copyright Biometrika Trustees. Reprinted with permission.

Table A30(a) Lower Critical Values of r for the Runs Test* ($\alpha = 0.05$)

n_1	$n_2=2$	3	4	5	6	7	8	9	10	11	12	13	14	15	16	17	18	19	20
2											2	2	2	2	2	2	2	2	2
3				2	2	2	2	2	2	2	2	2	2	3	3	3	3	3	3
4			2	2	2	3	3	3	3	3	3	3	3	4	4	4	4	4	4
5			2	2	3	3	3	3	3	4	4	4	4	4	4	4	5	5	5
6		2	2	3	3	3	3	4	4	4	4	5	5	5	5	5	5	6	6
7		2	2	3	3	3	4	4	5	5	5	5	5	6	6	6	6	6	6
8		2	3	3	4	4	5	5	5	5	6	6	6	6	6	7	7	7	7
9		2	3	3	4	4	5	5	5	6	6	6	7	7	7	7	8	8	8
10		2	3	3	4	5	5	5	6	6	7	7	7	7	8	8	8	8	9
11		2	3	4	4	5	5	6	6	7	7	7	8	8	8	9	9	9	9
12	2	2	3	4	4	5	6	6	7	8	8	9	9	9	9	9	9	10	10
13	2	2	3	4	5	5	6	6	7	7	8	8	9	9	9	10	10	10	10
14	2	2	3	4	5	5	6	7	7	8	8	8	8	8	10	10	10	11	11
15	2	3	3	4	5	6	6	7	7	8	8	9	9	10	10	11	11	11	12
16	2	3	4	4	5	6	6	7	8	8	9	9	10	10	11	11	11	12	12
17	2	3	4	4	5	6	7	7	8	9	9	10	10	11	11	11	12	12	12
18	2	3	4	5	5	6	7	8	8	9	9	10	10	11	11	12	12	13	13
19	2	3	4	5	6	6	7	8	8	9	10	10	11	11	12	12	13	13	13
20	2	3	4	5	6	6	7	8	9	9	10	10	11	12	12	13	13	13	14

*Any value of r that is equal to or smaller than that shown in the body of this table for given values of n_1 and n_2 is significant at the 0.05 level. Tabled values are appropriate for one-tailed test at stated significance level or two-tailed test at twice the significance level.

Source: Adapted from Swed, F. S. and Eisenhart, C. (1943). "Tables for Testing Randomness of Grouping in a Sequence of Alternatives," *Annals of Mathematical Statistics*, **14**, 66–87. Used by permission of the Institute of Mathematical Statistics.

Table A30(b) Upper Critical Values of r for the Runs Test* ($\alpha = 0.05$)

	Upper Critical Value																		
n_1	$n_2 = 2$	3	4	5	6	7	8	9	10	11	12	13	14	15	16	17	18	19	20
2																			
3																			
4				9	9														
5			9	10	10	11	11												
6			9	10	11	12	12	13	13	13	13								
7				11	12	13	13	14	14	14	14	15	15	15					
8				11	12	13	14	14	15	15	16	16	16	16	17	17	17	17	17
9					13	14	14	15	16	16	16	17	17	18	18	18	18	18	18
10					13	14	15	16	16	17	17	18	18	18	19	19	19	20	20
11					13	14	15	16	17	17	18	19	19	19	20	20	20	21	21
12					13	14	16	16	17	18	19	19	20	20	21	21	21	22	22
13						15	16	17	18	19	19	20	20	21	21	22	22	23	23
14						15	16	17	18	19	20	20	21	22	22	23	23	23	24
15						15	16	18	18	19	20	21	22	22	23	23	24	24	25
16							17	18	19	20	21	21	22	23	23	24	25	25	25
17							17	18	19	20	21	22	23	23	24	25	25	26	26
18							17	18	19	20	21	22	23	24	25	25	26	26	27
19							17	18	20	21	22	23	23	24	25	26	26	27	27
20							17	18	20	21	22	23	24	25	25	26	27	27	28

*Any value of r that is equal to or greater than that shown in the body of this table for given values of n_1 and n_2 is significant at the 0.05 level. Tabled values are appropriate for one-tailed test at stated significance level or two-tailed test at twice the significance level.

Source: Adapted from Swed, F. S. and Eisenhart, C. (1943). "Tables for Testing Randomness of Grouping in a Sequence of Alternatives," *Annals of Mathematical Statistics*, **14**, 66–87. Used by permission of the Institute of Mathematical Statistics.

Table A31 Critical Values for the Durbin–Watson Test: $\alpha = 5\%$

n	$p=1$		$p=2$		$p=3$		$p=4$		$p=5$	
	d_L	d_U	d_L	d_U	d_L	d_U	d_L	d_U	d_L	d_U
15	1.08	1.36	0.95	1.54	0.82	1.75	0.69	1.97	0.56	2.21
16	1.10	1.37	0.98	1.54	0.86	1.73	0.74	1.93	0.62	2.15
17	1.13	1.38	1.02	1.54	0.90	1.71	0.78	1.90	0.67	2.10
18	1.16	1.39	1.05	1.53	0.93	1.69	0.82	1.87	0.71	2.06
19	1.18	1.40	1.08	1.53	0.97	1.68	0.86	1.85	0.75	2.02
20	1.20	1.41	1.10	1.54	1.00	1.68	0.90	1.83	0.79	1.99
21	1.22	1.42	1.13	1.54	1.03	1.67	0.93	1.81	0.83	1.96
22	1.24	1.43	1.15	1.54	1.05	1.66	0.96	1.80	0.86	1.94
23	1.26	1.44	1.17	1.54	1.08	1.66	0.99	1.79	0.90	1.92
24	1.27	1.45	1.19	1.55	1.10	1.66	1.01	1.78	0.93	1.90
25	1.29	1.45	1.21	1.55	1.12	1.66	1.04	1.77	0.95	1.89
26	1.30	1.46	1.22	1.55	1.14	1.65	1.06	1.76	0.98	1.88
27	1.32	1.47	1.24	1.56	1.16	1.65	1.08	1.76	1.01	1.86
28	1.33	1.48	1.26	1.56	1.18	1.65	1.10	1.75	1.03	1.85
29	1.34	1.48	1.27	1.56	1.20	1.65	1.12	1.74	1.05	1.84
30	1.35	1.49	1.28	1.57	1.21	1.65	1.14	1.74	1.07	1.83
31	1.36	1.50	1.30	1.57	1.23	1.65	1.16	1.74	1.09	1.83
32	1.37	1.50	1.31	1.57	1.24	1.65	1.18	1.73	1.11	1.82
33	1.38	1.51	1.32	1.58	1.26	1.65	1.19	1.73	1.13	1.81
34	1.39	1.51	1.33	1.58	1.27	1.65	1.21	1.73	1.15	1.81
35	1.40	1.52	1.34	1.58	1.28	1.65	1.22	1.73	1.16	1.80
36	1.41	1.52	1.35	1.59	1.29	1.65	1.24	1.73	1.18	1.80
37	1.42	1.53	1.36	1.59	1.31	1.66	1.25	1.72	1.19	1.80
38	1.43	1.54	1.37	1.59	1.32	1.66	1.26	1.72	1.21	1.79
39	1.43	1.54	1.38	1.60	1.33	1.66	1.27	1.72	1.22	1.79
40	1.44	1.54	1.39	1.60	1.34	1.66	1.29	1.72	1.23	1.79
45	1.48	1.57	1.43	1.62	1.38	1.67	1.34	1.72	1.29	1.78
50	1.50	1.59	1.46	1.63	1.42	1.67	1.38	1.72	1.34	1.77
55	1.53	1.60	1.49	1.64	1.45	1.68	1.41	1.72	1.38	1.77
60	1.55	1.62	1.51	1.65	1.48	1.69	1.44	1.73	1.41	1.77
65	1.57	1.63	1.54	1.66	1.50	1.70	1.47	1.73	1.44	1.77
70	1.58	1.64	1.55	1.67	1.52	1.70	1.49	1.74	1.46	1.77
75	1.60	1.65	1.57	1.68	1.54	1.71	1.51	1.74	1.49	1.77
80	1.61	1.66	1.59	1.69	1.56	1.72	1.53	1.74	1.51	1.77
85	1.62	1.67	1.60	1.70	1.57	1.72	1.55	1.75	1.52	1.77
90	1.63	1.68	1.61	1.70	1.59	1.73	1.57	1.75	1.54	1.78
95	1.64	1.69	1.62	1.71	1.60	1.73	1.58	1.75	1.56	1.78
100	1.65	1.69	1.63	1.72	1.61	1.74	1.59	1.76	1.57	1.78

Table A31 (continued)   Critical Values for the Durbin–Watson Test: $\alpha = 2.5\%$

	$p = 1$		$p = 2$		$p = 3$		$p = 4$		$p = 5$	
n	d_L	d_U	d_L	d_U	d_L	d_U	d_L	d_U	d_L	d_U
15	0.95	1.23	0.83	1.40	0.71	1.61	0.59	1.84	0.48	2.09
16	0.98	1.24	0.86	1.40	0.75	1.59	0.64	1.80	0.53	2.03
17	1.01	1.25	0.90	1.40	0.79	1.58	0.68	1.77	0.57	1.98
18	1.03	1.26	0.93	1.40	0.82	1.56	0.72	1.74	0.62	1.93
19	1.06	1.28	0.96	1.41	0.86	1.55	0.76	1.72	0.66	1.90
20	1.08	1.28	0.99	1.41	0.89	1.55	0.79	1.70	0.70	1.87
21	1.10	1.30	1.01	1.41	0.92	1.54	0.83	1.69	0.73	1.84
22	1.12	1.31	1.04	1.42	0.95	1.54	0.86	1.68	0.77	1.82
23	1.14	1.32	1.06	1.42	0.97	1.54	0.89	1.67	0.80	1.80
24	1.16	1.33	1.08	1.43	1.00	1.54	0.91	1.66	0.83	1.79
25	1.18	1.34	1.10	1.43	1.02	1.54	0.94	1.65	0.86	1.77
26	1.19	1.35	1.12	1.44	1.04	1.54	0.96	1.65	0.88	1.76
27	1.21	1.36	1.13	1.44	1.06	1.54	0.99	1.64	0.91	1.75
28	1.22	1.37	1.15	1.45	1.08	1.54	1.01	1.64	0.93	1.74
29	1.24	1.38	1.17	1.45	1.10	1.54	1.03	1.63	0.96	1.73
30	1.25	1.38	1.18	1.46	1.12	1.54	1.05	1.63	0.98	1.73
31	1.26	1.39	1.20	1.47	1.13	1.55	1.07	1.63	1.00	1.72
32	1.27	1.40	1.21	1.47	1.15	1.55	1.08	1.63	1.02	1.71
33	1.28	1.41	1.22	1.48	1.16	1.55	1.10	1.63	1.04	1.71
34	1.29	1.41	1.24	1.48	1.17	1.55	1.12	1.63	1.06	1.70
35	1.30	1.42	1.25	1.48	1.19	1.55	1.13	1.63	1.07	1.70
36	1.31	1.43	1.26	1.49	1.20	1.56	1.15	1.63	1.09	1.70
37	1.32	1.43	1.27	1.49	1.21	1.56	1.16	1.62	1.10	1.70
38	1.33	1.44	1.28	1.50	1.23	1.56	1.17	1.62	1.12	1.70
39	1.34	1.44	1.29	1.50	1.24	1.56	1.19	1.63	1.13	1.69
40	1.35	1.45	1.30	1.51	1.25	1.57	1.20	1.63	1.15	1.69
45	1.39	1.48	1.34	1.53	1.30	1.58	1.25	1.63	1.21	1.69
50	1.42	1.50	1.38	1.54	1.34	1.59	1.30	1.64	1.26	1.69
55	1.45	1.52	1.41	1.56	1.37	1.60	1.33	1.64	1.30	1.69
60	1.47	1.54	1.44	1.57	1.40	1.61	1.37	1.65	1.33	1.69
65	1.49	1.55	1.46	1.59	1.43	1.62	1.40	1.66	1.36	1.69
70	1.51	1.57	1.48	1.60	1.45	1.63	1.42	1.66	1.39	1.70
75	1.53	1.58	1.50	1.61	1.47	1.64	1.45	1.67	1.42	1.70
80	1.54	1.59	1.52	1.62	1.49	1.65	1.47	1.67	1.44	1.70
85	1.56	1.60	1.53	1.63	1.51	1.65	1.49	1.68	1.46	1.71
90	1.57	1.61	1.55	1.64	1.53	1.66	1.50	1.69	1.48	1.71
95	1.58	1.62	1.56	1.65	1.54	1.67	1.52	1.69	1.50	1.71
100	1.59	1.63	1.57	1.65	1.55	1.67	1.53	1.70	1.51	1.72

Table A31 (continued) Critical Values for the Durbin–Watson Test: $\alpha = 1\%$

n	$p = 1$		$p = 2$		$p = 3$		$p = 4$		$p = 5$	
	d_L	d_U	d_L	d_U	d_L	d_U	d_L	d_U	d_L	d_U
15	0.81	1.07	0.70	1.25	0.59	1.46	0.49	1.70	0.39	1.96
16	0.84	1.09	0.74	1.25	0.63	1.44	0.53	1.66	0.44	1.90
17	0.87	1.10	0.77	1.25	0.67	1.43	0.57	1.63	0.48	1.85
18	0.90	1.12	0.80	1.26	0.71	1.42	0.61	1.60	0.52	1.80
19	0.93	1.13	0.83	1.26	0.74	1.41	0.65	1.58	0.56	1.77
20	0.95	1.15	0.86	1.27	0.77	1.41	0.68	1.57	0.60	1.74
21	0.97	1.16	0.89	1.27	0.80	1.41	0.72	1.55	0.63	1.71
22	1.00	1.17	0.91	1.28	0.83	1.40	0.75	1.54	0.66	1.69
23	1.02	1.19	0.94	1.29	0.86	1.40	0.77	1.53	0.70	1.67
24	1.04	1.20	0.96	1.30	0.88	1.41	0.80	1.53	0.72	1.66
25	1.05	1.21	0.98	1.30	0.90	1.41	0.83	1.52	0.75	1.65
26	1.07	1.22	1.00	1.31	0.93	1.41	0.85	1.52	0.78	1.64
27	1.09	1.23	1.02	1.32	0.95	1.41	0.88	1.51	0.81	1.63
28	1.10	1.24	1.04	1.32	0.97	1.41	0.90	1.51	0.83	1.62
29	1.12	1.25	1.05	1.33	0.99	1.42	0.92	1.51	0.85	1.61
30	1.13	1.26	1.07	1.34	1.01	1.42	0.94	1.51	0.88	1.61
31	1.15	1.27	1.08	1.34	1.02	1.42	0.96	1.51	0.90	1.60
32	1.16	1.28	1.10	1.35	1.04	1.43	0.98	1.51	0.92	1.60
33	1.17	1.29	1.11	1.36	1.05	1.43	1.00	1.51	0.94	1.59
34	1.18	1.30	1.13	1.36	1.07	1.43	1.01	1.51	0.95	1.59
35	1.19	1.31	1.14	1.37	1.08	1.44	1.03	1.51	0.97	1.59
36	1.21	1.32	1.15	1.38	1.10	1.44	1.04	1.51	0.99	1.59
37	1.22	1.32	1.16	1.38	1.11	1.45	1.06	1.51	1.00	1.59
38	1.23	1.33	1.18	1.39	1.12	1.45	1.07	1.52	1.02	1.58
39	1.24	1.34	1.19	1.39	1.14	1.45	1.09	1.52	1.03	1.58
40	1.25	1.34	1.20	1.40	1.15	1.46	1.10	1.52	1.05	1.58
45	1.29	1.38	1.24	1.42	1.20	1.48	1.16	1.53	1.11	1.58
50	1.32	1.40	1.28	1.45	1.24	1.49	1.20	1.54	1.16	1.59
55	1.36	1.43	1.32	1.47	1.28	1.51	1.25	1.55	1.21	1.59
60	1.38	1.45	1.35	1.48	1.32	1.52	1.28	1.56	1.25	1.60
65	1.41	1.47	1.38	1.50	1.35	1.53	1.31	1.57	1.28	1.61
70	1.43	1.49	1.40	1.52	1.37	1.55	1.34	1.58	1.31	1.61
75	1.45	1.50	1.42	1.53	1.39	1.56	1.37	1.59	1.34	1.62
80	1.47	1.52	1.44	1.54	1.42	1.57	1.39	1.60	1.36	1.62
85	1.48	1.53	1.46	1.55	1.43	1.58	1.41	1.60	1.39	1.63
90	1.50	1.54	1.47	1.56	1.45	1.59	1.43	1.61	1.41	1.64
95	1.51	1.55	1.49	1.57	1.47	1.60	1.45	1.62	1.42	1.64
100	1.52	1.56	1.50	1.58	1.48	1.60	1.46	1.63	1.44	1.65

Source: Durbin, J. and Watson, G. S. (1951). "Testing for serial correlation in least square regression II," *Biometrika*, **38**, 159–178. Copyright Biometrika Trustees. Reprinted with permission.

Index

Index of Examples and Data

*T1.9 = Table 1.9.
F3.2 = Figure 3.2.

689